Nematode Parasites of Vertebrates

Their Development and Transmission

Dedication

This book is dedicated to Professor Alain G. Chabaud of the Muséum National d'Histoire Naturelle, Paris, for his many important contributions to our understanding of the nematode parasites of vertebrates

NEMATODE PARASITES OF VERTEBRATES

Their Development and Transmission
2nd Edition

R.C. Anderson

Department of Zoology
University of Guelph
Guelph, Ontario
Canada

CABI *Publishing*

CABI *Publishing* is a division of CAB *International*

CABI Publishing
CAB International
Wallingford
Oxon OX10 8DE
UK

Tel: +44 (0)1491 832111
Fax: +44 (0)1491 833508
Email: cabi@cabi.org

CABI Publishing
10 E 40th Street
Suite 3203
New York, NY 10016
USA

Tel: +1 212 481 7018
Fax: +1 212 686 7993
Email: cabi-nao@cabi.org

A catalogue record for this book is available from the British Library, London, UK.

Library of Congress Cataloging-in-Publication Data
Anderson, R. C. (Roy Clayton), 1926–
    Nematode parasites of vertebrates : their development and transmission / R.C.
Anderson. --2nd ed.
      p. cm.
    Includes bibliographical references.
    ISBN 0-85199-421-0 (alk. paper)
    1. Nematoda. 2. Vertebrates--Parasites. I. Title.
QL391.N4A46 2000
592'.57165--dc21

99-42444
CIP

ISBN 0 85199 421 0

Typeset by AMA DataSet Ltd, UK
Printed and bound in the UK by Biddles Ltd, Guildford and King's Lynn

Contents

List of Figures

Preface to the Second Edition

This edition includes relevant information in some 450 articles appearing from 1989 to 1998, with a few from early 1999. Some articles overlooked or unavailable for the first edition have been included, e.g. the discovery of transmammary transmission in *Toxocara cati* and an important early work on *Haemonchus contortus* in the UK. In addition, inconsistencies in the nomenclature used in the camallanoids and capillariines in the first edition have been eliminated. *Teladorsagia circumcincta* replaces *Ostertagia circumcincta*, *Litomosoides sigmodontis* replaces *L. carinii* and *Contracaecum rudolphi* replaces *C. spiculigerum*.

The number of species covered has been increased by 34, giving an overall total of 595. In addition to the species described in detail, many other species or genera not yet investigated are mentioned if they are related or relevant to the understanding of studied species, e.g. *Ascarophis* spp., members of the little understood Muspiceoidea, and some newly recognized species of *Trichinella*. Since the first edition there have been significant contributions to our knowledge of rhabditoids, trichostrongyloids, metastrongyloids, ascaridoids, dracunculoids and trichinelloids.

New theoretical considerations include comments on the loss of heteroxeny in some lungworms (metastrongyloids) and the substitution of vertebrate intermediate hosts for terrestrial gastropods which have allowed the adaptation of some species to marine mammals. The appearance of secondary monoxeny in ascaridoid parasites of: (i) predators and (ii) non-predators by different routes is outlined and helps to explain the host distribution of ascaridoids and the characteristics of their development and transmission. Finally, new observations of oviposition in the genus *Filaria* fill the gap between *Parafilaria* and *Stephanofilaria*, perhaps leading to the microfilaria and its relationship to biting insects in the evolution of filarioids.

Molecular studies that have been helpful in defining species in *Trichinella* and in the discovery of sibling species in ascaridoids (especially of marine mammals) and other families are covered.

Some illustrations photographed directly from published articles did not reproduce well in the first edition. These have been replaced by accurate pen and ink copies prepared by Uta Strelive. A number of reviewers expressed appreciation of the illustrations in the first edition but wished there were more of them. Accordingly the number has been increased from 33 to 43. Illustrations of transmission and development usually provide accurate depictions of first and third-stage infective larvae in addition to depicting the behaviour of the nematode in the tissues of its host. However, additional independent illustrations of infective larvae have been provided since these stages are liable to be encountered in the field by researchers.

There are two departures from the CIH Keys. Firstly, the Nematoda is regarded as a phylum and thus the subclasses become classes. In addition, the Dioctophymatoidea and the Trichinelloidea are placed in separate suborders of the Enoplida. It is anticipated that important cladistic studies now being undertaken in France will eventually necessitate changes in the higher categories of the bursate nematodes. It would be premature to include these changes in this book devoted to only a minority of species.

The aim of the book remains 'to summarize and synthesize our knowledge of the basic features of the development and transmission of parasitic nematodes of vertebrates and to place this information in the context of the modern classification as found in the *CIH Keys to the Nematode Parasites of Vertebrates*'. The book also aims to place the information in a historical context and the material is closely referenced to allow readers readily to find the relevant primary literature among the approximately 3200 references given in the book.

R.C.A.

Acknowledgements to the Second Edition

I am indebted to a number of individuals who made the preparation of the second edition possible. Bernadette Ardelli, a PhD candidate, found time from her research to put the data into the computer. Her skill and professionalism are warmly acknowledged. Uta R. Strelive again prepared illustrations with her customary skill and devotion to excellence. Erica Floto and Pooi-Leng Wong collected reprints, made photostats and arranged and checked references.

Dr František Moravec (Czech Republic) kindly advised on the nomenclature of the camallanoids and the capillariids. Dr Dwight Bowman (USA) provided very helpful notes on moults in the eggs of ascaridoids. Dr Lena Measures (Canada), Dr Murray Lankester (Canada) and Dr David Spratt (Australia) allowed me to read and use unpublished manuscripts. Technical advice was obtained from Dr Odile Bain, Dr Marie-Claude Durette-Desset (France) and Dr Ian Beveridge (Australia).

Publishers and editors allowed me to reproduce illustrations in the following journals: *Annales de Parasitologie Humaine et Comparée, Canadian Journal of Zoology, Environmental Biology of Fishes, Folia Parasitologica, Journal of Parasitology* and *Systematic Parasitology*. An illustration of *Anisakis* transmission was kindly provided by the Institut Maurice-Lamontagne, Mon-Joli, Quebec. I am grateful to Drs W.B. Scott and E.J. Crossman for permission to illustrate fishes based on their book, *The Freshwater Fishes of Canada* (Fisheries Research Board of Canada).

I extend my appreciation to the Department of Zoology for its support. I also acknowledge with gratitude the continued funding I have received from the National Science and Engineering Research Council of Canada.

The continued support and advice of Mr Tim Hardwick (Book Publisher, CABI *Publishing*) and the excellent production and editorial work of Ms Rachel Robinson and Ms Emma Critchley of CABI are gratefully acknowledged.

Preface to the First Edition

Several monographs exist with detailed information on development and transmission of medically important nematodes. As well, there are several textbooks with information on parasitic nematodes of interest to practising physicians and veterinarians and to students of these professions. Chandler *et al.* (1950, in *Introduction to Nematology*, ed. J.R. Christie) provided a useful summary of the characteristics of the transmission of a few major groups but much of the material is now out of date. Chabaud (1965, in *Traité de Zoologie*) accompanied his important systematic analyses with a few brief comments on transmission. There is, however, no book which treats all the parasitic nematodes of vertebrates which have been studied and relates the information to the latest concepts about systematics and relationships.

The aim of this book is to summarize and synthesize our knowledge of the more basic features of the development and transmission of the parasitic nematodes of vertebrates and to place this information in the context of the modern classification of the nematodes as found in the *CIH Keys to the Nematode Parasites of Vertebrates*. The arrangement of the orders and superfamilies follows that provided by Prof. A.G. Chabaud in No. 1 of the CIH Keys. The arrangement of the various families and subfamilies follows that found in each of the keys to genera (Nos 2–10). However, within each family or subfamily, genera and species are placed in alphabetical order with a few exceptions, including the genera *Brugia* and *Wuchereria*, which are combined in a heading so that the closely related *Wuchereria bancrofti* and *Brugia malayi* would not be treated distantly from each other in the family Onchocercidae. In some instances it has been desirable to treat two species at the same time (e.g. *Protostrongylus stilesi* and *P. rushi*; *P. kamenskyi* and *P. pulmonalis*). The development and transmission of species of *Rhabdias* or *Strongyloides* (Rhabditoidea) are so similar that species in each genus are discussed collectively. Except for these few exceptions, each species is reviewed separately.

An attempt has been made to generalize, starting mainly with superfamilies and following with families, subfamilies and sometimes genera. This scheme works well when dealing with well-studied relatively homogeneous groups like the bursate superfamilies, the oxyuroids, dracunculoids, acuarioids and filarioids. It is, however, much more difficult, and often impossible, to generalize at the higher taxonomic levels when dealing with such superfamilies as the Cosmocercoidea, Seuratoidea, Thelazioidea and Habronematoidea. Although there are morphological and taxonomic reasons which justify the contents of those superfamilies, each contains biologically diverse and probably distantly related species which preclude much generalization.

Considerable documentation has been provided to enable the reader readily to find and consult the primary literature (a book is a poor substitute for the original literature) but it will be appreciated that it has been necessary to be highly selective as concerns the references to intensively studied species. An entire career could be devoted to the study of a single group or even a single medically important species. For example, fairly recent compilations list 379 articles on bovine ostertagiasis, 383 articles on bovine dictyocauliasis (one species only) and 5449 references to animal and human onchocerciasis. Out of this mass of literature, an attempt has been made to select those articles which seem best to describe the most basic features of development and transmission of the species considered.

It is hoped this book will be of practical use to parasitologists, physicians, veterinarians, zoologists, and wildlife and fisheries biologists and that it will encourage a fresh appreciation of the astonishing diversity of 'life styles' even within superfamilies often regarded as biologically homogeneous. For example, the trichostrongyloids range from the typical gut parasite with the usual free-living larvae, to species which are transmitted by emesis and others which live in bile ducts and mammary glands of their hosts. The lungworms (metastrongyloids) range from typical inhabitants of the lungs which use molluscs as intermediate hosts, to forms in which first-stage larvae passed in faeces or vomit are the infective stage. In the dracunculoids of marine fish are found species in which first-stage larvae are microfilarioid and occur in the blood, where they are available to blood-sucking crustacean intermediate hosts – a remarkable convergence with the distantly related filarioids of terrestrial vertebrates. In the filarioids transmission ranges from the Filariidae, which release eggs into the environment by means of a break in the skin, to louse-transmitted, ephemeral Onchocercidae, which flood the skin of shorebirds with long-lived microfilariae and then disappear.

R.C.A.

Acknowledgements to the First Edition

I am indebted to a number of individuals who made the preparation of this book possible. I acknowledge the extraordinary editorial skills of Dr Cheryl Bartlett (University College of Cape Breton) who, with help from Mr Bernard MacLennon, reviewed each chapter three times, provided much helpful advice and eliminated numerous errors. Pooi-Leng Wong put the material into the computer, was largely responsible for the format and used her extensive knowledge of nematode systematics to detect and correct spelling and other errors in various drafts; her dedication throughout is gratefully acknowledged. Uta R. Strelive prepared the illustrations with her usual concern for detail and excellence. Mrs Erica Floto located and ordered obscure articles not available at Guelph and made countless photocopies. I am grateful to the staff of the Commonwealth Institute of Parasitology for help in obtaining articles. I am particularly grateful to Dr Ralph Lichtenfels of the United States Department of Agriculture and Dr Marie-Claude Durette-Desset of the Muséum National d'Histoire Naturelle, Paris, for photocopies of a number of articles. Dr Bob Kabata (Pacific Biological Station) and Professor Eugene Balon (University of Guelph) translated some Russian articles. Comments on certain technical matters were gratefully received from Dr Frank Moravec (Czechoslovak Academy of Science), Dr W.M. Hominick (Imperial College, England) and Dr Franz Schulte (Berlin). Grazia Bishop (St Albans, UK) kindly helped with the citing of Russian journals and I am most grateful to her.

The publishers of the following kindly allowed me to use illustrations from their publications: Akademia Kiad, *Annales de Parasitologie Humaine et Comparée*, *Canadian Bulletin of Fisheries and Aquatic Sciences*, *Canadian Journal of Zoology*, International Institute of Parasitology, CRC Press Inc., *Journal of Helminthology*, *Journal of Parasitology*, *Journal of Wildlife Diseases*, *Parasitology*, *Proceedings of the Helminthological Society of Washington* and *Transactions of the American Microscopical Society*.

I acknowledge the generous support I have received over a period of many years from the National Science and Engineering Research Council of Canada.

The support and advice of Mr Tim Hardwick (Book Publisher, CABI) in this endeavour was greatly appreciated, and I am most grateful for the excellent editorial and production work of Ms Pippa Smart and Ms Lindsay Gallaher who helped in many ways, especially in working out the format for the book.

Chapter 1

Introduction

The study of the development and transmission of parasitic nematodes started in the middle of the 19th century with early investigations on the transmission of *Trichinella spiralis* by Herbst, Virchow, Leuckart, Zenker and others. This was followed a few years later (1865–1866) by the discovery of alternation of generations in *Rhabdias bufonis* in amphibians by Metchnikoff and Leuckart. During the same period Leuckart and Marchi discovered that *Mastophorus muris* of the stomach of mice developed in flour beetles (*Tenebrio*) and Metchnikoff and Leuckart showed that *Camallanus lacustris* of freshwater fish developed in copepods. The latter finding apparently led directly to Fedchenko's important discovery in 1871 of the transmission of the human guinea worm by means of copepod intermediate hosts. In the 1870s Manson reported the development of *Wuchereria bancrofti* in mosquitoes and in 1890 Hamann discovered that the acuarioid *Echinuria uncinata* of ducks and geese developed in cladocerans (*Daphnia*) and Cori showed that the ascaridoid *Porrocaecum ensicaudatum* of passerine birds developed in earthworms. At the turn of the century the free-living stages of hookworms were known and Looss discovered skin penetration. At the same time Maupas and Seurat discovered the basic pattern of moults in free-living as well as parasitic nematodes and provided a theoretical template for the comparative study of nematode development from the free-living to the parasitic forms.

The early workers, therefore, provided a sound basis for the further investigation of the nematodes of vertebrates and since their early work some 594 species have been investigated, with representatives in all 27 superfamilies found in vertebrates. Our information on some species is limited to a few observations but many species, especially those found in humans (about 35 species) and domesticated animals, as well as in some wild animals, have been intensively studied. The following pages will review and attempt to synthesize this extensive and truly remarkable body of information based on the experimental study of nematodes for a century and a half. The study of transmission is important not only because of its relevance to human and animal medicine, but also because an understanding of transmission can help to explain how nematodes became parasitic in the first place and how they managed to radiate in some host groups (and not in others) during the course of evolution.

General Systematic Arrangement and Distribution of Nematodes

It has been estimated that about 16,000–17,000 nematode species have been described and that at least 40,000 species exist. Estimates of 500,000 to a million species have no basis in fact. Currently there are about 2271 described genera in 256 families. About 33% of all the nematode genera which have been described occur as parasites of vertebrates, equal to the percentage of genera known in marine and freshwater (Anderson, 1984). These figures are undoubtedly biased by the fact that there have been many more systematic parasitologists than specialists on the systematics of aquatic nematodes. Nevertheless, the variety and numbers of nematodes which have adopted a parasitic existence in vertebrates is impressive.

The classification used herein follows Chitwood (1950) and the CIH Keys (Anderson *et al.*, 1974–1983) except that the Nematoda is recognized as a phylum instead of a class of the disputed Aschelminthes (for recent discussions of the problem see d'Hondt, 1997, and Garey and Schmidt-Rhaesa, 1998) and the dioctophymatoids and the trichinelloids are placed in separate suborders of the Enoplida. Proposals (Inglis, 1983; Adamson, 1987) which change the names of some higher taxa do not depart fundamentally from those used in the keys which are used extensively to classify parasitic nematodes by specialists and non-specialists alike. Phylogenetic studies using molecular methods (Blaxter, 1993; Vanfleteren *et al.*, 1994; Blaxter *et al.*, 1998) have not yet seriously challenged the system used in the CIH Keys although one can anticipate clarification as the current system is tested by cladistic and molecular methods. In the CIH system (Fig. 1.1) the class Adenophorea consists of two major orders. Species of the order Chromadorida are abundant in marine habitats and did not contribute any members to parasitism. The order Enoplida includes nematodes found in marine and

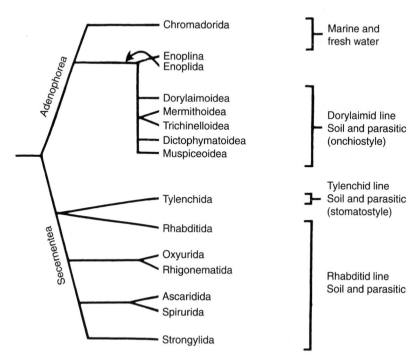

Fig. 1.1. Arrangement of the higher taxa of the Nematoda based on the *CIH Keys to the Nematode Parasites of Vertebrates* edited by Anderson *et al.* (1974–1983). (After R.C. Anderson, 1988 – courtesy *Journal of Parasitology.*)

fresh water as well as soil. Species of the suborder Dorylaimina of the Enoplida live mainly in soil and possess an onchiostyle, a tooth-like spear in the buccal cavity used in some species to kill and digest small organisms. The dorylaimines gave rise to four parasitic superfamilies, three of which are found in vertebrates (Trichinelloidea, Dioctophymatoidea, ?Muspiceoidea) and one in invertebrates, mainly insects (Mermithoidea). The superfamilies of the Enoplida found in vertebrates consist of few genera (e.g. *Capillaria*, *Trichinella*, *Trichuris*) and, except for the Capillariinae, few species. The affinities of these genera are generally revealed by the presence in the first-stage larvae of the onchiostyle characteristic of the free-living dorylaimids; an onchiostyle has not been found in the Muspiceoidea which has, however, been incompletely studied.

The class Secernentea includes the tylenchids (Tylenchida), which live in soil and feed mainly on plant cells and fungi; some authors regard the aphelenchids as a distinct order (Siddiqi, 1985; Hunt, 1993). A number of species have become parasites of invertebrates, especially insects, mites and leeches. The tylenchids possess a modified buccal cavity known as a stomatostyle, which they use to pierce cells and suck fluids. Although very successful as parasites of various insects, the tylenchids never invaded vertebrates if we ignore *Myoryctes wiesmanni* Ebert, 1863 said to have been found in a frog in Europe.

The order Rhabditida includes numerous soil-dwelling saprophytic nematodes which feed on bacteria and are among the most abundant metazoan inhabitants of soil enriched with decomposing organic matter. B.G. Chitwood (1937, 1950) regarded members of the family Rhabditidae as the most primitive of the class and this view has prevailed. In addition to the parasitic Rhabditida the soil-dwelling rhabditids apparently gave rise to four major orders parasitic in vertebrate animals, namely the Strongylida, Oxyurida, Ascaridida and Spirurida; they also gave rise to the Rhigonematida of diplopods (Adamson and Van Waerebeke, 1985). The Oxyurida is the only order with numerous representatives in both vertebrates and invertebrates and it has been suggested that those in vertebrates were derived from ancestors parasitizing insects.

A striking characteristic of the nematode parasites of vertebrates is the dominance of the secernenteans over the adenophoreans and it is generally agreed that the rhabditid-line gave rise to about 92% of all the parasites found in vertebrate animals. Also, since the parasitic secernenteans and the few parasitic adenophoreans can be traced back to soil ancestors, it has been hypothesized that nematode parasitism could not have arisen until animals invaded land, first the invertebrates and later the tetrapods (Anderson, 1984). This explains why nematode parasites are practically non-existent in such important marine and freshwater invertebrate taxa as the molluscs, polychaetes and crustaceans in contrast to the relative richness of the nematode fauna in their terrestrial counterparts, the earthworms, insects, diplopods and terrestrial slugs and snails. The reasons why aquatic nematodes failed to give rise to parasites is not clear but it has been suggested that the aquatic environment is not conducive to contacts between nematodes and potential hosts because of the dispersal effects of water (Inglis, 1965); the necessity for decaying organic matter in soil which preadapted nematodes for a parasitic life and provided large populations (Osche, 1954, 1963); the loss of kinocilia, which made it difficult or impossible for nematodes to move effectively in water (Anderson, 1984); and the lack of a dispersal stage like the dauer larva (see below) in the marine and freshwater nematodes (Anderson, 1984). Another possibility is that the Chromadorida in particular are physiologically too specialized and well adapted to marine environments to make the transfer to parasitic environments, in contrast to the rhabditids, which are more generalized morphologically and physiologically.

Nematode parasites occur in fish, however, and it is necessary to account for their origins in the light of the hypothesis that nematodes did not become parasites until the appearance of land animals. Fish nematodes are, except for some capillariines, all members of secernentean superfamilies well represented in terrestrial hosts. With the possible exception of a mere 20 species of oxyuroids (cf. the 497 species in the terrestrial vertebrates – Anderson and Lim, 1996), the nematodes of fish utilize intermediate and/or paratenic hosts in transmission. Although many fish nematodes undoubtedly await discovery, it seems clear that the nematode fauna in fish is, unlike that in terrestrial hosts, restricted to a few major taxa and not highly diverse and rich in species, with the possible exception of the dracunculoids and the camallanoids. For these reasons, it has been suggested that the ancestors of fish nematode parasites originated in early terrestrial vertebrates and transferred later to fish after they had acquired heteroxeny and paratenesis (Anderson, 1984, 1988). In the aquatic milieu these heteroxenous nematodes, which were probably not then highly specific, found new intermediate and paratenic hosts (for example, aquatic insects and crustaceans instead of terrestrial insects; fish rather than amphibians and reptiles) and were able to become established in fish. New forms could not arise in the aquatic ecosystem to supplement the acquired nematode fauna. Thus the fish nematode fauna lacks major diversity, is limited in species and has obvious affinities with parasites found in terrestrial hosts (Anderson, 1996).

It is worth noting here that the nematode fauna of cetaceans and pinnipeds is, like that of fish, restricted, with obvious affinities with species in terrestrial mammals and dominated by heteroxenous forms. One may assume that practically all the monoxenous forms that lived in terrestrial ancestors of marine mammals could not adapt to the aquatic ecosystem and, with a few exceptions, became extinct when their hosts adopted an aquatic existence.

Development of Nematodes

The most important early work on nematode development was conducted by Maupas (1899, 1900). Although professionally an archivist with the National Library in Algiers, Maupas found time to conduct highly original studies at home on ciliates, rotifers and nematodes and pioneered the methods for their cultures. Maupas noted that female rhabditid nematodes commonly found in soil laid eggs which developed into tiny first-stage larvae which hatched and immediately commenced to feed on bacteria by means of pharyngeal pumping; rhabditids have an oesophagus with a bulb containing valves that allows them to ingest food. In a few hours the larvae entered a lethargus, a period of 1–2 h, during which they remained motionless. The larvae then moulted and recommenced feeding. The process of feeding, lethargus and moulting occurred four times and at the final moult all the adult characteristics had appeared, since development of the reproductive organs and secondary sexual characteristics, which began during development of the late third stage, culminated in the late fourth stage (see, for example, the development of *Rhabditis orbitalis* in Chapter 2). Maupas suspected that all nematodes passed through five stages, separated by four moults, and almost a century of research has borne out this prediction. Maupas also observed that when his cultures became exhausted of nutrients all stages disappeared except some at the early third stage that did not shed the second-stage cuticle. These arrested, non-feeding larvae survived without moisture and food for prolonged periods but, when transferred to fresh cultures, they exsheathed and started to feed and grow to the fourth and adult stages. Such larvae are now known as **dauer larvae** (from Fuchs, 1937: *die dauer*, 'the enduring').

Dauer larvae do not engage in pharyngeal pumping (Swanson and Riddle, 1981). The mouth is plugged and the body appears thin compared with second- and third-stage larvae. The cuticle of the dauer larva differs in that the outer cortex is thicker and there is an additional striated underlayer not found in other stages (Cassada and Russell, 1975). The factors involved in the formation of dauer larvae are not fully understood. They form, as indicated above, when bacterial food sources become scarce and perhaps during crowding in response to a pheromone (Golden and Riddle, 1982). When transferred to new cultures they exsheath, if they have not already done so, and begin pharyngeal pumping as early as 1–2 h (Cassada and Russell, 1975) as noted earlier by Maupas (1889, 1900). Kimura *et al.* (1997) cloned and sequenced a gene (*daf-2*) in *Caenorhabditis elegans* that when damaged can block or enhance the ability to switch to the dauer stage. During times of ample food *C. elegans* maintains high levels of an insulin-like hormone which binds to *daf-2*. This, in turn, may trigger *daf-23* (which encodes a P13 kinase) to activate the second messenger molecules passing signals to burn fuel. When the worm population increases, the food supply decreases; increased pheromone concentrations trip the worms' chemosensory alarms and insulin levels decrease – an indication of reduced glucose availability. The two signals push the nematodes into the dauer stage (for review of Kimura *et al.* see Roush, 1997). The relevance of these findings to the cessation of development, usually at the early third stage, in the Secernentea in general may be clarified in the future (Rajan, 1998).

Dauer larvae generally remain motionless in culture unless disturbed. They react rapidly to stimulation such as touch or vibration and they have the ability to stand on their tails on the substrate and wave their bodies in the air. The dauer larva is a superb dispersal stage and it readily attaches to the surface of animals (especially insects) it encounters in the environment and thus can be carried to new food sources. For example, dauer larvae of rhabditids are invariably found in parallel rows under the elytra of dung beetles, which carry them to fresh dung as the need arises. Osche (1954, 1963) has discussed the phoretic associations of dauer larvae with other organisms.

The Rule of the Infective Third Stage

Chabaud (1954, 1955), in a series of influential articles, analysed information on the transmission and development of the nematodes of vertebrates. He emphasized that in the Secernentea (the name 'Phasmidia' was used but has been dropped because it was preoccupied) found in vertebrates the infective stage was almost always at the beginning of the third larval stage (immediately or shortly after the second moult) whether or not the parasite was monoxenous, with free-living stages, or heteroxenous, with early larval stages taking place in an intermediate host. This rule of the infective third stage is strong evidence that secernentean nematodes were derived from free-living rhabditids with a dauer larva. The infective third stage in secernentean nematodes of vertebrates is apparently, therefore, homologous to the early third stage and the dauer larvae of free-living rhabditids. This is most clearly suggested by the long-lived, sheathed, resistant, early third-stage larva which is the infective stage in many strongyloids (Strongyloidea) and trichostrongyloids (Trichostrongyloidea). In addition, the free-living stages of the parasitic Rhabditida (e.g. *Rhabdias*, Strongyloides) as well as the hookworms (Ancylostomatoidea), strongyloids and trichostrongyloids are remarkably similar morphologically to homologous stages of the free-living nematodes such as in *Rhabditis*.

The adenophorean nematodes in vertebrates do not follow the rule of the infective third stage. They presumably evolved from dorylaimines which inhabited soil (Steiner, 1917) and lacked the equivalent of a dauer larva. Thus, the trichinelloids (Trichinelloidea) infect the definitive host in the first stage even when they use hosts such as oligochaetes or fish in transmission (i.e. some Capillariinae). *Trichinella spiralis* grows dramatically in the striated muscles of the host but nevertheless remains in the first stage. In the heteroxenous Dioctophymatoidea the infective stage is either the third (*Dioctophyme*, *Soboliphyme*) or the fourth (*Eustrongylides*) larval stage.

Principles of Transmission

Monoxeny

A number of major groups of nematodes are **monoxenous** (*mono* Gr. = one, xenus Gr. = host) in that they infect the host directly without the intervention of an intermediate host (Rhabditida; Strongylida except the Metastrongyloidea; Cosmocercoidea; Heterakoidea). In **primary monoxeny** there is no reason to believe there was ever during the course of evolution an intermediate host involved in transmission (Anderson, 1988). Fülleborn (1920) proposed that percutaneous transmission in monoxenous nematodes was the primitive mode of infecting the host and, in fact, this mode of transmission occurs commonly in the more primitive nematodes including the Rhabditida (e.g. *Rhabdias*, *Strongyloides*) and the Strongylida (some Ancylostomatoidea and Trichostrongyloidea). Adamson (1986, 1989) postulated that only those groups that arose from ancestors infecting the host percutaneously developed tissue parasitism and the use of intermediate hosts. Sukhdeo *et al.* (1997) used nucleotide sequence data to explore the phylogenetic relationships of nine selected taxa of the Strongylida along with *Strongyloides stercoralis*, *Ascaris suum* and *Contracaecum rudolphi* (syn. *C. spiculigerium*). The results in the Strongylida produced two clades, one in which skin penetration was lost but tissue migration retained as the oral infection route evolved and another in which both skin penetration and tissue migration were lost and infection is by the oral route. The authors concluded that ancestral Strongylida possessed skin-penetrating and tissue-migrating larvae which were modified over the course of time as the oral route evolved. Often the same species can infect the host orally as well as through the skin (e.g. *Ancylostoma caninum*, *Nippostrongylus brasiliensis*) and circumstances and host behaviour probably determine which route is more important in these species under natural conditions.

Percutaneous transmission invariably involves a larval migration in the body. Larvae generally enter the lymph and blood vessels, reach the heart and lungs and are coughed up and swallowed, but some may wander for prolonged periods in the host's tissues (especially in the immune host) and contribute to the evolution of **transplacental** and **transmammary transmission**, as in *Strongyloides* spp., *Ancylostoma caninum* and *Uncinaria lucasi*. In *U. lucasi*, selection has exploited fully the tendency of larvae of hookworms to persist in the tissues of the immune host and transmammary transmission has become crucial for the survival of this nematode in fur seals, which spend many months at sea – well beyond the life span of adult worms.

Secondary monoxeny assumes the loss of one of the hosts during the course of evolution and a return to direct transmission. Fülleborn (1927) interpreted the tissue phase in *Ascaris lumbricoides* as an event that once took place in an intermediate host that has been lost during the course of evolution.

Fülleborn's interpretation as it applies to *Ascaris* spp. and other species in non-predators (e.g. *Baylisascaris laevi*, *Parascaris equorum*, *Toxocara pteropodis*, *T. vitulorum*) is misleading because in these ascaridoids it is the predatory definitive host that has been eliminated, not the intermediate host; and in species such as *Ascaris lumbricoides* of humans the latter serve as both intermediate and definitive host. What is novel in the relationship is the maturation in the gut of the host and not the tissue migration.

Secondary monoxeny is also attained in ascaridoids of predators but by eliminating the intermediate host, as in *Toxocara canis* and *T. cati*; thus the definitive host serves also as intermediate host. Secondary monoxeny by either route is important in these ascaridoids because it is associated with **visceral larval migrans**, which may evolve into transplacental transmission (as found, for example, in *Toxocara canis*) and transmammary transmission (as found in *Toxocara vitulorum*, *T. cati* and *T. pteropodis*).

Read and Skorping (1995) suggested that tissue migration is a trait which is selectively advantageous for some nematodes. Taxa which undertake migrations in the tissue during their development tend to grow faster and larger than close congenitors that develop directly in the gut without a tissue migration.

Secondary monoxeny also occurs rarely in the lungworms (Metastrongyloidea), i.e. *Filaroides hirthi*, *Osterus osleri* and *Andersonstrongylus captivensis*.

Seclusion and heteroxeny

Chabaud (1954) suggested that the evolutionary trend in parasitic nematodes was to isolate free-living stages from the rigours and uncertainties of the external environment, a process he referred to as **seclusion**.

Seclusion has been achieved in a number of ways by the nematode parasites of vertebrates including:

- the development of a resistant, ensheathed third-stage larva in the monoxenous bursate nematodes;
- the development of a resistant egg with ample nutrients so that all development can take place in the egg before hatching, as in the Amidostomatidae, *Nematodirus* of the Trichostrongyloidea, and in the Syngamidae of the Strongyloidea;

- the penetration into and sequestering of infective larvae in the soft bodies of invertebrates such as earthworms, as in *Stephanurus* and other Syngamidae;
- confining all development to a resistant egg which will not hatch until it attains the definitive host, as in the Oxyuroidea and the Heterakoidea;
- by adopting an intermediate host for the development and transmission of the larval stages.

As indicated earlier, **heteroxeny** (*hetero* Gr. = different), or the use of intermediate hosts, was discovered in the nematode parasites of vertebrates during the latter half of the 19th century. In the Secernentea, development in the intermediate host usually proceeds from the first to the early third stage, which is infective to the definitive or final host. Heteroxeny is very widespread within the nematode parasites of vertebrates. It has been adopted by the Metastrongyloidea of the order Strongylida; the Seuratoidea, Ascaridoidea and Subuluroidea of the order Ascaridida; the entire Spirurida; and the Dioctophymatoidea of the order Enoplida.

In the lungworms (Metastrongyloidea) – the only heteroxenous bursate group – intermediate hosts are generally terrestrial snails and slugs and, rarely, earthworms (*Metastrongylus* only). The Spirurida use a wide range of aquatic or terrestrial arthropods as intermediate hosts. It is a primitive feature of the Ascaridoidea and perhaps the Cucullanidae (Seuratoidea) that intermediate hosts are vertebrates, although the system has been modified in some ascaridoids during the course of evolution (Anderson, 1984, 1988). The Dioctophymatoidea use aquatic and terrestrial oligochaetes.

Such a widespread phenomenon as heteroxeny must provide major advantages for the transmission of vertebrate nematodes (Chabaud, 1957; Inglis, 1965). Some of the more obvious seem to be as follows:

- The preinfective stages are secluded from the uncertainties of the external environment.
- The intermediate host can greatly extend the life of larvae in time and space.
- The intermediate host can channel the parasite to the final host and take some of the chance out of transmission, because it is either a desired food item of the final host or a biting arthropod attracted to the final host and capable of serving as a vector.
- The intermediate host provides a nutrient-rich environment and larvae are capable of expanded growth, which can influence the rate of maturation in the final host and perhaps enhance the ability of adults to produce progeny (see also the section on precocity below).

Paratenesis

The term **paratenic host** was defined by Baer (1951) as one in which the infective stage of a parasite persists without essential development and usually without growth. The process of infective larvae passing from one paratenic host to another was defined by Beaver (1969) as **paratenesis**. Anderson (1984) expanded the definition to include the passage of a larval stage to a paratenic host and then to an intermediate host where development takes place, as, for example, in some ascaridoids in which copepod paratenic hosts transfer second-stage larvae to fish intermediate hosts. The expanded definition of a paratenic host is '*an organism which serves to transfer a larval stage or stages from one host to another but in which little or no development takes place*'. Primitively the paratenic host is an ecological rather than a physiological necessity in transmission. Infective larvae may be acquired by hosts which for various reasons cannot serve as paratenic hosts in the strict sense since they are not in the food chain of the definitive host. Such hosts have been referred to as '**paratenic trap hosts**' by Sharpilo *et al.* (1995).

Paratenesis in its simplest form is well demonstrated in the transmission of the metastrongyloid lungworms of mustelids which attain the infective stage in terrestrial gastropods. Mustelids probably do not acquire lungworms from eating gastropods. However, when an infected gastropod is eaten by a shrew or a rodent, the infective larvae invade the tissues and become slightly encapsulated, usually in the liver of these paratenic hosts. Thus, in the metastrongyloids the paratenic host is placed between the intermediate host and the final host. This pattern of paratenesis is also widespread in the order Spirurida. In addition, as indicated above, in the transmission of some ascaridoids a paratenic host occurs at the beginning of the developmental sequence in the form of an invertebrate, such as a crustacean in species in aquatic hosts and an earthworm in species in terrestrial hosts. In each group where it is found, paratenesis must have had a prolonged period of evolution, probably in response to shifts in food preferences of final hosts away from intermediate hosts to other animals which themselves continued to consume the intermediate hosts and were capable of becoming paratenic hosts. The phenomenon was probably crucial for the survival of many nematode parasites during the evolution of the Carnivora from unspecialized ancestors (Anderson, 1982).

There is an additional important consequence of paratenesis. The process adds to the number of hosts encountered by a parasite and there is always the possibility for the parasite to take advantage of its location by commencing to develop in the paratenic host, thus turning the latter into a new intermediate host through the process of precocity. This has apparently happened commonly in the Ascaridoidea.

Precocity and Capture

The rule of the infective third-stage larva is an extremely important generalization for our understanding of the transmission and development of the secernentean nematodes of vertebrates. There are, however, exceptions in which development proceeds beyond the early third stage and even up to the fifth or adult stage in the intermediate host. The phenomenon is called **precocity** and is defined as *growth and/or development beyond the expected in the intermediate host* (Anderson, 1988; Anderson and Bartlett, 1993). This strategy enables the parasite to compensate for host behaviour or other factors that might restrict transmission to reduced limits of space and time, since it accelerates gamete production when the parasite finally reaches the definitive host. Precocity manifests itself in four obvious ways:

- the unusual size (even gigantism) of the infective larvae;
- the precocious development of the genital primordia in the infective larvae;
- development to the fourth stage in the intermediate host;
- development to the subadult (fifth) stage in the intermediate host.

The possible role of precocity in nematode transmission can be most readily appreciated in those few species known to attain the fifth or subadult stage in the intermediate host, a phenomenon referred to as **extreme precocity** (Anderson, 1988). Extreme precocity has been known for many years although its significance has only recently been appreciated. Leuckart (1876) reported that *Hedruris androphora* (Habronematoidea: Hedruridae) of the intestine of the smooth newt (*Triturus vulgaris*) developed to the subadult stage in the haemocoel of the aquatic isopod *Asellus aquaticus*, an observation confirmed by Petter (1971) (similar observations have been made on *Hedruris ijima* of amphibians). European smooth newts are mainly terrestrial and come to water only for brief periods to spawn. Immature adult worms acquired as first-stage larvae in eggs the previous year are presumably present in the isopods and they would mature rapidly in the newts and produce eggs before the host returned to land. Thus infected isopods would be available the following spring when the newts returned to spawn.

Another example of extreme precocity is *Rhabdochona rotundicaudatum* (Thelazioidea: Rhabdochonidae) of the intestine of the common shiner (*Notropis cornutus*). Shiners spawn in a specific region of the stream and remain there for only a few weeks, during which they acquire subadult worms from the infected local mayfly larvae (*Ephemera simulans*), which emerge from their burrows at the time fish are present. The worms are ready to produce eggs almost as soon as they reach the gut of the fish. The latter excrete eggs on to the spawning area and these reinfect the local mayfly population before fish disperse in the stream.

Subadult worms of the genus *Ascarophis* (Habronematoidea: Cystidicolidae) have been reported in the haemocoel of marine decapods and amphipods. *Rabbium paradoxus* (Seuratoidea: Seuratidae), found as subadult worms in worker ants, is clearly another example of extreme precocity since other species in the genus are parasites of reptiles. The relationship of extreme precocity to transmission in these species is unknown but one might predict that it is related in some way to the behaviour of the definitive host.

Extreme precocity is not confined to nematodes which use crustaceans and aquatic insects as intermediate hosts. It is also found in some species which use vertebrate intermediate hosts and it may, in fact, result in **capture** – the transfer of a parasite from a vertebrate definitive host to another unrelated vetebrate and its subsequent speciation (*sensu* Chabaud, 1965).

Truttaedactnitis truttae (Seuratoidea: Cucullanidae) occurs mainly in the intestine of salmonid fishes in Europe but the larval stages and adults are found in lamprey (*Lampetra* spp.), which apparently serve as intermediate hosts. *T. pybusae* in North America, however, develops to the third stage in the liver of brook lamprey ammocoetes. During transformation of the ammocoete into the adult form, the larvae move to the intestine, moult and mature. Transformed lamprey will pass eggs of the nematode and there is no evidence that teleosts are involved in transmission. Thus, the intermediate host has apparently also become a final host. The significance of extreme precocity in this species is that it shows one of the ways in which a parasite from one vertebrate might establish a totally independent existence in a quite unrelated vertebrate that once served as an intermediate host.

The reports of *Hexametra angusticaecoides* (Ascaridoidea: Angusticaecinae) in chameleons and *Orneoascaris chrysanthemoides* in frogs are regarded as examples of extreme precocity of ascaridoids of snakes in their intermediate hosts. Finally, Anderson

(1988) raised the possibility that *Hysterothylacium haze* (Ascaridoidea: Anisakidae) of the body cavity and tissues of the yellowfin goby (*Acanthogobius flavimanus*) was a capture by means of extreme precocity of an ascaridoid of some predaceous fishes. Another possibility is that the yellowfin goby is an intermediate host for *H. haze* which occurs in some predaceous fishes serving as the definitive host.

Precocity is rarely as extreme as the examples given above but various forms of it are nevertheless widespread among secernenteans. An obvious example in the Strongylida is the parasitic infective larva of *Uncinaria lucasi*, which is passed in the milk of the northern fur seal mother to her pup. This larva appears post-partum and is substantially larger (938 μm) than the infective third-stage larva (645–795 μm) which invades the skin; the prepatent period is about 2 weeks, somewhat shorter than that in related species in terrestrial mammals.

Examples of precocity in the Spirurida and the Ascaridoidea are numerous and only a few illustrative examples will be mentioned. Compared with infective larvae in the more primitive monoxenous Rhabditida and Strongylida, infective larvae of hetero-xenous groups are relatively enormous. In the Spiruroidea and the Habronematoidea they are usually about 2–4 mm in length but in *Hartertia gallinarum* they are reported to be 20–30 mm and in *Spirura rytipleurites* 12 mm in length. Also, in some species (*Paracuaria adunca*, *Skrjabinoclava* spp., *Spinitectus* spp. and *Spirura* spp.) the repro-ductive systems are well defined in the infective third-stage larvae. Bartlett *et al.* (1989) and Wong *et al.* (1989) showed that the prepatent period tended to be shorter in acuarioids with precociously developed genital primordia than in those in which it was less developed. Bartlett *et al.* (1989) related the unusually short prepatent period of *Skrjabinocerca prima* in avocets (*Recurvirostra americana*) to the fact that the host leaves the nesting and transmission area as soon as the young are fledged.

Precocious growth of larvae in the invertebrate or vertebrate intermediate hosts is well established in ascaridoids such as *Amplicaecum robertsi*, *Anisakis* spp., *Hexametra* spp., *Pseudoterranova decipiens* and *Raphidascaris acus*. Indeed in some ascaridoids prac-tically all the growth takes place in intermediate hosts. Also in *R. acus* the third-stage larva moults to the fourth stage in the fish intermediate host.

Precocity is not confined to the secernenteans. The first-stage larva of *Trichinella spiralis* grows precociously in the muscles of the host and the reproductive system is highly developed. Thus *T. spiralis* matures with great rapidity in the intestine of the host. Also the infective stage of *Eustrongylides* spp. is a precociously developed fourth-stage larva in fish (cf. *Dioctophyme renale*).

Arrested Development

Arrested development (**hypobiosis**) is a well-studied phenomenon mainly in the Trichostrongyloidea of ungulates and lagomorphs and to a lesser extent in the Ancylostomatoidea. When arrest occurs an unusually high percentage of third- or fourth-stage larvae remain in the tissues (especially the mucosae) long after the normal prepatent period. Arrested larvae can, when they finally develop, greatly extend the patent periods of infections. Arrest is most important in trichostrongyloids and hookworms, because it allows species with limited adult life spans to survive in the host for prolonged periods when conditions are unsuitable for the development, survival and transmission of free-living larval stages. Arrested larvae tend to continue their development to egg-producing adulthood during parturition or lactation, when young animals are available with little or no immunity.

Extrinsic factors inducing arrest in infective larvae have been identified as cold, fluctuating temperatures, or hot dry conditions. Immunity has been shown to affect the number of larvae arresting and an important genetic component has been revealed. Arrest in the Strongylida is discussed more fully in Chapter 6.

The cessation of development in the definitive host is common in various groups of nematodes other than the Strongylida but its exact relationship to seasonal or immunological arrest is not always clear. Arrest should probably be distinguished from the normal cessation of development of larvae in intermediate and paratenic hosts in the ascaridoids and the Spirurida (cf. Michel, 1974). Some examples of cessation of development that seem appropriate are outlined below.

Physaloptera maxillaris of the stomach of skunks in temperate climates overwinters in the third stage apparently because of lack of food ingested by the host. Ascaridoids in the stomach of pinnipeds also appear to persist in the stomach as third-stage larvae when the host is not feeding.

White suckers (*Catostomus commersoni*) acquire larval *Philometroides huronensis* (Dracunculoidea) in summer and autumn but inseminated females remain near the bases of the fins until spring, when they enter the fin and mature. Larval *Philonema onchorhynchi* remain arrested in the tissues of sockeye salmon (*Onchorhynchos nerki*) for at least 2 years before reaching maturity in the spawning fish. Renewal of development of most dracunculoids in fish is probably related to temperature and/or hormonal factors which correlate larval production with the availability of intermediate hosts.

Postcyclic Parasitism

Bozhkov (1969) and Odening (1976) designated as **postcyclic transmission** those situations in which adult gut-dwelling parasites are transmitted to an unsuitable host by cannibalism and predation. The phenomenon is rather common in fish. For example, pike cannot serve as a definitive host of *Camallanus lacustris* but this nematode can be found frequently in the gut of pike which prey on fish usually harbouring this nematode (Moravec, 1978, 1994). Similarly, *Rhabdochona* spp. of cyprinids may be found in the gut of larger predatory fish. The acquired worms in the postcyclic host may live out their lives and even continue to pass eggs for a time. In assessing the nematode fauna of cold-blooded vertebrates, especially fish, it is important to consider the possibility of postcyclic transmission as a source of some parasites found.

Modes of Infecting the Definitive Host

Vertebrates acquire their nematode parasites in a variety of ways. The penetration of infective third-stage larvae into the skin of the host is found in the Rhabditida (*Rhabdias* spp., *Strongyloides* spp.), some of the hookworms (Ancylostomatoidea) and the Cosmocercoidea (*Cosmocerca communtata, Cosmocercoides variabilis*). However, most hookworms can also infect the host by the oral route. Skin penetration is rare in the

Strongyloidea (*Stephanurus dentatus*), and in the Trichostrongyloidea it is confined to species in amphibians (*Oswaldocruzia pipiens*) and rodents (*Hassalstrongylus musculi, Neoheligmonella* spp., *Nippostrongylus brasiliensis*). Most Strongylida infect the host by the oral route, i.e. ingesting infective larvae contaminating herbage or other food and faeces. Larvae of the syngamids (Strongyloidea) will sequester themselves in earthworm paratenic hosts. As noted earlier, skin penetration has probably led to prenatal (transplacental) and transmammary (lactational) transmission in *Strongyloides* spp. and to transmammary transmission in *Uncinaria lucasi* and other hookworms.

In the Oxyuroidea, Heterakoidea and the Trichinelloidea (except *Trichinella* spp. and *Aonchotheca philippinensis*) infection of the host is usually through the ingestion of eggs containing infective larvae. However, earthworm paratenic hosts are used by both the Heterakoidea and the Capillariinae (Trichinelloidea). Coprophagy is probably important in the transmission of pinworms (Oxyuroidea) to lagomorphs and rodents. Scavenging and cannibalism are important in the transmission of *Trichinella* spp. and *Calodium hepaticum*.

Ollulanus tricuspis (Trichostrongyloidea) is transmitted among felines and pigs by emesis or vomiting.

Although the lungworms (Metastrongyloidea) are normally heteroxenous, there are three species that have apparently abandoned intermediate hosts and are capable of infecting the definitive host in the first stage. In *Oslerus osleri* of canines, first-stage larvae are transmitted during regurgitative feeding. *Filaroides hirthi* of dogs and *Andersonstrongylus captivensis* of skunks are apparently transmitted through faeces contaminated with first-stage larvae; the latter species may also be transmitted by regurgitative feeding.

Autoinfection occurs in diverse taxa of monoxenous nematodes, including *Strongyloides stercoralis* (Rhabditoidea), *Ollulanus tricuspis* (Trichostrongyloidea), *Probstmayria vivipara* (Cosmocercoidea), *Gyrinicola batrachiensis* (Oxyuroidea) and *Aonchotheca philippinensis* (Trichinelloidea).

In the heteroxenous groups (Metastrongyloidea, Spirurida, Dioctophymatoidea) transmission is generally achieved through the ingestion of larvae in the tissues of intermediate and paratenic hosts, with the exception of the Filarioidea and some dracunculoids of fish which are transmitted by blood-sucking arthropod vectors.

Transplacental transmission appears in *Protostrongylus stilesi* (Metastrongyloidea) and in *Toxocara canis* (Ascaridoidea) and *Setaria marshalli* (Filarioidea).

Transmammary transmission is reported in *Toxocara canis, T. cati, T. pteropodis* and *T. vitularum*.

Trichinella spp. and some avian filarioids (*Eulimdana* of shorebirds) are unusual in that adults are ephemeral, i.e. they flood the tissues with long-lived progeny and then die and are eliminated by the host.

References (Introduction)

Adamson, M.L. (1986) Modes of transmission and evolution of life histories in zooparasitic nematodes. *Canadian Journal of Zoology* 64, 1375–1384.

Adamson, M.L. (1987) Phylogenetic analysis of the higher classification of the Nematoda. *Canadian Journal of Zoology* 65, 1478–1482.

Adamson, M.L. (1989) Constraints in the evolution of life histories in zooparasitic nematodes. In: Ko, R.C. (ed.) *Current Concepts in Parasitology*. Hong Kong University Press, Hong Kong, pp. 221–253.

Adamson, M.L. and Van Waerebeke, D. (1985) The Rhigonematida (Nematoda) of diplopods: reclassification and its cladistic representation. *Annales de Parasitologie Humaine et Comparée* 60, 685–702.

Anderson, R.C. (1982) Host–parasite relations and evolution of the Metastrongyloidea (Nematoda). *Deuxième Symposium sur la Specificité Parasitaire des Parasites de Vertébrés, 13–17 Avril 1981. Mémoires Muséum Nationale d'Histoire Naturelle, Nouvelle Série, Série A, Zoologie* 123, 129–133.

Anderson, R.C. (1984) The origins of zooparasitic nematodes. *Canadian Journal of Zoology* 62, 317–328.

Anderson, R.C. (1988) Nematode transmission patterns. *Journal of Parasitology* 74, 30–45.

Anderson, R.C. (1996) Why do fish have so few roundworm (nematode) parasites? *Environmental Biology of Fishes* 46, 1–5.

Anderson, R.C. and Bartlett C.M. (1993) The significance of precocity in the transmission of the nematode parasites of vertebrates. *Canadian Journal of Zoology*, 71, 1917–1922.

Anderson, R.C. and Lim, L.H.S. (1996) *Synodontisia moraveci* n.sp. (Oxyuroidea: Pharyngodonidae) from *Osteochilus melanopleurus* (Cyprinidae) of Malaysia, with a review of pinworms in fish and a key to species. *Systematic Parasitology* 34, 157–162.

Anderson, R.C., Chabaud, A.G. and Willmott, S. (eds) (1974–1983) *CIH Keys to the Nematode Parasites of Vertebrates, Nos 1–10*. Commonwealth Agricultural Bureaux, Farnham Royal, UK.

Baer, J.G. (1951) *Ecology of Animal Parasites*. University of Illinois Press, Urbana, Illinois.

Bartlett, C.M., Anderson, R.C. and Wong, P.L. (1989) Development of *Skrjabinocerca prima* (Nematoda: Acuarioidea) in *Hyalella azteca* (Amphipoda) and *Recurvirostra americana* (Aves: Charadriiformes), with comments on its precocity. *Canadian Journal of Zoology* 67, 2883–2892.

Beaver, P.C. (1969) The nature of visceral larva migrans. *Journal of Parasitology* 55, 3–12.

Blaxter, M.L. (1993) Hemoglobins: divergent nematode globins. *Parasitology Today* 9, 353–360.

Blaxter, M.L., De Ley, P., Garey, J.R., Liu, L.X., Scheldeman, P., Vierstraete, A., Vanfleteren, J.R., Mackey, L.Y., Dorris, M., Frisse, L.M., Vida, J.T. and Thomas, W.K. (1998) A molecular evolutionary framework for the phylum Nematoda. *Nature* 392, 71–75.

Bozhkov, D. (1969) Postcyclic parasitism and postcyclic hosts in helminths. *Izvestiya Zoologicheskii Inst.* Muz BAN 29, 183–189. (In Russian.)

Cassada, R.C. and Russell, R.L. (1975) The dauerlarva, a post-embryonic developmental variant of the nematode *Caenorhabditis elegans*. *Developmental Biology* 46, 326–342.

Chabaud, A.G. (1954) Sur le cycle évolutif des spirurides et nématodes ayant une biologie comparable. Valeur systematique des caractères biologique. *Annales de Parasitologie Humaine et Comparée* 29, 42–88.

Chabaud, A.G. (1955) Essai d'interpretation phylétique des cycles évolutifs chez les nématodes parasites de vertébrés. *Annales de Parasitologie Humaine et Comparée* 30, 83–126.

Chabaud, A.G. (1957) Spécificité parasitaire chez les nématodes parasites de vertébrés. In: *Premier Symposium sur la Spécificité Parasitaire des Parasites de Vertébrés*. Institut de Zoologie, Université de Neuchatel. Imprimerie Paul Attinger S.A., Neuchatel, pp. 230–243.

Chabaud, A.G. (1965) Spécificité parasitaire. I. Chez les nématodes parasites de vertébrés. In: Grassé, P.P. (ed.) *Traité de Zoologie. Anatomy. Systématique, Biologie*, Vol. 4, Part 2. *Némathelminthes (Nématodes)*, pp. 548–564.

Chitwood, B.G. (1937) A revised classification of the Nematoda. *30 year Jubileum K.I. Skrjabin, Moscow*, pp. 69–80.

Chitwood, B.G. (1950) An outline classification of the Nematoda. In: Chitwood, B.G. and Chitwood, M.B. (eds) *Introduction to Nematology*. University Park Press, Baltimore, Maryland, pp. 12–25.

D'Hondt, J.L. (1997) Qu'est-il advanu des anciens 'Aschelminthes'. *Bulletin de la Societé Biologique de France* 122, 355–364.

Fuchs, A.G. (1937) Neue parasitische und halbparasitischa Nematoden bei Borkenkäfern und einige andere Nematoden. *Zoologische Jahrbücher (Systematik)* 71, 123–190.

Fülleborn, F. (1920) Ueber die Anpassung der Nematoden an den Parasitismus und den Infectionsweg bei Askaris und anderen Fadenwürmern des Menschen. *Archiv für Schiffs- und Tropen-Hygiene, Pathologie und Therapie Exotischer Krankheiten* 24, 340–347.

Fülleborn, F. (1927) Ueber das Verhalten der Larven von *Strongyloides stercoralis*, Hakenwürmern und *Ascaris lumbricoides* im Körper des Wirtes und ein Versuch, es biologisch zu deuten. Beihefte (2). *Archiv für Schiffs- und Tropen-Hygiene, Pathologie und Therapie Exotischer Krankheiten* 31, 151–202.

Garey, J.R. and Schmidt-Rhaesa, A. (1998) The essential role of 'minor' phyla in molecular studies of animal evolution. *American Zoologist* 38, 907–917.

Golden, J.W. and Riddle, D.L. (1982) A pheromone influences larval development in the nematode *Caenorhabditis elegans*. *Science* 218, 578–580.

Hunt, D.J. (1993) *Aphelenchida, Longidoridae and Trichodoridae; their Systematics and Bionomics.* CAB International, Wallingford, UK, 352 pp.

Inglis, W.G. (1965) Patterns of evolution in parasitic nematodes. In: Taylor, A.E.R. (ed.) *Evolution of Parasites*. Third Symposium of the British Society of Parasitology. Blackwell Scientific Publications, Oxford, UK, pp. 79–124.

Inglis, W.G. (1983) An outline classification of the phylum Nematoda. *Australian Journal of Zoology* 31, 243–255.

Kimura, K.D., Tissenbaum, H.A., Liu, Y. and Ruvkun, G. (1997) Daf-2 an insulin receptor-like gene that regulates longevity and diapause in *Caenorhabditis elegans*. *Science* 277, 897–898.

Leuckart, R. (1876) *Die menschlichen Parasiten und die von ihnen herrührenden Krankheiten*, Vol. 2. Leipzig, pp. 513–882.

Maupas, E.F. (1899) Le mue et l'enkystement chez les nématodes. *Archives de Zoologie Expérimentale et Générale* 7, 563–628.

Maupas, E.F. (1900) Modes et formes de reproduction des nématodes. *Archives de Zoologie Expérimentale et Générale* 8, 463–624.

Michel, J.F. (1974) Arrested development of nematodes and some related phenomena. *Advances in Parasitology* 12, 279–366.

Moravec, F. (1978) Occurrence of the endoparasitic helminths in a pike (*Esox lucius*) from the Mácha Lake fishpond system. *Věstnik, Československé Společnosti Zoologické* 43, 174–193.

Moravec, F. (1994) *Parasitic Nematodes of Freshwater Fish in Europe*. Academia, Prague, and Kluwer, Dordrecht, The Netherlands. 473 pp.

Moravec, F. (1998) *Nematodes of Freshwater Fishes of the Neotropical Region*. Academia, Prague.

Odening, K. (1976) Concepts and terminology of hosts in parasitology. *Advances in Parasitology* 14, 1–93.

Osche, G. (1954) Über Verhalten und Morphologie der Dauerlarve freilebender Nematoden. *Zoologischer Anzeiger* 152, 65–73.

Osche, G. (1963) Morphological, biological and ecological considerations in the phylogeny of parasitic nematodes. In: Dougherty, E.C. (ed.) *The Lower Metazoa, Comparative Biology and Phylogeny*. University of California Press, Berkeley, and Cambridge University Press, London, pp. 283–302.

Petter, A.J. (1971) Redescription d'*Hedruris androphora* Nitzsch, 1821 (Nematoda, Hedruridae) et étude de son developpement chez l'hôte intermediare. *Annales de Parasitologie Humaine et Comparée* 46, 479–495.

Rajan, T.V. (1998) A hypothesis for the tissue specificity of nematode parasites. *Experimental Parasitology* 89, 140–142.

Read, A.F. and Skorping, A. (1995) The evolution of tissue migration by parasitic nematode larvae. *Parasitology* 111, 359–371.

Roush, W. (1997) Worm longevity gene cloned. *Science* 277, 897–898.

Sharpilo, V.P., Tkach, V.V. and Lisitsyna, O.I. (1995) Paratenic parasitism and 'trap hosts'. In: *Proceedings of the Jubilee Conference of the Ukrainian Society of Parasitologists,* Kiev, May 16–17, 1995, pp. 111–118. (In Russian.)

Siddiqi, M.R. (1985) *Tylenchida, Parasites of Plants and Insects.* Commonwealth Institute of Parasitology of the Commonwealth Agricultural Bureaux, Farnham Royal, UK, 644 pp.

Steiner, G. (1917) Ueber die Verwandtschafts verhältnisse und die systematische Stellung der Mermithiden. *Zoologischer Anzeiger* 48, 263–267.

Sukhdeo, S.C., Sukhdeo, V.K., Black, M.B. and Vrijènhoek R.C. (1997) The evolution of tissue migration in parasitic nematodes (Nematoda: Strongylida) inferred from a protein-coding mitochondrial gene. *Biological Journal of the Linnean Society* 61, 281–298.

Swanson, M.M. and Riddle, D.L. (1981) Critical periods in the development of the *Caenorhabditis elegans* dauer larva. *Developmental Biology* 84, 27–40.

Vlanfleteren, J.R., Van de Peey, Y., Blaxter, M.L., Tweedi, A.R., Trotman, C., Liu, L., Van Hauwaert, M.L. and Moens, L. (1994) Molecular genealogy of some nematode taxa as based on cytochrome C and globin amino acid sequences. *Molecular Phylogenetics and Evolution* 3, 92–101.

Wong, P.L., Anderson, R.C. and Bartlett, C.M. (1989) Development of *Skrjabinoclava inornatae* (Nematoda: Acuarioidea) in fiddler crabs (*Uca* spp.) (Crustacea) and Western Willets (*Catoptrophorus semipalmatus inornatus*) (Aves: Scolopacidae). *Canadian Journal of Zoology* 67, 2893–2901.

Part I

Class Secernentea

Chapter 2
Order Rhabditida
2.1
The Superfamily Rhabditoidea

There is only one superfamily associated with vertebrates. This superfamily is divided into several families (Anderson and Bain, 1982). The transmission of members of the families Rhabditidae, Rhabdiasidae and Strongyloididae have been studied. Most members of the Rhabditidae occur in soil where they feed on bacteria. Among this group **phoresy** is commonly used to aid in dispersal to new habitats suitable for development. Dauer larvae attach to invertebrates, especially insects, which carry them to fresh environments where they can initiate new colonies. In some species dauer larvae invade the tissues of soft-bodied invertebrates such as earthworms. The latter can, when they die and decompose, form the substrate for colonies of nematodes initiated by dauer larvae which have persisted in the tissues of the living host. For a review of such associations see Osche (1963) and Poinar (1983). Recent studies have revealed a new form of phoresy in the genus *Rhabditis* (*Pelodera*). A special third-stage larva, distinct from the dauer larva, is formed which invades the orbit and skin of rodents. The larva gains nourishment from the rodent but without growing or developing significantly. Larvae eventually leave the rodent and establish colonies in suitable places in the host's environment. The rodent acts, therefore, as a dispersal agent.

Members of the Rhabdiasidae and Strongyloididae have both parasitic and free-living stages. The parasitic form is highly modified and the affinities of these forms with the rhabditoids were not appreciated until free-living generations were discovered and studied. It is in the free-living stages that these unusual parasites revert to the primitive condition and reveal their taxonomic affinities to the rhabditoids.

Family Rhabditidae

Rhabditis (*Pelodera*)

Adults and larvae of members of the subgenus *Pelodera* of the genus *Rhabditis* (some authors regard *Pelodera* as a genus) have been reported in rotting vegetation in saltwater habitats (*R. commandorica*), wet soil in freshwater habitats (*R. punctata*), manure

(*R. strongyloides*) and rodents' nests (*R. cutanea, R. nidicolis, R. orbitalis*) (see Sudhaus *et al.*, 1987). The latter four species have unusual associations with mammals. *R. strongyloides* will multiply in hair soiled with manure on the body of cattle and dogs and one strain can result in a dermatitis superficially resembling mange (Sudhaus and Schulte, 1988). *R. strongyloides* is, however, a common free-living species found in soil enriched with manure and its association with the bodies of animals is fortuitous. *R. nidicolis* has been found in the nests of field voles (*Microtus agrestis*) but its relationship to the vole is still unknown (Sudhaus and Schulte, 1986).

R. orbitalis Sudhaus and Schulte, 1986
R. cutanea Sudhaus, Schulte and Homminick, 1987

The biology of *R. orbitalis* and *R. cutanea* is intimately related to that of the rodents with which they are associated (Sudhaus *et al.*, 1987; Schulte, 1989). Third-stage larvae of *R. orbitalis* have been reported frequently in the conjunctival sac of mice, voles and lemmings (Muridae and Arvicolidae) (Rausch, 1952; Stammer, 1956; Dollfus *et al.*, 1961; Osche, 1963; Poinar, 1965; Kinsella, 1967; Cross and Santana, 1974; Prokopic *et al.*, 1974; Cliff *et al.*, 1978; Cliff and Anderson, 1980; Schulte, 1989; Casanova *et al.*, 1996; Sainsbury *et al.*, 1996). Third-stage larvae of *R. cutanea* were found coiled in the hair follicles of wood mice (*Apodemus* spp.) (Hominick and Aston, 1981; Sudhaus *et al.*, 1987). Larvae of both species developed in culture or in moist enriched conditions in nests and runways of their rodent hosts. During this development there were the usual larval stages resulting in third-stage larvae which rapidly transformed into fourth-stage larvae and breeding fifth-stage adults (Figs 2.1 and 2.2). However, some third-stage larvae in the colony arrested soon after they were formed, giving rise to dauer larvae (Fig. 2.3E) which retained the shed cuticle of the second stage. Dauer larvae are adapted to survive conditions of food deprivation and to some degree, lack of moisture. When conditions suitable for development are attained, dauer larvae rapidly developed to fourth-stage larvae and reproducing adults.

The peculiar feature of *R. orbitalis* and *R. cutanea* is that, in addition to the third-stage larva (Fig. 2.3F) which proceeds directly to the fourth stage and a dauer larva (Fig. 2.3E) – both typical of rhabditids in general – they produced another kind of third stage (Fig. 2.3D) known as the infective larva (Schulte, 1988). This larva was longer and more slender than the dauer larva, highly active, and sensitive to heat and vibrations in the environment. Unlike most free-living nematodes these larvae did not require films of moisture in which to move. On a dry surface, they attached the head end to the substrate and pulled the rest of the body forward, sometimes even somersaulting. They attached the tail to the substrate and waved back and forth or even propelled themselves from the substrate. It is these unusual larvae which occurred in the orbits or hair follicles of rodents with which they are associated (Sudhaus *et al.*, 1987; Schulte, 1989). The existence of three types of third-stage larvae (the third which develops to the fourth stage; the dauer larvae; and the parasitic larvae) in the life cycle of *R. orbitalis* is referred to as **larval triphenism**.

According to Schulte (1989), parasitic larvae of *R. orbitalis* obtained nourishment by endosmosis from the lachrymal secretions and increased in size. The intestinal cells became densely packed with lipid droplets and the phasmids enlarged, perhaps to excrete salts acquired from orbital fluid. It is assumed that, after obtaining nutrients

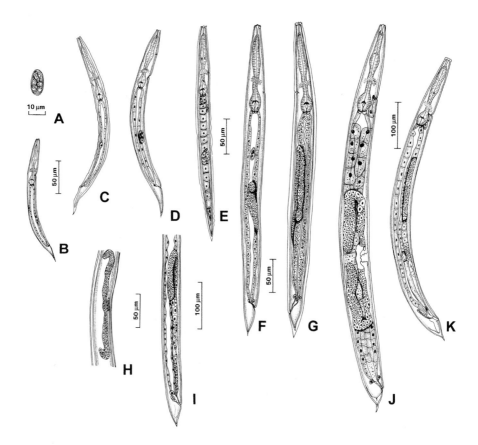

Fig. 2.1. Developmental stages of *Rhabditis* (*Pelodera*) *orbitalis*: (A) egg; (B) first-stage larva; (C) moulting first-stage larva; (D) second-stage larva; (E) moulting second-stage larva; (F) developing third-stage larva (male); (G) moulting third-stage larva (male); (H) genital primordium of moulting female third-stage larva; (I) caudal extremity of male fourth-stage larva; (J) female fourth-stage larva; (K) moulting male fourth-stage larva. (After G.M. Cliff and R.C. Anderson, 1980 – courtesy *Canadian Journal of Zoology*.)

and growing somewhat, larvae leave the orbit and enter the nest or runways of the host, where (if conditions are suitable) they can establish free-living generations. The behaviour of the infective third-stage larvae of *R. cutanea* seems to be similar to that of *R. orbitalis* (see Sudhaus *et al.*, 1987) although this species has not been investigated as thoroughly as the latter species. In superficially invading the host, the third-stage larvae of *R. orbitalis* and *R. cutanea* find a protected environment with food which they can store and use for subsequent development. For these nematodes the rodent acts as an agent of dispersal in situations where places for free-living development may be sparse. At the same time the association with the rodent is essentially harmless.

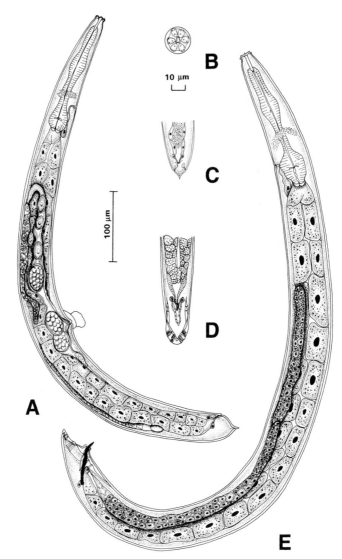

Fig. 2.2.   Adult *Rhabditis (Pelodera) orbitalis*: (A) female; (B) cephalic extremity; (C) caudal extremity of female, ventral view; (D) caudal extremity of male, ventral view; (E) male. (After G.M. Cliff and R.C. Anderson, 1980 – courtesy *Canadian Journal of Zoology*.)

Schulte and Poinar (1991) crossed Californian and European populations of *R. orbitalis* and concluded that the species was holarctic. They suggested that the association of the nematodes with rodents evolved among colonies of lemmings.

Hass (1990) observed that insemination was more efficient in *R. strongyloides*, *R. punctata* and *R. nidicolis* than in *R. orbitalis* and *R. cutanea*. Sperm loss during coitus occurred in all species and in some cases could be attributed to changes in the position of the spicules while inserted in the vagina (*R. cutanea*, *R. strongyloides*). The number of

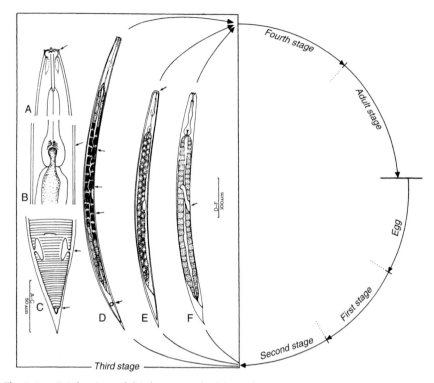

Fig. 2.3. Triphenism of third stage in the life cycle of *Rhabditis orbitalis*. Infective larva: (A) anterior region, ventral view; (B) pharyngeal area, ventral view; (C) caudal region with enlarged phasmids and a mucron on tail tip; (D) infective larva *in toto*, lateral view; (E) dauer larva *in toto*, lateral view; (F) usual third-stage larva *in toto*, lateral view. (After F. Schulte, 1989 – courtesy *Proceedings of the Helminthological Society of Washington*.)

sperm emitted by males greatly exceeded the number of eggs subsequently laid by females.

Family Rhabdiasidae

Parasitic adults of members of the Rhabdiasidae are common lung parasites of amphibians and reptiles throughout the world. The biology of some species of *Rhabdias* has been studied in some detail; the genus contains about 42 nominal species (Baker, 1987). A few observations have been made of the development of some species of the related genera *Entomelas* and *Pneumonema* of reptiles.

Rhabdias

Metchnikoff (1865, 1866) and Leuckart (1865a,b) discovered that eggs of *Rhabdias bufonis*, passed in faeces of the toad host, hatched and larvae produced a free-living generation of adult worms with morphological characters of free-living Rhabditidae. These authors believed that the parasitic adult was parthenogenetic but later studies have shown that the parasite is a **protandrous hermaphrodite** (Boviri, 1911; Schleip, 1911; Goodey, 1924a,b; Schaake, 1931; Dreyfus, 1937a,b; Whicker and Lanter, 1968; Baker, 1979a,b). Eggs (Fig. 2.4A) deposited in lungs by the hermaphrodite pass up the respiratory system, are swallowed and passed in faeces of the host. First-stage larvae (Fig. 2.4C) which hatch from eggs are **rhabditiform**. In some species (e.g. *R. fuscovenosa* of snakes), first-stage larvae generally develop directly into second- and then infective third-stage larvae, a process known as **homogonic** development (Railliet, 1899; Goodey, 1924a; Travassos, 1926; Chu, 1936; Williams, 1960; Baker, 1979a; Kuzmin, 1996; Spieler and Schierenberg, 1995). Other species (e.g. *R. bufonis*, *R. americanus*, *R. brachylaimus*, *R. multiproles*, *R. ranae*) have heterogonic free-living stages which result in reproducing adults that eventually lead to infective larvae (Chu, 1936; Yuen, 1965; Kloss, 1971; Baker, 1979a). Both **homogonic** and **heterogonic** development may occur in a single species although one may predominate (Chu, 1936; Williams, 1960). In heterogonic development, growth to the adult stage in the external environment is extremely rapid. Baker (1979a) noted that in *R. americanus* of toads and *R. ranae* of frogs, mature adults appeared in cultures within 27 h at room temperature (~22°C). Adults are typically rhabdititoid in morphology. The female is amphidelphic and produces a few large eggs which develop to ensheathed infective third-stage larvae (Fig. 2.4B). The female then becomes senescent and the larvae consume her internal organs and eventually break free of the maternal cuticle. This process, known as **matricidal endotoky**, is common among free-living rhabditids. Most infective larvae appear in cultures about 6 days after eggs are passed from the host.

Infective larvae of *R. americanus* and *R. ranae*, and presumably other species in amphibians, infect the host by skin penetration. Species in reptiles, however, probably infect the host orally (Maupas in Seurat, 1916; Chu, 1936; Chabaud *et al.*, 1961; Baker, 1979a) although this has not been firmly established. Larvae which penetrate the skin of amphibians lose their sheaths during the process (Fülleborn, 1928; Schaake, 1931; Williams, 1960; Baker, 1979a). Larvae then migrate via fascia and, over a period of several days, to the body cavity of the host (Baker, 1979a), where they grow to late third and fourth stages (Fig. 2.4D,E) and to the subadult stage of the parasitic hermaphrodite. These subadults must invade the lungs if they are to mature and produce eggs (Baker, 1979a). It is common to find mature hermaphrodites in lungs of wild-caught frogs and toads as well as numerous subadults in the body cavity. Massive numbers of worms generally are not found in the lungs of wild-caught amphibians. This raises the possibility that there may be some regulation of the number of nematodes that invade the lungs at any one time from a pool of subadults remaining in the body cavity. In Baker's (1979a) experiments, which lasted from 25 days (*R. ranae* in *Rana sylvatica*) to 60 days (*R. americanus* in *Bufo americanus*), more subadult worms always occurred in the body cavity of the host than mature adults in the lungs. *R. americanus* first appeared in lungs 18 days after toads were infected but gravid worms were found only after 30 days. *R. ranae* in frogs developed more rapidly and gravid hermaphrodites were found in lungs as early as 9 days postinfection.

Tadpoles of toads and frogs were infected experimentally with *R. americanus* and *R. ranae*, respectively (Baker, 1979a). Thus, the absence of *Rhabdias* in wild tadpoles has

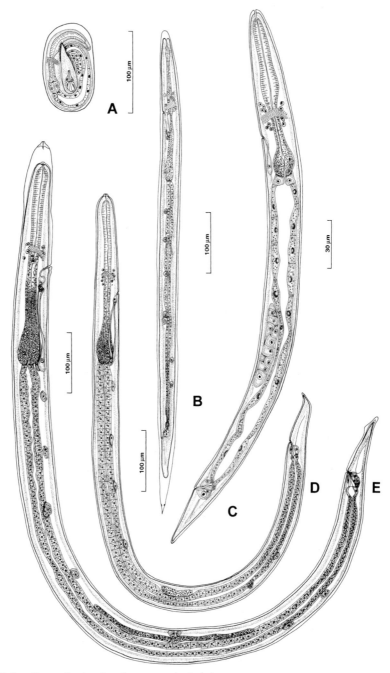

Fig. 2.4. Stages in the development of *Rhabdias americanus*: (A) egg; (B) ensheathed third-stage infective larva, lateral view; (C) first-stage larva, lateral view; (D) parasitic third-stage larva 4 days postinfection, lateral view; (E) parasitic fourth-stage larva 6 days postinfection, lateral view. (After M.R. Baker, 1979 – courtesy *Canadian Journal of Zoology*.)

an ecological rather than a physiological basis. Baker (1979a) was also able to transmit the frog species to toads and the toad species to frogs.

The possible role of paratenic hosts in transmission of species of *Rhabdias* has not been fully elucidated but infective larvae of *R. bufonis* and *R. americanus* will invade and survive in tissues of snails (Fülleborn, 1928; Baker, 1979a). Chu (1936) found that infective larvae of *R. fuscovenosa* from snakes failed to penetrate snails, tadpoles, frogs and toads.

There have been few epizootiological studies of species of *Rhabdias*. In England the lowest prevalences and intensities of *Rhabdias* spp. in various amphibians were observed in summer (Lees, 1962). Baker (1979a) carried out field studies of two species in southern Ontario, Canada. Prevalence and intensity of *R. ranae* in *Rana sylvatica* were lowest in summer and highest in spring and early autumn. Many subadult worms were found in the body cavity of frogs in late summer and early autumn. In contrast, none occurred in the body cavity in spring and few in late autumn. Baker concluded that there was only one generation, and possibly two, of the parasite annually and that transmission occurred mainly in spring and autumn. Young-of-the-year frogs acquired infections the same summer as they transformed and the parasite survived winter as adults in the frogs.

For information on the transmission and development of rhabdiasid nematodes in palearctic hosts see Kuzmin (1996, 1997).

The authorities and hosts of the species of *Rhabdias* mentioned above are as follows (summarized from Baker, 1987).

- *R. americanus* Baker, 1978
  Host: *Bufo americanus*
- *R. brachylaimus* (Linstow, 1903)
  Host: *Bufo melanostictus*
- *R. bufonis* (Schrank, 1788)
  Hosts: *Bombina bombina*, *Bufo* spp., *Hyla arborea*, *Pelobates fuscus*, *Rana* spp.
- *R. fuelleborni* Travassos, 1921
  Hosts: *Bufo marinus*, *Bufo* spp., *Thoropa miliaris*
- *R. fuscovenosa* (Raillet, 1895)
  Hosts: *Natrix tigrina*, and numerous other snakes
- *R. gemellipara* Chabaud, Brygoo and Petter, 1961
  Hosts: *Brookesia ebenaui*, *Camaeleo* spp., *Zonosaurus madagascariensis*
- *R. multiproles* Yuen, 1965
  Host: *Rana* spp.
- *R. ranae* Walton, 1929
  Hosts: *Acris gryllus*, *Hyla* spp., *Pseudacris triseriata*, *Rana* spp.

Entomelas

E. entomelas (Dujardin, 1845)

E. entomelas is commonly found in the blindworm (*Anguis fragilis*) in Europe. Seurat (1920), Travassos (1930) and Moravec (1974) showed that development of this species is heterogonic. Moravec (1974) described the free-living males and females. He assumed that they would eventually give rise to infective larvae which, he felt, probably invaded snails and earthworms which serve as paratenic hosts, as suggested earlier by Seurat (1920).

Pneumonema

P. tiliquae Johnston, 1916

Ballantyne and Pearson (1963) and Ballantyne (1991a) investigated *P. tiliquae* of the blue-tongued lizard (*Tiliqua scincoides*) of Australia. They showed that the parasite was a protandrous hermaphrodite which passed eggs containing a fully developed larva. In the external environment, development was heterogonic and the infective larvae exhibited matricidal endotoky. Ballantyne and Pearson (1963) believed that the presence of snails and slugs stimulated exsheathing of infective larvae. Larvae apparently entered snails and travelled to the pericardium. It was possible to infect lizards by feeding them snails and slugs containing larvae or by oral inoculation of infective larvae. There was no evidence that larvae would invade the skin of the lizard host. In the lizard, larvae fed and grew in the small intestine before penetrating the gut and invading the body cavity where two moults occurred. Subadults then moved into the lungs and matured. Development of the gonad ceased if adult worms remained in the body cavity (Ballantyne, 1991b).

Family Strongyloididae

Strongyloides

Members of the genus are extremely widespread parasites of the gut mucosa of tetrapods throughout the world; over 40 species are described. One prominent species in humans and dogs, *Strongyloides stercoralis*, and several in domesticated animals have been the subject of considerable research by numerous parasitologists. For a review of work prior to about 1940 the reader is referred to Chandler *et al.* (1950), who called attention to the early contributions of Grassi (1878), Leuckart (1883), van Durme (1902), Leichtenstern (1905), Looss (1905), Fülleborn (1914), Sandground (1926), Nishigori (1928), Kreis (1932), Faust (1933), Lucker (1934, 1942), Graham (1936, 1938, 1939a,b, 1940), Beach (1935, 1936) and Chitwood and Graham (1940). For an account of strongyloidiasis in humans, refer to Grove (1989, 1996).

The development and transmission of the various species of *Strongyloides* are similar and will be treated collectively. The parasitic form of *Strongyloides* is a delicate, thread-like female which produces eggs lacking vitelline membranes. Eggs are generally passed in faeces of the host although in some species first-stage larvae may be passed. In the external environment, they complete their development to rhabditiform first-stage larvae. These larvae either develop into infective third-stage larvae, sometimes referred to as **filariform** or **strongyliform**, which invade the host, or develop into a single free-living generation consisting of males and females. The progeny of this generation give rise to infective larvae. Numerous authors have reported this pattern and described in detail the various developmental stages (e.g. Griffiths, 1940a,b; Basir, 1950; Reesal, 1950, 1951a,b; Premvati, 1958a–c; Little, 1962, 1966a,b; Abadie, 1963; Wertheim and

Lengy, 1965; Miyamoto, 1970; Wertheim, 1970; Lyons *et al.*, 1973; Mirck, 1975a,b: Bartlett, 1995). In some species that are usually heterogonic, strains may exist that are entirely homogonic.

In recent years we have arrived at a better understanding of the events which take place in the life cycle of species of *Strongyloides* and the following summary probably applies to many of the species reported in the genus (see Georgi, 1982).

The parasitic form of species of *Strongyloides* produces only genotypically female eggs by mitotic (apomictic) parthenogenesis (Triantaphyllou and Moncol, 1977; Moncol and Triantaphyllou, 1978; Georgi, 1982; Viney, 1994). Eggs embryonate in the intestine and hatch into rhabditiform first-stage larvae, which typically pass out in the faeces and undergo the free-living phase of development. In the external environment rhabditiform larvae either develop into female infective larvae (the homogonic pathway) or grow and moult four times to produce a single free-living generation consisting of both males and females (the heterogonic pathway). Recent research indicates that two classes of amphidial neurons (ASF and ASI) acting together control the direction of the free-living pathway (Ashton *et al.*, 1998). Females resulting from heterogonic development produce eggs by meiotic parthenogenesis but development of the eggs must be initiated by the penetration of the sperm, although the sperm and egg pronuclei do not fuse (but see Viney *et al.*, 1993; Viney, 1994). Apparently nuclei of the second maturation division of the egg recombine to give the diploid chromosome number. Larvae arising from these eggs develop only into infective larvae, which will invade and mature as females in the host. For a pictorial outline of the above sequence of events see Fig. 2.5, taken from Georgi (1982). It is not known what stimuli acting during oogenesis or embryogenesis determine whether larvae will develop into males or females during heterogonic development, although intra-host factors have been cited as possibilities, e.g. duration and intensity of infection, species, age and immunological status. Such factors may modify gene activity or sex-hormone balance in the embryo. It is also not known what factors determine whether larvae will develop into free-living rhabditiform females or into strongyloid infective larvae but it is believed the relative proportions of the two may depend on environmental conditions such as pH, pO_2, pCO_3, consistency of substrate, temperature and level of nutrition.

It has been widely accepted for many years that infective larvae of *Strongyloides* spp. invade the skin and enter blood or lymphatic vessels and are carried to the heart and then to the lungs. In lungs, larvae leave blood vessels, enter air spaces and then move up the trachea and are eventually swallowed by the host (Looss, 1905; Fülleborn and Schilling-Torgau, 1911; Fülleborn, 1914). In the intestine, worms grow rapidly and moult twice to attain the adult stage in about 2 weeks. Some of the evidence for lung migration in *Strongyloides ratti* was recently questioned by Wilson (1983) and Tindall and Wilson (1988, 1990a,b), and some authors (Tada *et al.*, 1979; Tanaka *et al.*, 1989; Bhopole *et al.*, 1992; McHugh *et al.*, 1994; Koga *et al.*, 1999) reported that infective larvae of *Strongyloides ratti* move in subcutaneous tissues to the head and from there to the mouth. Schad *et al.* (1989), after experiments with puppies, questioned whether the classical route was followed in *S. stercoralis*. Takamure (1995) reported that larvae of *S. venezuelensis* of rats migrated within subcutaneous tissues and muscles for 42 h postinfection and then accumulated in the lungs from 45 h postinfection; presumably larvae reached the lungs in blood. Larvae left the lungs, travelled up the trachea and were swallowed.

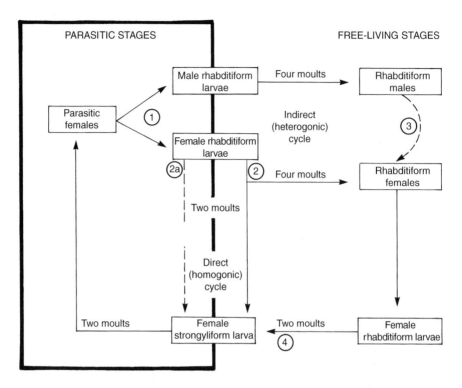

Fig. 2.5. The free-living and parasitic generations in the life cycle of *Strongyloides* spp. (1) Parasitic females produce only genotypically female eggs by mitotic partheno-genesis but both male and female rhabditiform larvae are produced. (2) Female rhabditiform larvae may develop into either free-living rhabditiform females or into strongyliform infective larvae. (3) Each egg of the free-living female must be penetrated by a sperm for development but in some species male and female pronuclei do not fuse. The females reproduce by meiotic parthenogenesis. (4) Rhabditiform larvae of the free-living generation develop only into strongyloid larvae. (Modified after J.R. Georgi, 1982 – courtesy CRC Press Inc.)

The prepatent period in infections with *Strongyloides* spp. seems to be variable among species and hosts. It has been reported to be as short as 3 days in *S. ransomi* (see Frickers, 1953) or as long as 17 days in *S. stercoralis*, although in the latter species it has been reported to vary between 8 and 17 days. Turner (1955) gives a prepatent period of 9 days in infections of *S. papillosus* and Griffiths (1940a,b) 7–10 days in *S. agoutii*. The peak of egg production seems generally to occur around the first 3–8 weeks of the patent period but low-grade infections may persist for months in individual animals.

Infections of *S. stercoralis* can persist in humans and canines through autoinfection, i.e. eggs hatch in the host and the rhabditiform larvae develop in the gut to the infective stage (Faust and De Groat, 1940; Faust, 1949). The latter then invade the intestinal mucosa or perianal skin, and undertake a lung migration before establishing themselves in the intestine, where they grow into the usual parthenogenetic females. Autoinfection accounts for the persistence of strongyloidiasis in the absence of opportunities for reinfection in the usual way. In the immunocompromised host, autoinfection can result in fatal disseminating strongyloidiasis (Purtilo *et al.*, 1974; Scowden *et al.*, 1978; Schad *et al.*, 1984; Genta, 1986; Davidson, 1992; Purvis *et al.*, 1992; Simpson *et al.*, 1993; Grove, 1996). Schad *et al.* (1993) proposed that in well established (both active and chronic) infections of *S. stercoralis* in puppies the rate of larval development is 'down-regulated' and most larvae do not attain infectivity internally but pass out as slender rhabditiform larvae. A few, however, which are wider and shorter, autoinfect. In the immunologically naïve host, development proceeds unhindered and many larvae become autoinfective; this results in a brief period of hyperinfection. As resistance develops, the proportion of larvae passed from the host as preinfective rhabditoid larvae increases and the adult population in the gut decreases.

Mansfield *et al.* (1996) and Schad *et al.* (1997) demonstrated that barren female *S. stercoralis* persisting in a chronically infected host can under certain conditions return to fecundity and their progeny can quickly initiate autoinfection.

Heavy infections of *Strongyloides papillosus* have been associated with sudden death (by ventricular fibrillation – cardiac arrest) in calves and lambs (Tsuji *et al.*, 1992; Taira, 1992), apparently related to the presence of live parasitic females in the small intestine (Nakamura *et al.*, 1994a). *S. papillosa* will produce a wasting condition in rabbits but cardiac functions are not affected (Nakamura *et al.*, 1994b).

Transplacental transmission has been proposed by some authors to account for early infection in young newborn animals (Enigk, 1952; Stone, 1964). In addition, the infective larvae of some species (e.g. *S. westeri*, *S. papillosus*, *S. ransomi*, *S. ratti*, *S. fulleborni*, *S. venezuelensis*) can remain in the superficial tissues of the resistant host and, in the case of the lactating host, be passed in the milk to the suckling young (Moncol and Batte, 1966; Supperer and Pfeiffer, 1967; Katz, 1969; Stewart *et al.*, 1969; Lyons *et al.*, 1969, 1973; Moncol and Grice, 1974; Zamirden and Wilson, 1974; Moncol, 1975; Wilson *et al.*, 1976; Nolan and Katz, 1981). In *S. westeri* of horses, the parturient mare does not shed eggs of the nematode but readily passes larvae to the foal, which will pass eggs 10–14 days later; for a review of the discovery see Lyons (1994). Larvae invading the mare probably continue to accumulate in her tissues (Georgi, 1982). There is no evidence that *S. stercoralis* can be transmitted lactationally (Mansfield and Schad, 1995).

The authorities and hosts of the species of *Strongyloides* mentioned above are as follows:

- *S. agouti* Griffiths, 1940
  Host: agouti (*Dasyprocta agouti*)
- *S. fulleborni* Linstow, 1905
  Hosts: non-human primates (e.g. chimpanzee, baboon)
- *S. papillosus* (Wedl, 1856)
  Hosts: sheep, goats, cattle, zebra, camels, rabbits, deer
- *S. procyonis* Little, 1966
  Host: raccoon (*Procyon lotor*)
- *S. ransomi* Schwartz and Alicata, 1930
  Host: swine
- *S. ratti* Sandground, 1925
  Host: rodent

- *S. robustus* Chandler, 1942
  Host: squirrel
- *S. stercoralis* (Bavay, 1876)
  Hosts: humans, dogs, cats, non-human primates
- *S. westeri* Ihle, 1917
  Host: equine
- *S. venezuelensis* Brumpt, 1934
  Host: Norway rat.

Parastrongyloides

P. trichosuri Mackerras, 1959
P. peramelis Mackerras, 1959
Mackerras (1959) studied development of *P. trichosuri* and *P. paremalis* of the small intestine of marsupials in Australia. The parasitic stages include both males and females. Development of larvae was apparently both homogonic and heterogonic and similar to that of members of the genus *Strongyloides*.

Family Cylindrocorporidae

Longibucca

L. lasiura McIntosh and Chitwood, 1934
L. lasiura (syn. *L. eptesica*) is a parasite of the stomach of bats (*Pipistrellus subflavus*, *Myotis lucifugus*, *M. ciliolabrum*, *Eptesicus fuscus* and *Lasionycteris noctivagans*) in North America (Elsea, 1953; Measures, 1994a). Members of the genus are ovoviviparous and it is possible that the worms are autoinfective, since various larval stages as well as adults are sometimes found in the bats. However, additional research is needed (Measures, 1994b).

References (Rhabditida)

Abadie, S.H. (1963) The life cycle of *Strongyloides ratti*. *Journal of Parasitology* 49, 241–248.

Anderson, R.C. and Bain, O. (1982) No. 9. Keys to genera of the Superfamily Rhabditoidea, Dioctophymatoidea, Trichinelloidea and Muspiceoidea. In: Anderson, R.C., Chabaud, A.G. and Willmott, S. (eds) *CIH Keys to the Nematode Parasites of Vertebrates*. Commonwealth Agricultural Bureaux, Farnham Royal, UK, pp. 1–26.

Ashton, F.T., Bhopale, V.M., Holt, D., Smith, G. and Schad, G.A. (1998) Developmental switching in the parasitic nematode *Strongyloides stercoralis* is controlled by the ASF and ASI amphidial neurons. *Journal of Parasitology* 84, 691–695.

Baker, M.R. (1979a) The free-living and parasitic development of *Rhabdias* spp. (Nematoda: Rhabdiasidae) in amphibians. *Canadian Journal of Zoology* 57, 161–178.

Baker, M.R. (1979b) Seasonal population changes in *Rhabdias ranae* Walton, 1929 (Nematoda: Rhabdiasidae) in *Rana sylvatica* of Ontario. *Canadian Journal of Zoology* 57, 179–183.

Baker, M.R. (1987) *Synopsis of the Nematoda Parasitic in Amphibians and Reptiles*. Memorial University of Newfoundland Occasional Papers in Biology, No. 11. 325 pp.

Ballantyne, R.J. (1991a) Life history and development of *Pneumonema tiliquae* (Nematoda: Rhabdiasidae). *International Journal for Parasitology* 21, 521–533.

Ballantyne, R.J. (1991b) Post-embryonic development of the reproductive system of *Pneumonema tiliquae* (Nematoda: Rhabdiasidae). *International Journal for Parasitology* 21, 535–547.

Ballantyne, R.J. and Pearson, J.C. (1963) The taxonomic position of the nematode *Pneumonema tiliquae*. *Australian Journal of Science* 25, 498.

Bartlett, C. (1995) Morphology, homogonic development, and lack of a free-living generation in *Strongyloides robustus* (Nematoda, Rhabditoidea), a parasite of North American sciurids. *Folia Parasitologica* 42, 102–114.

Basir, M.A. (1950) The morphology and development of the sheep nematode, *Strongyloides papillosus* (Wedl, 1856). *Canadian Journal of Research*, D 28, 173–196.

Beach, T.D. (1935) The experimental propagation of *Strongyloides* in culture. *Proceedings of the Society for Experimental Biology and Medicine* 32, 1484–1486.

Beach, T.D. (1936) Experimental studies on human and primate species of *Strongyloides*. V. The free-living phase of the life cycle. *American Journal of Hygiene* 23, 243–277.

Bhopale, V.M., Smith, G., Jordan, H.E. and Schad, G.A. (1992) Development biology and migration of *Strongyloides ratti* in the rat. *Journal of Parasitology* 78, 861–868.

Boviri, T. (1911) Ueber das Verhalten der Geschlechtschromosomen bei Hermaphroditismus. Beobachtungen an *Rhabditis nigrovenosa*. *Verhandlungen der Physikalisch-Medizinischen Gesellshaft in Würzburg* 41, 83–97.

Casanova, J.C., Arrizabalaga, A., Spakulova, M. and Morand, S. (1996) The first record of *Rhabdias (Pelodera) orbitalis* (Nematoda: Rhabditidae), a larval parasite in the eyes of the rodent *Chionomys nivalis* on the Iberian Peninsula. *Helminthologica* 33, 227–229.

Chabaud, A.G., Brygoo, E.R. and Petter, A.J. (1961) Description et caractères biologiques de deux nouveaux *Rhabdias* Malgaches. *Annales de Parasitologie Humaine et Comparée* 36, 752–763.

Chandler, A.C., Alicata, J.E., Chitwood, B.G. and Chitwood, M.B. (1950) Life history (Zooparasitica) II. Parasites of vertebrates. In: Christie, B.G. (ed.) *Nematology. Section II.* Chitwood Publisher, Monumental Printing Company, Baltimore, Maryland, pp. 267–301.

Chitwood, B.G. and Graham, G.L. (1940) Absence of vitelline membranes on developing eggs in parasitic females of *Strongyloides ratti*. *Journal of Parasitology* 26, 183–190.

Chu, T. (1936) Studies on the life history of *Rhabdias fuscovenosa* var. *catanensis* (Rizzo, 1902). *Journal of Parasitology* 22, 140–160.

Cliff, G.M. and Anderson, R.C. (1980) Development of *Pelodera strongyloides* (Schneider, 1860) Schneider, 1866 (Nematoda: Rhabditidae) in culture. *Journal of Helminthology* 54, 135–146.

Cliff, G.M., Anderson, R.C. and Mallory, F.F. (1978) Dauerlarvae of *Pelodera strongyloides* (Schneider, 1860) (Nematoda: Rhabditidae) in the conjunctival sacs of lemmings. *Canadian Journal of Zoology* 56, 2117–2121.

Cross, J.H. and Santana, F.J. (1974) *Pelodera strongyloides* (Schneider) in the orbit of Taiwan rodents. *Chinese Journal of Microbiology* 7, 137–138.

Davidson, R.A. (1992) Infection due to *Strongyloides stercoralis* in patients with pulmonary disease. *Southern Medical Journal* 85, 28–31.

Dollfus, R.F., Desportes, C., Chabaud, A.G. and Campana-Rouget, Y. (1961) Station expérimentale de parasitologie de Richelieu (Intre-et-Loire). Contributions de la faune parasitaire régionale. Chapitre II. Liste des parasites par ordre systématique. E. Nématodes. *Annales de Parasitologie Humaine et Comparée* 36, 303–311.

Dreyfus, A. (1937a) Sobre o mechanismo de formação dos espermatozoides nas zonas testiculares da forma parasita de *Rhabdias fulleborni* Trav. *Revista de Biologia e Hygiene, Sao Paulo* 8, 5–9.

Dreyfus, A. (1937b) Hermaphroditismo alternante proterogynice em *Rhabdias fulleborni* Trav. *Memoires de l'Institut Butantan* 11, 289–297.

Durme, P. van (1902) Quelques notes sur les embryons de *Strongyloides intestinalis* et leur pénétration par la peau. *Thompson Yates Laboratories Report* 4, 471–474.

Elsea, J.R. (1953) Observations on the morphology and biology of *Longibucca eptesica* n.sp. (Nematoda: Cylindrocorporidae), parasitic in the bat. *Proceedings of the Helminthological Society of Washington* 20, 65–76.

Enigk, K. (1952) Pathogenität und Therapie des Strongyloidenbefalles der Haustiere. *Monatshefte für Praktische Tierheilkunde* 4, 97–112.

Faust, E.C. (1933) Experimental studies on human and primate species of *Strongyloides*. II. The development of *Strongyloides* in the experimental host. *American Journal of Hygiene* 18, 114–132.

Faust, E.C. (1949) *Human Helminthology. A Manual for Physicians, Sanitarians and Medical Zoologists.* Lea and Febiger, Philadelphia.

Faust, E.C. and De Groat, A. (1940) Internal autoinfection in strongyloidiasis. *American Journal of Tropical Medicine* 20, 359–375.

Frickers, J. (1953) Strongyloidosis bij het varken. *Tijdschrift voor Diergeneeskunde* 78, 279–299.

Fülleborn, F. (1914) Untersuchungen über den Infektionsweg bei *Strongyloides* und *Ancylostomum* und die Biologie dieser Parasiten. *Archiv für Schiffs und Tropenhygiene* 18, 26–80.

Fülleborn, F. (1928) Ueber den Infektionsweg bei *Rhabdias bufonis* (*Rhabdonema nigrovenosum*) des Frosches nebst Versuchungen über die Lymphzirkulation des letzteren. *Zentralblatt für Bakteriologie und Parasitenkunde und Infektionskrankheiten. Abteilung I. Originale* 109, 444–462.

Fülleborn, F. and Schilling-Torgau, V. (1911) Untersuchungen über den Infektionsweg bei *Strongyloides* und *Ancylostomum*. Vorläufige Mitteilung. *Archiv für Schiffs und Tropenhygiene* 15, 569–571.

Genta, R.M. (1986) *Strongyloides stercoralis*: immunobiological considerations on an unusual worm. *Parasitology Today* 2, 241–246.

Georgi, J.R. (1982) Strongyloidiasis. In: *CRC Handbook Series on Zoonoses: Parasitic Zoonoses. Section C*, Vol. 2, pp. 257–267.

Goodey, T. (1924a) The anatomy and life history of the nematode *Rhabdias fuscovenosa* (Railliet) from the grass snake *Tropidonotus natrix*. *Journal of Helminthology* 2, 51–64.

Goodey, T. (1924b) Two new species of the nematode genus *Rhabdias*. *Journal of Helminthology* 2, 203–208.

Graham, G.L. (1936) Studies on *Strongyloides*. I. *S. ratti* in parasitic series, each generation in the rat established with a single homogonic larva. *American Journal of Hygiene* 24, 71–87.

Graham, G.L. (1938) Studies on *Strongyloides*. II. Homogonic and heterogenic progeny of the single, homogonically derived *S. ratti* parasite. *American Journal of Hygiene* 27, 221–234.

Graham, G.L. (1939a) Studies on *Strongyloides*. IV. Seasonal variation in the production of heterogenic progeny by singly established *S. ratti* from a homogonically derived line. *American Journal of Hygiene* 30, 15–27.

Graham, G.L. (1939b) Studies on *Strongyloides*. V. Constitutional differences between a homogonic and a heterogenic line of *S. ratti*. *Journal of Parasitology* 25, 365–375.

Graham, G.L. (1940) Studies on *Strongyloides* VII. Length of reproductive life in a homogonic line of singly established *S. ratti*. *Revista de Medicina Tropical Parasitologia Bacteriologia Clinica Laboratoria* 6, 89–103.

Grassi, G.B. (1878) *L'Anguillula intestinali*. Nota preventiva. *Gazzetta Medica Italiana Lombardia. Milano*, v. 38, 7.5 v. 5:471–474.

Griffiths, H.J. (1940a) Experimental studies on *Strongyloides agoutii* in the guinea pig. *Canadian Journal of Research* 18, 307–324.

Griffiths, H.J. (1940b) Studies on *Strongyloides agouti* sp.nov. from the agouti (*Dasyprocta agouti*). *Canadian Journal of Research* 18, 173–190.

Grove, D. (ed.) (1989) *Strongyloidiasis. A Major Roundworm Infection of Man.* Taylor and Francis, London.

Grove, D.J. (1996) Human strongyloidiasis. *Advances in Parasitology* 38, 251–309.

Hass, B. (1990) A quantitative study of insemination and gamete efficiency in different species of the *Rhabditis strongyloides* group (Nematoda). *Invertebrate Reproduction and Development* 18, 205–208.

Hominick, W.M. and Aston, A.J. (1981) Association between *Pelodera strongyloides* (Nematoda: Rhabditidae) and wood mice, *Apodemus sylvaticus. Parasitology* 83, 67–75.

Katz, F.F. (1969) *Strongyloides ratti* (Nematoda) in newborn offspring of inoculated rats. *Proceedings of the Pennsylvania Academy of Science* 43, 221–225.

Kinsella, J.M. (1967) Helminths of microtines in western Montana. *Canadian Journal of Zoology* 43, 269–274.

Kloss, G.R. (1971) Alguns *Rhabdias* (Nematoda) de *Bufo* no Brasil. *Papeis Avulsos do Departamento de Zoologia* 24, 1–52.

Koga, M., Nig, A. and Tada, L. (1999) *Strongyloides ratti*: migration study of third-stage larvae in rats by whole-body autoradiography after ^{35}S methionine labeling. *Journal of Parasitology* 85, 405–409.

Kreis, H.A. (1932) Studies on the genus *Strongyloides* (nematodes). *American Journal of Hygiene* 16, 450–491.

Kuzmin, Y.I. (1996) The free-living development and types of life cycles of rhabdiasid nematodes (Nematoda: Rhabdiasidae) from the palearctic region. *I.I. Schmalhausen Institute of Zoology* 16/17, 133–138. (In Ukrainian.)

Kuzmin, Y.I. (1997) The life cycle and new data on distribution of *Rhabdias sphaerocephala* (Nematoda, Rhabdiasidae). *Vestnik Zoologii* 31, 49–57. (In Ukrainian.)

Lees, E. (1962) The incidence of helminths parasites in a particular frog population. *Parasitology* 52, 95–102.

Leichtenstern, O. (1899) Zur Lebensgeschichte der *Anguillula intestinalis. Zentralblatt für Bakteriologie und Parasitenkunde und Infektionskrankheiten. Abteilung I. Originale* 25, 226–231.

Leichtenstern, O. (1905) Studien über *Strongyloides stercoralis* (Bavay) (*Anguillula intestinalis*), und *stercoralis* nebst Bemerkungen über *Ancylostomum duodenale. Arbeiten aus dem Kaiserlichen Gesundheitsamte* 22, 309–350.

Leuckart, K.G.F.R. (1865a) Zur Entwickelungsgeschichte des *Ascaris nigrovenosa.* Zugleich eine Erwiderung gegen Herrn Candidat Mecznikow. *Archiv für Anatomie, Physiologie und Wissenschaftliche Medizin. Berlin* pp. 641–658.

Leuckart, K.G.F.R. (1865b) Helminthologische Experimentaluntersuchungen. Vierte Reihe. *Nachrichten von der K. Gesellschaft der Wissenschaften und der Georg-Augusts Universität. Göttingen* 8, 219–232.

Leuckart, K.G.F.R. (1883) Ueber die Lebensgeschichte der sog. *Anguillula stercoralis* und deren Beziehungen zu der sog. *Ang. intestinalis.* Berichte über die Verhandlungen der Königlichen Sächsischen Gesellschaft der Wissenschaften zu Leipzig. *Mathematisch-Physische Klasse* 34, 85–107.

Little, M.D. (1962) Experimental studies on the life cycle of *Strongyloides. Journal of Parasitology* (Suppl.) 48, 41.

Little, M.D. (1966a) Comparative morphology of six species of *Strongyloides* (Nematoda) and redefinition of the genus. *Journal of Parasitology* 52, 69–84.

Little, M.D. (1966b) Seven new species of *Strongyloides* (Nematoda) from Louisiana. *Journal of Parasitology* 52, 85–97.

Looss, A. (1905) Die Wanderung der *Ancylostomum* und *Strongyloides* Larven von der Haut nach dem Darm. *Comptes Rendus 6th Congress International Zoology* (Berne, 14–19 Aout, 1904) pp. 225–233.

Lucker, J.T. (1934) Development of the swine nematode *Strongyloides ransomi* and the behavior of its infective larvae. *USA Department of Agriculture Technical Bulletin* No. 437, 1–30.

Lucker, J.T. (1942) The dog *Strongyloides* with special reference to occurrence and diagnosis of infections with the parasite. *Veterinary Medicine* 37, 128–137.

Lyons, E.T. (1994) Vertical transmission of nematodes: emphasis on *Uncinaria lucasi* in Northern fur seals and *Strongyloides westeri* in equids. *Journal of the Helminthological Society of Washington* 61, 169–178.

Lyons, E.T., Drudge, J.H. and Tolliver, S.C. (1969) Parasites from mare's milk. *Blood Horse* 95, 2270–2271.

Lyons, E.T., Drudge, J.H. and Tolliver, S.C. (1973) On the life cycle of *Strongyloides westeri* in the equine. *Journal of Parasitology* 59, 780–787.

Mackerras, M.J. (1959) *Strongyloides* and *Parastrongyloides* (Nematoda: Rhabdiasoidea) in Australian marsupials. *Australian Journal of Zoology* 7, 87–104.

Mansfield, L.S. and Schad, G.A. (1995) Lack of transmammary transmission of *Strongyloides stercoralis* from a previously hyperinfected bitch to her pups. *Journal of the Helminthological Society of Washington* 62, 80–83.

Mansfield, L.S., Niamatali, S., Bhopale, V., Volk, S., Smith, G., Lok, J.B., Genta, R.M. and Schad, G.A. (1996) *Strongyloides stercoralis*: maintenance of exceedingly chronic infections. *American Journal of Tropical Medicine and Hygiene* 55, 617–624.

McHugh, T.D., Jenkins, T., Greenwood, R. and McLaren, D.J. (1994) The migration and attrition of *Strongyloides ratti* in naive and sensitized rats. *Journal of Helminthology* 68, 143–148.

Measures, L.N. (1994a) Seasonal dynamics of the bat stomach worm, *Longibucca lasiura* (Nematoda: Rhabdtoidea) in Alberta. *Canadian Journal of Zoology* 72, 791–794.

Measures, L.N. (1994b) Synonomy of *Longibucca eptesica* with *Longibucca lasiura* (Nematoda: Rhabdtoidea) and new host and geographic records. *Journal of Parasitology* 80, 486–489.

Metchnikoff, I. (1865) Ueber die Entwicklung von *Ascaris nigrovenosa*. *Archiv für Anatomie, Physiologie und Wissenschaftliche Medizin, Leipzig*, pp. 409–420.

Metchnikoff, I. (1866) On the development of *Ascaris nigrovenosa*. *Quarterly Journal of the Microscopical Society of London* 6, 25–32.

Mirck, M. (1975a) On the life cycle of *Strongyloides westeri* Ihle, 1917 (Nematoda: Strongyloididae) in Shetland ponies in the Netherlands. Preliminary communication. *Tropical and Geographical Medicine* 27, 231.

Mirck, M. (1975b) On the life cycle of *Strongyloides westeri* Ihle, 1917 (Nematoda: Strongyloididae) in Shetland ponies in the Netherlands. Preliminary communication. *Tropical and Geographical Medicine* 27, 442.

Miyamoto, K. (1970) Experimental strongyloidiasis in the rat inoculated with *Strongyloides fuelleborni*. I. Migration of the larvae. *Japanese Journal of Parasitology* 19, 487–493.

Moncol, D.J. (1975) Supplement to the life history of *Strongyloides ransomi* Schwartz and Alicata, 1930 (Nematoda: Strongyloididae) of pigs. *Proceedings of the Helminthological Society of Washington* 42, 86–92.

Moncol, D.J. and Batte, E.G. (1966) Trans-colostral infection of newborn pigs with *Strongyloides ransomi*. *Veterinary Medicine/SAC* 61, 583–586.

Moncol, D.J. and Grice, M.J. (1974) Transmammary passage of *Strongyloides papillosus* in the goat and sheep. *Proceedings of the Helminthological Society of Washington* 41, 1–4.

Moncol, D.J. and Triantaphyllou, A.C. (1978) *Strongyloides ransomi*: factors influencing the *in vitro* development of the free-living generation. *Journal of Parasitology* 64, 220–225.

Moravec, F. (1974) Description of the free-living males and females of *Entomelas entomelas* (Dujardin, 1845) (Nematoda: Rhabdiasidae). *Scripta Facultatis Scientiarum Naturalium Universitatis Purkynianae Brunensis* 4, 97–100.

Nakamura, Y., Tsuji, N. and Taira, N. (1994a) Wasting condition under normal cardiac rhythms in rabbits following *Strongyloides papillosus* infection. *Journal of Veterinary Medical Science* 56, 1005–1007.

Nakamura, Y., Tsuji, N., Taira, N and Hirose, H. (1994b) Parasitic females of *Strongyloides papillosus* as a pathogenic stage for sudden cardiac death in infected lambs. *Journal of Veterinary Medical Science* 56, 723–727.

Nishigori, M. (1928) The factors which influence the external development of *Strongyloides stercoralis* and on auto-infection with this parasite. *Taiwan Igakkai Zasshi* 277, 31–33. (In Chinese.)

Nolan, T.J. and Katz, F.F. (1981) Transmammary transmission of *Strongyloides venezuelensis* (Nematoda) in rats. *Proceedings of the Helminthological Society of Washington* 48, 8–12.

Osche, G. (1963) Morphological, biochemical, and ecological considerations in the phylogeny of parasitic nematodes. In: Dougherty, E.C. (ed.) *The Lower Metazoa, Comparative Biology and Phylogeny.* University of California Press, Berkeley, pp. 283–302.

Poinar, G.O. Jr (1965) Life history of *Pelodera strongyloides* (Schneider) in the orbits of murid rodents in Great Britain. *Proceedings of the Helminthological Society of Washington* 32, 148–151.

Poinar, G.O., Jr. (1983) *The Natural History of Nematodes.* Prentice-Hall, Englewood Cliffs, New Jersey, 323 pp.

Premvati (1958a) Studies on *Strongyloides* of primates. I. Morphology and life history of *Strongyloides fulleborni* von Linstow, 1905. *Canadian Journal of Zoology* 36, 65–77.

Premvati (1958b) Studies on *Strongyloides* of primates. II. Factors determining the direct and the indirect mode of life. *Canadian Journal of Zoology* 36, 185–195.

Premvati (1958c) Studies on *Strongyloides* of primates. III. Observations on the free-living generations of *S. fulleborni. Canadian Journal of Zoology* 36, 447–452.

Prokopic, J.M., Barus, V. and Hodokova, Z. (1974) Preliminary report on the incidence of larvae of the family Rhabditidae (Nematoda) in the eyes of rodents. *Folia Parasitologia* 21, 189–192.

Purtilo, D.T., Meyers, W.M. and Conner, D.H. (1974) Fatal strongyloidiasis in immuno-suppressed patients. *American Journal of Medicine* 56, 488–493.

Purvis, R.S., Beightler, E.L., Diven, D.G., Sanchez, R.L. and Tyring, S.K. (1992) *Strongyloides stercoralis* hyperinfection. *International Journal for Dermatology* 31, 160–164.

Railliet, A. (1899) Évolution sans hétérogonie d'un angiostome de la couleuvre a collier. *Compte Rendu de l'Academie Sciences Paris* 129, 1271–1273.

Rausch, R. (1952) Studies on the helminth fauna of Alaska XI. Helminth parasites of microtine rodents. Taxonomic considerations. *Journal of Parasitology* 38, 418–442.

Reesal, M.R. (1950) Observations on the path of larvae of *Strongyloides agoutii* in the guinea pig and the effectiveness of the method of inoculation. *Journal of Parasitology* 36, 39.

Reesal, M.R. (1951a) Observations on the biology of the infective larvae of *Strongyloides agoutii. Canadian Journal of Zoology* 29, 109–115.

Reesal, M.R. (1951b) Observations on the development of *Strongyloides agoutii* of the agouti in the guinea pig. *Canadian Journal of Zoology* 29, 116–120.

Sandground, J.H. (1926) Biological studies on the life cycle in the genus *Strongyloides* Grassi, 1879. *American Journal of Hygiene* 6, 337–388.

Sainsbury, A.W., Bright, P.W., Morris, P.A. and Harris, E.A. (1996) Ocular disease associated with *Rhabditis orbitalis* nematodes in a common dormouse (*Muscardinus avellanarius*). *Veterinary Record* 139, 192–193.

Schaake, M. (1931) Infektionsmodus und Infektionsweg der *Rhabdias bufonis* Schrank (Angiostomum nigrovenosum) und die Metamorphose des Genitalapparates der hermaphroditischen Generation. *Zeitschrift für Parasitenkunde* 3, 517–648.

Schad, G.A., Hellman, M.E. and Muncey, D.W. (1984) *Strongyloides stercoralis*: hyperinfection in immunosuppressed dogs. *Experimental Parasitology* 57, 287–296.

Schad, G.A., Aikens, L.M. and Smith, G. (1989) *Strongyloides stercoralis*: is there a canonical migratory route through the host? *Journal of Parasitology* 75, 740–749.

Schad, G.A., Smith, G., Meygeri, Z., Bhopale, V.M., Niamatali, S. and Maze, R. (1993) *Strongyloides stercoralis*: an initial autoinfective burst amplifies primary infection. *American Journal of Tropical Medicine and Hygiene* 48, 716–725.

Schad, G.A., Thompson, F., Talham, G., Holt, D., Nolan, T., Ashton, F.T., Lange, A.M. and Bhopale, V.M. (1997) Barren female *Strongyloides stercoralis* from the occult chronic infections are rejuvenated by transfer to parasite-naive recipient hosts and give rise to an autoinfective burst. *Journal of Parasitology* 83, 785–791,

Schleip, W. (1911) Das Verhalten des Chromatins bei *Angiostomum* (*Rhabdonema*) *nigrovenosum*. Ein Beitrag zur Kenntnis der Beziehungen zwischen Chromatin und Geschlechts-bestimmung. *Archiv für Zellforschung, Leipzig* 7, 87–138.

Schulte, F. (1988) Larvaler 'Triphänismus' und die Evolution alternativer Entwicklungszyklen bei einem saprobionten Nematoden: *Rhabditis* (*Pelodera*) *orbitalis* (Nematoda) als Larvalparasit im Auge von Mäusen. *Zeitschrift für Zoologische Systematic Evolution-forschung.* 16, 237–249.

Schulte, F. (1989) Life history of *Rhabditis* (*Pelodera*) *orbitalis*, a larval parasite in the eye orbits of arvicolid and murid rodents. *Proceedings of the Helminthological Society of Washington* 56, 1–7.

Schulte, F. and Poinar, G.O. Jr (1991) On the geographic distribution and parasitism of *Rhabditis* (*Pelodera*) *orbitalis* (Nematoda: Rhabditidae). *Proceedings of the Helminthological Society of Washington* 58, 82–84.

Scowden, E.B., Schaffner, W. and Stone, W.J. (1978) Overwhelming strongyloidiasis. An unappreciated opportunistic infection. *Medicine (Baltimore)* 57, 527–544.

Seurat, L.G. (1916) Contributions a l'étude des formes larvaires des nématodes parasites hétéroxènes. *Bulletin Scientifique de la France et de la Belgique, Paris* 49, 297–377.

Seurat, L.G. (1920) Histoire naturelle des nématodes de la Bérbérie. Première partie. Morphologie, développement, éthologie et affinités des nématodes. *Université d'Alger, Faculté des Sciences, Fondation Joseph Azoubib. Travaux du Laboratoire de Zoologie Generale*, 221 pp.

Simpson, W.G., Gerhardstein, D.C. and Thompson, J.R. (1993) Disseminated *Strongyloides stercoralis* infection. *Southern Medical Journal* 86, 821–825.

Spieler, M. and Schierenberg, E. (1995) On the development of the alternating free-living and parasitic generations of the nematode *Rhabdias bufonis*. *Invertebrate Reproduction and Development* 28, 193–203.

Stammer, H.J. (1956) Die Parasiten deutscher Kleinsäuger. Verhandlungen der Deutschen Zoologischen Gesellschaft. *Zoologischer Anzeiger, Supplementband* 19, 362–390.

Stewart, T.B., Smith, W.N. and Stone, W.M. (1969) Strongyloidosis: prenatal infection of pigs with *Strongyloides ransomi*. *Georgia Veterinarian* 21, 13–18.

Stone, W.M. (1964) *Strongyloides ransomi* prenatal infection in swine. *Journal of Parasitology* 50, 568.

Sudhaus, V.W. and Schulte, F. (1986) Auflösung des Artenkomplexes *Rhabditis* (*Pelodera*) '*strongyloides*' (Nematoda) und Beschreibung zweier neuen kryptischen Arten mit Bindung an Nagetieren. *Zoologische Jahrbücher (Systematik)* 113, 409–428.

Sudhaus, W. and Schulte, F. (1988) *Rhabditis* (*Pelodera*) *strongyloides* (Nematoda) als Verursacher von Dermatitis, mit systematischen und biologischen Bemerkungen über verwandte Arten. *Zoologische Jahrbücher (Systematik)* 115, 187–205.

Sudhaus, W., Schulte, F. and Hominick, W.M. (1987) A further sibling species of *Rhabditis* (*Pelodera*) *strongyloides* (Nematoda): *Rhabditis* (*P.*) *cutanea* n.sp. from the skin of wood mice (*Apodemus sylvaticus*). *Revue Nématologie* 10, 319–326.

Supperer, R. and Pfeiffer, H. (1967) Zum Problem der pränatalen Strongyloidesinvasion beim Schwein. *Wiener Tierärztliche Monatsschrift* 54, 101–103.

Tada, P.A.G., Mimori, T. and Nakai, M. (1979) Migration route of *Strongyloides ratti* in albino rats. *Japanese Journal of Parasitology* 28, 219–227. (In Japanese.)

Taira, N., Ura, S., Nakamura, Y. and Tsuji, N. (1992) Sudden death in calves infected with *Strongyloides papillosus. Japan Agricultural Research Quarterly* 26, 203–209.

Takamure, A. (1995) Migration route of *Strongyloides venezuelensis* in rodents. *International Journal for Parasitology* 19, 757–760.

Tanaka, M., Mimori, T., Minematsu, T. and Tada, I. (1989) The acquirement of growth ability for third-stage larvae of *Strongyloides ratti* during the head-passage in the rat. *International Journal of Parasitology* 19, 757–760.

Tindall, N.R. and Wilson, P.A.G. (1988) Criteria for a proof of migration routes of immature parasites inside hosts exemplified by studies of *Strongyloides ratti* in the rat. *Parasitology* 96, 551–563.

Tindall, N.R. and Wilson, P.A.G. (1990a) A basis to extend the proof of migration routes of immature parasites inside hosts: estimated time of arrival of *Nippostrongylus brasiliensis* and *Strongyloides ratti* in the gut of the rat. *Parasitology* 100, 275–280.

Tindall, N.R. and Wilson, P.A.G. (1990b) An extended proof of migration routes of immature parasites inside hosts: pathways of *Nippostrongylus brasiliensis* and *Strongyloides ratti* in the rat are mutually exclusive. *Parasitology* 100, 281–288.

Travassos, L. (1926) Entwicklung des *Rhabdias fulleborni* n.sp. *Deutsche Tropenmedizin Zeitschrift* 30, 594–602.

Travassos, L. (1930) Pesquizas helminthologicas realisados em Hamburgo. VII. Notas sobre os Rhabdiasoidea Railliet, 1916 (Nematoda). *Memorias do Instituto Oswaldo Cruz* 24, 161–181.

Triantaphyllou, A.C. and Moncol, D.J. (1977) Cytology, reproduction, and sex determination of *Strongyloides ransomi* and *S. papillosus. Journal of Parasitology* 63, 961–973.

Tsuji, N., Itakisashi, T., Nakamura, Y., Taira, N., Kubo, M., Ura, S. and Genno, A. (1992) Sudden cardiac death in calves with experimental infections of *Strongyloides papillosus. Journal of Veterinary Medical Science* 54, 1137–1143.

Turner, J.H. (1955) Preliminary report of experimental strongyloidiasis in lambs. *Proceedings of the Helminthological Society of Washington* 22, 132–133.

Viney, M.E. (1994) A genetic analysis of reproduction in *Strongyloides ratti. Parasitology* 109, 511–519.

Viney, M.E., Matthews, B.E. and Walliker, D. (1992) On the biological and biochemical nature of cloned populations of *Strongyloides ratti. Journal of Helminthology* 66, 45–52.

Viney, M.E., Matthews, B.E. and Walliker, D. (1993) Mating in the nematode parasite *Strongyloides ratti*: proof of genetic exchange. *Proceedings of the Royal Society of London B* 254, 213–219.

Wertheim, G. (1970) Growth and development of *Strongyloides venezuelensis* Brumpt, 1934 in the albino rat. *Parasitology* 61, 381–388.

Wertheim, G. and Lengy, J. (1965) Growth and development of *Strongyloides ratti* Sandground, 1925 in the albino rat. *Journal of Parasitology* 51, 636–639.

Whicker, M.H. and Lanter, F.H. (1968) A preliminary report on gametogenesis and fertilization in the heterogenic life cycle of *Rhabdias ranae* Walton, 1929. *Bulletin of the Association of Southeastern Biology* 15, 59.

Williams, R.W. (1960) Observations on the life history of *Rhabdias sphaeorcephala* Goodey, 1924 from *Bufo marinus* L., in the Bermuda Islands. *Journal of Helminthology* 34, 93–98.

Wilson, P.A.G. (1983) Roundworm juvenile migration in mammals: the pathways of skin-penetrators reconsidered. In: Meerovitch, E. (ed.) *Aspects of Parasitology*, pp. 459–485. Institute of Parasitology, McGill University, Montreal.

Wilson, P.A.G., Gentle, M. and Scott, D.S. (1976) Milk-borne infection of rats with *Strongyloides ratti* and *Nippostrongylus brasiliensis*. *Parasitology* 72, 355–360.

Yuen, P.H. (1965) Some studies on the taxonomy and development of some rhabdisoid and cosmocercoid nematodes from Malayan amphibians. *Zooligischer Anzeiger* 174, 275–298.

Zamirden, M. and Wilson, P.A.G. (1974) *Strongyloides ratti*: relative importance of maternal sources of infection. *Parasitology* 69, 445–453.

Chapter 3

Order Strongylida (the Bursate Nematodes)

The order includes the bursate nematodes, divided into five well-defined superfamilies. The Diaphanocephaloidea is a peculiar minor superfamily with two genera in lizards and snakes. The cephalic extremity is in the form of two lateral jaw-like structures. The hookworms (Ancylostomatoidea) and the strongyles (Strongyloidea) have large, globular buccal capsules which enable them to attach to the intestinal mucosa and suck the blood of the host. Buccal capsules are absent in trichostrongyles (Trichostrongyloidea) and lungworms (Metastrongyloidea) and the buccal cavity is markedly reduced in size. The trichostrongyles are mainly gut parasites; a few species occur in other parts of the body. The lungworms are usually found within the lungs but many species are associated with the vascular system distant from the lungs.

The diaphanocephaloids, hookworms, strongyles and trichostrongyles are monoxenous, although paratenic hosts are frequently used in the transmission of some species. The lungworms are almost entirely heteroxenous and most make use of earthworms and gastropods as intermediate hosts in which development from the first to the third larval stage takes place.

Transmission in the monoxenous superfamilies usually involves the development in the external environment of eggs to the first stage which hatches, and moults to the second stage. The first and second stages may feed on bacteria in the environment (e.g. hookworms and strongyles) or they may have sufficient nutrient reserves in their cells to develop and persist for considerable periods without feeding (e.g. *Dictyocaulus*). Second- and third-stage larvae that feed in the environment have an oesophageal bulb with a prominent valve which is absent or much reduced in species that do not feed. In some species (e.g. *Dictyocaulus*) development to the first stage and hatching take place in the definitive host. In other species embryonated eggs are passed by the host and the entire development to the third stage occurs in the egg (e.g. Amidostomatidae, *Nematodirus*, Syngaminae). In one peculiar genus (*Ollulanus*) the entire development to the third stage takes place *in utero*. The third and infective stage, which generally retains the cuticle of the second stage (often with a long attenuated tail), does not feed and typically enters the host through the skin (many hookworms) or *per os* (most strongyles and trichostrongyles). The third stage is homologous to the dauer larva of the free-living rhabditids from which the Strongylida presumably evolved. It is the infective and dispersal stage and generally resistant to adverse environmental conditions. Larvae which penetrate skin exsheath before or during their invasion of the skin. Larvae which are ingested by the definitive host exsheath in some region of the gut (the rumen, abomasum or intestine, depending on the species). It is believed that larvae respond to factors in the gut and release a fluid which acts on a specific region of the cuticle, causing a break which allows larvae to escape (Sommerville, 1957).

Most of the Strongylida have a tissue phase in the definitive host. In the diaphanocephaloids, hookworms and strongyles which have large buccal capsules in the adult form, a provisional or temporary buccal capsule is formed during development to the fourth stage. During development of the fourth stage the adult buccal capsule is formed beside and beneath the provisional buccal capsule, which is shed at the final moult. A reduced provisional buccal capsule appears rarely in the fourth stage of some trichostrongyloids, the adults of which lack buccal capsules. Buccal capsules are absent in the metastrongyloids.

3.1

The Superfamily Diaphanocephaloidea

The diaphanocephaloids are curious bursate nematodes of the digestive tract of terrestrial snakes and, rarely, lizards (Lichtenfels, 1980a). The buccal cavity is large and complex and the oral opening is dorsoventrally expanded, giving the buccal region a characteristic bivalved appearance. The oral opening is sometimes surrounded by a delicate cuticular membrane reminiscent of the corona radiata found in the superfamily Strongyloidea. The superfamily consists of one family (Diaphanocephalidae), two genera (*Diaphanocephalus* and *Kalicephalus*), six subgenera and about 33 species. Schad and Kuntz (1964) reported that species in the genus *Kalicephalus* lack host specificity and in an individual host there is marked intolerance between species (i.e. species space themselves apart in the gut).

Knowledge of the development and transmission of the diaphanocephalids is limited to a few observations on four species of *Kalicephalus*. Female worms deposit thin-shelled, smooth, oval eggs in an advanced stage (morula) of segmentation. Development of the eggs and larval stages is rapid in faecal cultures and water, and essentially similar to that of the hookworms (see section 3.2). The first and second stages are **rhabditiform** and the third stage **strongyliform**, as in most bursate nematodes with free-living stages (for definitions of these forms see section 3.2 on the Ancylostomatoidea). The tail in all three stages is conical and pointed and the infective larva retains the cuticle of the second stage.

Snakes can be infected experimentally by oral inoculation of third-stage larvae (Schad, 1956). Larvae often invade the stomach or intestinal wall and become encapsulated in the mucosa, where development to the fourth stage, with its provisional buccal capsule, takes place. The prepatent period is about 2–3 months in some species.

How snakes become infected under natural conditions is a mystery. Snakes test their environment with their tongues and Schad (1956) suggested larvae might attach to the latter. There is no evidence that skin penetration occurs. There is also the possibility that infective larvae could invade soft-bodied animals such as snails, slugs and amphibians cohabiting with snakes. If larvae persist in their tissues, these animals could serve as paratenic hosts.

Family Diaphanocephalidae

Kalicephalus

K. costatus costatus (Rudolphi, 1819)

K. c. costatus (syn. *K. philodryadus* according to Schad, 1962) was collected by Ortlepp (1923a) from the stomach and intestine of *Philodryas serra* from South America. Eggs were 78–92 × 46–50 μm in size. At 25°C, eggs in tap water developed to the first stage and hatched in 24 h. First-stage larvae were about 312 μm in length. The tail was 'filiform' and 60 μm long. The genital primordium consisted of a single cell. In 30 h larvae moulted to the second stage. The second moult occurred 4–5 days after cultures had been made. The exsheathed larvae were 480–512 μm long. Attempts to infect a grass snake (*Tropidonotus natrix*) orally were unsuccessful.

K. agkistrodontis Harwood, 1932
K. parvus Ortlepp, 1923
K. rectiphilus Harwood, 1932

These three species were studied by Schad (1956). The size and morphology of the eggs and the various larval stages as well as the rate of development were similar in the three species and the data on only one species will be given. Eggs of *K. parvus* were 80–100 × 40–50 μm in size. The length of first-, second- and third-stage larvae were, respectively: 300–380 μm, 460–500 μm and 470–590 μm. Eggs hatched in tap water at room temperature in 20–26 h. The first moult commenced in 72 h and the second in 120 h. In 144 h cultures contained sheathed third-stage larvae.

Schad (1956) collected female specimens of *K. parvus* from *Coluber c. coluber*, cultured their eggs to the third stage and then inoculated them orally into gopher snakes (*Pituophis catenifer*). Larvae remained ensheathed in the stomach lumen for about 10 days but by 14 days some had invaded the stomach wall and become encapsulated in the mucosa. In about 109 days development towards the fourth-stage commenced but fourth stage larvae were not collected from the snakes. Eggs appeared in faeces of *P. catenifer* in 115 days.

The development of *K. agkistrodontis* differed from *K. parvus* in that parasitic stages were found only in the lumen of the gut, and fourth-stage larvae as well as immature adult worms were found as early as 23 days. A gravid female worm was found in an experimentally infected snake 58 days postinfection.

The early parasitic stages of *K. rectiphilus* were encapsulated in the intestine, and not the stomach as in *K. parvus*. Development was highly variable and 33 days postinfection third-, fourth- and early fifth-stage worms were found encapsulated in the posterior part of the duodenum.

K. willeyi Linstow, 1904

Ahmed and Saleh (1967) maintained adults and cultured larvae of *K. willeyi* collected from *Vipera russelli*. Adults were maintained for 36 days at 32°C and larvae developed to the third stage in Hedon-Fleig medium. The authors claimed that, in culture, first-stage larvae invaded female worms through the vulva and lived on the contents of the uterus as they developed.

3.2

The Superfamily Ancylostomatoidea

Hookworms occur in the small intestine of mammals. Like species of the Strongyloidea, they have large, highly cuticularized buccal capsules provided with teeth or cutting plates (Chabaud, 1974) which they use to attach themselves to the intestinal mucosa of the host. However, unlike species of strongyles, the buccal region is never hexagonal in transverse section and a corona radiata is absent. Also many hookworms, unlike strongyles, infect the host mainly through skin penetration. The Ancylostomatoidea has a single family, the Ancylostomatidae, divided into the Ancylostomatinae and the Bunostominae based on the position of the duct of the dorsal oesophageal gland and other characters described by Lichtenfels (1980c). Fourth-stage and adult hookworms suck blood, and infection by the three species that are well-known pathogens in humans results in anaemia caused by blood loss, particularly in malnourished individuals. For excellent reviews on hookworms in humans, refer to Chandler (1929), Komiya and Yasuraoha (1966), Miller (1979) and Schad and Warren (1990).

Female worms deposit oval, thin-shelled eggs into the gut lumen (Fig. 3.1A). Eggs are generally in the four- to six-cell stage when passed in faeces of the host. Under suitable conditions of moisture and temperature (e.g. 23–33°C) the egg embryonates in about 24 h to a first-stage larva about 250–300 μm in length with a pointed tapered tail, an elongate buccal cavity and an oesophagus usually with a valved bulb (Fig. 3.1B–F). This larva is generally referred to as **rhabditiform**, mainly because the oesophagus resembles that of the free-living rhabditids. Larvae feed on bacteria and within 2 days moult to the second stage (Fig. 3.1H), which is about 400–430 μm in length but otherwise similar to the first stage. Within 4–5 days after eggs are passed from the host, the second-stage larvae become lethargic and development proceeds to the third and infective stage (Fig. 3.1H). The oesophagus loses its valved bulb and the tail becomes relatively short and blunt. The larva retains the cuticle of the second stage with its attenuated tail. Such larvae are often referred to as **strongyliform**. Unlike the first two larval stages, the third stage does not feed and its survival depends on stored nutrients in its tissues. Strongyliform larvae are active, 500–700 μm in length, long-lived and often resistant to adverse environmental conditions. Some tend to be thermotactic, thigmotactic, phototactic and negatively geotactic. They readily contaminate the food and environment of the host (see for example Harada, 1954a).

Many hookworms (unlike strongyles) infect the host by penetrating the skin after shedding their sheaths, an important discovery made by Looss (1898) when he accidentally spilt water containing larvae on his hand. Enzymes released by the larvae aid their passage through the skin (Bruni and Passalacqua, 1954; Lewert and Lee, 1954). In

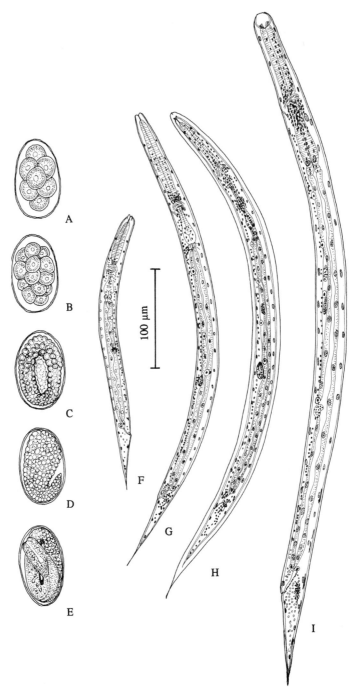

Fig. 3.1. Stages in the development of *Uncinaria stenocephala*: stages in the development of the egg; (F) first-stage larva; (G) second-stage larva; (H) sheathed infective third-stage larva; (I) male fourth-stage larva. (After H.C. Gibbs, 1961 – courtesy *Canadian Journal of Zoology*.)

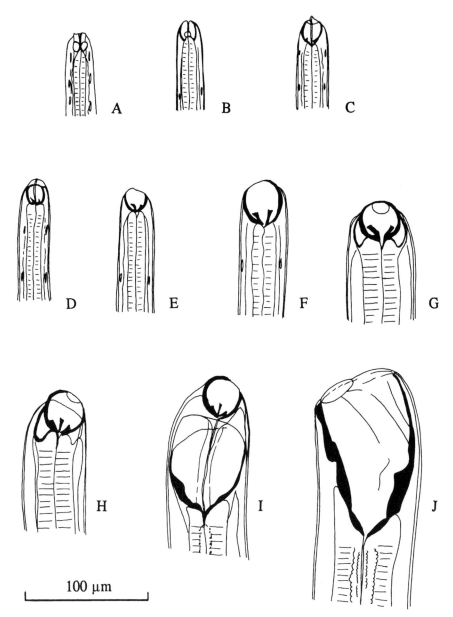

Fig. 3.2.   Development of the provisional and adult buccal capsule in *Uncinaria stenocephala*: (A) parasitic third stage showing beginning of provisional buccal capsule; (B) later stage of A; (C) parasitic third stage with fully developed provisional buccal capsule; (D) early third stage moulting; (E) early fourth stage; (F) fourth stage with fully developed provisional buccal capsule; (G) late fourth stage, beginning of adult buccal capsule (ventral); (H) lateral view of G; (I) completion of development of adult buccal capsule; (J) adult buccal capsule (lateral). (After H.C. Gibbs, 1961 – courtesy *Canadian Journal of Zoology*.)

a few hours larvae enter subcutaneous tissues, invade lymphatic capillaries and are carried to regional lymph nodes, the larger lymph channels leading to the thoracic duct, and the general circulation, which carries them to the heart and lungs. In lungs, larvae enter alveoli and begin to develop to the fourth stage (Fig. 3.1I). Development continues as larvae migrate from the lungs to the trachea and intestine. Arriving as fourth-stage larvae in the intestine, they have acquired a spherical provisional or temporary buccal capsule which enables them to attach to the gut wall (Fig. 3.2A–F). During development of the fourth stage (Fig. 3.2G–J) the buccal capsule of the adult parasite forms posterior and lateral to the provisional buccal capsule, the cuticular lining of which is shed at the final moult. Worms reach the adult stage in 2–7 weeks, depending on the species.

Although skin penetration appears to be the main route for some hookworms to infect the host, third-stage larvae reaching the intestine by the oral route nevertheless will grow to adulthood, although they may undergo a brief sojourn in the mucosa before maturing in the lumen of the intestine (Leichtenstern, 1886; Fülleborn, 1926, 1927; Foster and Cross, 1934).

The phenomenon of arrest, widely known and studied in trichostrongyles of herbivores, is also reported in hookworms such as *Ancylostoma caninum* of canids, *Uncinaria lucasi* of fur seals and *Ancylostoma duodenale* of humans. In the resistant host, unsheathed infective third-stage larvae invade and persist in the tissues and can eventually be mobilized in the female host for transmammary and prenatal infection of the young. There is evidence that environmental factors such as temperature, acting on the free-living stages of hookworms, may also induce arrest among worms in the host (Schad *et al.*, 1982; Gibbs, 1986; Prociv and Luke, 1995).

Family Ancylostomatidae

Subfamily Ancylostomatinae

The duct of the dorsal oesophageal gland is usually in a dorsal gutter on the inner surface of the buccal capsule.

Ancylostoma

The 14 or so members of the genus are distinguished by a cephalic extremity bent dorsally, the oral opening lined with teeth and a large globular buccal capsule. Species occur in the intestine of Carnivora, Primates, Edentata, Rodentia and Suidae. The genus contains several species important in animal and human health.

A. braziliense de Faria, 1910
A. ceylanicum (Looss, 1911)
These two species, mainly from dogs and felines, were regarded as conspecific until Biocca (1951) showed they were distinct species. *A. braziliense* was found in dogs and felines (*Felis catis, F. pardus, F. serval*) in Brazil and is known from dogs in various warm regions of the world. The eggs and larval stages are similar to those of *A. duodenale.*

A. braziliense is a common cause of **cutaneous larva migrans**, called 'creeping eruption', resulting from the reaction to third-stage larvae migrating in the skin of humans (Dove, 1932; Beaver, 1956; Faust *et al.,* 1975).

A. ceylanicum is a parasite of the intestine of cats, dogs and humans in southeast Asia. Yoshida (1971a,b) confirmed the conclusion of Biocca (1951) and added new distinguishing characters to separate *A. braziliense* and *A. ceylanicum.* He also showed that the infective larva of *A. ceylanicum* was significantly longer (712.1 ± 15.0 µm) than that of *A. braziliense* (662.1 ± 17.0 µm) but was indistinguishable from the third-stage larva of *A. duodenale* (719.3 ± 23.1 µm). Wijers and Smit (1966) infected eight volunteers cutaneously with *A. ceylanicum.* All acquired itchy papular eruptions at the site of larval penetration and abdominal discomfort began 15–20 days postinfection. The prepatent period was 21 days. Bearup (1967) gave the prepatent period as 24 days and Yoshida *et al.* (1972) as 18–26 days in humans. Rep (1965) and Yoshida (1968) reported prepatent periods of 14–17 days in dogs and cats.

Rep *et al.* (1968a,b) showed that monospecific infections of either *A. braziliense* or *A. ceylanicum* always resulted in the production of eggs, whereas heterospecific infections consisting of males of one species and females of the other species did not result in the production of eggs.

Yoshida *et al.* (1974) studied the development of *A. braziliense* and *A. ceylanicum* in puppies infected by oral inoculation of infective larvae. Development of the two species was identical. Larvae exsheathed in the intestine and as early as 3 h postinfection invaded the intestinal mucosa, where over a period of 1–2 days (on the second and third days postinfection) they developed to the fourth stage, which then migrated into the lumen of the intestine. Early fifth-stage worms were found along with fourth-stage larvae 6 days postinfection. Immature adults gradually increased in numbers and by 10 days all worms found were adult. Eggs first appeared in faeces on the 14th day. Of the larvae given, 69–90% ($\bar{x}$ = 78%) of *A. ceylanicum* and 84–97% ($\bar{x}$ = 90%) of *A. braziliense* reached sexual maturity in the intestine of the host.

Yoshida *et al.* (1974) infected puppies by cutaneous exposure to the two hookworm species. One puppy was examined 24 h after infection with *A. ceylanicum* and a few third-stage larvae were found in the skin at the site of exposure. In 2 days larvae were found in the lungs. In 3 days larvae were found in the lungs, trachea, stomach and intestine. The data indicated that third-stage larvae developed during the lung–trachea migration and the late third-stage larvae moulted to the fourth stage in the intestine. There was no evidence that larvae invaded the mucosae. In 7 days early fifth-stage worms were present, and eggs appeared in faeces after 14 days.

A. braziliense seems to behave somewhat differently after cutaneous penetration. Third-stage larvae underwent a rapid lung–trachea migration and reached the intestine as third-stage larvae, which invaded the mucosa. Fourth-stage larvae were found on and after the 3rd day of infection and within 7 days a few early fifth-stage worms were found. Eggs appeared in faeces after 14 days.

A. caninum (Ercolani, 1859)

A. caninum is a common cosmopolitan hookworm of the intestine of dogs and other canids. It was once thought that hookworms occurring in cats belonged to a distinct strain (Scott, 1928, 1929; McCoy, 1931a; Foster and Cort, 1932; Foster and

Daengsvang, 1932) until Biocca (1954) showed that the cat hookworm was a distinct species called *A. tubaeforme* (see also Burrows, 1962).

According to McCoy (1930) eggs of *A. caninum* hatched in 6–12 days at 12°C, in 4–5 days at 15°C, in 1.5–2.0 days at 17°C, in 1 day at 23°C, in 10–12 h at 30°C and in 9 h at 37°C. First-stage larvae were 300–340 μm in length, with conical, pointed tails. Second-stage larvae were 400–430 μm in length. Infective larvae were about 630 μm in length and appeared in cultures in 22 days at 15°C, in 9 days at 17°C, in 4–5 days at 23°C, in 58–66 h at 30°C and in 47 h at 37°C. Larvae failed to develop to the infective stage at 15°C and most died at temperatures above 37°C.

Balasingam (1964b) concluded that 25–35°C was suitable for the development of eggs, with the maximum hatch taking place at 25°C. Infective larvae appeared in cultures at 25–35°C, the optimum temperature being 30°C. Eggs and larvae of *A. caninum* were readily killed by freezing temperatures. The free-living stages were not as tolerant of cold as those of *Uncinaria stenocephala*.

Dogs can become infected with *A. caninum* by several routes. Leuckart (1866) and Leichtenstern (1886) first showed that *A. caninum* could be transmitted by oral inoculation of infective larvae and the oral route is still regarded as important in canine infections. Larvae exsheathed in the stomach and intestine and they invaded the gastric glands or the glands of Lieberkühn of the small intestine. They remained there for a few days and then emerged into the lumen, where they moulted to the fourth stage as early as 3 days postinfection. Worms developed to the adult stage rapidly and eggs first appeared in faeces in 15–26 days. Oral transmission and direct development in the intestine were confirmed by Foster and Cross (1934). After Looss (1898) reported skin penetration by *A. duodenale*, it was discovered that dogs could also be infected with *A. caninum* by this route. The lung–trachea migration takes at least 2–7 days.

Foster (1932, 1935) and Clapham (1962) showed that transplacental transmission occurred in pregnant bitches which were infected cutaneously or orally. Worms did not mature in the puppies until they were born. Puppies passed eggs 10–12 days after birth and mortality in infected puppies was high. Finally, Stone and Girardeau (1966) found larvae in the colostrum of a bitch; the discovery of the milk transmission of *Uncinaria lucasi* predated this discovery by several years. A prerequisite for prenatal and trans-mammary transmission is the existence of migrating third-stage larvae in the tissues of the pregnant bitch. Arrested unsheathed third-stage larvae of *A. caninum* were first noted in dogs by Scott (1928). Arrest is especially likely in older dogs which display resistance to infection and is most prominent in females (Miller, 1965). Enigk and Stoye (1968) gave infective larvae to bitches at various times between mating and parturition. In five of six bitches, larvae appeared in the milk; numbers of larvae decreased rapidly and disappeared in 20 days. Prenatal infection occurred in four of seven puppies from one bitch and three of five from another. Stone and Peckham (1970) collected infective larvae from the milk of lactating bitches and gave them to puppies. All nine experimental puppies became infected and passed eggs in their faeces 12–16 days postinfection. Arrested larvae were found in muscle fibres (Lee *et al.*, 1975), where they survived for at least 240 days (Stoye, 1973). Stoye and Schmelzle (1986) confirmed that larvae reached the milk of bitches through a somatic rather than a blood route. Larvae occurred in mammary glands only during lactation, especially during the first week after parturition when they were available to suckling puppies.

Schad (1983) reported that third-stage larvae subjected to precipitous changes in temperatures tended to arrest in the tissues of dogs. Schad and Page (1982) orally infected dogs with 'chilled arrest prone' larvae and removed those adult worms resulting from the infection with anthelmintics. Larvae that had invaded the tissues were not stimulated to return to the gut and develop by the removal of adult worms. Nevertheless, arrested larvae spontaneously developed during a period of 2–3 months.

Finally, Little (1959) and Miller (1970a) showed that larvae given to insects and mice sequestered themselves in the tissues and remained alive for long periods; larvae persisted for the life of the mouse. Harada (1954b) reported that infective larvae attached to the pulvillus and abdomen of bluebottle and house flies. These observations raise the possibility that insects, rodents and perhaps other animals could serve as paratenic hosts in the transmission of hookworms to free-ranging canids.

Adult worms are voracious blood suckers and in the living dog the sucking movements of the oesophagus were as high as 120–150 per minute (Wells, 1931). Sucking continued even during copulation as the blood passed through the gut of the male worm and from under the folds of the bursa clasping the female. Sucking even continued when the body was transected behind the oesophagus. Wells (1931) concluded that each worm would remove 0.84 cm^3 of arterial blood in 24 h.

A. duodenale (Dubini, 1843)

A. duodenale is referred to as the 'Old World hookworm' but its original distribution was probably the North Temperate zone of the eastern hemisphere (Faust, 1949). Human migrations have extended the distribution of the parasite through the Middle East into the Mediterranean regions of Europe and Africa and into tropical regions of the world, including southeast Asia, the south Pacific and Indonesia. It was introduced to the Americas by European exploration and colonization. It is essentially a subtropical species and does not occur in the northern hemisphere above 52° latitude; free-living stages require temperatures in excess of 22°C (Miller, 1979).

The history of the discovery of *A. duodenale* and the elucidation of its development and transmission by Dubini, Leuckart, Wucherer, Grassi, Parona, Perroncito, Bentley and especially Looss (who discovered skin penetration and the lung–tracheal migration) is one of the most interesting chapters in medical science and zoology. For an account of these contributions refer to Grove (1990) and translations of some of the classic papers in Kean *et al.* (1978).

According to Chandler (1929), eggs were 50–80 × 36–42 μm in size and were in the two- to eight-cell stage when passed in faeces of the host. It has been estimated that a single female can produce 10,000–25,000 eggs each day. Development to the first stage (250–300 μm in length) under favourable soil conditions and at a temperature of 25°C required 1–2 days. The second stage (500–600 μm in length) was reached in 3 days and the infective stage in 5–8 days. The optimum temperature for development of the free-living stages ranged from 21 to 27°C. Infective larvae were 500–700 μm in length. The tail was conical and sharply pointed. The cuticle of the second-stage had a slender, tapered tail extending a short distance posterior to the tail of the infective larvae. The wall of the buccal cavity was thin and delicate (cf. *Necator americanus*). Infective larvae were thermotactic and tended to occur in the upper region of the soil, where they moved vertically in films of moisture. They survived best in moist, shaded areas with humus and vegetation. The life span of the larvae varied from 2 to 12 months, depending on environmental conditions.

Miller (1979) has noted that there is no suitable non-human host in which detailed aspects of the development of *A. duodenale* can be studied. Although it is possible to infect chimpanzees, there are ethical and economic factors which make this primate unsuitable for experimentation. Infective larvae exsheath and penetrate intact skin, hair follicles or pores, especially if they are in soil adhering to the body. Reaction to invading larvae takes place within minutes as an intense burning and itching. The resultant pruritus, known as **ground itch** or **hookworm dermatitis**, is common on the feet (between the toes) in endemic regions. Larvae evidently travel in the usual way via the lymphatics and blood vessels to the heart and lungs, where they begin to develop to the fourth stage before migrating to the intestine (Looss, 1911). Humans can also acquire infection by ingesting larvae. The latter exsheath in the gut and develop directly to adulthood without a migration to the lungs.

The prepatent period of *A. duodenale* has been reported as 31–41 days by Brumpt (1952), 45–78 days by Looss (1911), 43–162 days by Yoshida *et al.* (1958), 88 days by Koike (1960) and 68–69 days by Akimoto (1966). Schad *et al.* (1973) reported arrested development in *A. duodenale* in West Bengal, India. Infective larvae acquired in the dry season did not develop until the next monsoon season, when conditions for free-living stages and transmission were favourable. Factors involved in their arrest are unknown. Prociv and Luke (1995) reported arrest in Australia.

Nawalinski and Schad (1974) reported the case of an individual infected percutaneously on 24 June. In 7–14 days a non-productive cough and pharyngitis developed, indicating normal larval migration to the lungs. However, eggs were not found in stools after 22 weeks. Four months later, when stool examinations were again initiated, eggs were found and egg output increased for 2 months. Unfortunately the arresting stage is not known but it is presumably the third (cf. *Ancylostoma caninum*).

A. duodenale has a maximum life span of 6–7 years but the maximum egg output occurs in the first 12–18 months (Kendrick, 1934) and the number of worms declines in the absence of reinfection within 2–3 years. In endemic areas, however, there is a continual turnover in the hookworm populations in humans (Hsieh, 1970; Miller, 1970b).

A. tubaeforme (Zeder, 1800)

A. tubaeforme (syn. *A. caninum longispiculum*) is a cosmopolitan hookworm of cats. Scott (1929, 1930), McCoy (1931a) and others noted that it was almost impossible to infect dogs with the feline species or to infect cats with the canine species. These characteristics were attributed to strain differences until Biocca (1954) showed convincingly that the species found commonly in cats was morphologically distinct from the species in dogs. Eggs were 55–75 × 34–45 μm in size and their development and free-living stages were similar to those of *A. caninum*. McCoy (1931a) gave the prepatent period in cats as 17–27 ($\bar{x}$ = 21.6) days and reported that a single worm produced about 2350 eggs per day.

Matthews (1975) used *in vitro* techniques to study skin penetration of infective larvae of *A. tubaeforme*. The author found no evidence of enzyme secretions from larvae and suggested that larvae are able to move through tissues mechanically without the aid of enzymes (cf. Lewert and Lee, 1954).

Okoshi and Murata (1967a,b) infected kittens by oral inoculation of larvae in gelatin capsules and by subcutaneous inoculation. The recovery of worms in relation to the number given was 50.2% by the oral method and 61.5% by the subcutaneous method. The average prepatent period was 19 days after oral infection and 22 days after subcutaneous infection. In oral infections, larvae developed directly to adults in the intestine, whereas in subcutaneous infection larvae migrated in the tissues and eventually to the intestine, where they developed to adulthood. Okoshi and Murata (1968a) confirmed earlier work that the cat was an unsuitable host for *A. caninum*. Onwuliri *et al.* (1981) gave the prepatent period as 18 days and reported that egg output reached its peak in about 4–5 weeks postinfection.

Okoshi and Murata (1968b) gave infective larvae orally to mice. Larvae persisted unchanged in the tissues, especially in the lungs, brain and gut for long periods. Most larvae given subcutaneously migrated to the lungs or the gut. Norris (1971) infected mice orally and percutaneously. Numerous larvae were found in the tissues (mainly the muscles) of mice infected orally and few were found in mice infected percutaneously. Larvae underwent a short pulmonary migration before settling mainly in the muscles, where they lived for up to 18 months. These observations raise the possibility that rodents might serve as paratenic hosts of hookworms of cats.

Nwosu (1978) studied the effects of temperature on the development of eggs of *A. tubaeforme*. Over 90% of eggs developed and hatched at 15–37°C but a maximum of 110 h was required for development at 15°C and only 15 h at 37°C.

Globocephalus

In members of this genus the anterior extremity is uniquely directed anterodorsally (Lichtenfels, 1980c). The host distribution of the genus includes pigs, primates, mustelids, opossums, rodents and deer. Seven or eight species occur in swine.

G. urosubulatus (Alessandrini, 1909)

G. urosubulatus is a parasite of the intestine of pigs in Europe, Africa and North and South America. According to Ehlert (1962) eggs passed in faeces of the host hatched in 22–26 h at 26°C. The second stage was formed in 20–26 h after hatching and the third stage was attained in 40–46 h after the moult to the second stage. Pigs were infected experimentally *per os* but the experiment did not rule out the possibility of percutaneous infection. The prepatent period was 26–47 days. The longest prepatent period was in older pigs and in pigs with heavy infections; it was suggested that development was delayed in these hosts. Swine may pass eggs for 14 months after infection.

Placoconus

Members of the genus occur in raccoons, mustelids and bears in the New World. The buccal capsule has five articulating plates and a dorsal gutter but the genus is closely related to *Uncinaria*.

P. lotoris (Schwartz, 1925)

P. lotoris is found in the posterior region of the small intestine of raccoons (*Procyon lotor*) and skunks (*Mephitis nigra*) in North America. The development and transmission of *P. lotoris* have been investigated by Balasingam (1964a). Eggs were 68–76 × 38–44 µm in size and were in the 8- to 16-cell stage when passed in faeces of the host. In faecal cultures eggs developed and hatched into first-stage larvae within 24 h at 22–24°C. First-stage larvae were 273–358 µm in length. The first moult occurred 24–48 h after larvae hatched. Second-stage larvae were 420–630 µm and third-stage larvae were 578–653 µm in length. The tail in all stages was conical and sharply pointed.

Infective larvae were given orally to raccoons and skunks. Larvae exsheathed in the stomach and invaded the gastric and duodenal glands. In 48 h the early fourth stage had appeared and larvae returned to the lumen and moulted to the subadult stage in 7–10 days. As worms developed to the late fourth stage and adults they gradually moved posteriorly in the small intestine, until in 20 days half the worms were in the posterior fourth of the small intestine. Eggs appeared in the faeces of infected animals 25–27 days postinfection.

Attempts to infect raccoons, skunks, cats, dogs and ferrets by placing larvae on the skin or by inoculating them subcutaneously were not successful. Infective larvae were given orally to cats, dogs and ferrets. Larvae were passed in faeces and any that established themselves in these hosts failed to mature.

Uncinaria

Members of the genus have globular buccal capsules with two ventrolateral and two dorsolateral cutting plates. Species in carnivores belong in the subgenus *Uncinaria*.

U. lucasi Stiles, 1901

U. lucasi is a common parasite of northern fur seals (*Callorhinus ursinus*) and northern sea-lions (*Eumetopias jubata*) which breed in the North Pacific. The transmission of this species, elucidated by Olsen (1958), Olsen and Lyons (1965) and Lyons and Keyes, (1978) revealed remarkable adaptations to the seasonal life cycle of fur seals on the Pribilof Islands, Alaska in the north Pacific. Fur seals have strong homing instincts and are highly social and gregarious. Breeding animals come together in rookeries from early June until late August. The harems then break up and the seals, including young-of-the-year, abandon the rookery and spend all their time at sea until their return to the rookery in spring. Adult populations of worms cannot survive in seals during their prolonged 9-month period at sea, presumably because of the innate short life span of the worms and the immune system of the host. *U. lucasi* displays remarkable mechanisms to overcome these special problems.

Several facts revealed by Olsen and his colleagues led to the current understanding of development and transmission: (i) adult worms did not occur in the intestines of adult or yearling seals returning to the rookery in early June, indicating that adult worms failed to survive the period at sea; (ii) adult worms only occurred in pups 2 weeks to 4–5 months of age on the rookery; (iii) the soil on the rookery was heavily contaminated with infective larvae prior to the arrival of the seals, indicating that larvae could survive the severe winter on the rookery; (iv) arrested third-stage larvae of *U. lucasi* were found in the tissues of all age groups of seals.

The life cycle is now understood to consist of three interrelated stages:

1.   The intestinal phase. Newly born pups acquired parasitic third-stage larvae from the milk of the mother. These larvae, found in the belly blubber and mammary glands of pregnant cows, were 796–975 µm in length. Vacuoles near the cephalic end could be seen and were regarded as the primordia of the provisional buccal capsule of the fourth-stage larvae; larvae were also larger than free-living third-stage larvae and unsheathed (see below). In the intestine of seal pups parasitic third-stage larvae developed to fourth-stage larvae and mature adults in about 2 weeks; the final moult occurred in 4–5 days. Large numbers of eggs were then passed in faeces of infected pups. Eggs were oval with clear, smooth shells; they were in the eight-cell stage when passed from the host, and 126–140 × 80–98 µm in size.

2.   The free-living phase. At 22°C first-stage rhabditiform larvae, 510–630 µm in length, were present in eggs in 24 h. In about 40 h larvae moulted within the eggs to the second stage, which were 600–690 µm in length. In 60 h the second moult occurred, resulting in strongyliform third-stage larvae (645–795 µm in length) with a short, blunt tail and enclosed in two shed cuticles. Hatching of third-stage larvae took place in 100 h at room temperature in the laboratory but was delayed for several months on the rookery. Newly hatched third-stage larvae appeared for the first time on the rookery around the end of August or the beginning of September (diurnal temperature of the soil ranged from 10 to 12°C). On hatching, third-stage larvae shed the first-stage cuticle but retained the second-stage cuticle with its delicate attenuated tail.

3.   The tissue phase. When adult and yearling seals returned to the rookery the follow- ing June, they were exposed to infective third-stage larvae which readily penetrated into their flippers and other parts of the body. The paths followed by larvae that had penetrated the skin of seals have not been determined. Larvae may enter blood vessels or lymphatics and become distributed in tissues of the body, including the blubber of the belly and the mammary glands. There is evidence that the parasitic third-stage larvae are present in the milk for only a short time postpartum. The precocious development of the larvae in the tissues of the female seals ensures that development to adulthood in the pups will be considerably accelerated. The reason worms do not mature in adult seals is unknown but it might be suspected that the seals are immune and their immunity may be continually stimulated by arrested larvae in tissue.

Lyons and Keyes (1978) failed to infect three pups with third-stage larvae from the belly tissues of bulls and bachelors but were successful in infecting two pups with larvae from pregnant cows. However, larvae were smaller (640.5–732.0 µm in length) in bulls, bachelors, 2-year-old males, male and female yearlings and pups in contrast with those ($\bar{x}$ = 938.1 µm) from pregnant females. This raised the possibility that hormones in pregnant cows elicit the precocious development to the parasitic third stage.

Heavy infections of fourth-stage and adult hookworms in pups resulted in blood loss and anaemia. Considerable mortality in the pups was attributed to hookworms.

U. stenocephala (Railliet, 1884)
U. stenocephala (syn. *Dochmoides stenocephala*) is a common parasite of the small intestine of canids (dogs, foxes, wolves) and a few other carnivores mainly in the

northern hemisphere. According to Gibbs (1961) eggs were 72–80 × 45–55 μm in size and passed from the host in the 8- to 16-cell stage (for development see Figs 3.1 and 3.2). Eggs developed from 7.5°C to 27°C but the rate of development was inversely proportional to temperature. At 20°C, regarded as the optimum temperature by Gibbs and Gibbs (1959), most first-stage larvae developed and hatched by 12 h and by 96 h infective larvae were present in cultures. First-stage larvae were 290–360 μm (Fig. 3.1F), second-stage larvae 430–530 μm (Fig. 3.1G) and infective larvae 500–580 μm (Fig. 3.1I) in length. The tails of the various larval stages were cone-shaped and sharply pointed but not markedly attenuated. Infective larvae remained viable in tap water at 5°C for 10 months. Gibbs and Gibbs (1959) and Balasingam (1964b) have shown that eggs and infective larvae of *U. stenocephala* were more resistant to freezing than those of *A. caninum* (which developed best at 23–30°C) and they concluded this parasite was adapted to cooler, northern climates.

Gibbs (1961) infected puppies by orally administering infective larvae contained in water in gelatin capsules. Larvae shed their sheaths in the stomach within 18 h and larvae invaded mucous glands of the pyloric region of the stomach (19.2%) and the mucosa of the duodenum (70.5%). A few larvae were found in the mucosa of the ileum and a few remained in the lumen of the gut. Two days postinfection, larvae were in the lumen of the small intestine and undergoing the third moult. In 7–9 days the final moult commenced and by 10 days all worms found were in the fifth stage. Eggs appeared in the faeces of the host in 15 days. Most mature worms were found in the third quarter of the small intestine.

Gibbs (1961) had limited success in infecting pups by placing larvae in a drop of water on the groin region. A few third-stage larvae were found in the lungs a short time after infection. In one experiment 3000 larvae were applied to the skin of puppies. Only ten larvae were found in the small intestine; these larvae developed normally. Attempts to infect puppies by subcutaneous inoculation of larvae were not successful. Gibbs (1961) concluded, therefore, that oral infection was the normal route for larvae to invade the host.

Subfamily Bunostominae

Members of the subfamily have a tooth-like dorsal cone supporting the duct of the dorsal oesophageal gland (Lichtenfels, 1980c).

Bunostomum

Members of the genus occur in the small intestine of Bovidae, Antilocapridae, Cervidae and Elephantidae.

B. phlebotomum (Railliet, 1900)

B. phlebotomum is a cosmopolitan parasite of the small intestine, mainly of cattle although it has been reported from zebu and, rarely, roe deer. Eggs passed in faeces of the host were 79–117 × 47–70 μm in size with an irregular, roughened surface and in the four- to eight-cell stage (Krug and Mayhew, 1949). Conradi and Barnette (1908) and Schwartz (1924) made preliminary observations of development of the eggs. According to Schwartz (1924) eggs reached the first stage and hatched in 24 h at 21–27°C. Eighteen hours later the second moult was in progress leading to the second stage, which was about 490 μm in length. Both first and second stages had long, attenuated tails. Third-stage larvae were 500–540 μm in length in 6-day-old cultures. These larvae retained both the first- and second-stage cuticles, as noted also by Supperer (1958). However, the first-stage cuticle was discarded when larvae encountered solid objects in the environment.

Schwartz (1924) and Sprent (1946b) showed that third-stage larvae were phototactic and thermotactic. Schwartz (1924) noted that larvae tended to climb the sides of vessels but he did not know if this was caused by phototropism or negative geotropism. Sprent (1946b) showed that larvae were not negatively geotactic and observed that larvae in culture when breathed on or warmed moved 'vigorously, the most obvious tendency being to stretch themselves out into space waving their bodies in the air, the long, tapering tail-sheath serving as an anchor.' Also larvae could withstand drying for 5 days on a slide at 25°C and 75% relative humidity and they survived in cultures for 3 months.

Schwartz (1924) failed to discover if larvae penetrated skin but Mayhew (1939) and Sprent (1946a) demonstrated that the host was usually infected by this route. Sprent (1946a) suggested that, since larvae tended to remain in dung, the host was probably most often infected from larvae leaving dung adhering to their bodies. Large numbers of larvae invading the skin of calves resulted in an irritating pruritus. Sprent (1946a) sectioned skin and various tissues after exposure to larvae. In 15 min larvae had penetrated the basement membrane between the epithelium and the dermis. Other larvae were in hair follicles. In 30 min larvae had reached the dermis and in 60 min some were found in cutaneous blood vessels. Moulting to the fourth stage took place in the lungs of calves 10 days later and the fourth-stage larvae migrated to the intestine via the trachea. Mayhew (1946, 1948, 1949) also showed that calves could be infected by the oral administration of larvae as well as by the cutaneous route. Sprent (1946a) found larvae in faeces of calves as early as 56 days after applying larvae to the skin. The prepatent period reported by Mayhew (1946) varied from about 60 to 72 days (excluding one animal in which eggs were first found at 115 days). Mayhew (1950) reported that adult hookworms lived for 8.5–24 months in three calves.

B. trigonocephalum (Rudolphi, 1808)

B. trigonocephalum (syn. *Monodontus trigonocephalum*) is mainly a parasite of the small intestine of sheep and goats but it has been reported in alpaca, llama, chamois and red deer.

According to Hesse (1923) eggs were 80–93 × 47–65 μm in size and were passed in faeces in the 8- to 16-cell stage. At 22–23°C first-stage larvae were formed and hatched in 24 h; larvae left the egg shell from a point lateral to one pole. Larvae were 360–600 μm in length and had long, filamentous tails. Second-stage larvae were 450–600 μm in length. The second moult occurred 24 h after the first moult and the third-stage larvae were 450–700 μm in length. According to Belle (1959) eggs failed to embryonate below 15°C or above 35°C and the most favourable temperatures for development were 20–30°C. Eggs were not resistant to unusually high or freezing temperatures or desiccation. Crofton (1965) reported that eggs developed only from 15 to 34°C.

Cameron (1923b, 1927c) studied the behaviour of infective larvae and reported that they were lethargic at 15°C but active at 22–40°C. Larvae died when frozen or dried. They were thermotactic, phototactic and negatively geotactic. Belle (1959) also noted that larvae were killed by freezing and Shorb (1942) and Kates (1943) reported that there was no evidence that larvae survived winter on pastures in Maryland, USA.

Beller (1928), Ortlepp (1939) and Lucker and Neumayer (1946) studied development of *B. trigonocephalum* in sheep. They infected sheep by orally administering infective larvae in water and by placing larvae in a small amount of water on the skin in a wool-free area, usually under a foreleg. The site of penetration of larvae was obviously irritated since the sheep responded by scratching the area with a hindfoot. One lamb that received 5000 larvae on its skin harboured 500 hookworms 4 months later. Far lighter infections resulted when 5000 or 50,000 larvae were given orally than when similar numbers of larvae were applied to the skin (see also Stoye, 1965). Exposure to extremely large numbers of larvae (over half a million) was followed by inhibition of the development of larvae (presumably because of stimulation of immune processes in the host). The prepatent period in moderate to heavy infections was 53–60 days. Heavy infections with *B. trigonocephalum* resulted in anaemia and depression of the growth rate of the host.

Gaigeria

Members of the genus occur in Bovidae and Antilocapridae in Africa and India.

G. pachyscelis Railliet and Henry, 1910

This is a common parasite of the small intestine of sheep and goats in the drier regions of South Africa. It also is found in the Congo, India and Indonesia. Ortlepp (1937) suggested it was introduced to South Africa by importation of Asian livestock and he recorded its presence in the Indian antelope *Boselaphus tragocemalus*.

Ortlepp (1934, 1937) carried out a detailed experimental study of *G. pachyscelis* in sheep in South Africa. Eggs were 108–115 × 58–61 μm in size and were passed in faeces of the host in the 16- or 32-cell stage. At 26°C most first-stage larvae had hatched in faecal cultures in 36–48 h. First-stage larvae were 237–262 μm in length with attenuated tails. The second moult occurred about 24 h after first-stage larvae had hatched. After 4–6 days of incubation second-stage larvae had moulted to third-stage larvae, which were 580–670 μm in length. Infective larvae retained the second-stage cuticle with its markedly attenuated tail. Third-stage larvae migrated upward in films of moisture and were positively phototactic and thermotactic. They died within an hour after being dried on a glass slide.

Numerous attempts by Ortlepp (1934, 1937) to infect sheep by the oral inoculation of infective larvae were unsuccessful. However, sheep readily became infected when larvae in small amounts of water were placed between their hooves or on skin behind the ears. Thirty-one of 34 sheep were successfully infected by the latter method. Larvae reached the lungs after penetrating the skin and remained there for 13–14 days, during which they reached the fourth stage; some fourth-stage larvae had already migrated to the intestine at this time. The provisional buccal capsule of the fourth stage had a dorsal tooth and two subventral lancets. By 21 days females in the intestine were 3.56 mm and males 3.25 mm in length. The final moult took place 7–8 weeks postinfection. The prepatent period was usually about 10 weeks but sometimes was almost 22 weeks.

Nascimento *et al.* (1988) infected goats and sheep with *G. pachyscalis*. The prepatent period in sheep was 62–77 day and in goats 63–78 days. Mean daily oviposition in ovines ranged from 1600 to 3500 eggs per female in sheep and 1600 to 2500 per female in goats.

Ansari and Singh (1980) reported that larvae invaded the skin and migrated to the lungs in mice, guinea pigs, rats and rabbits. However, larvae developed only to the fourth stage in guinea pigs and rabbits.

Arantes *et al.* (1983) placed 500 larvae on the inguinal region of each of 12 goats and concluded (from necropsy data) that 7.6% of the larvae developed successfully in the host.

Ansari (1981) found ten infective larvae in colostrum of one of six ewes infected experimentally with larvae of *G. pachyscelis*. The ewe had been given 10,000 infective larvae. There was no sound evidence for prenatal infection.

Necator

Members of the genus occur in humans, apes, Suidae and, rarely, rhinoceroses, pangolins and dogs. *N. americanus* is an important human pathogen. The buccal cavity is relatively small in this species and the oral opening is provided with cutting plates rather than teeth, as in *Ancylostoma duodenale*.

N. americanus Stiles, 1902

N. americanus is known as the New World hookworm. Its original distribution in humans was apparently south of the Tropic of Cancer in Africa, southern Asia, the East Indian Archipelago and the Pacific Islands (Faust, 1949). The parasite was introduced to the Americas by slaves and colonists from Africa and other regions of the world. It was once common in the southeastern USA.

Eggs were $50–80 \times 36–42$ µm in size and in the four-cell stage when passed in faeces of the host (Chandler, 1929). A single female produced 6000–11,000 eggs per day (Stoll, 1923). Development of the eggs and the free-living stages was similar to that of *A. duodenale* but the optimum temperature was 25–28°C (Brumpt, 1958). Also, eggs of *N. americanus* died at temperatures below 7°C, unlike those of *A. duodenale*.

The infective larva of *N. americanus* can be distinguished from that of *A. duodenale* in that the walls of the buccal cavity are more prominent, the genital primordium is more anterior and smaller, the oesophagus is broader, and there is a slight constriction of the intestine at its junction with the oesophagus (Chandler, 1929).

According to Miller (1979), who reviewed the subject, experimental percutaneous infections of humans have usually been markedly successful whereas oral infections have been much less successful. Mizuno and Ito (1963) and Nagahana *et al.* (1963) showed that oral infection was successful only if larvae were allowed to invade the buccal epithelium. Smith (1904), in a lucid account of the biology of *N. americanus*, stressed the importance of infected mud between the toes as a source of infection, leading initially to pruritus or 'ground itch' resulting from the penetration of larvae in the skin.

The prepatent periods reported in humans infected with *N. americanus* varied from 44 to 100 days (Smith, 1904; Looss, 1911; Nagahana *et al.*, 1963).

According to Kendrick (1934) worms can live for 5–6 years and Palmer (1955) reported one infection which persisted for 15 years in the absence of reinfection. Beaver (1988) described the egg production in a light 17- or 18-year-old infection in a volunteer.

Orihel (1971) pointed out that *N. americanus* has been reported from non-human primates on a number of occasions. He found infections in chimpanzees (*Pan troglodytes*) and African pata monkeys (*Erythrocebus patas*) from the primate colony of the Delta Regional Primate Research Centre in Louisiana, USA. Laboratory-bred and feral chimpanzees and laboratory-bred pata monkeys were successfully infected with larvae of *N. americanus* cultured from human faeces. The prepatent period was 42–54 days.

Hoagland and Schad (1978) compared the development and transmission of *N. americanus* and *Ancylostoma duodenale* in the light of *r* and *K* selection and the constraints of body size on reproduction.

3.3

The Superfamily Strongyloidea

Members of the superfamily have large, complex buccal capsules, often with a **corona radiata** (a series of leaf-like structures on the border of the labial region), and are mainly gut parasites although a few species occur in the respiratory or urinary systems (Lichtenfels, 1980b). The superfamily is divided into the families Strongylidae (including the large strongyles of equines), Chabertiidae (including the nodular worms, *Oesophagostomum* spp.), Syngamidae (including the gapeworms of birds and a single species in swine, *Stephanurus dentatus*) and Deletrocephalidae (of *Rhea americana*). Their oval, smooth-shelled, poorly developed eggs pass out in waste products of the host and, like those of hookworms, embryonate rapidly in the external environment under suitable conditions of moisture and temperature. In the Strongylidae, Chabertiidae and Stephanurinae, the first-stage larva hatches and develops rapidly to the infective stage. Prior to, and during, moulting, larvae are lethargic. The small first-stage rhabditiform larva has a long filamentous tail, and a narrow, elongate buccal cavity. The second stage is slightly larger than the first but morphologically similar to it. During development to the strongyliform third stage the tail becomes blunt and conical. The stoma is closed and the larva retains the shed cuticle of the second stage with its attenuated tail.

In the gapeworms (Syngaminae), development to the third stage takes place in the egg. The oesophagus in the first two larval stages lacks the valved bulb and the attenuated tail used in feeding and movement by larvae of the strongylids, the chabertiids and *Stephanurus dentatus* (Stephanurinae).

The usual route by which strongyles infect the final host is oral; members of the superfamily are predominantly parasites of herbivorous hosts such as horses, ruminants, ratite birds and certain Australian marsupials (Lichtenfels, 1980b). The infective larvae of some species, such as the gapeworms and the swine kidney worm (Syngamidae), sequester themselves in earthworm and gastropod paratenic hosts, where they are available to hosts such as swine, anseriforme and passeriforme birds and small non-herbivorous mammals.

Family Strongylidae

Members of the family have round or oval oral openings with well-developed corona radiata or lips and are distinguished from members of the family Chabertiidae by details in the structure of the ovejector and the dorsal ray (Lichtenfels, 1980b). Members of the family are found in the gut of a diverse group of hosts. Although richly represented in equines, representatives can also be found in elephants, hyracoids, ostriches, rhinoceroses, tapirs, tortoises, warthogs and Australian marsupials.

The development and transmission of only a few important species in equines have been investigated. It can probably be assumed, however, that development outside the host of all strongylids is basically similar. Smooth, thin-shelled, oval-shaped eggs are passed in the faeces of the host in an early stage of segmentation. In suitable conditions of temperature and humidity, eggs embryonate to rhabditiform first-stage larvae, which hatch in 1–2 days. Larvae grow and moult to second-stage larvae, which develop and moult to early strongyliform third-stage larvae in about 5–6 days. First- and second-stage larvae have long slender tails. The third-stage larva retains the cuticle of the second stage as a tight-fitting sheath with its attenuated, filamentous tail.

Although sheathed infective larvae do not feed, they live for several weeks or months at warm temperatures and can survive winter and dry conditions for long periods. Infective larvae are active and crawl over vegetation when moisture conditions are suitable. Negatively geotactic and mildly phototactic, the larvae readily become available to their herbivorous hosts.

It can probably be assumed that infective larvae always exsheath in the small intestine and, if the behaviour of species which have been studied is any indication, all strongylids undergo a tissue phase before returning to the large intestine, where they mature.

A summary of the development of the three important large strongylids (*Strongylus* spp.) is given below in some detail as well as that of *Triodontophorus tenuicollis*, all of horses. However, about 50 lesser-known strongylids of horses have been described, including species in the genera *Cylicostephanus* (= *Cylicocercus*), *Cylicocyclus*, *Cylicodontophorus*, *Gyalocephalus* and *Poteriostomum* (Poluszynski, 1930; Lucker, 1934, 1936a, 1938). In addition, Lucker (1936) noted that, in culture, rhabditiform first-stage larvae hatched from eggs of *Cyathostomum catenatum* (= *Cylicocercus catinatus*) and *Gyalocephalus capitatus* in 24–48 h. The first moult took place in about 72 h. Moulted exsheathed strongyliform larvae were present in the culture in 5 days. Eggs of *Cylicostephanus goldi* (= *Cylicocercus goldi*) hatched in 24 h and sheathed strongyliform larvae first appeared 4 days later.

Limited information is available on stages of some smaller strongylids found in nodules and in the lumen of the colon and caecum of horses (Cuille *et al.*, 1913; Ihle and van Oordt, 1923; Gibson, 1953; Müller, 1953). Worms were referred to the genus *Trichonema*, which Lichtenfels (1975) and others regard as a synonym of *Cyathostomum*. Poluszynki (1930) gave the length of the larvae of '*Trichonema* spec.' as 850 (760–960) μm with the sheath and 520 (470–550) μm without the sheath. Third- and fourth-stage larvae (the latter with well-defined provisional buccal capsules with a single tooth) were found in nodules in the gut wall. Fourth-stage larvae evidently migrated from the mucosa to the lumen of the gut, where they grew and moulted to the adult stage. Worms were said to mature in 20–30 days and the prepatent period was 3–4 months, according to Wetzel (1954). There is evidence that inhibited fourth-stage larvae persist in nodules for prolonged periods and only enter and mature in the lumen of the gut when adults have been eliminated or reduced in numbers, as, for example, when anthelmintics are administered to the host (Gibson, 1953). This may be a common phenomenon in the superfamily.

Subfamily Strongylinae

Strongylus

Members of the genus are blood-sucking parasites of the colon and caecum of equines (asses, donkeys, horses and zebras) and are often referred to as **palisade worms**. The development and transmission of three species have been investigated in some detail. When the infective larvae are ingested with food of the host, the third and fourth stages undertake curious migrations in the tissues before returning to the gut, where they mature. The migration routes followed by the three species are dissimilar. The need to migrate in the tissues of the host may be a legacy from skin-penetrating ancestors (*Stephanurus dentatus*, the syngamid strongylid of swine, infects the host orally or by skin penetration). Since the three strongylids which have been studied occur together in the same host, it is tempting to think that differences in their behaviour and sites of development may be adaptations that minimize competition during a critical period in their lives.

During development three vesicular or bulb-like structures are reported to appear at the anterior end of the third-stage larva (Hobmaier, 1925; Wetzel, 1941b; Wetzel and Kersten, 1956; McCraw and Slocombe, 1974) and it has been suggested that these structures aid in movement through tissues, since they are said to inflate and deflate.

S. edentatus (Looss, 1900)

S. edentatus is a common parasite of the caecum and colon of horses. Infective larvae were 790 (740–840) μm in length including the tail of the sheath and 520 (480–550) μm without the sheath (Poluszynki, 1930). Infective larvae exsheathed and invaded mainly the terminal portion of the jejunum and ileum, according to Rooney (1970), and the right ventral colon and caecum, according to McCraw and Slocombe (1974). Larvae then entered veins and within 2 days migrated to the liver via the hepatic portal system (Wetzel and Kersten, 1956; McCraw and Slocombe, 1974, 1978). Larvae grew in the liver and moulted to the fourth stage in 11–18 days. In 30 days postinfection larvae migrated extensively in the liver and by 42 days were noted in the hepatorenal ligament, which McCraw and Slocombe (1974) regarded as the most important path of larvae *en route* to the large intestine. They noted the proximity of the right lobe of the liver to the base of the caecum and the short length of the hepatorenal ligament. Migrating larvae of *S. edentatus* were generally found in many other regions of the body, including retroperitoneal sites in the flanks and kidney and on the omentum, diaphragm and pancreas. However, the fate of these worms is uncertain and McCraw and Slocombe (1978) found no evidence to suggest that larvae distant from the caecum and colon ever reached the gut. There was evidence that fourth-stage larvae moulted to subadults in 13–15 weeks. Worms which had reached the subserosa of the gut wall in 3–5 months were in haemorrhagic nodules, which eventually opened into the lumen and released the subadult worms. The prepatent period was about 11 months (Wetzel, 1942).

S. equinus Müller, 1782

S. equinus is a cosmopolitan parasite of the caecum and colon of equines and is generally found in smaller numbers than *S. edentatus* and *S. vulgaris*. Infective larvae were 980 (920–1020) μm including the tail of the sheath and 720 (640–770) μm without the sheath (Poluszynski, 1930). Infective larvae exsheathed in the small intestine of foals and invaded the wall of the large intestine, where they became encapsulated and developed to the early fourth stage in about 2 weeks (Wetzel and Enigk, 1938; Wetzel, 1941; Enigk, 1970; McCraw and Slocombe, 1985). Fourth-stage larvae emerged from nodules and entered the peritoneal cavity. By 19 days larvae had invaded the liver, where they remained for at least 12 weeks. However, in about 4 weeks larvae began to move extensively in, and on, the liver and some began to invade the pancreas (especially 12–17 weeks postinfection) which is closely associated with the right lobe of the liver. In about 15 weeks the worms moulted to the subadult stage and over a period of several weeks returned to the gut wall from the pancreas. In the gut they elicited nodules which eventually burst, releasing worms into the lumen of the large intestine. Wetzel (1942) gave the prepatent period as 261 days. According to Enigk (1970) and Ershov (1970) the prepatent period was about 8.5 months.

As in *S. edentatus*, some subadult *S. equinus* wandered to regions of the body remote from the large intestine (e.g. retroperitoneal spaces in the flanks, perirenal fat, diaphragm and omentum) but the fate of these worms was not known.

S. vulgaris (Looss, 1900)

S. vulgaris is a ubiquitous, cosmopolitan parasite of the caecum and right ventral colon of wild and domestic horses, wild asses and zebras. The parasite attaches itself to a plug of mucosa drawn into its sizable buccal capsule. Excellent reviews of the pathogenesis and development of *S. vulgaris* have been published by Soulsby (1965), Enigk (1970), McCraw and Slocombe (1976) and Ogbourne and Duncan (1977) and the brief account given below has relied heavily upon these publications.

Hatching and larval development take place at temperatures of 8–39°C. At 30°C development to the infective stage required 3–4 days compared with 16–20 days at 12°C. Infective larvae were 1020 (930–1090) μm in length including the tail of the sheath and 740 (670–790) μm without the sheath (Poluszynski, 1930). Under field conditions infective larvae emerged from faeces in greatest numbers when the latter were moistened by rain (Ogbourne, 1972, 1973) and larvae migrated into the soil and on to blades of grass. Infective larvae remained alive under dry conditions for months and even years. They were also markedly resistant to freezing temperatures. Medica and Sukhdeo (1997) suggested that fatty acids provide the energy for activity of the larvae.

Larvae on herbage ingested by horses exsheathed in less than 1.5 h in the small intestine as a result of the stimulation of duodenal contents and coliform bacteria (Poynter, 1954, 1956; Rupasinghe, 1975). The larva emerged from the sheath through a cap which detached from the anterior end. The subsequent fate of exsheathed larvae now seems well established through the work of several parasitologists (Enigk, 1950, 1951, 1970; Skalinskii, 1952, 1954; Ershov, 1960, 1970; Drudge *et al.*, 1966; Duncan and Pirie, 1972; Duncan, 1973). The conclusions, based on the results of study of naturally and experimentally infected foals, conform to what Georgi (1973) referred to as the Kikuchi–Enigk model. According to Duncan and Pirie (1972) exsheathed third-stage larvae penetrated the intestinal wall 1–3 days postinfection, grew and moulted to the fourth stage by 7 days, and then invaded the lumen of submucosal arterioles. Larvae migrated in the arterioles to the caecal and colic arteries, which they reached in 14 days, and the anterior mesenteric artery, which they reached by 21 days. During this initial migratory period larvae were only 1–2 mm in length. They grew markedly in the arteries and by 120 days postinfection were 10–18 mm in length and had moulted to the fifth stage, although they retained the fourth-stage cuticle. In 3–4 months the young worms migrated back to the intestinal wall by way of the lumen of arteries. On arrival at the serosal surface, pea-sized nodules formed around the worms in the arteries. Such nodules were common 4 months postinfection. Larvae eventually escaped from the nodules and entered the lumen of the intestine, where they reached maturity in 6–8 weeks. The prepatent period was, therefore, 6–7 months.

S. vulgaris is associated with thrombi and other lesions in the wall of the arteries, especially the mesenteric artery, which interfere with the blood supply to important regions of the alimentary tract. It is regarded as the major helminth pathogen of horses.

Triodontophorus

The genus is closely related to *Strongylus* but is distinguished by details of the cephalic end and the spicules. About five species are known from horses (Levine, 1968). Members of the genus suck blood; they are usually found attached to the gut wall and engorged with blood.

T. tenuicollis Boulenger, 1916

T. tenuicollis is regarded as the most important member of the genus found in horses, where it occurs in the dorsal colon. It is frequently associated with ulcers and several worms may be associated with an individual ulcer (Ransom and Hadwen, 1918). *In utero* eggs were in the one- to 16-cell stage and 98–105 × 54–58 μm in size. Ortlepp (1925) cultured eggs collected from female worms. First-stage larvae were fully formed within 24 h at 26°C and most eggs hatched within 40 h. The first-stage larvae were 485–512 μm long with long slender tails. The first moult occurred in 24–30 h after hatching and second-stage larvae were 600–700 μm in length. Larvae moulted the second time in 4 days. Larvae were 500–530 μm in length, excluding the sheath, and the tail was conical. They were negatively geotactic and were resistant to desiccation and marked variations in temperatures (−8°C to 60°C).

Fourth-stage larvae, found in the lumen of the colon of horses, were briefly described by Ortlepp (1925).

Family Chabertiidae

Chabertiids are medium-sized strongyles found in artiodactyls, macropodids, rodents and primates.

Subfamily Chabertiinae

Castorstrongylus

C. castoris Chapin, 1925

C. castoris is a parasite of the colon of beaver (*Castor canadensis*) in North America. Romanov (1969) reported that eggs were 94–124 × 48–59 μm in size and in the eight-cell stage when oviposited. First-stage larvae appeared in eggs in 5–6 days at 18–25°C and two moults occurred, in 8–10 and 15–16 days. Third-stage larvae which hatched retained the cuticles of the two previous stages. Infective larvae were 700–780 µm in length with the sheath and 480–550 µm without the sheath. Twelve immature worms were found in a beaver given 30 infective larvae 78 days previously.

Chabertia

C. ovina (Fabricius, 1788)

C. ovina is known as the large-mouthed bowel worm of camels, cattle, chamois, deer, gazelles, goats and sheep. It is readily recognized because the oral opening is directed anteroventrally. Adult worms attach to the intestine by drawing a portion of the mucosa into the buccal capsule, where it is believed to be predigested before passing into the oesophagus (Wetzel, 1931). Gordon and Graham (1933) and Ross and Kauzal (1933) believed that *C. ovina* adults also ingested blood. Eggs deposited by female worms were in the 16-cell stage when deposited by female worms but usually in the morula stage when passed in faeces (Crofton, 1963). Eggs, which were 90–100 × 53–59 µm in size, hatched in 24–28 h and first-stage larvae were 250 µm in length (Threlkeld, 1948). They completed the first moult in 36 h when they were 350 µm in length. Second-stage larvae grew to about 650 µm by the 6th day and grew and moulted to the third stage by the 7th day, when they were about 750 µm in length; larvae retained the second-stage cuticle.

According to Crofton (1965) hatching of eggs took place above 6°C and below 36°C. At 36°C it occurred in 17 h, at 26°C in 24 h and at 16°C in 48 h. Crofton (1963) pointed out that *C. ovina* is usually found in greatest numbers in temperate and cool temperate climates. Both eggs and larvae can withstand freezing temperatures (Ross and Kauzal, 1933; Andrews, 1934) and infective larvae can develop from eggs kept at 4–6°C (Ross and Gordon, 1936).

Andrews (1934) gave 1000 infective larvae to a lamb and then additional doses of 1000 larvae 73, 79 and 96 days later. Eggs appeared in the faeces of the lamb 52 days after the first feeding. The lamb passed eggs for 102 days. Threlkeld (1947, 1948) gave infective larvae to several lambs and examined them at intervals. By 90 h larvae had exsheathed and were found attached to, or within, the wall of the colon. By 96 h larvae were found in the lumen of the intestine. In 6 days larvae had developed to the fourth stage. In 23 days a larva was undergoing the final moult and in 25 days all worms were subadults; males were 3 mm and females 4 mm in length. Nematodes were noted *in copula* in 38 days and eggs were found in the faeces of the infected lambs 47–54 days postinfection. According to Threlkeld (1948) the activity of *C. ovina* from 25 days postinfection until maturity was reached was coincident with severe diarrhoea and discharge of bloody mucus.

Cyclodontostomum

This genus was previously assigned to the Ancylostomatoidea but is now recognized as belonging to the Strongyloidea (Lichtenfels, 1980b).

C. purvisi Adams, 1933

C. purvisi is a parasite of the small and large intestine of various species of *Rattus* in Malaysia. Varughese (1973) collected specimens from the large intestine and caecum of Malayan giant rats (*R. mulleri, R. sabanus* and *R. whiteheadi*) and established the parasite in laboratory mice and rats. Eggs were 80×50 μm in size and developed into first-stage larvae, which hatched within 24 h at 26–28°C. First-stage larvae were 294–438 μm in length with fairly long, thin tails. The second-stage larva was not found in the culture. Sheathed infective larvae, reached in 4–5 days after hatching of the first-stage larvae, were 397–454 μm in length. Oral administration of larvae resulted in infections in laboratory mice and rats ('other routes failed to result in infection'). In the gut of the rodent the third-stage larva moulted in 5–10 days and the subadult stage was attained in 16–23 days.

Ternidens

T. deminutus (Railliet and Henry, 1905)

T. deminutus is a parasite mainly of the large intestine of African and Asian non-human primates, including chimpanzees, gorillas, baboons, macaques and various monkeys. It occurs commonly or sporadically in humans in enzootic regions of Africa (Sandground, 1929, 1931; Amberson and Schwarz, 1952). Eggs of *T. deminutus* are larger ($70–94 \times 40–60$ μm) than those of other nematodes (e.g. hookworms) found in the faeces of humans in Africa. According to Goldsmid (1971) eggs were usually passed from the host in the eight-cell stage. Eggs hatched when the first stage was reached in about 48–72 h at 29°C. The larva moulted to the second stage on or about the 2nd or 3rd day and to the ensheathed third stage in 8–10 days. Infective larvae had only limited ability to withstand desiccation (Blackie, 1932) and they had a tendency to ascend vegetation (Sandground, 1931). Larvae survived for up to 10 h at −5°C and for 12 weeks at 5 ± 1°C (Goldsmid, 1971).

The mode of transmission of *T. deminutus* is still not understood. It seems evident that, like most other strongylids, the infective larvae do not penetrate skin. Attempts to infect humans and baboons by oral inoculation and skin penetration of infective larvae have been largely unsuccessful (Blackie, 1932; Goldsmid, 1971, 1974). The possibility that a paratenic host ('intermediate host') is involved in transmission has been suggested by Amberson and Schwartz (1952) and others. Goldsmid (1982), in a useful review, asks, 'why has *T. deminutus* infection in man been recorded only in Africa and mainly in Zimbabwe?' The possibility that there is an ecological association between savannah-inhabiting monkeys and humans has been raised, as well as the possibility that some local customs, such as insect-eating, may predispose certain Africans to infections. Perhaps insects and other terrestrial invertebrates consumed by humans and other primates are capable of harbouring infective larvae. It is worth noting that *Syngamus trachea* can use earthworms and molluscan paratenic hosts in transmission and *Cyathostoma lari* larvae have been found in the dung beetle, *Aphodius fossor*, as well as in earthworms. It is likely that infective larvae of many strongyles are capable of sequestering themselves in tissues of a variety of terrestrial invertebrates that may or may not be involved in transmission under natural conditions.

Boch (1956) reported fourth-stage *T. deminutus* in nodules in the gut wall of *Macacus cynomolgus* and suggested that the presence of mature worms in the gut lumen inhibited the emergence of fourth-stage larvae from nodules.

Subfamily Oesophagostominae

Oesophagostomum

Members of the genus, known as nodular worms, are common parasites of the large intestine of pigs, ruminants, primates and rodents. Those in domestic animals are considered significant pathogens. The buccal capsule is relatively reduced and thin-walled. The cuticle of the cephalic end is inflated and a transverse groove is usually present near the excretory pore; this groove tends to offset the cephalic end from the rest of the body of the parasite. The genus is rich in species and is divided into subgenera (Lichtenfels, 1980b). For a review of sporadic reports in humans, see Polderman and Blotkamp (1995).

Marotel (1908) recognized that the three morphological stages found in nodules in cattle in France belonged to *Oesophagostomum radiatum*. Veglia (1924, 1928) provided the first sound account of the development and transmission of a species of the genus *Oesophagostomum*. Since then about seven other species parasitic in livestock have been studied. Eggs are oval with thin shells and are laid usually in the 16- to 32-cell stage (Fig. 3.3A). In faeces, eggs develop rapidly into rhabditiform first-stage larvae (Fig. 3.3B) which hatch as early as 24 h at optimum temperatures. First-stage larvae escape from one pole of the egg. Larvae feed on bacteria in the environment and moult to the second-stage (Fig. 3.3C) about 24 h after hatching. Both first- and second-stage larvae have long attenuated tails, rhabditiform oesophagi and a limited number of intestinal cells. Second-stage larvae moult to infective larvae (Fig. 3.3D) as early as 3–5 days from the time of hatching. The third-stage larva has a club-shaped oesophagus (strongyliform), a delicate tubular buccal cavity and a sharp, conical tail. The larva, however, retains the cuticle of the second stage with its long filamentous tail. Infective larvae can live for long periods and, in some species, are capable of surviving winter.

The definitive host usually becomes infected by ingesting third-stage larvae, although calves have apparently been infected experimentally through the skin (Mayhew, 1939). In the small intestine larvae exsheath and within about 3 days invade the mucosa, where they become surrounded by the transparent capsule in which they are coiled (Fig. 3.3G). In the capsules the larvae develop to the fourth stage (Fig. 3.3E–I) and then enter the lumen, where they shed the cuticle of the third stage. The fourth-stage larva has an oval, thin-walled provisional buccal capsule. Cervical papillae appear and the oesophagus takes on the form of that in adults. Sexes can be differentiated by the shape of the tail, which in the male is directed dorsally and is shorter than that in the female. Fourth-stage larvae move from the small intestine posteriorly to the caecum and colon, where they moult and reach adulthood. The final moult occurs about 2 weeks postinfection but the females do not produce and deposit eggs until about 1 month postinfection, when eggs appear in the faeces of the host.

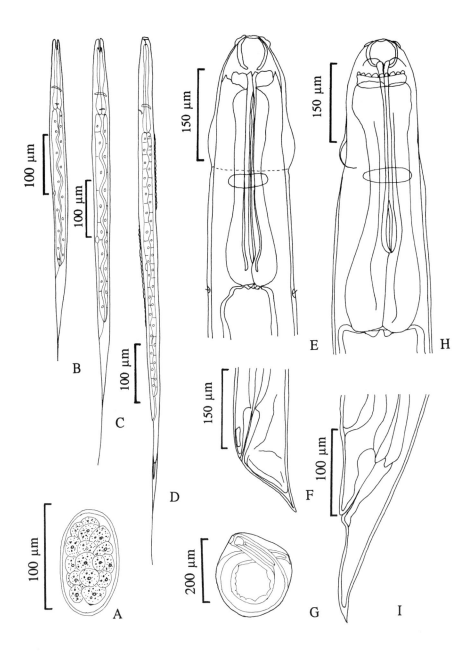

Fig. 3.3. Developmental stages of *Oesophagostomum venulosum*: (A) unembryonated egg; (B) first-stage larva; (C) second-stage larva; (D) infective third-stage larva; (E) late fourth-stage larva, anterior end, dorsal view; (F) late fourth-stage larva, posterior end; (G) encapsulated late third-stage parasitic larva (4 days postinfection); (H) late fourth-stage larva, anterior end; (I) late female fourth-stage larva, posterior end. (After A. Goldberg, 1951 – courtesy *Proceedings of the Helminthological Society of Washington.*)

O. asperum (Railliet and Henry, 1913)

O. asperum is a parasite of the caecum and colon of goats. According to Rao and Venkataratnam (1977a) eggs were 95–112 × 60–68 µm in size. When cultured in aerated water at room temperature (13.6–37.4°C) first-stage larvae hatched in about 60 h. First-stage larvae were 390–461 µm in length and the intestine was composed of eight to ten cells. The first moult commenced in 70 h when second-stage larvae were 687–722 µm in length. The second moult occurred in 122–150 h in faecal culture. Infective larvae were 895–920 µm in length; the tail of the sheath was 117–128 µm in length. Infective larvae were active in water for 20 days and survived for 45 days.

Rao and Venkataratnam (1977b) gave five kids infective larvae orally. Fourth-stage larvae were collected from the large intestine of two kids 22 and 29 days postinfection. Immature and mature fifth-stage worms were found in one kid 8 weeks postinfection. The prepatent period was said to be 48 days.

O. columbianum Curtice, 1890

O. columbianum is a common parasite of the large intestine of sheep, goats, alpaca and certain wild African antelope. According to Veglia (1924, 1928) eggs embryonated into first-stage larvae in 15–20 h and a day later moulted to the second stage. Infective larvae appeared in about 3 days after the second moult. Larvae migrated on to vegetation. When ingested by the host, larvae exsheathed and entered the mucosa and submucosa of the small and large intestine, where they became encapsulated. In 4 days larvae moulted to the fourth stage, which entered the lumen of the large intestine 5–8 days post-infection. Fourth-stage larvae moulted to adults 27 days postinfection and eggs appeared in faeces at 40 days. Some adults lived for up to 21 months.

Agrawal (1966) showed that eggs would develop and hatch, and larvae develop, at temperatures from 15 to 37°C but the optimum was 30°C. At room temperature infective larvae lived for 103 days.

Fotedar and Wali (1982) gave the size of eggs as 80–90 × 50–60 µm and reported that they were laid in the four- to eight-cell stage. First-stage larvae were 300–350 µm, second-stage larvae 650–700 µm and third-stage larvae 700–800 µm in length. The optimum temperature for growth was 28–30°C. Infective larvae remained alive for 3 weeks.

Attempts to infect lambs cutaneously with *O. columbianum* were unsuccessful and fourth-stage larvae given orally to lambs failed to result in infection (Shelton and Griffiths, 1968a). Shelton and Griffiths (1968b) reported that subcutaneous inoculation of infective larvae seemed to sensitize lambs and subsequent oral infections were more severe because larvae were retained longer than normal in the gut wall.

Dash (1973) reported two histotropic phases in infections of *O. columbianum* in lambs. In first infections, third-stage larvae became encapsulated mainly in the wall of the anterior small intestine. The third moult occurred 5 and 10 days postinfection, after which larvae returned to the lumen and migrated to the large intestine. Some larvae entered the wall of the large intestine and were arrested at the mid-fourth stage whereas other larvae grew to adulthood in the lumen of the large intestine. Few larvae developed to adulthood after second infections. Some were arrested in the first histotropic phase in the small intestine and others in the second histotropic phase in the large intestine. Dash (1973) suggested that the second histotropic phase is abnormal and indicative of a poorly adapted host–parasite relationship, as also suggested by Taylor and Michel

(1953), Dobson (1966) and Shelton and Griffiths (1968a,b). Perhaps *O. columbianum* is derived from species in wild hosts such as African antelopes.

O. dentatum (Rudolphi, 1803)

O. dentatum is a cosmopolitan parasite of the large intestine of pigs and peccaries. Goodey (1924, 1926) described the various larval stages, including the fourth. First-stage larvae were 435 μm in length. Sheathed infective larvae were 660–720 μm in length (including the sheath with its tail which was 180 μm in length). Infective larvae were 'positively geotactic', resistant to desiccation for only 1–2 days and 'very active' at 37°C. Spindler (1936) reported that infective larvae under grass lived for 14 months and larvae on bare experimental plots survived for about 9 months. Larvae in soil moved towards moisture (see also Myasnikova, 1946). Kaarma (1970) noted that eggs did not hatch at 1–11°C but did so at 15–18°C and 25–26°C. First-stage larvae appeared in eggs within 24 h and hatched in 48 h. Infective larvae appeared in 7 days at 25–26°C and in 10 days at 15–18°C.

Kotlán (1948) infected pigs and reported a histotropic phase in which infective larvae invaded the intestinal mucosa and developed to the fourth stage in 4 days and then re-entered the lumen of the intestine, where growth and the final moult took place in as little as 10 days. Most larvae moulted to the adult stage in 20–30 days and worms were depositing eggs 49 days postinfection.

McCracken and Ross (1970) studied pure infections of *O. dentatum* in experimentally infected pigs. According to these authors 90% of larvae were in the mucosa and in the third stage (1.1 mm in length) in 3–5 days. By 7 days most worms were in the fourth stage and only 20% were in the gut wall; the majority (80%) were in the lumen of the large intestine. Male larvae were about 2 mm and female larvae about 4 mm in length. In 14 days all worms were in the fourth stage and found in the lumen of the intestine, and by 28 days all worms were in the adult stage. The prepatent period was 35 days. Lesions in infected pigs were most obvious in the caecum and were first observed 48 h postinfection. By 4 days numerous distinct nodules became apparent in the caecum and mid-colon. After 3 weeks lesions in the gut disappeared. Talvik *et al.* (1997) reported the prepatent period as 17–18 days (see also *O. quadrispinulatum*) and discussed why the figure is smaller than those reported elsewhere.

Rose and Small (1980) reported on the development and survival of *O. dentatum* in the climatic and environmental conditions prevailing in England.

Christensen *et al.* (1995) infected pigs with 2000, 20,000 and 200,000 eggs and examined them 3–47 days later. The prepatent period in the first two groups was 19 days. In the third heavily infected group the prepatent period was 27–33 days and worms consisted mainly of immature stages. Christensen *et al.* (1996) implanted into pigs stunted worms from heavily infected pigs and found that the worms grew larger than if they had remained in pigs with high populations of worms; apparently the 'stunted' worms consisted mainly of fourth-stage larvae (70%). The fate of stunted adults transplanted into the pigs is uncertain. The authors suggest that stunted growth may be reversible (however, the validity of their conclusion might be questionable if larvae were retarded at the fourth stage rather than the adult stage – R.C.A.).

Christensen (1997) studied infections in pigs with differing female–male sex ratios. Unusually high female–male sex ratio reduced fecundity, as based on eggs excreted by pigs. Nevertheless, even when males were scarce, females managed to get fertilized.

Christensen *et al.* (1997) studied worm population in pigs given trickle infections (100, 1000 or 10,000 larvae each week). Eggs per female (determined by faecal egg counts) were higher in the two larger doses, and females increased in length over time but male lengths remained constant from the 8th week. Only very small numbers of worms eventually were found in the intestine in pigs injected percutaneously or intravenously with 10,000 third-stage larvae. The results suggest that subcutaneous infection of swine is of little importance in the epizootiology of *O. dentatum* in pigs (Nosal *et al.*, 1998). Barnes (1997) attempted to model the population dynamics in 'trickle' or single infections.

O. kansuensis Hsiung and Kung, 1955
Twenty lambs were infected by Xiao and Kong (1987) with third-stage larvae of *O. kansuensis*. In 24 h most larvae had exsheathed and 90% entered the large intestine, where they developed. In 2 days most larvae had reached the fourth stage. Young adults appeared in 24 days and adults in 30 days. The prepatent period was 60–61 days in spring and summer and about 3 months or more in winter. The authors claim there is no histotrophic phase or nodule formation and ulcers were not noted. Fourth-stage larvae moved back and forth in the caecum and colon.

O. quadrispinulatum (Marcone, 1901)
O. quadrispinulatum (syn. *O. longicaudum*) is found, along with *O. dentatum*, in pigs (Taffs, 1969). This species is considered to be more pathogenic than *O. dentatum* and produces larger nodules in the host. *O. quadrispinulatum* tends to occupy the caecum and proximal part of the colon, in contrast with *O. dentatum*, which is usually found more distally (Jacobs, 1967; Kendall *et al.*, 1977). The two species can be separated by the shape of the oesophagus and the length of the tail in the female but otherwise they are morphologically similar (Kendall *et al.*, 1977).

Sonntag (1991) reported that most eggs occurred in the centre of the faeces of pigs. Eggs developed into third-stage larvae from 10°C to 40°C and development time was positively correlated with temperature. However, more third-stage larvae resulted from eggs kept at 20–25°C than at other temperatures. Larvae did not develop at 4°C and −5°C; eggs tolerated these latter temperatures for only 5–10 days. Rate of development was independent of humidity but was significant for their subsequent survival. The optimum humidity for the survival of larvae was 75–100%. Larvae migrated from faeces at 15°C and 25°C with an ambient humidity of 100%; they started to leave faeces in 5–6 days.

According to Spindler (1933) infective larvae were 645–650 μm in length, including the sheath. He infected pigs and 48 h postinfection found encapsulated third-stage larvae in the colon. In 17 days he found fourth-stage larvae and early fifth-stage worms in the lumen of the caecum and colon and fourth-stage worms emerging from nodules. By 35 days the nodules, free of the parasite, were almost totally resolved. The prepatent period was 50–53 days.

Nickel and Haupt (1969) claimed that *O. quadrispinulatum* matured in pigs as early as 3 weeks postinfection. Kendall *et al.* (1977) recovered fourth-stage larvae as early as 4 days postinfection and reported adults in 14 days and eggs in the faeces as early as 33 days. They noted considerable variability in development. For example, third-stage larvae were recovered for up to 72 days postinfection. Talvic *et al.* (1997)

stated that the prepatent period of *O. quadrispiculatum* was 17–19 days (see also *O. dentatum*).

Barutzki and Gothe (1998) noted that first- and second-stage larvae of *O. quadrispinulatum* generally remained in faeces whereas infective larvae tended to leave and move into the surrounding habitat. After a period of 1–4 weeks out of faeces, third-stage larvae were said to return to helminth-free faeces, travelling 80–150 cm.

O. radiatum (Rudolphi, 1803)

O. radiatum is a common cosmopolitan parasite of the large intestine of cattle, zebra and water buffalo. Development and transmission have been studied by Anantaraman (1942), Andrews and Maldonado (1941) and Roberts *et al.* (1962). Eggs were 88–95 × 44–55 μm in size. They embryonated to first-stage larvae and hatched in 12–14 h. First-stage larvae were 400–430 μm and second-stage larvae 630–700 μm in length. The infective stage appeared in culture in 4 days and was 820 μm long. According to Andrews and Maldonado (1941) infective larvae appeared in cultures in about 6 days at 25–30°C. According to Roberts *et al.* (1962), who were unable to infect calves percutaneously (cf. Mayhew, 1939), the third moult was completed in the gut wall in 5–7 days and in 7–14 days fourth-stage larvae entered the lumen of the gut, where the final moult occurred in 17–22 days. The prepatent period was 32–42 days and egg production peaked usually in 6–10 weeks, followed by a rapid decline in egg counts and elimination of worms from the host. According to Anantaraman (1942) infective larvae survived intermittent desiccation if alternated with wetting.

Andrews and Maldonado (1941) infected calves. Third-stage larvae exsheathed and invaded the ileum, caecum and colon in 24–48 h postinfection. Larvae developed in the gut wall and moulted to the fourth stage in 8–9 days. In 10 days fourth-stage larvae entered the lumen of the gut and reached adulthood in the caecum and colon. The prepatent period was 37–41 days.

O. venulosum (Rudolphi, 1809)

O. venulosum is mainly a parasite of sheep and goats but has been found in deer, bighorn sheep, chamois and other ruminants throughout the world. Development was concisely described by Goldberg (1951) (Fig. 3.3). Eggs passed in faeces of the host were 87–105 × 55–64 μm in size and usually in the 16- to 32-cell stage. In culture (charcoal–faeces) first-stage larvae hatched close to one end of the egg in about 24 h. Larvae were on average 554 μm in length 10–12 h after hatching. The first moult occurred in about 24 h after hatching. Second-stage larvae were on average 730 μm in length with 16–20 intestinal cells. At room temperature third-stage larvae were present 3–5 days posthatching. The mouth was closed by the collapse of the buccal tube and three lips were present. Larvae averaged 899 μm in length, including the sheath; without the sheath, larvae averaged 727 μm in length.

Goldberg (1951) gave infective larvae to ten lambs and a goat and examined them at various times thereafter. Three days postinfection, third-stage larvae (about 905 μm in length) were found coiled and encapsulated in the mucosa of the small intestine; the area forming the provisional buccal capsule was clear. Larvae were quiescent. On day 4 postinfection most larvae were found in the lumen of the intestine near the mucosa. Nearly all larvae (96%) had the well-developed provisional buccal capsule of the fourth stage and some were exsheathing. The data showed that development from the third to the fourth stage occurred while larvae were encapsulated in the intestinal wall. After the fourth stage was reached worms emerged into the lumen of the gut and shed the third-stage cuticle. By 5 days postinfection most larvae (95%) had migrated to the caecum and the first metre of the colon. Approximately 98% of the larvae were in the fourth stage. The fourth and final moult usually occurred 13–16 days postinfection, when worms were little more than a third the length of mature adults. Most worms were mature by 31 days postinfection and the average prepatent period was 28 days.

Family Syngamidae

Syngamids, along with the Deletrocephalidae of *Rhea*, have a hexagonal oral opening. Corona radiata and lips are absent. Unlike the deletrocephalids, syngamids lack a dorsal gutter and a perioral groove (Lichtenfels, 1980b). Syngamids occur mainly in the respiratory system of birds and mammals, but one species (*Stephanurus dentatus*) is associated with the urinary system of swine and another (*Archeostrongylus italicus*) occurs in the intestine of porcupines (*Hystrix cristata*). The development and transmission of species in the Stephanurinae and Syngaminae have been investigated.

Subfamily Stephanurinae

The subfamily contains the monotypic genus *Stephanurus* with the single species occurring in swine.

Stephanurus

S. dentatus Diesing, 1839

S. dentatus occurs in capsules in perirenal fat and in the walls of the ureter and adjacent tissues of swine. Capsules containing adult worms communicate by channels with the ureters or the renal pelvis, and eggs of the parasite can pass into the urine and leave the body of the host (Fig. 3.4). According to Alicata (1935b) eggs were 91–114 × 53–65 μm in size and composed of 32–64 cells when passed in urine. Eggs developed to first-stage larvae and hatched in 1–2 days. First-stage larvae moulted in 20 h at 22–24°C when they were 530 μm in length. In 70 h second-stage larvae moulted to the third stage and were 518–610 μm in length (excluding the sheath, which is the second-stage cuticle with its attenuated tail).

Infective larvae survived in soil with faeces in the laboratory for 108 days (Ross and Kauzal, 1932) but Spindler (1934) reported that larvae survived for only 76 days under field conditions and were not resistant to low temperatures, dry conditions or exposure to sunlight. Larvae were negatively phototactic and positively thermotactic. However, during early morning when dew was present larvae were found on blades of grass.

Tromba (1955) gave eggs to earthworms (*Eisenia foetida*) and subsequently found unsheathed larvae in the lumen of the intestine of most of them. Some larvae were noted in masses of amoebocytes called 'brown bodies'. Earthworms harboured larvae for up to 35 days and pigs given earthworms containing larvae became infected. Batte *et al.*

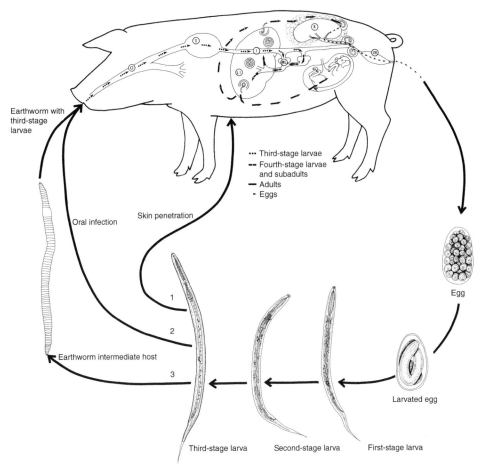

Fig. 3.4.   Development and transmission of *Stephanurus dentatus.* I = intestine; K = kidney; LI = liver; ML = mesenteric lymph node; O = oesophagus; S = stomach; UB = urinary bladder. (Original, U.R. Strelive.)

(1960) confirmed the observations of Tromba (1955). Sinha (1967) infected the earthworms *Eutypheus waltoni* and *Pheretima* sp.

The behaviour of the parasite in swine has been investigated by Bernard and Bauche (1914), Schwartz and Price (1929, 1931, 1932), Ross and Kauzal (1932), Spindler (1934), Batte *et al.* (1960) and Waddell (1969), who reported that swine could become infected by ingesting infective larvae or by the penetration of larvae through the skin, especially abraded skin. They also showed that larvae developed for a time in the liver before invading the ureters and kidneys.

Lichtenfels and Tromba (1972) carried out a careful study in which the larval stages were identified. They gave 2000–20,000 infective larvae to 6–8-week-old pigs which were subsequently examined for worms at various intervals. Larvae exsheathed in the gut and most larvae entered lymphatic vessels and were found in mesenteric lymph nodes as early as 1 day postinfection. Larvae moulting to the fourth stage were found in the nodes as early as 3 days. They appeared in the liver in 10 days and were in the early fourth stage at this time. Moulting fourth-stage larvae and subadult worms were present in the liver 31 days postinfection. The authors concluded that only larvae which had developed in lymph nodes 'contributed substantially to the fourth-stage and adult worms found in the host following infection'.

According to Schwartz and Price (1929, 1931, 1932) worms which had completed their development in the liver perforated the liver capsule and invaded the peritoneal cavity about 77 days postinfection. Worms migrated to the perirenal regions in about 107–113 days and invaded perirenal fat. In the latter, worms burrowed towards the ureters, which they perforated and entered and within which they became encapsulated. Some worms wandered from the ureters into the kidney pelvis and under the kidney capsule. According to Batte *et al.* (1960) worms may spend 4–9 months in the liver before moving to the ureters and the prepatent period was 9–16 months. Batte *et al.* (1966) reported that sows may pass eggs for at least 3 years following an initial infection. These authors also gave larvae to pregnant gilts (young females) and demonstrated transplacental transmission.

Ross and Kauzal (1932) allowed larvae to penetrate the skin of guinea pigs and reported that larvae remained in the skin for 7 days and underwent the third moult. Fourth-stage larvae appeared in the liver in 3–40 days and underwent the fourth moult in 2 months. In about 6 months they migrated directly to the ureters and produced eggs.

In pigs, aberrant fourth-stage and subadult *S. dentatus* may wander in tissues and be found encapsulated in the pancreas, heart, lungs, spleen and skeletal muscles. They may rarely invade the central nervous system and cause paralysis. The parasite has been reported sporadically in cattle, which are liable to ingest larvae crawling on vegetation.

S. dentatus is regarded as an important pathogen of swine, resulting in destruction of liver tissue, thrombosis of hepatic vessels, cirrhosis, peritonitis and cystitis. Infection significantly reduces weight gains in piglets. Also, the parasite may transfer to cattle and cause liver damage.

Steward and Tromba (1957) took advantage of the long prepatent period and the rapid growth of pigs to control the infection. They recommended that breeding be confined to gilts and that the latter should be marketed after weaning their first litters. There would be, therefore, little or no opportunity for transmission to occur between the sow and her young and in a short time the parasites could be eliminated from the swine.

Subfamily Syngaminae

Members of the subfamily occur in the respiratory system of their hosts. Two genera occur in mammals (*Mammomonogamus* and *Rodentogamus*) and three in birds (*Syngamus, Boydinema* and *Cyathostoma*). Species in mammals produce eggs devoid of opercula whereas eggs of the species in birds are operculate. Eggs pass from the respiratory system to the glottis, are swallowed and passed in faeces of the host. Development to the third and infective stage takes place in the egg and the first two larval stages lack the valved oesophageal bulb and the filamentous tail found in free-living stages of most species in other families and subfamilies of the Strongyloidea. The third-stage larva may remain in the egg or hatch spontaneously before being ingested by a paratenic or final host. Development and transmission of species found in birds have been studied rather well but those in mammals have received little attention. Species of the Syngaminae in birds are known as gapeworms since they partially block the respiratory system of birds, causing them to gape or gasp for breath.

Cyathostoma

Members of the genus lack a collar around the oral opening. Males and females are not always found in couples, as in *Syngamus trachea*. Some members of the genus are infective in the egg stage but other species require the intervention of earthworm paratenic hosts.

C. bronchialis (Mühlig, 1884)

C. bronchialis is a parasite of the respiratory system, including the air sacs of ducks, geese and swans. Eggs are ovoid with an indistinct operculum and 74–83 × 49–62 μm in size. Eggs require about 2 weeks to embryonate to the third stage. Cram (1927) found only a single worm in ducks and geese given fully embryonated eggs.

Vasilev and Denev (1972) reported larvae of *C. bronchialis* in various earthworms (*Allolobophora caliginosa, Bimastus tenuis, Lumbricus polyphemus, L. rubellus, L. terrestris, Octolasium complanatum, O. lacteum* and *O. rebelli*). The authors infected galliforme birds by feeding them earthworms but believed that geese were the most suitable hosts.

Fernando *et al.* (1973) were unable to infect geese (*Anser cygnoides, Branta canadensis*) by the oral inoculation of embryonated eggs or hatched third-stage larvae (290–354 μm). However, geese fed naturally infected earthworms (*Allolobophora* sp. and *Lumbricus* sp.) from a waterfowl park became infected. Third-stage larvae invaded the gut wall and entered the peritoneal cavity and air sacs. They migrated to the lungs, where the third moult occurred in 24 h and the fourth moult in about 4 days postinfection. Adult worms migrated to the trachea as single worms in 6 days and were found *in copula* at 7 days. Worms were fully grown by 13 days, when eggs were first found in tracheal mucus.

C. lari Blanchard, 1849

C. lari is a common red parasite of the nasal and orbital cavities of gulls (Laridae) in Europe. Pemberton (1959) allowed eggs from worms found in *Larus ridibundus* to incubate in water at room temperature. He reported larvae in the eggs in 7 days and a moult in 10 days; the second moult was evidently overlooked. Hatching commenced in 13 days and continued for 2 months. Threlfall (1965) also studied development of eggs at different temperatures and suggested that a period of low temperature was necessary for development. Barus (1970) reported that eggs were 85–97 × 41–47 μm in size and had polar opercula. He studied development at different temperatures (20–28°C) and reported that the third stage was reached in 8–14 days. First-stage larvae were 213–248 μm in length and second-stage larvae 255–270 μm. Third-stage larvae, which hatched through one of the opercula, were 248–297 μm in length. The tail of the third-stage larva ended in a ventral finger-like projection.

Pemberton's (1959) attempts to infect chickens with eggs and larvae were unsuccessful. Eggs were given to *Lumbricus terrestris* which were fed 3 weeks later to

3-week-old chickens and, at the same time, two chickens were given eggs and larvae; at necropsy a single worm was found in a chicken given earthworms. Threlfall (1965) kept eggs at room temperature for 23 days and then used them to inoculate orally three 15-day-old fowl chicks and two herring gull chicks. All birds became infected. Gull chicks placed on litter in a pen previously used by infected birds also became infected. *Eiseniella tetraedra* was found naturally infected and *Eisenia foetida* was experimentally infected. A dung beetle, *Aphodius fossor*, was also infected. The evidence seems to indicate that birds can acquire the parasite from ingesting eggs or invertebrate paratenic hosts.

Bakke (1975) found *C. lari* commonly in gulls in Norway. He noted that birds were not infected in spring (April) when they first arrived, but prevalence and intensity peaked in July and decreased as the migration approached in September–October. He believed that birds were infected directly by the ingestion of eggs and larvae on the nesting sites and indirectly by the ingestion of earthworms on farmland. He concluded that reinfection of gulls in spring occurred from the ingestion of earthworms in which larvae had wintered (cf. *Porrocaecum ensicaudatum* in passeriformes).

Mammomonogamus

Members of the genus occur in mammals in the tropics, including cattle, sheep, goats, deer, felids and elephants. The worms occur in the larynx (*M. laryngeus*), nasal cavities (*M. ierei*, *M. nasicola*), pharynx (*M. indicus*) and middle ear (*M. auris*). Two species (*M. laryngeus*, *M. nasicola*) have been reported in humans but are basically parasites of ungulates (Freitas *et al.*, 1995; Nosanchuk *et al.*, 1995). Daoning *et al.* (1997, 1998) described a new species (*M. gangquiensis*) in a human in China and reported three cases of *M. laryngeus* in humans. Fengyi (1997) reported a case in Shanghai.

M. ierei (Buckley, 1934)
Pairs of this species (syn. *Syngamus ierei*) were found in the nares of domestic cats in Trinidad (Buckley, 1934). According to Buckley (1934) eggs were 92 × 49 μm in size with pitted shells. When passed from the host, eggs were in the four- to six-cell stage. At 26–30°C the first-stage larva (320–400 μm in length) was attained in 3–4 days and the second stage (450 μm) in 5–6 days. Third-stage larvae (450 μm) appeared after 8 days and hatched with the first- and second-stage cuticles. The first sheath was lost a day or two after hatching and the second in about 12 days. Attempts to infect cats by oral inoculation of larvae were unsuccessful.

Syngamus

S. trachea (Montagu, 1811)
S. trachea, a red parasite, is a common cosmopolitan inhabitant of the trachea of wild and domesticated galliformes, some passeriformes and rarely other birds. Males are considerably smaller (2.0–6.0 mm in length) than females (5.0–20.0 mm) according to Cram (1927). The mature male remains attached by means of its bursa to the vulvar region of the female. Eggs, which were 85–90 × 50 μm, passed out of the vulva and

under the margin of the male bursa. Eggs were usually in the 16-cell stage when passed from the host and developed into third-stage larvae (about 308 µm in length, with a conical tail) in about 1 week at 25°C. Ortlepp (1923b) claimed that there was only one moult in the egg but Wehr (1937) observed a moult in the egg at 5 days and another at 7 days at 24–30°C. Some eggs hatched after about 10 days whereas others did not, according to Wehr (1937). Ortlepp (1923b) reported that eggs hatched at 25°C but not at 20°C.

Birds have been infected experimentally by oral inoculation of eggs or free third-stage larvae (Ortlepp, 1923b; Morgan and Clapham, 1934; Wehr, 1937). However, several authors have shown that infective larvae invaded earthworms (*Eisenia foetida, Helodrilus caliginosus, Lumbricus terrestris*) and became encapsulated in the muscles and that birds could readily be infected by feeding on infected earthworms (Walker, 1886; Seurat, 1916; Waite, 1920; Clapham, 1934; Taylor, 1935a, 1938; Wehr, 1937; Nolst, 1971, 1973; Bates, 1972). *E. foetida*, in particular, flourishes in soil contaminated by poultry faeces and is probably an important paratenic host in places where poultry are raised in semi-natural conditions. Larvae do not change morphologically in the earthworms and can persist for 3.5 years (Taylor, 1938).

According to Nolst (1971, 1973) eggs hatched in the second half of the midgut of *E. foetida*. Larvae then penetrated the wall of the posterior intestine and entered the coelom. From the coelom, larvae invaded the body muscles associated with the posterior intestine and dispersed into the muscles of all segments of the host. Larvae did not change morphologically in the earthworm paratenic host and eventually each became encapsulated. According to Nolst (1971) larvae which occurred in the first 45 segments of *E. foetida* were in the external muscle layer of the body wall. In other segments worms were found in the internal muscle layer, the skin, the coelom and the nervous system; in the rectum worms were found in the transverse muscles.

Other invertebrates may also serve as paratenic hosts of *S. trachea* under natural conditions. Taylor (1935a, 1938) experimentally infected terrestrial snails and slugs. Clapham (1939a,b) infected fly larvae (*Lucilia sericata* and *Musca domestica*). Rizhikov (1941) infected chickens by feeding them *Lymnaea stagnalis* which had been exposed to larvae of *S. trachea* and suggested that ducks and corvids known to eat snails may acquire infections from such paratenic hosts. Enigk and Dey-Hazra (1970a) concluded that in northwest Germany *S. trachea* survives cold seasons in paratenic hosts, since eggs will not embryonate below 15°C. Madsen (1952) reported that Danish wild birds which consumed earthworms were more often infected with gapeworms than other birds. There is limited experimental evidence that birds given larvae in earthworms developed infections of higher intensity than those given eggs containing larvae (Wehr, 1937).

A number of authors have followed development in the definitive host (Ortlepp, 1923b; Clapham, 1935a; Wehr, 1937; Barus and Blažek, 1965; Enigk and Dey-Hazra, 1970a; Fernando *et al.*, 1971b). The consensus is that infective larvae exsheathed in the duodenum and invaded the duodenal wall. Some larvae entered the peritoneal cavity but most entered veins in the gut wall. Larvae appeared in the liver as early as 2 h post-infection (Fernando *et al.*, 1971b) and in the lungs in 4 h. The third moult occurred in 2 days in the lungs, according to Fernando *et al.* (1971b), and the fourth some time between 4 and 7 days, according to Wehr (1937), who found subadults in 7 days and noted that fourth-stage larvae could be distinguished readily by the presence of the provisional buccal capsule. Wehr (1937) also observed worms *in copula* in the lungs of chicks 9 days postinfection but Fernando *et al.* (1971) found copulating worms in bronchi 5 days postinfection. Some worms moved in pairs to the trachea as early as 7 days postinfection, however. The prepatent period was 12–17 days (Wehr, 1971). The maximum longevity in chickens was 147 days, in turkeys 224 days and in guinea fowl 98 days (Wehr, 1939). Barus (1966) reported that infections lasted for 23–92 days in chickens and 48–126 days in turkeys.

3.4

The Superfamily Trichostrongyloidea

The Trichostrongyloidea is by far the largest superfamily among the bursate nematodes. Divided into 14 families and 24 subfamilies by Durette-Desset and Chabaud (1977, 1981) and Durette-Desset (1983) the group is distinguished from the hookworms and the strongyles by the fact that the buccal capsule is absent or greatly reduced and lips and **corona radiata** are vestigial or absent. The lateral lobes of the bursa are highly developed although the dorsal lobe may be considerably reduced. The host range of the trichostrongyloids is broad. Found in all terrestrial vertebrate groups, they are extremely diverse as parasites of mammals, especially bats, rodents and ruminants. They also occur in monotremes and Australian marsupials. They are much less common and diverse in amphibians, reptiles and birds.

Trichostrongyloids are essentially parasites of the stomach and intestine of their hosts. Nevertheless, species of the subfamily Dictyocaulinae (with *Dictyocaulus*) are widespread and successful parasites of the trachea and bronchi of ruminants and equines. Also, some rare genera occur in bile ducts (*Hepatojarakus*), nasal cavities (*Nasistrongylus*) and mammary glands (*Mammanidula*) of their hosts.

Herbivorous hosts usually acquire their trichostrongyloids by ingesting infective larvae contaminating their food. Infective larvae of species in ruminants, like those of the strongyles, have a tendency to climb in films of moisture on to vegetation, where they become available to the host.

Exsheathment, the process whereby an ingested ensheathed larva escapes from the cuticle of the second stage in the gut of the host, has been extensively studied in the Trichostrongyloidea (Sommerville, 1957; Rogers and Sommerville, 1963, 1968). The initial stimulus comes from the host and, depending on the species of nematode, may include dissolved gaseous carbon dioxide, undissociated carbonic acid or hydrochloric acid, and/or pepsin at a low pH. The larva responds to these stimuli by producing an exsheathing fluid, which arises from an area near the excretory gland and which attacks an encircling area of the sheath near the cephalic end. Typically the anterior end of the sheath detaches from the main body of the sheath like a cap, thus allowing the larva to escape.

In oral infections, exsheathed larvae may invade the gut mucosa and develop to the fourth stage before re-entering the lumen, where they mature. In species found in such hosts as rodents both oral and percutaneous infection may occur. In percutaneous infection larvae migrate by way of the lymph and blood to the heart and lungs. In lungs the parasite moults to the fourth stage which migrates up the trachea to the throat and gut, where it matures (e.g. *Nippostrongylus brasiliensis*).

There are some oddities in the development and transmission of the trichostrongyloids, however. In *Dictyocaulus* spp. infective larvae invade the gut wall and migrate by way of the lymphatics and blood to the lungs, where they mature. Larvae of *Hepatojarakus malayae* apparently follow the hepatic portal system to the liver, where they mature. *Ollulanus tricuspis* of the stomach of felines and pigs is autoinfective and is transmitted entirely by emesis. Undoubtedly other interesting surprises await discovery. Perhaps *Mammanidula asperocutis* of the mammary glands of *Sorex* spp. is transmitted through the milk.

The phenomenon of **arrest** or **inhibition** during development in the definitive host is an important part of the biology of many trichostrongyloids, especially those in lagomorphs and ruminants, and has been extensively investigated. Michel (1974) and Gibbs (1986) have published major reviews. In many trichostrongyloids third-stage larvae ingested by the host exsheath and invade the gut mucosa where they develop rapidly to the early fourth stage and then return to the gut lumen, where they mature in a few days. Some members of the genus *Trichostrongylus* (e.g. *T. colubriformis* and *T. vitrinus*) may remain as third-stage larvae in the mucosa. When arrest occurs, an unusually high percentage of larvae remain in the mucosa for prolonged periods without further development. Arrested development plays an important role in the transmission of some trichostrongyloids because it allows species with limited adult life spans to survive in the arrested stage during periods when external conditions are unsuitable for the development, survival and transmission of larval stages (Anderson *et al.*, 1965; Blitz and Gibbs, 1972a,b; Michel, 1974). For example, in temperate regions where transmission cannot occur in winter, larvae acquired in autumn develop to the fourth stage in the mucosa and remain arrested until spring, when conditions are again favourable for transmission, whereupon they invade the lumen of the gut and grow to egg-producing adults. Arrested development also occurs in areas where there are periods of extreme dryness unsuitable for transmission (Williams and Bilkovich, 1971; Shimshony, 1974; Baker *et al.*, 1981, 1984; Williams *et al.*, 1983; Gatongi *et al.*, 1998).

Arrested trichostrongyloid larvae tend to develop to adulthood in ewes during parturition and lactation. Thus, transmission is enhanced at a time when young animals with little or no immunity are available for infection for the first time (Kassai and Aitkin, 1967; Kassai, 1968; Dineen and Kelly, 1973; Gibbs, 1982; Gibbs and Barger, 1986). The synchronized maturation of arrested larvae in spring in lactating ewes results in the so called **spring rise** in numbers of eggs passed by the host (Taylor, 1935b), a phenomenon also referred to as the **periparturient rise** by Salisbury and Arundel (1970).

Some factors responsible for arrest have been recognized. Experiments have shown that an important genetic component is involved and it is possible, by experimental, selection to increase the percentage of larvae in a population that are liable to arrest (Watkins and Fernando, 1984, 1986a,b). Waller and Thomas (1975) noted earlier that *Haemonchus contortus* in northern England always arrests and that it is an obligatory part of its development. Seasonal factors are also important in some species: **seasonal arrest** has been likened to diapause in insects and apparently occurs in response to such stimuli as low and/or fluctuating temperatures in autumn or the presence of hot, dry conditions in summer acting upon the free-living stages, especially the third larval stage (Armour *et al.*, 1969; Fernando *et al.*, 1971; Blitz and Gibbs, 1972a; Hutchinson *et al.*, 1972; McKenna, 1973; Armour and Bruce, 1974; Armour, 1978; Horak, 1981). Increasing moisture and variations in photoperiod (Gibbs, 1973; Connan, 1975; Cremers and Eysker, 1975) have also been suggested as possible stimuli initiating arrest.

In addition to factors extrinsic to the host, factors within the host have been suggested as causes of arrest, the most important being immunity (Dunsmore, 1961; Ross, 1963; Donald *et al.*, 1964; Dineen *et al.*, 1965a,b). Larvae entering a host with an established population of adult worms are most likely to arrest (Fox, 1976; Gibson and Everett, 1976; Michel, 1978; Behnke and Parish, 1979; Snider *et al.*, 1981; Adams, 1983; Smith *et al.*, 1984). The number of arresting worms in a host may be related to the number of adult worms present (Dineen, 1978) in that, as adults become senescent and die, there is a relaxation of immunity and some arrested larvae then leave the gut wall to replace the lost adults. Thus, the host immune system might regulate the number of adult worms.

The possible effects of environmental factors (photoperiod, temperature, diet) on host physiology have been investigated to determine if the latter could be related to arrest but results have been inconclusive (Gibbs, 1986). Parasite interactions have also been suggested as possible initiators of arrest, including the intensity of larvae and/or adult worms in the host (crowding effects) (Schad, 1977) and the possible presence of pheromones or other parasite-related chemicals which might cause invading larvae to arrest in the gut wall (Michel, 1963; Gibbs, 1986).

Factors responsible for the cessation of arrest and the resumption of development are not well understood. As noted earlier, loss or decline in immunity in the host may result in the cessation of arrest. However, there is little evidence that a decline in immunity can break arrest initiated by environmental factors (e.g. seasonal arrest) (Gibbs, 1986) and it has been suggested that seasonally arrested larvae behave as though they are in diapause, in which case an 'internal clock' would break arrest at a predetermined time after its induction by some stimulus (Armour and Bruce, 1974; Horak, 1981).

Family Strongylacanthidae

Strongylacantha

S. glycirrhiza Beneden, 1873
This is a parasite of the intestine of rhinolophoid bats (Chiroptera). Seurat (1920) reported that eggs were 111×56 µm in size and laid in the four- or eight-cell stage. Eggs embryonated rapidly at room temperature and first-stage larvae hatched. Larvae were 410–435 µm in length; they moulted twice without growing significantly and retained both shed cuticles. Larvae did not feed but their cells were packed with nutrient granules. The buccal cavity of third-stage larvae lacked hooks found in the adult and the tail was conical and rather thick.

Family Amidostomatidae

Members of the family are parasites of the gizzard of aquatic birds including geese, ducks, swans, coots and, rarely, charadriiformes. The family is divided into the Amidostomatinae, with a long, well-developed buccal cavity, and the Epomidiostomatinae in which the buccal cavity is greatly reduced. Eggs are large, oval and smooth shelled and are passed from the host in the morula stage with about 32 cells. Development of eggs in the external environment is extremely rapid at a range of temperatures and leads to the third-stage larva enclosed in the shed cuticles of the first and second stages. The third-stage larva hatches from the egg and in a day or two the first-stage cuticle is lost but that of the second stage is retained as a sheath. Free-living third-stage larvae require several days before they are infective. Eggs and larvae are not resistant to desiccation and their survival depends on the presence of moisture. Larvae are reported to remain alive in water for prolonged periods and can rise in the water column or stay in one place by vigorous swimming. Larvae tend to be negatively geotactic and will ascend vegetation. Birds usually become infected by ingesting larvae on food (geese in particular are grazers) but one species (*Amidostomum anseris*) can also infect geese by penetrating the skin and undertaking a lung–trachea migration before establishing itself in the gizzard. The role of skin penetration in transmission under natural conditions is unknown. When larvae reach the gut in oral infection, they shed the second-stage cuticle and invade the epithelium at the junction of the proventriculus and gizzard, where development to the fourth and adult stages takes place. The prepatent period is about 2–3 weeks.

Subfamily Amidostomatinae

Amidostomum

A. acutum (Lundahl, 1848)

A. acutum (syns *A. anatinum*, *A. boschadis*, *A. skrjabini*: see Czaplinski, 1962; Kobulei and Rhizhikov, 1968; Lomakin, 1988) is a common, cosmopolitan parasite of the gizzard of geese and ducks. According to Zajicek (1964) eggs developed in tap water, soil or faeces at 18–24°C and third-stage larvae hatched from eggs after about 50 h of incubation. Leiby and Olsen (1965) removed eggs from the uteri of identified female worms from ducks in Colorado, USA (mixed infections of two species occurred in the ducks) and cultured them in water. Eggs were 70–78 × 44–56 μm in size. Only eggs in the morula stage developed. At 22–24°C, first-stage larvae (501–517 μm) in length, appeared in the eggs in 13–14 h. These larvae entered a lethargus in 19–21 h and moulted in 22–24 h. The second moult took place in 31–36 h and the third-stage larvae, surrounded by the first- and second-stage cuticles, became active in the egg in 32–39 h. Eggs hatched in 52–56 h at 18–20°C and in 24–28 h at 36°C. The free third-stage larva shed the first-stage cuticle in about 27 h. Third-stage larvae required 5 days at 18–20°C to become infective. The ensheathed infective larvae were 532–642 μm in length with a tapered tail ending in a rounded point. The third stage was slightly shorter than the second stage, because of its shorter tail.

According to Leiby and Olsen (1965) infective larvae exsheathed in 20 h in the gut of ducklings and domestic chicks. Larvae penetrated the epithelium near the junctions of the proventriculus and intestine with the gizzard and remained under the epithelium in these regions until maturity. The third moult occurred in 2–3 days and the fourth in

7–8 days postinfection. The fourth-stage larva had a well-developed buccal capsule. The prepatent period was 14–21 days.

A. anseris (Zeder, 1800)

A. anseris (syn. *A. nodulosum*) is a cosmopolitan parasite mainly of wild and domesticated geese, although it has been reported in ducks and some other birds. Herman and Wehr (1954) and Cowan and Herman (1955) called attention to infections in Canada geese (*Branta canadensis*) in North America and reported severe disease signs in heavily infected birds. Cowan (1955) noted eggs were in the morula stage when passed by the host and that they hatched in 24–48 h in tap water at room temperature (cf. Bausov, 1969; Wang, 1983). Larvae shed the cuticle of the first larval stage at hatching. The prepatent period in goslings given infective larvae was 14–25 days. The parasites did not mature simultaneously and infections persisted in the birds for 18 months. Geller (1962) concluded that the worms sucked blood and that eggs did not develop beyond the morula stage in the host because of the latter's elevated temperature, which inhibited development of the eggs.

Enigk and Dey-Hazra (1968, 1969, 1970b) reported that eggs reached the infective stage in 23 days at 20°C. The larvae were able to swim and could keep themselves in the water column and even rise to the surface. Both eggs and larvae were highly susceptible to desiccation. Thirty-one of 43 domestic goslings were successfully infected by placing infective larvae on the skin. The prepatent period in these birds was 15–18 days, similar to that when birds were infected orally. Tracheotomy revealed that larvae which invaded the skin migrated to the lungs and appeared in the mucus of the trachea 16–32 h later.

Stradowski (1974) reported that the eggs of *A. anseris* could survive freezing but most third-stage larvae were killed. Goslings were given eggs and third-stage larvae of known age (Stradowski, 1971). Only 3.3% of larvae grew and established themselves in goslings given eggs containing fully developed third-stage larvae but 55% of larvae given 1 h after hatching established themselves in the gizzard. If larvae were given at 24 h and 1 week after hatching, 60.3% and 68%, respectively, established themselves in the birds. Only 3% of larvae given to goslings 9 weeks posthatching invaded the gizzard.

Few worms established themselves in ducks aged 6 months to 2 years and none reached maturity (Stradowski, 1972). However, 21-day-old ducks became infected, although only 17.3% of 30 larvae given were recovered from the gizzards of the birds 35 days postinfection. The prepatent period was prolonged, i.e. 33–67 days. Thus, ducks are not as suitable hosts for *A. anseris* as geese. Stradowski (1977) reported the prepatent period in domestic geese as 14.1–31.1 days, depending on the age (2–3 days and 1–3 years) of the birds, with the shortest period in the juveniles. In geese 2–3 months and 1–2 years of age the patent periods were 174 and 129 days, respectively; after egg production had ceased, worms lived for only an additional 7–8 days.

Phuc and Varga (1975) failed to infect domestic chicks with *A. anseris*. Attempts to infect ducklings (500 larvae *per os*) were successful but only 15% of the larvae reached adulthood; the worms were stunted and development was delayed. Egg output in infected ducklings was only 350 per day. In goslings, in contrast, a mean of 70% of the larvae reached adulthood, the earliest prepatent period was 14 days, and egg output reached an average of 1850 per day. These results suggest that *A. anseris* is more adapted to geese than to ducks. However, Wang (1983) reported that in ducklings the fourth stage was reached in 3 days and the fifth in 7–8 days. The prepatent period was 16–18 days.

Yaron (1968) gave pigeons (*Columba livia*) infective larvae but the worms failed to reach adulthood in this host.

A. fulicae (Rudolphi, 1819)

A. fulicae (syn. *A. raillieti*; see Pavlov, 1960) is a parasite of coots (*Fulica americana, F. atra*) in North America and Europe.

Barus (1964) reported that eggs (from *F. atra*) were 110–111 × 50–68 μm in size and in the morula stage when passed from the host; Leiby and Olsen (1965) gave the size of the eggs (from *F. americana*) as 100–117 × 67–74 μm and confirmed they were passed from the host in the morula stage. Barus (1964) cultured eggs in saline at 20–27°C. Infective larvae were 630–690 μm in length without the cuticle of the second stage, and 690–780 μm with the sheath. Larvae could be distinguished from those of *A. anseris* and *A. acutum* by certain morphometric differences and the shape of the tail.

According to Leiby and Olsen (1965) eggs kept in water at 22–24°C contained first-stage larvae in 12–14 h. The first lethargus occurred in 19–20 h and the first moult was completed in 21–24 h; second-stage larvae were 625–648 μm in length. The second moult commenced in 27–29 h and was completed 1–2 h later. Eggs hatched in 36–40 h. Eggs kept at 18–20°C and 36–38°C hatched in 43–47 h and 24–30 h, respectively. Eggs kept at 6°C for 35 days failed to hatch. Following hatching, third-stage larvae required 9 days at 18–20°C to become infective. Infective third-stage larvae were 612–672 μm in length (presumably without the sheath). The tail was conical with a rounded extremity.

Domestic ducklings and chicks were given *per os* ensheathed third-stage larvae which had hatched from eggs 9–12 days earlier. At necropsy 28 h postinfection, exsheathed larvae were found in the epithelial lining of the gizzard near its junction with the proventriculus. The authors claimed that the third moult occurred in 4–5 days and the fourth 12–14 days postinfection and that the prepatent period was 21–27 days. (These data are at variance with earlier comments on the lack of infectivity of larvae. They may be based on the results of infecting domestic chicks and examining them at 9, 10 and 12 days; see Table 6 of Leiby, 1963 – R.C.A.) The fourth–stage larva was said to have a well-developed buccal capsule.

Subfamily Epomidiostomatinae

Epomidiostomum

E. uncinatum (Lundahl, 1848)

E. uncinatum (syn. *E. anatinum*; see Czaplinski, 1962) is a cosmopolitan parasite of the gizzard of ducks and geese. Leiby and Olsen (1965) collected eggs from ducks in Colorado and cultured them in water. Eggs from female worms were in the one-cell or the morula stage and were 70–93 × 44–56 μm in size. Only eggs in the morula stage developed in water. At 22–24°C first-stage larvae (maximum length 398 μm) appeared in 15–17 h. The first lethargus occurred in 26–29 h and the moult was completed in 28–30 h (second-stage larvae were 406–436 μm in length). The second moult was completed in 39–41 h and hatching took place in 43–49 h. At 18–20°C and 36°C eggs hatched in 56–59 h and 34–36 h, respectively. Eggs kept at 6°C developed but did not hatch. Larvae and eggs were not resistant to desiccation. Third-stage larvae hatched, retaining both the first- and second-stage cuticles, but the first was lost in 1–2 days. Larvae required 4 days after hatching to become infective and were 408–457 μm in length.

Third-stage larvae, which were 4–13 days old and had developed in water at 18–20°C, readily infected ducklings and chicks (Table 8, in Leiby, 1963). They exsheathed and penetrated the lining of the gizzard within 28 h postinfection. The third moult took place in 3–4 days and the final moult in 7–8 days postinfection. Eggs appeared in the faeces of the birds in 16–24 days. The fourth-stage larva had a wide, shallow buccal capsule.

Wang (1983, English summary only) reported that first-stage larvae developed in eggs in 24 h and hatched in 29 h at 28–30°C. They moulted to the second stage in 2 days and to the third in 5 days and were infective by 6 days. (These reports are at variance with other reports on this and other species in the family – R.C.A.) In ducklings, larvae moulted to the fourth stage in 2–3 days and to adults in 14–15 days.

Family Trichostrongylidae

Subfamily Libyostrongylinae

Libyostrongylus

L. douglasi (Cobbold, 1882)

The development and transmission of the 'wire worm' of the proventriculus of the ostrich (*Struthio camelus*) were described by Theiler and Robertson (1915) in South Africa. Eggs were ovoid and 59–74 × 36–44 μm in size. Eggs required 3–4 days to pass from the stomach of the host and appear in the faeces. During this period eggs developed from the one-cell or two-cell stage to the morula stage. Eggs developed at 27–28°C and first-stage larvae hatched and developed to the infective stage in 98 h. At 32–33°C, 35°C and 36°C comparable rates of development were 67 h, 60 h and 59 h, respectively. Development of eggs and hatching were retarded at 6–9°C and larvae failed to reach the third stage.

Ova and first-stage larvae kept at room temperature in dry conditions for 2–3 years nevertheless developed normally when moistened. Larvae failed to reach the third stage in water and winter temperatures greatly retarded development of eggs and larvae but did not stop it.

First-stage larvae were about 240 μm in length when they hatched from eggs. The tail was slender and ended in a minute knob on a slender appendix. The oesophagus had a valved bulb. The earliest second-stage larvae were 600 μm in length. The tail of this stage was conical with a rounded tip. Second-stage larvae grew and moulted to the third stage, which was about 745 μm in length including the sheath and 530 μm without it. (The lengths of larvae given by the authors were frequently contradictory – R.C.A.) Infective larvae were distinguished by the presence of a knob on the end of the tail with numerous minute spines ('prickly'). They were markedly active at warm temperatures and in moist conditions, and negatively geotactic in films of moisture and when not exposed to direct sunlight. They were negatively heliotactic and died after prolonged exposure to sunlight. Larvae tended to climb to the tips of blades of grass and also readily disappeared into the soil in bright sunlight. They avoided decomposing faeces and were resistant to desiccation; some survived drying for up to 9 months. The larvae exsheathed in artificial digestive juices and dilute solutions of various acids.

Attempts to infect ostriches by placing numerous larvae on the skin were unsuccessful. Infective larvae given *per os* exsheathed in the proventriculus of the birds and apparently invaded the mucosa where, in 5–6 (average 4.3) days, they moulted to the fourth stage. In 18–21 days most larvae had moulted to the fifth stage but the third- and fourth-stage cuticles sometimes persisted around the worms for 27 days. Adult worms were found on the surface of the mucosa and in the lumen of glands. Worms grew after the final moult and by 33 days males were about 4.2 mm and females 5.4 mm in length. Sperm was noticed in males in 23 days and eggs appeared in the ovijector of females in 33 days postinfection. The authors concluded that one ostrich with an undetermined number of adult worms passed 3.5 million eggs in 1 day.

The authors investigated the use of various drugs and chemicals to kill free-living and parasitic stages; they also investigated husbandry practices which would eliminate or control the infection in ostriches. They concluded succinctly that 'the solution of the wire worm problem does not lie in the drenching of birds with drugs, but in the rearing of birds free of the parasite. That this can be done, our experiments clearly show.'

Paralibyostrongylus

P. hebrenicutus (Lane, 1923)

This is a parasite of *Atherurus africanus* and can be transmitted to guinea pigs and rabbits. Cassone *et al.* (1992) followed its development in guinea pigs. First-stage larvae were 335–420 µm and third-stage larvae 595–745 µm in length. The host can be infected by inoculation subcutaneously or orally. Larvae ensheathed soon after invading the host. Following subcutaneous inoculation larvae reached the lungs (presumably by way of the lymphatics and the right heart) in 8 h and were found in the gastric mucosa by day 3. Following oral inoculation, the infective larvae had invaded the gastric crypts in 24 h. Regardless of the route of infection, the third and fourth moults occurred in the mucosa by 19 days and eggs appeared in females on day 28.

Obeliscoides

Two clearly defined subspecies of the genus are recognized, namely: *O. cuniculi cuniculi* (Graybill, 1923) from the stomach of the cottontail rabbit (*Sylvilagus floridanus*) and less commonly from groundhogs (*Marmota monax*); and *O. c. multistriatus* (Measures and Anderson, 1983a) from the stomach of the snowshoe hare (*Lepus americanus*) (see Measures and Anderson, 1983a,b, 1984). Unfortunately, the type host of *O. c. cuniculi* is the domestic rabbit (*Oryctolagus cuniculus*) but Graybill (1923, 1924) thought that the worms in the domestic rabbit originated from cottontail rabbits. There is no evidence that hybrids between the two subspecies occur in nature. Experimental crosses of male *O. c. cuniculi* and female *O. c. multistriatus* produced viable progeny to the F_3 generation with systematic characters (in the synlophe) intermediate between the two subspecies (Measures and Anderson, 1984). The reciprocal mating, however, failed to result in progeny and it was concluded that the two subspecies are incipient species.

Alicata (1932) first described development and transmission of *O. cuniculi* in domestic rabbits. He did not give the source of the parasite he studied but we shall assume that he was dealing with *O. c. cuniculi* originating from cottontail rabbits. Similarly, we shall assume that Sollod *et al.* (1968) were dealing with the cottontail subspecies. Worley (1963) derived his material from cottontail rabbits in Montana and we may assume he studied *O. c. cuniculi*.

Studies of arrest by members of the Ontario Veterinary College (Fernando *et al.*, 1971; Hutchinson *et al.*, 1972; Hutchinson and Fernando, 1974, 1975; Watkins and Fernando, 1984, 1986a, b) were based on parasites maintained for many years in white rabbits but originating from snowshoe hares and presumably belonging to *O. c. multistriatus*. It is also assumed that field studies on snowshoe hares reported by Gibbs *et al.* (1977) refer to *O. c. multistriatus*.

O. cuniculi cuniculi (Graybill, 1923)

According to Alicata (1932) eggs were 75–91 × 42–53 μm in size and were passed in faeces in the 32-cell stage. In tap water, eggs embryonated to the first stage and hatched in about 30 h. Newly hatched first-stage larvae were 320–330 μm in length with conical, sharply pointed tails. Second-stage larvae (471–750 μm in length) were formed in 65 h. The first two larval stages fed constantly but before the second lethargus the buccal cavity closed. In 6 days, faecal cultures contained exsheathed third-stage larvae 653–710 μm in length; the tail of the infective larva was shorter than that of the second stage.

Infective larvae withstood 2°C to –4°C for 30 days but died when kept at –18°C for 3 days. Larvae were not markedly resistant to desiccation and died after a few hours of exposure to air-drying at room temperature. Larvae were phototactic but repelled by strong light. Infective larvae placed on the skin of rabbits failed to exsheath or penetrate. Rabbits were readily infected orally and larvae reached adulthood in the stomach in 16–20 days.

Sollod *et al.* (1968) reported that infective larvae exsheathed within 1 h post-infection in rabbits and the third-stage larvae invaded the gastric mucosa within 24 h. The third moult occurred in the mucosa and was completed by 5 days. The final moult probably occurred as worms migrated from the mucosa, starting on day 5 and continuing for at least 14 days, when 35% of all worms were on the surface of the gastric mucosa. The prepatent period was 19–25 days.

Worley (1963) transmitted *O. c. cuniculi* to New Zealand white rabbits (*Oryctolagus cuniculus*) and Dutch rabbits. Stable infections lasted for 4–8 months after oral inoculations of 80–10,000 larvae. The prepatent period was 16–23 days. The patent period averaged 138 days in four rabbits given 90–160 larvae and exceeded 196 days in two rabbits given 1000–10,000 larvae. Adult worms adhered closely to mucus coating the stomach and some worms were found in gastric crypts.

O. cuniculi multistriatus Measures and Anderson, 1983

Gravid females in saline discharged eggs in the morula stage (Measures and Anderson, 1983b). Eggs (Fig. 3.5A,B) were 91–111 × 41–51 μm in size and were generally

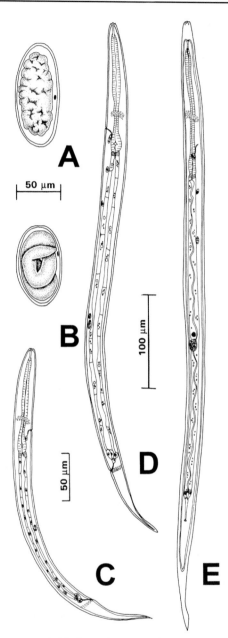

Fig. 3.5. Free-living stages of *Obeliscoides cuniculi multistriatus*: (A) newly laid egg; (B) larvated egg; (C) first-stage larva; (D) second-stage larva; (E) ensheathed third-stage infective larva. (After Measures and Anderson, 1983 – courtesy *Canadian Journal of Zoology*.)

distributed throughout the fecal pellets. At 22°C first-stage larvae appeared in eggs in 11 h and hatching took place in 30–36 h. First-stage larvae (Fig. 3.5C) were 369–399 μm in length. The first moult occurred in 84 h, when larvae were

415–473 μm in length. The second moult occurred at 6–9 days, when larvae were 529–589 μm in length (Fig. 3.5D). In 7–8 days infective larvae (Fig. 3.5E) were 503–582 μm without the sheath.

Infective larvae were given orally to New Zealand white rabbits (*Oryctolagus cuniculus*) (see Measures and Anderson, 1983c). Larvae exsheathed in the stomach within 24 h. The third moult occurred in 3 days, when male larvae were 0.9–1.2 mm in length, and the fourth moult at 8–11 days, when male larvae were about 4.3–4.7 mm in length; comparable figures for females were 1.0–1.2 mm and 6.8–8.8 mm. Males were mature in 16 days and copulation occurred in 15–16 days. The prepatent period in rabbits was 16–22 days.

Patent infections were established in groundhogs (*Marmota monax*), snowshoe hares (*Lepus americanus*) and cottontail rabbits (*Sylvilagus floridanus*) but attempts to infect mice, rats, hamsters, gerbils and guinea pigs were unsuccessful (Measures and Anderson, 1983c).

Gibbs *et al.* (1977) noted in Maine that the greatest intensity of *O. c. multistriatus* occurred from spring to summer (March to July) in snowshoe hares and most nematodes were adult. The percentage of fourth-stage larvae increased from late summer to winter and by December fourth-stage larvae constituted 60% of the total worm population. After February, numbers of fourth-stage larvae declined and the numbers of adults increased.

Stockdale *et al.* (1970), Fernando *et al.* (1971b) and Hutchinson *et al.* (1972) showed that infective larvae of *O. c. multistriatus* stored at 4°C (late autumn temperature in Ontario) for 2–8 weeks before being given orally to rabbits resulted in increasing proportions of larvae arresting in the fourth stage in the stomach. Prolonged storage at 4°C (for more than 10 weeks) resulted in a decrease in the percentage of larvae arresting. Storage at 5°C resulted initially in significant arrest but prolonged storage was followed by acclimation and the elimination of the tendency to arrest, and raising the temperature of storage did not significantly increase the percentage of larvae arresting in rabbits. Maintenance of larvae at 15°C followed by a sudden decrease to 5°C resulted in sudden arrest, which was not reversed even after 25 days of storage at 5°C.

Watkins and Fernando (1984) showed that arrested development has a genetic basis. They increased the propensity of *O. c. multistriatus* to arrest in response to cold treatment from 15% to over 90% in five generations by a process of selection (the transfer of arrested fourth-stage larvae from rabbit to rabbit). In subsequent generations the tendency to arrest remained high as long as selection pressure was maintained, and the isolate retained this tendency without prior cold treatment of infective larvae. In the absence of selection for arrest, however, the isolate reverted to one with a lower tendency to arrest. In subsequent articles Watkins and Fernando (1986a,b) reported that selection of arrested larvae was carried out by inoculating rabbits with single doses of cold-treated infective third-stage larvae to determine worm population structure throughout the infection and to assess the ability to arrest of the progeny produced during those infections. They concluded from these studies that the emergence from arrest was independent of the presence of adult worms in the stomach and that internal control of resumption of development may have a genetic basis, i.e. that larval development is controlled by polygenes and that several genotypes can exhibit the same phenotype. Arrested larvae of *O. c. multistriatus* act as a reservoir for replenishment of the adult population and increase the duration of the infection. In natural infections overwintering arrested larvae can apparently transmit genes for arrest to their progeny but the genes remain unexpressed until acted upon by the appropriate environmental stimuli.

Subfamily Cooperiinae

Cooperia

Members of *Cooperia* are cosmopolitan parasites of the small intestine – mainly of Bovidae but they also occur in Giraffidae, Camelidae and Cervidae.

C. curticei (Railliet, 1893)

C. curticei is a parasite of sheep, goats, deer and mouflon. Andrews (1939) reported that eggs were 70–82 × 35–41 μm in size and were in the 16- to 32-cell stage when passed in faeces of the host. At room temperature eggs hatched in about 20 h, the first-stage larva being about 300 μm long. The first moult took place about 30 h after hatching. Both first- and second-stage larvae had sharply pointed tails. In about 90 h after hatching larvae reached the infective stage, which was 711–860 μm in length including the sheath.

Ahluwalia and Charleston (1974) reported that eggs and larvae of *C. curticei* developed to the third stage in a temperature range of 10–37°C but not at 6°C and 42°C. There was little development of eggs in winter (June to August) in New Zealand. The most rapid development was from January to March, when in 9–10 days 90% of larvae on experimental plots had reached the third stage.

Two days after Andrews (1939) infected lambs orally, third-stage larvae were found exsheathed in the lumen of the small intestine. Three days later many third-stage larvae were in the crypts between villi in the anterior part of the small intestine. The third moult occurred 4 days postinfection in the crypts. The anterior ends of the larvae remained in the crypts but the posterior ends protruded between villi into the gut lumen. By day 5 most larvae were in the lumen of the intestine, when they were about 1.6 mm in length. The final moult took place 9–10 days postinfection in the lumen of the small intestine. Eggs appeared in faeces of lambs 15 days postinfection.

Sommerville (1960) reported that growth of *C. curticei* commenced 48–72 h after infection before the third moult and that fourth-stage larvae increased markedly in length until 6 days postinfection when the final lethargus commenced. The final moult usually occurred in 8 days and growth of the adults ceased in 12–14 days. Some worms failed to develop past the fourth moult and persisted in the gut for up to 22 days.

C. fuelleborni Hung, 1926

C. fuelleborni occurs in various antelopes in Africa. Anderson (1986) collected worms from impala (*Aepyceros melampus*) in Natal Province, South Africa, and reported that infective larvae found in culture in 6–7 days at 26°C were 546–655 μm in length, including the sheath, which had a tail 46–75 μm in length. Eggs were reported elsewhere to be 65–72 × 35 μm in size. According to Anderson (1986) fourth- and fifth-stage parasites occurred in the anterior region of the small intestine and few larvae were found in scrapings of the mucosa.

C. oncophora (Railliet, 1898)

C. oncophora is a parasite mainly of cattle but it also occurs in sheep, dromedaries, vicunas and alpacas and in wild ruminants such as bighorn sheep and pronghorns. Isenstein (1963) reported that eggs were 84.3 ± 4.7 μm (SD) $\times 42.4 \pm 2.5$ μm in size and were passed in faeces in the morula stage. Eggs cultured at 26–30°C hatched within 16.5 h. First-stage larvae were on average 375 μm in length. The first moult occurred 28 h after hatching and second-stage larvae were on average 808 μm in length. Both first- and second-stage larvae had long, conical, sharply pointed tails. The first moult took place in 28 h. Third-stage larvae were 863–1046 μm in length with the sheath. The tail of the larva in the sheath was slightly tapered and ended bluntly.

Calves were given infective larvae orally. Exsheathed third-stage larvae were found in the abomasum and small intestine 13 h postinfection. In about 3 days all worms were found in the small intestine and were undergoing the third moult; by 4 days worms were in the fourth stage. In 10 days postinfection larvae were moulting to the adult stage, or had already completed the fourth moult; males were 4.05 mm and females 4.9 mm in length. In 11 days nearly all worms recovered were in the fifth stage; males were 4.7 mm and females 6.5 mm in length. The prepatent period in the calves was 17–22 days. Two calves were each given 30,000 infective larvae: at necropsy on 25 and 31 days, 75–88% of the worms recovered were in the first 5.5 m of the small intestine.

Michel *et al.* (1970) showed that in Britain, from September to December, the parasitic development of an increasing proportion of *C. oncophora* in calves was arrested at the early fourth stage. Larvae of the same population developed normally when ingested by calves in spring .

Nilsson and Sorelius (1973) reported that *C. oncophora* is one of the two most common gastrointestinal parasites in young cattle in Sweden. Infective larvae over-wintered on pasture and were available in large numbers in May when cattle were turned out to pasture.

C. pectinata Ransom, 1907

C. pectinata, a widespread parasite of cattle, sheep, pronghorn and various wild ruminants, was studied by Herlich (1965a,b). The infective larvae were 760–937 μm in length including the sheath. In calves, third-stage larvae established themselves in the small intestine – mainly the duodenum – and underwent the third moult 3 days postinfection. Fourth-stage larvae were present in 6 days. In 8 days some worms were undergoing the fourth moult; by 10 days 94% of the worms found were adults and the remainder were fourth-stage larvae. There was no evidence that worms penetrated tissues but they were intertwined around villi. In 12 days each female contained 204 eggs but the prepatent period was 14–17 days. Egg counts reached their highest level in 20–67 days postinfection; they remained high for about 3 weeks and then generally declined to zero (or nearly zero) 15 weeks postinfection.

C. punctata (Linstow, 1907)

C. punctata is a common parasite of cattle and various wild ruminants. According to Stewart (1954) eggs were on average 79×36 μm in size and in the 16-cell stage when laid. At 20–26°C eggs hatched in 20 h, when larvae were about 318 μm long. Second-stage larvae appeared in cultures in 30 h after hatching at 28°C but were first seen 15 h later at 20–26°C. The newly moulted second-stage larvae averaged 619 μm in length. Infective third-stage larvae appeared in 75 h after hatching at 28°C and after 96 h at 20–26°C. The average length of the larvae was 798 µm (with the sheath). Alicata (1961) showed that most infective larvae of *C. punctata* died during the first month on pasture in Hawaii but a few survived if moisture conditions were suitable.

Stewart (1954) gave infective larvae orally to calves. Some third-stage larvae, recovered 1 day postinfection, were found exsheathed in the abomasum (they were only 772 µm in length, having cast the sheath). Early fourth-stage larvae (1.2 mm) were noted 3 days postinfection. Most worms found at 8 days were in the early adult stage and a few were still moulting. By 10 days eggs were forming in the uteri.

Alicata and Lynd (1961) studied the pathogenesis of *C. punctata* in calves and reported that eggs were found in the faeces 12 days postinfection, but apparently they did not examine the faeces earlier than that. Highest egg output occurred 17 day postinfection and then the number of eggs passed declined markedly during the following 9 weeks.

Bailey (1949) gave the prepatent period as 11–16 days and the patent period as up to 9 months. Mayhew (1962) gave the patent period as 4.5–63.5 months.

Cooperioides

C. hamiltoni (Mönnig, 1932)
This is a parasite of the small intestine of impala (*Aepyceros melampus*) in Natal, South Africa. According to Anderson (1992) first-stage larvae were 567–648 µm in length. At 26°C third-stage larvae appeared in culture in 12–16 days.

C. hepaticae Ortlepp 1938
This is a parasite of the impala (*Aepyceros melampus*) in Natal, South Africa. First-stage larvae were 546–685 µm and third-stage larvae 562–675 µm in length (Anderson, 1992). At 26°C third-stage larvae appeared in culture in 12–16 days.

Impalaia

I. tuberculata Mönnig, 1923
Anderson (1995) found this species in the impala (*Aepyceros melanepus*) in Natal, South Africa. The larvae are easy to distinguish in the intestine because they are tightly coiled. Early fourth-stage larvae were 1.3–1.8 mm in length.

Subfamily Graphidiinae

Graphidium

G. strigosum (Dujardin, 1845)
G. strigosum is a parasite of the stomach (mainly the fundus) of palaearctic Leporidae including *Lepus europaeus*, *L. capensis* and *Oryctolagus cuniculus*. Broekhuizen and Kemmers (1976) concluded that the wild rabbit (*O. cuniculus*) was the original and best

adapted host of *G. strigosum* since it tolerates the parasite well, in contrast to *L. europaeus* in which it causes serious stomach lesions. The spread of the parasite in hare populations is apparently dependent on the presence of rabbits.

According to Wetzel and Enigk (1937) eggs were 90–100 × 50–58 μm in size and in the morula stage when passed in faeces of the host. Development to the first-stage larva and hatching occurred in 8–10 h at 20°C when larvae were 310–380 μm in length. The first moult occurred 24–30 h after the first moult, when larvae were 480–530 μm in length. The second moult occurred in a further 2–3 days. Ensheathed infective third-stage larvae were 680–740 μm in length; they climbed to the tips of vegetation at dusk but moved downward when subjected to heat and bright sunlight.

Attempts to infect *O. cuniculus* by skin penetration were unsuccessful. In oral infections, larvae exsheathed in the stomach and commenced to grow immediately. The third moult occurred in 2–3 days, when worms were 1.3 mm in length. The final moult was not recorded but rabbits passed eggs 13 days postinfection. Fourth-stage larvae were found coiled in the ducts of the gastric glands of the fundus. In 9 days these larvae re-entered the lumen and reached adulthood. Adult worms occurred in mucus, with the head buried in the stomach grooves but unattached to the mucosa (cf. *Obeliscoides cuniculi*).

The relationship of *G. strigosum* and *Trichostrongylus retortaeformis* in wild rabbits has been reported by Dudzinski and Mykytowycz (1963), Dunsmore and Dudzinski (1968), Boag (1987) and Boag and Kolb (1989).

Hyostrongylus

Members of the genus are parasites of Ethiopian Leporidae, Suidae and a few Bovidae.

H. rubidus (Hassall and Stiles, 1892)

H. rubidus is a small, reddish cosmopolitan stomach parasite of pigs. Alicata (1935) was the first to study the development and transmission. Eggs were thin, transparent and oval, with unequal poles (one less convex than the other), and 60–76 × 31–38 μm in size. Eggs were in the 'early tadpole' stage when passed in the faeces of the host. First-stage larvae were 290–315 μm at hatching in 39 h at 22–24°C. The first moult occurred in 74 h when larvae were 546–554 μm in length. The first- and second-stage larvae had long filamentous tails and tubular buccal cavities. Second-stage larvae moulted in 122 h. Elsewhere in the article (p. 56) Alicata (1935) stated the first moult occurred in 103–113 h and the second in 105–113 h. The infective larvae had a conical tail with a minute digitiform, terminal process and were 715–735 μm in length.

Alicata (1935) noted that infective larvae were not resistant to desiccation or freezing and he was unable to demonstrate skin penetration. Rose and Small (1982) reported that eggs of *H. rubidus* developed completely at 10–27°C but no development occurred at 4°C. Infective larvae survived on herbage for up to 10 months in southern England but were rapidly killed by freezing and desiccation.

Alicata (1935) gave infective larvae orally to guinea pigs. In 15 min larvae had exsheathed and were found attached to the gastric mucosa, where they penetrated the epithelial folds and caused ulcers. The third moult occurred in 5 days. Fourth-stage larvae had a small provisional buccal capsule; males were 0.92–1.12 mm and females

0.89 mm in length. The final moult occurred in 13 days. Males were fully developed in 17 days and females in 19 days. Kendall and Small (1974) gave the prepatent period as 17 days in pigs. Egg output increased rapidly by 24 days and then declined by day 30; these pigs had been infected daily with large numbers of eggs. Taffs (1966) gave the prepatent period in pigs as 3–4 weeks.

Kotlan (1949) noted that larvae developing in the gastric mucosa of pigs resulted in destruction of the epithelium and the formation of lentil-sized nodules. According to Gerber (1968) and Boch *et al.* (1968) larvae in the gastric mucosa did not develop uniformly and fourth-stage larvae as well as subadults were found after the end of the prepatent period. Soulsby (1965) reported that adults produced a chronic catarrhal gastritis, leading to the formation of a diphtheritic membrane as well as ulceration in pigs.

Subfamily Ostertagiinae

Ostertagia

Members of *Ostertagia* are found in the abomasum of cattle, sheep, goats and holarctic Cervidae (Cervinae, Odocoileinae). The name of the genus is derived from Ostertag, who described worms found in Germany in 1890. Early observations on the development of *O. ostertagi* were made by Stadelmann (1892) and Stödter (1901). The earliest modern study was undertaken by Threlkeld (1934).

Eggs are passed from the host in the morula stage. Their development, as well as that of the free-living stages, is typical of most members of the family. First- and second-stage larvae are rhabditiform with conical pointed tails and valved oesophagi. Third-stage larvae are strongyliform. Sheathed infective larvae crawl on herbage and are available to grazing hosts. Larvae exsheath in the abomasum and then invade the gastric pits and glands, mainly of the fundic and pyloric region, where they elicit nodules and swellings containing one or more larvae (Sommerville, 1953, 1954). In a few days third-stage larvae undergo the third moult. In the absence of arrest fourth-stage larvae typically leave the mucosa as early as 4 days postinfection and enter the lumen of the abomasum, where they mature. Some worms may undergo the fourth moult before returning to the lumen. Worms may be found free in the abomasum or attached to the mucosa. The prepatent period is about 3 weeks. Variability in development and the presence of arrested larvae can prolong the patent period considerably.

Species of *Ostertagia* are well known for their propensity to arrest in the early fourth stage. The phenomenon has been intensively studied in cattle and sheep and has also been reported in wild deer (*Odocoileus virginianus*) in Canada (Baker and Anderson, 1974; Belem *et al.*, 1993).

Several species of the Ostertaginae are known to cause disease in animals but only the well studied *O. ostertagia* and the related *Teladorsagia circumcincta* are considered below.

O. ostertagi (Stiles, 1892)

Commonly known as the medium stomach worm, *O. ostertagi* is a cosmopolitan parasite of cattle. Armour and Ogbourne (1982) and Myers and Taylor (1989) regarded it as one of the most important helminth pathogens of cattle in the temperate world; it is also known to cause problems in subtropical climates where there is winter rainfall.

According to Threlkeld (1946) eggs were 70–84 × 40–50 μm in size. Eggs hatched after 24 h in culture and first-stage larvae were 300–500 μm in length. The first moult occurred 2 days after hatching and the second-stage larvae were on average 750 μm in length. Infective larvae present in cultures in 5–6 days were 850–900 μm in length. According to Pandey (1972) eggs hatched at temperatures from 4 to 35°C and infective larvae developed in fecal cultures from 10 to 35°C; the optimum temperature was 25°C.

Threlkeld (1946) orally infected calves and examined them at various times. His report of moults was confusing. Moults reported in 72 and 96 h in the abomasum may both have been the third. The moult observed in 10 days was presumably the fourth moult. Immature adults were found attached to the mucosa and coiled within nodules 'or lesions' within the mucosa. Mature worms were noted in 21 days and eggs were present in faeces in 23 days.

Douvres (1956) reported two moults in calves: one in 3 days (the third) and one in 7–10 days (the fourth). Mature worms were found 21 days postinfection in one calf.

Rose (1969) reported that in calves the third moult occurred on day 4 and the fourth in 12 days. Spermatogenesis was completed in 15 days and females contained ova in 16 days.

Michel (1969) showed that eggs of *O. ostertagi* deposited on herbage during April–June in the UK first gave rise to third-stage larvae in July or August. Thereafter little or no development took place. Thus, in western Europe there was probably only one and at most two generations per year. In addition, third-stage larvae can survive the winter and infect calves in the spring. This seasonal pattern of transmission has been reported in many other regions (Armour and Ogbourne, 1982). Gettinby *et al.* (1979) have developed a formula to predict the timing and magnitude of the seasonal increase in numbers of third-stage larvae available to cattle. Smith and Grenfell (1985) reviewed the population dynamics of *O. ostertagi* and Gibbs (1988) reviewed the epizootiology of bovine ostertagiasis in north temperate regions.

The possible role of earthworms in bringing larvae (which may be as deep as 15 cm in soil) to the surface of pastures has been noted by Armour and Ogbourne (1982), who cited the observations of Oakley (1981) on larval *Dictyocaulus viviparus*.

Armour and Ogbourne (1982) reviewed the history of knowledge of arrest in *O. ostertagi* and only a few highlights derived from their review will be outlined herein. Anderson *et al.* (1965) reported a significant increase in arrested fourth-stage larvae in cattle in autumn in the UK and suggested this was the result of physiological changes in the host or in the parasite during autumn. Armour and Bruce (1974) and Michel (1974) showed that most larvae conditioned to cold and arrested in calves did not resume development until 4 months. The seasonal arrest of larvae ingested during autumn grazing is known to be common in the northern hemisphere. Elsewhere arrested fourth-stage larvae accumulate at different seasons. In temperate Argentina and Australia they accumulate in spring whereas in New Zealand they accumulate in autumn. In the southern USA larvae are most common in late winter and early spring. In these countries attempts to induce arrest by cold-conditioning were unsuccessful. Studies in Australia by Smeal and Donald (1981) indicated that the propensity to arrest is genetically determined and independent of cold temperatures. It was concluded by Armour and Ogbourne (1982) 'that when arrested development of *O. ostertagi* occurs during a particular season it is a heritable trait of the larva itself and may occur in response to various adverse environmental stimuli, one of which in the northern hemisphere appears to be declining temperature or cold condition.'

Satrija and Nansen (1992a) studied the effects on calves of concurrent infections of *O. ostertagia* and *Cooperia oncophora* designed to mimic natural infections. *C. oncophora* had, apparently, no effect on infections with *O. ostertagi*.

Teladorsagia

T. circumcincta (Stadelmann, 1894)

T. circumcincta, which was originally placed in the genus *Ostertagia*, is the most important species of the genus found in sheep (Crofton, 1963) but it occurs in a variety of other ruminants, including deer, llamas, pronghorns, mouflon, bighorn sheep and antelope. Gasnier and Cabaret (1996) used isoenzyme electrophoresis to show that goat and sheep strains ('lines') of *T. circumcincta* exist and Gasnier *et al.* (1997) showed that the two lines could be separated morphologically in features of the dorsal ray.

Threlkeld (1934) was the first to investigate the development of this species. Eggs were 80–100 × 40–50 µm in size. In faeces taken from sheep, eggs were in the 8- to 16-cell stage as well as in the morula stage. Eggs hatched in 18–24 h (temperature not given) and first-stage larvae were 350 µm in length. Threlkeld did not accurately report moulting but stated that ensheathed infective larvae were 880–1000 µm in length; Dikmans and Andrews (1933) gave the length as 797–866 µm including the sheath.

Crofton (1965) reported that a strain of *T. circumcincta* developed and hatched at 4–34°C. At the higher temperatures hatching occurred in 17 h. At 20°C, 13°C and 6°C the times required were, respectively, 24, 13 and 7 days. Crofton and Whitlock (1965) reported that a strain in New York hatched only at 10–38°C and that at 10°C the New York strain needed twice the time to hatch as the British strain.

Threlkeld (1934) infected ten lambs and examined them for worms at various times. Unsheathed third-stage larvae (802–935 µm in length) were found in the lumen of the abomasum in 3 days, but in 3.7–4.0 days (90–96 h) most larvae were in the abomasal wall in 'small white raised areas' each containing larvae (802–1300 µm in length). In 7–9 days larvae were apparently in the fourth stage and females were 7 mm and males 5 mm in length. In 15–16 days larvae of all stages were found and 'the percentage of adults was comparatively small considering the number of infective larvae given to the lambs.'

Sommerville (1953, 1954) infected lambs and examined them 3–84 days later. In 3 days all larvae in the abomasum were exsheathed and in the third stage. A few similar larvae were found with their heads in gastric pits. In 4 days all larvae were in the fourth stage, thus the third moult took place 72–96 h postinfection. Most larvae were coiled in gastric pits and glands. At later times even adult worms occurred in the mucosa (7–35 days).

Dunsmore (1960, 1963) noted an increase in numbers of arrested larvae in lambs with increased doses of larvae and also that when an adult population was removed by anthelmintics it was replaced by previously arrested larvae in the mucosa. Dunsmore (1965) also reported a correlation between the time of parturition and the 'spring rise' in eggs of *O. circumcincta* in lambing ewes.

Armour *et al.* (1966) reported that most larvae were in gastric glands in 4 days and that these larvae grew and moulted to young adults by day 8, at which time they had emerged from the gastric glands. Many fourth-stage larvae arrested in the gastric glands but by 60 days most had resumed their development in the glands or on the mucosal surface.

Smith (1989) provided a critical review and analysis of the structure and behaviour of four models to depict the results of trickle infections of *T. circumcincta* in sheep.

Giangasperó *et al.* (1992) reported summer arrest of *T. circumcincta* (also *Marshallagia marshalli*) in Awassi sheep in northwest Syria characterized by cold winters and hot summers. Rickard (1993) reported parasitic gastritis associated with inhibited larvae of *Telodorsagia* sp.

Longistrongylus

L. sabie Mönnig, 1932

Anderson (1995) found this species in the small intestine of impala (*Aepyceros melanepus*) in Natal, South Africa. Third-stage larvae were 632–718 μm in length. Male early fourth-stage larvae were 2.2–4.1 mm and early female larvae 1.6–5.3 mm in length.

Subfamily Trichostrongylinae

Trichostrongylus

Members of the genus are small reddish worms which occur in birds and ruminants throughout the world and in primitive rodents. Some ruminant species are transmissible to rodents and lagomorphs, and zoonotic infections of some species have been reported in humans. Species inhabit the abomasum or the anterior part of the small intestine of ruminants and lagomorphs and the caecum of rodents and birds. Like most trichostrongyloids, there are free-living preparasitic stages and the infective larvae tend to crawl on vegetation, where they are available to the host; skin penetration is unknown in the genus. Infective larvae are resistant to drying conditions. They tolerate low temperatures but are generally killed by freezing temperatures. Some species are reported to invade and remain for a period of time in the mucosa of the host before entering the lumen, where they mature.

T. axei (Cobbold, 1879)

T. axei is a common cosmopolitan parasite of the abomasum of ruminants, including cattle, sheep and deer. It also occurs in the stomach and first part of the intestine of horses. Barrett *et al.* (1970) found *T. axei* in *Spermophilis richardsoni* in Montana and transmitted it experimentally to this ground squirrel. Eggs were deposited in the morula stage and were 79–92 × 31–41 μm in size (Shorb, 1940). First-stage larvae were 289–388 μm and infective larvae 619–762 μm in length, including the sheath (Keith, 1953). Ciordia and Bizzell (1963) considered 25°C to be the optimal temperature for the development of *T. axei* and four other species of trichostrongyloids. According to

Douvres (1957) larvae in the third moult were found in calves in 4–6 days. Fourth-stage larvae were found in 7 days postinfection and the fourth moult occurred in 10 days. Worms were adults in 15 days. Mayhew (1962) estimated that a single population of *T. axei* in sheep produced eggs for up to 1 year and 3 months.

Lyons *et al.* (1996) reported that *T. axei* had been maintained in domestic rabbits (*Oryctolagus cuniculi*) by serial passage since 1953. One strain was from cattle and one from horses (see also Lyons *et al.*, 1987).

Ross *et al.* (1967) reported a histotrophic migration in association with the development of plaque lesions in calves recently infected with *T. axei*.

T. colubriformis (Giles, 1852)

T. colubriformis (syn. *T. instabilis*) is a cosmopolitan parasite of the small intestine of cattle and other ruminants. It is transmissible to rabbits (Purvis and Sewell, 1971). Mönnig (1926) reported that eggs were 79–101 × 39–47 μm in size and were passed in the 24- to 32-cell stage in faeces of the host. Eggs developed and hatched in 19 h at 18–21°C. The first moult occurred in 25–28 h after hatching, when the larva was about 585 μm in length. In 60 h the third stage was formed; it was 674–749 μm in length and the tip of the tail had two small tubercles. Mönnig (1926) noted that embryonated eggs were markedly resistant to drying.

Nakamura (1937) reported that development of eggs to the infective stage took 2.7–3.3 days at 28°C. Ciordia and Bizzell (1963) reported that 4 days were required for development to the third stage in faeces at 25°C. According to Wang (1967) eggs developed when maintained at 10–30°C but not at 5°C or 35°C; the optimal temperatures were 15–30°C and the resulting infective larvae that developed at these temperatures were 625–687 μm in length (averages at three different temperatures). Andersen *et al.* (1966) reported that the optimal survival temperature for free-living stages was 4°C and larvae stored at this temperature for 308 days remained infective. Waller and Donald (1970) noted that eggs survived desiccation down to a relative humidity of 75%. Beveridge *et al.* (1989) compared the effects of temperature and relative humidity on the development and survival of larval *Trichostrongylus* spp., including *T. colubriformis*, and related the data to Australian conditions.

Douvres (1957) orally infected calves and a guinea pig and examined them for developing worms at various times postinfection. Most worms were found in washings of the small intestine. The third moult occurred in 3–4 days and the fourth 6–10 days postinfection. Worms were adult by 15 days. Bizzell and Ciordia (1965a) reported that the prepatent period in calves was 18–21 days and the average patent period was 47 days.

Rahman and Collins (1990a) infected 12 goats with *T. colubriformis*. Most worms established themselves within the first 3 m of the intestine. Worms invaded the mucosa and developed for 11 days before returning to the lumen as young adults. In 21 days eggs were noted in uteri of females.

Eysker (1978) and Ogunsusi and Eysker (1979) demonstrated arrested development in the third stage.

T. retortaeformis (Zeder, 1800)

This is a parasite of the small intestine of European rabbits (*O. cuniculus*) and hares (*Lepus europaeus*). It has been reported sporadically in other hosts and was introduced with rabbits to Australia, where it was studied by Dunsmore (1966). According to Mönnig (1930) eggs were 85–91 × 46–56 μm in size.

Crofton (1946a,b) carried out detailed studies on the behaviour of infective larvae on vegetation in the UK. He concluded that larvae were 'distributed in that portion of the herbage where there is least climatic change. The type, height and density of the herbage modify the effect of climatic factors which influence the extent of the movement'. Eggs failed to develop and hatch when maximum temperatures did not exceed 10°C. Eggs developed and hatched in 8 days or less at maximum temperatures of 15.5–26.6°C. Grass blades reduced the rate of loss of moisture from faecal pellets and so larvae were able to develop. Eggs passed by the host in autumn survived winter and hatched in spring but eggs passed in the colder winter periods died. Migration of larvae on to vegetation ceased at temperatures below 2.7°C and high rates of evaporation reduced the number of larvae found on grass blades. Some infective larvae survived cold periods but many died during periods of high evaporation.

According to Prasad (1959) the optimum temperature for development in faecal cultures was 25°C when the infective stage was reached in 3–5 days. Gupta (1961a) reported that ova hatched and larvae developed to the infective stage at 5–30°C but 35°C was lethal.

Bailey (1968) concluded that the host provided a stimulus for exsheathment and that this depended upon the concentration of undissociated carbonic acid, pH and temperature.

T. rugatus Mönnig, 1925

This is a parasite of the small intestine of sheep and goats in Africa, Australia and Malaysia. It has been reported in bighorn sheep in the USA (Becklund and Senger, 1967). Beveridge and Barker (1983) infected 16 sheep and examined them for worms at intervals from 2 to 20 days. The third moult occurred in 4–6 days and the fourth in 10 days postinfection. The prepatent period was 16 and 18 days in four sheep. Growth of larvae in lambs was relatively slow and constant between days 2 and 10 but was followed by a period of rapid growth. The parasites were found mainly in the first 6 m of the gut and 71% occurred in the first 3 m. It was estimated that each female produced 24–35 eggs each day.

T. sigmodontis Baylis, 1945

This is a parasite of the caecum and lower small intestine of cotton rats (*Sigmodon hispidus*), rice rats (*Oryzomys palustris*) and nutria (*Myocastor coypus*) in North America. According to Thatcher and Scott (1962) eggs passed by the host in the 16-cell to morula stage were 66–80 × 35–38 μm in size. At 20–25°C eggs hatched in 36–48 h and first-stage larvae were 310–380 μm in length. The first moult apparently occurred in 2–3 days and by 4 days infective larvae, 650–700 μm in length, appeared in cultures. Cotton rats, rice rats, gerbils (*Meriones unguiculatus*) and hamsters (*Mesocricetus auratus*) were infected orally. The prepatent period was 9–12 days.

T. skrjabini Kalantarian, 1930

This is a parasite of the abomasum and small intestine of sheep, mouflon and roe deer in the CIS. Altaev (1961) found that the optimal temperature for development of eggs and larvae in faecal cultures was 23–25°C. Eggs hatched in 42 h at these temperatures and the infective stage was reached in 4 days. Larvae did not develop in agar, saline or water. Poorly embryonated eggs did not resist desiccation but more developed eggs (10–24 h) were more resistant. First- and second-stage larvae were killed when desiccated for 1 h whereas infective larvae survived at room temperature for 9 months.

T. tenuis (Mehlin, 1846)

T. tenuis (syn. *T. pergracilis*) is reported as a parasite of the caeca mainly of grouse and geese which consume terrestrial vegetation. It is well known as a parasite of *Lagopus lagopus* and *L. mutus* in northern Europe. The status of reports of *T. tenuis* in the New World requires clarification since the species in the bobwhite quail (*Colinus virginianus*) formerly identified as *T. tenuis* (syn. *pergracilis*) is now recognized as a distinct North American species, *T. cramae* Durette-Desset *et al.* (1993). The assessment of the species reported in domestic geese in North America (Cram and Cuvillier, 1934; Cram and Wehr, 1934) should be clarified since the worms may have been misidentified as *T. tenuis*. However, *T. tenuis* could have been introduced into the USA, since Cram and Cuvillier (1934) reported that 'a flock of pheasants which had been imported from England and were held in captivity suffered from the effects of heavy infestations with *Trichostrongylus tenuis*, whereas pheasants reared on the premises showed no clinical evidence of disease'. The nematode has been reported in *L. lagopus* and *L. mutus* in Norway (Holstadt *et al.*, 1994); intensities were higher in *L. lagopus* than in *L. mutus*.

The first observations on the development and transmission of *T. tenuis* were contained in a report of a Committee of Inquiry on Grouse Disease published in 1911 in the UK (Lovat, 1911). In 1912 an abridged, popular version was edited by Leslie and Shipley (1912). Within this latter volume the problem of grouse diseases consists of chapters (IV, V, VII and VIII) contributed, respectively, by Wilson, Wilson and Leslie, Shipley and Shipley and Leiper (Wilson *et al.*, 1912). The authors related the prevalence, intensity and pathogenicity of *T. tenuis* to 'grouse disease' and well-known fluctuations in red grouse (*Lagopus lagopus scoticus*) on grouse moors in the UK. New investigations have confirmed most of the basic findings of the 1911 report, and added new dimensions to our understanding of the disease and how it might be managed on moorlands for grouse shooting.

The 1911 report noted that adults of *T. tenuis* were extremely thin, transparent and difficult to see with the naked eye even though males were 8 mm and females 10 mm in length. The cephalic end of the worms penetrated deeply into the caecal mucosa. Eggs were about 75×46 μm in size and were passed in faeces of grouse in the morula stage with 64 cells. In moist faeces and at suitable temperatures eggs reached the first stage, which hatched in about 36–48 h; larvae were about 360 μm in length. The first moult occurred in 36–48 h and the second in 8–16 days, depending on the temperature. Heather (*Calluna vulgaris*) is the staple food of adult red grouse and it was demonstrated experimentally that sheathed infective larvae crawled to the tips of moist heather and accumulated there in drops of water during misty weather. It was also shown that larvae could withstand desiccation on the tips of heather. Two adult grouse were given infective larvae in water. One bird died of aspiration pneumonia but the other became emaciated and died and many *T. tenuis* were found in its caeca. The authors concluded, as did Cobbold earlier, that *T. tenuis* was the causal agent of a disease that regularly killed grouse, mainly in spring and they speculated on the factors responsible for its seasonality.

Prevalence in adult grouse is 100% and chicks acquire infections soon after their diet turns from insects to vegetation. The prepatent period is about 7–8 days. Birds do not develop protective immunity and since worms are fairly long-lived intensities can be extremely high (up to 9000–10,000 individuals) (Hudson and Dobson, 1989; Shaw and Moss, 1989b).

Watson *et al.* (1987, 1988) showed that *T. tenuis* threaded themselves into the caecal mucosa with anterior and posterior ends protruding into the lumen. As noted earlier by Wilson (in Leslie and Shipley, 1912), the worms caused trauma, atrophy and flattening of the epithelial cells, which would probably interfere with the normal digestion of heather and other plant material.

McGladdery (1984) confirmed earlier observations that free-living third-stage larvae were negatively geotactic and positively phototactic. They readily ascended heather in moist conditions. Shaw *et al.* (1989) showed that eggs and early free-living stages were not resistant to desiccation or extreme cold. Eggs failed to develop at low temperatures and most third-stage larvae failed to survive winter on grouse moors. Connan and Wise (1994) reported that eggs of *T. tenuis* are rather sensitive to cold but that infective larvae are capable of withstanding winter temperatures on Yorkshire grouse moors. The larvae may, however, be susceptible to desiccation, especially on the tips of heather.

Shaw (1988) provided evidence that arrest occurs in third-stage larvae. Recently acquired third-stage larvae apparently retained the second-stage cuticle for a few days and could be distinguished from established exsheathed larvae. In August and September most larvae in grouse were in the fourth stage but some third-stage larvae were sheathed, indicating that transmission was taking place. In winter (November and December) ensheathed larvae were absent but exsheathed larvae were present, suggesting that few new larvae were being ingested during this period and that the exsheathed third-stage larvae were arrested. In spring (March and April) the proportion of third-stage larvae decreased significantly in the parasite population and sheathed larvae were absent, indicating that overwintering third-stage larvae had developed to fourth and adult stages when the output of eggs in the faeces of the birds increased. Shaw (1988) suggested that the more or less synchronized development of larvae in grouse resulting eventually in the 'spring rise' in worm egg production may contribute to the disease and mortality occurring in grouse in spring. Delahay *et al.* (1995) reported that short-term energy unbalance, weight loss and loss of condition occurred in grouse at the onset of the development of fourth-stage larvae to adults and up to the onset of patency.

Shaw and Moss (1989a) reported that prevalences of *T. tenuis* in red grouse in Scotland were high even at low grouse densities and that prevalence, intensity and aggregation were higher in older than younger grouse. Populations of adult worms could apparently survive for over 2 years. The output of eggs from the worms decreased with age of the worm population, especially in winter (November–December). Moss *et al.* (1993) analysed the numbers of eggs of *T. tenuis* passed in faeces of red grouse over 8 years and proposed that rainfall in the previous summer explained much of the year-to-year variation in egg numbers, probably because transmission is greater in wet than in dry summers. The data apparently do not support the hypothesis that cyclic-type population changes in grouse were caused by *T. tenuis*.

Hudson and Dobson (1997) examined and quantified transmission rates and density-dependent reductions in egg production of *T. tenuis*. The sustained cycles observed in long-term dynamics of the grouse populations suggest that density-dependent reductions in worm fecundity and establishment are absent or only operating at levels undetectable in field studies.

Moss *et al.* (1990) decided that there was no relationship between intensity of *T. tenuis* and fox predation (cf. Hudson, 1986a). Shaw (1990) noted that captive red grouse infected with *T. tenuis* started to lay later in spring and laid fewer eggs at a slower rate than uninfected hens. Hens infected in March exhibited inappetence a week postinfection and gained less weight than controls. This suggested the possible significance of developing larvae to the health of the birds, as noted also in spring, when arrested larvae proceed to develop (Shaw, 1988). Hudson *et al.* (1992) reported that grouse killed by predators had more caecal worms than birds that were shot. More birds were heavily infected in places where predator control was intense, leading to the possibility that predators select heavily infected over less heavily infected grouse. Dogs found fewer birds treated with oral anthelminthics than untreated birds. A modified model taking predation into account was proposed. Dobson and Hudson (1995) claimed that models demonstrated that small numbers of predators selectively removing heavily infected individuals may allow the size of the red grouse population to increase, since the predators reduce the regulatory role of the parasites.

Hudson (1986b) compared the productivity of red grouse treated with levamisole hydrochloride with untreated birds in the field. Treatment significantly increased the production of young birds and breeding success was related to decreased intensities of *T. tenuis*.

Several attempts have been made to model the red grouse – *T. tenuis* system. Watson *et al.* (1984) modelled changes in breeding density emphasizing overwinter survival, whereas Potts *et al.* (1984) and Hudson *et al.* (1985) modelled cyclic changes in bag records emphasizing winter survival and dispersal. The models support the conclusion that *T. tenuis* infections in red grouse reduce breeding production and that this is reflected in cyclic changes in the numbers of grouse shot in the autumn. Hudson *et al.* (1985) emphasized the importance of humidity in the survival of free-living stages and showed by modelling that the 'interaction of a long-lived free-living stage and parasite induced reduction in fecundity would be of prime importance in producing the cycles in grouse density'.

For a general review of the relationship of *T. tenuis* to red grouse, the reader is referred to Hudson and Dobson (1989).

Moss *et al.* (1996) pointed out that red grouse have unstable population dynamics and that if some cocks were removed during the increase phase it might prevent the usual cyclic decline. The results indicate that age structural changes and associated behaviour may result in cycles by affecting recruitment.

Watson and Shaw (1991) pointed out that *T. tenuis* is uncommon in ptarmigan (*Lagopus mutus*) yet cyclic-type declines also occur in this species in Scotland.

T. vitrinus Looss, 1905

T. vitrinus is a cosmopolitan parasite of the duodenum of sheep, goats, cattle and other ruminants; it is frequently associated with *T. colubriformis* and it has been reported in pigs and rabbits. According to Shorb (1939) eggs were 93–118 × 41–52 μm in size. Infective larvae were 622–796 μm in length. The tail was blunt with two small terminal tubercles (Dikmans and Andrews, 1933). According to Crofton (1965) eggs hatched in

19 h at 36°C, 24 h at 28°C, 48 h at 18°C and 7 days at 9–10°C. Rose and Small (1984) studied the survival of eggs and larvae under field conditions in England. Eggs and larvae developed from April to March but development was slow from October to March and mortality was high. In dry periods, mortality of preinfective stages was high, especially in short herbage. Infective larvae survived for up to 16 months on grass plots and survived severe winter cold. In the laboratory, eggs and larvae developed from 4 to 27°C and were killed by continuous freezing.

Eysker (1978) and Ogunsusi and Eysker (1979) reported arrested development in the third stage.

Subfamily Haemonchinae

Haemonchus

Members of the genus are found in the abomasum of ungulates and one well-known species (*H. contortus*) is regarded as one of the most pathogenic helminths in domesticated animals. The genus was revised by Gibbons (1979) who placed some 15 described species or subspecies and varieties, including *H. placei* and *H. contortus cayugensis*, into synonymy with *H. contortus* (Rudolphi, 1803) which she regarded as a polymorphic species. Whitlock and Le Jambre (1981) reported bionomic and genetic differences which they believed justified the recognition of *H. placei* and *H. c. cayugensis* as distinct from *H. contortus*. They suggested that the latter represented a complex of sibling species. Jacquiet *et al.* (1997) used morphology of spicules to separate species. The general biology of *H. contortus* and the reported differences in biology between this species and *H. placei* are outlined below.

H. bedfordi Le Roux, 1926
The infective larvae of *H. bedfordi* from the impala (*Aepyceros melampus*) in South Africa were 615–718 μm in length (Anderson, 1995). The author contrasted larvae of *H. bedfordi* with those of *H. contortus* and *H. placei*.

H. contortus (Rudolphi, 1803)
The twisted stomach or wire worm is a red, blood-sucking parasite found in sheep, goats, cattle, bison and deer. White-tailed deer (*Odocoileus virginianus*) have been infected experimentally with *H. contortus* from sheep (Foreyt and Trainer, 1970) and McGhee *et al.* (1981) transferred *H. contortus* between white-tailed deer, lambs and calves. Morphologically the parasites from deer were indistinguishable from those in cattle and sheep. Conder *et al.* (1992) infected jirds (*Meriones unguiculatus*); growth was slow and incomplete in jirds but they did invade the stomach and some persisted for 14 days.

Ransom (1906) made some of the earliest observations on *H. contortus*. He noted that eggs in faeces were in 'various stages' of segmentation and that they developed and first-stage larvae (350 μm in length) hatched within 2 days at 16–20°C. Most larvae reached the sheathed infective stage in 10–14 days. Eggs and newly hatched larvae were not resistant to cold and desiccation, but infective larvae were highly resistant to drying and cold, including freezing temperatures. Infective larvae climbed to the tips of blades of grass in moist conditions and Ransom concluded that the definitive host became infected by ingesting larvae on herbage.

Veglia (1916) carried out a detailed study of *H. contortus* with special reference to the development and characteristics of the free-living stages. Eggs were 66.5–79.0 × 43.3–46.6 µm in size and with 24–26 blastomeres in fresh faeces; eggs deposited by females were usually in the four-cell stage and developed as they passed through the gut of the host. The optimal temperature for development was 20–30°C. In liquid cultures first-stage larvae hatched from eggs in 14–17 h at 26°C. First-stage larvae were 340–350 µm in length and had a valved oesophageal bulb and attenuated, sharply pointed tails. The buccal cavity was tubular. The genital primordium consisted of two cells. About 1 h after hatching, larvae commenced to feed. In 10–12 h, when larvae were 400–450 µm in length, the first lethargus and moult occurred and larvae recommenced feeding. Gamble *et al.* (1989) studied the second moult in *H. contortus* and suggested that the second cuticle is digested by 44–kDa zinc metalloprotease, possibly released from oesophageal glands. Prominent lateral alae made their appearance. The second lethargus and moult commenced 40 h after the first moult or 60–65 h after larvae had hatched. Infective larvae in faecal cultures were 754–756 µm in length (apparently this included the sheath) in 65 h at 26°C. The larva retained the second-stage cuticle with its pointed, attenuated tail. The tail of the larva within the sheath was conical and much less attenuated than that of the second stage. Prominent lateral alae were present. The genital primordium had 16 nuclei and the buccal cavity was elongated but closed anteriorly.

Veglia (1916) made extensive observations on the effects of various environmental factors on the behaviour of larvae and their survival. For example, larvae died at high temperatures (70–85°C) and in decomposed faeces. Eggs failed to develop when kept at 4°C and would not develop when returned to warmer temperatures. Mature larvae left faecal material and were highly resistant to desiccation, apparently in part because of the presence of lipid in the intestinal cells and because they tended to clump together. Infective larvae moved upwards in films of moisture, especially at night and at times of diffuse light, and were negatively heliotactic.

Attempts to infect lambs percutaneously were unsuccessful but after oral inoculation with infective larvae lambs passed eggs on average 15 days postinfection. Larvae exsheathed in the mouth and abomasum. In the latter they immediately began to feed as they lodged between villi. In 2 days larvae were 655–840 µm in length. The third lethargus started in 30–36 h postinfection and lasted for 12 h, during which the moult occurred; larvae (still 750–850 µm in length) escaped from the third cuticle through a longitudinal slit. A provisional buccal capsule was present in the early fourth stage which enabled larvae to attach to the mucosa and suck blood for the first time. Larvae caused small haemorrhages as early as 3 days postinfection. By day 7, sexes were easily distinguishable and males were 2.7–3.0 µm and females 3.7–4.0 µm in length. The fourth moult occurred in 9–11 days and was preceded by a lethargus of about 24 h. The adult buccal capsule was more globular than that of the fourth stage. Eggs were deposited by females in about 15 days.

Coyne and Smith (1992) noted that eggs did not develop at 5°C and the optimum temperature for full development was 20°C. They produced a model to estimate development and mortality rates from their experimental data.

Rahman and Collins (1990b) infected goats with 40,000 larvae and examined them at various days postinfection. More worms established themselves in the fundic than in the middle or pyloric third of the abomasum. Emergence into the lumen started between 7 and 11 days postinfection and by 4 days all worms had moulted to the fourth stage. By day 18 some (13.2%) females had eggs. In 21 days over half the females were gravid.

A number of authors have reported on the effects of environmental factors on the development and survival of eggs and larvae of *H. contortus*. For example, Rees (1950) emphasized the necessity for a favourable combination of temperature, humidity and light to initiate vertical migration of larvae on grass blades. Silverman and Campbell (1958) concluded that in Scotland *H. contortus* requires about an optimum of 2 weeks in summer for development of the egg to the third-stage larva. They noted that embryonated eggs were more resistant to adverse conditions than unembryonated eggs. For example, rapid dessication of faeces destroyed most eggs unless they were all embryonated. Rose (1963b, 1964) compared larval survival and availability on herbage of different heights and lushness. Narain and Chaudhry (1971) and Sood and Kaur (1975) restudied the effects of temperature on development and Todd *et al.* (1976) reported that desiccation protected free larvae from death at storage below freezing temperatures but was harmful at temperatures above freezing. Gibson and Everett (1976) concluded that in southern England the climate was not particularly favourable for the development and survival of the free-living stages of *H. contortus*. Hsu and Levine (1977) related humidity and temperature to development of the free-living stages of *H. contortus* (and *Trichostrongylus colubriformes*). Jasmer *et al.* (1987) compared cold hardiness in *H. contortus* and *Teladorsagia circumcincta*.

H. contortus is noted for a marked propensity to arrest in the early fourth stage and the process is considered as the primary means of surviving winter in temperate climates. According to Blitz and Gibbs (1971b) arrested male larvae were at the stage equivalent to 3–4 days postinfection and the genital primordium was not differentiated whereas females were slightly more advanced and the genital primordium had started to develop. Cylindrical crystals of unknown significance were common in intestinal cells of arrested larvae; these crystals disappeared with development of larvae. Arrested larvae occurred on and not in the mucosa.

Dineen *et al.* (1965a), Soulsby (1966) and Michel (1968) suggested that arrest was related to high levels of resistance in lambs. Dineen *et al.* (1965a) showed that the percentage of arrested larvae of *H. contortus* increased in lambs given repeated doses of infective larvae.

Gibbs (1967), Connan (1968) and Muller (1968) showed that arrested larvae of *H. contortus* acquired in fall matured in spring and the worms contributed to the spring rise in eggs passed in faeces of sheep. Arrested larvae, transferred in winter to parasite-free lambs, failed to develop for 10 weeks and egg counts in the faeces of the host rose markedly as usual in April (Blitz and Gibbs, 1971a,b). In addition, a high percentage of infective larvae exposed to environmental conditions (14.25–12.50 photoperiod and mean temperature of 17°C) in September in Quebec, Canada, arrested (Blitz and Gibbs, 1972a). It was concluded that arrested larvae mature spontaneously in spring as if they have been in an environmentally induced diapause (Blitz and Gibbs, 1972b). Jacquiet *et al.* (1995) showed that in dry periods *H. contortus* is able to survive because adult worms can maintain their ability to produce eggs for up to 50 weeks.

Waller and Thomas (1975) studied *H. contortus* in lambs in northeast England. In late June small numbers of adult *H. contortus* were present. Subsequently the number of worms rose rapidly and the percentage of arrested worms increased monthly until it was 100% in September, October and November. There were no differences in worm numbers or percentage arrested in experimental and tracer lambs, suggesting that autumn climatic conditions and host immunity were not responsible for arrest in this strain of *H. contortus*.

Connan (1975) reported that chilling infective larvae (4°C) had no effect on the percentage of larvae arresting and that exposing larvae to autumn climatic conditions in East Anglia, England, was not necessary to induce a high percentage of larvae to arrest. Apparently a suitable stimulus was provided by keeping the culture in the dark at 25°C for 12 days. Mansfield *et al.* (1977) also concluded that temperature had no effects on inducing arrest in *H. contortus*. Eysker (1981) reported that exposure of infective larvae to 15–16°C induced arrest in sheep and that exposure to lower temperatures was much less effective. Connan (1978) reported that the number of larvae arresting in experimentally infected lambs varied positively with the water content of faeces in which larvae had been cultured and the time larvae were in the cultures (6 or 12 days). Two groups of sheep were allowed to graze on infected pasture: (i) from 10 to 27 September and (ii) from 22 November to 9 December. Both groups began to pass eggs between February and May and there was in both groups a sharp rise in egg counts in April, presumably because of the simultaneous maturation of arrested larvae.

H. placei (Place, 1893)

H. placei is regarded as the main species of *Haemonchus* found in cattle and is distinguished from *H. contortus* of ovines by spicule differences and lengths of infective larvae (Roberts *et al.*, 1954; Bremner, 1956). Field evidence suggested that cattle were resistant to infection with *H. contortus* (Roberts and Bremner, 1955).

Bremner (1956) infected calves with *H. placei* and examined them at various times. In 36 h exsheathed third-stage larvae (620–760 μm in length) were found in the abomasum. Many larvae had invaded pits of the gastric glands and some had entered gland tubules. In 48 h 22.4% of larvae were in the fourth stage (620–860 μm in length) and other larvae were moulting third stage. In 76 h many larvae (720–1200 μm in length) had embedded their anterior extremities within gastric pits. In 10 days numerous blood clots were associated with fourth-stage larvae. In 13 days 66% of worms were adults, 9% were moulting and 25% were in the fourth stage. In 15 days all worms were adults but females had no eggs.

Bremner (1956) concluded that most *H. placei* underwent the fourth moult in 11–15 days (cf. 9 and 11 days in *H. contortus*). Eggs did not appear in faeces until 26–28 days postinfection (cf. 11 days in *H. contortus*).

Herlich *et al.* (1958) compared the morphology of worms from cattle and sheep sharing pasture contaminated with *H. placei* and *H. contortus*. Spicules were longer in *H. placei* ('the bovine strain') than in *H. contortus* ('the ovine strain'). In *H. placei* the vulva process was typically knob-like whereas in *H. contortus* it was linguiform. *H. placei* recovered from experimentally infected lambs were shorter and fewer in number than those recovered from similarly infected calves whereas the opposite was true for the ovine strain. The prepatent period of *H. placei* was 24–32 days (cf. 18–22 days in *H. contortus*).

Le Jambre (1979) hybridized *H. contortus* from sheep and *H. placei* from cattle. The vulvar morph-type of *H. placei* and the inability of the eggs of this species to hatch and develop at 11°C were inherited as dominant traits in the hybrids. The third-stage larvae from crossing male *H. placei* with female *H. contortus* were similar in size to those of *H. placei* whereas larvae from the reciprocal mating were intermediate in size. Hybrids of thiabendazole-resistant female *H. contortus* and non-resistant male *H. placei* were resistant. F_1 males of male *H. contortus* and female *H. placei* hybrids were sterile, due to meiotic disturbances which stopped spermatogenesis during metaphase I, but were fertile in the reciprocal cross until the F_2 generation. Female hybrids from these generations had low levels of fertility when backcrossed to males of either parent species. Chromosomes in eggs of unfertilized females did not undergo meiosis but increased by endomitosis and in faeces exhibited uneven cytoplasmic division and abnormal shape.

Family Dictyocaulidae

The family consists of the Mertensinematinae of amphibians and reptiles and Dictyocaulinae of the respiratory system of Bovidae, Cervidae, Equidae and Suidae.

Subfamily Dictyocaulinae

The subfamily contains *Bronchonema* and the well-studied and important genus *Dictyocaulus* of ruminants and horses.

Dictyocaulus

Three well-known species of the genus have been investigated – two in ungulates and one in equines. Members of the genus are medium-sized and occur in the bronchi and trachea, where they are associated with bronchitis, giving rise to a clinical syndrome (including coughing) known in cattle and sheep in the UK as **husk** (for a brief history of husk see Allan and Johnson, 1960). Eggs are oval and thin-shelled, and embryonate to the first-stage *in utero*. After oviposition most eggs hatch in the air passages but some remain unhatched, at least until they reach the large intestine. First-stage larvae passed in faeces of the host are rather stout with slightly tapered tails with rounded ends. The cells of the larvae (except in *D. arnfieldi*) are packed with nutrient granules which obscure anatomical details but there is no evidence that the oesophagus has a valved bulb or that the larva feeds (Daubney, 1920). Under suitable conditions of temperature and moisture the first-stage larva enters a lethargus and moults to the second stage in a few hours. The second stage is similar in morphology to the first-stage larva and its cells are also packed with granules. The larva moults to the third stage in a few more hours and the third-stage larva retains for a short time both shed cuticles but that of the first stage is eventually cast. The third stage is typically strongyliform and the tail is rather short and sharply conical. The granules which obscured the internal anatomy of the first two larval stages largely disappear. The sheath fits closely to the body and barely extends beyond

the tail end of the enclosed larva. Third-stage larvae are negatively geotactic and fairly resistant to cold and freezing conditions and moderate drying.

The definitive host becomes infected by ingesting infective third-stage larvae contaminating its food. Larvae migrate on to vegetation but many apparently remain in faeces. The dispersal of larvae from faeces by means of exploding sporangia of fungus (*Pilobolus* spp.) may be important in the transmission of larvae to cattle which avoid feeding near faecal material (Robinson, 1962; Robinson *et al.*, 1962; Bizzell and Ciordia, 1965b; Taranik *et al.*, 1978; Doncaster, 1981; Rodriguez Diego *et al.*, 1981, 1983; Jørgensen *et al.*, 1982; Boon *et al.*, 1983; Roque *et al.*, 1983). In the definitive host, third-stage larvae invade the small intestine, enter the lymphatics and travel to the mesenteric lymph nodes, where they remain for 6–7 days and develop to the fourth and fifth stages without significant growth (Hobmaier and Hobmaier, 1929d; Kauzal, 1933; Anderson and Verster, 1971; Verster *et al.*, 1971). The small fifth-stage worms as well as some fourth-stage larvae then travel via the thoracic duct to the anterior vena cava and from there to the heart and the pulmonary arteries. In lungs the worms grow markedly, mature and produce larvated eggs in 3–4 weeks postinfection. The patent period in infections is variously reported and may be greatly extended by the presence of arrested immature worms (late fourth and early fifth stages) in the lungs (Taylor and Michel, 1952; Michel, 1974; Gupta and Gibbs, 1975).

An effective live vaccine produced by attenuating infective larvae with ionizing radiation has been developed. The attenuated larvae invade lymph nodes and stimulate an immune response but usually fail to reach the lungs, or are expelled rapidly if they do so (Jarrett *et al.*, 1959; Poynter *et al.*, 1960; Jarrett and Sharp, 1963; Sharma *et al.*, 1988).

D. arnfieldi (Cobbold, 1884)

D. arnfieldi is a parasite of equines and has also been reported from the Indian tapir (*Tapirus indicus*). LaPage (1956) regarded the donkey as the usual host. According to Cameron (1926) eggs were 'almost round' and about 90 μm in diameter, and larvated when oviposited. The first-stage larva was 430 μm in length with a spear-shaped tail. The oesophagus was clearly visible (cf. *Dictyocaulus filaria* and *D. viviparus*) and club-shaped; the cells of the larvae were not packed with nutrient granules. There was no evidence of a valved oesophageal bulb but the terminal end of the oesophagus was expanded. The first-stage larva moulted in 24 h after hatching but retained the moulted cuticle. The tail of the second stage was conical with a blunt tip. The second moult took place 12–24 h after the first. The third-stage larva was morphologically similar to the first- and second-stage larvae. It generally cast the first-stage cuticle but retained that of the second stage. Larvae did not survive long in culture.

Soliman (1960), who described adults in detail, gave the length of first-stage larvae as 290–390 μm. He also noted that the internal anatomy of the larva was not obscured by nutrient granules as in other species of *Dictyocaulus*. He reported that adults 'lie with their heads facing the terminal branches of the trachea and with their tails pointing towards the main trunk of the bronchial tree'. They 'face the stream of the inflammatory exudate upon which they feed'.

D. eckerti Skrjabin, 1931

Schneider *et al.* (1996) and Epe *et al.* (1997) have confirmed by molecular methods that *D. eckerti* of deer is a distinct species from *D. filaria* and *D. viviparus*. The results

support morphological criteria (i.e. buccal capsule structure) used to distinguish the species in deer (Durette-Desset *et al.*, 1988). This helps to explain why the results of some experiments transferring worms from deer to livestock and vice versa have often been somewhat ambiguous. For a review of these and newer studies see Bienioschek *et al.* (1996) who infected a fallow deer with *D. eckerti* which first passed larvae on day 24 and for 25 days thereafter.

D. filaria (Rudolphi, 1809)

D. filaria is a common cosmopolitan parasite of sheep and goats as well as wild antelope and deer. It has been well studied by several early parasitologists including Maupas and Seurat (1912), who observed two moults leading to the infective larvae, and Romanovitch and Slavin (1914) and Guberlet (1919) – Guberlet was influenced by the work of Ransom (1906) on *Haemonchus contortus* – who orally infected lambs and later recovered adult worms in the lungs. Daubney (1920) described in detail the development of eggs and larvae. Hobmaier and Hobmaier (1929d) discovered that eggs hatched in the bronchi and that infective larvae entered the lymph nodes and developed to the fourth and fifth stages before migrating to the lungs via the lymphatic system.

According to Daubney (1920) eggs were 116–138 × 68–90 μm in size. First-stage larvae were 550–585 μm in length and stout with rounded, slightly tapered tails. Daubney (1920) concluded that the first-stage larva did not feed, a conclusion already reached by Maupas and Seurat (1912). Daubney (1920) apparently studied larval development over a range of temperatures and combined the results in his account. 'Under suitable conditions of temperature and moisture' the first moult began 'six to forty-eight hours' after hatching but the first-stage cuticle was usually retained. The second-stage larva was about the same length as the first stage but the tail was more slender and pointed, with a slight constriction anterior to the tip. Granules in the cells were less dense than in the first stage. The second moult occurred 12–48 h after the first moult and the second-stage sheath was also retained. The infective third-stage larva was slightly longer than the second stage and the tail was similar in morphology to that of the second stage. The granules in the cells were much reduced in older larvae and the oesophagus was typically strongyliform. The first-stage cuticle was cast in 12–48 h after reaching the third stage

Kauzal (1933) studied larval development at 18–23°C and reported the first moult in 48 h and the second in 6–7 days. Larvae reached the infective stage in 19 days at 4–5°C and remained active for 100 days whereas they degenerated after 36 days at room temperature. He related this cold-hardiness to the fact that husk is more common in cooler, elevated environments in Australia.

Deorani (1965b) studied development of eggs and larvae and described them in detail. The embryonated eggs were 119–138 × 57–88 μm in size and first-stage larvae were 548–582 μm in length. Second-stage larvae were 531–565 μm and third-stage larvae 514–548 μm in length. He also reported that all larval stages were sensitive to desiccation.

Gerichter (1951) studied development of *D. filaria* in Israel and reported that at 21–25°C 90% of larvae hatched in 48 h. the first lethargus occurred in 24 h and the first moult was completed in an additional 24 h. The second-stage larva remained unchanged for about 24 h and then entered the second lethargus, which was completed in a few hours. The third stage shed the first cuticle within 24 h after the second moult and became infective on day 5. At 30°C hatching commenced in 24 h (60%); all had hatched in 48 h and the infective stage was reached in 4 days. At 37°C development proceeded to the second stage and ceased.

Rose (1955) studied the survival of larvae on pastures and reviewed earlier studies, including those of Soliman (1952, 1953a). Infective larvae persisted for 23 weeks from January to June but only 7 weeks from June to August in the UK. The rate of development of eggs and larvae varied with the season; it was 1 week in June and August. Infective larvae survived for 48 weeks in wet faeces at low temperatures; they were not markedly resistant to desiccation but tolerated freezing temperatures for several days. As Kauzal (1933) had observed earlier, eggs and larvae developed at low temperature (20 days at 5°C).

Gallie and Nunns (1976) reported that in England development to the infective stage took 4–9 days in late spring and summer, 1.5–4.0 weeks in autumn and 5.5–7 weeks in winter, and that the proportion of first-stage larvae developing to the infective stage was 10–28% in autumn and winter and 2–25% in spring and summer. Infective larvae survived from autumn until spring in sufficient numbers to perpetuate transmission but not to cause clinical disease.

Anderson and Verster (1971) and Verster *et al.* (1971) carried out a detailed study of the migration of *D. filaria* from the gut to the lungs of sheep. Third-stage larvae arrived in the right colic mesenteric lymph nodes of lambs 18 h postinfection and remained there for 6 days, during which time they developed to the fourth stage (third moult in 3 days) and the fifth stage (fourth moult in 5 days). The genital system developed rapidly in the fourth stage. Fifth-stage worms were astonishingly tiny (males 576–656 μm and females 576–672 μm in length). Worms were first found in lung tissue on day 4. The numbers of worms found in the lungs increased markedly up to 28 days and decreased to zero in the mesenteric lymph nodes by 21 days. From 12 days the number of worms in the lung parenchyma decreased whereas those in the trachea and bronchi increased. By 28 days about 82% of all worms recovered were found in the trachea and bronchi. Lymph drained from the thoracic duct of one lamb contained fifth-stage worms on day 7 and showed that they entered the circulatory system by way of the thoracic duct. The uterus of female worms in the trachea contained embryonated eggs on day 28.

Kassai *et al.* (1972) also confirmed that worms reached the lungs by migrating from lymph nodes through the thoracic duct and reported substantial larval migration 8–9 days postinfection. Some worms reached the lungs when migration through the thoracic duct was prevented from the fifth day. (However, Verster *et al.* (1971) reported that some worms migrated as early as day 4 – R.C.A.)

Kauzal (1933) noted that lambs ceased passing larvae in 12–14 weeks. Goldberg (1952) reported that the prepatent period in 10 lambs was 32–57 days and the average patent period was 45 days. Pulatov (1984) reported a prepatent period in goats of 28–41 days and a patent period of 96–124 days.

D. viviparus (Bloch, 1782)

D. viviparus is a widespread parasite of cattle and deer. Strains of the species probably exist; Presidente and Knapp (1973) failed to infect calves with *D. viviparus* isolated from black-tailed deer (*Odocoileus hemionus columbianus*) in western North America (see

D. eckerti). An excellent review including 383 references has been published by Jørgensen and Ogbourne (1985).

According to Daubney (1920) eggs were 75–96 × 51–58 µm in size and first-stage larvae 300–360 µm in length. Soulsby (1965), who gave an extensive account of the biology of *D. viviparus*, reported that eggs were 82–88 × 32–38 µm in size and first-stage larvae 390–450 µm in length. Development and morphology of the larval stages were similar to those of *D. filaria*.

The optimal temperature for development was about 27°C according to Daubney (1920) but eggs developed normally and third-stage larvae appeared in 7–10 days in cultures at 8°C. First-stage larvae failed to survive freezing temperatures (−9°C) for 48 h but 60% of second-stage larvae survived and continued to develop to third-stage larvae at higher temperatures.

Third-stage larvae migrated upwards in culture as long as films of moisture were present and they migrated in darkness and diffuse sunlight but avoided direct sunlight. Third-stage larvae were capable of surviving desiccation for a considerable period.

Soliman (1952, 1953a) carried out a number of studies on the development and behaviour of larvae and the effects of environmental factors on them.

Michel and Rose (1954) investigated the dispersal and survival of larvae on pastures in the UK. In May, larvae survived for only 4 weeks but they survived for 13 weeks from October to January. Most larvae remained in faeces and most infective larvae were recovered when faeces were thinly spread on the vegetation (cf. dispersal of larvae by fungi).

The course of the infection of *D. viviparus* in calves is similar to that in *D. filaria*. Jarrett *et al.* (1957) and Jarrett and Sharp (1963) showed that third-stage larvae invaded the mesenteric lymph nodes in the region of the ileum, caecum and the first part of the colon and remained there for 3–8 days. In 7 days fourth-stage larvae first reached the lungs but large numbers only arrived in 13–15 days. Jarrett *et al.* (1957) noted lung lesions as early as 7 days. The prepatent period was reported as 21–30 days (Soliman 1953b; Rubin and Lucker, 1956; Jarrett and Sharp, 1963).

The patent period was 27–72 days (Rubin and Lucker, 1956). Taylor and Michel (1952) reported that the presence of arrested larvae in the lungs may extend the patent period for up to 150 days.

Family Molineidae

Subfamily Ollulaninae

Ollulanus

O. tricuspis Leuckart, 1865
O. tricuspis (syn. *O. skrjabini*, *O. suis*) is a curious tiny, autoinfective trichostrongyloid found with its anterior end buried in the gastric crypts mainly of domestic cats but also of other felines such as lions, cheetahs, tigers and wild cats (Hasslinger, 1985; Rickard and Foreyt, 1992). It occurs less commonly in dogs and foxes. It also occurs in pigs (Mason, 1975; Volkov, 1983; Voronkova *et al.*, 1985; Voronkova, 1986).

Adults are unusually small, the males 0.7–0.8 mm and females 0.8–1.0 mm in length. The male has a well-developed bursa (Cameron, 1923a). The head end of the female is typically curved ventrally and the tail ends in three to six cusps (Hasslinger and Wittmann, 1982; Blanchard *et al.*, 1985). According to Cameron (1927b) the female is monodelphic and the vulva is in the posterior region of the body. Eggs are few in number (one to three) and enormous in size in relation to the size of the female (Cameron, 1927b). The egg embryonates *in utero* into a first-stage larva, 350 μm in length, with a round, blunt tail and a rhabditiform oesophagus. This larva hatches *in utero*. The second-stage larva apparently also develops *in utero* and is distinguished from the first-stage larva by the presence of three cusps on the tail; this larva was about 340 μm in length. Third-stage larvae were apparently found in the uterus of the female and in the lumen of the stomach. Those found in the uterus were about 400 μm in length whereas those found in the stomach were about 500 μm in length. The tail of the third-stage larva was short and had at least three terminal cusps, and the oesophagus was typically strongyliform. Fourth-stage larvae were usually found on the surface of the gastric mucosa; females were about 625 μm and males 650 μm in length.

Cameron (1927b) concluded from his experiments that transmission occurred by emesis or vomition. The parasite elicits a catarrhal condition in the stomach which encourages vomiting. Hungry cats readily consume the vomit of other cats and in doing so acquire the infection. Autoinfection can result in the build-up of a sizable population of worms in the host. Collins and Charleston (1972) reported 4500 worms in one cat and fatal infections have been reported in tiger cubs (Lensink *et al.*, 1979). Hasslinger and Trah (1981) found an average intensity of 1500 in 542 cats and 11,028 worms in one animal. Presumably, populations that build up in the stomach of cats are controlled by regular vomition (Tiberio *et al.*, 1983).

Pigs, which are capable of vomiting, also probably acquire infections and presumably control superinfections in the same way as cats and other carnivores.

Subfamily Molineinae

Hepatojarakus

Members of the genus occur in oriental and Australasian rodents, often in sites other than the intestine. Species have a reduced corona radiatum.

H. malayae Yeh, 1955

H. malayae is a curious trichostrongyloid found in the bile duct of rats (*Rattus argentiventer, R. cremoriventer, R. exulans, R. rajah, R. tiomanicus, R. whiteheadi*) in Malaysia. Balasingam (1965) infected rats by giving them infective larvae *per os*. The development and transmission were subsequently studied by Ow-Yang (1974) in *R. tiomanicus*. Eggs were 60–63 × 37–39 μm in size. They were in the 8- to 16-cell stage when released by females but 16–32 cells when passed in faeces. Eggs developed to the first larval stage, which hatched in 13–15 h at 28°C. Larvae were 142–198 μm in length. The first moult occurred in about 13.5–15.5 h and second-stage larvae were 298–355 μm in length. The second moult started in about 27–30 h. Infective third-stage larvae were 355–497 μm in length and most escaped from the second-stage cuticle through a cap-like break in the anterior region of the second-stage cuticle 3.5–4.9 days after hatching. The tail of the third-stage larva was conical and ended in a rounded point.

Third-stage larvae were not resistant to desiccation and survived up to 32 days on moist filter paper. Larvae tended to extend themselves 'in an upright position' at the edge of moist filter paper and readily responded to CO_2 and changes in light intensity.

Infective larvae were given orally to rats and also inoculated into the body cavity. The prepatent period was 14–19 days by both methods of infection. Attempts to infect rats by placing larvae on the skin were not successful. Larvae penetrated the fundic wall of the stomach and presumably followed the hepatic portal system to the liver, where they appeared in 18 h postinfection. The third moult took place in the liver parenchyma in 60–72 h, when larvae were 745–933 µm in length and the final moult took place 8–11 days postinfection; males (7.2–8.5 mm in length) usually moulted before females (15.0–17.9 mm in length). Worms were observed penetrating hepatic ducts 16 days postinfection. Worms eventually entered the lumen of the hepatic ducts and matured.

Molineus

Members of the genus are parasites of the upper part of the small intestine of carnivores and primates (Cebidae). Mature *M. torulosus* (Molin, 1861) have been reported in macroscopic nodules in the small intestine of *Cebus* spp. (Macchioni and Pierotti, 1972; Brack *et al.*, 1973; Brack, 1976). Larger nodules had an ulcerated opening which communicated with the gut lumen.

M. barbatus (Chandler, 1942)

M. barbatus is a parasite of raccoons (*Procyon lotor*) and skunk (*Mephitis mephitis*) in North America. According to Gupta (1961b, 1963) eggs were 54–61 × 38–45 µm in size and were passed in faeces of the host in the morula stage. First-stage larvae (230–270 µm in length) hatched in 8 days at 10°C, 4–5 days at 15°C, 24 h at 20°C, 22 h at 25°C and 17–18 h at 30–35°C. The tail was conical and relatively short. At 22–24°C the first moult occurred in 43 h after hatching, when the larva was 570 µm in length. The second moult occurred 90 h after hatching and the infective larva was 520 µm excluding the second-stage cuticle. Infective larvae were killed rapidly by desiccation. They were thermotactic, phototactic and negatively geotactic.

Two of seven ferrets (*Putorius putorius*) injected subcutaneously with larvae developed transient (13 days) low grade infections. The prepatent period was 10–11 days. Five of seven ferrets given larvae orally became infected and eggs appeared in their faeces in 8–10 days but the duration of the infections was brief.

Balasingam (1963) infected dogs and cats orally and by subcutaneous inoculation and reported a prepatent period of 8 days and a patent period of 'more than a month'.

Oswaldocruzia

Members of the genus are found in the intestine of amphibians and reptiles throughout the world.

O. filiformis (Goeze, 1782)

O. filiformis is a common parasite of amphibians and reptiles (*Anguis, Bombina, Bufo, Colubes, Coronella, Eremiae, Hyla, Lacerta, Natrix, Ophisaurus, Pelobates, Rana, Salamandra, Tachydromus, Talescopus, Trituris, Vipera*) in Europe, the eastern CIS and the Canary Islands (Baker, 1987).

Hendrikx (1981, 1983) reported on the development and transmission of *O. filiformis* found commonly in *Bufo bufo* in The Netherlands. Unfortunately, it is impossible to interpret most of the findings in these articles because of confusion in the identity of larval stages. Hendrikx and Moppes (1983) acknowledged this fact (p. 534) and provided a reinterpretation of earlier findings.

Hendrikx (1983) successfully infected toads orally with third-stage larvae and questioned whether percutaneous infection occurred as reported in *Oswaldocruzia pipiens* (see below). According to the reinterpretation by Hendrikx and Moppes (1983) infective larvae exsheathed in the stomach of toads in 18 h postinfection and entered the mucosa, where they grew to about 1.0 mm in length. Larvae then entered the lumen of the stomach and migrated to the intestine where they moulted to the fourth stage when males were 1.0–1.5 mm and females 1.5–2.0 mm in length.

Hendrikx and Moppes (1983) stated that 'L$_4$ stages develop from day 9 in the lumen of the intestine' and 'the young L$_4$ stages are lying deeply between the villi close to the epithelium'. Also, 'late L$_4$ stages are situated more superficially in the lumen of the intestine as are L$_5$ (*sic*) and adults' and 'egg production by females was observed on day 29 after infection'.

Griffin (1988) cultured the free-living stages of *O. filiformis* in tap water at various temperatures from 6 to 32°C. Eggs did not develop at 4°C but developed and hatched at 6 ± 1°C. The maximum temperature for development was 32°C. Increases in temperature up to 28°C increased the rate of development and reduced the hatching time.

O. pipiens Walton, 1929

O. pipiens is a widely reported species found in the intestine of *Anolis carolinensis, Bufo* spp., *Eumeces fasciatus, Eurycea lucifuga, Hyla* spp., *Leiolopisma laterale, Pseudacris triseriata, Rana* spp., *Sceloporus undulatus* and *Terrapene* spp. in North America (Baker, 1987).

Baker (1978) studied the development and transmission of this species in *Rana sylvatica* and *Bufo americanus* in Ontario, Canada. Eggs were deposited by females in the 8- to 16-cell stage (Fig. 3.6A). Eggs in faeces developed into first-stage larvae (Fig. 3.6B) which hatched in 24 h. First-stage larvae were 297–317 μm in length with a pointed, conical tail and a well-developed oesophagus with a valved bulb. Infective third-stage larvae (Fig. 3.6C) were present in faecal cultures in water in 3–4 days. Larvae migrated out of faecal matter into the surrounding water. Third-stage larvae were 562–631 μm in length without the second-stage sheath and 644–716 μm in length including the sheath. The tail was conical and sharply pointed. The genital primordium consisted of only a few cells.

Infective larvae were placed on moist filter paper in dishes and frogs and toads were restrained on the paper for 4 h before being transferred to terraria at 14–18°C. In *R. sylvatica*, third-stage larvae (Fig. 3.6D–F) were attached to the mucosa of the cardiac

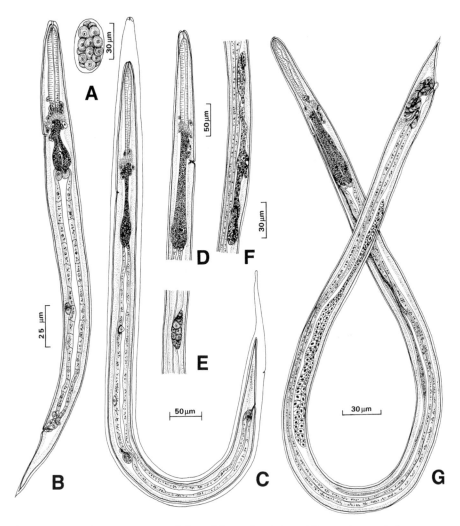

Fig. 3.6. Development stages of *Oswaldocruzia pipiens*: (A) unembryonated egg; (B) first-stage larva; (C) ensheathed third-stage infective larva; (D) parasitic third-stage larva, anterior end; (E) third-stage larva, genital primordium; (F) genital primordium of female fourth-stage larva; (G) male fourth-stage larva. (After M.R. Baker, 1978 – courtesy *Canadian Journal of Zoology*.)

region of the stomach 1–3 days postinfection. Fourth-stage (Fig. 3.6G) and moulting fourth-stage larvae were found in the lumen of the intestine 5–10 days postinfection; adult males and subgravid females appeared in 12 days. In *B. americanus* examined 14 days postinfection, females without eggs were present. In 18 days, females with eggs were found. Toads were infected with larvae derived from worms in frogs and the parasite was successfully transmitted to tadpoles of *R. sylvatica* although it was never found in wild-caught tadpoles.

Examination of *R. sylvatica* collected during spring and summer indicated that young frogs acquired infections soon after their transformation from tadpoles. Peak prevalences and intensities occurred in spring (May–June) and early autumn (September). All worms collected in April were mature, showing that worms overwintered in the frogs. Young-of-the-year frogs contained worms in August. Apparently the later summer period, when the marsh shrank in size, was ideal for the transmission of *O. pipiens* to young frogs which had just transformed and moved on to the shoreline of the marsh; these frogs presumably retained their infections until the following spring.

Subfamily Nematodirinae

Nematodirus

About 40 species have been described in the genus and Rossi (1983) pointed out that, apart from those in livestock, most species occur in holarctic Cervidae and Bovidae. *Nematodirus* species are essentially parasites of mammals in temperate or cold, often elevated, environments (Rossi, 1983) and their development and transmission reflect this fact. Eggs are remarkable for their large size and their development to the infective third stage within the egg, as noted by Ransom (1911), Railliet and Henry (1912) and Maupas and Seurat (1913). Development in the egg is slow compared with that of other trichostrongyloids and the third-stage larva is unusually long. Larvae in eggs are remarkably resistant to freezing temperatures and desiccation. During hatching the larva sheds the first-stage cuticle (which is usually left behind in the egg shell) but retains the second-stage cuticle with its markedly attenuated tail. The eggs of some species (e.g. *N. spathiger* and probably *N. helvetianus*) hatch during the summer in the usual way when conditions of moisture and temperature are favourable (Kates, 1950; Goldberg and Rubin, 1956). In dry conditions the eggs remain dormant. In northern England eggs of *N. filicollis* and especially *N. battus* remain dormant after a prolonged period of development in summer. Eggs of these species apparently undergo mass hatching in spring in response to a rise in temperature, presumably following sensitization by exposure to low temperatures in winter (Gibson, 1959; Thomas, 1959b; Thomas and Stevens, 1960; Mitchell *et al.*, 1985). The results of a study in one part of the world might not be wholly applicable to another region where environmental conditions are different. Rickard *et al.* (1987, 1989), using 22 groups of three to four tracer lambs allowed to graze for 28 days on pasture with *Nematodirus* species in Oregon, found that *N. battus* and *N. filicollis* were transmitted during the entire year, with major peaks in late fall and winter. They suggested transmission of *N. battus* and *N. filicollis* was mainly dependent on precipitation whereas that of *N. spathiger* was more dependent on temperature in the study area.

N. abnormalis May, 1920

N. abnormalis is a cosmopolitan parasite of sheep, goats, camels and antelopes. It usually occurs in small numbers with other species but in South Australia it is frequently the dominant species or may occur monospecifically (Beveridge and Ford, 1982). It is the dominant species in sheep in Turkey (Onar, 1975).

According to Onar (1975) eggs were lenticular in shape and 190–246 × 80–118 μm in size. Eggs were in the six- to eight-cell stage when passed in faeces of the host. Eggs kept at 6°C for 5 months did not develop beyond the morula stage. At 15°C first-stage larvae appeared in eggs on day 16; the second stage was reached in 28 days and the third stage in 36 days. Free larvae were noted in cultures in 48 days. At 20°C free larvae were observed in 22 days and at 25°C they were noted in 16 days. At 30°C free larvae were noted in 10 days and at 35°C in 8 days. Infective third-stage larvae were 909–1072 μm in length. The tail in lateral view had a deep cleft 'but arising within the cleft from the ventral lobe is a tubercular or stump-like process' which did not project beyond the tips of the lobes.

Neiman (1977) reported hatching in 38 days at 8–10°C and in 4 days at 35–36°C but few larvae hatched at higher temperatures. Eggs frozen for 3.5 months developed normally when transferred to warm temperatures. Eggs passed in faeces on pastures from January to March hatched in mid-April. Rabbits were successfully infected. Goat and infected kids passed eggs 17–18 days postinfection.

According to Beveridge *et al.* (1985) *N. abnormalis* developed in the anterior 4 m of the small intestine of lambs. The third and fourth moults took place 4–6 and 12–14 days, respectively, postinfection. Lambs were given 60,000 larvae and a large proportion remained inhibited at the early fourth stage. The main effects of the infection were caused by coiling around villi and deformation of the epithelium. There was no evidence of penetration into the lamina propria.

N. battus Crofton and Thomas, 1951

N. battus is a parasite of sheep, goats and cattle and frequently occurs along with other members of the genus in these hosts. According to Thomas (1959a) eggs were 152–182 × 67–77 μm in size, passed in faeces of the host in the seven- to eight-cell stage, and brown in colour. At 20–22°C eggs developed to first-stage larvae in 9–10 days. The first moult took place in 12–13 days, and larvae were in the third stage in 28–30 days. At 15°C the third stage was not reached until 50 days. Hatching of the eggs never occurred in cultures. According to Thomas (1957) infective larvae were 963–1136 μm in length including the sheath and 774–928 μm without the sheath. The tail of the third-stage larva tapered to a sharp point and there were two 'notches' on the dorsal surface – a small anterior one and a larger posterior one about 37 μm from the tip of the tail.

In lambs the fourth stage was reached within 5 days postinfection, when larvae were 1.4–4.0 mm in length. In 12 days immature adult worms were present. Eggs were present in the faeces 2 weeks postinfection.

According to Mapes and Coop (1972) the third moult in lambs occurred in 2–4 days and the fourth 8–10 days postinfection. The period of maximum larval penetration into the intestinal mucosa occurred in about 4 days and most larvae had returned to the mucosal surface by day 6 while they were still in the fourth stage. A few larvae moulted to the fifth stage while deep in the mucosa.

N. filicollis (Rudolphi, 1802)

N. filicollis is a cosmopolitan parasite of sheep, goats, alpaca and various Cervidae. Boulenger (1915) published an account reportedly of this species but there is some question as to the species he studied. Thomas (1957) believed Boulenger worked with a mixed infection and that the third-stage larva described by the latter was characteristic of *N. spathiger* and not *N. filicollis*. According to Thomas (1957) the eggs of *N. filicollis* were 134–168 × 71–87 µm in size. Thomas (1959a) reported that at 21°C eggs embryonated to the first stage in 8–9 days. The first-stage cuticle became visible in 11–12 days and larvae had reached the second stage in 12–16 days. In 24–27 days the second-stage cuticle was visible and the larva was in the infective third stage. At 28°C development was completed in 20 days and at 21°C development required 40–45 days. Infective third-stage larvae were 752–1018 µm in length with the sheath and 496–672 µm without the sheath. Thus, the infective larva of *N. filicollis* is the shortest of any members of the genus in livestock. The tail of the infective larva ended bluntly and had a deep notch, which divided it into a stout ventral lobe and a slightly smaller dorsal lobe (Thomas, 1957).

Larvae are markedly resistant to freezing temperatures (Poole, 1956; Smith and Archibald, 1965). In lambs the prepatent period was said to be 2–3 weeks.

N. helvetianus May, 1920

N. helvetianus is mainly a parasite of cattle but has been reported in sheep, goats and camels. According to Herlich (1954) eggs passed in faeces of the host were in the two- to eight-cell stage and 185–245 × 92–113 µm in size. Zviaguintzev (1934) reported that eggs were highly resistant to cold and survived in various stages at winter temperatures in Moscow. Although the optimum temperature for development was 28°C, eggs developed from 3 to 29°C. At 34–35°C half the eggs died. Herlich (1954) noted that eggs in the eight-cell stage kept at −10°C for 21 days developed normally when transferred to 28°C. At 20–22°C about 75% of the eggs reached the infective stage; hatching began 17–20 days after incubation and continued sporadically for 121 days. Optimal development occurred at 28–29°C, when 85–90% of the eggs reached the infective stage.

First-stage larvae appeared in eggs after 64 h of incubation at 28°C. These larvae were 990 µm in length with attenuated tails. The first moult occurred in the eggs in 4 days, when larvae were 1.07 mm in length. The second moult started on day 6 and was completed 16 h later; larvae began hatching 8–10 days after incubation and most were free by the end of 20 days. Larvae with the sheath were 0.97–1.24 mm in length. (According to the illustration in Herlich (1954) the tail of the sheath extended about 160 µm beyond the tail of the third-stage larva – R.C.A.) The tail of the third-stage larva ended in a digitiform appendage flanked by a dorsal and a ventral lobe. Rose (1966) studied the effects of temperature on eggs and larvae and found infective larvae on vegetation and well above ground level.

Herlich (1954) found sheathed infective larvae in the omasum and abomasum of a calf given larvae orally 5 h before necropsy. Exsheathed larvae were found in the small intestine of a calf given larvae 4 days earlier. Fourth-stage larvae were present in 8 days and by 15 days adult worms were present on the mucosa and coiled among villi. The prepatent period was 21–26 days and the patent period 12–132 days. Smith (1974) reported arrested fourth-stage larvae in calves in October and November in eastern Canada.

N. spathiger (Railliet, 1896)

N. spathiger is a common parasite of sheep, goats, cattle and numerous other domesticated and wild ruminants. Eggs were 181–230 × 91–107 µm in size according to Shorb (1939) and were in the two- to eight-cell stage when passed in the faeces of the host. According to Kates and Turner (1955) the early second stage was noted in eggs cultured for 5–6 days at 24°C (75°F). From 6 to 14 days second-stage larvae 'transformed into infective or third-stage larvae and the sheath of the second moult was formed'. Eggs started to hatch after 14 days in culture and hatching continued for a 'week or more'. According to Thomas (1957) the infective larva of *N. spathiger* was 976–1130 μm including the sheath and 723–810 μm without the sheath. The tail was bifid and with a terminal rod-shaped structure (cf. the larva described by Boulenger (1915) as *N. filicollis*).

Eggs containing larvae survived freezing temperatures of −8°C to −10°C for 440 days and free infective larvae survived for 150–230 days (Zurliiski, 1978). Infective larvae frozen for 150 days or dried at 15–20°C for 166 days were infective to lambs. Under field conditions in Montana, Marquardt *et al.* (1959) reported that the most rapid development occurred from April to September and the largest numbers of larvae occurred on pasture in spring. They regarded sunlight and high temperature at the soil surface as detrimental to the larvae.

The prepatent period in lambs was 2 weeks (Kates and Turner, 1955). Most larvae had apparently undergone the third moult prior to day 5 and the fourth moult did not take place until after 10 days postinfection.

Family Herpetostrongylidae

Paraustrongylus

P. ratti Obendorf, 1979
P. ratti is a parasite of the intestine of *Rattus fuscipes* in Australia. Beveridge and Durette-Desset (1993) described the larval stages. The infective larvae were 500–550 μm in length, with an elongate tail with a dorsal and ventral spike. Male fourth-stage larvae were 1.09–1.76 mm and females 1.4–2.00 mm in length.

Family Ornithostrongylidae

Subfamily Ornithostrongylinae

Ornithostrongylus

Members of the genus are found in the intestine of neotropical birds.

O. quadriradiatus (Stevenson, 1904)
O. quadriradiatus (syn. *O. crami*) is a cosmopolitan blood-sucking parasite of the duodenum of pigeons and doves (*Columba livia*, *Streptopelia turtur*, *Zenaida macroura*) (Wehr, 1930, 1971; Komarov and Beaudette, 1931; Cram, 1933). It is regarded as an important pathogen which results in a sudden acute enteritis, with heavy mortality.

Adult worms are slender, spirally coiled, red and deeply embedded in the mucosa and difficult to detect. According to Cram and Cuvillier (1931) and Cuvillier (1937) eggs were 64–88 × 32–49 μm in size and in the 32- to 64-cell stage when passed in the faeces of the host. Eggs developed into first-stage larvae which hatched within 19 h, when they were about 290–390 μm in length. The first moult occurred in 27 h, when larvae were 528 μm in length. The second moult usually occurred on day 3 after eggs were passed from the host. Third-stage larvae were 530–640 μm in length, including the second-stage cuticle (without the sheath, 585 μm). The tail of third-stage larvae ended in a terminal point and a pair of lateral, asymmetrical subterminal spines. Larvae survived for 5–6 weeks in culture and tended to migrate from faecal material into places where there was water. They also tended to be negatively geotactic.

Pigeons were infected *per os*. The third moult took place in 2 days and fourth-stage larvae were present in the birds by 40 h. The final moult occurred 24–48 h after the third moult and eggs appeared in the faeces 5–6 days postinfection.

According to Cuvillier (1937) fourth-stage larvae were usually found on the surface of the mucosa of the proventriculus and duodenum but some larvae had penetrated deeply into the glands. In later stages few worms were found in the proventriculus. Infections could be retained for 5 months; the author noted that third-stage larvae were found as late as 10 days postinfection. Thus, variability in development of larvae could extend the patent period considerably. Eggs were not resistant to freezing or desiccation and first- and second-stage larvae died after a few minutes of drying. Third-stage larvae were only moderately resistant to desiccation. The parasite could not be transmitted to chickens, turkeys or ducks.

Family Heligmosomidae

The related families Heligmosomidae and Heligmonellidae have complex cuticular ridges, collectively known as the **synlophe**, which attach the nematode to villae as they coil about these structures in the intestine (Durette-Desset, 1971).

In the Heligmosomidae the axis of orientation of the ridges is subfrontal and in the Heligmonellidae the axis is generally oblique (Durette-Desset, 1983). The female in the heligmosomids has a terminal caudal spine lacking in females of the heligmonellids and there are differences in details of the bursal rays. Durette-Desset (1971), in her classic thesis, used the structure of the synlophe to relate genera of these nematodes to the evolution and dispersal of their hosts during the course of evolution.

Bansemir and Sukhdeo (1996) suggested from experimental results that *Heligmosomoides polygyrus* in the intestine of mice tended to select regions in the intestine with the longest villi.

Heligmosomids are common parasites of soricoid insectivores and oriental, holarctic and neotropical rodents. The family contains the genus *Heligmosomoides*, with the well-studied species *H. polygyrus* of mice.

Heligmosomoides

H. polygyrus polygyrus (Dujardin, 1845)
H. p. polygyrus (syn. *Nematospiroides dubius* Baylis according to Durette-Desset, 1971) is a parasite of the duodenum (Panter, 1969) of wild house mice (*Mus musculus*),

white-footed mice (*Peromyscus maniculatus*) and field mice (*Apodemus sylvaticus*) (Elton *et al.*, 1931; Spurlock, 1943; Ehrenford, 1954). It is readily maintained in mice in the laboratory and has been the subject of various studies.

Spurlock (1943) showed that various inbred strains of mice could easily be infected orally. According to Ehrenford (1954) eggs were 70–84 × 37–53 μm in size and were passed in faeces in the 8- to 16-cell stage. At 23–28°C hatching took place in 26 h and first-stage larvae were 300–600 μm in length. Ehrenford (1954) noted a moult at 48 h after hatching and assumed this was the final moult leading to the sheathed infective larva. He claimed 4–6 days were required for the development of the infective larvae, which were 480–563 μm in length. Larvae migrated to the edge of the culture dishes and assumed a 'vertical stance using the tail as a foot.' They also remained motionless for long periods unless stimulated by mechanical vibration, air currents or heat.

Infection was by oral ingestion and larvae were unable to penetrate the skin. In the gut of mice, larvae exsheathed and were found 24 h postinfection in the small intestine. From 24 to 48 h postinfection, larvae penetrated the intestinal mucosa and came to lie close to the longitudinal muscle layer of the gut wall; they could be seen grossly from the outside of the intestine. The third moult occurred 2 days postinfection and the final moult (the fourth, but regarded by Ehrenford as the third) 6–8 days postinfection, after which worms returned to the intestinal lumen and matured. Eggs appeared in faeces 9 days postinfection and were produced for up to 8 months. Bansemir and Sukhdeo (1994) suggested that *H. polygyrus* feed on host tissue and not on host ingestion or blood.

Fahmy (1956) collected worms from *A. sylvaticus* and *M. musculus* in Scotland. Eggs kept at 22°C hatched in 23–24 h. Eggs kept at 26°C hatched after 19.5–20 h. First-stage larvae were 343–365 μm in length. The tail was conical but sharply pointed. Larvae examined 24 h after hatching were 418–440 μm in length. Sheathed third-stage larvae, present in 48–56 h, were 443–461 μm long (presumably excluding the sheath). Eggs were found 12 days postinfection in faeces of mice given larvae orally.

Bryant (1973) restudied *H. p. polygyrus* and reported for the first time the two moults in the free-living stages. At 20°C eggs hatched in 36–37 h. The first moult occurred in 28–29 h and the third 17–20 h after the first moult. In mice, the third moult occurred in the intestine in 70–96 h and the fourth in 144–166 h postinfection. The first eggs were detected in the faeces of mice in 10 days.

Dobson (1962) showed that the mouse (*M. musculus*) is a more suitable host for *H. p. polygyrus* than the rat. Development was slower in the rat than in the mouse and worms lived further down the intestine of the rat.

Bawden (1969) noted that more larvae of *H. p. polygyrus* established themselves in mice on inadequate diets than on ample diets but the 'low plan' diet reduced the ability of the adult worms to survive.

Lewis and Bryant (1976) reported on the distribution of *H. p. polygyrus* in the intestine of mice and Behnke and Parish (1979) studied the arrest of larvae in immune mice. Hernandez and Sukhdea (1995) stressed the significance of self- and allogrooming in the transmission of *H. polygyrus*. Kavaliers and Colwell (1995) suggested that female mice can discriminate between the odours of parasitized and non-parasitized male mice.

Sukhdeo and Mettrick (1983) transplanted adult worms into different regions of the intestine of mice. Worms consistently migrated anteriorly to the duodenum. Liu (1965) reported that, in mice, larvae were in the gastrointestinal mucosa as early as 4 h after oral inoculation. They moved into the deepest portion of the gastric mucosa of the fundic region. Larvae left the gastric mucosa within 36 h postinfection and entered the intestine. Larvae encapsulated in the muscularis externa on day 3. Worms vacated their capsules by day 8 and emerged into the lumen of the intestine.

Brown *et al.* (1994a) studied the epizootiology of *H. polygrus* in a wild population of *A. sylvaticus* in England. The parasite had an overdispersed distribution, with higher prevalences in males and heavier mice. Brown *et al.* (1994b) concluded that eggs per gram of faeces fluctuated according to a 24 h cycle and that the origin of this cycle lay in the pattern of egg release by worms. Prevalence in intensity peaked in spring and declined in autumn. Infected mice moved further and faster than uninfected mice but the reasons for this are still unclear.

N'Zobadila *et al.* (1996a,b) contrasted the morphogenesis of *H. p. polygyrus* in *Apodemus flavicollis* and *M. musculus* with *Heligmosomum mixtum* of *Clethrionomys glareolus* and *Heligmosomoides laevis* of *Microtus arvalis*. The results revealed the close affinities of *Heligmosomoides* and *Heligmosomum*.

H. kurilensis kobayashii (Nadtochii, 1966)
Asakawa (1987) studied the development of this species, collected from *Apodemus speciosus* from Japan, in mice and jirds (*Meriones unguiliculatus*). Larvae hatched after incubation for 24 h at 20°C. The first moult occurred between 24–48 h after hatching and infective larvae appeared in 5 days. In mice infected orally the fourth-stage larvae appeared in the mucosa of the small intestine and in about 10 days the adult worms appeared. In jirds, adults elicited capsules under the serosa of the upper small intestine and died.

Family Heligmonellidae

The family includes parasites of talpoid insectivores, lagomorphs and rodents.

Subfamily Nippostrongylinae

The subfamily includes the genus *Hassalstrongylus* of cricetid rodents and *Nippostrongylus* of oriental and Australian murids and oriental dermapterans. *N. brasiliensis* is a well-studied species found worldwide in rats.

Hassalstrongylus

Members of the genus are found mainly in cricetid rodents.

H. musculi (Dickmans, 1935)
H. musculi (syn. *Longistriata musculi*) is a parasite of the intestine of *M. musculus* and *Oryzomys palustris* in North America (Durette-Desset, 1972). According to Schwartz and Alicata (1935) eggs were 61–68 × 38 μm in size and well segmented when passed in

faeces of the host. Eggs hatched in about 24 h in cultures at 28°C. First-stage larvae were 296–311 μm in length. The second moult was not observed but infective larvae apparently appeared in cultures in 4 days. In a charcoal–faeces culture the infective larvae lost the second-stage cuticle. Infective larvae were 610–677 μm length, apparently without the second-stage cuticle. The tail of the third-stage larva had two subventral processes about 10 μm from the tip.

Schwartz and Alicata (1935) successfully infected mice by the oral inoculation of larvae and by placing larvae on the skin. The route of migration of larvae from the skin to the intestine was not determined but there was no evidence the lungs were involved. Regardless of the method of infecting mice, the prepatent period was always 7 days. The patent period in mice infected percutaneously was about 2 weeks. Mice infected orally passed more eggs than mice infected percutaneously and the patent period was longer (32 and 63 days).

Neoheligmonella

N. dossoi Durette-Desset and Cassone, 1986
N. tranieri Durette-Desset and Cassone, 1986

Both species were found in the small intestine of the rodent *Uranomys ruddi* (Muridae) of the Ivory Coast, Africa, and have been studied by Durette-Desset and Cassone (1986, 1987a,b) and Durette-Desset *et al.* (1989). Eggs of *N. dossoi* were 65 × 40 μm and those of *N. tranieri* were 60 × 40 μm in size. Eggs were in the morula stage *in utero*. The preparasitic larval stages of the two species were indistinguishable. First- and second-stage larvae have attenuated tails and were on average 263 μm and 499 μm in length, respectively. Infective larvae were on average 725 μm in length. The first-stage larva lacked cuticular 'crêtes' but the second- and third-stage larvae had alae. In culture, infective larvae appeared both sheathed and unsheathed. The authors suggested that in the rainy season larvae tended to be unsheathed and highly infective whereas during the dry season larvae tended to be sheathed.

U. ruddi were infected both orally and by skin penetration. Larvae acquired orally invaded gastric glands, entered blood vessels leading to the hepatic portal system and passed to the lungs by way of the heart in 12 h. The third moult occurred in the lungs and larvae passed up the respiratory system, were swallowed and reached the intestine, where they underwent the final moult in 5–6 days. The prepatent period was 15–17 days.

N. pseudospira Durette-Desset, 1970

This species was studied by N'Zobadila and Durette-Desset (1992) in the rodent *Arvicanthis niloticus* from Mali. After oral infection larvae were found in the stomach for 12 h. They appeared in the liver in 8–12 h and in the lungs from 12 h to 3 days. The larvae moulted in the lungs and migrated to the gut on day 3 as sheathed fourth-stage larvae. The larvae shed the third cuticle in the second quarter of the intestine. The final moult occurred in the first half of the intestine and the prepatent period was 11 days.

Nippostrongylus

N. brasiliensis (Travassos, 1914)

N. brasiliensis (syn. *Heligmosomum muris, Nippostrongylus muris*) is a cosmopolitan parasite of the small intestine of rats (*Rattus assimilis, R. conatus, R. norvegicus, R. rattus*) and mice (*M. musculus*); the parasite is also transmissible to other rodents (Kassai, 1982). Adult worms inhabit the duodenum, jejunum and, less commonly, the upper ileum, where they burrow among villi and crypts of the mucosa. Jenkins (1974) regarded the duodenum as the favoured site where worms could be found after they had been expelled from other sites in the intestine. Because of the ease with which it can be maintained in the laboratory for a variety of studies, *N. brasiliensis* has been the subject of much research. Kassai (1982) compiled and summarized 684 references dealing with all aspects of the biology of the parasite and Ogilvie and Jones (1971) reviewed studies on immunity. Haley (1961) provided a useful review of earlier work.

Yokogawa (1922) was the first to investigate the development and transmission. He reported that eggs were $54–62 \times 31–34\,\mu m$ in size and that they commenced to segment in the uterus. Eggs were passed in the faeces of the host in the 16- to 20-cell stage. Most eggs developed to first-stage larvae and hatched in 20–24 h at 'temperatures favourable for development'. First-stage larvae were 280–300 µm in length, with tapered, sharply pointed tails. Two days after hatching, larvae commenced to moult. Yokogawa (1922) observed only one moult and concluded that the infective stage was the second stage. These infective larvae were 620–750 µm in length without the sheath and 770–820 µm with the sheath. Lucker (1936b) reported that first-stage larvae were 270–550 µm in length and that growth culminated in a moult in the first 36–48 h; the moult was preceded by an unusually early separation of the first cuticle, which was fine and unstriated. Second-stage larvae were 470–750 µm in length and when fully grown moulted the second time to form the infective larvae. Both first- and second-stage larvae had elongate buccal cavities and attenuated tails. The anterior part of the buccal cavity of the third stage was compressed. According to Yokogawa (1922) third-stage larvae were unsheathed in culture. Chandler (1932) noted that the anterior end of the sheath was usually split and the larva could extend or retract its body into the sheath. According to Africa (1931a) larvae remained in their sheaths for several weeks but the latter eventually turned brown and contracted. Weinstein and Jones (1956) gave the average length of the exsheathed larvae as 635 µm and Haley (1961) gave a range of 500–700 µm and an average of 632 µm. Weinstein (1996) reported that the free-living larvae of *N. brasiliensis* concentrated vitamin B_{12} in their tissues as they fed on bacteria in rat faeces. The infective larvae contained the highest amount reported in any animal tissue. It is suggested this store of vitamin would be available to the rapidly growing and differentiating worms in the rat host. The author states that 'the intimate relationship between host and faeces and the free-living larvae of nematode parasites is an ancient one' and that 'little is known of the evolutionary selections and adaptations that have occurred between particular hosts and their bacterial flora and the concomitant co-evolution of fecal nematode cycles'.

Africa (1931a) reported that infective larvae tolerated cold (0–7°C) for 13 days but mortality was greater than at room temperatures. According to Boardman (1933) infective larvae were not resistant to desiccation and died in a few minutes when exposed to low humidities. Larvae were also reported to be negatively heliotactic and positively phototactic and thermotactic (Africa, 1931a; Boardman, 1933; Cunningham, 1956) but Parker and Haley (1960) concluded that the positive response to light was a response to heat produced by light.

Yokogawa (1922) infected rats by feeding them bread contaminated with infective larvae but only a small percentage of the larvae attained adulthood in the intestine. Larvae placed on the skin penetrated rapidly into the tissues and reached the lungs apparently by way of the blood. A moult to the fourth stage took place in the lungs and larvae then passed up the trachea, were swallowed and matured in 5–6 days in the intestine. The prepatent period was 6–7 days. Schwartz and Alicata (1934b) confirmed the findings of Yokogawa (1922) and concluded that ingested larvae only developed to maturity if they had undertaken a migration to the lungs.

Taliaferro and Sarles (1939), Gharib (1955) and Lee (1972) showed that infective larvae penetrated the skin of young mice directly and sometimes within 5 min. Lewert and Lee (1954) found larvae in the dermis within 15 min. Gharib (1953, 1955) observed that larvae crawled about on the surface of the skin. Some held their anterior ends perpendicular to the skin and then bent over, forcing the head into the skin. After remaining in the stratum corneum for a short time, larvae moved deeper into the epidermis and then migrated to the loose subcutaneous connective tissue. Lee (1972) concluded that larvae in the epidermis moved between cells and that in the dermis and subcutaneous tissues they moved through the ground substance and the collagen of the connective tissue. As noted earlier by Lewert and Lee (1954) there was histolysis of collagen and ground substance around larvae but Lee (1972) thought that this might not be the result of enzymatic activity.

Gharib (1953, 1955) reported larvae in lymph nodes (precrural and axilla). Clarke (1967, 1968) examined blood and lymph nodes of rats heavily infected percutaneously and concluded that larvae migrated mainly by the blood, although some were also found in lymph nodes. Twohy (1956) reported that, after subcutaneous inoculation, infective larvae appeared in the lungs as early as 11 h, and by 15 h most larvae which migrated had reached the lungs. However, in 18–19 h some 30% of the larvae inoculated were still at the site of the inoculation and would presumably not have migrated. Inoculation of larvae was more successful than allowing a similar number of larvae to penetrate the skin. Tindall and Wilson (1990a,b) also confirmed that *N. brasiliensis* penetrated skin and migrated to the lungs.

Larvae remained in the lungs for 35–50 h, according to Yokogawa (1922). According to Twohy (1956) there was logarithmic growth in 19–32 h postinfection when the third moult occurred. There was no further growth until worms had migrated via the trachea to the intestine. Yokogawa (1922) and Twohy (1956) found fourth-stage larvae in the intestine in 41.5–45 h after percutaneous infection; by 59 h most larvae had left the lungs. In the intestine fourth-stage larvae grew rapidly; the final moult began in 90–108 h postinfection and lasted for 12–15 h. Eggs generally appeared in the faeces of rats in 6 days. Simaren and Ogunkoya (1970) reported that the prepatent period was 6 days in rats infected orally or subcutaneously. The patent period was usually 11–19 days (Africa, 1931b) but might be longer in infections of high intensity than in those of low intensity (Haley and Parker, 1961). Simaren and Ogunkoya (1970) reported the patent period as 6–12 days in oral infections and 6–15 days in subcutaneous infections; however, more worms were found in rats infected subcutaneously than in rats infected orally.

According to Phillipson (1969) copulation did not occur until at least 104 h after subcutaneous inoculation of larvae into rats, and most worms had mated by 120 h. Each female in a mixed infection of males and females could produce about 1000 fertile eggs every 24 h. Female worms recovered from rats on day 10 and transplanted to uninfected rats produced about 3500 eggs without an additional insemination. Young female worms after insemination could produce about 1000–1500 eggs per day and it was estimated that females could store 4000–4500 sperm. Thus a female would probably require fertilization at least every other day.

Although Wilson *et al.* (1976a) suggested that transmammary transmission may occur in *N. brasiliensis*, their experiment did not adequately rule out the possibility of skin invasion (Wilson *et al.*, 1976b). There is, therefore, no evidence that transmammary transmission occurs in *N. brasiliensis*. Kassai (1982) has provided a generalized chart of the development of *N. brasiliensis* (Fig. 3.7).

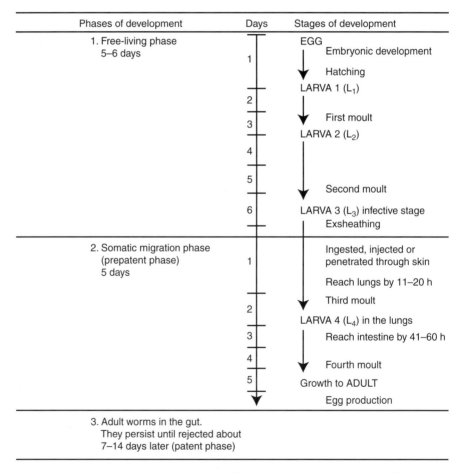

Phases of development	Days	Stages of development
1. Free-living phase 5–6 days	1	EGG Embryonic development Hatching LARVA 1 (L_1)
	2	
	3	First moult LARVA 2 (L_2)
	4	
	5	
	6	Second moult LARVA 3 (L_3) infective stage Exsheathing
2. Somatic migration phase (prepatent phase) 5 days	1	Ingested, injected or penetrated through skin Reach lungs by 11–20 h
	2	Third moult LARVA 4 (L_4) in the lungs
	3	Reach intestine by 41–60 h
	4	Fourth moult
	5	Growth to ADULT Egg production
3. Adult worms in the gut. They persist until rejected about 7–14 days later (patent phase)		

Fig. 3.7. Development and transmission of *Nippostrongylus brasiliensis*. (Slightly modified after T. Kassai, 1982 – courtesy International Institute of Parasitology and Akadémia Kiad, Budapest.)

3.5

The Superfamily Metastrongyloidea

The Metastrongyloidea is a moderately sized superfamily of bursate nematodes consisting of about 181 species classified into 46 genera and seven families (Anderson, 1978, 1982). The superfamily is confined to mammals and its species are most common in Artiodactyla (14 genera), Carnivora (14), Marsupialia (seven) and Cetacea (Odontoceti) (six). A number of genera occur in the Insectivora (five), Rodentia (three) and Primates (ten). Hares and rabbits (Lagomorpha) share *Protostrongylus* with the artiodactyls. Cattle and horses are devoid of metastrongyloids, their place being taken by *Dictyocaulus* spp. (Trichostrongyloidea) in these hosts.

Most metastrongyloids as adults occupy lungs of the host (e.g. *Protostrongylus*) but there are numerous important exceptions. Some members of the Protostrongylidae (*Elaphostrongylus* and *Parelaphostrongylus*) are associated with veins distant from the lungs and some members of the Angiostrongylidae occupy pulmonary and mesenteric arteries. Many members of the Pseudaliidae of Odontoceti, as well as all members of the Skrjabingylidae, occupy frontal sinuses.

Some species living in lungs deposit unembryonated eggs in the lung tissues or air spaces (e.g. Protostrongylidae, *Aelurostrongylus abstrusus*) whereas species living outside the lungs deposit unembryonated eggs that are carried to the lungs in blood (e.g. *Angiostrongylus* spp., *Elaphostrongylus* spp., *Parelaphostrongylus* spp.). Eggs develop in the lungs to first-stage larvae. Many metastrongyloids are ovoviviparous in that eggs develop into first-stage larvae *in utero* and female worms deposit these fully developed larvae into the lungs (e.g. Crenosomatidae, some Angiostrongylidae, many Filaroididae). Larvae of these species leave the lungs via the bronchial escalator and are swallowed and passed in the host's faeces.

In some metastrongyloids the lungs have been bypassed entirely. For example, in sinus worms (Skrjabingylidae, presumably some Pseudaliidae) larvae leave the sinuses, enter the throat and are swallowed. In addition, in certain Angiostrongylidae living in mesenteric arteries (*Angiostrongylus costaricensis*, *A. siamensis*) eggs embryonate in the intestinal wall and first-stage larvae apparently migrate into the lumen of the gut and are then passed in faeces of the host.

With the peculiar exception of the Metastrongylidae of Suidae which use earthworm intermediate hosts, a general feature of transmission in the Metastrongyloidea is the use of gastropods (usually terrestrial) in which development occurs to the third and infective stage; this was first discovered by Hobmaier and Hobmaier (1929c) in *Muellerius capillaris*. Many hosts (notably the artiodactyls) acquire their lungworm fauna from the accidental or deliberate ingestion of gastropods containing infective larvae. Species in Carnivora frequently utilize vertebrate paratenic hosts (e.g. amphibians, reptiles, rodents and shrews) to channel infective larvae to the definitive host. There are, however, some unusual exceptions to this general pattern of transmission in the Angiostrongylidae and the Filaroididae. In some species first-stage larvae are directly infective to the definitive host; these highly specialized species (*Andersonstrongylus captivensis*, *Filaroides hirthi*, *Oslerus osleri*) occur in skunks and canids. One species (*Filaroides decorus*) in sea-lions utilizes coprophagic fish as an intermediate hosts, and *Otostrongylus circumlitus* of seals uses American plaice, at least in Canadian waters.

Some recent observations seem to shed some light on these apparent anomalies in the transmission of the lungworms. It has been shown recently that first-stage larvae of *Angiostrongylus cantonensis* and *A. vasorum* can develop to the infective third stage in amphibians although the latter are not likely to be important in the transmission of these nematodes under natural conditions. This shows that some lungworms of carnivores have first-stage larvae with an innate ability to reach the infective stage in vertebrates. This may explain the fact that *Andersonstrongylus captivensis* (Angiostrongylidae), *Filaroides hirthi* and *Oslerus osleri* (Filaroididae) infect their hosts in the first stage, i.e. the mustelid or canid host serves as both intermediate and final hosts. The unusual ability of larvae to attain the infective third-stage in vertebrates explains how *Filaroides decorus* of sea-lions and *Otostrongylus circumlitus* (Crenosomatidae) of seals were able to stay with their hosts as the latter evolved from terrestrial into marine mammals. These highly adaptable parasites were able to replace the molluscan intermediate hosts with marine fish and thus ensure their survival in the new environment. Possibly lungworms of the Odontoceti survived the evolution of their terrestrial hosts into the marine environment by similar methods. Lungworms which failed to have an innate ability to reach the infective stage in vertebrates would presumably have been unable to survive in a marine host and would have become extinct.

There is considerable diversity in the migratory behaviour and development of metastrongyloids in the definitive host. Some infective larvae penetrate the gut and migrate in the lymph vessels and nodes to the heart, then pass to the lungs where they develop to adulthood (e.g. *Cystocaulus ocreatus*, *Muellerius capillaris*). Other species follow the hepatic portal system to the heart and lungs (e.g. *Crenosoma* spp.). Infective larvae of at least one species (e.g. *Aelurostrongylus pridhami*) penetrate the gut, enter the body cavity, cross the diaphragm and invade the lungs directly from the pleural cavity. Species like *Parelaphostrongylus tenuis* and *Angiostrongylus cantonensis*, living distant from the lungs, migrate as third-stage larvae to the central nervous system, where they develop to adulthood before moving to the definitive site in the host.

A number of lungworms undertake their migration to the definitive sites in the fifth or subadult stage. For example, species of *Skrjabingylus* as well as *Filaroides martis* reach the subadult stage in the gut wall and then undertake remarkably complex, highly directed migrations to the frontal sinuses and lungs, respectively. Subadult *Angiostrongylus vasorum* migrate from visceral lymph nodes to the hepatic portal system and then travel to the liver, heart and finally to the pulmonary arteries where they mature.

Family Metastrongylidae

The three cosmopolitan species attributed to the family belong in the genus *Metastrongylus* (syn. *Choerostrongylus*) which is characterized by a pair of massive trilobed lips, long filiform spicules and an atypical bursa. The development and transmission of *M. apri*, *M. pudendotectus* and *M. salmi* of domestic pigs and wild boars have been investigated. The three species often occur together in the bronchi and bronchioles (mainly of the diaphragmatic lobe) of the same individual host (Ewing and Todd, 1961a,b). Members of the genus are peculiar in that they produce fully larvated eggs with thick, sculptured shells which pass through the respiratory system and out in the faeces of the host. The first-stage larva has a ventrally bent tail with a rounded, bulbous terminal end. Eggs remain viable for prolonged periods in moist conditions (Kates, 1941; Rose, 1959). Hobmaier and Hobmaier (1929a,b) first showed that eggs hatched in earthworms and larvae invaded the tissues and developed to the infective stage; there were two moults. These observations were confirmed and expanded by Schwartz and Alicata (1934a), Alicata (1935c), Schwartz and Porter (1938), Rose (1959), Tromba (1959) and Wang (1974). Nguyen Duc and Bùi (1995) claimed that eggs (20–80%) of *Metastrongylus* (species not indicated) hatched in sandy soil in 36 h but less commonly (10–15%) hatched in water. They reported the sites at which larvae occurred in the earthworms *Ocnerodrilus* sp., *Pheretima campanulata* and *Lampito mauriti* in Vietnam.

Swine acquire the lungworms when they ingest infected earthworms, which are frequently abundant in their environment. Third-stage larvae invade the intestine of swine and enter the lymphatic vessels. They tend to accumulate in lymph nodes for a short period of time but eventually reach the lungs by way of the right heart. The worms moult twice in the lymphatics or lungs as early as 3 days postinfection. The prepatent period is about 3–6 weeks.

Ewing *et al.* (1982) noted that *M. apri* and *M. pudendotectus* are usually found together in swine and provided evidence that these nematodes will rarely develop if they infect the host singly, but they both develop and mature when mixed in swine. They concluded that *M. apri* is a facultative mutualist and *M. pudendotectus* an obligate mutualist. Leignel *et al.* (1997) used sequencing of ribosomal DNA internal transcribed spacer 2 (ITS2) and random amplified polymorphic DNA assay (RAPD) to distinguish *Metastrongylus asymmetricus*, *M. confusus*, *M. pudendotectus* and *M. salmi*. The ITS2 sequences were similar in *M. salmi* and *M. confusus* but RAPD analysis showed that the four species were distinct.

Rose (1973) reviewed the information on the lungworms of domestic pigs as well as those found in sheep. Humbert and Drouet (1993) followed seasonal variations in egg production of *Metastrongylus* spp. in wild boar populations in the Chambord game reserve in France. In the CIS, Anisimava and Maklakova (1992) regarded the earthworms *Lumbricus rubellus* and *Nicodrilus caliginosus* as the most important hosts of *Metastrongylus* spp.

Metastrongylus

M. apri (Gmelin, 1790)

Earthworms reported to serve as intermediate hosts of *M. apri* by various authors (Schwartz and Alicata, 1934a; Dunn, 1955; Dayton, 1957; Refuerzo and Reyes, 1959; Rose, 1959; Reyes and Refuerzo, 1967; Wang, 1974; Kontrimavichus *et al.*, 1976) are as follows: *Allolobophora caliginosa, A. chloritica, A. longa, A. terrestris, Bumastus tenuis, Dendrobaena octaedra, D. rubida, D. subrubicunda, Diplocardia* sp., *Eisenia austriaca, E. bonnberti, E. foetida, E. lonnbergi, E. nordenskioldi, E. rosea, E. veneta, Eiseniella tetraedra, Helodrilus caliginosus, H. foetidus, Lampito mauritii, Lumbricus custaneus, L. rubellus, L. terrestris, Octalasium lacteum, Perionyx excavatus, Pheretima aspergillum, P. bahli, P. houlleti* and *P. hupeiensis.*

According to Schwartz and Alicata (1934a) first-stage larvae were about 275–305 μm in length. They invaded various tissues of earthworms such as *H. foetidus*. According to Schwartz and Porter (1938) the vast majority of larvae occurred in the calciferous glands but a few were found in the oesophagus, heart, dorsal blood vessel and crop. Schwartz and Alicata (1934a) reported two moults in rapid succession, the first as early as 9 days and the second 10 days postinfection. Infective larvae were about 685 μm in length and they retained the sheath of the second moult (that of the first was apparently discarded). Rose (1959) reported that at 22–23°C the second stage was reached in 6 days and the third in 9 days. At 9 days the third stage was granular in appearance and only lost this appearance in 16 days. Wang (1974) reported that development to the infective stage required 8 days at 24–30°C and 1 month at 14–21°C.

Infective larvae in swine apparently penetrated the intestine and reached the lymphatic system. They tended to accumulate in mesenteric lymph nodes, especially those of the large intestine near the caecum (Schwartz and Alicata, 1934a; Dunn, 1956). Larvae reached lungs apparently by following lymphatic vessels which led to the blood and right heart.

Düwel and Schleich (1971) infected guinea pigs with *M. apri* larvae from *Eisenia foetida*.

M. pudendotectus (Wostokov, 1905)

This species often occurs with *M. apri* (see Ewing and Todd, 1961b) and most experimental studies have not separated the two species. Known intermediate hosts are *Allolobophora caliginosa, Eisenia lonnbergi, E. rosea, Lumbricus rubellus, L. terrestris* and *Octalasium lacteum.* In swine the prepatent period is said to be 3–6 weeks by some authors (Hobmaier and Hobmaier, 1929a,b; Schwartz and Alicata, 1934a). Tiunov (1965) determined the minimum prepatent period as 21–22 days although most pigs passed eggs 30–35 days postinfection. According to the same author, single infections lasted for only 4–6 months in swine.

M. salmi (Gedoelst, 1923)

Known intermediate hosts of *M. salmi* are *H. caliginosus, L. terrestris* and *P. excavatus* (see Alicata, 1935c; Bhattacharyya *et al.*, 1971). First-stage larvae were 275–295 μm in length. Larvae developed rapidly in earthworm intermediate hosts and the cuticle of the first stage was usually shed before the cuticle of the second stage became completely detached. The second cuticle was retained, however, by the infective third-stage larvae, which were 550–630 μm in length (Schwartz and Alicata, 1934a; Alicata, 1935). Development in swine was probably similar to that of *M. apri*, which is a much commoner species and therefore more thoroughly studied than *M. salmi*.

Family Protostrongylidae

The family consists of 13 clearly related genera distributed mainly among Bovidae (sheep, goats, African antelopes) and Cervidae (deer). About half the known species are members of the genus *Protostrongylus*, some members of which occur in lagomorphs. Since protostrongylids are predominantly parasites of ruminants it is speculated that they are secondarily adapted to lagomorphs (Anderson, 1982). The family is distinguished from other metastrongyloids by unusually complex gubernacula and telamons (Boev, 1975).

Most protostrongylids occur in lungs, bronchi, bronchioles, alveolar ducts and alveoli (e.g. *Cystocaulus, Muellerius, Neostrongylus, Protostrongylus, Varestrongylus*). A few species are associated with veins distant from the lungs (e.g. *Elaphostrongylus, Parelaphostrongylus*). It is characteristic of all species in the family that female worms deposit unembryonated eggs, which develop into first-stage larvae in the lungs of the host. Eggs of species living distant from the lungs are deposited in veins and carried by the blood to the lungs via the right heart. Eggs are filtered from the blood and come to lie in capillaries, where they embryonate and become encapsulated by host response. First-stage larvae hatch from eggs, leave the area of inflammation, enter the airways and pass up the respiratory system and out in faeces of the host. First-stage larvae are generally long lived and resistant to drying and freezing temperatures.

First-stage larvae of protostrongylids fall sharply into two types. Larvae of species of *Protostrongylus* have rather long, tapered tails ending in a sharp point. In most other species which have been studied (*Cystocaulus, Elaphostrongylus, Muellerius, Parelaphostrongylus, Varestrongylus*) the tail has a more or less prominent posteriorly directed dorsal spine a short distance from the tail tip; this spine is said to be minute and difficult to distinguish in *Neostrongylus linearis*. The spine is shed with the cuticle of the first moult in the gastropod intermediate host. In addition, all first-stage larvae in the family seem to have well-developed lateral alae.

Hobmaier and Hobmaier (1929c) were the first to show that protostrongylid first-stage larvae (of *Muellerius capillaris*) would invade the foot of terrestrial gastropods and develop to infective third-stage larvae. The larvae moult twice and, in the foot, they retain the two shed cuticles of the first and second stage. These loose cuticles are rapidly discarded after the larvae are freed from snail tissue. Protostrongylids are not highly specific in their use of intermediate hosts and many species have been shown experimentally to develop in a wide range of terrestrial snails and slugs and even in aquatic species. Nevertheless, in any one region, only a few intermediate host species are probably important in natural transmission to the definitive host. The latter usually becomes infected by the accidental ingestion of gastropods crawling on vegetation. Some authors have reported that infective larvae leave the foot of snails and slugs and survive for considerable periods on vegetation, where they are transmitted to herbivores.

Other authors dispute this and the evidence seems to favour the opinion that the host normally becomes infected by ingesting gastropods.

Concurrent infections of two or more species may occur in an individual final host. In addition, first-stage larvae of species of the same genus, and even of different genera, may be morphologically indistinguishable. Since eggs of protostrongylids only embryonate in tissues of the final host, there is no obvious practical way to separate species for experimental work in many concurrent infections. Thus, in the following account, *Protostrongylus kamenskyi* and *P. pulmonalis* of lagomorphs are considered together, as are *P. stilesi* and *P. rushi* of Rocky Mountain bighorn sheep.

Cystocaulus

C. ocreatus (Railliet and Henry, 1907)

This is a parasite of the alveoli, alveolar ducts and bronchi of sheep and goats in Europe, Africa and Asia. According to Gerichter (1951) first-stage larvae were 390–420 μm in length; according to Davtyan (1949) they were 380–480 μm in length. The tail tip was bent dorsally and ended in a proximal tubular piece with tiny spines and a long, sharp, dagger-like piece. There was a curved dorsal spine on the tail as in *M. capillaris*. First-stage larvae invaded and developed in a variety of gastropods as follows: *Abida frumentum, Agriolimax agrestis, A. schulzi, Cathaica semenovi, Cepaea hortensis, C. vindobonensis, Chondrula septemdentata, Cochlicella acuta, Ena asiatica, E. eleonorae, Euparypha pisana, Fruticicola rubens, Helicella barbesiana, H. candicans, H. joppensis, H. obvia, H. vestalis, Helix cavata, H. pometia, Jaminia potaniniana, Levantina caecariana, L. hierosolima, Limax flavus, L. maximus, Monacha syriaca, Monachoides umbrosa, Parachondrula aptycha, Retinella nittelina, Theba carthusiana, T. pisana* and *Zebrina detrita* (see Boev, 1940, 1975; Joyeux and Gaud, 1946; Davtyan, 1948; Gerichter, 1951; Kassai, 1957; Zdarska, 1960). Manga-González and Morrondó-Pelayo (1990) showed that numerous species of snails were suitable intermediate host of combinations of *C. ocreatus* and *Muellerius capillaris* (i.e. *Cernuella cespitum arigonis, Helicella itala, H. ordunensis, Cochlicella barbara, Cepaea nemoralis* and *Helis aspersa*). Mixtures of larvae from *C. ocreatus* and *Neostrongylus linearis* developed in *C. cespitum arigonis, H. ordunensis* and *C. nemoralis*.

According to Zdarska (1960) first-stage larvae reached the infective stage in gastropods in 28 days at 26–28°C. Kassai (1957) claimed that the infective stage was reached in 20–46 days at 18–22°C. Gerichter (1951) stated that the first moult occurred 13–14 days and the second 15–17 days postinfection; an additional 1–2 days were required for larvae to become infective. At this time they were 750–790 μm in length and the tip of the tail had a button-like appendage.

According to Davtyan (1949) and Sagoyan (1950) infective larvae invaded the small intestine and colon of the final host and were found in mesenteric lymph nodes 36–40 h postinfection. Larvae apparently reached lungs by way of the hepatic portal system or lymph vessels. The parasite moulted twice in lung interstitial tissue (6–7 and 13 days after infection) and then moved into alveolar ducts. Worms matured by 28–30 days and unembryonated eggs were laid. The prepatent period was 40–44 days; Gerichter (1951) gave the prepatent period as 42 days.

Some authors believe that infective larvae leave molluscs and that sheep and goats become infected when they ingest free larvae on vegetation. Other authors (Azimov *et al.*, 1973) question this and believe that the only route of infection is by the ingestion of gastropods, since the latter do not normally shed larvae.

Elaphostrongylus

Members of the genus have been reported from red deer (*Cervus elaphus*), sika deer (*C. nippon*), reindeer (*Rangifer tarandus*) and moose (*Alces alces*) in Europe and the eastern CIS. Pryadko and Boev (1971) and Kontrimavichus *et al.* (1976) recognized a single species consisting of: (i) *E. cervi cervi* in *Cervus elaphus* in Scotland and presumably western Europe; (ii) *E. cervi panticola* in *C. elaphus*, *C. elaphus sibiricus*, *C. elaphus brauneri*, *C. nippon* and *Alces alces* in eastern Europe and the Far East (e.g. Altai); and (iii) *E. cervi rangiferi* in *Rangifer tarandus* in Europe and the CIS. Stéen *et al.* (1989) described a new species (*E. alces*) in moose in Sweden and distinguished it from *E. cervi* and *E. rangiferi* on the basis of some differences in morphology and the location in the host. In moose, worms are found epidurally rather than in the subdural and subarachnoid spaces as in reindeer. Halvorsen *et al.* (1989) provided some support for the proposal of Stéen *et al.* (1989) when they were unable to transmit the parasite from red deer and moose to reindeer. Stéen *et al.* (1997) transferred *E. rangiferi* to moose and *E. alces* to reindeer. Both species matured in the alternative host but produced fewer larvae than in the usual host, to which they were well adapted. Reindeer infected with *E. alces* developed patent infections after 39–130 days. In moose the prepatent period of *E. alces* was 39–73 days. *E. rangiferi* infections were patent in moose after 133 days. Lankester (1977) transmitted *E. rangiferi* from caribou (*R. t. caribou*) in Newfoundland to *Alces alces*, raising the possibility that this is a single species (see Kutzer and Prosl, 1975) and perhaps distinct strains. Gibbons *et al.* (1991) restudied specimens and concluded that three species should be recognized, namely *E. cervi*, *E. rangiferi* and *E. alces*. We prefer to use subspecies designations at this time to distinguish *E. cervi* from western Europe and *E. cervi* from the Far East.

E. alces Stéen, Chabaud and Rehbinder, 1989

E. alces occurs in the epidural spaces of the spinal canal and beneath fascia of the muscles and chest of moose in Sweden. *E. alces* females are larger than other species of the genus and have a bottle-shaped oesophagus and an oval bursa with long rays. According to Lankester *et al.* (1998) first-stage larvae were 417 (377–445) µm in length and third-stage larvae from the snail *Arianta arbustorum* were 714 (675–756) µm in length (smaller than that of *E. cervi*). In moose the prepatent period is said to be 39–73 days (Stéen *et al.*, 1997); the variability may be related to the apparent heavy doses of uncertain numbers used to infect the animals.

E. cervi cervi Cameron, 1931
E. cervi panticola Lubimov, 1945

As indicated above, these two subspecies occur normally in *Cervus* spp. in Europe and the Far East (Altai). *E. c. cervi* was first reported from Scotland and is presumably the subspecies reported in central Europe in *Cervus elaphus*. It is undoubtedly the subspecies introduced to New Zealand with importation of *C. elaphus* from Britain for the purpose of game ranching (Mason *et al.*, 1976; Mason and McCallum, 1976; Watson, 1984).

The development and transmission of *E. c. panticola* have been studied in considerable detail in the Altai region of Kazakhstan of the CIS where it occurs in red deer (*C. elaphus*), Siberian maral deer (*C. e. sibericus*) and sika deer (*C. nippon*) on game farms; it has also been reported in moose (*Alces alces*) in this region.

E. c. panticola occurs in the skeletal muscles and meninges of the host (Liubimov, 1948, 1959a,b; Pryadko *et al.*, 1963, 1964). According to Panin (1964a,b) eggs developed and larvae hatched near the sites on the meninges or muscles where adult worms were present. First-stage larvae, which were 321–396 μm in length, eventually were carried to the lungs with blood, coughed up, swallowed and passed in the faeces. They were resistant to drying and remained alive in dried faeces for more than a year. They also survived the severe winter temperatures of the Altai. Panin (1964a,b) found natural infections in the following molluscs on deer pastures in the Altai, Kazakhstan: *Agriolimax agrestis* (prevalence 5.9%), *Bradybaena fruticum* (8.5%), *Cochlicopa lubrica* (1.3%), *C. lubricella* (5.7%), *C. pseudonitens* (6.3%), *Discus ruderatus* (3.2%), *Euconulus fulvus* (2.1%), *Perforatella bicallosa* (5.9%), *Perpolita petronella* (0.1%), *Succinea altaica* (8.0%), *S. granulosa* (3.7%), *Vitrina rugulosa* (1.1%), *Zenobiella aculeata* (10.8%), *Z. nordenskioldi* (23.8%) and *Zonitoides nitidus* (5.5%). Řezáč *et al.* (1994) in the Czech Republic successfully infected *Arianta arbustorum* and *Helix pomtia* with larvae of *E. cervi*. They concluded that larvae invaded the superficial furrows of the foot.

Experimental infections were produced in the above terrestrial species as well as in the aquatic snail *Radix ovata*. In *Z. nordenskioldi* the first moult took place 18–20 days and the second 30–32 days postinfection, apparently at ambient temperatures (Panin and Rusikova, 1964). The infective larvae were 910–1110 μm in length and could survive in molluscs for up to 2 years.

Deer became heavily infected on new pastures that had high populations of terrestrial gastropods which had become infected throughout spring and summer (see also Prosl and Kutzer, 1980a). Development required 119–124 days and large numbers of larvae were passed 5–6 months postinfection. The central nervous system was involved in a high percentage of Siberian maral deer, many of which displayed neurological signs; it is apparently not known if the neuropil is involved in the early development of the worm in the central nervous system (cf. *Parelaphostrongylus tenuis*). In sika deer most worms localized in the muscles and neurological disease was much less prevalent than in maral deer. Neurological signs reported in infected deer included loss of herding instinct, lowered head, circling, ataxia, aggressiveness and general paralysis. These clinical signs may have been complicated by concurrent neural infections of the filarioid *Setaria altaica* (Liubimov, 1948, 1959a,b; Shol, 1964). Prosl and Kutzer (1980b) reported, however, that even large numbers of *E. c. cervi* in the meninges of the brain and spinal cord failed to produce neurological signs in red deer in central Europe. The route followed by larvae to the central nervous system has not been determined in red deer but Olsson *et al.* (1998) showed that larvae in guinea pigs migrated via the body cavity to muscles of the lateral body wall and entered the vertebral canal perhaps along spinal nerves (cf. *Parelaphostrongylus tenuis*).

Pusterla *et al.* (1997) reported neurological signs in goats in Switzerland which were probably caused by *E. cervi* (cf. *Parelaphostrongylus tenuis* and *E. rangiferi*).

Mason (1989, 1994, 1995) reviewed the present status of *E. c. cervi* in New Zealand, where the parasite was first noted in 1975 in red deer and wapiti from Fiordland. Watson (1983, 1986) in New Zealand infected red deer calves with 200 infective larvae each and reported the prepatent period as 107–125 days. Two 18-month-old calves given 400–500 infective larvae started to pass first-stage larvae 86 and 98 days later. Apparently, adult worms can produce eggs for 6 years (Watson, 1986).

Gajadhar and Tessaro (1995) showed that *E. cervi* developed in the North American molluscs *Triodopsis multilineata* and *Deroceras reticulatus*. Two mule deer (*Odocoileus hemionis*) of western North America were given larvae orally. One animal passed larvae beginning 121 days postinoculation and the other had numerous first-stage larvae in its lungs and faeces 128 days postinfection when it was killed. Both mule deer developed progressive neurological disease beginning on day 104 post-infection and had to be killed a few weeks later. A red deer infected at the same time remained normal.

E. rangiferi Mitskevich, 1960

This is a parasite of the meninges and muscles of reindeer (*Rangifer tarandus tarandus*) in Scandinavia and the CIS. It was introduced to caribou (*Rangifer tarandus caribou*) in Newfoundland, Canada, with an importation of reindeer from Norway early in this century (Lankester and Northcott, 1979; Lankester and Fong, 1989). It can be transmitted experimentally to moose (*Alces alces*) (see Lankester, 1977) and occurs naturally in moose in Newfoundland (Lankester and Fong, 1996).

Roneus and Nordkvist (1962) carried out a detailed histological study of the infection in sick reindeer in Sweden and noted that the parasites 'were most often found in the subarachnoid space and only occasionally in the subdural space. In no instance was there evidence that the nematode had penetrated into the adjacent nervous tissues'. The parasite was also found in the fascia of the superficial muscles, particularly beneath the latissimus dorsi and the deep pectoral. The authors also noted groups of eggs in the leptomeninges, in a section of the pituitary and also in the epimysium and fascia.

Mitskevich (1957, 1958) reported that eggs developed and hatched near the place where they were deposited. The larvae, she felt, were carried by the blood or lymph to the heart and lungs whereupon they were coughed up, swallowed and passed in faeces. First-stage larvae were 288–403 µm in length, had a dorsal spine on the tail, were resistant to adverse environmental conditions and were capable of surviving for up to 27 months.

According to Mitskevich (1957, 1958), who first studied *E. rangiferi* in the CIS, the parasite attained the infective stage in the terrestrial gastropods *Trichida hispida* and *Succinea putris* and various species of the freshwater genera *Coretus*, *Galba*, *Lymnaea* and *Radix*. Terrestrial hosts were the most suitable although species of *Agriolimax* were refractory. The infective stage was reached in 27–35 days in suitable molluscs kept at room temperature. Infective larvae were 933 µm in length. Lankester (1977) and Lankester and Northcott (1979) infected *Mesodon thyroidus* and *Triodopsis albolabris* with larvae from caribou in Newfoundland. First-stage larvae were 381–490 µm and third-stage larvae 937–1041 µm in length. Lankester and Fong (1996) suggest that *Deroceras laeve* may be the main host in Newfoundland.

Skorping and Halvorsen (1980) successfully infected several species of snails in Norway with larvae from the faeces of reindeer. The larvae attained the infective stage in the foot of *Arion silvaticus, Deroceras laeve, Discus ruderatus, Euconulus fulvus* and *Trichida hispida* in about 20 days at 20°C. Development in *Arianta arbustorum, Deroceras reticulatum* and *Succinea pfeifferi* was considerably slower. In *Arion hortensis, A. subfuscus, Clausilia bidentata, Cochlicopa lubrica, Nesovitrea* spp. and *Vitrina pellucida* only a few larvae developed and at a slow rate. *Punctum pygmaeum* and *Vertigo lilljeborgi* were refractory to infection. Skorping (1985) experimentally infected the aquatic snail *Lymnaea stagnalis*.

Halvorsen and Skorping (1982) studied the effects of temperature on the growth and development of *E. rangiferi* in the gastropods *Arianta arbustorum* and *Euconulus fulvus*. In the former snail the most rapid development to the third stage was 11 days at 28°C; at 12°C it was 75 days. In *E. fulvus* development was 11.5 days at 24°C and 48.5 days at 12°C.

According to authors in the CIS, infective larvae invaded the intestinal wall of reindeer and were carried by the blood to the central nervous system (especially the brain) and the intermuscular connective tissues, where they matured and deposited eggs (Kontrimavichus *et al.*, 1976). It is not clear how the blood route was determined. Handeland (1994) and Handeland *et al.* (1994) apparently gave doses of '300' to '1000' infective larvae to 12 reindeer calves. The doses were described thus: 'the number of snails necessary to give the predetermined infective dose (crushed snails in water) was decided on the basis of the L3 average of 18 (SD 15) found in 50 randomly picked snails'. The authors describe the various clinical signs in 11 animals and at necropsy listed and described lesions found in the gut wall, liver, lungs, kidneys, heart and central nervous system. They believed that the presence of larvae in these organs indicated that the worms travelled by the blood (i.e. hepatic portal system to the heart and lungs). However, their results could be reinterpreted by assuming that the large doses given might result in the abnormal dispersal of larvae, probably accompanied by strong host reaction. This could account for lesions in the superficial tissue of the liver and kidney after larvae had entered the body cavity. Also, larvae in the lungs were in capsules and clearly not migrating. There is, therefore, little reason to believe the worms reach the central system by the route postulated by the authors (see also Ollson *et al.*, 1998, concerning *E. cervi*). Handeland and Skorping (1992a, 1993) and Handeland *et al.* (1993) repeated experiments similar to the above and reported cerebrospinal nematodiasis in goats and sheep given heavy doses of larvae.

Lankester (1977) reported a 74-day prepatent period but it is now known that the experimental infections may have been contaminated with another protostrongylid (Lankester and Fong, 1989). There appears to be no evidence that *E. rangiferi* develops in the neuropil of reindeer and the neurological signs (ataxia, lumbar paresis, paralysis and torticollis) reported in infected hosts are apparently the result of damage to meninges and perhaps nerve roots (see for example Roneus and Nordkvist, 1962; Lankester and Northcott, 1979). Lankester (1977), found worms in the ventricles of the brain of a moose calf infected with the *E. rangiferi* from caribou in Newfoundland. The relationship between the various signs observed in reindeer and the muscle phase of the infection is not clear.

According to Mitskevich (1963, 1964) and Polyanskaya (1964) in the CIS, most infections of reindeer take place after a new generation of molluscs appears in July and

August. Young reindeer became infected at the age of 4–5 months. Wissler and Halvorsen (1976) and Halvorsen *et al.* (1980) discussed the relationship between transmission, gastropod density and movements of reindeer in Norway. They concluded that most infections of reindeer take place in late summer or autumn on summer ranges, probably on the calcium-rich bogs above the timberline. Higher than normal summer temperatures may have induced increased development of larvae in intermediate host which precipitated the outbreaks (Handeland and Slettbakk, 1994). Halvorsen *et al.* (1985) reported that *E. rangiferi* has an adult longevity of several years. Also, an initial period of high larval production was followed by a greatly reduced output of larvae. In male reindeer the larval output was high in autumn and early winter but in females it was high in late winter and spring. This sexual difference was regarded as an adaptation to maximize transmission throughout the year.

Muellerius

M. capillaris (Mueller, 1889)

M. capillaris (syn. *Synthetocaulus capillaris*) is a cosmopolitan parasite of the lung parenchyma of sheep and goats. It frequently occurs together with *Cystocaulus ocreatus*. As indicated previously, it was the first metastrongyloid studied in which gastropods were shown to be intermediate hosts (Hobmaier and Hobmaier, 1929c). Numerous gastropods have been shown experimentally to be suitable intermediate hosts but probably only a few are significant in field transmission in any one locality. For example, in the USA, Mapes and Baker (1950) noted that *Deroceras reticulatum* was commonly (31%) infected on pastures whereas *Arion circumscriptus* in the same area was not infected. Sultanov *et al.* (1975) concluded that *Helicella candaharica* was the only mollusc found naturally infected with *M. capillaris* in Uzbekistan, CIS, although other species could be readily infected in the laboratory. Trushin (1976) showed that in the Kalinin province of the CIS, freshwater snails played no role in the transmission of *M. capillaris*. The following gastropods have, nevertheless, been shown to be suitable intermediate hosts experimentally and under field conditions: *Agriolimax agrestis, A. reticulatum, A. laevis, Anguispira alternata, Anisus contortus, A. leucostoma, Arianta arbustorum, Arion circumscriptus, A. hortensis, A. subfuscus, A. empiricorum, Bradybaena fruticum, Cepaea hortensis, C. nemoralis, C. vindobonensis, Cernuella cespitum, C. virgata, Cingulifera planospira, Chondrula septemdentata, Cocclicella acuta, Cochlicopa lubrica, Deroceras reticulatum, Euparypha pisana, Fruticicola hispida, Galba corvus, Goniodiscus rotundatus, Gyraulus albus, G. laevis, Helicella acuta, H. barbesiana, H. candaharica, H. candicans, H. diecta, H. obvia, H. vestalis, H. virgata, Helicolimax pellucidus, Helicigona arbustarum, Helix cavata, H. pometia, Hyalina cellaria, Levantina caesareana, L. hierosolima, Limax flavus, Lymnaea stagnalis, Monacha syriaca, Monachoides umbrosa, Milax sowerbyi, Planorbis planorbis, P. coreus, Perforatella bidentata, Physa fontinalis, Praticollella griseola, Pseudotrichia rubigibosa, Radix peregra, Retinella nittellina, R. petronella, Succinea putris, S. pfeiffei, Theba pisana, Trichia hispida, Zebrina cylindrica, Z. detrita, Zonitoides arboreus* and *Z. nitidus* (Hobmaier and Hobmaier, 1929c; Pavlov, 1937; Williams, 1942; Joyeux and Gaud, 1946; Gerichter, 1948, 1951; Mapes and Baker, 1950; Rose, 1957; Egorov, 1960; Boev, 1975; Sultanov *et al.*, 1975; Zdzitowiecki, 1976; Sauerlander, 1979). Manga-González and Morrondó-Pelayo

(1990) combined larvae of *M. capillaris* with *Cytocaulus acreatus* in snails (see also *C. ocreatus*). Solomon *et al.* (1996) showed that development of larvae of *Muellerius capillaris* from the Nubian ibex ceased to develop in aestivating *Trochoidea seetzanii* and *Theba pisana* in Israel. Reactivation of the snails caused the larvae to continue to develop to the infective third stage (cf. Lankester and Anderson (1968) regarding *Parelaphostrongylus tenuis* in aestivation and its effect on the development of larvae in gastropods).

According to Gerichter (1951) first-stage larvae were 300–320 μm long; the tail was undulating and sharply pointed and had a posteriorly directed spine a short distance from its tip. In gastropods held at the 'optimum temperature' of 20–30°C, the first moult occurred in 28 days and was followed in 2 days by the second moult. An additional 2–3 days were required for larvae to become infective. Rose (1957) noted that the infective stage was reached in 92 days at 5°C, 37 days at 10°C, 18 days at 15°C, 13 days at 20°C and only 8 days at 25°C. Rose (1957) found various stages of *M. capillaris* in naturally infected slugs from sheep pasture and concluded that gastropods could be repeatedly infected (see also Lankester and Anderson, 1968).

Larvae released from snails penetrated the intestine of the final host and attained the mesenteric lymph nodes in 18–72 h (Svarc, 1968). Larvae were apparently carried in lymph to the right heart and appeared in lungs as early as 96 h although migration to lungs was only completed in 4–5 days (Boev, 1975). In lungs, worms left the blood and established themselves in alveoli and bronchioles. Eggs were deposited in the lung parenchyma where they developed into first-stage larvae and elicited granulomas (Rose, 1958). Gerichter (1951) gave the prepatent period as 42 days ('six weeks') and Sultanov *et al.* (1975) as 38–48 days.

Neostrongylus

N. linearis (Marotel, 1913)

N. linearis occurs in the small bronchi of sheep, goats, tur (*Capra cylindricornis*) and chamois (*Rupicapra rupicapra*) in Europe and Israel. According to Müller (1934) first-stage larvae were 300–350 μm in length and had a minute dorsal spine on the tail. Larvae invaded and developed to the third stage in *Arianta arbustorum, Arion hortensis, A. subfuscus, Cepaea hortensis, C. nemoralis, Cernuella cespitum, C. virgata, Cochlicella ventricosa, Derocercas agrestis, Fruticicola strioleta, Helicella apicina, H. aspersa, H. neglecta, H. variabilitis* and *H. pometia* (Müller, 1934; Rojo-Vazquez and Cordero del Campillo, 1974; Marcos Martinez, 1977). Zdarska (1960 in Boev, 1975) found larvae of *N. linearis* in naturally infected *Cepaea vindobonensis, Helicella candicans* and *Zebrina detrita*. Manga-González and Morrondó-Pelayo (1994) concluded that in Spain the molluscs susceptible to infection with *N. linearis* were, in decreasing order, *Oestophorella buvinieri, Oestophora barbula, C. nemoralis* and *H. aspersa*.

Müller (1934) reported that larvae reached the infective stage in 10–14 days at room temperature. Rojo-Vazquez and Cordero del Campillo (1974) reported that at 18°C the first moult took place in 10 days and the second in 14 days in *H. neglecta* and *H. variabilitis*. Comparable rates of development in *C. nemoralis* and *H. aspersa* were 13–14 and 13–15 days, respectively, for the first moult and 18–21 and 19 days for the second. Third-stage larvae were about 500 μm long. The prepatent period in lambs given infective larvae by Rojo-Vazquez and Cordero del Campillo (1974) was 60–85 days. Castanon-Ordonez (1982) infected 18 lambs and examined them at various times later. Larvae were found in the wall of the caecum and colon 2 days postinfection. Larvae followed blood to the liver and most had reached the lungs by day 6. Two moults occurred in the lungs and the fourth stage was observed 8 days and the subadults 10 days postinfection.

Parelaphostrongylus

The genus consists of three species in North American cervids. *P. odocoilei* is a western species found in muscles of mule deer and woodland caribou. *P. tenuis* is a parasite of the cranial veins and meninges of white-tailed deer in eastern North America. *P. andersoni* appears to have a continental distribution and occurs in the muscles of white-tailed deer, as well as woodland and barren-ground caribou.

P. andersoni Prestwood, 1972

Adult *P. andersoni* occurs adjacent to, or partially within, blood vessels in muscles of the hindbody of white-tailed deer (*Odocoileus virginianus*) in the southeastern USA, Michigan and southern British Columbia (Anderson and Prestwood, 1981; Pybus and Samuel, 1981; Pybus *et al.*, 1990). The parasite has also been reported in woodland and barren-ground caribou in Canada, including Newfoundland, and may have a wider distribution in this host in North America and perhaps Eurasia (Lankester and Hauta, 1989; Lankester and Fong, 1996). *P. andersoni* has been transmitted experimentally to mule deer (*Odocoileus h. hemionus*) (Pybus and Samuel, 1984a). It may occur in moose in Newfoundland (Lankester and Fong, 1996).

Female worms deposit eggs in the veins with which they are associated. Eggs are carried to the lungs, where they lodge in alveolar capillaries and become surrounded by inflammatory cells which eventually lead to granulomas. Eggs embryonate and larvae break out of nodules, pass up the respiratory tract and are passed out of the host in the mucus which coats the faecal pellets.

First-stage larvae were 308–382 µm in length and indistinguishable from those of *P. tenuis*, which may occur together with *P. andersoni* in the same host in eastern North America (Pybus *et al.*, 1990). Larvae invaded and developed to the infective stage in the following gastropods: *Deroceras laeve*, *Mesodon perigraptus*, *M. thyroides*, *Triodopsis albolabris*, *T. multilineata* and *T. vannostrandi* (Prestwood, 1972; Nettles and Prestwood, 1976; Anderson and Prestwood, 1981). The infective stage, which was reached in about 3 weeks, depending on temperature, was 0.99–1.20 mm in length. Prestwood (1972) noted a slight protuberance on the dorsal aspect of the tail which helped to distinguish it from larvae of *P. tenuis*. Ballantyne and Samuel (1984) contrasted the morphology of the infective larva of *P. andersoni* with that of the other two species in the genus. Lankester and Hauta (1989) found considerable variation in the shape of the tail of third-stage *P. andersoni* and advised caution in the use of this feature to distinguish species. According to Prestwood (1972) experimentally infected fawns started to pass first-stage larvae in 60–69 days. Muscles infected with *P. andersoni* included the longissimus dorsi, gluteal, thigh and psoas (Nettles and Prestwood, 1976).

The parasites were usually associated with extravasation of blood; occasionally whitish streaks and fibrous areas were noted.

Prestwood and Nettles (1977) noted that the results of low-level infections of white-tailed deer with *P. andersoni* contrasted greatly with those of moderate or heavy single infections (200–5000 larvae). Moderate infections (200 infective larvae) were characterized by: (i) an absolute eosinophilia, which peaked 2–3 weeks postinfection and declined gradually; (ii) viable adult worms in muscles; (iii) large numbers of eggs and larvae in lungs; (iv) large numbers of motile larvae in faeces; and (v) susceptibility to reinfection (Nettles and Prestwood, 1976). Postmortem features of these infections resembled those seen in naturally infected deer. Repeated administration of low numbers of infective larvae, on the other hand, resulted in: (i) a sustained leucocytosis with absolute eosinophilia, which declined only after the administration of infective larvae was stopped; (ii) postmortem evidence that some larvae failed to negotiate the caecal wall; (iii) reduced numbers of eggs reaching lungs; (iv) reduced viability of eggs reaching lungs due to encapsulation by host reaction; and (v) resistance to subsequent challenge with infective larvae.

P. odocoilei (Hobmaier and Hobmaier, 1934)

P. odocoilei occurs mainly in the skeletal muscles of mule and black-tailed deer (*Odocoileus hemionus*) in western North America (Hobmaier and Hobmaier, 1934; Brunetti, 1969; Pybus *et al.*, 1984). It has been reported in mountain goat (*Oreamus americanus*) (see Pybus *et al.*, 1984) and in caribou (*Rangifer tarandus*) (see Gray and Samuel, 1986). It has been transmitted experimentally to moose (*Alces alces*) (see Platt and Samuel, 1978; Pybus and Samuel, 1980). It is apparently not transmissible to white-tailed deer (*Odocoileus virginianus*) (see Platt and Samuel, 1978; Pybus and Samuel, 1984a).

According to Hobmaier and Hobmaier (1934), first-stage larvae were 378 μm in length and had a prominent dorsal spine on the tail; Brunetti (1969) gave the length of the larva as 367.5 μm.

Suitable experimental intermediate hosts were *Agriolimax agrestis*, *A. campetris*, *Deroceras laeve*, *Epigramophora arrosa*, *Euconulus fulvus*, *Helix aspersa*, *Planorbis* sp., *Triodopsis multilineata*, *Vitrina limpida* and *Zonitoides arboreus* (see Hobmaier and Hobmaier, 1934; Brunetti, 1969; Platt and Samuel, 1978, 1984). Third-stage larvae were 624 μm in length according to Hobmaier and Hobmaier (1934).

The behaviour of larvae in the definitive host early in infection has not been investigated. There is, however, no evidence that any part of development involves the central nervous system as in *Parelaphostrongylus tenuis* (see Pybus and Samuel, 1984b). Adult worms were found associated mainly with small veins or lymphatics in the muscles. Eggs oviposited in veins passed to the lungs, where they were filtered from the blood and became surrounded by granulomas. Eggs oviposited in lymphatics were found in superficial and deep inguinal, aortic, ruminal, tracheal and bronchial lymph nodes (Brunetti, 1969). However, Pybus and Samuel (1984b) concluded that most eggs were deposited in blood vessels. Lesions associated with infections were described by Pybus and Samuel (1984b).

Platt and Samuel (1978) reported that the prepatent period was 49–52 days in mule deer (*O. h. hemionus*), 58–72 days in black-tailed deer (*O. h. columbianus*) and 68–72 days in moose (*Alces alces*). Brunetti (1969) gave the prepatent period as 2.5 months in

O. h. columbianus. Platt and Samuel (1978) and Pybus and Samuel (1984a) followed larval output in experimentally infected fawns.

P. tenuis (Dougherty, 1945)

P. tenuis, the so-called meningeal or brain worm, is an extremely common and widely distributed neurotropic nematode of white-tailed deer (*O. virginianus*) in most regions of eastern North America. Prevalences as high as 86% have been reported in deer populations. Intensities ranged from one to 150 but the higher figure was exceptional and usually only one to nine worms were found in individual deer. Reviews of most known aspects of *P. tenuis* have been published by Anderson (1971), Anderson and Lankester (1974), Anderson and Prestwood (1981), Lankester (1987) and Lankester and Samuel (1998). More recent reports of *P. tenuis* in the USA are New York (Garner and Porter, 1991), Minnesota (Pitt and Jordan, 1994) and Warsaw Island, Georgia (Davidson *et al.*, 1996).

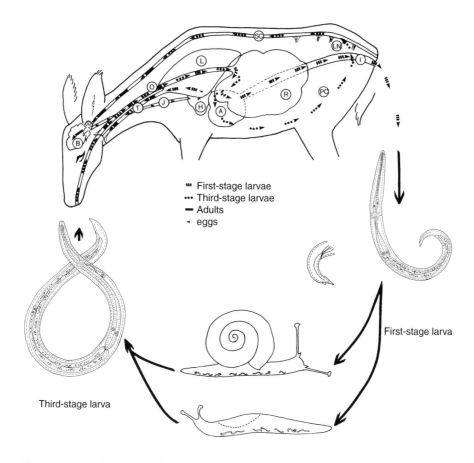

Fig. 3.8. Development and transmission of *Parelaphostrongylus tenuis*. A = abomasum; B = brain; H = heart; I = intestine; J = jugular vein; L = lung; LN = lumbar nerve; O = oesophagus; PC = peritoneal cavity; R = rumen; SC = spinal cord; T = trachea. (Original, U.R. Strelive.)

Adult *P. tenuis* occur in the cranial subdural space and the venous sinuses of the cranium (Fig. 3.8). Anderson (1963b) reported that the round, unsegmented eggs were oviposited by gravid female worms into the venous circulation and were carried as emboli to the right heart and lungs, where they lodged in small capillaries and became incorporated into granulomas. Occasionally females deposited eggs on the meninges or in the wall of the venous sinuses. These eggs developed normally into first-stage larvae but the fate of the larvae was not determined. Larvae that developed in lungs escaped from their granulomas and passed up the respiratory tract and out in faeces of the host. Larvae, found only in the mucus which coats the faecal pellet (Lankester and Anderson, 1968), were 310–380 μm in length and had a prominent spine on the tail. These larvae were long lived and resistant to drying and freezing (Lankester and Anderson, 1968; Shostak and Samuel, 1984). Although larvae will retain their viability for long periods in frozen deer faeces, they are not well adapted to survive variable winter conditions under the snow in the northern mixed biomes of northeastern North America where white-tailed deer reach their northern limits of distribution (Forrester and Lankester, 1998). In field experiments only 27% of larvae survived from December to April in this region of North America. Nevertheless, meningeal worm still manages to infect 82% of the deer herd and 79% of fawns are infected before they move to summer range (see below).

A number of terrestrial gastropods have been infected experimentally but only a few species may be significant in natural transmission. Larvae attained the infective stage in the foot of *Anguispira alternata*, *Arion circumscriptus*, *Deroceras laeve* (syn. *D. gracile*), *D. reticulatum*, *Discus cronkhitei*, *Haplotrema concavum*, *Mesodon thyroidus*, *Stenotrema fraternum*, *Triodopsis albolabris*, *T. notata* and *Zonitoides arboreus* (Anderson, 1963a; Lankester and Anderson, 1968; Kearney and Gilbert, 1978). *A. alternata* was a poor host because larvae were reluctant or unable to penetrate its tissues. Also, *A. circumscriptus* and *D. reticulatum* (introduced to North America) were poor experimental hosts in comparison with native species like *Deroceras laeve* and *T. albolabris*. The infective larvae were 900–1080 μm in length. Studies have shown that infective larvae can survive winter in gastropods. Also, development is retarded in aestivating and hibernating gastropods but continues when the latter become active again. Snails do not acquire an immunity to infection and it is possible to superimpose one infection on another (Lankester and Anderson, 1968).

Experimental studies were supported by field studies on Navy Island, Ontario, Canada, where *D. laeve* and *Z. nitidus* were the most common and most often infected gastropods (Lankester and Anderson, 1968). Larvae were also found in *A. alternata*, *A. circumscriptus*, *Cionella lubrica*, *D. reticulatum* and *Succinea ovalis* but prevalence was low and except for the last two slugs, these species were too uncommon to be significant in transmission.

The most important intermediate host in Ontario, the native slug *D. laeve*, is an annual species which, in Ontario at least, dies after laying eggs in June and July (Lankester and Anderson, 1968). *Zonitoides arboreus* and *Z. nitidus* live for 2–3 years. On Navy Island, Ontario, prevalences of larval *P. tenuis* in *D. laeve* in a damp forested habitat rose to 25% in spring. The slug was undoubtedly an important source of infection until mid-July, when the adults died and were replaced by juveniles. Prevalence in *Z. nitidus* (4.3%) did not fluctuate greatly and this snail was probably an important host throughout spring, summer and autumn. On Navy

Island, intensity in gastropods was low, e.g. 4.7 (1–97) in *D. laeve* and 1.4 (1–5) in *Z. nitidus*.

Most attempts (Parker, 1966; Gleich, 1972; Gleich and Gilbert, 1976; Gleich *et al.*, 1977; Upshall *et al.*, 1986) to find infected gastropods in enzootic areas have not been highly successful. Evidently, however, transmission to deer can occur without high prevalences and intensities in gastropods, because the large food consumption of deer ensures that they will eventually ingest infected snails and slugs on vegetation. It is known that patent infections can be produced in deer with extremely small numbers of infective larvae. Lankester and Peterson (1996) examined terrestrial molluscs on deer range in northeastern Minnesota, some in a deer yard. Ten (0.08%) of 12,096 snails and slugs were infected with 3.2 ± 2.5 infected larvae of *P. tenuis*. Prevalence was greater in a traditional deer wintering yard (seven of 4401, 0.16%) where deer aggregated for almost 5 months at a density of $50~\text{km}^{-2}$ than on summer range (three of 7695, 0.04%) where they occurred on $4~\text{km}^{-2}$. Despite such low densities of infected gastropods, their ingestion by chance remained a tenable explanation for the high prevalence of *P. tenuis* in the deer. These results accord with the observation of infection in deer on the same range (see Slomke *et al.*, 1995, below). There is much evidence that fawns are infected a few months after birth in enzootic areas (Peterson and Lankester, 1991). Maze and Johnston (1986), however, found relatively high prevalences of *P. tenuis* in *Discus cronkhitei*, *Triodopsis albolabris* and *Ventridens intertextus* in Elk State Forest, Pennsylvania. Most infected gastropods came from an old reclaimed strip mine, which may have been a focus of infection for animals in the area.

Slomke *et al.* (1995) examined 379 deer in northeastern Minnesota from 1991–1993. Small numbers of adult *P. tenuis* were found in 82% but over a third (118 of 311) were not passing larvae in their faeces. Most occult infections were sterile because only one sex was present. Intensities of worms per deer (3.2 ± 2.2) did not increase with age after 1 year but most old deer eventually became infected. Worms first appeared in the cranium of fawns in September, but larvae were not noted in the faeces until mid-December and most fawns were not positive until January (i.e. 4–5 month prepatent period). *P. tenuis* is apparently a long-lived species and since intensity did not increase with age, it is likely deer are immune to additional infection. Fecundity of female worms was greatest in fawns and larval output in all deer passing larvae increased in spring.

In deer, infective larvae released from snail tissues penetrated the gastrointestinal wall, crossed the peritoneal cavity and followed lumbar and perhaps other nerves to the vertebral canal (Anderson and Strelive, 1969). About 10 days were required for this journey. Migratory third-stage larvae invaded dorsal horns of grey matter, where they developed for about 1 month (Anderson, 1963a, 1965). By 40 days postinfection most worms, now subadults, had left the neural parenchyma and entered the spinal subdural space, where they matured and migrated to the cranium. Some worms remained in the subdural space. Others invaded cranial venous sinuses (especially the intercavernous) and the females oviposited eggs into the blood. The prepatent period was 82–91 days.

P. tenuis is generally clinically silent in white-tailed deer. It has, however, occasioned a great deal of research because it causes neurological disease in a variety of wild, semidomesticated and domesticated ungulates. For reviews, refer to Anderson (1971, 1972), Anderson and Lankester (1974), Anderson and Prestwood (1981), Lankester (1987) and Lankester and Samuel (1998). Neurological disease ('moose sickness') was known in moose for almost half a century before its cause was determined (Anderson, 1964). The disease in moose occurs in places where the range of this cervid overlaps that of white-tailed deer in eastern North America (Anderson, 1972). The disease was considered responsible for declines in moose populations in the 1940s and 1950s in Maine, Minnesota, Nova Scotia and Ontario when extremely high deer populations intermingled with high moose populations (but see Whitlaw and Lankester, 1994a,b). Infected moose displayed a variety of neurological signs associated with traumatic damage inflicted on the neuropil by the activities of small numbers of *P. tenuis*. Signs included lumbar weakness, ataxia, torticollis, arching, blindness, fearlessness, depression, paresis, paraplegia and death. The disease in moose was related to the spread northward in recent times of white-tailed deer into moose range, probably because of human activities (agriculture, forestry, fires) which provided suitable habitat for deer; climatic factors may also have been involved (Anderson, 1972).

P. tenuis is highly pathogenic in caribou and reindeer (Anderson and Strelive, 1967a; Anderson, 1970; Trainer, 1973) and may have been a factor in the failure of some attempts to introduce these animals into certain areas in North America (Dauphine, 1975; Bergerud and Mercer, 1989). The disease has been noted in reindeer in a zoo park as well (Nichols *et al.*, 1986).

Wapiti (*Cervus elaphus*) and red deer and their hybrids are susceptible to neurological disease (Anderson and Strelive, 1966; Woolf *et al.*, 1977; Olsen and Woolf, 1978, 1979) but can tolerate the infection and even pass a few larvae (Anderson and Strelive, 1966; Pybus *et al.*, 1989). It is unknown if *P. tenuis* can persist in wapiti in the absence of the normal white-tailed deer host.

Raskevitz *et al.* (1991) studied the presence of gastropods on range and range utilization by white-tailed deer and wapiti in an area in Oklahoma where these ungulates were sympatric. Wapiti spent most time in open fields and meadows, where gastropods were relatively scarce. Deer spent most time in forested areas, where most gastropods were found. The authors suggested that the reason a viable wapiti herd existed in the area was 'at least partially because of the habitat preference by wapiti and the reduced availability of infected gastropods in the selected areas preferred by the wapiti'. Such factors may have allowed deer and wapiti to be sympatric in eastern North America prior to settlement.

Mule and black-tailed deer (*O. hemionus*) and hybrids with white-tailed deer developed severe neurological disease when infected by *P. tenuis* (see Anderson *et al.*, 1966; Nettles *et al.*, 1977; Tyler *et al.*, 1980). The disease has been noted in both captive and free-living mule deer.

Neurological disease associated with *P. tenuis* has been reported in free-ranging and captive fallow deer (*Dama dama*) in contact with white-tailed deer (Kistner *et al.*, 1977; Nettles *et al.*, 1977; Woolf *et al.*, 1977).

Non-cervid ungulates are also susceptible to neurological disease when infected with *P. tenuis*. The disease has been experimentally produced in sheep and goats (Anderson and Strelive, 1966, 1969, 1972) and has been reported on several occasions in these animals on farms in the USA where the animals share range with infected deer (Mayhew *et al.*, 1976; Jortner *et al.*, 1985; O'Brien *et al.*, 1986). It has recently been diagnosed in a calf in Michigan (Yamini *et al.*, 1997) and a heifer in Virginia (Duncan and Patton, 1998). In addition, the parasite has been shown experimentally to be highly pathogenic in pronghorn antelope (G.V. Tyler and C.P. Hibler, personal communication, 1980) and bighorn sheep (Pybus *et al.*, 1996).

The disease has been diagnosed in llamas, guanocos and alpacas (Brown *et al.*, 1978; Baumgartner *et al.*, 1985; Rickard *et al.*, 1994; Scarrat *et al.*, 1996) and eland (*Taurotragus oryx*) (A.K. Prestwood, personal communication, 1980). It was diagnosed in sable antelope (*Hippotragus niger*) and suspected in bongo antelope (*Tragelaphus eurycercus*) and scimitar-horned oryx (*Oryx dammah*) (Nichols *et al.*, 1986) held in a zoo park surrounded by infected white-tailed deer. It has been diagnosed in black buck antelopes (*Antilope cervicapra*) in a zoo park in Louisiana (Oliver *et al.*, 1996).

Pneumostrongylus

P. calcaratus Mönnig, 1932

P. calcaratus is a common parasite of the lungs of African antelope, especially impala (*Aepyceros melampus*). It is usually found in firm, tan-grey nodules along the lobar borders of the lungs (Gallivan *et al.*, 1989). In primary infections adults, larvae and eggs were found in alveoli and bronchioles within the nodule. First-stage larvae were 314–341 μm in length and the tail had a dorsal spine.

Anderson (1974, 1976) exposed slugs (*Urocyclus* (*Elisolimax*) *flavescens*) to larvae from lung lesions in impala and recovered larval stages. Third-stage larvae appeared 13–14 days after exposure of the slugs and were 604–789 μm in length. *U. flavescens* was observed on various trees, mainly during summer, in the Nyala Game Ranch but were most commonly found on *Ziziphus mucronata*, a tree regularly browsed by impala and other antelopes. In the coastal area of Natal where the slugs were common, four of ten collected in autumn and three of 34 collected in spring had larval stages of lungworms. Only five naturally infected slugs had more than ten larvae and the highest number found in a single slug was 93. Attempts to infect two impala were unsuccessful. Anderson (1982) gave third-stage larvae to a goat (killed 17 days postinfection) and two lambs (killed at 10 and 51 days postinfection). Worms were found only in the lamb killed at 51 days postinfection.

Protostrongylus

Members of the genus occur in lagomorphs (the first five species below) and ungulates (the last six species below).

P. boughtoni Goble and Dougherty, 1943

This is a parasite of the bronchi, bronchioles and alveoli of snowshoe hare (*Lepus americanus*) in Canada. First-stage larvae were 253–307 μm in length and developed to the infective stage in the terrestrial gastropod *Vallonia pulchella* in 28–30 days at 18°C (Kralka and Samuel, 1984a,b). The first moult took place 14–18 days and the second 28–30 days postinfection. Ova were noted in the lungs of one hare infected 17 days previously; in other hares the prepatent period was 25–27 days. Experimentally infected hares passed larvae for 41–104 days. In 12–23 days there was a rapid rise in larval output, followed by a marked decline in the number of larvae passed. The parasite was transmitted successfully to domestic rabbits (*Oryctolagus cuniculus*). Kralka and Samuel (1984a,b) demonstrated that larvae only rarely left snails and that free larvae are not likely to be significant in transmission.

Kralka and Samuel (1990) reported that *Vertigo gouldi* was the major intermediate host in boreal forest habitats in north central Alberta, Canada, where prevalence in hares was 100%. Juveniles became infected within a month of birth and intensities increased to relatively high numbers within 3 months and then declined. There was no evidence of transplacental transmission.

P. cunicularum (Joyeux and Gaud, 1946)

This is a parasite of the bronchi of the wild rabbit (*O. cuniculus*) in Europe. First-stage larvae were 315–370 μm in length. According to Joyeux and Gaud (1946) *Helicella rugosiuscula* was a suitable intermediate host. Infective larvae were 550–750 μm in length. The prepatent period was 26–37 days.

P. kamenskyi Schulz, 1930
P. pulmonalis (Frolich, 1802)

P. kamenskyi and *P. pulmonalis* (syn. *P. terminalis*) occur together in the lungs of hares (*Lepus* spp.) and rabbits (*Sylvilagus nuttali*) in Europe and Asia. First-stage larvae were 340–350 μm in length and similar in both species. Several authors studied mixed infections (Ryzhikov *et al.*, 1956; Boev, 1975). Development to the infective stage occurred in the gastropods *Pupilla muscorum* and *Vallonia tenuilabris*, which were considered natural intermediate hosts in the CIS (Ryzhikov *et al.*, 1956). Larvae developed also in *Succinea elegans* and *Vertigo alpestris*. The first moult occurred in 9–12 days, the second in 20–22 days and larvae were infective in 30–36 days postinfection. The infective larvae were 500–640 μm in length. Some larvae were said to leave the foot of gastropods and could be found on vegetation (cf. Kralka and Samuel, 1990). The prepatent period in hares was 19–22 days.

P. tauricus Schulz and Kadenazii, 1949

This is a common parasite of the bronchioles of hares (*Lepus europaeus*) in Europe. First-stage larvae were 340–360 μm in length. Boev (1975) listed the following gastropods as suitable intermediate hosts: *Helicella krynizkyi*, *H. obvia*, *Pupilla muscorum*, *Vallonia costatus* and *V. enniensis*. According to Kadenatsii (1958) larval development in molluscs was completed in 20–25 days at optimal temperatures (presumably greater than 15°C). The author believed that larvae which left molluscs in mucus were important in transmission to hares (cf. the observations of Kralka and Samuel (1984a,b) on *P. boughtoni*). Babos (1961) claimed that the first moult occurred in molluscs on day 8 and the second on day 28 postinfection. Infective larvae were 500–540 μm in length. Larvae given to hares appeared in lungs in 12–48 h. The first eggs were deposited in lungs 22 days postinfection and 11–12 days were required for larvae to develop in the eggs. Rodonaya (1977) reported that *Helicella derbentina* was a suitable intermediate host in Georgia, CIS. Development was completed in this gastropod in 25–30 days; attempts to infect *Enomphalia ravergieri*, *Helix lucorum* and *Vallonia pulchella* failed. The prepatent period was 40 days according to Rodonaya (1977) and 25–30 days according to Kadenatsii (1958). Infected hares passed larvae for 8–9 months.

P. davtiani (Savina, 1940)

P. davtiani occurs in the bronchi, alveolar ducts and alveoli of sheep, goats (including *Capra aegagrus*), mouflon (*Ovis ophion, O. musimon*) and argali (*O. ammon*) in Europe and Asia. First-stage larvae were 290–350 μm in length. The species was said to develop in a wide range of gastropod intermediate hosts (Boev, 1975). In Uzbekistan, CIS, *Subzebrinus sogdianus* and *Xeropicta candaharica* were natural intermediate hosts and *Subzebrinus albiplicatus* and *S. pfeifferi* were infected experimentally (Kulmamatov, 1981). Larvae reached the infective stage in molluscs in 20–41 days at 22–30°C. The prepatent period in lambs was 28–32 days.

P. hobmaieri (Schulz, Orlov, and Kutass, 1933)

P. hobmaieri occurs in the bronchi of sheep, mouflon (*O. musimon*), wild goats (*Capra caucasica, C. sibirica*) and chamois (*Rupicapra rupicapra*) in Asia and Europe. Its biology has been reviewed by Boev (1975). First-stage larvae were 260–320 μm in length. Gastropod intermediate hosts were *Euconulus fulvus, Pupilla muscorum, P. sterri, P. triplicata, Vallonia costata, V. pulchella, Vertigo alpestris, V. antivertigo, V. pygmea, V. ronnebyansis* and *Zonitoides nitidus* (see also Politov, 1973). At room temperature the first moult occurred 12–16 days postinfection and the second moult 1–2 days later. In the lungs of the final host, larvae hatched from eggs 28–30 days postinfection and appeared in faeces in 33–36 days.

P. rufescens (Leuckart, 1865)

P. rufescens (syn. *P. kochi*) occurs in the smaller bronchi of sheep, goats, wild goats (*Capra aegagrus* and *C. sibirica*) and mouflon (*O. musimon*) in Europe, Asia, Africa, North America and Australia. According to Gerichter (1951) the first-stage larvae were 370–400 μm in length. Hobmaier and Hobmaier (1930) successfully infected three species of the genus *Helicella*. Gerichter (1951) infected *Helicella barbesiana, H. vestalis* and *Monacha syriaca. P. rufescens* is more specific in its use of intermediate hosts than *Cystocaulus ocreatus* and *Muellerius capillaris*, according to Gerichter (1951). Boev (1975) reviewed CIS and other work on this lungworm and listed the following as suitable intermediate hosts: *Agriolimax kervillei, Albida frentum, Cepaea vindobonensis, Cochlicella contermina, Cochlicopa lubrica, Enomphalia strigella, Euconulus fulvus, Eulota fruticum, Goniodiscus ruderatus, Helicella candicans, H. candaharica, H. obvia, Helix aspersa, H. promatia, Monachoides incarnata, Pupilla muscorum, Succinea putris, Vallonia costata, V. pulchella, Theba corthusiana, Zenbrina detrita* and *Zenobiella rubiginosa*. Soltys (1964) reported that *Helicella obvia* was found naturally infected in Poland and *Cepaea vindobonensis* and *Succinea putris* were infected experimentally.

At 20–30°C larvae in gastropods attained the third stage in 46–49 days, according to Gerichter (1951), and third-stage larvae were 620–700 μm in length. According to Davtyan (1949) some infective larvae leave infected gastropods and so sheep may acquire infections by ingesting larvae on vegetation, but other authors (e.g. Matekin *et al.*, 1954) question this possibility.

P. stilesi Dikmans, 1931
P. rushi Dikmans, 1937

These two species occur together in Rocky Mountain bighorn sheep (*Ovis canadensis*) and less commonly in mountain goats (*Oreamus americanus*) in North America. A third species, *P. frosti* Honess, 1942, occurs uncommonly in the lung parenchyma of bighorn sheep but nothing is known of its biology (Forrester, 1971; Hibler *et al.*, 1982). *P. stilesi* occurs in the lung parenchyma whereas *P. rushi* occurs in bronchi and bronchioles. *P. stilesi* has been associated with a fatal pneumonia in lambs of bighorn sheep and is considered the most important species. Monson and Post (1972) transmitted *P. stilesi* to bighorn–mouflon hybrids.

First-stage larvae are more numerous in the core of the faecal pellet (cf. *Parelaphostrongylus tenuis*) than on the surface mucus (Forrester and Lankester, 1997b). Larvae of *P. stilesi* and *P. rushi* developed to the infective stage in *Euconulus fulvus*, *Pupilla blandi*, *P. muscorum* and *Vallonia cyclophorella* as well as in species of *Pupoides* and *Vertigo* (see Forrester, 1971; Hibler *et al.*, 1982) (Fig. 3.9). Development to the infective stage required 45–60 days, depending on the temperature (Hibler *et al.*, 1982). Adult bighorn became infected by accidentally ingesting snails on vegetation. The infective larvae migrated to the lungs, presumably by way of the blood or lymphatic

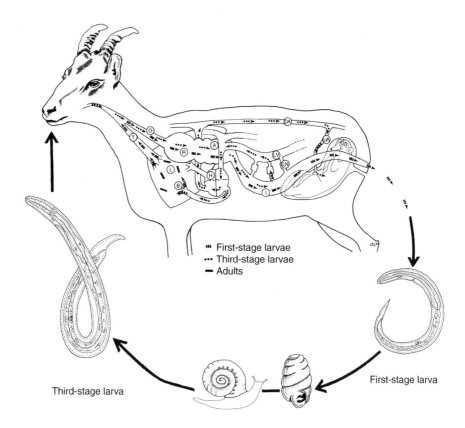

First-stage larvae

■■ First-stage larvae
••• Third-stage larvae
— Adults

Third-stage larva

Fig. 3.9. Development and transmission of *Protostrongylus stilesi*. A = abomasum; B = bronchi; DA = dorsal aorta; H = heart; I = intestine; L = lung; LN = lymph node; LV = lymph vessel; O = oesophagus; R = rumen; T = trachea; UA = umbilical artery. (Original, U.R. Strelive.)

system, and developed to adulthood in about 20 days. Eggs deposited in the lungs developed into first-stage larvae, which appeared in faeces in 28–35 days postinfection (Hibler *et al.*, 1982). Robb and Samuel (1990) found *Protostrongylus*-type infective larvae (presumably *P. stilesi* and *P. rushi*) on winter range in Alberta, Canada, in *Euconulus fulvus* (6.8% prevalence), *Vertigo gouldi* (5.6%), *V. modesta* (3.2%), *Columella* spp. (2%), immature pupillids (2.8%), *Catinella* sp. (31.6% – only 19 examined), *Discus cronkhitei* (0.9%) and *Vitrina alaskana* (0.8%). The predominant intermediate hosts were considered to be the first three. Prevalence of larvae was highest in September and April, when the bighorns used to range.

There was considerable circumstantial evidence that one or both species of *Protostrongylus* were transplacentally transmitted (Pillmore, 1956, 1959; Rufi, 1961; Forrester and Senger, 1964; Howe, 1965). Hibler *et al.* (1972, 1974, 1982) finally demonstrated that bighorn sheep infected with *P. stilesi* became immune to subsequent higher levels of infection and that infective larvae acquired by these animals accumulated in their lungs. These larvae were found in lungs of pregnant ewes until the last 6 weeks of pregnancy, when they apparently crossed the placenta and lodged in the liver of the fetus. At birth, larvae left the liver and entered the lungs of newborn lambs. They gradually migrated towards the dorsal aspect of the diaphragmatic lobes and most were found in this site in 3.5–5 weeks. In this location larvae developed to adulthood in about 20 days and females commenced to deposit eggs. A marked inflammatory reaction was directed against eggs and first-stage larvae that had hatched from them; it was estimated that 30 million larvae may occur in the lungs. Larvae were often aspirated into the anterior and ventral lobes of the lungs and were rapidly surrounded by inflammatory cells, which formed granulomas. The latter became numerous and coalesced. Bacteria (*Corynebacterium* spp., *Neisseria* sp., *Pasteurella* spp., *Staphylococcus* spp. and *Streptococcus* spp.) and myxoviruses invaded the diseased tissue. In about 6 weeks suppurative bronchopneumonia developed and consolidation of large areas of the lungs took place. Lambs frequently died in 2–5 days following consolidation. Treatment of ewes with cambendazole, dichlorvos and thiabendazole eliminated third-stage larvae from the lungs of ewes and the fetal liver. For an account of these experiments refer to Hibler *et al.* (1977, 1982).

Hibler *et al.* (1982) also noted the gregarious nature of bighorn sheep and their tendency to develop repetitive patterns of habitat use. Thus, they tend to fertilize with their faeces restricted lambing, bedding and resting areas, where the vegetation becomes lush and the mollusc populations are high. Such places are ideal sites for the transmission of lungworms.

Boag and Wishart (1982), Samson and Holmes (1985) and Robb and Samuel (1990) have provided evidence that bighorn in Alberta acquire *Protostrongylus* spp. from ingesting snails on their return to winter range in late summer and early autumn. Arnett *et al.* (1993) studied larvae in the faeces of bighorn in southwestern Montana. Numbers of larvae in faeces of ewes and rams declined from late autumn until early spring, whereas numbers of larvae in lamb faeces increased as winter progressed.

P. skrjabini (Boev, 1936)

This nematode occurs in bronchi and bronchioles of sheep, goats and argali (*Ovis ammon*) in the CIS and China. Experimentally, *P. skrjabini* developed in *Helicella candaharica* and *Subzebrinus albiplicatus* and reached the infective stage in 25–30 days at 15–25°C. The prepatent period in kids and lambs was 30 days (Azimov *et al.*, 1976).

Varestrongylus

V. alpenae (Dikmans, 1935)
This species has been found uncommonly in the small bronchioles of white-tailed deer (*Odocoileus virginianus*) in eastern and central North America (Dikmans, 1935; Dougherty, 1945; Cheatum, 1948; Prestwood and Pursglove, 1974; Anderson and Prestwood, 1981; Gray *et al.*, 1985). O'Roke and Cheatum (1950) reported that *V. alpenae* developed in various terrestrial gastropods. Unfortunately *Parelaphostrongylus tenuis* and *P. andersoni* were unknown at the time these studies were conducted; the former species in particular is very widespread in deer in eastern North America (Anderson and Prestwood, 1981). It is possible that larvae collected by early workers belonged, in fact, to one or both of these species. Gray *et al.* (1985) reported larvae (296 ± 15 µm) in faeces of white-tailed deer in central and eastern Saskatchewan. The larvae developed to the infective stage (492–600 µm long) in *Triodopsis albolabris* and *T. multilineata*. The infective larvae were given to mule (*Odocoileus hemionus*) and white-tailed deer fawns. Mule deer passed larvae in 43–49 days and white-tailed deer in 54–55 days postinfection. Adult nematodes identified as *V. alpenae* were recovered from the lung parenchyma of two of the mule deer fawns.

V. capreoli (Stroh and Schmid, 1938)
This is a parasite of the smaller bronchi of roe deer (*Capreolus capreoli*), mouflon (*Ovis musimon*), moose (*Alces alces*) and goats in Europe and Asia. First-stage larvae were 285–341 µm in length and had the typical dorsal spine on the tail. According to Stroh and Schmid (1938) larvae developed in species of *Agriolimax* and *Cepaea*. Zdarska (1960 in Boev, 1975) experimentally infected *Cepaea hortensis*, *Clausilia pila*, *Cochlicopa lubrica*, *Discus rotundatus* and *Monachoides incarnata*. Naturally infected molluscs were *Arion rufus*, *A. subfuscus*, *Deroceras reticulatum*, *Monachoides umbrosa*, *Succinea putris*, *Trichia unidentata* and *Vitrina sellucida*.

V. pneumonicus Bhalerao, 1932
This is a parasite of the bronchi and bronchioles of wild and domesticated sheep and goats in Asia and parts of the CIS. First-stage larvae were similar in morphology to those of *Muellerius capillaris* and 240–280 µm in length. Larvae invaded the following snails and grew and moulted twice to the infective stage: *Agriolimax agrestis*, *A. schulzi*, *Cochlicopa lubrica*, *Ena eleonorae*, *Euconulus fulvus*, *Fruticola lantzi*, *Macrochlamys cassida*, *M. kazachstani*, *M. schmidti*, *M. turanica*, *Phenacolimax rugulosa*, *Subzebrinus labiellus*, *Succinea evoluta*, *S. martensiana*, *Vallonia pulchella* and *Zonitoides nitidus* (see Boev, 1940, 1952; Boev and Vol'f, 1940; Bhalerao, 1945).

According to Boev (1975) worms matured in sheep in 73–92 days postinfection but larvae were still absent from lungs at that time. Bhalerao (1932), on the other hand, found mature worms in lungs in 10 weeks. Deorani (1965a) cultured eggs in saline at 30–32°C; active larvae appeared in the eggs in 16–18 h and some hatched in 24–26 h.

V. sagittatus (Mueller, 1890)

This is a common lung parasite of red deer (*Cervus elaphus*), maral deer (*C. elaphus canadensis*) and fallow deer (*Dama dama*) in Europe and part of Asia. First-stage larvae were 233–305 µm in length (Panin, 1967). The tail had a well-developed dorsal spine. First-stage larvae were highly resistant to desiccation and variations in temperature, including freezing. The most suitable natural intermediate hosts on deer farms were *Bradybaena fruticum* (prevalence 3.8%), *Succinea altaica* (4.8%), *Zenobiella aculeata* (5.1%) and *Z. nordenskoldi* (10.5%). The first moult occurred 13–15 days and the second 18–20 days after larvae had invaded the molluscs. They were infective 22–24 days postinfection and 576–640 µm in length. The prepatent period was 134 days and adult worms remained viable in deer for several years.

Infections were present in young marals 4–6 months old and in animals up to 14 years of age. Infections peaked in winter and spring, dropped in summer and were minimal in autumn (Pryadko *et al.*, 1965).

Family Crenosomatidae

Crenosomatids are ovoviviparous nematodes which occupy the bronchi, sinuses and veins of insectivores, carnivores and, rarely, marsupials. They have well-developed bursae with highly developed dorsal rays and the vulva is in the mid-region of the body. There are only five genera in the family and the transmission of species has been investigated in only three of them (*Crenosoma*, *Otostrongylus* and *Troglostrongylus*).

Crenosoma

Species of *Crenosoma* inhabit the bronchi and bronchioles of their hosts. The cuticle is thrown into a series of crenellated, transverse folds, especially evident in the anterior part of the body. Each fold has prominent longitudinal striations. First-stage larvae of *Crenosoma* spp. pass in the usual way up the respiratory tract, are swallowed and passed in faeces. Terrestrial gastropods serve as intermediate hosts of species in terrestrial hosts. Infective larvae from gastropods invade the hepatic portal system of the final host and reach the lungs a few hours after infection. One species in seals reaches the infective stage in fish.

C. mephitidis Hobmaier, 1941

This is a parasite of skunk (*Mephitis mephitis*, *M. occidentalis* and *Spilogale gracilis*) in North America (Hobmaier, 1941a; Craig, 1972; Stockdale *et al.*, 1974). It developed to the infective stage in species of the gastropod genera *Agriolimax*, *Epiphragmophora*, *Helix*, *Limax* and *Milax* (Hobmaier, 1941a) as well as in *Anguispira alternata*, *Mesodon thyroides*, *Triodopsis albolabris* and *T. tridentata* (see Craig, 1972). First-stage larvae were 270–305 µm in length according to Hobmaier (1941a) and 233–348 µm according to Craig (1972). The first moult took place in *Mesodon thyroides* 6–7 days and the second 12 days postinfection (Craig, 1972). Third-stage larvae were 525–560 µm in length according to Hobmaier (1941a) and 532–642 μm according to Craig (1972). The tails of both the first- and third-stage larvae ended in a tiny curved mucron (Craig, 1972).

Hobmaier (1941a) claimed that the garter snake *Thamnophis sirtalis* was a paratenic host, as larvae were noted in tubercles covering the outside of the stomach and intestine and 'in and under the mucous membrane of both organs', and that 'frogs, toads and lizards, could be artificially infected' but no details were given. Craig (1972) noted that *C. mephitidis* became encapsulated in voles (*Microtus pennsylvanicus*).

Stockdale *et al.* (1974) concluded from experiments that infective larvae invaded the stomach of the final host and entered the hepatic portal system. They migrated from the liver to the lungs by way of the posterior vena cava, heart and pulmonary arteries (cf. *C. vulpis*). The third and fourth moults occurred 4 and 7 days, respectively, postinfection.

According to Hobmaier (1941a), puppies and foxes were infected experimentally and the prepatent period was 19 days in these animals. Craig (1972) infected a dog and found mature worms 21 days later.

C. petrowi Morozov, 1939

Larvae (227–283 μm in length) from faeces of the black bear (*Ursus americanus*) in Canada developed to the third infective stage (555–567 μm) in *Mesodon thyroidus* (Addison and Fraser, 1994). At 23.5°C the first and second moults in the snail occurred on days 6–7 and 9–11, respectively. The prepatent period in five infected bears was 19–25 days. The nematodes failed to develop in skunk (*Mephitis mephitis*), raccoons (*Procyon lotor*) and red foxes (*Vulpes vulpes*).

C. striatum (Zeder, 1800)

This is a common parasite of the European hedgehog (*Erinaceus europaeus*). Petrov (1941) showed that *Agriolimax agrestis*, *Arion circumscriptus* and *Succinea putris* could serve as intermediate hosts. Lammler and Saupe (1968) found that *Arianta arbustorum* and *Cepaea hortensis* were the most suitable of all the snails they infected, including aquatic species (e.g. *Australorbis glabratus*, *Bithynia leachi*, *Lymnaea* spp. and *Planorbis* spp.). Two hedgehogs were infected with larvae from snails but attempts to infect a variety of other mammals (e.g. cats, dogs, mice, rats) failed, indicating that *C. striatum* is specific to hedgehogs. According to Barus and Blazek (1971) first-stage larvae were 255–287 μm in length. The larvae developed in *Lymnaea peregra*, *Oxychilus glaber*, *Succinea putris*, *Monachoides umbroa*, *Arion circumscriptus*, *Limax tenellus* and *Milax rusticus*. At 20°C the first moult occurred in 8–10 days and the second in 12–18 days, when larvae were 423–511 μm in length. The prepatent period in four hedgehogs given infective larvae was 19–21 days.

Panebianco (1957) carried out a histological study of *C. striatum* in hedgehogs.

C. vulpis (Dujardin, 1845)

C. vulpis is a cosmopolitan parasite of the bronchi of canids (*Alopex lagopus*, *Canis familiaris*, *C. lupus*, *Nyctereutes procyonoides*, *Urocyon cinereoergenteus*, *Vulpes fulva* and *V. vulpes*). It has also been reported in various mustelids (*Lutra lutra*, *Martes* spp., *Meles meles*) but some of these reports need confirmation.

Wetzel and Muller (1935a,b), Petrov and Gagarin (1938), Wetzel (1940), Petrov (1941) and Stockdale and Hulland (1970) reported the following terrestrial gastropods as suitable intermediate hosts: *Agriolimax agrestis, Arianta arbustorum, Arion circumscriptus, A. hortensis, A. intermedius, Cepaea hortensis, C. nemoralis, Fruiticicola fruticulum, Helix pomatia, Mesodon thyroides, Succinea putris, Triodopsis albolabris, Zonitoides excavata* and *Z. nitida.*

Wetzel (1940) reported that first-stage larvae were 246–308 µm in length. According to Wetzel and Muller (1935a,b) the infective stage was reached in 17 days in molluscs. Third-stage larvae were 458–549 µm in length.

In the final host, third-stage larvae penetrated the intestine and reached the lungs within 20 h by way of the visceral lymphatics, thoracic duct, vena cava and the right heart (Wetzel and Muller, 1935a,b). The third moult occurred in lungs 3 days postinfection and the final moult on the day 7. The prepatent period was 18–21 days.

On the other hand, Stockdale and Hulland (1970) marshalled convincing evidence that third-stage larvae reached the heart by way of the hepatic portal system. In the liver, larvae broke out of the venules of the portal triads and entered the parenchyma. They then apparently entered the hepatic vein and were carried to the heart and lungs. In the latter the third moult occurred about 4 days postinfection. The final moult took place on or shortly before day 8. During the prepatent period, larvae and later the subadults seemed to move up the bronchial tree from the bronchioles to the larger bronchi. Females deposited larvae 19 days postinfection. Shaw *et al.* (1996) described the clinical signs of eosinophilic bronchitis in dogs infected with *C. vulpis.*

Wetzel (1941a) reported that infected dogs and foxes passed larvae for 240–290 days after a single infection. Attempts to infect cats were unsuccessful. He did not find encapsulated or other larvae in rats and mice given infective larvae and concluded that canids probably acquired the parasite by ingesting snails and slugs.

Otostrongylus

O. circumlitus (Railliet, 1899)

O. circumlitus is a large lungworm found in the main bronchi and bronchioles of seals. It has a circumpolar distribution in *Phoca vitulina, Phoca hispida, Erignathus barbatus, Halichoerus grypus, Phoca largha* and *Histriophoca fasciata* and also occurs in *Mirouga angustirostris* of California (see Bergeron *et al.*, 1997a,b; Gosselin *et al.*, 1998). Eggs were 72–129 × 51–76 µm in size, with thin shells. First-stage larvae were 373–469 µm in length in two species of seal and remained alive in sea water at 5°C for 3–6 months (Bergeron *et al.*, 1997). Larvae failed to develop in various amphipods and mysids. However, larvae from *P. hispida* invaded the intestinal wall of American plaice (*Hippoglossoides platessoides*), goldfish (*Carassius auratus*) and sculpins (*Myxocephalus scorpioides*). Plaice were the most suitable hosts since complete development to the third stage occurred in this host. After 30 days, second-stage larvae were in the circular muscle layers of the gut. The second moult occurred at 56 days and the third-stage larvae (785–1007 µm) were found under the serosa. Grey seals (*H. grypus*) given infective larvae from plaice infected with *O. circumlitus* from *M. angustrirostris* of California passed first-stage larvae in 35 days; the prepatent period in harbour seals (*P. vitulina*) was 31–33 days (L. Measures, personal communication).

Troglostrongylus

T. brevior Gerichter, 1949

This species was found in the bronchi of felines (*Catolynx chaus* and *Felis ocreata*) near the Dead Sea. First-stage larvae from faeces were 300–310 µm in length; the tail had a deep dorsal incision and a shallower ventral one near its tip. The following gastropods were infected: *Chondrula septemdentata*, *Helicella barbesiana*, *H. ustalis*, *Limax flavus*, *Monacha syriaca*, *Retinella nitellina* and *Theba pisana* (see Gerichter, 1949).

Gerichter (1949) studied the rate of development of *T. brevior* in gastropods over a wide temperature range (4–27°C). He stated that the species reached the infective stage in 40 days in gastropods held at 4–8°C. The most rapid development was in *Helicella* spp. (8 days at 22–27°C). Domestic kittens given infective larvae in snails passed larvae 28 days later. Attempts to determine if mice could serve as paratenic hosts were inconclusive. Infective larvae were given to four mice which were examined 10–120 days later but only a single larva was found in one mouse 120 days later.

Family Angiostrongylidae

The angiostrongylids are characterized by a typical bursa and a posterior vulva. The family is well represented in marsupials, insectivores, carnivores and rodents. Transmission of most species studied involves gastropod intermediate hosts. Paratenic hosts are generally used in species in carnivores. One species, *Andersonstrongylus captivensis*, is unusual in that first-stage larvae are directly infective to the final host (cf. certain species in the family Filaroididae). The family contains a number of species associated with the vascular system, two of which are of exceptional interest because they are also neurotropic.

Andersonstrongylus

A. captivensis Webster, 1978

A. captivensis is a parasite of the lung parenchyma of skunk (*Mephitis mephitis*). Webster (1980) fed ground skunk lungs containing gravid female worms and larvae to several uninfected skunks. The latter passed first-stage larvae 20–24 days later. Webster (1980) suggested that transmission could be by cannibalism, coprophagy or regurgitative feeding to young. Webster (1981) has discussed the systematic relationship of *A. captivensis* to *Angiostrongylus milksi* (Whitlock, 1956) found in dogs.

Angiostrongylus

The genus *Angiostrongylus* contains species found in rodents, shrews and carnivores. Some species (e.g. *A. cantonensis*, *A. dujardini*, *A. mackerrasae*, *A. malaysiensis*, *A. sandarsae* and *A. vasorum*) inhabit the pulmonary arteries and the right heart. These species release eggs which embryonate in the lungs. *A. cantonensis* and *A. mackerrasae* develop from the third to the early adult stage in the brain, after which they migrate via the venous system to the heart and pulmonary arteries. Some other species

(*A. costaricensis, A. siamensis*) occur in mesenteric veins, and larvae and eggs pass through the gut wall and into the intestine. Three species (*A. andersoni, A. blarinae* and *A. michiganensis*) occupy lungs (bronchioles).

A. andersoni (Petter, 1972)

A. andersoni (syn. *Morerastrongylus andersoni*) occurs in large abscesses in the lungs of the African rodent *Taterillus nigeriae*. Eggs embryonated in alveoli and first-stage larvae passed in faeces attained the third stage in *Arion hortensis, Deroceras reticulatum, Helix aspersa, Lymnaea stagnalis* and *Planorbarius corneus* (see Petter, 1974). According to Petter and Cassone (1975), larvae were ingested by the aquatic snails *L. stagnalis* and *P. corneus* and were found eventually in various regions of the body. In *L. stagnalis* at 22°C, the first moult occurred in 8 days and the second in 11 days (Petter, 1974). In the final host, third-stage larvae entered the body cavity and reached the liver (probably by direct penetration of the organ) in 14 h. They passed from there to the lungs, which they reached in 24 h (Petter, 1974). The third moult took place 2–3 days and the fourth 5–6 days postinfection. The prepatent period was 24 days.

A. cantonensis (Chen, 1935)

This is a neurotropic parasite of the pulmonary arteries and rarely the right heart of rats. It has occasioned an extensive literature because it is transmissible to humans. Reviews of facts concerning *A. cantonensis* have been published by Alicata (1965b, 1969, 1988), Anderson (1968), Rosen (1967) and Alicata and Jindrak (1970) and the reader is referred to these for more information.

Mackerras and Sandars (1954, 1955) described the development and neurotropic behaviour in rats of a species of *Angiostrongylus* that they identified as *A. cantonensis* (Chen, 1935). However, Bhaibulaya (1975) concluded that the species studied by Mackerras and Sandars in Australia was in fact *A. mackerrasae* Bhaibulaya, 1968 which behaves in rats like *A. cantonensis*. The description of the development in rats given below refers, therefore, to *A. cantonensis* as studied by Jindrak (1968), Wallace and Rosen (1969) and Bhaibulaya (1975). *A. mackerrasae* will be briefly discussed later.

Adults live in the pulmonary arteries and unsegmented eggs are deposited into the blood; they lodge as emboli in smaller vessels of the lungs, and embryonate in about 6 days (Weinstein *et al.*, 1963). First-stage larvae pass through the respiratory tract, into the gut, and out in the faeces. Terrestrial, aquatic and amphibious gastropods serve as intermediate hosts. In the gastropod the first moult takes place in 7–9 days and the moult to the infective stage takes place 12–16 days postinfection at 21–26°C; larvae develop in the viscera in addition to the foot.

Knowledge of the route taken by larvae to the brain of rats, where they develop to subadults, is based largely on the results of heavily dosing rats with larvae in the hope that some larvae will be found at necropsy. It might be worthwhile, therefore, to confirm current concepts with additional experiments using smaller numbers of larvae. According to Jindrak (1968), Wallace and Rosen (1969) and Bhaibulaya (1975), most infective larvae penetrated the stomach of rats and entered the hepatic portal system and the mesenteric lymphatic system. They were then carried to the heart and lungs. Larvae entered alveoli, invaded the pulmonary veins, returned to the left heart and were distributed around the body by the arterial circulation. Larvae reached the central nervous system (mainly the cerebrum) 2–3 days postinfection. Those in the neural parenchyma grew and the third moult took place 4–6 days and the fourth moult 7–9 days postinfection. In 12–14 days postinfection, young adults invaded the subarachnoid space of the brain, where they spent the next 2 weeks. In 28–33 days postinfection, worms invaded the cerebral vein and travelled to the heart and pulmonary arteries, where they matured. Oviposition of single-celled eggs took place 33 days postinfection. Eggs carried in blood to the lungs embryonated in about 1 week into first-stage larvae. The latter appeared in faeces of rats 42–45 days postinfection.

The parasite has been found in wild rats (*Bandicota indica*, *Melomys littoralis*, *Rattus annandalei*, *R. argentiventer*, *R. assimilis*, *R. bowersi*, *R. conatus*, *R. coxinga*, *R. diardi*, *R. exulans*, *R. jalorensis*, *R. losea*, *R. rattus*, *R. muelleri* and *R. norvegicus*), mainly in the tropical belt, latitude 23° N to 23° S. It has been reported in Africa, Australia, China, Taiwan, Malaysia, Caroline Islands, Cook Islands, Hawaiian Islands, Loyality Islands, Orchid Islands, Mariana Islands, New Caledonia, New Hebrides Islands, Tahiti, Thailand, Ryukyu Islands, Okinawa, Mauritius, Sri Lanka, Philippines, Solomon Islands, Egypt (Yousif and Ibrahim, 1978), India and probably Indonesia (Alicata and Jindrak, 1970). It occurs in Cuba (Pascual *et al.*, 1981). It has been suggested that *A. cantonensis* is a recent immigrant to the Pacific islands from eastern Asia and that its spread may have been associated with introductions of infected rats and molluscs (e.g. *Achatina fulica*) (Alicata, 1966). In Nigeria the intermediate host is *Archachatina marginata* (Udonsi and Oranusi, 1991).

A. cantonensis has been experimentally transmitted to white rats, tree shrews (*Tupaia glis*), mongoose (*Herpestes urva*), white mice, rabbits, guinea-pigs, primates (*Macaca* spp.) and slow lorises (*Nycticebus coucang*) but failed to mature in any of these species except rats (Loison *et al.*, 1962; Weinstein *et al.*, 1962, 1963; Alicata 1963a, 1964a; Courdurier *et al.*, 1964; Guilhon, 1965, 1967; Heyneman and Lim, 1965, 1966a,b, 1967a,b,c; Nishimura, 1965, 1966; Wood, 1965; Shoho, 1966; Lee, 1967). Wood's (1965) report of the transmission of *A. cantonensis* to the mongoose requires confirmation. Infections in captive tamarins (*Sanguinus* spp.) were reported by Carlisle *et al.* (1998).

The infection is largely silent in rats with infections of low to moderate intensity but circling, cannibalism and paraplegia may occur in heavy experimental infections. When lungs become involved, coughing, sneezing and raspy breathing may be evident. Death may be caused in rats by massive damage to the lungs or by blockage of pulmonary arteries. In mice, ataxia, lethargy, ocular abnormalities, twitching, circling, collapse and death have been noted following infection. In experimentally infected rabbits, photophobia, bloody tears, paraplegia, displacement of the eyeballs, ocular lesions, abnormal positions of the head and respiratory abnormalities are reported following infection (Shoho, 1966). Neurological signs in experimentally infected monkeys (*Macaca* spp.), slow lorises (*Nycticebus coucang*) and guinea pigs are rare or absent (Loison *et al.*, 1962; Weinstein *et al.*, 1962, 1963; Alicata *et al.*, 1963; Lim, 1970). Tree shrews (*Tupaia glis*) usually die when heavily infected (Lim, 1970).

A. cantonensis develops to the intective stage in a wide range of aquatic and terrestrial gastropods. Infection is by penetration of the body surface or by ingestion. The following gastropods are considered natural intermediate hosts: *Achatina fulica*, *Bellamya ingallsiana*, *Bradybaena similaris*, *Cipangopaludina chinensis*, *Deroceras laeve*, *Euglandina rosea*, *Girasia peguensis*, *Indoplanorbis exustus*, *Laevicaulus alte*, *Macrochlamys resplendens*, *Microparmarion malayanus*, *Opeas javanicum*, *Pupina complanata*, *Pila ampullacea, P. scutata, Quantula striata, Subulina octona, Vaginalus plebeius* and *Veronicella alte.*

Experimentally, the following gastropods have been infected: *Biomphalaria* spp., *Bithynia* sp., *Bradybaena oceania, Bulinis* spp., *Deroceras reticulatum, Drepanotrema simmonsi, Euglandina rosea, Euhadra hickonis, Ferrissia tenuis, Fossaria ollula, Fruticola despecta, Helicina orbiculata, Helisoma* sp., *Indoplanorbis exustus, Lenistes carinatus, Limax arborum, L. flavus, L. maximus, L. marginalis, Lymnaea* spp., *Marisa cornuarietis, Mesodon thyroidus, Onchidium* sp., *Physa acuta, Planorbis planorbis, Plesiophysa hubendicki, Segmentina hemisphaerula, Semisalcospira libertina, Stagnicola elodes* and *Succinea lauta* as well as the bivalves *Crassostrea virginica* and *Mercenaria mercenaria* (see Lim *et al.*, 1962, 1963, 1965; Alicata 1963b,c,d, 1965b, 1966; Richards, 1963; Chiu, 1964; Cheng and Alicata, 1965; Cheng and Burton, 1965; Harinasuta *et al.*, 1965; Lim and Heyneman, 1965; Punyagupta, 1965; Cheng, 1966; Knapp, 1966; Wallace and Rosen, 1966; Ash, 1967, 1976; Knapp and Alicata, 1967; Richards and Merritt, 1967; Chang *et al.*, 1968; Yousif and Lammler, 1975; Brockelman *et al.*, 1976; Hori *et al.*, 1976; Sauerlander, 1976; Katakura, 1979; Higa *et al.*, 1986).

Freshwater prawns (especially *Macrobrachium lar*), land crabs (*Ocypode ceratophthalma* and *Cardiosoma hirtipes*), coconut crabs (*Birgus latro*), planarians (*Geoplana septemlineata*), frogs and toads may serve as paratenic hosts since they feed on gastropods and infective larvae are able to survive for several days in their alimentary tract or tissues. Fish may serve as paratenic hosts under certain conditions. Human infections frequently occur through the agency of these paratenic hosts which are often eaten raw (accidentally or deliberately) or whose juices are used in the preparation of local dishes (Alicata, 1962, 1963b, 1964a,b, 1965a,b; Alicata and Brown, 1962, 1963; Cheng and Burton, 1965; Cheng, 1966; Wallace and Rosen, 1966, 1967; Ash, 1967, 1968, 1976; Alicata and Jindrak, 1970; Ko, 1978). Infective larvae can survive several days in water and the possibility that larvae which have left molluscs can reach humans through drinking water has been raised. Also it has been shown that larvae may leave living molluscs and contaminate vegetables (Heyneman and Lim, 1967b,c).

According to Oku *et al.* (1979, 1980) first-stage larvae given to tadpoles of the frogs *Xenopus laevis* and *Rana chensinensis* developed to the infective stage and were infective to rats.

A high percentage of susceptible molluscs harbours infective larvae in endemic areas (see especially Lim *et al.*, 1963, 1965; Lim and Heyneman, 1965; Heyneman and Lim, 1965, 1967a). In addition, individual snails may carry large numbers of infective larvae. Knapp (1966) reported 59,400 larvae in eight *Achatina fulica* from Hawaii. Whether or not rats feed on all species of snails reported to harbour larvae has not been reported. The same may be said of prawns and land crabs harbouring infective larvae presumably as a result of feeding on infected molluscs. Lim (1966) and Lim and Heyneman (1965), however, have shown experimentally that rats preferred *Microparmarion malayanus* and *Macrochlamys resplendens* to six other species. These preferred molluscs had the highest prevalences of infection in nature and were considered the most important species in maintaining the parasite in all habitats studied except rice fields in Malaysia. In rice fields, *Pila scutata, Bellamya ingallsiana* and *Indoplanorbis exustus* were considered the most important hosts. Field studies in Malaysia have also demonstrated that *M. resplendens* and *M. malayanus* feed on rodent faeces in rat runways and that rat nests frequently contain broken shells of these and other species of susceptible snails.

Heyneman and Lim (1965) found differences in the prevalence of infection in rats and molluscs in three different habitats in Malaysia; the highest prevalences occurred in areas where farms and forests merged. Tree-dwelling rats (*Chiropodomys gliroides* and *Rattus canus*), which are rarely if ever exposed to infection, died when infected experimentally. Some ground-dwelling rats (*R. rajah* and *R. sabanus*) were resistant to infection. 'Selection pressure favouring resistant animals doubtless exerts a strong evolutionary influence on continually exposed animals' (Heyneman and Lim, 1965).

Ow-Yang and Lim (1965) and Heyneman and Lim (1965) showed that infection of rats with small numbers of larvae protected them from intense infections that would otherwise have been lethal. Heyneman and Lim (1966a) found that a Malaysian strain of *A. cantonensis* developed more successfully in rats than a Hawaiian strain.

A. cantonensis in humans produces the condition known as eosinophilic meningitis. Infective larvae invade the central nervous system and some succeed in growing to the adult stage without producing progeny. Neurological signs and symptoms evidently vary but neck stiffness, drowsiness, headache, vomiting, vertigo, facial paralysis and paraesthetic skin areas have been reported (Franco *et al.*, 1960; Rosen *et al.*, 1961, 1962, 1967). Fever is rare in adults but has been reported in infants with typical signs. A consistent feature, considered diagnostic, is pleocytosis with a high percentage of eosinophils. Onset is sudden or insidious and most patients recover within a few days or weeks. Recurrences are common. Only a few cases are fatal and paralysis of limbs is unreported.

A. costaricensis Morera and Cespedes, 1971

A. costaricensis was first discovered in the small branches of the mesenteric artery of humans in Costa Rica, where it was associated with granulomatous lesions accompanied by massive eosinophilic infiltrations involving all layers of the intestine and regional lymph nodes (Morera and Cespedes, 1971). Encapsulated eggs and larvae were present in the lesions. Morera (1973) exposed the veronicellid slug *Vaginulus plebeius* to first-stage larvae (260–269 µm in length). Larvae invaded the slug and developed, the first moult occurring in 4 days and the second in 11–13 days. Third-stage larvae were 460–482 µm in length. In cotton rats (*Sigmodon hispidus*) infective larvae penetrated the gut in the ileocaecal region by 24 h. Most larvae were found in lymphatics of the intestinal wall and mesentery where the third moult occurred in 3–4 days, when larvae were 925 µm (females) and 825 µm (males) in length. The final moult occurred in 5–7 days and by 10 days the worms had migrated to the mesenteric arteries and its branches in the intestinal wall. Oviposition began 18 days postinfection, when females were 11.2 mm and males 9.5 mm in length. Eggs developed in the gut wall, first-stage larvae migrated across the gut wall into the intestinal lumen and first appeared in faeces of the host in 24 days. Monge *et al.* (1978) infected *Biomphalaria glabrata* and collected infective larvae 30 days later which were given orally to cotton rats. Larvae appeared in the faeces of the rats in 22 days and adult worms were found in the mesenteric arteries.

Mota and Lenzi (1995) massively infected mice and suggested that infective larvae reached the gut vessels after a pulmonary circulation. (This conclusion requires confirmation – R.C.A.)

The parasite is now known to be widely distributed in the arteries of the caecum and in branches of the cranial mesenteric artery of rodents in Central and South America, i.e. *Tylomys watsoni, Liomys salvini, L. adspersus, Proechimys semispinosus,*

Peromyscus nudipes, Zygodontomys microtenus, Orzomys albigularis, O. caliginosus, O. nigripes, O. ratticeps, O. fulvescens, Sigmodon hispidus, Rattus rattus, R. norvegicus (see Morera, 1973, 1978, 1985; Morera *et al.*, 1982; Tesh *et al.*, 1973). The parasite was also found aberrantly in the coati (*Nasua narica*) and in marmosets (*Saguinus mystax* and *Callithrix penicillata*) (Monge *et al.*, 1978; Sly *et al.*, 1982; Brack and Shröpel, 1995). According to Morera (1973, 1985) the cotton rat (*S. hispidis*) is the main definitive host and *V. plebeius* the main intermediate host in Costa Rica, where prevalences in the slug range from 28 to 75%. Intermediate hosts in southern Brazil include *Phyllocaulus variegatus, P. soleformis, Bradybaena similaris, Belocaulus angustipes, Limax maximus, L. flavus* (Graff-Teixera *et al.*, 1989, 1993; Rambo *et al.*, 1997). Definitive hosts reported in Brazil are *O. nigripes* and *O. ratticeps* (Santos *et al.*, 1996). Hata (1994, 1996) and Hata and Kojima (1994) have successfully cultured *A. costaricensis* in defined media. Thiengo *et al.* (1997) redescribed *A. costaricensis* from Brazil.

A. dujardini Drozdz and Doby, 1970
This is a parasite of the heart and pulmonary arteries of the European rodents *Apodemus sylvaticus* and *Clethrionomys glareolus*. Drozdz *et al.* (1971) concluded that the most suitable intermediate hosts were the aquatic snails *Biomphalaria glabrata, Hygromia limbata, Lymnaea corvus, L. stagnalis, L. peregra, Planorbis planorbis, Planorbarius corneus* and *Retinella incerta*. The terrestrial gastropods *Agriolimax laevis, A. reticulatus, Cepaea nemoralis* and *Helix aspersa* were not as suitable as hosts, since larvae became encapsulated in them. Larvae were apparently ingested by the gastropods. The first moult took place 12–14 days and the second 17–19 days postinfection, at room temperature. The infective stage was attained in 18 days.

Drozdz and Doby (1970) gave larvae to rodents and examined the latter at various times. They concluded that larvae penetrated the intestine and entered the liver. Some larvae reached the lungs as early as 28 h postinfection from the liver. Development was rapid in the lungs and fourth-stage larvae appeared by 72 h. In the following 3 days, larvae developed into adults, a process completed by 7 days in the lungs, pulmonary arteries and the right heart. Twelve days postinfection, worms were mature; in 16 days unsegmented eggs appeared in the lungs. In 24–26 days larvae appeared in the faeces.

A. mackerrasae Bhaibulaya, 1968
This species has been reported in the pulmonary arteries and right ventricle of *Rattus fuscipes* and *R. norvegicus* in Queensland, Australia. It is apparently the species studied by Mackerras and Sandars (1955) in their classic work on the neurotropic behaviour of these kinds of nematodes in rats (Bhaibulaya, 1968). *A. mackerrasae* has much longer spicules (400–560 μm), the posterior end of the female tail ends in a minute mucron and the posterior lateral ray is longer than in *A. cantonensis*. Also, the length of the vagina and the distance between the anus and vulva differ slightly in the two species.

The development of *A. mackerrasae* in gastropods and rats was similar to that reported in *A. cantonensis* and the larval stages were indistinguishable. Mackerras and Sandars (1955) reported that the first moult in gastropods occurred 12–13 days and the second 17–18 days postinfection at 21–26°C.

Conclusions about the migration of *A. mackerrasae* in rats were based on the results of finding larvae in rats given 500–1000 infective larvae. Both Mackerras and Sandars (1954) and Bhaibulaya (1975) concluded that worms entered the hepatic portal system and became blood-borne. According to Bhaibulaya (1975), third-stage larvae penetrated the wall of the gut and reached the liver. Larvae then reached the lungs via the heart. In lungs, larvae were dispersed by the arterial system and most eventually reached the brain (mainly the cerebrum and olfactory tracts). Conclusions regarding the migration to the brain of the host probably should be reinvestigated using smaller numbers of infective larvae.

In the brain, the third moult occurred 6–10 days postinfection and larvae then migrated to the subarachnoid space where they grew and moulted 10–11 days postinfection. Young adults subsequently migrated from the brain to the pulmonary arteries, where they were found in 22–26 days. In 35 days, females were mature and eggs were released; the latter were filtered from the blood in the lungs. Eggs embryonated in the lungs and first-stage larvae appeared in faeces of infected rats 40–42 days postinfection.

A. malaysiensis Bhaibulaya and Cross, 1971

A. malaysiensis occurs in the pulmonary arteries and right heart of *Rattus jalorensis* in Malaysia. According to Bhaibulaya and Cross (1971), third-stage larvae found in the gastropods *Bradybaena similaris*, *Microparmarion malayanus* and *Laevicaulus alte* were given to *R. jalorensis* and adult worms were recovered for identification. Sawabe and Makiya (1995) reported *A. malaysiensis* in *Rattus norvegicus* in Kitakyushi City, Japan. The infectivity of *A. malaysiensis* in *Biomphalaria glabrata* was lower than that of *A. cantonensis* and suggested that larvae of the former species are more likely to be destroyed in the gut of the snail than larvae of the latter lungworm.

A. siamensis Ohbayashi, Kamiya and Bhaibulaya, 1979

A. siamensis is a parasite of the mesenteric arteries of *Rattus berdmorei*, *R. rattus*, *R. sabanus* and *R. surifer* in Thailand (Ohbayashi *et al.*, 1979; Kamiya *et al.*, 1980). Kamiya *et al.* (1980) and Katakura *et al.* (1981) successfully infected *Biomphalaria glabrata*. Third-stage larvae from snails which had been kept at 25–30°C for 35 days were given orally to laboratory mice, cotton rats (*Sigmodon hispidus*), laboratory rats and gerbils (*Meriones unguiculatus*); all became infected. The prepatent period was stated to be 'earlier than day 29 postinfection'.

According to Kudo *et al.* (1983) third-stage larvae invaded the wall of the caecum and colon of mice and migrated to the marginal and intermedial sinuses of the mesenteric lymph nodes and lymphatic vessels peripheral to the nodes, where the third (30–72 h postinfection) and fourth (4–7 days) moults occurred. Young adult worms then moved to the colon via the lymph vessels and on day 6 moved into arterioles of the colon and mesentery. This migration was completed by day 10. Worms reached the mesenteric arteries and their branches in the lower portion of the small intestine and caecum. Oviposition commenced on day 22 and eggs released into blood lodged mainly in capillaries of the lower portion of the small intestine, caecum and upper colon. Larvae invaded the lumen of the intestine and appeared in faeces 31 days postinfection.

A. vasorum (Baillet, 1866)

This small species occurs in the pulmonary arteries and less commonly in the right ventricle of dogs and foxes (*Cerdocyon thous*, *Ducicyon azarae*, *D. vetulus*, *Fennecus zerda*, *Vulpes vulpes*) in certain areas in Europe, North and South America, and Africa

(Uganda) (Prestwood *et al.*, 1981; Bolt *et al.*, 1994). The unembryonated eggs are carried to the capillaries of the lungs, where they develop into first-stage larvae which hatch and are eventually passed in faeces of the host. Guilhon (1960, 1963) showed that *Arion ater* and *A. rufus* were suitable intermediate hosts in France, and he massively infected dogs by forcing them to consume infected slugs.

Rosen *et al.* (1970) carried out a series of experiments to elucidate the development of *A. vasorum* in more detail. They successfully infected the aquatic gastropod *Biomphalaria glabrata* (aquatic) and a number of terrestrial species found in the Pacific (e.g. *Bradybaena similaris*, *Laevicaulus alte*, *Prosoples javanicum* and *Subulina octona*). In *B. glabrata*, infective larvae (510–610 μm in length) were recovered in 16 days at 24°C. Large numbers of infective larvae were found in the slug *L. alte* but relatively few in the other gastropods. Sauerlander and Eckert (1974) successfully infected *Achatina fulica*. According to Prestwood *et al.* (1981) the terrestrial slug *Deroceras laeve*, which is extremely common in North America, is a suitable intermediate host.

Rosen *et al.* (1970) infected several dogs and examined them at various times postinfection. Infective larvae appeared in the visceral lymph nodes, where they underwent two moults. The tiny subadults then migrated via the hepatic portal system to the liver, heart and pulmonary arteries, where they matured. The prepatent period was 49 days.

Lima *et al.* (1994) reported *A. vasorum* in the fox *Dusicyon vetulus* in Brazil. Larvae developed in *Biomphalaria glabrata* and two crossbred dogs were infected; the prepatent period was 42–47 days.

Bolt *et al.* (1993) reported that frogs (*Rana temporaria*) could harbour third-stage larvae and serve as paratenic hosts of *A. vasorum*. They claimed to have found third-stage larvae in frogs given first-stage larvae 30 days earlier, and a fox was successfully infected with the larvae (cf. Oku *et al.* (1979, 1980) who reported that first-stage larvae of *A. cantonensis* reach the infective stage in *Xenopus* and *Rana chansinensis*).

For a review of *A. vasorum* with special reference to pathology and clinical manifestations, see Prestwood *et al.* (1981) and Bolt *et al.* (1994).

Aelurostrongylus

A. abstrusus (Raillet, 1898)

A. abstrusus is a common parasite of terminal and respiratory bronchioles and alveolar ducts of domestic cats. The female is oviparous and eggs embryonate in alveolar ducts, filling and distending them and adjacent alveoli. First-stage larvae pass up the bronchial escalator and are passed in faeces in the usual way. Cameron (1927a) concluded that first-stage larvae would develop to an infective stage in mice, which served as intermediate hosts. However, Hobmaier and Hobmaier (1935a,b), Gerichter (1949), Mackerras (1957) and Hamilton and McCaw (1967) showed that terrestrial gastropods were intermediate hosts and Cameron's observations could not be duplicated (see also Baudet, 1933).

Hobmaier and Hobmaier (1935a) infected *Agriolimax agrestis*, *A. columbianus*, *Helminthoglypta arosa*, *H. californiensis*, *H. nickleana* and *Helix aspersa*. The first moult occurred 1 week postinfection and the second in 2–3 weeks at an unspecified temperature. They also gave third-stage larvae to frogs, toads, lizards, snakes, sparrows, chickens, ducklings and small rodents and recovered viable larvae in their tissues some time later. They successfully infected cats with larvae from both snails and paratenic hosts. Cats frequently vomited shortly after receiving larvae. In cats, larvae (which were highly active), entered the mucous membranes of the oesophagus, stomach and upper intestine during the first day postinfection. At the end of this time some larvae were already in the lungs. The third moult took place in the lungs 5–6 days and the final moult 8–9 days postinfection. Females began to oviposit after the fourth week and larvae appeared in the faeces after the fifth week.

Mackerras (1957) infected *Agriolimax laevis* with *A. abstrusus* and successfully transferred the parasite to cats; third-stage larvae were 520–530 μm in length. Larvae in cats migrated rapidly to the lungs, where they moulted twice. The prepatent period was given as 39 days. In mice, third-stage larvae congregated mainly on the gastrosplenic ligament, where they were surrounded by small yellow capsules; they remained viable for at least 12 weeks.

Gerichter (1949) infected the gastropods *Agriolimax* sp., *Chondrula septemdentata*, *Helicella barbesiana*, *H. vestalis*, *Helix cavata*, *Levantina hierosolyma*, *L. sesareana*, *Limax flavus*, *Monacha syriaca*, *Retinella nitellina* and *Theba pisana*. Development to the second stage took 11 days (apparently at 30°C). The second moult took place on day 15 and the infective stage was reached on day 18 postinfection. Third-stage larvae were 460–510 μm in length. Cats passed larvae 39 days after being infected.

In a detailed study of the pathogenesis of *A. abstrusus*, Stockdale (1970c) reported that *A. abstrusus* in the cat caused focal interstitial pneumonia during the first 25 days. Worms started to lay eggs as early as 25 days postinfection. The eggs became centres of extensive granulomas.

A. falciformis (Schlegel, 1933)

This ovoviviparous species occurs in the lung parenchyma of badgers (*Meles meles*) in Europe. Larvae passed in faeces of infected badgers developed to the infective third stage in the gastropods *Arion hortensis*, *Cepaea hortensis*, *C. nemoralis*, *Deroceras agrestis*, *Euomphalia strigella*, *Fruticola hispida* and *Succinea putris* (Wetzel, 1937, 1938). A 2-year-old badger given snails with infective larvae passed first-stage larvae 21 days later. Schlegel (1933, 1934) described lesions in lungs associated with *A. falciformis*.

A. pridhami Anderson, 1962

A. pridhami occurs in the lung parenchyma (alveoli, alveolar ducts, respiratory and terminal bronchioles) of wild mink (*Mustella vison*) in North America (Anderson, 1962; Stockdale, 1970a,b). The female is ovoviviparous and first-stage larvae (sometimes still in the egg) are deposited into air spaces. Larvae passed in faeces of mink infected *Ampullaria cuprina*, *Anguispira alternata*, *Armiger crista*, *Deroceras laeve* (syn. *D. gracile*), *Discus cronkhitei*, *Gyraulus deflexus*, *Physa integra*, *Succinea ovalis* and *Zonitoides arboreus*. Stockdale (1970b) infected *Mesodon thyroides* and *Triodopsis albolabris*. Larvae invaded the epidermis of the foot or were ingested and then migrated to the foot. At 22–23°C, the first moult took place on days 5–6 and the second on day 13. Third-stage larvae were 360–392 μm in length and had lateral alae and pointed tails.

Stockdale (1970b) studied the migration of larvae to the lungs and concluded that they migrated through the stomach wall into the peritoneal cavity, crossed the diaphragm and invaded the visceral pleura of the lungs within 24 h postinfection. The third moult took place in the lungs about day 3 and the final moult by about day 7 postinfection. The prepatent period was 21–28 days (Anderson, 1962; Stockdale, 1970b).

A. pridhami occurred together with *Filaroides martis* in mink; larval stages were identical in morphology. However, the prepatent period of *F. martis* was longer (41–53 days) and mink invariably vomited when given infective larvae of *A. pridhami* but not when given infective larvae of *F. martis* (see Anderson, 1962; Stockdale, 1970b).

Infective larvae of *A. pridhami* became encapsulated in the liver of mice, passerine birds, frogs and fish. Presumably, paratenic hosts are important in the transmission of this species to mink.

Didelphostrongylus

D. hayesi Prestwood, 1976
D. hayesi is a parasite of the alveoli of subpleural regions of the lungs of the opossum (*Didelphis marsupialis*). Prestwood (1976) exposed *Mesodon perigraptus* and *Triodopsis albolabris* to larvae, and third-stage larvae recovered from snails were given orally to young opossums. The prepatent period was 22 days. First-stage larvae (228–253 µm in length) had sharply pointed tails, as did third-stage larvae (330–396 µm in length).

Trilobostrongylus

T. bioccai Anderson, 1963
T. bioccai occurs free in large, thin-walled capsules under the pleura of the lungs of fisher (*Martes pennanti*) in North America. The female is ovoviviparous and first-stage larvae developed to the infective stage in the terrestrial molluscs *Deroceras laeve* (syn. *D. gracile*), *Discus cronkhitei* and *Zonitoides arboreus* as well as in the aquatic snail *Physa integra*. In the snails, the first moult occurred on day 11 and the second on day 16 at 22–23°C (Anderson, 1963a). Delicate lateral alae were present on both first- and second-stage larvae.

Family Filaroididae

The Filaroididae contains abursate species. The family is probably derived from the Angiostrongylidae which has some genera with markedly reduced bursae (e.g. *Andersonstrongylus*, *Madafilaroides*). The transmission of species of the Filaroididae is unusually varied. *Filaroides martis* and *Oslerus rostratus* have the typical type of transmission that has come to be expected in the metastrongyloids of carnivores (i.e. development in gastropods and the involvement of paratenic hosts). The infective stage of *Oslerus osleri* and *Filaroides hirthi* of canids, however, is the first stage and the parasites are monoxenous. In addition, *O. osleri* is transmitted to pups during **regurgitative feeding** whereas *F. hirthi* is transmitted through faeces contaminated with larvae. *Andersonstrongylus captivensis* of the family Angiostrongylidae can also be transmitted by giving the final host (skunk) first-stage larvae; this fact probably also indicates the close affinities between the Angiostrongylidae and the Filaroididae. It is worth noting that Craig (1972) was not successful in infecting terrestrial gastropods with *Filaroides mephitidis* of skunks. Anderson (1963a) noted that *F. canadensis* of the otter (*Lutra canadensis*) failed to develop in *Physa integra*, which was a suitable intermediate host of *F. martis*; first-stage larvae of *F. canadensis* were found alive and undeveloped in tissues and tissue spaces of snails 35 days after exposure.

One species (*Filaroides decorus*) of the California sea-lion reaches the infective stage in coprophagous fish which serve as intermediate hosts.

Filaroides

F. decorus Dougherty and Herman, 1947

This is a parasite of the lungs of the California sea-lion (*Zalophus californianus*). According to Dailey (1970), first-stage larvae were passed in faeces of the sea-lion. Faeces were readily consumed by the opaleye (*Girella nigricans*), a fish of the family Girellidae, and first-stage *F. decorus* invaded the tissues of the fish and attained the infective stage. By 3 days postinfection, larvae were in the mucosa and submucosa of the intestine. Five to 12 days postinfection, larvae were found under the longitudinal muscle layer of the intestine. The first moult apparently occurred 12–15 days and the second 24–36 days postinfection. Third-stage larvae were in the serosa or the mesenteric fat. A young experimentally infected sea-lion began to pass larvae 21 days after being given infected *G. nigricans*. The larval stages were typical of those of the members of the Angiostrongylidae and Filaroididae.

Apparently, sea-lions acquire infections near rookeries, where there are ample coprophagic fish liable to acquire larvae from eating faeces. Under these conditions, sea-lions may acquire massive infections of *F. decorus*.

F. hirthi Georgi and Anderson, 1975

Georgi (1976, 1987) and Georgi *et al.* (1976, 1977, 1979) found *F. hirthi* commonly in lung parenchyma of beagle dogs in kennels in several locations in the USA and determined its mode of transmission. Female worms produced the usual thin-shelled eggs containing first-stage larvae; the latter were shed in faeces. Pups became infected early in life, apparently during the latter part of the nursing period. Dogs could be infected successfully by feeding them ground lungs with the parasite and by oral inoculation of larvae collected from faeces. It was concluded that transmission is mainly through faecal contamination. Also, there was evidence (larvae in lymph nodes long after exposure to infection) of autogenous reinfection by some larvae. Georgi *et al.* (1979) detected first-stage larvae in lungs of infected dogs as early as 6 h after oral inoculation. This was evidence that larvae reached the lungs by the hepatic portal system and/or the mesenteric lymph drainage. There were four moults in the lungs, mainly at days 1, 2, 6 and 9. The prepatent period was about 5 weeks. Georgi *et al.* (1979) produced normal infections by injecting eggs and first-stage larvae into the jugular vein.

F. martis (Werner, 1782)

F. martis (syn. *F. bronchialis*) is a common parasite of wild mink (*Mustela vison*) in which it occurs in peribronchial nodules at the hilus of the lobes of the lungs. Petrov and

Gagarin (1938) and Dubnitski (1955) infected *Agriolimax reticulatus*, *Arion intermedius*, *A. agrestis*, *Eulota fruiticola*, *Succinea putris*, *Zonitoides excavata* and *Z. nida* with *F. martis* in the CIS. Anderson (1962) successfully infected terrestrial gastropods (*Anguispira alternata*, *Derocercas laeve* (syn. *D. gracile*), *Discus cronkhitei*, *Succinea ovalis* and *Zonitoides arboreus*) as well as aquatic snails (*Ampullaria cuprina*, *Armiger crista*, *Gyraulus deflexus* and *Physa integra*) by exposing them to large numbers of larvae from female worms (Fig. 3.10). Larvae were apparently ingested by the gastropods but they also penetrated the foot of terrestrial gastropods. At 22–23°C, moults occurred at 5–6 and 13 days postinfection. Anderson (1962) gave infective larvae to mink, which passed first-stage larvae 41–53 days later. Infective larvae given to mice became encapsulated in the liver.

Stockdale and Anderson (1970) and Ko and Anderson (1972) showed that *F. martis* moulted twice in the gastric mucosa of mink, once 3–4 days postinfection and

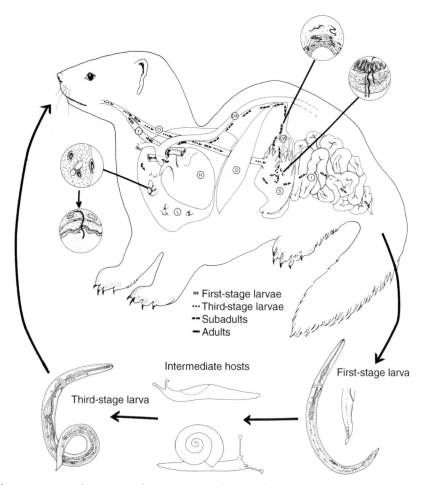

" First-stage larvae
··· Third-stage larvae
-- Subadults
— Adults

Intermediate hosts

Third-stage larva

First-stage larva

Fig. 3.10. Development and transmission of *Filaroides martis*. CA = coeliac artery; D = diaphragm; DA = dorsal aorta; H = heart; I = intestine; L = liver; O = oesophagus; S = stomach; T = trachea. (Original, U.R. Strelive.)

again at about 5 days. Worms left the stomach as minute subadults (males $\bar{x}$ = 1.4 mm; females $\bar{x}$ = 1.5 mm) on day 9 and migrated along the adventitia or adipose tissue on the coeliac artery to the dorsal aorta, which was reached as early as day 10. Worms then migrated through the diaphragm. Most apparently followed the dorsal aorta to the base of the heart, where they transferred to the adventitia of the pulmonary arteries, which they followed to the lungs. Worms reached the lungs in about 15–21 days and they became encapsulated in connective tissue surrounding bronchi and the pulmonary arteries. Female worms eventually pushed their tails through the bronchial epithelium into the lumen of the bronchi and released first-stage larvae on the bronchial escalator.

Oslerus

O. osleri (Cobbold, 1876)

O. osleri (syn. *Filaroides osleri*) is a widely distributed parasite of canids, especially wild canids such as coyotes, foxes and dingos. It is less commonly found in domestic dogs. The parasite occurs in nodules on the trachea and bronchi. Urquhart *et al.* (1954), Dorrington (1964, 1965, 1968), Polley and Creighton (1977) and Dunsmore and Spratt (1979) have determined its mode of transmission. Female worms produce numerous larvated eggs, which are released into the respiratory tract of the host, and eggs and larvae can be found in faeces by standard techniques. Larvae in faeces are generally dead or inactive and degenerate in appearance. However, young dogs and related canids can be infected by oral inoculation of free first-stage larvae or larvated eggs from the lungs of an infected host. Dorrington (1968) thought that the bitch might transmit the infections to pups by licking and cleaning soon after birth. Dunsmore and Spratt (1979) noted that wild canids feed pups by regurgitation. They suggested that this was the likely mode of transmission and might be why the infection is more common in wild canids such as coyotes and dingos than in dogs. Dunsmore and Spratt (1979) found fourth-stage larvae in a dingo 14 days postinfection and immature fifth-stage worms 34 days postinfection. They found adult worms in nodules 70 days postinfection and moderately sized nodules in a dingo 18 weeks postinfection.

O. rostratus Gerichter, 1949

O. rostratus (syn. *Anafilaroides rostratus*) inhabits the lung parenchyma of felines, including the domestic cat and *Lynx rufus* (Gerichter, 1949; Klewer, 1958). Gerichter (1949) noted eggs and free larvae in uteri of female worms and described first-stage larvae as 300–320 μm in length. Some development of the larvae took place in tissues of the gastropods *Monacha syriaca* and *Theba pisana* but the results were inconclusive because the snails were kept at 4.8–12.5°C, which was probably too low a temperature to support development. Klewer (1958) found *O. rostratus* in lynx and evidently infected the slug *Limax maximus*. In the foot of the latter, the first moult took place 6–7 days and the second by 10 days postinfection.

Seneviratna (1959) infected *Achatina fulica*, *Helix aspersa*, *Laevicaulus alte* and *Mariella dussumieri*. He claimed that the first moult was completed in 14–18 days and the second in 20–28 days in gastropods kept at 24–30°C. Larvae developed in the foot but they were sometimes noted in other parts of the body when the third stage was reached. Mice were shown to be suitable paratenic hosts. Seven-day-old chicks given larvae were fed to a cat a week later; the cat was infected when examined several months later. Seneviratna's (1959) attempts to follow development in cats were inconclusive. Several cats given larvae vomited after receiving larvae and he failed to determine the stage found in the lungs of some of them. He claimed that all moults in the cat were completed by day 46 postinfection and he first observed larvae on day 78.

Family Skrjabingylidae

The family contains the single genus *Skrjabingylus*, the five species of which occur in the frontal sinuses of Mustelidae. Members of the genus are easily recognized on the basis of their location in the host, their red colour, the modified bursa with lateral fleshy lobes and a median vulva. These nematodes are associated frequently with cavitations of bones surrounding the frontal sinuses (Hansson, 1968; King, 1977; Lewis, 1978).

First-stage larvae formed *in utero* pass from the sinuses to the back of the throat of the final host and are swallowed and eliminated with faecal material. Petrov and Gagarin (1938) showed that *S. petrowi* developed in the snail *Succinea ovalis*. Hobmaier (1941b) confirmed the role of terrestrial snails in the development of *S. chitwoodorum* and reported encapsulated larvae in snakes and frogs which he regarded as actual or potential paratenic hosts. Lankester and Anderson (1971) showed that infective larvae of *S. chitwoodorum* and *S. nasicola* invaded the gut wall or lymph nodes of the final host and moulted into subadults, which entered the body cavity and followed nerves to the central nervous system. The nematodes entered the spinal subarachnoid space and then moved to the front of the brain. They followed olfactory nerves which pass through the cribriform plate of the ethmoid bone and they reached the nasal sinuses as early as 6 days postinfection. Presumably all species in the genus behave similarly.

Skrjabingylus

S. chitwoodorum Hill, 1939

This is a common parasite of skunk (*Mephitis mephitis* and *Spilogale putorius*) in North America. Hobmaier (1941b), Lankester and Anderson (1971) and Lankester (1983) reported the following terrestrial gastropods as suitable experimental intermediate hosts: *Agriolimax agrestes, Limax maximus, L. cinereus, L. flavus, L. niger, Mesodon thyroides, Milax* sp. and *Triodopsis albolabris*. According to Hobmaier (1941b) *Epiphragmophora* sp. and *Helix pomatia* were less suitable hosts than those listed above.

First-stage larvae were 428–545 μm in length and the tail ended in a cone-shaped structure with six circular grooves (Lankester, 1983) (Fig. 3.11). The first moult in gastropods occurred 10–14 days and the second 15–18 days postinfection at 21°C. Third-stage larvae were 776–922 μm in length and had wide lateral alae and long excretory glands; the genital primordia were of two sizes, a smaller one for the male and a larger for the female (Lankester, 1983). Hobmaier (1941b) reported encapsulated third-stage larvae in garter snakes (*Thamnophis sirtalis*) and larvae given to frogs were subsequently found encapsulated in the tissues of the host. Lankester (1983) gave infective larvae to *Bufo americanus, Rana pipiens, T. sirtalis*, the fish *Ictaluris nebulosis* and white mice and recovered some of them from the tissues of these hosts several days later.

Skunk invariably vomited for up to 30 min after being given infective larvae. Third-stage larvae in the gut and lymph nodes moulted 1–2 and 3–4 days postinfection (Lankester and Anderson, 1971). Four to five days postinfection, the tiny fifth-stage worms (715–830 μm) migrated into the body cavity and into the abdominal wall.

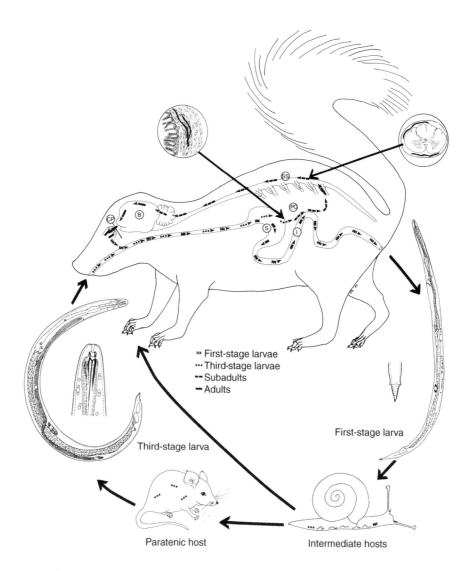

First-stage larvae
Third-stage larvae
Subadults
Adults

Third-stage larva

First-stage larva

Paratenic host   Intermediate hosts

Fig. 3.11. Development and transmission of *Skrjabingylus chitwoodorum*. B = brain; CP = cribiform plate; I = intestine; PC = peritoneal cavity; S = stomach; SS = subdural space. (Original, U.R. Strelive.)

Moving beneath the perineurium of nerves, the worms entered the vertebral column and invaded the spinal subarachnoid space, in which they migrated anteriorly to the frontal sinuses. They arrived in the sinuses as early as 6 days postinfection; females were 1.9 mm and males 1.2 mm in length at this time. The prepatent period was 31–33 days (Lankester, 1970).

S. lutrae Lankester and Crichton, 1972

This is a parasite of the otter (*Lutra canadensis*) in North America. Larvae developed to the infective stage in the snail *Mesodon roemori* (see Lankester and Crichton, 1972).

S. nasicola (Leuckart, 1842)

S. nasicola is a common, cosmopolitan parasite of members of the genus *Mustela* (i.e. *M. altaica, M. frenata, M. erminea, M. eversmanni, M. itatsi, M. lutreola, M. nivalis, M. putorius, M. rixosa, M. sibirica* and *M. vison*). The parasite developed to the infective stage in the following gastropods: *Agriolimax reticulatus, Cochlicopa lubrica, Euparypha pisana, Helicella arenosa, Mesodon thyroidus* and *Zenobiella rubiginosa* (Dubnitski, 1956; Theron, 1975; Lankester, 1983). Dubnitski (1956) reported moulting times of 7–8 and 15–17 days (temperature not specified). According to Lankester (1983) the first moult in gastropods occurred 10–14 days and the second 15–18 days postinfection at 21°C. According to Theron (1975) the worms moulted at 12 and 16 days in *Helicella arenosa* at 20°C; comparable figures in *Euparypha pisana* were 18 and 23 days.

First-stage larvae were 353–407 μm in length and similar in morphology to those of *S. chitwoodorum* (Lankester, 1983). Third-stage larvae were 825–980 μm in length. Although otherwise similar to those of *S. chitwoodorum*, the wall of the buccal capsule of *S. nasicola* was heavily cuticularized and asymmetrical in lateral view since it was extended laterally. Genital primordia were small and of the same size and shape in all larvae.

Mink (*Mustela vison*) invariably vomited shortly after oral inoculation of infective larvae. Vomiting often persisted for 3 h. The neurotropic behaviour and migration of *S. nasicola* were similar to those of *S. chitwoodorum* (see Lankester and Anderson, 1971) and the prepatent period was 30–31 days (Lankester, 1983).

The role of paratenic hosts in the transmission of *S. nasicola* has been the subject of several articles. Hansson (1967) infected mustelids by feeding them rodents (*Apodemus* spp., *Clethrionomys glareolus*) and shrews (*Sorex araneus*) naturally infected in Sweden. Duncan (1976) questioned whether shrews were likely to be involved, mainly because mustelids do not normally feed on them. He felt that stoats and weasels may well ingest molluscs to acquire their infections. Gamble and Riewe (1982) suggested that rodents (especially myomorphs), shrews, amphibians and reptiles are paratenic hosts of *S. nasicola*. Weber (1986) studied the possible role of rodents in the transmission of *S. nasicola*.

Hansson (1974) stated that *Mustela* spp. were infected in the cold months of the year and intensity was highest in winter and spring rather than in autumn.

S. petrowi Bagaenow and Petrov, 1941

This is a common parasite of pine marten (*Martes martes*), stone marten (*M. foina*) and sable marten (*M. zibellina*) in the CIS (Kontrimavichus *et al.*, 1976). Petrov and Gagarin (1938) reported that this species developed in the snail *Succinea ovalis*.

Family Pseudaliidae

The pseudaliids (seven genera) of toothed whales (Odontoceti) have highly modified bursae with unique fusion of rays. Species occur in the lungs, cranial sinuses, middle ear, eustachian tubes and circulatory system of their hosts. One genus (*Stenuroides*) occurs in the lungs of the mongoose (*Herpester ichneumon*). Little is known of their transmission (Conlogue *et al.*, 1985; Moser and Rhinehart, 1993; Balbuena *et al.*, 1994). Lungworms have not been reported in the filter-feeding baleen whales. Perhaps these whales are descendants of terrestrial ancestors which lacked lungworms, or their feeding habits precluded transmission of lungworms.

Halocercus

H. lagenorhynchi Baylis and Daubney, 1925

Woodard *et al.* (1969) reported worms in a 21-year-old dolphin (*Tursiops truncatus*) born in captivity and a 2.5-month-old calf with an unusually heavy infection, which raised the possibility of transplacental or transmammary transmission. Dailey *et al.* (1990, 1991) found adult *H. lagenorhynchi* in the lungs of *Tursiops truncatus* from Florida. Four immature *T. truncatus*, ranging in size from a full-term fetus to specimens 134 cm in length, were infected with numerous sexually mature worms. Female worms contained larvated eggs. These observations are evidence of prenatal infection but do not indicate how the female whale acquires larvae that can transfer to the fetus.

References (Strongylida)

Adams, D.B. (1983) Observations on the self cure reaction and other forms of immunological responsiveness against *Haemonchus contortus* in sheep. *International Journal for Parasitology* 13, 571–578.

Addison, E.M. and Fraser, G.A. (1994) Life cycle of *Crenosoma petrowi* (Nematoda: Metastrongyloidea) from black bears (*Ursus americanus*). *Canadian Journal of Zoology* 72, 300–302.

Africa, C.M. (1931a) Studies on the activity of the infective larvae of the rat strongylid, *Nippostrongylus muris*. *Journal of Parasitology* 17, 196–206.

Africa, C.M. (1931b) Studies on the host relations of *Nippostrongylus muris*, with special reference to age resistance and acquired immunity. *Journal of Parasitology* 18, 1–13.

Agrawal, V. (1966) The effects of temperature on the survival and development of the free-living stages of *Oesophagostomum columbianum* (Curtice, 1890) (Nematoda). *Transactions of the American Microscopical Society* 85, 99–106.

Ahluwalia, J.S. and Charleston, W.A.G. (1974) Studies on the development of the free-living stages of *Cooperia curticei*. *New Zealand Veterinary Journal* 22, 191–195.

Ahmed, Z. and Saleh, M. (1967) Survival and propagation of *Kalicephalus willeyi* in vitro. *Revista di Parasitologia* 28, 123–127.

Akimoto, H. (1966) Epidemiological studies on hookworm infection. Analysis of hookworm infection in Japan and experimental infection to man by the cutaneous route. *Japanese Journal of Parasitology* 15, 1–29. (In Japanese.)

Alicata, J.E. (1932) Life history of the rabbit stomach worm, *Obeliscoides cuniculi*. *Journal of Agricultural Research* 44, 401–419.

Alicata, J.E. (1935) Early developmental stages of nematodes occurring in swine. *USA Department of Agriculture Technical Bulletin* No. 489. 96 pp.

Alicata, J.E. (1961) Survival of the infective larvae of *Cooperia punctata* of cattle on pasture in Hawaii. *Proceedings of the Helminthological Society of Washington* 28, 181–183.

Alicata, J.E. (1962) *Angiostrongylus cantonensis* (Nematoda: Metastrongylidae) as a causative agent of eosinophilic meningoencephalitis of man in Hawaii and Tahiti. *Canadian Journal of Zoology* 40, 5–8.

Alicata, J.E. (1963a) Incapability of vertebrates to serve as paratenic hosts for infective larvae of *Angiostrongylus cantonensis*. *Journal of Parasitology* 49, 48.

Alicata, J.E. (1963b) The incidence of *Angiostrongylus cantonensis* (Chen) among rats and molluscs in New Caledonia and nearby islands and its possible relation to eosinophilic meningitis in Noumea, New Caledonia. *South Pacific Commission Technical Paper* No. 139, 1–9.

Alicata, J.E. (1963c) Morphological and biological differences between the infective larvae of *Angiostrongylus cantonensis* and those of *Anafilaroides rostratus*. *Canadian Journal of Zoology* 41, 1179–1183.

Alicata, J.E. (1963d) Experimental work on eosinophilic meningitis. *Plantation Health* 28, 17–23.

Alicata, J.E. (1964a) Parasitic infections of man and animals in Hawaii. *Technical Bulletin Hawaii Agricultural Experimental Station* 61, 1–139.

Alicata, J.E. (1964b) Land crabs as probable paratenic hosts for the infective larvae of *Angiostrongylus cantonensis*. *Journal of Parasitology* 50, 39.

Alicata, J.E. (1965a) Notes and observations on marine angiostrongylosis and eosinophilic meningoencephalitis in Micronesia. *Canadian Journal of Zoology* 43, 667–672.

Alicata, J.E. (1965b) Biology and distribution of the rat lungworm, *Angiostrongylus cantonensis* and its relationship to eosinophilic meningoencephalitis and other neurologic disorders of man and animals. *Advances in Parasitology* 3, 223–248.

Alicata, J.E. (1966) The presence of *Angiostrongylus cantonensis* in islands of the Indian Ocean and probable role of the giant African snail, *Achatina fulica*, in dispersal of the parasite to the Pacific Islands. *Canadian Journal of Zoology* 44, 1041–1049.

Alicata, J.E. (1969) Present status of *Angiostrongylus cantonensis* infection in man and animals in the tropics. *Journal of Tropical Medicine and Hygiene* 72, 53–63.

Alicata, J.E. (1988) *Angiostrongylus cantonensis* (eosinophilic meningitis): historical events in its recognition as a new parasitic disease of man. *Journal of the Washington Academy of Sciences* 78, 38–46.

Alicata, J.E. and Brown, R.W. (1962) Observations on the method of human infection with *Angiostrongylus cantonensis* in Tahiti. *Canadian Journal of Zoology* 40, 755–760.

Alicata, J.E. and Brown, R.W. (1963) Observations on the cause and transmission of eosinophilic meningitis in Hawaii and the South Pacific. *Proceedings of the Hawaiian Academy of Science*, 1962–1963, pp. 14–15.

Alicata, J.E. and Jindrak, K. (1970) *Angiostrongylosis in the Pacific and Southeast Asia*. Charles C. Thomas, New York.

Alicata, J.E. and Lynd, F.T. (1961) Growth rate and other signs of infection in calves experimentally infected with *Cooperia punctata*. *American Journal of Veterinary Research* 22, 704–707.

Alicata, J.E., Loison, G. and Cavallo, A. (1963). Parasitic meningoencephalitis experimentally produced in a monkey with larvae of *Angiostrongylus cantonensis*. *Journal of Parasitology* 49, 156–157.

Allan, D. and Johnson, A.W. (1960) A short history of husk. *Veterinary Record* 72, 42–44.

Altaev, A.K. (1961) Observations on the development of free-living larvae of *Trichostrongylus skrjabini*. *Trudi Gel'mintologicheskoi Laboratorii, Akademiya Nauk SSSR* 11, 10–21. (In Russian.)

Amberson, J.M. and Schwarz, E. (1952) *Ternidens deminutus* Railliet and Henry, a nematode parasite of man and primates. *Annals of Tropical Medicine and Parasitology* 46, 227–237.

Anantaraman, M. (1942) The life-history of *Oesophagostomum radiatum*, the bovine nodular worm. *Indian Journal of Veterinary Science and Animal Husbandry* 12, 87–132.

Andersen, F.L., Wang, G.T. and Levine, N.D. (1966) Effect of temperature on survival of the free-living stages of *Trichostrongylus colubriformis. Journal of Parasitology* 52, 713–721.

Anderson, I.G. (1986) Observations on the life-cycle and larval morphogenesis of *Cooperia fuelleborni* (Nematoda: Trichostrongyloidea) parasitic in impala, *Aepyceros melampus. Journal of Helminthology* 60, 113–122.

Anderson, I.G. (1995) Observations on the life-cycles and larval morphogenesis of *Haemonchus bedfordi, Impalaia tuberculata* and *Longistrongylus sabie* (Nematoda: Trichostrongyloidea) parasitic in impala, *Aepyceros melampus. South African Journal of Zoology* 30, 37–45.

Anderson, I.G.H. (1974) Preliminary note on the life cycle of the lungworm, *Pneumostrongylus calcaratus*, Mönning, 1932. *Journal of the South African Veterinary Association* 45, 219–220.

Anderson, I.G.H. (1976) Studies on the life cycle of the lungworm, *Pneumostrongylus calcaratus*, Mönning, 1932. *Journal of the South African Veterinary Association* 47, 23–27.

Anderson, I.G.H. (1982) The life cycle of the lungworm, *Pneumostrongylus calcaratus. Journal of the South African Veterinary Association* 53, 109–114.

Anderson, I.G.H. (1992) Observations on the life-cycles and larval morphogenesis of, and transmission experiments with *Cooperioides hamiltoni* and *Cooperioides hepaticae* (Nematoda: Trichostrongyloidea) parasitic in impala, *Aepyceros melampus. South African Journal of Zoology* 27, 81–88.

Anderson, R.C. (1962) The systematics and transmission of new and previously described metastrongyles (Nematoda: Metastrongylidae) from *Mustela vison. Canadian Journal of Zoology* 40, 893–920.

Anderson, R.C. (1963a) Further studies on the taxonomy of metastrongyles (Nematoda: Metastrongyloidea) of Mustelidae in Ontario. *Canadian Journal of Zoology* 41, 801–809.

Anderson, R.C. (1963b) The incidence, development, and experimental transmission of *Pneumostrongylus tenuis* Dougherty (Metastrongyloidea: Protostrongylidae) of the meninges of the white-tailed deer (*Odocoileus virginianus borealis*) in Ontario. *Canadian Journal of Zoology* 41, 775–792.

Anderson, R.C. (1964) Neurologic disease in moose infected experimentally with *Pneumostrongylus tenuis* from white-tailed deer. *Pathologia Veterinaria* 1, 289–322.

Anderson, R.C. (1965) The development of *Pneumostrongylus tenuis* in the central nervous system of white-tailed deer. *Pathologia Veterinaria* 2, 360–379.

Anderson, R.C. (1968) The pathogenesis and transmission of neurotropic and accidental nematode parasites of the central nervous system of mammals and birds. *Helminthological Abstracts* 37, 191–210.

Anderson, R.C. (1970) Neurologic disease in reindeer (*Rangifer tarandus tarandus*) introduced into Ontario. *Canadian Journal of Zoology* 49, 159–166.

Anderson, R.C. (1971) Lungworms. In: Davies, J.W. and Anderson, R.C. (eds) *Parasitic Diseases of Wild Mammals*. Iowa State University Press, Ames, pp. 81–126.

Anderson, R.C. (1972) The ecological relationships of meningeal worm and native cervids in North America. *Journal of Wildlife Diseases* 8, 304–309.

Anderson, R.C. (1978) No. 5. Keys to genera of the Superfamily Metastrongyloidea. In: Anderson, R.C., Chabaud, A.G. and Willmott, S. (eds) *CIH Keys to the Nematode Parasites of Vertebrates*. Commonwealth Agricultural Bureaux, Farnham Royal, UK, pp. 1–40.

Anderson, R.C. (1982) Host–parasite relations and evolution of the Metastrongyloidea (Nematoda). In: *Deuxième Symposium sur la Spécificité Parasitaire des Parasites des Vertébrés*

13–17 Avril 1981. Mémoires du Muséum National D'Histoire Naturelle Série A, Zoologie, 123, 129–133.

Anderson, R.C. and Lankester, M.W. (1974) Infections and parasitic diseases and arthropod pests of moose in North America. In: Bedard, J. (ed.) *Moose Ecology. Naturaliste Canadien* 101, 23–50.

Anderson, R.C. and Prestwood, A.K. (1981) Lungworms. In: Davidson, W.R., Hayes, F., Nettles, V. and Kellog, F. (eds) *Disease and Parasites of White-tailed Deer.* Miscellaneous Publication No. 7 of Tall Timber Research Station, Tallahassee, Florida pp. 266–317.

Anderson, R.C. and Strelive, U.R. (1966) Experimental cerebrospinal nematodiasis (*Pneumostrongylus tenuis*) in sheep. *Canadian Journal of Zoology* 44, 889–894.

Anderson, R.C. and Strelive, U.R. (1967a) The experimental transmission of *Pneumostrongylus tenuis* to caribou (*Rangifer tarandus terraenovae*). *Canadian Journal of Zoology* 46, 503–510.

Anderson, R.C. and Strelive, U.R. (1967b) The penetration of *Pneumostrongylus tenuis* into the tissues of white-tailed deer. *Canadian Journal of Zoology* 45, 285–289.

Anderson, R.C. and Strelive, U.R. (1969) The effect of *Pneumostrongylus tenuis* (Nematoda: Metastrongyloidea) on kids. *Canadian Journal of Comparative Medicine* 33, 280–286.

Anderson, R.C. and Strelive, U.R. (1972) Experimental cerebrospinal nematodiasis in kids. *Journal of Parasitology* 58, 816.

Anderson, R.C., Lankester, M.W. and Strelive, U.R. (1966) Further experimental studies of *Pneumostrongylus tenuis* in cervids. *Canadian Journal of Zoology* 44, 851–861.

Anderson, N., Armour, J., Jennings, F.W., Ritchie, J.S.D. and Urquhart, G.M. (1965) Inhibited development of *Ostertagia ostertagi. Veterinary Record* 77, 146–147.

Anderson, P.J.S. and Verster, A. (1971) Studies on *Dictyocaulus filaria.* II. Migration of the developmental stages in lambs. *Onderstepoort Journal of Veterinary Research* 38, 185–190.

Andrews, J.S. (1934) Egg production by *Nematodirus* spp. (Trichostrongylidae) and by *Chabertia ovina* (Strongylidae) following repeated experimental infections of sheep with these nematodes. *Proceedings of the Helminthological Society of Washington* 1, 51.

Andrews, J.S. (1939) Life history of the nematode *Cooperia curticei* and development of resistance in sheep. *Journal of Agricultural Research* 58, 771–785.

Andrews, J.S. and Maldonado, J.R. (1941) The life history of *Oesophagostomum radiatum,* the common nodular worm of cattle. *Research Bulletin, Puerto Rico Agricultural Experiment Station* No. 2, 14 pp.

Anisimava, A.I. and Maklakova, L.P. (1992) Infection of earthworms with metastrongylid larvae in different biocoenoses. *Vestsi Akadèmii Navuk BSSR. Sevyya Biyalagichnykh Navuk* 3–4, 85–88.

Ansari, M.Z. (1981) A preliminary note on the prenatal and transcolostral infection of *Gaigeria pachyscelis* in lambs. *Indian Journal of Animal Health* 20, 155–157.

Ansari, M.Z. and Singh, K.S. (1980) Studies on the biology of the infective larvae of *Gaigeria pachyscelis* Railliet and Henry, 1910 in abnormal hosts. *Indian Journal of Animal Sciences* 50, 632–637.

Arantes, I.G., do Nascimento, A.A. and da C. Pereira, O. Jr (1983) *Gaigeria pachyscelis* Railliet and Henry, 1910 (Nematoda: Ancylostomatidae). Estudo morfobiológico de larvas e de adultos jovens em parasitismo em caprinos. *Arquivos do Instituto Biológico São Paulo* 50, 21–37.

Armour, J. (1978) Arrested development in cattle nematodes with special inference to *Ostertagia ostertagi.* In: Borgsteede, F.H.M. (ed.) *Facts and Reflections III. Workshop on Arrested Development of Nematodes in Sheep and Cattle.* Central Veterinary Institute, Lelystad, The Netherlands, pp. 77–88.

Armour, J. and Bruce, R.G. (1974) Inhibited development of *Ostertagia ostertagi* infections – a diapause phenomenon in a nematode. *Parasitology* 69, 161–174.

Armour, J. and Ogbourne, C.P. (1982) *Bovine Ostertagiasis: a Review and Annotated Bibliography.* Miscellaneous publication No. 7 of the Commonwealth Institute of Parasitology. Commonwealth Agricultural Bureaux, Farnham Royal, UK, 93 pp.

Armour, J., Jarrett, W.F. and Jennings, F.W. (1966) Experimental *Ostertagia circumcincta* infection in sheep: development and pathogenesis of single infection. *American Journal of Veterinary Research* 27, 1267–1278.

Armour, J., Jennings, F.W. and Urquhart, G.M. (1969) Inhibition of *Ostertagia ostertagi* at the early fourth larval stage. II. The influence of environment on host or parasite. *Research in Veterinary Science* 10, 238–244.

Arnett, E.B., Irby, L.R. and Cook, J.G. (1993) Sex and age specific lungworm infection in Rocky Mountain bighorn sheep during winter. *Journal of Wildlife Diseases* 29, 90–93.

Asakawa, M. (1987) Genus *Heligmosomoides* Hall, 1916 (Heligmosomidae: Nematoda) from the Japanese wood mice *Apodemus* spp. III. The life cycle of *Heligmosomoides kurilensis dobayashii* (Nadtochii, 1966) in ICR mice and preliminary experimental infection to jirds. *Journal of the College of Dairying, Japan. Natural Science* 12, 131–140.

Ash, L.R. (1967) Observations on the role of molluscs and other invertebrates in the transmission of *Angiostrongylus cantonensis* infection to man in New Caledonia. *South Pacific Commission Seminar on Helminthiasis and Eosinophilic Meningitis, Noumea, New Caledonia*, pp. 1–9.

Ash, L.R. (1968) The occurrence of *Angiostrongylus cantonensis* in frogs of New Caledonia with observations on paratenic hosts of metastrongyles. *Journal of Parasitology* 54, 432–436.

Ash, L.R. (1976) Observations on the role of molluscs and planarians in the transmission of *Angiostrongylus cantonensis* infection to man in New Caledonia. *Revista de Biologia Tropical, Costa Rica* 24, 163–174.

Azimov, D.A., Ubaidullaev, Y.V. and Zimin, Y.M. (1973) Protostrongylid diseases of sheep and goats. *Veterinariya* 9, 66–68. (In Russian.)

Azimov, D.A., Sultanov, M.A., Kulmamatov, E.N., Zimin, Y.M. and Isakova, D.T. (1976) The ontogenesis of *Protostrongylus skrjabini* (Boev, 1936). *Doklady Akademii Nauk Uzbekskoi SSR (Uzbekiston SSR Fanlar Akademiyasining, Dokladlari)* 10, 65–67. (In Russian.)

Babos, S. (1961) Zur Kenntnis der Protostrongylosen der Leporiden, unter besonderer Berücksichtigung der in Ungarn vorkommenden *Protostrongylus* Arten. *Helminthologia* 3, 1–4.

Bailey, M.A. (1968) The role of the host in initiation of development of the parasitic stages of *Trichostrongylus retortaeformis* (Nematoda). *Comparative Biochemistry and Physiology* 26, 897–906.

Bailey, W.S. (1949) Studies on calves experimentally infected with *Cooperia punctata* (v. Linstow, 1907) Ransom, 1907. *American Journal of Veterinary Research* 10, 119–129.

Baker, M.R. (1978) Development and transmission of *Oswaldocruzia pipiens* Walton, 1929 (Nematoda: Trichostrongylidae) in amphibians. *Canadian Journal of Zoology* 56, 1026–1031.

Baker, M.R. (1987) *Synopsis of the Nematoda Parasitic in Amphibians and Reptiles.* Memorial University of Newfoundland, Occasional Papers in Biology, No. 11, 325 pp.

Baker, M.R. and Anderson, R.C. (1974) Seasonal changes in abomal worms (*Ostertagia* spp.) in white-tailed deer (*Odocoileus virginianus*) at Long Point, Ontario. *Canadian Journal of Zoology* 53, 87–96.

Baker, N.F., Fisk, R.A., Bushnell, R.B. and Oliver, M.N. (1981) Seasonal occurrence of infective nematode larvae on irrigated pasture grazed by cattle in California, USA. *American Journal of Veterinary Research* 42, 1188–1191.

Baker, N.F., Fisk, R.A. and Rimbey, C.W. (1984) Seasonal occurrence of infective nematode larvae in California, U.S.A. high sierra pastures grazed by cattle. *American Journal of Veterinary Research* 45, 1393–1397.

Bakke, T.A. (1975) Studies on the helminth fauna of Norway XXIX: the common gull, *Larus canus* L., as final host for *Syngamus* (*Cyathostoma*) *lari* (Blanchard, 1849) (Nematoda, Strongyloidea). *Norwegian Journal of Zoology* 23, 37–44.

Balasingam, E. (1963) Experimental infection of dogs and cats with *Molineus barbatus* Chandler, 1942, with a discussion on the distribution of *Molineus* spp. *Canadian Journal of Zoology* 41, 599–602.

Balasingam, E. (1964a) Studies on the life cycle and developmental morphology of *Placoconus lotoris* (Schwartz, 1925) Webster, 1956 (Ancylostomidae: Nematoda). *Canadian Journal of Zoology* 42, 869–902.

Balasingam, E. (1964b) Comparative studies on the effects of temperature on free-living stages of *Placoconus lotoris*, *Dochmoides stenocephala* and *Ancylostoma caninum*. *Canadian Journal of Zoology* 42, 907–918.

Balasingam, E. (1965) Studies on *Hepatojarakus malayae* Yeh, 1955 (Trichostrongylidae: Nematoda). *Medical Journal of Malaya* 20, 68–69.

Balbuena, J.A., Aspholm, P.E., Andersen, K.I. and Bjørge, A. (1994) Lung-worms (Nematoda: Pseudaliidae) of harbour porpoises (*Phocoena phocoena*) in Norwegian waters: patterns of colonization. *Parasitology* 108, 343–349.

Ballantyne, R.J. and Samuel, W.M. (1984) Diagnostic morphology of the third-stage larvae of three species of *Parelaphostrongylus* (Nematoda, Metastrongyloidea). *Journal of Parasitology* 70, 602–604.

Bansemir, A.D. and Sukhdeo, M.V.K. (1994) The food resource of adult *Heligmosomoides polygyrus* in the small intestine. *Journal of Parasitology* 80, 24–28.

Bansemir, A.D. and Sukhdeo, M.V.K. (1996) Villus length influences habitat selection by *Heligmosomoides polygyrus*. *Parasitology* 113, 311–316.

Barnes, E.H. (1997) Population dynamics of the parasitic stages of *Oesophagostomum dendatum* in pigs in single and trickle infections. *International Journal of Parasitology* 27, 1595–1604.

Barrett, R.E., Billeb, B.R. and Worley, D.E. (1970) Infectivity of *Trichostrongylus axei* and *T. colubriformis* for Richardson's squirrel (*Spermophilus richardsoni*). *Journal of Parasitology* 56, 752.

Barus, V. (1964) Über die exogene Phase des Entwicklungszyklus von *Amidostomum fulicae* (Rudolphi, 1819) (Nematoda, Amidostomatidae). *Zeitschrift für Parasitenkunde* 24, 112–120.

Barus, V. (1966) The longevity of the parasitic stages and the dynamics of egg production of the nematode *Syngamus trachea* (Montagu, 1811) in chickens and turkeys. *Folia Parasitologica* 13, 274–277.

Barus, V. (1970) Beitrag zur Morphologie der Larven der Nematodenart *Cyathostoma lari* Blanchard, 1849, während der exogenen Entwicklungsphase. *Zeitschrift für Parasitenkunde* 34, 151–157.

Barus, V. and Blažek, K. (1965) Revision der exogen und endogenen phase des Entwicklungszyklus und der Pathogenität von *Syngamus* (*Syngamus*) *trachea* (Montagu, 1811) Chapin, 1925 im Organismus des Endwirtes. *Ceskoslovenská Parasitologie* 12, 47–70.

Barus, V. and Blazek, T. (1971) The life cycle and the pathogenicity of the nematode *Crenosoma striatum*. *Folia Parasitologica* 18, 215–226.

Barutzki, D. and Gothe, R. (1998) Untersuchungen zur migration und translation von *Oesophagostomum quadrispinulatum* larven aus dem kot. *Journal of Veterinary Medicine B* 45, 223–233.

Bates, H. (1972) Does *Syngamus trachea* (Nematoda) have a direct life cycle? *Bulletin of the Georgia Academy of Science* 30, 127–128.

Batte, E.G., Harkema, R. and Osborne, J.C. (1960) Observations on the life cycle and pathogenicity of the swine kidney worm (*Stephanurus dentatus*). *Journal of the American Veterinary Medical Association* 136, 622–625.

Batte, E.G., Moncol, D.J. and Barber, C.W. (1966) Prenatal infection with the swine kidney worm (*Stephanurus dentatus*). *Journal of the American Veterinary Medical Association* 149, 758–765.

Baudet, E.A.R.F. (1933) *Aelurostrongylus abstrusus* de lungworm van der kat. *Tijdschrift voor Diergeneeskunde* 60, 982–984.

Baumgartner, W., Zajac, A.M., Hull, B.L., Andrews, F. and Garry, F. (1985) Parelaphostrongylosis in llamas. *Journal of the American Veterinary Medical Association* 187, 1243–1245.

Bausov, I.A. (1969) Observations on the biology of *Amidostomum anseris*. *Uchenye Zapiski Kurskogo Gosudarstvennogo Pedagicheskogo Instituta* 59, 100–109. (In Russian.)

Bawden, R.J. (1969) Some effects of the diet of mice on *Nematospiroides dubius* (Nematoda). *Parasitology* 59, 203–213.

Bearup, A.J. (1967) *Ancylostoma braziliense* (Correspondence). *Tropical and Geographical Medicine* 19, 161–162.

Beaver, P.C. (1956) Parasitological reviews. Larva migrans. *Experimental Parasitology* 5, 587–621.

Beaver, P.C. (1988) Light, long-lasting *Necator* infection in a volunteer. *American Journal of Tropical Medicine and Hygiene* 39, 369–372.

Becklund, W.W. and Senger, C.M. (1967) Parasites of *Ovis canadensis* in Montana, with a checklist of the internal and external parasites of the Rocky Mountain bighorn sheep in North America. *Journal of Parasitology* 53, 157–165.

Behnke, J.M. and Parish, H.A. (1979) *Nematospiroides dubius*: arrested development of larvae in immune mice. *Experimental Parasitology* 47, 116–127.

Belem, A.M.G., Couvillion, C.E., Siefker, C. and Griffin, R.N. (1993) Evidence for arrested development of abomasal nematodes in white-tailed deer. *Journal of Wildlife Diseases* 29, 261–265.

Belle, E.A. (1959) The effect of microenvironment on the free-living stages of *Bunostomum trigonocephalum*. *Canadian Journal of Zoology* 37, 289–298.

Beller, K.F. (1928) Infektionswege und Entwicklung des Hakenwurms vom Schafe (*Bunostomum trigonocephalum* Rud. 1808). *Zeitschrift für Infektionskrankheiten, Parasitäre Krankheiten und Hygine der Haustiere* 29, 232–251.

Bergeron, E., Measures, L.N. and Huot, J. (1997a) Lungworm (*Otostrongylus circumlitus*) infections in ringed seals (*Phoca hispida*) from eastern Arctic Canada. *Canadian Journal of Fisheries and Aquatic Sciences* 54, 2443–2448.

Bergeron, E., Measures, L.N. and Huot, J. (1997b) Experimental transmission of *Otostrongylus circumlitus* (Railliet, 1899) (Metastrongyloidea: Crenosomatidae), a lungworm of seals in eastern arctic Canada. *Canadian Journal of Zoology* 75, 1364–1371.

Bergerud, A.T. and Mercer, W.E. (1989) Caribou introduction in eastern North America. *Wildlife Society Bulletin* 17, 111–120.

Bernard, P.N. and Bauche, J. (1914) Influence du mode de pénétration cutanée ou buccale du *Stephanurus dentatus* sur les localisations de ce nématode dans l'organisme du porc et sur son évolution. *Annales de l'Institut Pasteur* 28, 450–469.

Beveridge, I. and Barker, I.K. (1983) Morphogenesis of *Trichostrongylus rugatus* and distribution during development in sheep. *Veterinary Parasitology* 13, 55–65.

Beveridge, I. and Durette-Desset, M.-C. (1993) Adult and larval stages of *Paraustrostrongylus ratti* (Nematoda: Trichostrongyloidea) from *Rattus fuscipes*. *Transactions of the Royal Society of South Australia Incorporated* 117, 27–36.

Beveridge, I. and Ford, G.E. (1982) The trichostrongyloid parasites of sheep in South Australia and their regional distribution. *Australian Veterinary Journal* 59, 177–179.

Beveridge, I., Martin, R.R. and Pullman, A.L. (1985) Development of the parasitic stages of *Nematodirus abnormalis* in experimentally infected sheep and associated pathology. *Proceedings of the Helminthological Society of Washington* 52, 119–131.

Beveridge, I., Pullman, A.L., Martin, R.R. and Barelds, A. (1989) Effects of temperature and relative humidity on development and survival of the free-living stages of *Trichostrongylus colubriformis, T. rugatus* and *T. vitrinus. Veterinary Parasitology* 33, 143–153.

Bhaibulaya, M. (1968) A new species of *Angiostrongylus* in an Australian rat, *Rattus fuscipes. Parasitology* 58, 789–799.

Bhaibulaya, M. (1975) Comparative studies on the life history of *Angiostrongylus mackerrasae* Bhaibulaya, 1968 and *Angiostrongylus cantonensis* (Chen, 1935). *International Journal for Parasitology* 5, 7–20.

Bhaibulaya, M. and Cross, J. (1971) *Angiostrongylus malaysiensis* (Nematoda: Metastrongylidae), a new species of rat lungworm from Malaysia. *Southeast Asian Journal of Tropical Medicine and Public Health* 2, 527–533.

Bhalerao, G.D. (1932) On some nematode parasites of goats and sheep of Muktesar. *Indian Journal of Veterinary Science* 2, 242–254.

Bhalerao, G.D. (1945) Interesting mode of the life-cycle of the lungworm *Varestrongylus pneumonicus. Current Science* 4, 106–107.

Bhattacharyya, H.M., Sinha, P.K. and Sarkar, P.B. (1971) Studies on the incidence of *Metastrongylus salmi* in West Bengal with observations on its life cycle. *Indian Veterinary Journal* 48, 993–996.

Bienioschek, S., Rehbein, S. and Ribbeck, R. (1996) Cross-infections between fallow deer and domestic ruminants with large lungworms (*Dictyocaulus* spp.). *Applied Parasitology* 37, 229–238.

Biocca, E. (1951) On *Ancylostoma braziliense* (de Faria, 1910) and its morphological differentiation from *A. ceylanicum* (Looss, 1911). *Journal of Helminthology* 25, 1–10.

Biocca, E. (1954) Ridesonizione di *Ancylostoma tubaeforme* (Zeder, 1800) parasitta del gatto, considerato erroneamente sinonimo di *Ancylostoma caninum* (Ercolani, 1859) parassita del cane. *Revista de Parassitologia* 15, 267–278.

Bizzell, W.E. and Ciordia, H. (1965a) The course of uncomplicated experimental infections with *Trichostrongylus colubriformis* in calves. *Journal of Parasitology* 51, 174.

Bizzell, W.E. and Ciordia, H. (1965b) Dissemination of infective larvae of trichostrongylid parasites of ruminants from feces to pasture by the fungus, *Pilobolus* spp. *Journal of Parasitology* 51, 184.

Blackie, W.K. (1932) A helminthological survey of southern Rhodesia. *Memoir Series, No. 5, London School of Hygiene and Tropical Medicine*, p. 91.

Blanchard, J.L., Hargis, A.M. and Prieur, D.J. (1985) Scanning electron microscopy of *Ollulanus tricuspis* (Nematoda). *Proceedings of the Helminthological Society of Washington* 52, 315–317.

Blitz, N.M. and Gibbs, H.C. (1971a) An observation on the maturation of arrested *Haemonchus contortus* larvae in sheep. *Canadian Journal of Comparative Medicine* 35, 178–180.

Blitz, N.M. and Gibbs, H.C. (1971b) Morphological characterization of the stage of arrested development of *Haemonchus contortus* in sheep. *Canadian Journal of Zoology* 49, 991–995.

Blitz, N.M. and Gibbs, H.C. (1972a) Studies on the arrested development of *Haemonchus contortus* in sheep. I. The induction of arrested development. *International Journal for Parasitology* 2, 5–12.

Blitz, N.M. and Gibbs, H.C. (1972b) Studies on the arrested development of *Haemonchus contortus* in sheep. II. Termination of arrested development and the spring rise phenomenon. *International Journal for Parasitology* 2, 13–22.

Boag, B. (1987) The helminth parasites of the wild rabbit *Oryctolagus cuniculus* and the brown hare *Lepus capensis* from the Isle of Coll, Scotland. *Journal of Zoology* 212, 352–355.

Boag, B. and Kolb, H.H. (1989) Influence of host age and sex on nematode populations in the wild rabbit (*Oryctolagus cuniculus* L.). *Proceedings of the Helminthological Society of Washington* 56, 116–119.

Boag, D.A. and Wishart, W.D. (1982) Distribution and abundance of terrestrial gastropods on a winter range of bighorn sheep in southwestern Alberta. *Canadian Journal of Zoology* 60, 2633–2640.

Boardman, E.T. (1933) A comparative study of the behaviour of the preparasitic larvae of four bursate nematodes. Unpublished dissertation, Welch Medical Library, Johns Hopkins University, Baltimore, Maryland.

Boch, J. (1956) Knötchenwurmbefall (*Ternidens deminutus*) bei Rhesusaffen. *Zeitschrift für Angewandte Zoologie* No. 2, pp. 207–214.

Boch, J., Gerber, H.C. and Hörchner, F. (1968) Ein Beitrag zur Kenntnis der Hyostrongylose des Schweines. *Berliner und Münchener Tierärztliche Wochenschrift* 81, 145–148.

Boev, S.N. (1940) Helminthiasis of lungs of sheep of eastern Kazakhstan. *Trudi Kazakhskogo Nauchno-Issledovatel'skogo Veterinarnogo Instituta, Kazgosizdat, Alma-Ata* 4, 283–302. (In Russian.)

Boev, S.N. (1952) Lung nematodes and nematode diseases of ruminants of Kazakhstan. Unpublished PhD thesis, Alma Ata, USSR. (In Russian.)

Boev, S.N. (1975) *Protostrongylids. Fundamentals of Nematology*, Vol. 25. Academy of Science of the USSR. (Translated by Amerind Publishing Co., New Delhi, 1984.)

Boev, S.N. and Vol'f, Z.V. (1940) Cycle of development of lung helminths from the subfamily Synthetocaulinae Skrj. 1932 from sheep of Kazakhstan. *Trudi Kazakhskogo Nauchno-Issledovatel'skogo Veterinarnogo Instituta, Kazgosizdat, Alma-ata* 4, 250–260. (In Russian.)

Bolt, G., Monrad, J., Frandsen, F., Henriksen, P. and Bietz, H.H. (1993) The common frog (*Rana temporaria*) as a potential paratenic and intermediate host for *Angiostrongylus vasorum*. *Parasitology Research* 79, 428–430.

Bolt, G., Monrad, J., Koch, J. and Jensen, A.L. (1994) Canine angiostrongylosis: a review. *Veterinary Record* 135, 447–452.

Boon, J.H., Cremers, H.J.W.M., Hendriks, J. and Vliet, G. van (1983) Lungworm infection in cattle, a persistent problem? *Tijdschrift voor Diergeneeskune* 108, 435–438.

Boulenger, C.L. (1915) The life history of *Nematodirus filicollis* Rud., a nematode parasite of the sheep's intestine. *Parasitology* 8, 133–155.

Brack, M. (1976) Pathologisch-Anatomische Aspekte des Befalls mit *Molineus torulosus* oder *Filaroides cebus* beim Kapuziner (*Cebus apella*). In: Ippen, R. and Schröder, H.D. (eds) *Erkrankungen der Zootiere, Verhandlungsbericht des XVIII Internationalen Symposiums über die Erkrankungen der Zootiere, 16–20 June, 1976, Innsbruck, Berlin, Germany (DDR)*. Akademie-Verlag, pp. 213–218.

Brack, M. and Schröpel, M. (1995) *Angiostrongylus costaricensis* in a black-eared marmoset. *Tropical and Geographical Medicine* 47, 136–138.

Brack, M., Myers, B.J. and Kuntz, R.E. (1973) Pathogenic properties of *Molineus torulosus* in capuchin monkeys, *Cebus apella*. *Laboratory Animal Science* 23, 360–365.

Bremner, K.C. (1956) The parasitic life-cycle of *Haemonchus placei* (Place) Ransom (Nematoda: Trichostrongylidae). *Australian Journal of Zoology* 4, 146–151.

Brockelman, C.R., Chusatayanond, W. and Baidikul, V. (1976) Growth and localization of *Angiostrongylus cantonensis* in the molluscan host, *Achatina fulica*. *Southeast Asian Journal of Tropical Medicine and Public Health* 7, 30–37.

Broekhuizen, S. and Kemmers, R. (1976) The stomach worm, *Graphidium strigosum* (Dujardin) Railliet and Henry, in the European hare, *Lepus europaeus* Pallas. In: Pielowski, Z. and Pucek, Z. (eds) *Ecology and Management of European Hare Populations*. Państwowe Wydawnictwo Rolnicze i Leśne, Warsaw, Poland, pp. 157–171.

Brown, E.D., Macdonald, D.W., Tew, T.E. and Todd, I.A. (1994a) *Apodemus sylvaticus* infected with *Heligmosomoides polygyrus* (Nematoda) in an arable ecosystem: epidemiology and effects of infection on the movements of male mice. *Journal of Zoology* 234, 623–640.

Brown, E.D., Macdonald, D.W., Tew, T.E. and Todd, I.A. (1994b) Rhythmicity of egg production by *Heligmosomoides polygyrus* in wild wood mice, *Apodemus sylvaticus. Journal of Helminthology* 68, 105–108.

Brown, T.T., Jordon, H.E. and Demorest, C.N. (1978) Cerebrospinal parelaphostrongylosis in llamas. *Journal of Wildlife Diseases* 14, 441–444.

Brumpt, L.C. (1952) Deductions cliniques tirées de cinquante cas de l'ankylostomose provoquée. *Annales de Parasitologie Humaine et Comparée* 27, 237–249.

Brumpt, L.C. (1958) Ankylostomose. *Revue de Praticien* 8, 279–289.

Brunetti, O.A. (1969) Redescription of *Parelaphostrongylus* (Boev and Schul'ts, 1950) in California deer, with studies on its life history and pathology. *California Fish and Game* 55, 307–316.

Bruni, A. and Passalacqua, A. (1954) Sulla presenza di una mesomucinasi (jaluronidasi) in *Ancylostoma duodenale. Bollettino della Società Italiana di Biologia Sperimentale* 30, 789–791.

Bryant, V. (1973) The life cycle of *Nematospiroides dubius*, Baylis, 1926 (Nematoda: Heligmosomidae). *Journal of Helminthology* 46, 263–268.

Burrows, R.B. (1962) Comparative morphology of *Ancylostoma tubaeforme* (Zeder, 1800) and *Ancylostoma caninum* (Ercolani, 1859). *Journal of Parasitology* 48, 715–718.

Buckley, J.J.C. (1934) On *Syngamus ierei* sp.nov. from domestic cats, with some observations on its life-cycle. *Journal of Helminthology* 12, 89–98.

Cameron, T.W.M. (1923a) On the morphology of *Ollulanus tricuspis* Leuckart, 1865, a nematode parasite of the cat. *Journal of Helminthology* 1, 157–160.

Cameron, T.W.M. (1923b) On the biology of the infective larva of *Monodontus trigonocephalum* (Rud.) of sheep. *Journal of Helminthology* 1, 205–214.

Cameron, T.W.M. (1926) On the morphology of the adults and the free-living larvae of *Dictyocaulus arnfieldi*, the lungworm of equines. *Journal of Helminthology* 4, 61–68.

Cameron, T.W.M. (1927a) Observations on the life history of *Aelurostrongylus abstrusus* (Railliet), the lungworm in cats. *Journal of Helminthology* 2, 55–66.

Cameron, T.W.M. (1927b) Observations on the life history of *Ollulanus tricuspis* Leuck., the stomach worm of the cat. *Journal of Helminthology* 5, 67–80.

Cameron, T.W.M. (1927c) On the parasitic development of *Monodontus trigonocephalum*, the sheep hookworm. *Journal of Helminthology* 5, 149–162.

Carlisle, M.S., Prociv, P., Genman, J., Pass, M.A., Campbell, G.L. and Mudie, A. (1998) Cerebrospinal angiostrongyliasis in five captive tamarans (*Sanguinus* spp.). *Australian Veterinary Journal* 76, 167–170.

Cassone, J., Vuong, P.N. and Durette-Desset, M.C. (1992) Life cycle of *Paralibyostrongylus hebrenicutus* (Nematoda: Trichostrongylidae). *Annales de Parasitologie Humaine et Comparée* 67, 33–41.

Castanon-Ordonez, L. (1982) Ciclo interno de *Neostrongylus linearis* (Marotel, 1913) Gebauer, 1932 (Nematoda, Protostrongylidae) en la oveja. *Hygia Pecoris* 4, 21–31.

Chabaud, A.G. (1974) No. 1. Keys to subclasses, orders and superfamilies. In: Anderson, R.C., Chabaud, A.G. and Willmott, S. (eds) *CIH Keys to the Nematode Parasites of Vertebrates*. Commonwealth Agricultural Bureaux, Farnham Royal, UK, pp. 6–17.

Chandler, A.C. (1929) *Hookworm Disease – Its Distribution, Biology, Epidemiology, Pathology, Diagnosis, Treatment and Control*. MacMillan Press, New York.

Chandler, A.C. (1932) Experiments on resistance of rats to superinfection with the nematode, *Nippostrongylus muris. American Journal of Hygiene* 16, 750–782.

Chang, P.K., Cross, J.H. and Chen, S.S.S. (1968) Aquatic snails as intermediate hosts for *Angiostrongylus cantonensis* on Taiwan. *Journal of Parasitology* 54, 182–183.

Cheatum, E.L. (1948) A contribution on the life cycle of the lungworm, *Leptostrongylus alpenae* (Nematoda: Metastrongyloidea) with observations on its incidence and biology. Unpublished PhD thesis, University of Michigan, Ann Arbor, Michigan.

Cheng, T.C. (1966) Perivascular leucocytosis and other types of cellular reactions in the oyster *Crassostrea virginica* experimentally infected with the nematode *Angiostrongylus cantonensis*. *Journal of Invertebrate Pathology* 8, 52–58.

Cheng, T.C. and Alicata, J.E. (1965) On the modes of infection of *Achatina fulica* by the larvae of *Angiostrongylus cantonensis*. *Malacologia* 2, 267–274.

Cheng, T.C. and Burton, R.W. (1965) The American oyster and clam as experimental intermediate hosts of *Angiostrongylus cantonensis*. *Journal of Parasitology* 51, 296.

Chiu, J.K. (1964) Snail hosts of *Angiostrongylus cantonensis* in Taipei, Taiwan. *Bulletin of the Institute of Zoology, Academia Sinica* 3, 55–62.

Christensen, C.M. (1997) The effect of three distinct sex ratios at two *Oesophagostomum dentatum* worm populations densities. *Journal of Parasitology* 83, 636–640.

Christensen, C.M., Barnes, E.H. Nansen, P, Roepstorff, A. and Slotved, H.C. (1995) Experimental *Oesophagostomum dentatum* infection in the pig: worm populations resulting from single infections with three doses of larvae. *International Journal for Parasitology* 25, 1491–1498.

Christensen, C.M., Barnes, E.H., Nansen, P. and Grondahl-Nielson, C. (1996) Growth and fecundity of *Oesophagostomum dentatum* in high-level infections and after transplantation into naive pigs. *Parasitology Research* 82, 364–368.

Christensen, C.M., Barnes, E.H. and Nansen, P. (1997) Experimental *Oesophagostomum dentatum* infections in the pig: worm populations at regular intervals during trickle infections with three dose levels of larvae. *Parasitology* 115, 545–552.

Ciordia, H. and Bizzell, W.E. (1963) The effects of various constant temperatures on the development of the free-living stages of some nematode parasites of cattle. *Journal of Parasitology* 49, 60–63.

Clapham, P.A. (1934) Experimental studies on the transmission of gapeworm (*Syngamus trachea*) by earthworms. *Proceedings of the Royal Society of Medicine, London, Series B* 115, 18–29.

Clapham, P.A. (1935a) On the experimental transmission of *Syngamus trachea* from starlings to chickens. *Journal of Helminthology* 13, 1–2.

Clapham, P.A. (1935b) Some helminth parasites from partridges and other English birds. *Journal of Helminthology* 13, 139–148.

Clapham, P.A. (1939a) Three new intermediary vectors for *Syngamus trachea*. *Journal of Helminthology* 17, 191–194.

Clapham, P.A. (1939b) On flies as intermediate hosts of *Syngamus trachea*. *Journal of Helminthology* 17, 61–64.

Clapham, P.A. (1962) Pre-partum infestation of puppies with *Ancylostoma caninum*. (Correspondence). *Veterinary Record* 74, 754–755.

Clarke, K.R. (1967) The migration route of the third-stage larvae of *Nippostrongylus brasiliensis* (Travassos, 1914). *Journal of Helminthology* 41, 285–290.

Clarke, K.R. (1968) The migration route of the third stage larvae of *Nippostrongylus brasiliensis* (Travassos, 1914). *Acta Leidensia* 36, 62–66.

Collins, G.H. and Charleston, W.A.G. (1972) *Ollulanus tricuspis* and *Capillaria putorii* in New Zealand cats (Correspondence). *New Zealand Veterinary Journal* 20, 82.

Conder, G.A., Johnson, S.S., Hall, A.D., Fleming, N.W., Mills, M.D. and Guimond, P.M. (1992) Growth and development of *Haemonchus contortus* in jirds, *Meriones unguiculatus*. *Journal of Parasitology* 78, 492–497.

Conlogue, C.T., Ogden, J.A. and Foreyt, W.J. (1985) Parasites of dall's porpoise (*Phocaenoides dalli* True). *Journal of Wildlife Diseases* 21, 160–166.

Connan, R.M. (1968) The post-parturient rise in fecal nematode egg counts of ewes: its aetiology and epidemiological significance. *World Review of Animal Production* 4, 53–58.

Connan, R.M. (1975) Inhibited development in *Haemonchus contortus*. *Parasitology* 71, 239–246.

Connan, R.M. (1978) Arrested development in *Haemonchus contortus*. In: *Facts and Reflections. III*. Workshop on 'Arrested development of nematodes in sheep and cattle', Central Veterinary Institute, Lelystad, The Netherlands, pp. 53–62.

Connan, R.M. and Wise, D.R. (1994) Further studies on the development and survival at low temperatures of the free living stages of *Trichostrongylus tenuis*. *Research in Veterinary Science* 57, 215–219.

Conradi, A.F. and Barnette, E. (1908) Hookworm disease of cattle (Uncinariasis). *South Carolina Agricultural Experimental Station Bulletin* No. 137, 23 pp.

Courdurier, J., Guilhon, J.C., Malarde, L., Laigret, J., Desmoulins, G. and Schollhammer, G. (1964) Réalisation du cycle d'*Angiostrongylus cantonensis* au laboratoire. *Bulletin de la Société de Pathologie Exotique* 57, 1255–1262.

Cowan, A.B. (1955) Some preliminary observations on the life history of *Amidostomum anseris* Zeder, 1800. *Journal of Parasitology* 41, 43.

Cowan, A.B. and Herman, C.M. (1955) Winter losses of Canada geese at Pea Island, North Carolina. *Proceedings of the Southeastern Association of Game and Fish Commissioners Meeting, Daytona Beach, Florida, October 2–5, 1955*, pp. 172–174.

Coyne, M.J. and Smith, G. (1992) The development and mortality of the free-living stages of *Haemonchus contortus* in laboratory culture. *International Journal of Parasitology* 22, 641–658.

Craig, R.E. (1972) Lungworms (Nematoda: Metastrongyloidea) of striped skunk (*Mephitis mephitis*). Unpublished MS thesis, University of Guelph, Guelph, Ontario.

Cram, E.B. (1927) Bird parasites of the nematode suborders Strongylata, Ascaridata and Spirurata. *USA National Museum Bulletin* No. 140.

Cram, E.B. (1933) Ornithostrongylosis, a parasitic disease of pigeons. *The Bureau Veterinarian* 9, 1 and 3.

Cram, E.B. and Cuvillier, E. (1931) *Ornithostrongylus quadriradiatus* of pigeons; observations on its life history, pathogenicity and treatment. *Journal of Parasitology* 17, 116.

Cram, E.B. and Cuvillier, E. (1934) Observations on *Trichostrongylus tenuis* infestation in domestic and game birds in the USA. *Parasitology* 26, 340–345.

Cram, E.B. and Wehr, E.E. (1934) The status of species of *Trichostrongylus* of birds. *Parasitology* 26, 335–339.

Cremers, H.J.W.M. and Eysker, M. (1975) Experimental induction of inhibited development in a strain of *Haemonchus contortus* in the Netherlands. *Tropical and Geographical Medicine* 27, 229.

Crofton, H.D. (1946a) The ecology of immature phases of trichostrongyle nematodes. I. The vertical distribution of infective larvae of *Trichostrongylus retortaeformis* in relation to their habitat. *Parasitology* 39, 17–25.

Crofton, H.D. (1946b) The ecology of immature phases of trichostrongyle nematodes. II. The effect of climatic factors on the availability of the infective larvae of *Trichostrongylus retortaeformis* to the host. *Parasitology* 39, 26–38.

Crofton, H.D. (1963) Nematode parasite populations in sheep and on pasture. *Technical Communication* No. 35, Commonwealth Bureau of Helminthology, Farnham Royal, UK, 104 pp.

Crofton, H.D. (1965) Ecology and biological plasticity of sheep nematodes. I. The effect of temperature on the hatching of eggs of some nematode parasites of sheep. *Cornell Veterinarian* 55, 242–250.

Crofton, H.D. and Whitlock, J.H. (1965) Ecology and biological plasticity of sheep nematodes. III. Studies on *Ostertagia circumcincta* (Stadelmann, 1894). *Cornell Veterinarian* 55, 259–262.

Cuille, J.M., Marotel, G. and Roquet, M.I.C. (1913) Nouvelle et grave entérite vermineuse du cheval; la cylicostomose larvaire. *Bulletin de la Société Vétérinaire de Lyon* 16, 172–184.

Cunningham, F.C. (1956) A comparative study of tropisms exhibited by different stages of *Nippostrongylus muris* (Nematoda: Trichostrongylidae). *Catholic University of America, Biological Studies* 36, 1–24.

Cuvillier, E. (1937) The nematode, *Ornithostrongylus quadriradiatus*, a parasite of the domesticated pigeon. *USA Department of Agriculture Technical Bulletin* No. 569, 1–36.

Czaplinski, B. (1962) Nematodes and acanthocephalans of domestic and wild Anseriformes in Poland. I. Revision of the genus *Amidostomum* Railliet and Henry, 1909. *Acta Parasitologica Polonica* 10, 125–164.

Dailey, M.D. (1970) The transmission of *Parafilaroides decorus* (Nematoda: Metastrongyloidea) in the California sea-lion (*Zalophus californianus*). *Proceedings of the Helminthological Society of Washington* 37, 215–222.

Dailey, M.D., Odell, D.K. and Walsh, M.T. (1990) Transmission of lungworms (Nematoda: Pseudaliidae) in the cetacean *Tursiops truncatus*. *Bulletin de la Société Francaise de Parasitologie, Seventh International Congress of Parasitology (Abstract), Paris, Aug. 20–24*, p. 288.

Dailey, M., Walsh, M., Odell, D. and Campbell, T. (1991) Evidence of prenatal infection in the bottlenose dolphin (*Tursiops truncatus*) with the lungworm *Halocercus lagenorhynchi* (Nematoda: Pseudaliidae). *Journal of Wildlife Diseases* 27, 164–165.

Daoning, L., Guiyan, L. and Ximei, Z. (1997) Three cases of *Mammomonogamus laryngeus* infections. *Chinese Journal of Parasitology and Parasitic Diseases* 15, 281–284.

Daoning, L., Zhuaya, L., Ai He, Y. and Wang-Ximei, Z. (1998) First case of infections by *Mammomonogamus gangquiensis*. *Academic Journal of Sun Yat Sen University Medical School* 19, 246–249.

Dash, K.M. (1973) The life cycle of *Oesophagostomum columbianum* (Curtice, 1890) in sheep. *International Journal for Parasitology* 3, 843–851.

Daubney, R. (1920) The life-histories of *Dictyocaulus filaria* (Rud.), and *Dictyocaulus viviparus* (Bloch). *Journal of Comparative Pathology and Therapeutics* 33, 225–266.

Dauphine, T.C. (1975) The disappearance of caribou reintroduced to Cape Breton Highlands National Park. *Canadian Field Naturalist* 89, 299–310.

Davidson, W.R., Doster, G.L. and Freeman, R.C. (1996) *Parelaphostrongylus tenuis* on Wassaw Island, Georgia: a result of translocating white-tailed deer. *Journal of Wildlife Diseases* 32, 701–703.

Davtyan, E.A. (1948) Susceptibility of molluscs to larval infections of some lung nematodes of sheep and goats. *Trudi Erevanskogo Zooveterinarnogo Instituta* 10, 105–119. (In Russian.)

Davtyan, E.A. (1949) Cycle of development of lung nematodes of sheep and goats of Armenia. *Zoologicheskii Sbornik, Akademiya Nauk Armyanskoi SSR, Erevan, USSR* 6, 185–266. (In Russian.)

Dayton, D.A. (1957) The earthworm *Eisenia lonnbergi*, a new intermediate host for swine lungworm. *Journal of Parasitology* 43, 282.

Delahay, R.J., Speakman, J.R. and Moss, R. (1995) The energetic consequences of parasitism: effects of a developing infection of *Trichostrongylus tenuis* (Nematoda) on red grouse (*Lagopus lagopus scoticus*) energy balance, body weight and condition. *Parasitology* 110, 473–482.

Deorani, V.P.S. (1965a) Studies on the embryonation of eggs of *Varestrongylus pneumonicus* (Protostrongylidae: Nematoda); and guinea pig infection with larvae from terrestrial snails. *Indian Journal of Helminthology* 17, 85–88.

Deorani, V.P.S. (1965b) Studies on the life-history and bionomics of *Dictyocaulus filaria* (Rudolphi, 1809) (Metastrongylidae: Nematoda). *Indian Journal of Helminthology* 17, 89–103.

Dikmans, G. and Andrews, J.S. (1933) A comparative morphological study of the infective larvae of the common nematodes parasitic in the alimentary tract of sheep. *Transactions of the American Microscopical Society* 52, 1–25.

Dikmans, G. (1935) Two new lungworms, *Protostrongylus coburni* n.sp. and *Pneumostrongylus alpenae* n.sp. from the deer, *Odocoileus virginianus* in Michigan. *Transactions of the American Microscopical Society* 54, 138–144.

Dineen, J.K. (1978) The nature and role of immunological control in gastrointestinal helminthiasis. In: Donald, A.D., Southcott, H. and Dineen, J.K. (eds) *The Epidemiology and Control of Gastrointestinal Parasites of Sheep in Australia*. Commonwealth Scientific and Industrial Research Organization, Melbourne, Australia, pp. 121–135.

Dineen, J.K. and Kelly, J.D. (1973) Immunological unresponsiveness of neonatal rats to infection with *Nippostrongylus brasiliensis*. The competence of neonatal lymphoid cells in worm expulsion. *Immunology* 25, 141–150.

Dineen, J.K., Donald, A.D., Wagland, B.M. and Offner, J. (1965a) The dynamics of the host–parasite relationship. III. The response of sheep to primary infection with *Haemonchus contortus. Parasitology* 55, 515–525.

Dineen, J.K., Donald, A.D., Wagland, B.M. and Turner, J.H. (1965b) The dynamics of the host–parasite relationship. II. The response of sheep to primary and secondary infection with *Nematodirus spathiger* infection in sheep. *Parasitology* 55, 163–171.

Dobson, C. (1962) Certain aspects of the host–parasite relationship of *Nematospiroides dubius* (Baylis). *Parasitology* 52, 41–48.

Dobson, C. (1966) The distribution of *Oesophagostomum columbianum* along the alimentary tract of sheep. *Australian Journal of Agricultural Research* 17, 765–777.

Dobson, A. and Hudson, P. (1995) The interaction between the parasites and predators of red grouse *Lagopus lagopus scoticus. Ibis (London)* 137, S 87–96.

Donald, A.D., Dineen, J.K., Turner, J.H. and Wagland, B.M. (1964) The dynamics of the host–parasite relationship. I. *Nematodirus spathiger* infection in sheep. *Parasitology* 54, 527–544.

Doncaster, C.C. (1981) Observations on relationships between infective juveniles of bovine lungworm, *Dictyocaulus viviparus* (Nematoda: Strongylidae) and the fungi, *Pilobolus kleinii* and *P. crystallinus* (Zygomycotina: Zygomycetes). *Parasitology* 82, 421–428.

Dorrington, J.E. (1964) Lungworm in dogs in the western Cape Province. *Journal of the South African Veterinary Medical Association* 35, 499–501.

Dorrington, J.E. (1965) Preliminary report on the transmission of *Filaroides osleri* (Cobbold, 1879) in dogs. *Journal of the South African Veterinary Medical Association* 36, 389.

Dorrington, J.E. (1968) Studies on *Filaroides osleri* infestation in dogs. *Onderstepoort Journal of Veterinary Research* 35, 225–285.

Dougherty, E.C. (1945) The nematode lungworms (suborder Strongylina) of North American deer of the genus *Odocoileus. Parasitology* 36, 199–208.

Douvres, F.W. (1956) Morphogenesis of the parasitic stages of *Ostertagia ostertagi*, a nematode parasite in cattle. *Journal of Parasitology* 42, 626–635.

Douvres, F.W. (1957) The morphogenesis of the parasitic stages of *Trichostrongylus axei* and *Trichostrongylus colubriformis*, nematode parasites of cattle. *Proceedings of the Helminthological Society of Washington* 24, 4–14.

Dove, W.E. (1932) Further studies on *Ancylostoma braziliense* and the etiology of creeping eruption. *American Journal of Hygiene* 15, 664–712.

Drozdz, J. and Doby, J.M. (1970) Evolution morphologique, migrations et chronologie du cycle de *Angiostrongylus* (*Parastrongylus*) *dujardini* Drozdz et Doby, 1970 (Nematoda: Metastrongyloidea) chez ses hôtes-définitifs. *Bulletin de la Société Scientifique de Bretagne* 45, 229–239.

Drozdz, J., Doby, J.M. and Mandahl-Barth, G. (1971) Etude des morphologie et évolution larvaires de *Angiostrongylus (Parastrongylus) dujardini* Drozdz et Doby, 1970 Nematoda: Metastrongyloidea. Infestation des mollusques hôtes intermédiaires. *Annales de Parasitologie Humaine et Comparée* 46, 265–276.

Drudge, J.H., Lyons, E.T. and Szanto, J. (1966) Pathogenesis of migrating stages of helminths, with special references to *Strongylus vulgaris*. In: Soulsby, E.J.L. (ed.) *Biology of Parasites. Emphasis on Veterinary Parasites*. Academic Press, New York and London, pp. 199–214.

Dubnitski, A.A. (1955) Intermediate hosts in the life cycle of development of the nematode *Filaroides bronchialis*. *Karakulevodstvo i Zverovodstvo* 8, 51–52. (In Russian.)

Dubnitski, A.A. (1956) Studies on the life cycle of the nematode *Skrjabingylus nasicola*, a parasite affecting the frontal sinuses of fur-bearers of the marten family. *Karakulevodstvo i Zverovodstvo* 9: 59–61. (In Russian.)

Dudzinski, M.L. and Mykytowycz, R. (1963) Relationship between sex and age of rabbits, *Oryctolagus cuniculus* (L.) and infection with nematodes *Trichostrongylus retortaeformis* and *Graphidium strigosum*. *Journal of Parasitology* 49, 55–59.

Duncan, J.L. (1973) The life cycle, pathogenesis and epidemiology of *S. vulgaris* in the horse. *Equine Veterinary Journal* 5, 20–25.

Duncan, N. (1976) Theoretical aspects concerning transmission of the parasite *Skrjabingylus nasicola* (Leuckart, 1842) to stoats and weasels, with a review of the literature. *Mammal Review* 6, 63–74.

Duncan, J.L. and Pirie, H.M. (1972) The life cycle of *Strongylus vulgaris* in the horse. *Research in Veterinary Science* 13, 374–379.

Duncan, R.B. and Patton, S. (1998) Natural occurring cerebrospinal parelaphostrongylosis in a heifer. *Journal of Veterinary Investigation* 10, 287–291.

Dunn, D.R. (1955) The culture of earthworms and their infection with *Metastrongylus* species. *British Veterinary Journal* 111, 97–101.

Dunn, D.R. (1956) The pig lungworm (*Metastrongylus* spp.). II. Experimental infection of pigs with *M. apri*. *British Veterinary Journal* 112, 327–337.

Dunsmore, J.D. (1960) Retarded development of *Ostertagia* species in sheep. *Nature, London* 186, 986–987.

Dunsmore, J.D. (1961) Effect of whole body irradiation and cortisone on the development of *Ostertagia* spp. in sheep. *Nature, London* 192, 139–140.

Dunsmore, J.D. (1963) Effect of removal of an adult population of *Ostertagia* from sheep on concurrently existing arrested larvae. *Australian Veterinary Journal* 39, 459–463.

Dunsmore, J.D. (1965) *Ostertagia* spp. in lambs and pregnant ewes. *Journal of Helminthology* 39, 159–184.

Dunsmore, J.D. (1966) Nematode parasites of free-living rabbits, *Oryctolagus cuniculus* (L.), in eastern Australia. I. Variations in the number of *Trichostrongylus retortaeformis* (Zeder). *Australian Journal of Zoology* 14, 185–199.

Dunsmore, J.D. and Dudzinski, M.L. (1968) Relationship of numbers of nematode parasites in wild rabbits, *Oryctolagus cuniculus* (L.), to host sex, age, and season. *Journal of Parasitology* 54, 462–474.

Dunsmore, J.D. and Spratt, D.M. (1979) The life history of *Filaroides osleri* in wild and domestic canids in Australia. *Veterinary Parasitology* 5, 275–286.

Durette-Desset, M.C. (1971) Essai de classification des nématodes héligmosomes. Corrélation avec la paléobiogéographie des hôtes. *Mémoires du Muséum National d'Histoire Naturelle Nouvelle Série A Zoologie* 49, 1–126.

Durette-Desset, M.C. (1972) Compléments morphologique à l'étude de quelques nématodes héligmosomes, parasites de rongeur américains. *Annales de Parasitologie Humaine et Comparée* 47, 243–249.

Durette-Desset, M.C. (1983) No. 10. Keys to the genera of the superfamily Trichostrongyloidea. In: Anderson, R.C. and Chabaud, A.G. (eds) *CIH Keys to the Nematode Parasites of Vertebrates.* Commonwealth Agricultural Bureaux, Farnham Royal, UK, pp. 1–68.

Durette-Desset, M.C. and Cassone, J. (1986) Sur deux nématodes trichostrongyloides parasites d'un muridé Africain. I. Description des adultes. *Annales de Parasitologie Humaine et Comparée* 61, 565–574.

Durette-Desset, M.C. and Cassone, J. (1987a) Sur deux nématodes trichostrongyloides parasites d'un muridé Africain. II. Chronologie des cycles, description des stades larvaires et des immatures. *Annales de Parasitologie Humaine et Comparée* 62, 133–158.

Durette-Desset, M.C. and Cassone, J. (1987b) Sur deux nématodes trichostrongyloides parasites d'un muridé Africain. III. Dualité physiologique des larves infestantes, liée aux rythmes saisonniers. *Annales de Parasitologie Humaine et Comparée* 62, 577–589.

Durette-Desset, M.C. and Chabaud, A.G. (1977) Essai de classification des nématodes Trichostrongyloidea. *Annales de Parasitologie Humaine et Comparée* 52, 539–558.

Durette-Desset, M.C. and Chabaud, A.G. (1981) Nouvel essai de classification de nématode Trichostrongyloidea. *Annales de Parasitologie Humaine et Comparée* 56, 297–312.

Durette-Desset, M.C., Chabaud, A.G. and Moore, J. (1993) *Trichostrongylus cramae* n.sp. (Nematoda), a parasite of bobwhite quail (*Colinus virginianus*). *Annales de Parasitologie Humaine et Comparée* 68, 43–48.

Durette-Desset, M.C., Hugonnet, L. and Chabaud, A.G. (1988) Redescription de *Dictyocaulus noerneri* Railliet et Henry, 1907 parasite de *Capreolus capreolus.* Comparison avec *D. viviparus* (Bloch, 1782). *Annales de Parasitologie Humaine et Comparée* 63, 285–295.

Durette-Desset, M.C., Vuong, P.N. and Cassone, J. (1989) Sur deux nématodes trichostrongyloides parasites d'un muridé Africain. IV. Migrations chez le vertébré. *Annales de Parasitologie Humaine et Comparée* 64, 120–133.

Düwel, D. and Schleich, H. (1971) Das Meerschweinchen – ein Modellwirt für Metastrongyliden. *Berliner und Münchener Tierärztliche Wochenschrift* 84, 405–408.

Egorov, Y.G. (1960) Biology of *Muellerius capillaris. Trudi nauchno-Issledovatelskogo Veterinarnogo Instituta, Minsk* 1, 160–170. (In Russian.)

Ehlert, H. (1962) Zur Kenntnis des Hakenwurmes des Schweines *Globocephalus urosubulatus* (Alessandrini, 1909) Cameron, 1924. Dissertation, Giessen, 87 pp.

Ehrenford, F.A. (1954) The life cycle of *Nematospiroides dubius* Baylis (Nematoda: Heligmosomidae). *Journal of Parasitology* 40, 480–481.

Elton, C., Ford, E.B., Baker, J.R. and Gardner, A.D. (1931) The health and parasites of wild mouse population. *Proceedings of the Zoological Society of London,* pp. 657–721.

Enigk, K. (1950) Zur Entwicklung von *Strongylus vulgaris* (Nematodes) im Wirtstiere. *Zeitschrift für Tropenmedizin und Parasitologie* 2, 287–306.

Enigk, K. (1951) Weitere Untersuchungen zur Biologie von *Strongylus vulgaris* (Nematodes) im Wirtstiere. *Zeitschrift für Tropenmedizin und Parasitologie* 2, 523–535. (English version in *Cornell Veterinarian* 63, 223–246.)

Enigk, K. (1970) The development of the three species of *Strongylus* of the horse during the prepatent period. In: Bryans, J.T. and Gerber, H. (eds) *Equine Infectious Diseases,* Vol. 2. *Proceedings of the Second International Conference on Equine Infectious Diseases, Paris, 1969,* S. Karger, Basel, pp. 259–268 (Discussion pp. 323–325).

Enigk, K. and Dey-Hazra, A. (1968) Die perkutane Infektion bei *Amidostomum anseris* (Strongyloidea, Nematoda). *Zeitschrift für Parasitenkunde* 31, 155–165.

Enigk, K. and Dey-Hazra, A. (1969) Zum Verhalten der exogen Entwicklungsformen von *Amidostomum anseris* (Strongyloidea, Nematoda). *Archiv für Geflügelkunde* 33, 259–273.

Enigk, K. and Dey-Hazra, A. (1970a) Zur Biologie und Pathogenität von *Syngamus trachea* (Strongyloidea, Nematoda). *Deutsche Tierärztliche Wochenschrift* 77, 489–528.

Enigk, K. and Dey-Hazra, A. (1970b) The behaviour of exogenous stages of *Amidostomum anseris* (Strongyloidea, Nematoda). In: Singh, K.S. and Tandan, B.K. (eds) *H.D. Srivastava Commemoration Volume.* Indian Veterinary Research Institute, Izatnagar, UP, pp. 603–619.

Enigk, K. and Stoye, M. (1968) Untersuchungen über den Infektionsweg von *Ancylostoma caninum. Medizinische Klinik* 63, 1012–1017.

Epe, C., Samson-Himmelstjarna, T. and Schneider, T. (1997) Differences in a ribosomal DNA sequence of lungworm species (Nematoda: Dictyocaulidae) from fallow deer, cattle, sheep and donkeys. *Research in Veterinary Science* 62, 17–21.

Ershov, V.S. (1960) *Parasitology and Parasitic Diseases of Livestock.* Israel Program for Scientific Translations, Jerusalem.

Ershov, V.S. (1970) The biology of horse strongyles and the allergic reactions caused by them. In: Bryans, J.T. and Gerber, H. (eds) *Equine Infectious Diseases. Proceedings of the Second International Conference on Equine Infectious Diseases, Paris, 1969*, Vol. 2. S. Karger, Basel, pp. 304–309 (Discussion pp. 323–325).

Ewing, M.S., Ewing, S.A., Keener, M.S. and Mulholland, R.J. (1982) Mutualism among parasitic nematodes: a population model. *Ecological Modelling* 15, 353–366.

Ewing, S.A. and Todd, A.C. (1961a) Metastrongylosis in the field: species and sex ratios of the parasites, preferential location in the respiratory apparatus of the host, and concomitant lesions. *American Journal of Veterinary Research* 22, 606–609.

Ewing, S.A. and Todd, A.C. (1961b) Association among members of the genus *Metastrongylus* Molin, 1861 (Nematoda: Metastrongylidae). *American Journal of Veterinary Research* 22, 1077–1080.

Eysker, M. (1978) Inhibition of the development of *Trichostrongylus* spp. as third-stage larvae in sheep. *Veterinary Parasitology* 4, 29–33.

Eysker, M. (1981) Experiments on inhibited development of *Haemonchus contortus* and *Ostertagia circumcincta* in sheep in the Netherlands. *Research in Veterinary Science* 30, 62–65.

Fahmy, A.M. (1956) An investigation on the life cycle of *Nematospiroides dubius* (Nematoda: Heligmosomidae) with special reference to the free-living stages. *Zeitschrift für Parasitenkunde* 17, 394–399.

Faust, E.C. (1949) *Human Helminthology. A Manual for Physicians, Sanitarians and Medical Zoologists.* Lea and Febiger, Philadelphia.

Faust, E.C., Beaver, P.C. and Jung, R.C. (1975) *Animal Agents and Vectors of Human Disease.* Lea and Febiger, Philadelphia.

Fengyi, Q. (1997) *Mammomonogamus laryngeus* infection case occurred in Shanghai. *Chinese Journal of Parasitology and Parasitic Diseases* 15, 198–200.

Fernando, M.A., Stockdale, P.H.G. and Remmler, O. (1971a) The route of migration, development, and pathogenesis of *Syngamus trachea* (Montagu, 1811) Chapin, 1925, in pheasants. *Journal of Parasitology* 57, 107–116.

Fernando, M.A., Stockdale, P.H.G. and Ashton, G.C. (1971b) Factors contributing to the retardation of development of *Obeliscoides cuniculi* in rabbits. *Parasitology* 63, 21–29.

Fernando, M.A., Hoover, I.J. and Ogungbade, S.G. (1973) The migration and development of *Cyathostoma bronchialis* in geese. *Journal of Parasitology* 59, 759–764.

Foreyt, W. and Trainer, D.O. (1970) Experimental haemonchosis in white-tailed deer. *Journal of Wildlife Diseases* 6, 35–42.

Forrester, D.J. (1971) Bighorn sheep lungworm–pneumonia complex. In: Davis, J.W. and Anderson, R.C. (eds) *Parasitic Diseases of Wild Mammals.* Iowa State University Press, Ames, pp. 158–173.

Forrester, D.J. and Senger, C.M. (1964) Prenatal infection of bighorn sheep with protostrongylid lungworms. *Nature, London* 201, 1051.

Forrester, S.G. and Lankester, M.W. (1997a) Extracting protostrongylid nematode larvae from ungulate feces. *Journal of Wildlife Diseases* 33, 511–516.

Forrester, S.G. and Lankester, M.W. (1997b) Extracting *Protostrongylus* spp. larvae from bighorn sheep feces. *Journal of Wildlife Diseases* 33, 865–872.

Forrester, S.G. and Lankester, M.W. (1998) Overwinter survival of first-stage larvae of *Parelaphostrongylus tenuis* (Nematoda: Protostrongylidae). *Canadian Journal of Zoology* 76, 704–710.

Foster, A.O. (1932) Prenatal infection with the dog hookworm, *Ancylostoma caninum. Journal of Parasitology* 18, 112–118.

Foster, A.O. (1935) Further observations on prenatal hookworm infection of dogs. *Journal of Parasitology* 21, 302–308.

Foster, A.O. and Cort, W.W. (1932) The effect of a deficient diet on the susceptibility of dogs and cats to non-specific strains of hookworms. *American Journal of Hygiene* 16, 582–601.

Foster, A.O. and Cross, S.X. (1934) The direct development of hookworms after oral infection. *American Journal of Tropical Medicine* 14, 565–573.

Foster, A.O. and Daengsvang, S. (1932) Viability and rate of development of the eggs and larvae of the two physiological strains of the dog hookworm, *Ancylostoma caninum. Journal of Parasitology* 18, 245–251.

Fotedar, D.N. and Wali, N. (1982) On the development of juvenile stages of some bursate nematode parasites of sheep and pathogenicity of *Oesophagostomum. Indian Journal of Helminthology* 34, 123–135.

Fox, J.C. (1976) Inhibited development of *Obeliscoides cuniculi* in rabbits: the effects of active and passive immunization and resumption of larval development. *Veterinary Parasitology* 1, 209–220.

Franco, R., Bories, S. and Couzin, B. (1960) A propos de 142 cas de meningite à éosinophiles observés à Tahiti et en Nouvelle-Caledonie. *Médecine Tropicale, Marseille* 20, 41–55.

Freitas, A.L., De Carli, G. and Blankenhein, M.H. (1995) *Mammomonogamus (Syngamus) laryngeus* infection: a new Brazilian human case. *Revista do Instituto de Medicina Tropical de São Paulo* 37, 177–179.

Fülleborn, F. (1926) Ueber das Verhalten der Hakenwurmlarven bei der Infektion per os. *Archiv für Schiffs- und Tropen-Hygiene, Pathologie und Therapie Exotischer Krankheiten* 30, 638–653.

Fülleborn, F. (1927) Durch Hakenwurmlarven des Hundes (*Uncinaria stenocephala*) beim Menschen erzeugte 'Creeping Eruption'. *Hamburgische Universität. Abhandlungen aus dem Gebiet der Auslandskunde. Band 26, Reihe D: Medizin u. Veterinärmedizin, Band 2: Arbeiten über Tropenkrankheiten und deren Grenzgebiete*, pp. 121–133.

Gajadhar, A.A. and Tessaro, S.V. (1995) Susceptibility of mule deer (*Odocoileus hemionus*) and two species of North American molluscs to *Elaphostrongylus cervi* (Nematoda: Metastrongyloidea). *Journal of Parasitology* 81, 593–596.

Gallie, G.J. and Nunns, V.J. (1976) The bionomics of the free-living larvae and the transmission of *Dictyocaulus filaria* between lambs in North-East England. *Journal of Helminthology* 50, 79–89.

Gallivan, G.J., Barker, I.K., Alves, R.M.R., Culverwell, J. and Girdwood, R. (1989) Observations on the lungworm, *Pneumostrongylus calcaratus*, in impala (*Aepyceros melampus*) from Swaziland. *Journal of Wildlife Diseases* 25, 76–82.

Gamble, R.L. and Riewe, R.R. (1982) Infestations of the nematode *Skrjabingylus nasicola* (Leuckart, 1842) in *Mustela frenata* (Lichtenstein) and *M. erminea* (L.) and some evidence of a paratenic host in the life cycle of this nematode. *Canadian Journal of Zoology* 60, 45–52.

Gamble, H.R., Lichtenfels, J.R. and Purcell, J.P. (1989) Light and scanning electron microscopy of the ecdysis of *Haemonchus contortus* infective larvae. *Journal of Parasitology* 75, 303–307.

Garner, D.L. and Porter, W.F. (1991) Prevalence of *Parelaphostrongylus tenuis* in white-tailed deer in northern New York. *Journal of Wildlife Diseases* 27, 594–598.

Gasnier, N. and Cabaret, J. (1996) Evidence for the existence of a sheep and a goat line of *Teladorsagia circumcincta* (Nematoda). *Parasitology Research* 82, 546–550.

Gasnier, N., Cabaret, J. and Durette-Desset, M.C. (1997) Sheep and goat lines of *Teladorsagia circumcincta* (Nematoda): from allozyme to morphological identification. *Journal of Parasitology* 83, 527–529.

Gatongi, P.M., Prichard, R.K., Ranjan, S., Gathuma, I.M., Manyua, W.K., Cheruiyot, H. and Scott, M.E. (1998) Hypobiosis of *Haemonchus contortus* in natural infections of sheep and goats in a semi-arid area of Kenya. *Veterinary Parasitology* 77, 49–61.

Geller, E.R. (1962) The impossibility of autoinfection with *Amidostomum* in geese. *Zoologicheski Zhurnal* 41, 993–996. (In Russian.)

Georgi, J.R. (1973) The Kikuchi–Enigk model of *Strongylus vulgaris* migrations in the horse. *Cornell Veterinarian* 63, 220–222.

Georgi, J.R. (1976) *Filaroides hirthi*: experimental transmission among beagle dogs through ingestion of first-stage larvae. *Science* 194, 735.

Georgi, J.R. (1987) Parasites of the respiratory tract. *Veterinary Clinics of North America: Small Animal Practice* 17, 1422–1442.

Georgi, J.R., Fahnestock, G.R., Bohm, M.F.K. and Adsit, J.C. (1979) The migration and development of *Filaroides hirthi* larvae in dogs. *Parasitology* 79, 39–47.

Georgi, J.R., Fleming, W.J., Hirth, R.S. and Cleveland, D.J. (1976) Preliminary investigation of the life history of *Filaroides hirthi* Georgi and Anderson, 1975. *The Cornell Veterinarian* 66, 309–323.

Georgi, J.R., Georgi, M.E. and Cleveland, D.J. (1977) Patency and transmission of *Filaroides hirthi* infection. *Parasitology* 75, 251–257.

Gerber, H.C. (1968) Untersuchungen über Vorkommen und Entwicklung von *Hyostrongylus rubidus* beim Schwein. Dissertation, Berlin, 29 pp.

Gerichter, C.B. (1948) Observation on the life history of lung nematodes using snails as intermediate hosts. *American Journal of Veterinary Research* 9, 109–112.

Gerichter, C.B. (1949) Studies on the nematodes parasitic in the lungs of Felidae in Palestine. *Parasitology* 39, 251–262.

Gerichter, C.B. (1951) Studies on the lung nematodes of sheep in the Levant. *Parasitology* 41, 166–183.

Gettinby, G., Bairden, K., Armour, J. and Benitez-Usher, C.A. (1979) A prediction model for bovine ostertagiasis. *Veterinary Record* 105, 57–59.

Gharib, H.M. (1953) Skin penetration and migration via lymphatics of experimental animals by infective larvae of *Nippostrongylus muris*. *Transactions of the Royal Society of Tropical Medicine and Hygiene* 47, 264.

Gharib, H.M. (1955) Observations on skin penetration by the infective larvae of *Nippostrongylus brasiliensis*. *Journal of Helminthology* 29, 33–36.

Giangasperó, M., Bahhady, F.A., Orita, G. and Gruner, L. (1992) Summer-arrested development of abomasal trichostrongylids in Awassi sheep in semi-arid areas of north-west Syria. *Parasitology Research* 78, 594–597.

Gibbons, L.M. (1979) Revision of the genus *Haemonchus* Cobb, 1898 (Nematoda: Trichostrongylidae). *Systematic Parasitology* 1, 3–24.

Gibbons, L.M., Halvorsen, O. and Stuve, G. (1991) Revision of the genus *Elaphostrongylus* Cameron (Nematoda; Metastrongyloidea) with particular reference to species of the genus occurring in Norwegian cervids. *Zoologica Scripta* 20, 15–26.

Gibbs, H.C. and Gibbs, K.E. (1959) The effects of temperature on the development of the free-living stages of *Dochmoides stenocephala* (Railliet, 1884) (Ancylostomatidae: Nematoda). *Canadian Journal of Zoology* 37, 247–257.

Gibbs, H.C. (1961) Studies on the life cycle and developmental morphology of *Dochmoides stenocephala* (Railliet, 1884) (Ancylostomatidae: Nematoda). *Canadian Journal of Zoology* 39, 325–348.

Gibbs, H.C. (1967) Some factors involved in the spring rise phenomenon in sheep. In: Soulsby, E.J.L. (ed.) *Reactions of the Host to Parasitism*. N.E. Elwert Universitas Verlagsbuchand, Marburg, pp. 160–179.

Gibbs, H.C. (1973) Transmission of parasites with reference to the strongyles of domestic sheep and cattle. *Canadian Journal of Zoology* 51, 281–289.

Gibbs, H.C. (1982) Mechanisms of survival of nematode parasites with emphasis on hypobiosis. *Veterinary Parasitology* 11, 25–48.

Gibbs, H.C. (1986) Hypobiosis in parasitic nematodes – an update. *Advances in Parasitology* 25, 129–174.

Gibbs, H.C. (1988) The epidemiology of bovine ostertagiasis in the north temperate regions of North America. *Veterinary Parasitology* 27, 39–47.

Gibbs, H.C. and Barger, I.A. (1986) *Haemonchus contortus* and other trichostrongylid infections in parturient, lactating and dry ewes. *Veterinary Parasitology* 22, 57–66.

Gibbs, H.C. and Everett, G. (1978) Further observations on the effect of different levels of larval intake on the output of eggs of *Ostertagia circumcincta* in lambs. *Research in Veterinary Science* 24, 169–173.

Gibbs, H.C., Crenshaw, W.J. and Mowatt, M. (1977) Seasonal changes in stomach worms (*Obeliscoides cuniculi*) in snowshoe hares in Maine. *Journal of Wildlife Diseases* 13, 327–332.

Gibson, T.E. (1953) The effect of repeated anthelmintic treatment with phenothiazine on the faecal egg counts of housed horses, with some observations on the life cycles of *Trichonema* spp. in the horse. *Journal of Helminthology* 27, 29–40.

Gibson, T.E. (1959) The survival of the free-living stages of *Nematodirus* spp. on pasture herbage. *The Veterinary Record* 71, 362–366.

Gibson, T.E. and Everett, G. (1976) The ecology of the free-living stages of *Haemonchus contortus*. *British Veterinary Journal* 132, 50–59.

Gleich, J.G. (1972) Terrestrial gastropods from central Maine; distribution, relative abundance, and relationship to parasitic nematodes, especially *Pneumostrongylus tenuis*. Unpublished MS thesis, University of Maine, Orono, Maine.

Gleich, J.G. and Gilbert, F.F. (1976) A survey of terrestrial gastropods from central Maine. *Canadian Journal of Zoology* 54, 620–627.

Gleich, J.G., Gilbert, F.F. and Kutscha, N.P. (1977) Nematodes in terrestrial gastropods from central Maine. *Journal of Wildlife Diseases* 13, 43–46.

Goldberg, A. (1951) Life history of *Oesophagostomum venulosum*, a nematode parasite of sheep and goats. *Proceedings of the Helminthological Society of Washington* 18, 36–47.

Goldberg, A. (1952) Experimental infection of sheep and goats with the nematode lungworm, *Dictyocaulus filaria*. *American Journal of Veterinary Research* 13, 531–536.

Goldberg, A. and Rubin, R. (1956) Survival on pasture of larvae of gastric-intestinal nematodes of cattle. *Proceedings of the Helminthological Society of Washington* 23, 65–68.

Goldsmid, J.M. (1971) Studies on the life cycle and biology of *Ternidens deminutus* (Railliet and Henry, 1905) (Nematoda: Strongylidae). *Journal of Helminthology* 45, 341–352.

Goldsmid, J.M. (1974) The intestinal helminthozoonoses of primates in Rhodesia. *Annales des Société Belge de Médecine Tropicale* 54, 87–101.

Goldsmid, J.M. (1982) *Ternidens* infection. In: Schulz, M.G. (ed.) *CRC Handbook Series in Zoonoses. Section C. Parasitic Zoonoses*, Vol. II. CRC Press, Boca Raton, Florida, pp. 269–288.

Goodey, T. (1924) The anatomy of *Oesophagostomum dentatum* (Rud.) a nematode parasite of the pig, with observations on the structure and biology of the free-living larvae. *Journal of Helminthology* 2, 1–14.

Goodey, T. (1926) Some stages in the development of *Oesophagostomum dentatum* from the pig. *Journal of Helminthology* 4, 191–198.

Gordon, H.M. and Graham, N.P.H. (1933) A few cases of infestation with *Chabertia ovina*. *Australian Veterinarian Journal* 9, 198–199.

Gosselin, J.F., Measures, L.N. and Hust, J. (1998) Lungworm (Nematoda: Metastrongyloidea) infections in Canadian phocids. *Canadian Journal of Fisheries and Aquatic Sciences* 55, 825–834.

Graeff-Teixeira, C., Thomé, J.W., Pinto, S.C.C., Camillo-Coura, L. and Lenzi, H.L. (1989) *Phyllocaulis variegatus* an intermediate host of *Angiostrongylus costaricensis* in South Brazil. *Memórias do Instituto Oswaldo Cruz* 84, 65–68.

Graeff-Teixeira, C., de Avilla-Pires, F.D., de Cassia, R., Machado, C., Camillo-Coura, L. and Lenzi, H.L. (1990) Identificacão de roedores silvestres como hospedeiros do *Angiostrongylus costaricensis* no sul do Brasil. *Revista do Instituto de Medicina Tropical São Paulo* 32, 147–150.

Graeff-Teixeira, C., Thiengo, S.C., Thomé, J.W., Medeiros, A.B., Camillo-Coura, L. and Agostini, A.A. (1993) On the diversity of mollusc intermediate hosts of *Angiostrongylus costaricensis* Morera and Cespedes, 1971 in southern Brazil. *Memórias do Instituto Oswaldo Cruz* 88, 487–489.

Gray, J.B. and Samuel, W.M. (1986) *Parelaphostrongylus odocoilei* (Nematoda: Protostrongylidae) and a protostrongylid nematode in woodland caribou (*Rangifer tarandus caribou*) of Alberta, Canada. *Journal of Wildlife Diseases* 22, 48–50.

Gray, J.B., Samuel, W.M., Shostak, A.W. and Pybus, M.J. (1985) *Varestrongylus alpenae* (Nematoda: Metastrongyloidea) in white-tailed deer (*Odocoileus virginianus*) of Saskatchewan. *Canadian Journal of Zoology* 63, 1449–1454.

Graybill, H.W. (1923) A new genus of nematodes from the domestic rabbit. *Parasitology* 15, 340–342.

Graybill, H.W. (1924) *Obeliscoides*, a new name for the nematode genus *Obeliscus*. *Parasitology* 16, 317.

Griffin, C.T. (1988) The effect of constant and changing temperatures on the development of the eggs and larvae of *Oswaldocruzia filiformis* (Nematoda: Trichostrongyloidea). *Journal of Helminthology* 62, 281–292.

Grove, D.I. (1990) *A History of Human Helminthology*. CAB International, Wallingford, UK.

Guberlet, J.E. (1919) On the life history of the lungworm, *Dictyocaulus* (*Strongylus*) *filaria* Rud. in sheep. *Journal of American Veterinary Medical Association* 55, 621–627.

Guilhon, J. (1960) Rôle des Limacides dans le cycle évolutif d'*Angiostrongylus vasorum* (Baillet, 1866). *Comptes Rendus des Séances de l'Academie des Sciences* 251, 2252–2253.

Guilhon, J. (1963) Récherches sur le cycle évolutif du strongyle des vaisseaux du chien. *Bulletin de l'Academie Veterinaire de France* 36, 431–442.

Guilhon, J. (1965) Transmission d'*Angiostrongylus cantonensis* (Chen, 1935) au chien. *Compte Rendu Hebdomadaire des Séances de l'Academie des Sciences, Paris* 261, 1089–1091.

Guilhon, J. (1967) Angiostrongyline nematodes and human eosinophilic meningitis. *South Pacific Commission Seminar on Helminthiasis and Eosinophilic Meningitis, Noumea, New Caledonia*, pp. 1–5.

Gupta, R.P. and Gibbs, H.C. (1975) Infection patterns of *Dictyocaulus viviparus* in calves. *Canadian Veterinary Journal* 16, 102–108.

Gupta, S.P. (1961a) The effects of temperature on the survival and development of the free-living stages of *Trichostrongylus retortaeformis* Zeder (Nematoda). *Canadian Journal of Zoology* 39, 47–53.

Gupta, S.P. (1961b) The life history of *Molineus barbatus* Chandler, 1942. *Canadian Journal of Zoology* 39, 579–588.

Gupta, S.P. (1963) Mode of infection and biology of infective larvae of *Molineus barbatus* Chandler, 1942. *Experimentally Parasitology* 13, 252–255.

Haley, A.J. (1961) Biology of the rat nematode, *Nippostrongylus brasiliensis* (Travassos, 1914). II. Preparasitic stages and development in the laboratory rat. *Journal of Parasitology* 47, 727–732.

Haley, A.J. and Parker, J.C. (1961) Effect of population density on adult worm survival in primary *Nippostrongylus brasiliensis* infection in the rat. *Proceedings of the Helminthological Society of Washington* 28, 176–180.

Halvorsen, O. and Skorping, A. (1982) The influence of temperature on growth and development of the nematode *Elaphostrongylus rangiferi* in the gastropods *Arianta arbustorum* and *Euconulus fulvus*. *Oikos* 38, 285–290.

Halvorsen, O., Andersen, J., Skorping, A. and Lorentzen, G. (1980) Infection in reindeer with the nematode *Elaphostrongylus rangiferi* Mitskevich in relation to climate and distribution of intermediate hosts. *Proceedings of the Second International Reindeer/Caribou Symposium, Roros, Norway*, pp. 449–455.

Halvorsen, O., Skorping, A. and Hansen, K. (1985) Seasonal cycles in the output of first-stage larvae of the nematode *Elaphostrongylus rangiferi* from reindeer, *Rangifer tarandus tarandus*. *Polar Biology* 5, 49–54.

Halvorsen, O., Skorping, A. and Bye, K. (1989) Experimental infection of reindeer with *Elaphostrongylus* (Nematoda: Protostrongylidae) originating from reindeer, red deer and moose. *Canadian Journal of Zoology* 67, 1200–1202.

Hamilton, J.M. and McCaw, A.W. (1967) The role of the mouse in the life cycle of *Aelurostrongylus abstrusus*. *Journal of Helminthology* 41, 309–312.

Handeland, K. (1994) Experimental studies of *Elaphostrongylus rangiferi* in reindeer (*Rangifer tarandus tarandus*): life cycle, pathogenesis, and pathology. *Journal of Veterinary Medicine, Series B* 41, 351–365.

Handeland, K. and Skorping, A. (1992a) Experimental cerebrospinal elaphostrongylosis (*Elaphostrongylus rangiferi*) in goats. II. Pathological findings. *Journal of Veterinary Medicine, Series B* 39, 713–722.

Handeland, K. and Skorping, A. (1992b) The early migration of *Elaphostrongylus rangiferi* in goats. *Journal of Veterinary Medicine, Series B* 39, 263–272.

Handeland, K. and Skorping, A. (1993) Experimental cerebrospinal elaphostrongylosis (*Elaphostrongylus rangiferi*) in goats. I. Clinical observations. *Journal of Veterinary Medicine, Series B* 40, 141–147.

Handeland, K. and Slettbakk, T. (1994) Outbreaks of clinical cerebrospinal elaphostrongylosis in reindeer (*Rangifer tarandus*) in Finn mark, Norway, and their relation to climatic conditions. *Journal of Veterinary Medicine. Series B* 41, 407–410.

Handeland, K., Skorping, A. and Slettbakk, T. (1993) Experimental cerebrospinal elaphostronglosis (*Elaphostrongylus rangiferi*) in sheep. *Journal of Veterinary Medicine. Series B* 40, 181–189.

Handeland, K., Skorping, A., Stven, S. and Slettbakk, T. (1994) Experimental studies of *Elaphostrongylus rangiferi* in reindeer (*Rangifer tarandus tarandus*): Clinical observations. *Rangifer* 14, 83–87.

Hansson, I. (1967) Transmission of the parasitic nematode *Skrjabingylus nasicola* (Leuckart, 1842) to species of *Mustela* (Mammalia). *Oikos* 18, 247–252.

Hansson, I. (1968) Cranial helminth parasites in species of Mustelidae. I. Frequency and damage in fresh mustelids from Sweden. *Oikos* 19, 217–233.

Hansson, I. (1974) Seasonal and environmental conditions affecting the invasion of mustelids by larvae of the nematode *Skrjabingylus nasicola*. *Oikos* 25, 61–70.

Harada, F. (1954a) Investigations of hookworm larvae. III. Biological observations of infective larvae in migration towards vegetables. *Yokohama Medical Bulletin* 5, 212–229. (In Japanese.)

Harada, F. (1954b) Investigations of hookworm larvae. IV. On the fly as a carrier of infective larvae. *Yokohama Medical Bulletin* 5, 282–286. (In Japanese.)

Harinasuta, C., Setasubun, P. and Radomyos, P. (1965) Observations on *Angiostrongylus cantonensis* in rats and mollusks in Thailand. *Journal of the Medical Association of Thailand* 48, 158–172.

Hasslinger, M.A. (1985) Der Magenwurm der Katze, *Ollulanus tricuspis* (Leuckart, 1865) – zum gegenwärtigen Stand der Kenntnis. *Tierärztliche Praxis* 13, 205–217.

Hasslinger, M.A. and Trah, M. (1981) Untersuchungen zur Verbreitung und zum Nachweis des Magenwurmes der Katze, *Ollulanus tricuspis* (Leuckart, 1865). *Berliner und Münchener Tierärztliche Wochenschrift* 94, 235–238.

Hasslinger, M.A. and Wittmann, F.X. (1982) The tail morphology of different developmental stages of *Ollulanus tricuspis* (Leuckart, 1865) (Nematoda: Trichostrongyloidea). *Journal of Helminthology* 56, 351–353.

Hata, H. (1994) Essential amino acids and other essential components for development of *Angiostrongylus costaricensis* from third stage larvae to young adults. *Journal of Parasitology* 80, 518–520.

Hata, H. (1996) In vitro cultivation of *Angiostrongylus costaricensis* eggs to first stage larvae in chemically defined medium. *International Journal for Parasitology* 26, 281–286.

Hata, H. and Kojima, S. (1994) *Angiostrongylus costaricensis*: culture of third-stage larvae to young adults in a defined medium. *Experimental Parasitology* 73, 354–361.

Hendrikx, W.M.L. (1981) Seasonal fluctuation, localisation and inhibition of development of parasitic stages of *Oswaldocruzia filiformis* (Goeze, 1782) Travassos, 1917 (Nematoda: Trichostrongylidae) in *Bufo bufo* L., 1785 under natural and experimental conditions in the Netherlands. *Netherlands Journal of Zoology* 31, 481–503.

Hendrikx, W.M.L. (1983) Observations on the routes of infection of *Oswaldocruzia filiformis* (Nematoda: Trichostrongylidae) in Amphibia. *Zeitschrift für Parasitenkunde* 69, 119–126.

Hendrikx, W.M.L. and Moppes, M.C. van (1983) *Oswaldocruzia filiformis* (Nematoda: Trichostrongylidae): morphology of developmental stages, parasitic development and some pathological aspects of the infection in amphibians. *Zeitschrift für Parasitenkunde* 69, 523–537.

Herlich, H. (1954) The life history of *Nematodirus helvetianus* May, 1920, a nematode parasitic in cattle. *Journal of Parasitology* 40, 60–70.

Herlich, H. (1965a) The development of *Cooperia pectinata*, nematode parasite of cattle. *American Journal of Veterinary Research* 26, 1026–1031.

Herlich, H. (1965b) The effects of the intestinal worms, *Cooperia pectinata* and *Cooperia oncophora*, on experimentally infected calves. *American Journal of Veterinary Research* 26, 1032–1036.

Herlich, H., Porter, D.A. and Knight, R.A. (1958) A study of *Haemonchus* in cattle and sheep. *American Journal of Veterinary Research* 19, 866–872.

Herman, C.M. and Wehr, E.E. (1954) The occurrence of gizzard worms in Canada geese. *Journal of Wildlife Management* 18, 509–513.

Hernandez, A.D. and Sukhdeo, M.V.K. (1995) Host grooming and the transmission strategy of *Heligmosomoides polygyrus*. *Journal of Parasitology* 81, 856–869.

Hesse, A.J. (1923) On the free-living larval stages of the nematode *Bunostomum trigonocephalum* (Rud.) a parasite of sheep. *Journal of Helminthology* 1, 21–28.

Heyneman, D. and Lim, B.L. (1965) Correlation of habitat to rodent susceptibility with *Angiostrongylus cantonensis* (Nematoda: Metastrongylidae). *Medical Journal of Malaya* 20, 67–68.

Heyneman, D. and Lim, B.L. (1966a) Comparative susceptibility of primitive and higher primates to infection with *Angiostrongylus cantonensis* (Nematoda: Metastrongylidae). *Medical Journal of Malaya* 21, 341.

Heyneman, D. and Lim, B.L. (1966b) Comparison of developmental rates of two strains of *Angiostrongylus cantonensis* in the same laboratory-bred final hosts. *Medical Journal of Malaya* 21, 343.

Heyneman, D. and Lim, B.L. (1967a) A summary of recent studies on *Angiostrongylus cantonensis* host–parasite relations in Malaysia. *South Pacific Commission Seminar on Helminthiases and Eosinophilic Meningitis, Noumea, New Caledonia*. pp. 1–7.

Heyneman, D. and Lim, B.L. (1967b) *Angiostrongylus cantonensis*: proof of direct transmission with its epidemiological complications. *Science* 158, 1052–1058.

Heyneman, D. and Lim, B.L. (1967c) Discovery of a mode of transmission of *Angiostrongylus cantonensis* in Malaya. *Medical Journal of Malaya* 21, 377.

Hibler, C.P., Lange, R.E. and Metzger, C.J. (1972) Transplacental transmission of *Protostrongylus* spp. in bighorn sheep. *Journal of Wildlife Diseases* 9, 389.

Hibler, C.P., Metzger, C.J., Spraker, T.R. and Lange, R.E. (1974) Further observations on *Protostrongylus* sp. infection by transplacental transmission in bighorn sheep. *Journal of Wildlife Diseases* 10, 39–41.

Hibler, C.P., Spraker, T.R. and Schmidt, R.L. (1977) Treatment of bighorn sheep for lungworms. *Transactions of the 1977 Desert Bighorn Council* pp. 12–14.

Hibler, C.P., Spraker, T.R. and Thorne, E.T. (1982) Protostrongylosis in bighorn sheep. In: Thorne, E.T., Kingston, N., Jolley, W.R. and Bergstrom, R.C. (eds) *Diseases of Wildlife in Wyoming*. Wyoming Game and Fish Department, USA, pp. 208–213.

Higa, H.H., Brock, J.A. and Palumbo, N.E. (1986) Occurrence of *Angiostrongylus cantonensis* in rodents, intermediate and paratenic hosts on the island of Oahu. *Journal of Environmental Health* 48, 319–323.

Hoagland, K.E. and Schad, G.A. (1978) *Necator americanus* and *Ancylostoma duodenale*: life history parameters and epidemiological implications of two sympatric hookworms of humans. *Experimental Parasitology* 44, 36–49.

Hobmaier, M. (1925) Ueber die Entstehung des Aneurysma verminosum equi. *Zeitschrift für Infektionskrankheiten, Parasitäre Krankheiten und Hygiene der Haustiere* 28, 165–177.

Hobmaier, A. and Hobmaier, M. (1929a) Die Entwicklung der Larve des Lungenwurmes *Metastrongylus elongatus* (*Strongylus paradoxus*) des Schweines und ihr Invasionsweg, sowie vorläufige Mitteilung über die Entwicklung von *Choerostrongylus brevivaginatus*. *Münchener Tierärztliche Wochenschrift* 80, 365–369.

Hobmaier, A. and Hobmaier, M. (1929b) Biologie von *Choerostrongylus* (*Metastrongylus*) *pudendotectus* (*brevivaginatus*) aus der Lunge des Schweines, zugleich eine vorläufige Mitteilung über die Entwicklung der Gruppe *Synthetocaulus* unserer Haustiere. *Münchener Tierärztliche Wochenschrift* 80, 433–436.

Hobmaier, A. and Hobmaier, M. (1929c) Ueber die Entwicklung des Lungenwurmes *Synthetocaulus capillaris*, in Nacht-, Weg-und Schnirkelschnecken. *Münchener Tierärztliche Wochenschrift* 80, 497–500.

Hobmaier, A. and Hobmaier, M. (1929d) Die Entwicklung des Lungenwurmes des Schafes, *Dictyocaulus filaria* Rud., ausserhalb und innerhalb des Tierkörpers. *Münchener Tierärztliche Wochenschrift* 45, 621–625.

Hobmaier, A. and Hobmaier, M. (1930) Life history of *Protostrongylus* (*Synthetocaulus*) *rufescens*. *Proceedings of the Society for Experimental Biology and Medicine* 28, 156–158.

Hobmaier, A. and Hobmaier, M. (1934) *Elaphostrongylus odocoilei* n.sp., a new lungworm in black-tail deer (*Odocoileus columbianus*). Description and life history. *Proceedings of the Society for Experimental Biology and Medicine* 31, 509–514.

Hobmaier, M. (1941a) Description and extramammalian life of *Crenosoma mephitidis* n.sp. (Nematoda) in skunks. *Journal of Parasitology* 27, 229–232.

Hobmaier, M. (1941b) Extramammalian phase of *Skrjabingylus chitwoodorum* (Nematoda). *Journal of Parasitology* 27, 237–239.

Hobmaier, M. and Hobmaier, A. (1935a) Intermediate hosts of *Aelurostrongylus abstrusus* of the cat. *Proceedings of the Society for Experimental Biology and Medicine* 32, 1641–1647.

Hobmaier, M. and Hobmaier, A. (1935b) Mammalian phase of the lungworm *Aelurostrongylus abstrusus* in the cat. *Journal of the American Veterinary Medical Association* 87, 191–198.

Holstadt, O., Karbol, G. and Skorping, A. (1994) *Trichostrongylus tenuis* from willow grouse (*Lagopus lagopus*) and ptarmigan (*Lagopus mutus*) in northern Norway. *Bulletin of the Scandinavian Society for Parasitology* 4, 9–13.

Horak, I.G. (1981) The similarity between arrested development in parasitic nematodes and diapause in insects. *Journal of the South African Veterinary Association* 52, 299–303.

Hori, E., Kano, R. and Ishigaki, Y. (1976) Experimental intermediate hosts of *Angiostrongylus cantonensis*: studies on snails and a slug. *Japanese Journal of Parasitology* 25, 434–440. (In Japanese.)

Howe, D.L. (1965) Etiology of pneumonia in bighorn sheep. *Federal Aid Division Quarterly Report of the Wyoming Department of Game and Fish*, p. 21.

Hsieh, H.C. (1970) Studies on endemic hookworm: 2. Comparison of the efficacy of anthelmintics in Taiwan and Liberia. *Japanese Journal of Parasitology* 19, 534–536. (In Japanese.)

Hsu, C.K. and Levine, N.D. (1977) Degree-day concept in development of infective larvae of *Haemonchus contortus* and *Trichostrongylus colubriformis* under constant and cyclic conditions. *American Journal of Veterinary Research* 38, 1115–1119.

Hudson, P.J. (1986a) *Red Grouse, the Biology and Management of a Wild Game Bird*. Game Conservancy Trust, Fordingbridge, UK.

Hudson, P.J. (1986b) The effect of a parasitic nematode on the breeding production of red grouse. *Journal of Animal Ecology* 55, 85–92.

Hudson, P.J. and Dobson, A.P. (1989) Population biology of *Trichostrongylus tenuis*, a parasite of economic importance for red grouse management. *Parasitology Today* 5, 283–291.

Hudson, J.P. and Dobson, A.P. (1997) Transmission dynamics and host–parasite interactions of *Trichostrongylus tenuis* in red grouse (*Lagopus lagopus scoticus*). *Journal of Parasitology* 83, 194–202.

Hudson, P.J., Dobson, A.P. and Newborn, D. (1985) Cyclic and non-cyclic populations of red grouse: a role for parasitism? In: Rollinson, D. and Anderson, R.M. (eds) *Ecology and Genetics of Host–Parasite Interactions*. Academic Press, London, pp. 77–89.

Hudson, P.J., Dobson, A.P. and Newborn, D. (1992) Do parasites make prey vulnerable to predation? Red Grouse and Parasites. *Journal of Animal Ecology* 61, 681–692.

Humbert, J.F. and Drouet, J. (1993) Studies on the egg output variations and determinism of *Metastrongylus* spp., lungworms of the wild boar (*Sus scrofa* L.). *Research and Reviews in Parasitology* 53, 27–31.

Hutchinson, G.W. and Fernando, M.A. (1974) Enzymes of glycolysis and the pentose phosphate pathway during development of the rabbit stomach worm *Obeliscoides cuniculi*. *International Journal for Parasitology* 4, 389–395.

Hutchinson, G.W. and Fernando, M.A. (1975) Enzymes of the tricarboxylic acid cycle in *Obeliscoides cuniculi* (Nematoda; Trichostrongylidae) during parasitic development. *International Journal for Parasitology* 5, 77–82.

Hutchinson, G.W., Lee, E.H. and Fernando, M.A. (1972) Effects of variations in temperature on infective larvae and their relationship to inhibited development of *Obeliscoides cuniculi* in rabbit. *Parasitology* 65, 333–342.

Ihle, J.E.W. and van Oordt, G.J. (1923) On some strongylid larvae in the horse, especially those of *Cylicostomum*. *Annals of Tropical Medicine and Parasitology* 17, 31–45.

Isenstein, R.S. (1963) The life history of *Cooperia oncophora* (Railliet, 1898) Ransom, 1907, a nematode parasite of cattle. *Journal of Parasitology* 49, 235–240.

Jacobs, D.E. (1967) The linear distribution of two *Oesophagostomum* species in the intestine of the pig. *Acta Veterinaria Scandinavica* 8, 287–289.

Jacquiet, P., Cabaret, J., Cheikh, D. and Thiam, A. (1995) Experimental study of survival strategy of *Haemonchus contortus* in sheep during the dry season in desert areas of the Mauritania. *Journal of Parasitology* 81, 1013–1015.

Jacquiet, P., Cabaret, J., Cheikh, D. and Thiam, E. (1997) Identification of *Haemonchus* species in domestic ruminants based on morphometrics of spicules. *Parasitology Research* 83, 82–86.

Jarrett, W.F.H. and Sharp, N.C.C. (1963) Vaccination against parasitic disease: reactions in vaccinated and immune hosts in *Dictyocaulus viviparus* infection. *Journal of Parasitology* 49, 177–189.

Jarrett, W.F.H., McIntryre, W.I.M. and Urquhart, G.M. (1957) The pathology of experimental bovine parasitic bronchitis. *Journal of Pathology and Bacteriology* 73, 183–193.

Jarrett, W.F.H., Jennings, F.W., McIntyre, W.I.M., Mulligan, W., Sharp, N.C.C. and Urquhart, G.M. (1959) Immunological studies on *Dictyocaulus viviparus* infection in calves – double vaccination with irradiated larvae. *American Journal of Veterinary Research* 20, 522–526.

Jasmer, D.P., Wescott, R.B. and Crane, J.W. (1987) Survival of third-stage larvae of Washington isolates of *Haemonchus contortus* and *Ostertagia circumcincta* exposed to cold temperatures. *Proceedings of the Helminthological Society of Washington* 54, 48–52.

Jenkins, D.C. (1974) *Nippostrongylus brasiliensis* the distribution of primary worm populations within the small intestine of neonatal rats. *Parasitology* 68, 339–345.

Jindrak, K. (1968) Early migration and pathogenicity of *Angiostrongylus cantonensis* in laboratory rats. *Annals of Tropical Medicine and Parasitology* 62, 506–517.

Jørgensen, R.J. and Ogbourne, C.P. (1985). *Bovine Dictyocauliasis: a Review and Annotated Bibliography*. Commonwealth Institute of Parasitology, St Albans, UK.

Jørgensen, R.J., Rønne, H., Helsted, C. and Iskander, A.R. (1982) Spread of infective *Dictyocaulus viviparus* larvae in pasture and to grazing cattle: experimental evidence of the role of *Pilobolus* fungi. *Veterinary Parasitology* 10, 331–339.

Jortner, B.S., Troutt, H.F., Collins, T. and Scarratt, K. (1985) Lesions of spinal cord parelaphostrongylosis in sheep. Sequential changes following intramedullary larval migration. *Veterinary Pathology* 22, 137–140.

Joyeux, C. and Gaud, J. (1946) Recherches helminthologique marocaines. *Archives de l'Institut Pasteur du Maroc* 3, 383–461.

Kaarma, A. (1970) Effect of temperature on the development of eggs and larvae of *Oesophagostomum dentatum* (Rudolphi, 1803). *Sbornik Nauchnykh Trudov, Estonskii Nauchno-Issledovatel'skii Institut Zhivotnovodstva i Veterinarii* 21, 56–61.

Kadenatsii, A.N. (1958) Biological investigation of *Protostrongylus tauricus*, pulmonary parasite of hares. In: S.N. Boev (ed.) *Contributions to Helminthology*. Kazakh Gosudar, Izdatel, Alma Ata. (Israel Program for Scientific Translations, Jerusalem, 1968.)

Kamiya, M., Oku, Y., Katakura, K., Kamiya, H., Ohbayashi, M., Abe, H. and Suzuki, H. (1980) Report on the prevalence and experimental infections of *Angiostrongylus siamensis* Ohbayashi, Kamiya et Bhaibulaya, 1969, parasitic in the mesenteric arteries of rodents in Thailand. *Japanese Journal of Veterinary Research* 28, 114–121.

Kassai, T. (1957) Schnecken als Zwischenwirte der Protostrongyliden. *Zeitschrift für Parasitenkunde* 18, 5–19.

Kassai, T. (1968) Immunological tolerance to *Nippostrongylus brasiliensis* infection in rats. In: Soulsby, E.J.L. (ed.) *The Reaction of the Host to Parasitism*. N.E. Elwert Universitas Verlagsbuchand, Marburg, Lahn, pp. 250–258.

Kassai, T. (1982) *Handbook of* Nippostrongylus brasiliensis *(Nematode)*. Commonwealth Agricultural Bureaux, Akadémiai Kiadó, Budapest, Hungary.

Kassai, T. and Aitken, I.D. (1967) Induction of immunological tolerance in rats to *Nippostrongylus brasiliensis* infection. *Parasitology* 57, 403–418.

Kassai, T., Shnain, A.H., Kadhim, J.K., Altaif, K.I. and Jabbir, M.H. (1972) Collection of lymph from sheep by cannulation of the thoracic duct and application of the method in hosts infected with *Dictyocaulus filaria*. *Magyar Allatorvosok Lapja* 27, 691–696. (In Hungarian.)

Katakura, K. (1979) The development of three species of *Angiostrongylus* in the intermediate hosts. *Japanese Journal of Parasitology* 27, 27. (Abstract of thesis.)

Katakura, K., Oku, Y., Kamiya, M. and Ohbayashi, M. (1981) Development of the mesenteric metastrongyloid *Angiostrongylus siamensis* in *Biomphalaria glabrata*, an experimental intermediate host. *Japanese Journal of Parasitology* 30, 23–30.

Kates, K.C. (1941) Observations on the viability of eggs of lungworms of swine. *Journal of Parasitology* 27, 265–275.

Kates, K.C. (1943) Overwinter survival on pasture of preparasitic stages of some nematodes parasitic in sheep. *Proceedings of the Helminthological Society of Washington* 10, 23–25.

Kates, K.C. (1950) Survival on pasture of free-living stages of some common gastro-intestinal nematodes of sheep. *Proceedings of the Helminthological Society of Washington* 17, 39–59.

Kates, K.C. and Turner, J.H. (1955). Observations on the life cycle of *Nematodirus spathiger*, a nematode parasite in the intestine of sheep and other ruminants. *American Journal of Veterinary Research* 16, 105–115.

Kauzal, G. (1933) Observations on the bionomics of *Dictyocaulus filaria* with a note on the clinical manifestations in artificial infections in sheep. *Australian Veterinary Journal* 9, 20–26.

Kavaliers, M. and Colwell, D.D. (1995) Odours of parasitized males induce adverse responses in female mice. *Animal Behaviour* 50, 1161–1169.

Kean, B.H., Mott, K.E. and Russell, A.J. (eds) (1978) *Tropical Medicine and Parasitology. Classic Investigations*, Vol II. Cornell University Press, Ithaca and London.

Kearney, S.R. and Gilbert, F.F. (1978) Terrestrial gastropods from Himsworth Game Preserve, Ontario and their significance in *Parelaphostrongylus tenuis* transmission. *Canadian Journal of Zoology* 56, 688–694.

Keith, R.K. (1953) The differentiation of the infective larvae of some common nematode parasites of cattle. *Australian Journal of Zoology* 1, 223–235.

Kendall, S.B. and Small, A.J. (1974) The biology of *Hyostrongylus rubidus*. VIII. Loss of worms from the pig. *Journal of Comparative Pathology* 84, 437–441.

Kendall, S.B., Small, A.J. and Phipps, L.P. (1977) *Oesophagostomum* species in pigs in England. I. *Oesophagostomum quadrispinulatum*: description and life-history. *Journal of Comparative Pathology* 87, 223–229.

Kendrick, J.F. (1934) The length of life and rate of loss of hookworms, *Ancylostoma duodenale* and *Necator americanus*. *American Journal of Tropical Medicine and Hygiene* 14, 363–379.

King, C.M. (1977) The effects of the nematode parasite *Skrjabingylus nasicola* on British weasels (*Mustela nivalis*). *Journal of Zoology London* 182, 225–249.

Kistner, T.P., Johnson, G.R. and Rilling, G.A. (1977) Naturally occurring neurologic disease in a fallow deer infected with meningeal worms. *Journal of Wildlife Diseases* 13, 55–58.

Klewer, H.L. (1958) The incidence of helminth lung parasites of *Lynx rufus rufus* (Schabes) and the life cycle of *Anafilaroides rostratus* (Gerichter). *Journal of Parasitology* 44, 29.

Knapp, S.E. (1966) Distribution of *Angiostrongylus cantonensis* larvae in the giant African snail, *Achatina fulica*. *Journal of Parasitology* 52, 502.

Knapp, S.E. and Alicata, J.E. (1967) Failure of certain clams and oysters to serve as intermediate hosts for *Angiostrongylus cantonensis*. *Proceedings of the Helminthological Society of Washington* 34, 1–3.

Ko, R.C. (1978) Occurrence of *Angiostrongylus cantonensis* in the heart of a spider monkey. *Journal of Helminthology* 52, 229.

Ko, R.C. and Anderson, R.C. (1972) Tissue migration, growth, and morphogenesis of *Filaroides martis* (Nematoda: Metastrongyloidea) in mink (*Mustela vison*). *Canadian Journal of Zoology* 50, 1637–1649.

Kobulei, T. and Rhzhikov, K.M. (1968) Revision of the genus *Amidostomum* (Nematoda: Strongylata). *Parazitologiya* 2, 306–311. (In Russian.)

Koike, Y. (1960) The mode of infection of hookworms: experimental studies on oral infection of the infective larvae to the human host. *Journal of the Chiba Medical Society* 36, 1133–1149.

Komarov, A. and Beaudette, F.R. (1931) *Ornithostrongylus quadriradiatus* in squabs. *Journal of American Veterinary Medical Association* 79, 393–394.

Komiya, Y. and Yasuraoha, K. (1966) The biology of hookworms. In: *Progress of Medical Parasitology in Japan*, Vol. III. Megiro Parasitological Museum, Tokyo, pp. 1–114.

Kontrimavichus, V.L., Delyamure, S.L. and Boev, S.N. (1976) Metastrongyloids of domestic and wild animals. *Fundamentals of Nematology*, Vol. 26. Akademiya Nauk SSR. (Translated by Oxonian Press Pvt. Ltd, New Delhi, 1985.)

Kotlán, A. (1948) Studies on the life history and pathological significance of *Oesophagostomum* spp. of the domestic pig. *Acta Veterinaria Hungarica* 1, 14–30.

Kotlán, A. (1949) On the histotropic phase of parasitic larvae of *Hyostrongylus rubidus*. *Acta Veterinaria Hungarica* 1, 76–82.

Kralka, R.A. and Samuel, W.M. (1984a) Experimental life cycle of *Protostrongylus boughtoni* (Nematoda: Metastrongyloidea), a lungworm of snowshoe hares, *Lepus americanus*. *Canadian Journal of Zoology* 62, 473–479.

Kralka, R.A. and Samuel, W.M. (1984b) Emergence of larval *Protostrongylus boughtoni* (Nematoda: Metastrongyloidea) from a snail intermediate host, and subsequent infection in the domestic rabbit (*Oryctolagus cuniculus*). *Journal of Parasitology* 70, 457–458.

Kralka, R.A. and Samuel, W.M. (1990) The lungworm *Protostrongylus boughtoni* (Nematoda, Metastrongyloidea) in gastropod intermediate hosts and the snowshoe hare *Lepus americanus*. *Canadian Journal of Zoology* 68, 2567–2575.

Krug, E.S. and Mayhew, R.L. (1949) Studies on bovine gastro-intestinal parasites. XIII. Species diagnosis of nematode infections by egg characteristics. *Transactions of the American Microscopical Society* 61, 234–239.

Kudo, N., Oku, Y., Kamiya, M. and Ohbayashi, M. (1983) Development and migration route of *Angiostrongylus siamensis* in mice. *Japanese Journal of Research* 31, 151–163.

Kulmamatov, E.N. (1981) Morphobiological characteristics of the nematode *Protostrongylus davtiani* (Savina, 1940) in Uzbekistan. *Legochnye gel'mintozy zhvachnykh zhivotnykh* (*Nauchnye Trudi VASKhNIL*), pp. 25–31. (In Russian.)

Kutzer, E. and Prosl, H. (1975) Zur Kenntnis von *Elaphostrongylus cervi* Cameron, 1931. I. Morphologie und Diagnose. *Wiener Tierärztliche Monatsschrift* 62, 258–266.

Lammler, G. and Saupe, E. (1968) Infektionversuche mit dem Lungenwurm des Igels, *Crenosoma striatum* (Zeder, 1800). *Zeitschrift für Parasitenkunde* 31, 87–100.

Lankester, M.W. (1970) The biology of *Skrjabingylus* spp. (Metastrongyloidea: Pseudaliidae) in mustelids (Mustelidae). Unpublished PhD thesis, University of Guelph, Guelph, Ontario.

Lankester, M.W. (1977) Neurologic disease in moose caused by *Elaphostrongylus cervi* Cameron, 1931 from caribou. *Proceedings of the North American Moose Conference and Workshop* 13, 177–190.

Lankester, M.W. (1983) *Skrjabingylus* Petrov, 1927 (Nematoda: Metastrongyloidea) emended with redescriptions of *S. nasicola* (Leuckart, 1842) and *S. chitwoodorum* Hill, 1939 from North American mustelids. *Canadian Journal of Zoology* 61, 2168–2178.

Lankester, M.W. (1987) Pests, parasites and diseases of moose (*Alces alces*) in North America. Proceedings of the Second International Moose Symposium. *Swedish Wildlife Research* (Suppl.) 1, 461–485.

Lankester, M.W. and Anderson, R.C. (1968) Gastropods as intermediate hosts of *Pneumostrongylus tenuis* Dougherty of white-tailed deer. *Canadian Journal of Zoology* 46, 373–383.

Lankester, M.W. and Anderson, R.C. (1971) The route of migration and pathogenesis of *Skrjabingylus* spp. (Nematoda: Metastrongyloidea) in mustelids. *Canadian Journal of Zoology* 49, 1283–1293.

Lankester, M.W. and Crichton, V.J. (1972) *Skrjabingylus lutrae* n.sp. (Nematoda: Metastrongyloidea) from otter (*Lutra canadensis*). *Canadian Journal of Zoology* 50, 337–340.

Lankester, M.W. and Fong, D. (1989) Distribution of elaphostrongyline nematodes (Metastrongyloidea: Protostrongylidae) in Cervidae and possible effects of moving *Rangifer* spp. into and within North America. *Alces* 25, 133–145.

Lankester, M.W. and Fong, D. (1996) Protostrongylid nematodes in caribou (*Rangifer tarandus caribou*) and moose (*Alces alces*) in Newfoundland. *Rangifer* (Special Issue) 10, 73–83.

Lankester, M.W. and Hauta, P.L. (1989) *Parelaphostrongylus andersoni* (Nematoda: Protostrongylidae) in caribou (*Rangifer tarandus*) of northern and central Canada. *Canadian Journal of Zoology* 67, 1966–1975.

Lankester, M.W. and Northcott, T.H. (1979) *Elaphostrongylus cervi* Cameron, 1931 (Nematoda: Metastrongyloidea) in caribou (*Rangifer tarandus caribou*) of Newfoundland. *Canadian Journal of Zoology* 57, 1384–1392.

Lankester, M.W. and Peterson, W.J. (1996) The possible importance of wintering yards in the transmission of *Parelaphostrongylus tenuis* to white-tailed deer and moose. *Journal of Wildlife Diseases* 32, 31–38.

Lankester, M.L. and Samuel, W.M. (1998) *Pests, Parasites and Diseases*. In: Franzman, A.W. and Schwartz, C.C. (eds) *Ecology and Management of the North American Moose*. Wildlife Management Institute, Smithsonian Institute Press, Washington DC, and London, pp. 479–518.

Lankester, M.W., Olsson, I.M.C., Stéen, M. and Gajadhar, A.A. (1998) Extra-mammalian larval stages of *Elaphostrongylus alces* (Nematoda: Protostrongylidae), a parasite of moose (*Alces alces*) in Fennoscandia. *Canadian Journal of Zoology* 76, 33–38.

LaPage, G. (1956) *Veterinary Parasitology*. Oliver and Boyd, Edinburgh, UK.

Lee, D.L. (1972) Penetration of mammalian skin by the infective larva of *Nippostrongylus brasiliensis*. *Parasitology* 65, 499–505.

Lee, K.T., Little, M.D. and Beaver, P.C. (1975) Intracellular (muscle fibre) habitat of *Ancylostoma caninum* in some mammalian hosts. *Journal of Parasitology* 61, 589–598.

Lee, S.H. (1967) Sex ratio and distribution of *Angiostrongylus cantonensis* in laboratory-infected white rats. *Journal of Parasitology* 53, 830.

Leiby, P.D. (1963) Taxonomic and biological studies on the nematodes *Amidostomum* (Strongyloidea) and *Epomidiostomum* (Trichostrongyloidea) occurring in the gizzards of waterfowl. Unpublished PhD thesis, Colorado State University, University Microfilm Inc., Ann Arbor, Michigan.

Leiby, P.D. and Olsen, O.W. (1965) Life history studies on nematodes of the genera *Amidostomum* (Strongyloidea) and *Epomidiostomum* (Trichostrongyloidea) occurring in the gizzards of waterfowl. *Proceedings of the Helminthological Society of Washington* 32, 32–49.

Leichtenstern, O. (1886) Fütterungsversuche mit Ankylostoma-Larven. Eine neue Rhabditis-Art in den Fäces von Ziegelarbeitern: Berichtigung. *Centralblatt für Kliniske Medicin* 7, 673–675.

Leignel, V., Humbert, J.F. and Elard, L. (1997) Study by ribosomal DNA ITS 2 sequencing and RAP analysis on the systematics of four *Metastrongylus* species (Nematoda: Metastrongyloidea). *Journal of Parasitology* 83, 606–611.

Le Jambre, L.F. (1979) Hybridization studies of *Haemonchus contortus* (Rudlophi, 1803) and *H. placei* (Place, 1893) (Nematoda: Trichostrongylidae). *International Journal for Parasitology* 9, 455–463.

Lensink, B.M., Rijpstra, A.C. and Erken, A.H.M. (1979) *Ollulanus* infections in captive Bengal tigers. *Zoologische Garten* 49, 121–126.

Leslie, A.S. and Shipley, A.E. (eds) (1912) *The Grouse in Health and Disease. Being the Popular Edition of the Report of the Committee of Inquiry on Grouse Disease.* Smith, Elder and Co., London.

Leuckart, R. (1866) Zur Entwicklungsgeschicht der Nematoden. *Archiv für Wissenschaftliche Heilkunde, Leipzig* 2, 235–250.

Levine, N.D. (1968) *Nematode Parasites of Domestic Animals and Man.* Burgess Publishing, Minneapolis, Minnesota.

Lewert, R.M. and Lee, C.L. (1954) Studies on the passage of helminth larvae through host tissues. I. Histochemical studies on extracellular changes caused by penetrating larvae. II. Enzymatic activity of larvae *in vitro* and *in vivo. Journal of Infectious Diseases* 95, 13–51.

Lewis, J.W. (1978) A population study of the metastrongylid nematode *Skrjabingylus nasicola* in the weasel *Mustela nivalis. Journal of Zoology, London* 184, 225–229.

Lewis, J.W. and Bryant, V. (1976) The distribution of *Nematospiroides dubius* within the small intestine of laboratory mice. *Journal of Helminthology* 50, 163–171.

Lichtenfels, J.R. (1975) Helminths of domestic equids. Illustrated keys to genera and species with emphasis on North American forms. *Proceedings of the Helminthological Society of Washington (Special Issue)* 42, 92 pp.

Lichtenfels, J.R. (1980a) No. 8. Keys to genera of the Superfamilies Ancylostomatoidea and Diaphanocephaloidea. In: Anderson, R.C., Chabaud, A.G. and Willmott, S. (eds) *CIH Keys to the Nematode Parasites of Vertebrates.* Commonwealth Agricultural Bureaux, Farnham Royal, UK, pp. 1–41.

Lichtenfels, J.R. (1980b) No. 7. Keys to genera of the Superfamily Strongyloidea. In: Anderson, R.C., Chabaud, A.G. and Willmott, S. (eds) *CIH Keys to the Nematode Parasites of Vertebrates.* Commonwealth Agricultural Bureaux, Farnham Royal, UK, pp. 1–41.

Lichtenfels, J.R. and Tromba, F.G. (1972) The morphogenesis of *Stephanurus dentatus* (Nematoda: Strongylina) in swine with observations on larval migration. *Journal of Parasitology* 58, 757–766.

Lim, B.L. (1966) Land molluscs as food of Malaysian rodents and insectivores. *Journal of Zoology, London* 148, 544–560.

Lim, B.L. (1970) Experimental infection of primitive primates and monkeys with *Angiostrongylus cantonensis. Southeast Asian Journal of Tropical Medicine and Public Health* 1, 361–365.

Lim, B.L. and Heyneman, D. (1965) Host–parasite studies of *Angiostrongylus cantonensis* (Nematoda, Metastrongylidae) in Malaysian rodents: natural infection of rodents and molluscs in urban and rural areas of central Malaya. *Annals of Tropical Medicine and Parasitology, Great Britain* 59, 425–433.

Lim, B.L., Ow-Yang, C.K. and Lie Kian Joe (1962) Prevalence of *Angiostrongylus cantonensis* in Malayan rats and some possible intermediate hosts. *Medical Journal of Malaya* 17, 89.

Lim, B.L., Ow-Yang, C.K. and Lie Kian Joe (1963) Further results of studies on *Angiostrongylus cantonensis* Chen, 1935 in Malaya. *Singapore Medical Journal* 4, 179.

Lim, B.L., Ow-Yang, C.K. and Lie Kian Joe (1965) Natural infection of *Angiostrongylus cantonensis* in Malaysian rodents and intermediate hosts and preliminary observations on acquired resistance. *American Journal of Tropical Medicine and Hygiene* 14, 610–617.

Lima, W.S., Guimardes, M.P. and Lemos, J.S. (1994) Occurrence of *Angiostrongylus vasorum* in the lungs of the Brazilian fox *Dusicyon vetulus*. *Journal of Helminthology* 68, 87.

Little, M.D. (1959) Insects as possible paratenic hosts of *Ancylostoma caninum*. *Journal of Parasitology* (Suppl.) 47, 263–267.

Liu, S.K. (1965) Pathology of *Nematospiroides dubius*. I. Primary infections in C_3H and Webster mice. *Experimental Parasitology* 17, 123–135.

Liubimov, M.P. (1948) New helminths in the brain of maral deer. *Trudi Gel'mintologicheskoi Laboratorii* 1, 198–201. (In Russian.)

Liubimov, M.P. (1959a) The seasonal dynamics of elaphostrongylosis and setariosis in *Cervus elaphus maral*. *Trudi Gel'mintologicheskoi Laboratorii* 9, 155–156. (In Russian.)

Liubimov, M.P. (1959b) New observations on the epizootiology, prophylaxis, and therapy of elaphostrongylosis in maral deer. *Sbornik Stakei po Pantovnomu Olenevodstvu, Gorno-Altai, USSR* pp. 164–214. (In Russian.)

Loison, G., Cavallo, A. and Vervent, G. (1962) Etude expérimentale chez le singe *Macacus rhesus* du rôle de *Angiostrongylus cantonensis* dans l'étiologie des meningites à éosinophiles observés chez l'hômme dans certains territoire du Pacifique Sud. *Bulletin de la Société de Pathologie Exotique* 55, 1108–1122.

Lomakin, V.V. (1988) *Amidostomum auriculatum* sp.nov. (Amidostomatidae, Strongylidae), a new species of nematode parasite in *Anas querquedula*. *Zoologicheskii Zhurnal* 67, 780–784. (In Russian.)

Looss, A. (1898) Zur Lebensgeschichte des *Ancylostoma duodenale*. Eine Erwiderung an Herrn Prof. Dr. Leichtenstern. *Centralblatt für Bakteriologie* 24, 442–449, 483–488.

Looss, A. (1911) The anatomy and life history of *Ancylostoma duodenale* Dub. *Records of the Egyptian Government School of Medicine* 4, 167–616.

Lovat, Lord (ed.) (1911) *The Grouse in Health and Disease. Report of the Committee of Inquiry on Grouse Disease.* Smith, Elder and Co., London.

Lucker, J.T. (1934) The morphology and development of the preparasitic larvae of *Poteriostomum ratzii*. *Journal of the Washington Academy of Science* 24, 302–310.

Lucker, J.T. (1936a) Comparative morphology and development of infective larvae of some horse strongyles. *Proceedings of the Helminthological Society of Washington* 3, 22–25.

Lucker, J.T. (1936b) Preparasitic moults in *Nippostrongylus muris*, with remarks on the structure of the cuticula of trichostrongyles. *Parasitology* 28, 161–171.

Lucker, J.T. (1938) Description and differentiation of infective larvae of three species of horse strongyles. *Proceedings of the Helminthological Society of Washington* 5, 1–5.

Lucker, J.T. and Neumayer, E.M. (1946) Experiments on the pathogenicity of hookworm (*Bunostomum trigonocephalum*) infections in lambs fed an adequate diet. *American Journal of Veterinary Research* 7, 101–122.

Lyons, E.T. (1994) Vertical transmission of nematodes: emphasis on *Uncinaria lucasi* in northern fur seals and *Strongyloides westeri* in equids. *Journal of the Helminthological Society of Washington* 61, 169–178.

Lyons, E.T. and Keyes, M.C. (1978) Observations on the infectivity of parasitic third-stage larvae of *Uncinaria lucasi* Stiles, 1901 (Nematoda: Ancylostomatidae) of northern fur seals, *Callorhinus ursinus* Linn., on St. Paul Island, Alaska. *Journal of Parasitology* 64, 454–458.

Lyons, E.T., Drudge, J.H., Tolliver, S.C. and Swerczek, T.W. (1987) Infectivity of *Trichostrongylus axei* for *Bos taurus* calves after 25 years of passage in rabbits, *Oryctolagus cuniculus*. *Proceedings of the Helminthological Society of Washington* 54, 242–244.

Lyons, E.T., Drudge, J.H., Tolliver, S.C. and Collins, S.S. (1996) Some aspects of experimental infections of *Trichostrongylus axei* in domestic rabbits (*Oryctologus cuniculus*). *Journal of the Helminthological Society of Washington* 62, 275–277.

Lyons, E.T., Delong, R.L., Melin, S.R. and Tollivier, S.C. (1997) Uncinariasis in northern fur seal and California sea-lion pups from California. *Journal of Wildlife Diseases* 33, 848–852.

Macchioni, G. and Pierotti, P. (1972) Nodular enteritis by *Molineus torulosus* (Molin, 1861) in *Cebus capucinus*. *Annali della Facoltà di Medicina Veterinaria di Pisa* 25, 274–285.

Mackerras, M.J. (1957) Observations on the life history of the cat lungworm, *Aelurostrongylus abstrusus* (Railliet, 1898) (Nematoda: Metastrongylidae). *Australian Journal of Zoology* 5, 188–195.

Mackerras, M.J. and Sandars, D.F. (1954) Life history of the rat lungworm and its migration through the brain of its hosts. *Nature, London* 173, 956–958.

Mackerras, M.J. and Sandars, D.F. (1955) The life history of the rat lung-worm, *Angiostrongylus cantonensis* (Chen) (Nematoda: Metastrongylidae). *Australian Journal of Zoology* 3, 1–25.

Madsen, H. (1952) A study on the nematodes of Danish gallinaceous birds. *Danish Review of Game Biology* 2, 1–126.

Manga-González, M.Y. and Morrondó-Pelayo, M.P. (1990) Joint larval development of *Cystocaulus ocreatus/Muellerius capillaris* and *Cystocaulus ocreatus/Neostrongylus linearis* (Nematoda) in six species of Helicidae (Mollusca) experimentally infected. *Angewandte Parasitologie* 31, 189–197.

Manga-González, M.Y. and Morrondó-Pelayo, M.P. (1994) Larval development of ovine *Neostrongylus linearis* in four experimentally infected mollusc species. *Journal of Helminthology* 48, 207–210.

Mansfield, M.E., Todd, K.S. and Levine, N.D. (1977) Developmental arrest of *Haemonchus contortus* larvae in lambs given larval inoculum exposed to different temperatures and storage conditions. *American Journal of Veterinary Research* 38, 803–806.

Mapes, C.R. and Baker, D.W. (1950) Studies on the protostrongyline lungworms of sheep. *Journal of the American Veterinary Medical Association* 116, 433–435.

Mapes, C.J. and Coop, R.L. (1972) The development of single infections of *Nematodirus battus* in lambs. *Parasitology* 64, 197–216.

Marcos Martinez, M. del R. (1977) Histopathologia de les relaciones *Neostrongylus linearis* (Marotel, 1913) Gebauer, 1932, *Cernuella* (*Xeromagna*) *cespitum arigonis* (Rossmassler, 1854) y *C.* (*C.*) *virgata* (Da Costa, 1778) en infestacion experimental. *Anales de la Facultad de Veterinaria de Leon* 21, 103–174.

Marotel, G. (1908) L'oesophagostomose nodulaire. *Journal de Médecine Vétérinaire et de Zootechnie, Lyon* 59, 522–534.

Marquardt, W.C., Fritts, D.H., Senger, C.M. and Seghetti, L. (1959) The effect of weather on the development and survival of the free-living stages of *Nematodirus spathiger* (Nematoda: Trichostrongylidae). *Journal of Parasitology* 45, 431–439.

Mason, P.C. (1975) New parasite records from the South Island (Correspondence). *New Zealand Veterinary Journal* 23, 69.

Mason, P.C. (1989) *Elaphostrongylus cervi* – a review. *Surveillance* 16, 3–10.

Mason, P.C. (1994) Parasites of deer in New Zealand. *New Zealand Journal of Zoology* 21, 39–47.

Mason, P.C. (1995) *Elaphostrongylus cervi* and its close relatives; a review of protostrongylids (Nematoda, Metastrongyloidea) with spiny-tailed larvae. *Surveillance (Wellington)* 22, 19–24.

Mason, P.C. and McAllum, H.J.F. (1976) *Dictyocaulus viviparus* and *Elaphostrongylus cervi* in wapiti. *New Zealand Veterinary Journal* 24, 23.

Mason, P.C., Kiddey, H.R., Sutherland, R.J., Rutherford, D.M. and Green, M.G. (1976) *Elaphostrongylus cervi* in red deer. *New Zealand Veterinary Journal* 24, 22–23.

Matekin, P.V., Turlygina, E.S. and Shalaeva, E.M. (1954) Biology of lung protostrongylids of sheep and goats in the context of epizootiology of protostrongylosis in central Asia. *Zoologicheskii Zhurnal* 33, 373–394. (In Russian.)

Matthews, B.E. (1975) Mechanism of skin penetration by *Ancylostoma tubaeforme* larvae. *Parasitology* 70, 25–38.

Maupas, E. and Seurat, L.G. (1912) Sur l'évolution du strongyle filaire. *Compte Rendu des Séances de la Société de Biologie* 73, 522–526.

Maupas, E. and Seurat, L.G. (1913) La mue et l'enkystement chez les *Strongyles* du tube digestif. *Compte Rendu des Séances de la Société de Biologie* 74, 34–38.

Mayhew, I.G., Lahunta, A., Georgi, J.R. and Aspros, D.G. (1976) Naturally occurring cerebrospinal parelaphostrongylosis. *The Cornell Veterinarian* 66, 56–72.

Mayhew, R.L. (1939) Studies on bovine gastrointestinal parasites. I. The mode of infection of the hookworm and nodular worm. *Cornell Veterinarian* 29, 367–376.

Mayhew, R.L. (1946) Infection experiments with the hookworm (*Bunostomum phlebotomum*) in calves. *Proceedings of the Society for Experimental Biology and Medicine* 63, 360–361.

Mayhew, R.L. (1948) Studies on bovine gastrointestinal parasites. XI. The life cycle of the hookworm (*Bunostomum phlebotomum*) in the calf. *American Journal of Veterinary Research* 9, 35–39.

Mayhew, R.L. (1949) Studies on bovine gastro-intestinal parasites. XII. Additional infection experiments with the hookworm (*Bunostomum phlebotomum*) in the calf. *Journal of Parasitology* 35, 315–321.

Mayhew, R.L. (1950) Studies on bovine gastro-intestinal parasites XV. The length of life of the adult nodular worm and hookworm in the calf. *The American Midland Naturalist* 43, 62–65.

Mayhew, R.L. (1962) Studies on bovine gastrointestinal parasites. XXV. Duration of egg production of *Cooperia punctata*, *Trichostrongylus axei* and *Oesophagostumum radiatum*. *Journal of Parasitology* 48, 871–873.

Maze, R.J. and Johnston, C. (1986) Gastropod intermediate hosts of the meningeal worm *Parelaphostrongylus tenuis* in Pennsylvania: observations on their ecology. *Canadian Journal of Zoology* 64, 185–188.

McCoy, O.R. (1930) The influence of temperature, hydrogen-ion concentration, and oxygen tension on the development of the eggs and larvae of the dog hookworm, *Ancylostoma caninum*. *American Journal of Hygiene* 11, 413–448.

McCoy, O.R. (1931a) The egg production of two physiological strains of the dog hookworm, *Ancylostoma caninum*. *American Journal of Hygiene* 14, 194–202.

McCoy, O.R. (1931b) Immunity reactions of the dog against hookworm (*Ancylostoma caninum*) under conditions of repeated infection. *American Journal of Hygiene* 14, 268–303.

McCracken, R.M. and Ross, J.G. (1970) The histopathology of *Oesophagostomum dentatum* infections in pigs. *Journal of Comparative Pathology* 80, 619–623.

McCraw, B.M. and Slocombe, J.O.D. (1974) Early development of and pathology associated with *Strongylus edentatus*. *Canadian Journal of Comparative Medicine* 38, 124–138.

McCraw, B.M. and Slocombe, J.O.D. (1976) *Strongylus vulgaris* in the horse: a review. *Canadian Veterinarian Journal* 17, 150–157.

McCraw, B.M. and Slocombe, J.O.D. (1978) *Strongylus edentatus*: development and lesions from ten weeks postinfection to patency. *Canadian Journal of Comparative Medicine* 42, 340–356.

McCraw, B.M. and Slocombe, J.O.D. (1985) *Strongylus equinus*: development and pathological effects in the equine host. *Canadian Journal of Comparative Medicine* 49, 372–383.

McGhee, M.B., Nettles, V.F., Rollor, E.A., Prestwood, A.K. and Davidson, W.R. (1981) Studies on cross-transmission and pathogenicity of *Haemonchus contortus* in white-tailed deer, domestic cattle and sheep. *Journal of Wildlife Diseases* 17, 353–364.

McGladdery, S.E. (1984) Behavioural responses of *Trichostrongylus tenuis* (Cobbold) larvae in relation to their transmission to the red grouse, *Lagopus lagopus scoticus. Journal of Helminthology* 58, 295–299.

McKenna, P.B. (1973) The effect of storage on the infectivity and parasitic development of third-stage *Haemonchus contortus* larvae in sheep. *Research in Veterinary Science* 14, 312–316.

Measures, L.N. and Anderson, R.C. (1983a) New subspecies of the stomach worm *Obeliscoides cuniculi* (Graybill), of lagomorphs. *Proceedings of the Helminthological Society of Washington* 50, 1–4.

Measures, L.N. and Anderson, R.C. (1983b) Development of free-living stages of *Obeliscoides cuniculi multistriatus* Measures and Anderson, 1983. *Proceedings of the Helminthological Society of Washington* 50, 15–24.

Measures, L.N. and Anderson, R.C. (1983c) Development of the stomach worm, *Obeliscoides cuniculi* (Graybill), in lagomorphs, woodchucks and small rodents. *Journal of Wildlife Diseases* 19, 225–233.

Measures, L.N. and Anderson, R.C. (1984) Hybridization of *Obeliscoides cuniculi* (Graybill, 1923) Graybill, 1924 and *Obeliscoides cuniculi multistriatus* Measures and Anderson, 1983. *Proceedings of the Helminthological Society of Washington* 51, 179–186.

Medica, D.L. and Sukhdeo, M.V.K. (1997) Role of lipids in the transmission of the infective stage (L3) of *Strongylus vulgaris* (Nematoda: Strongylida). *Journal of Parasitology* 83, 775–779.

Michel, J.F. (1963) The phenomena of host resistance and the course of infection of *Ostertagia ostertagi* in calves. *Parasitology* 53, 63–84.

Michel, J.F. (1968) Immunity to helminths associated with the tissues. In: Taylor, A.E.R. (ed.) *Immunity to Parasites*. Blackwell, Oxford, pp. 67–89.

Michel, J.F. (1969) Observations on the epidemiology of parasitic gastro-enteritis in calves. *Journal of Helminthology* 43, 111–133.

Michel, J.F. (1974) Arrested development of nematodes and some related phenomena. *Advances in Parasitology* 12, 279–366.

Michel, J.F. (1978) Topical themes in the study of arrested development. In: Borgsteede, F.H.M. (ed.) *Facts and Reflections. III. Workshop on Arrested Development of Nematodes of Sheep and Cattle.* Central Veterinary Insitute, Lelystad, The Netherlands, pp. 7–17.

Michel, J.F. and Rose, J.H. (1954) Some observations on the free living stages of the cattle lungworm in relation to their natural environment. *Journal of Comparative Pathology* 64, 195–205.

Michel, J.F., Lancaster, M.B. and Hong, C. (1970) Observations on the inhibition of development of *Cooperia oncophora* in calves. *British Veterinary Journal* 126, 35–37.

Miller, T.A. (1965) Influence of age and sex on susceptibility of dogs to primary infection with *Ancylostoma caninum. Journal of Parasitology* 51, 701–704.

Miller, T.A. (1970a) Potential transport hosts in the life cycle of canine and feline hookworms. *Journal of Parasitology* (Suppl.) 56, 238.

Miller, T.A. (1970b) Studies on the incidence of hookworm infection in East Africa. *East African Medical Journal* 47, 354–363.

Miller, T.A. (1979) Hookworm infection in man. *Advances in Parasitology* 17, 315–384.

Mitchell, A., Bairden, K., Duncan, J.L. and Armour, J. (1985) Epidemiology of *Nematodirus battus* infection in eastern Scotland. *Research in Veterinary Science* 38, 197–201.

Mitskevich, V.Y. (1957) The causative agent of elaphostrongylosis of reindeer and its cycle of development. *Tezisy Dokladov Nauchnoi Konferentsii Vsesoynznogo Obshcestva Gel'mintologov (Posviash. 40 g. Oktiabr. Sotsial. Revoliuts). Part 1*, pp. 206–207. (In Russian.)

Mitskevich, V.Y. (1958) On the interpretation of the developmental cycle of the nematode *Elaphostrongylus rangiferi* n.sp. from reindeer. *Doklady Akademii Nauk SSSR* 119, 621–624. (In Russian.)

Mitskevich, V.Y. (1963) Elaphostrongylosis in reindeer. In: *Helminths of Man, Animals and Plants and their Control.* Papers presented to academician K.I. Skrjabin on his 85th birthday, Moscow. Izdatel'stvo Akademii Nauk SSSR, pp. 421–423. (In Russian.)

Mitskevich, V.Y. (1964) Life history of *Elaphostrongylus rangiferi* Miz., 1958. In: Boev, S.N. (ed.) *Parasites of Farm Animals in Kazakhstan,* Vol. 3. Alma-Ata Izdatel'stvo Akademii Nauk SSSR, 49–60. (In Russian.)

Mizuno, T. and Ito, S. (1963) The mode of infection of hookworms. XVII. Experimental infection in man with infective larvae. *Medicine and Biology, Tokyo* 67, 175–177. (In Japanese.) (Abstracted in *Helminthological Abstracts,* 1965, Vol. 34, No. 2408.)

Monge, E., Arroyo, R. and Solano, E. (1978) A new definitive natural host of *Angiostrongylus costaricensis* (Morera and Cespedes, 1971). *Journal of Parasitology* 64, 34.

Mönnig, H.O. (1926) The life histories of *Trichostrongylus instabilis* and *T. rugatus* of sheep in South Africa. *11–12th Annual Report of the Director of Veterinary Education and Research, Union of South Africa,* pp. 231–251.

Mönnig, H.O. (1930) Studies on the bionomics of the free-living stages of *Trichostrongylus* spp. and other parasitic nematodes. *16th Report of the Director of Veterinary Services, Department of Agriculture, Union of South Africa* 16, 175–198.

Monson, R.A. and Post, G. (1972) Experimental transmission of *Protostrongylus stilesi* to bighorn–mouflon sheep hybrids. *Journal of Parasitology* 58, 29–33.

Morera, P. (1973) Life history and redescription of *Angiostrongylus costaricensis* Morera and Céspedes, 1971. *American Journal of Tropical Medicine and Hygiene* 22, 613–621.

Morera, P. (1978) Definitive and intermediate hosts of *Angiostrongylus costaricensis.* In: *4th International Congress of Parasitology, 19–26 August 1978, Warsaw,* p. 17.

Morera, P. (1985) Abdominal angiostrongyliasis: a problem of public health. *Parasitology Today* 1, 173–175.

Morera, P. and Cespedes, R. (1971) *Angiostrongylus costaricensis* n.sp. (Nematoda: Metastrongyloidea), a new lungworm occurring in man in Costa Rica. *Revista de Biologia Tropical* 18, 173–185.

Morera, P., Perez, F., Mora, F. and Castro, L. (1982) Visceral larva migrans-like syndrome caused by *Angiostrongylus costaricensis. American Journal of Tropical Medicine and Hygiene* 31, 67–70.

Morgan, D.O. and Clapham, P.A. (1934) Some observations on gapeworm in poultry and game birds. *Journal of Helminthology* 12, 63–70.

Moser, M. and Rhinehart, H. (1993) The lungworm, *Halocercus* spp. (Nematoda: Pseudaliidae) in cetaceans from California. *Journal of Wildlife Diseases* 29, 507–508.

Moss, R., Trenholm, I.B., Watson, A. and Parr, R. (1990) Parasitism, predation and survival of hen red grouse *Lagopus lagopus scoticus* in spring. *Journal of Animal Ecology* 59, 631–642.

Moss, R., Watson, A., Trenholm, J.B. and Parr, R. (1993) Caecal thread worms *Trichostrongylus tenuis* in red grouse *Lagopus lagopus scoticus* effects of weather and host density upon estimated worm burdens. *Parasitology* 107, 199–209.

Moss, R., Watson, A. and Parr, R. (1996) Experimental prevention of a population cycle in red grouse. *Ecology* 77, 1512–1530.

Mota, E.M. and Lenzi, H.L. (1995) *Angiostrongylus costaricensis* life cycle: a new proposal. *Memórias do Instituto Oswaldo Cruz* 90, 707–709.

Muller, G.L. (1968) The epizootiology of helminth infestation in sheep in the south western districts of the Cape. *Onderstepoort Journal of Veterinary Research* 35, 159–194.

Müller, B. (1953) Über die Entwicklung und Differenzierung der *Trichonema*-Larven in der Darmwand des Pferdes. *Archiv für Experimentelle Veterinärmedizin* 7, 58–84, 153–175.

Müller, F.R. (1934) Ein Beitrag zur Entwicklung des Lungenwurmes *Neostrongylus linearis* Marotel, 1913. *Sitzungsberichte der Gesellschaft Naturforschender Freunde zu Berlin* 2, 158–161.

Myasnikova, E.A. (1946) The biology of *Oesophagostomum dentatum* (Rud., 1803). *Collected Papers on Helminthology Dedicated by his Pupils to K.I. Skrjabin in his 40th Year of Scientific, Educational and Administrative Achievement*, pp. 192–198. (In Russian.)

Myers, G.H. and Taylor, R.F. (1989) Ostertagiasis in cattle. *Journal of the Veterinary Diagnostic Investigation* 1, 195–200.

Nagahana, M., Tanabe, K., Yoshida, Y., Kondo, K., Ishikawa, M., Okada, S., Sato, K., Okamoto, K., Ito, S. and Fukutome, S. (1963) Experimental studies on the oral infection of *Necator americanus*. III. Experimental infection of three cases of human beings with *Necator americanus* larvae through the mucous membrane of the mouth. *Japanese Journal of Parasitology* 12, 162–167. (In Japanese.)

Nakamura, Y. (1937) Development of *Trichostrongylus instabliis* Section II. The larva. *Keio Igaku Tokyo* 17, 797–814.

Narain, B. and Chaudhry, H.S. (1971) Effect of temperature on the free-living stages of *Haemonchus contortus*. *Indian Biologist* 3, 54–56.

Nascimento, A.A., do, Arantes, I.G. and Artigas, P. de Toledo (1988) *Gaigeria pachyscelis* Railliet and Henry, 1910 (Nematoda, Ancyclostomatoidea) parasito de ovinos e caprinos. Oviposicão díaria média e período pré-patente. *Revista Centro de Ciencias Rurais, Santa Maria* 18, 177–181.

Nawalinski, T.A. and Schad, G.A. (1974) Arrested development in *Ancylostoma duodenale*: course of a self-induced infection in man. *American Journal of Tropical Medicine and Hygiene* 23, 894–898.

Neiman, P.K. (1977) The biology of *Nematodirus*, parasite of goats in southern Kirgiziya. *Byulleten Vsesoyuznogo Instituta Gel'mintologii im. K.I. Skrjabina* 21, 50–53. (In Russian.)

Nettles, V.F. and Prestwood, A.K. (1976) Experimental *Parelaphostrongylus andersoni* infection in white-tailed deer. *Veterinary Pathology* 13, 381–393.

Nettles, V.F., Prestwood, A.K., Nichols, R.G. and Whitehead, C.J. (1977) Meningeal worm-induced neurologic disease in black-tailed deer. *Journal of Wildlife Diseases* 13, 137–143.

Nguyen Duc, T. and Bùi, L. (1995) Persistence and development of larvae of *Metastrongylus* in the external environment and in the intermediate host. *Khoa Hoc Ky Thuat Thu Y* 2, 6–13.

Nichols, D.K., Montali, R.J., Phillips, L.G., Alvarado, T.P., Bush, M. and Collins, L. (1986) *Parelaphostrongylus tenuis* in captive reindeer and sable antelope. *Journal of the American Veterinary Medical Association* 188, 619–621.

Nickel, E.A. and Haupt, W. (1969) Ein Beitrag zum Verlauf der parasitischen Phase der Knötchenwurmentwicklung beim Schwein. *Archiv für Experimental Veterinärmedizin* 23, 1203–1210.

Nilsson, O. and Sorelius, L. (1973) Trichostrongyle infections of cattle in Sweden. *Nordisk Veterinaer Medicin* 25, 65–78.

Nishimura, K. (1965) Failure of infective larvae of *Angiostrongylus cantonensis* to establish prenatal infection in rats. *Journal of Parasitology* 51, 540.

Nishimura, K. (1966) *Angiostrongylus cantonensis* infection in albino rats. *Japanese Journal of Parasitology* 15, 116–123. (In Japanese.)

Nolst, A.M. (1971) Infestation d'*Eisenia foetida* Sav. par *Syngamus trachea* (Montagu, 1811) Chapin, 1925. *Annales de Parasitologie Humaine et Comparée* 46, 257–264.

Nolst, A.M. (1973) Etude, sur coupes histologiques, de la migration des larves de *Syngamus trachea* (Montagu, 1811) Chapin, 1925 au sein d'*Eisenia foetida* Sav. *Annales de Parasitologie Humaine et Comparée* 48, 559–566.

Norris, D.E. (1971) The migratory behaviour of the infective-stage larvae of *Ancylostoma braziliense* and *Ancylostoma tubaeforme* in rodent paratenic hosts. *Journal of Parasitology* 57, 998–1009.

Nosal, P., Christensen, C.M. and Nansen, P. (1998) A study on the establishment of *Oesophagostomum dentatum* in pigs following percutaneous exposure to third-stage larvae. *Parasitological Research* 84, 773–776.

Nosanchuk, J.S., Wade, S.E. and Landolf, M. (1995) Case report of and description of parasite in *Mammomonogamus laryngeus* (human syngamosis) infection. *Journal of Clinical Microbiology* 33, 998–1000.

Nwosu, A.B.C. (1978) Investigations into the free-living phase of the cat hookworm life cycle. *Zeitschrift für Parasitenkunde* 56, 243–249.

N'Zobadila, G. and Durette-Desset, M.C. (1992) Cycle biologique de *Neoheligmonella pseudospira* (Nematoda: Nippostrongylinae), parasite d'un muridé africain. *Annales de Parasitologie Humaine et Comparée* 67, 116–125.

N'Zobadila, G., Boyer, J. and Durette-Desset, M.C. (1996a) Morphogenèse d'*Heligmosomoides polygyrus polygyrus* (Dujardin, 1845) (Trichostrongylina, Heligmosomoidea) chez *Apodemus flavicollis* en France. Comparaison avec les espèces proches: *Heligmosomoides laevis* (Dujardin, 1845) et *Heligmosomum mixtum* Schulz, 1954, parasites d'Arvicolidae. *Bulletin du Muséum National d'Histoire Naturelle Paris, 4° Série* 18, 367–385.

N'Zobadila, G., Boyer, J., Vuong, P.N. and Durette-Desset, M.C. (1996b) Chronologie du cycle et étude des pseudo-kystes d'*Heligmosomoides polygyrus polygyrus* (Dujardin, 1845) (Trichostrongylina, Heligmosomoidea) chez *Apodemus flavicollis* en France. Comparaison avec les espèces proches. *Parasite* 3, 237–246.

Oakley, G.A. (1981) Survival of *Dictyocaulus viviparus* in earthworms. *Research in Veterinary Science* 30, 255–256.

O'Brien, T.D., O'Leary, T.P., Leininger, J.R., Sherman, D.M., Stevens, D.L. and Wolf, C.B. (1986) Cerebrospinal parelaphostrongylosis in Minnesota. *Minnesota Veterinarian* 26, 18–22.

Ogbourne, C.P. (1972) Observations on the free-living stages of strongylid nematodes of the horse. *Parasitology* 64, 461–477.

Ogbourne, C.P. (1973) Survival on herbage plots of infective larvae of strongylid nematodes of the horse. *Journal of Helminthology* 47, 9–16.

Ogbourne, C.P. and Duncan, J.L. (1977) *Strongylus vulgaris in the Horse: its Biology and Veterinary Importance*. Commonwealth Agricultural Bureaux, Farnham Royal, UK, pp. 1–40.

Ogilvie, B.M. and Jones, V.E. (1971) *Nippostrongylus brasiliensis*: a review of immunity and the host/parasite relationship in the rat. *Experimental Parasitology* 29, 138–177.

Ogunsusi, R.A. and Eysker, M. (1979) Inhibited development of trichostrongylids of sheep in northern Nigeria. *Research in Veterinary Science* 26, 108–110.

Ohbayashi, M, Kamiya, M. and Bhaibulaya, M. (1979) Studies on the parasite fauna of Thailand. I. Two new metastrongylid nematodes, *Angiostrongylus siamensis* sp.n. and *Thaistrongylus harinasutai* gen. et. sp.n. (Metastrongyloidea: Angiostrongylidae) from wild rats. *Japanese Journal of Veterinary Research* 27, 5–10.

Okoshi, S. and Murata, Y. (1967a) Experimental studies on ancylostomiasis in cats. IV. Experimental infection of *Ancylostoma tubaeforme* and *A. caninum* to cat. *Japanese Journal of Veterinary Science* 29, 251–258.

Okoshi, S. and Murata, Y. (1967b) Experimental studies on ancylostomiasis in cats. V. Visceral migration of larvae of *Ancylostoma tubaeforme* and *A. caninum* in cats. *Japanese Journal of Veterinary Science* 29, 315–327.

Okoshi, S. and Murata, Y. (1968a) Experimental studies on ancylostomiasis in cats. VII. Experimental infection and visceral migration of larvae of *Ancylostoma tubaeforme* and *A. caninum* in mice and chickens. *Japanese Journal of Veterinary Science* 30, 97–107.

Okoshi, S. and Murata, Y. (1968b) Experimental studies on ancylostomiasis in cats. VI. Visceral migration of larvae of *Ancylostoma tubaeforme* and *A. caninum* in dogs. *Japanese Journal of Veterinary Science* 30, 43–51.

Oku, Y., Katakura, K., Nagarsuka, J. and Kamiya, M. (1979) Possible role of tadpoles of *Rana chensininsis* as an intermediate host of *Angiostrongylus cantonensis*. *Japanese Journal of Veterinary Research* 27, 1–4. (In Japanese.)

Oku, Y., Katakura, K. and Kamiya, M. (1980) Tadpole of the clawed frog, *Xenopis laevis*, as an experimental intermediate host of *Angiostrongylus cantonensis*. *American Journal of Tropical Medicine and Hygiene* 29, 316–318.

Oliver, J.L. III, Trosclair, S.R., Morris, J.M., Paulsen, D.B., Duncan, D.E., Kim, D-Y., Vicek, T.J. and Nasarre, M.C.A. (1996) Neurologic disease attributable to infection with *Parelaphostrongylus tenuis* in blackbuck antelope. *Journal of the American Veterinary Medical Association* 209, 140–142.

Olsen, A. and Woolf, A. (1978) The development of clinical signs and the population significance of neurologic disease in a captive wapiti herd. *Journal of Wildlife Diseases* 14, 263–268.

Olsen, A. and Woolf, A. (1979) A summary of the prevalence of *Parelaphostrongylus tenuis* in a captive wapiti population. *Journal of Wildlife Diseases* 15, 33–35.

Olsen, O.W. (1958) Hookworms, *Uncinaria lucasi* Stiles, 1901, in fur seals, *Callorhinus ursinus* (Linn.), on the Pribilof Islands. *Transactions of the Twenty-third North American Wildlife Conference, March 3, 4 and 5, 1958*, pp. 152–175.

Olsen, O.W. and Lyons, E.T. (1965) Life cycle of *Uncinaria lucasi* Stiles, 1901 (Nematoda: Ancylostomatidae) of fur seals, *Callorhinus ursinus* Linn., on the Pribilof Islands, Alaska. *Journal of Parasitology* 51, 689–700.

Olsson, I.-M.C., Lankester, M.W., Gajadhar, A.A. and Stéen, M. (1998) Tissue migration of *Elaphostrongylus* spp. in guinea pigs (*Caira porcellus*). *Journal of Parasitology* 84, 968–975.

Onar, E. (1975) Observations on *Nematodirus abnormalis* (May, 1920): isolation, eggs and larvae, pre-parasitic development. *British Veterinary Journal* 131, 231–239.

Onwuliri, C.O.E., Nwosu, A.B.C. and Anya, A.O. (1981) Experimental *Ancylostoma tubaeforme* infection of cats: changes in blood values and worm burden in relation to single infections of varying size. *Zeitschrift für Parasitenkunde* 64, 149–155.

Orihel, T.C. (1971) *Necator americanus* infection in primates. *Journal of Parasitology* 57, 117–121.

O'Roke, E.C. and Cheatum, E.L. (1950) Experimental transmission of the deer lungworm *Leptostrongylus alpenae*. *Cornell Veterinarian* 40, 315–323.

Ortlepp, R.J. (1923a) Observations on the nematode genera *Kalicephalus*, *Diaphanocephalus*, and *Occipitodontus* g.n., and on the larval development of *Kalicephalus philodryadus* sp.n. *Journal of Helminthology* 1, 165–189.

Ortlepp, R.J. (1923b) The life-history of *Syngamus trachea* (Montagu) v. Siebold, the gape-worm of chickens. *Journal of Helminthology* 1, 119–140.

Ortlepp, R.J. (1925) Observations on the life history of *Triodontophorus tenuicollis* a nematode parasite of the horse. *Journal of Helminthology* 3, 1–14.

Ortlepp, R.J. (1934) Preliminary note on the life-history of *Gaigeria pachyscelis* (Raill and Henry, 1910), a hookworm of sheep. *Onderstepoort Journal of Veterinary Science and Animal Industry* 3, 347–349.

Ortlepp, R.J. (1937) Observations on the morphology and life-history of *Gaigeria pachyscelis* Raill, and Henry, 1910: a hookworm parasite of sheep and goats. *Onderstepoort Journal of Veterinary Science and Animal Industry* 8, 183–212.

Ortlepp, R.J. (1939) Observations on the life history of *Bunostomum trigonocephalum*, a hookworm of sheep and goats. *Onderstepoort Journal of Veterinary Science and Animal Industry* 12, 305–318.

Ow-Yang, C.K. (1974) On the life history of *Hepatojarakus malayae* (Nematoda: Trichostrongylidae), a parasite of the feral rat in Malaysia. *Journal of Helminthology* 48, 293–310.

Ow-Yang, C.K. and Lim, B.L. (1965) Observations on the presence of immunity in rats to *Angiostrongylus cantonensis* (Chen). *Medical Journal of Malaya* 20, 70.

Palmer, E.D. (1955) Course of egg output over a 15 year period in a case of experimentally induced necatoriasis americanus, in the absence of hyperinfection. *American Journal of Tropical Medicine and Hygiene* 4, 756–757.

Pandey, V.S. (1972) Effect of temperature on development of the free-living stages of *Ostertagia ostertagi*. *Journal of Parasitology* 58, 1037–1041.

Panebianco, F. (1957) Crenosomiasi nel' *Erinaceous europaeus*. Studio istopathologico. *Atti della Societa Italiana delle Scienze Veterinarie* 11, 731–733.

Panin, V.Y. (1964a) Life cycle of *Elaphostrongylus panticola* Liubimov, 1945. In: Boev, S.N. (ed.) *Parasites of Farm Animals in Kazakhstan*, Vol. 3. Alma-Ata: Izdatel'stvo Akademii Nauk Kazakhstan SSSR, 34–48. (In Russian.)

Panin, V.Y. (1964b) Role of terrestrial molluscs in spreading elaphostrongylosis in deer. In: Boev, S.N. (ed.) *Parasites of Farm Animals in Kazakhstan*, Vol. 3. Alma-Ata: Izdatel'stvo Akademii Nauk Kazakhstan SSR, 79–83. (In Russian.)

Panin, V.Y. (1967) Unravelling the cycle of development of *Bicaulis sagittatus* (Mueller, 1891) – parasite of deer. *Izvestiya Akademii Nauk Kazakhskoi SSR Seriya Biologicheskaya* 1, 50–56. (In Russian.)

Panin, V.Y. and Rusikova, D.I. (1964) Susceptibility of molluscs to infection with larvae of *Elaphostrongylus panticola* Liubimov 1945. In: Boev, S.N. (ed.) *Parasites of Farm Animals in Kazakhstan*, Vol. 3. Alma-Ata: Izdatel'stvo Akademii Nauk Kazakhstan SSR, pp. 84–89. (In Russian.)

Panter, H.C. (1969) Host–parasite relationships on *Nematospiroides dubius* in the mouse. *Journal of Parasitology* 55, 33–37.

Parker, G.R. (1966) Moose disease in Nova Scotia; gastropod nematode relationships. Unpublished MS thesis, Acadia University, Wolfville, Nova Scotia, Canada.

Parker, J.C. and Haley, A.J. (1960) Phototactic and thermotactic responses of filiform larvae of the rat nematode, *Nippostrongylus muris*. *Experimental Parasitology* 9, 92–97.

Pascual, J.E., Bouli, R.P. and Aguiar, H. (1981) Eosinophilic meningitis in Cuba, caused by *Angiostrongylus cantonensis*. *American Journal of Tropical Medicine and Hygiene* 30, 960–962.

Pavlov, P. (1937) Recherches expérimentales sur le cycle évolutif de *Synthetocaulus capillaris*. *Annales de Parasitologie Humaine et Comparée* 15, 500–503.

Pavlov, A.V. (1960) On the identity of *Amidostomum fulicae* (Rudolphi, 1819) and *A. raillieti* (Skrjabin, 1915). *Trudi Gelmintologicheskoi Laboratorii. Akademii Nauk SSSR* 10, 166–172. (In Russian.)

Pemberton, R.T. (1959) Life-cycle of *Cyathostoma lari* Blanchard 1849 (Nematoda, Strongyloidea). *Nature, London* 184, 1423–1424.

Peterson, W. J. and Lankester, M.L. (1991) Aspects of the epizootiology of *Parelaphostrongylus tenuis* in a white-tailed deer population. *Alces* 27, 183–192.

Petrov, A.M. (1941) On the study of the life cycles of nematodes, representatives of the genus *Crenosoma striatum* Molin, 1860. *Comptes Rendus (Doklady) de l'Academie des Sciences de l'URSS* 30, 574–575. (In Russian.)

Petrov, A.M. and Gagarin, V.G. (1938) Study of the cycle of development of pathogens of filariasis of the lungs (*Filaroides bronchialis*) and skrjabingylosis of the frontal sinus (*Skrjabingylus petrowi*) of fur animals. *Trudi Vsesoyuznogo Inta Gel'mintologii* 3, 127–133. (In Russian.)

Petter, A.J. (1972) Description d'une nouvelle espèce d'*Aelurostrongylus* parasite de Rongeur africain. *Annales de Parasitologie Humaine et Comparée* 47, 131–137.

Petter, A.J. (1974) Le cycle évolutif de *Morerastrongylus andersoni* (Petter, 1972). *Annales de Parasitologie Humaine et Comparée* 49, 69–82.

Petter, A.J. and Cassone, J. (1975) Mode de pénétration et localisation des larves de *Morerastrongylus andersoni* (Petter, 1972) (Metastrongyloidea, Nematoda) chez l'hôte intermédiaire. *Annales de Parasitologie Humaine et Comparée* 50, 469–475.

Phillipson, R.F. (1969) Reproduction of *Nippostrongylus brasiliensis* in the rat intestine. *Parasitology* 59, 961–971.

Phuc, D.V. and Varga, I. (1975) Experimental infection of chickens, ducklings and goslings with larvae of *Amidostomum anseris* (Zeder, 1800). *Acta Veterinaria Academiae Scientiarum Hungaricae* 25, 231–239.

Pillmore, R.E. (1956) Investigations of the life history and ecology of the lungworm *Protostrongylus stilesi*. *Federal Aid Division Quarterly Report of the Colorado Department of Game and Fish*, pp. 47–70.

Pillmore, R.E. (1959) The evidence for prenatal lungworm infection of bighorn lambs. *Journal of Colorado and Wyoming Academy of Science* 4, 61.

Pitt, W.C. and Jordan, P.A. (1994) A survey of the nematode parasite *Parelaphostrongylus tenuis* in the white-tailed deer, *Odocoileus virginianus*, in a region proposed for caribou, *Rangifer tarandus caribou*, reintroduction in Minnesota. *Canadian Field-Naturalist* 108, 341–346.

Platt, T.R. and Samuel, W.M. (1978) *Parelaphostrongylus odocoilei*: life cycle in experimentally infected cervids including the mule deer, *Odocoileus h. hemionus*. *Experimental Parasitology* 46, 330–338.

Platt, T.R. and Samuel, W.M. (1984) Mode of entry of first-stage larvae of *Parelaphostrongylus odocoilei* (Nematoda: Metastrongyloidea) into four species of terrestrial gastropods. *Proceedings of the Helminthological Society of Washington* 51, 205–207.

Polderman, A.M. and Blotkamp, J. (1995) *Oesophagostomum* infections in humans. *Parasitology Today* 11, 441–481.

Politov, Y.A. (1973) The life-cycle of *Protostrongylus hobmaieri* (Shults, Orlov and Kutass, 1933). *Trudi Kazakhskogo Nauchno-Issledovatel'skogo Veterinarnogo Instituta (Khimioprofilaktika, Patogenez i Epizootologiya gel'Mintozov sel'Skokhozyaistvennykh Zhivotnykh)* 15, 367–372. (In Russian.)

Polley, L. and Creighton, S.R. (1977) Experimental direct transmission of the lungworm *Filaroides osleri* in dogs. *Veterinary Record* 100, 136–137.

Poluszynski, G. (1930) Morphologisch-biologische Untersuchungen über die freilebenden Larven einiger Pferdestrongyliden. Vorläufige Mitteilung. *Tierärztliche Rundschau* 36, 871–873.

Polyanskaya, M.V. (1964) Study of the etiology of the so-called cerebral disease of reindeer (elaphostrongylosis of the cerebrum of reindeer). *Materialy Nauchnoi Konferntsii Vsesoynznogo Obshchestva Gel'mintologov, Moscow, USSR*, Part 2, 73–76. (In Russian.)

Poole, J.B. (1956) Reaction to temperature by infective larvae of *Nematodirus filicollis*, Trichostrongylidae (Nematoda). *Canadian Journal of Comparative Medicine and Veterinary Science* 20, 169–172.

Potts, G.R., Tapper, S.C. and Hudson, P.J. (1984) Population fluctuations in red grouse: analysis of bag records and a simulation model. *Journal of Animal Ecology* 53, 21–36.

Poynter, D. (1954) Second ecdysis of infective nematode larvae parasitic in the horse. *Nature, London* 173, 781.

Poynter, D. (1956) Effect of a coliform organism (*Escherichia*) on the second ecdysis of nematode larvae parasitic in the horse. *Nature, London* 177, 481–482.

Poynter, D., Jones, B.V., Nelson, A.M.R., Peacock, R., Robinson, J., Silverman, P.H. and Terry, R.J. (1960) Recent experiences with vaccination. *Veterinary Record* 72, 1078–1086.

Prasad, D. (1959) The effects of temperature and humidity on the free-living stages of *Trichostrongylus retortaeformis*. *Canadian Journal of Zoology* 37, 305–316.

Presidente, P.J.A. and Knapp, S.E. (1973) Susceptibility of cattle to an isolate of *Dictyocaulus viviparus* from black-tailed deer. *Journal of Wildlife Diseases* 9, 41–43.

Prestwood, A.K. (1972) *Parelaphostrongylus andersoni* sp.n. (Metastrongyloidea: Protostrongylidae) from the musculature of white-tailed deer (*Odocoileus virginianus*). *Journal of Parasitology* 58, 897–902.

Prestwood, A.K. (1976) *Didelphostrongylus hayesi* gen. et. sp.n. (Metastrongyloidea: Filaroididae) from the opossum, *Didelphis marsupialis*. *Journal of Parasitology* 62, 272–275.

Prestwood, A.K. and Nettles, V.F. (1977) Repeated low level infection of white-tailed deer with *Parelaphostrongylus andersoni*. *Journal of Parasitology* 63, 974–987.

Prestwood, A.K. and Pursglove, S.R. Jr (1974) *Leptostrongylus alpenae* in white-tailed deer (*Odocoileus virginianus*) of Georgia and Louisiana. *Journal of Parasitology* 60, 573.

Prestwood, A.K., Greene, C.E., Mahaffey, E.A. and Burgess, D.E. (1981) Experimental canine angiostrongylosis: I. Pathologic manifestations. *Journal of the American Animal Hospital Association* 17, 491–497.

Prociv, P. and Luke, R.A. (1995) Evidence for larval hypobiosis in Australian strains of *Ancylostoma duodenale*. *Transactions of the Royal Society of Tropical Medicine and Hygiene* 89, 379.

Prosl, H. and Kutzer, E. (1980a) Zur Biologie und Bekämpfung von *Elaphostrongylus cervi*. *Zeitschrift für Jagdwissenschaft* 26, 198–207.

Prosl, H. and Kutzer, E. (1980b) Zur Pathologie des Elaphostrongylusbefalles beim Rothirsch (*Cervus elaphus hippelaphus*). *Monatshefte für Veterinärmedizin* 35, 151–153.

Pryadko, E.I. and Boev, S.N. (1971) Systematics, phylogeny and evolution of elaphostrongyline nematodes of deer. *Izvestiya Akademii Nauk Kazakhsoki SSR Seriya Biologicheskaya* 5, 41–48. (In Russian.)

Pryadko, E.I., Visokov, S.N. and Frolov, V.S. (1963) Epizootiology of elaphostrongylosis in deer. In: Boev, S.N. (ed.) *Parasites of Farm Animals in Kazakhstan*, Vol. 3. Alma-Ata: Izdatel'stvo Akademii Nauk Kazakhstan SSR, 74–85. (In Russian.)

Pryadko, E.I., Shol, V.A., Beisova, T.K. and Teterin, V.I. (1964) Helminths of *Cervus elaphus sibiricus* and *C. nippon* and their distribution in deer farms in the Altai region of the Kazakh SSR. In: Boev, S.N. (ed.) *Parasites of Farm Animals in Kazakhstan*, Vol. 3. Alma-Ata: Izdatel'stvo Akademii Nauk Kazakhstan SSR, 61–70. (In Russian.)

Pryadko, E.I., Teterin, V.I. and Shol, V.A. (1965) Helminth infections in pantin deer of different ages and different seasons of year. *Izvestiya Akademii Nauk Kazakhskoi SSR, Series Biol.* 4, 57–64.

Pulatov, G.S. (1984) Details in the biology of *Dictyocaulus* in goats. *Uzbekskii Biologicheskii Zhurnal* 6, 53–54. (In Russian.)

Punyagupta, S. (1965) Eosinophilic meningoencephalitis in Thailand: summary of nine cases and observations on *Angiostrongylus cantonensis* as a causative agent and *Pila ampullacea* as a new intermediate host. *American Journal of Tropical Medicine and Hygiene* 14, 370–374.

Purvis, G.M. and Sewell, M.M.H. (1971) The host–parasite relationship between the domestic rabbit and *Trichostrongylus colubriformis* (Correspondence). *Veterinary Record* 89, 151–152.

Pusterla, N., Caplazi, P. and Braun, V. (1997) Cerebrospinal nematodiasis in seven goats. *Schweizer Archiv für Tierheilkunde* 139, 282–287.

Pybus, M.J. and Samuel, W.M. (1980) Pathology of the muscleworm, *Parelaphostrongylus odocoilei* (Nematoda: Metastrongyloidea) in moose. *Proceedings of the North American Moose Conference and Workshop* 16, 152–170.

Pybus, M.J. and Samuel, W.M. (1981) Nematode muscle worm from white-tailed deer of southeastern British Columbia. *Journal of Wildlife Management* 45, 537–543.

Pybus, M.J. and Samuel, W.M. (1984a) *Parelaphostrongylus andersoni* (Nematoda: Protostrongylidae) and *P. odocoilei* in two cervid definitive hosts. *Journal of Parasitology* 70, 507–515.

Pybus, M.J. and Samuel, W.M. (1984b) Lesions caused by *Parelaphostrongylus odocoilei* (Nematoda: Metastrongyloidea) in two cervid hosts. *Veterinary Pathology* 21, 425–431.

Pybus, M.J., Foreyt, W.J. and Samuel, W.M. (1984) Natural infections of *Parelaphostrongylus odocoilei* (Nematoda: Protostrongylidae) in several hosts and locations. *Proceedings of the Helminthological Society of Washington* 51, 338–340.

Pybus, M.J., Samuel, W.M. and Crichton, V. (1989) Identification of dorsal-spined larvae from free-ranging wapiti (*Cervus elaphus*) in southwestern Manitoba, Canada. *Journal of Wildlife Diseases* 25, 291–293.

Pybus, M.J., Samuel, W.M., Welch, D.A. and Wilke, C.J. (1990) *Parelaphostrongylus andersoni* (Nematoda: Protostrongylidae) in white-tailed deer from Michigan. *Journal of Wildlife Diseases* 26, 535–537.

Pybus, M.J., Samuel, W.M., Welch, D.A., Smits, J. and Haigh, J.C. (1992) Mortality of fallow deer (*Dama dama*) experimentally infected with meningeal worm, *Parelaphostrongylus tenuis*. *Journal of Wildlife Diseases* 28, 95–101.

Pybus, M.J., Groom, S. and Samuel, W.M. (1996) Meningeal worm in experimentally infected bighorn and domestic sheep. *Journal of Wildlife Diseases* 32, 614–618.

Rahman, W.A. and Collins, G.H. (1990a) The establishment and development of *Trichostrongylus colubriformis* in goats. *Veterinary Parasitology* 35, 195–200.

Rahman, W.A. and Collins, G.H. (1990b) The establishment and development of *Haemonchus contortus* in goats. *Veterinary Parasitology* 35, 189–193.

Railliet, A. and Henry, A. (1912) Observations sur les Strongylides du genre *Nematodirus*. *Bulletin de la Société de Pathologie Exotique* 5, 35–39.

Rambo, P.R., Agostini, A.A. and Graeff-Teixeira, C. (1997) Abdominal angiostrongylosis in southern Brazil – prevalence and parasitic burden in mollusc intermediate hosts from eighteen endemic foci. *Memórias do Instituto Oswaldo Cruz* 92, 9–14.

Ransom, B.H. (1906) The life history of the twisted wireworm (*Haemonchus contortus*) of sheep and other ruminants. *Bureau of Animal Industry Circular* No. 93, United States Department of Agriculture, pp. 1–7.

Ransom, B.H. (1911) The nematodes parasitic in the alimentary tract of cattle, sheep and other ruminants. *Bureau of Animal Industry Bulletin* No. 127, United States Department of Agriculture, 132 pp.

Ransom, B.H. and Hadwen, S. (1918) Horse strongyles in Canada. *Journal of the American Veterinary Medical Association* 53, 202–214.

Rao, S.H. and Venkataratnam, A. (1977a) Studies on the life history of *Oesophagostomum asperum* Railliet and Henry, 1913 in goat. I. Free-living larval stages. *Indian Veterinarian Journal* 54, 14–20.

Rao, S.H. and Venkataratnam, A. (1977b) Studies on the life-history of *Oesophagostomum asperum* Railliet and Henry, 1913 in goats. II. Parasitic stages. *Indian Veterinarian Journal* 54, 102–107.

Raskevitz, R.F., Kocan, A.A. and Shaw, J.H. (1991) Gastropod availability and habitat utilization by wapiti and white-tailed deer on range enzootic for meningeal worm. *Journal of Wildlife Diseases* 27, 92–101.

Rees, G. (1950) Observations on the vertical migrations of the third-stage larva of *Haemonchus contortus* (Rud.) on experimental plots of *Loinum perenne* S$_{24}$, in relation to meteorological and micrometeorological factors. *Parasitology* 40, 127–143.

Refuerzo, P.G. and Reyes, P.V. (1959) Studies on *Metastrongylus apri*. I. The earthworm host in the Philippines. *Philippine Journal of Animal Industry* 19, 55–62.

Rep, B.H. (1965) The pathogenicity of *Ancylostoma braziliense*. II. Cultivation of hookworm larvae. *Tropical and Geographical Medicine* 17, 329–337.

Rep, B.H., Vetter, J.C.M. and Eijsker, M. (1968a) Cross breeding experiments in *Ancylostoma braziliense* de Faria, 1910 and *A. ceylanicum* Looss, 1911. *Tropical and Geographical Medicine* 20, 367–378.

Rep, B.H., Vetter, J.C.M., Eijsker, M. and Van Joost, K.S. (1968b) Pathogenicity of *Ancylostoma ceylanicum*. V. Blood loss of the host and sexual behaviour of the hookworms. *Tropical and Geographical Medicine* 20, 177–186.

Reyes, P.V. and Refuerzo, P.G. (1967) Studies on *Metastrongylus apri*. II. Other earthworm hosts in the Philippines. *Philippine Journal of Animal Industry* 22, 103–107.

Řezáč, P., Palkovič, L., Holasová, E. and Bušta, J. (1994) Modes of entry of the first-stage larvae of *Elaphostrongylus cervi* (Nematoda: Protostrongylidae) into pulmonate snails *Arianta arbstorum* and *Helix pomatia*. *Folia Parasitologica* 41, 209–214.

Richards, C.S. (1963) *Angiostrongylus cantonensis*: intermediate host studies. *Journal of Parasitology* 49, 46–47.

Richards, C.S. and Merritt, J.W. (1967) Studies on *Angiostrongylus cantonensis* in molluscan intermediate hosts. *Journal of Parasitology* 53, 382–388.

Rickard, L.G. (1993) Parasitic gastritis in a llama (*Lama glama*) associated with inhibited larval *Telodorsagia* spp. (Nematoda: Trichostrongyloidea). *Veterinary Parasitology* 45, 331–335.

Rickard, L.G. and Foreyt, W.J. (1992) Gastrointestinal parasites of cougars (*Felis concolor*) in Washington and the first report of *Ollulanus tricuspis* in a sylvatic felid from North America. *Journal of Wildlife Diseases* 28, 130–133.

Rickard, L.G., Hoberg, E.P., Zimmerman, G.L. and Erno, J.K. (1987) Late fall transmission of *Nematodirus battus* (Nematoda: Trichostrongyloidea) in western Oregon. *Journal of Parasitology* 73, 244–247.

Rickard, L.G., Hoberg, E.P. and Bishop, J.K. (1989) Epizootiology of *Nematodirus battus, N. filicollis*, and *N. spathiger* (Nematoda: Trichostrongyloidea) in western Oregon. *Proceedings of the Helminthological Society of Washington* 56, 104–115.

Rickard, L.G., Smith, B.B., Gentz, E.J., Frank, A.A., Pearson, E.G., Walker, L.L. and Pybus, M.J. (1994) Experimentally induced meningeal worm *Parelphostrongylus tenuis* infection in the llama (*Lama lama*): clinical evaluation and implications for parasite translocation. *Journal of Zoo and Wildlife Medicine* 25, 390–402.

Rizhikov, K.M. (1941) Freshwater mollusc, *Lymnaea stagnalis* L., as a reservoir host of the nematode *Syngamus trachea* Mont. *Comptes Rendus (Doklady) de l'Académie des Sciences de USSR* 31, 831–832. (In Russian.)

Robb, L.A. and Samuel, W.M. (1990) Gastropod intermediate hosts of lungworms (Nematoda: Protostrongylidae) on a bighorn sheep winter range: aspects of transmission. *Canadian Journal of Zoology* 68, 1976–1982.

Roberts, F.H.S. and Bremner, K.C. (1955) The susceptibility of cattle to natural infestations of the nematode *Haemonchus contortus* (Rudolphi, 1803) Cobb, 1898. *Australian Veterinary Journal* 31, 133–134.

Roberts, F.H.S., Turner, H.N. and McKevett, M. (1954) On the specific distinctness of the ovine and bovine 'strains' of *Haemonchus contortus* (Rudolphi) Cobb (Nematoda: Trichostrongylidae). *Australian Journal of Zoology* 2, 275–295.

Roberts, F.H.S., Elek, P. and Keith, R.K. (1962) Studies on resistance in calves to experimental infections with the nodular worms *Oesophagostomum radiatum* (Rudolphi, 1803) Railliet, 1898. *Journal of Agricultural Research* 13, 551–573.

Robinson, J. (1962) *Pilobolus* spp. and the translation of the infective larvae of *Dictyocaulus* from faeces in pastures. *Nature, London* 193, 253–354.

Robinson, J., Poynter, D. and Terry, R.J. (1962) The role of the fungus *Pilobolus* in the spread of the infective larvae of *Dictyocaulus viviparus*. *Parasitology* 52, 17–18.

Rodonaya, T.E. (1977) The biology of the lungworm *Protostrongylus tauricus* from the European hare. *Parazitologicheskii Sbornik, Tiblisi* 4, 91–101. (In Russian.)

Rodriguez Diego, J., Roque, E. and Izquierdo, F. (1981) The role of *Pilobolus* spp. (Zygomycetes, Mucorales) in the spreading of the infective larvae of *Dictyocaulus viviparus* (Bloch, 1782) in Cuba (Abstract). *9th International Conference of the World Association for the Advancement of Veterinary Parasitology, 13–17 July 1981, Budapest*, p. 175.

Rodriguez Diego, J., Roque, E., Izquierdo, F. and Svarc, R. (1983) The role of *Pilobolus* spp. in dissemination of the invasive larvae of *Dictyocaulus viviparus* (Bloch, 1782) in Cuba. *Helminthologia* 20, 33–38.

Rogers, W.P. and Sommerville, R.I. (1963) The infective stage of nematode parasites and its significance in parasitism. *Advances in Parasitology* 1, 109–177.

Rogers, W.P. and Sommerville, R.I. (1968) The infectious process, and its relation to the development of early parasitic stages of nematodes. *Advances in Parasitology* 6, 327–348.

Rojo-Vazquez, F.A. and Cordero del Campillo, M. (1974) Le cycle biologique de *Neostrongylus linearis* (Marotel, 1913) Gebauer, 1932. *Annales de Parasitologie Humaine et Comparée* 49, 685–699.

Romanov, V.A. (1969) Study of the life-cycle of *Castorstrongylus castoris* Chapin, 1925 (Nematoda, Strongylidae), a parasite of European beavers (In Russian). *Problemy Parazitologie* Part 1, pp. 201–203.

Romanovitch, M.I. and Slavin, A.P. (1914) Étude sur l'évolution du *Dictyocaulus filaria* (*Strongylus filaria*) et l'infestation des mouton. *Compte Rendu des Séances de la Société de Biologie* 77, 444–445.

Roneus, O. and Nordkvist, M. (1962) Cerebrospinal and muscular nematodiasis (*Elaphostrongylus rangiferi*) in Swedish reindeer. *Acta Veterinaria Scandinavica* 3, 201–225.

Rooney, J.R. (1970) *Autopsy of the Horse*. Williams and Wilkins, Baltimore, Maryland.

Roque, E., Svarc, R. and Rodriguez Diego, J. (1983) Developmental synchronism of invasive larvae of *Dictyocaulus viviparus* with the growth of fungi *Pilobolus* spp. in bovine fecal material in Cuba. *Helminthologia* 20, 131–136.

Rose, J.H. (1955) Observations on the bionomics of the free-living larvae of the lungworm *Dictyocaulus filaria*. *Journal of Comparative Pathology* 65, 370–381.

Rose, J.H. (1957) Observation on the larval stages of a *Muellerius capillaris* within the inter-mediate hosts *Agriolimax agrestis* and *A. reticulatus*. *Journal of Helminthology* 31, 1–16.

Rose, J.H. (1958) Site of development of the lungworm *Muellerius capillaris* in experimentally infected lambs. *Journal of Comparative Pathology* 68, 359–362.

Rose, J.H. (1959) *Metastrongylus apri* the pig lungworm. Observations on the free-living embryonated egg and the larva in the intermediate host. *Parasitology* 49, 439–447.

Rose, J.H. (1963a) Experimental infections of calves with the nematode parasite, *Ostertagia ostertagi*. *Veterinary Record* 75, 129–132.

Rose, J.H. (1963b) Observations on the free-living stages of the stomach worm *Haemonchus contortus*. *Parasitology* 53, 469–481.

Rose, J.H. (1964) Relationship between environment and the development and migration of the free-living stages of *Haemonchus contortus*. *Journal of Comparative Pathology* 74, 163–172.

Rose, J.H. (1966) Investigations into the free-living phase of the life cycle of *Nematodirus helvetianus*. *Parasitology* 56, 679–691.

Rose, J.H. (1969) The development of the parasitic stages of *Ostertagia ostertagi*. *Journal of Helminthology* 43, 173–184.

Rose, J.H. (1973) Lungworms of the domestic pig and sheep. *Advances in Parasitology* 2, 559–599.

Rose, J.H. and Small, A.J. (1980) Observations on the development and survival of the free-living stages of *Oesophagostomum dentatum* both in their natural environment out-of-doors and under controlled conditions in the laboratory. *Parasitology* 81, 507–517.

Rose, J.H. and Small, A.J. (1982) Observations on the development and survival of the free-living stages of *Hyostrongylus rubidus* both in their natural environments out-of-doors and under controlled conditions in the laboratory. *Parasitology* 85, 33–43.

Rose, J.H. and Small, A.J. (1984) Observations on the bionomics of the free-living stages of *Trichostrongylus vitrinus*. *Journal of Helminthology* 58, 49–58.

Rosen, L. (1967) Biology and distribution of *Angiostrongylus cantonensis*. *South Pacific Commission Seminar on Helminthiases and Eosinophilic Meningitis, Noumea, New Caledonia*, 3 pp.

Rosen, L., Laigret, J. and Bories, S. (1961) Observations on an outbreak of eosinophilic meningitis on Tahiti, French Polynesia. *American Journal of Hygiene* 74, 26–42.

Rosen, L., Chappell, R., Laqueur, G.L., Wallace, G.D. and Weinstein, P.P. (1962) Eosinophilic meningoencephalitis caused by a metastrongylid lungworm of rats. *Journal of the American Medical Association* 179, 620–624.

Rosen, L., Loison, G., Laigret, J. and Wallace, G.D. (1967) Studies on eosinophilic meningitis 3. Epidemiologic and clinical observations on Pacific islands and the possible etiological role of *Angiostrongylus cantonensis*. *American Journal of Epidemiology* 48, 818–824.

Rosen, L., Ash, L.R. and Wallace, G.D. (1970) Life history of the canine lungworm *Angiostrongylus vasorum* (Baillet). *American Journal of Veterinary Research* 31, 131–143.

Ross, I.C. and Gordon, H.M. (1936) *The Internal Parasites and Parasitic Diseases of Sheep. Their Treatment and Control.* Angus and Robertson, Sydney.

Ross, I.C. and Kauzal, G. (1932) The life cycle of *Stephanurus dentatus* Diesing, 1839: the kidney worm of pigs with observations on its economic importance in Australia and suggestions for its control. *Bulletin of the Council for Scientific and Industrial Research, Australia* No. 58, 80 pp.

Ross, I.C. and Kauzal, G. (1933) Preliminary note on the pathogenic importance of *Chabertia ovina* (Fabricius, 1788). *Australian Veterinarian Journal* 9, 215–218.

Ross, J.G. (1963) Experimental infection of calves within the nematode parasite, *Ostertagia ostertagia*. *Veterinary Record* 75, 129–132.

Ross, J.G., Purcell, D.A., Dow, C. and Todd, J.R. (1967) Experimental infections of calves with *Trichostrongylus axei*; the course and development of infection and lesions in low level infections. *Research in Veterinary Science* 8, 206–210.

Rossi, P. (1983) Sur le genre *Nematodirus* Ransom, 1907 (Nematoda: Trichostrongyloidea). *Annales de Parasitologie Humaine et Comparée* 58, 557–581.

Rubin, R. and Lucker, J.T. (1956) The course and pathogenicity of initial infections with *Dictyocaulus viviparus*, the lungworm of cattle. *American Journal of Veterinary Research* 17, 217–226.

Rufi, V.G. (1961) Life cycles of lungworms. *Federal Aid Division Quarterly Report of the Wyoming Department of Fish and Game.*

Rupasinghe, D. (1975) Developmental, physiological and morphological observations on the free-living and parasitic stages of some strongylid nematodes of the horse. Unpublished PhD thesis, University of London.

Ryzhikov, K.M., Gubanov, N.M. and Fedorov, K.P. (1956) The life cycle of protostrongylids of mountain hares (*Lepus timidus*) under conditions existing in Yakutia. *Uchenie Zapiski. Moskovski Gosudarstvenni Pedagogicheski Instituto im V.I. Lenina* 96, 137–145. (In Russian.)

Sagoyan, I.S. (1950) Experimental cystocauliasis of sheep. *Trudi Armyanskogo Nauchno-Issledovatel'skogo Veterinarnogo Instituta, Arm G1Z, Erevan, USSR* 7, 127–139. (In Russian.)

Salisbury, J.R. and Arundel, J.H. (1970) Peri-parturient deposition of nematode eggs by ewes and residual pasture contamination as sources of infection for lambs. *Australian Veterinary Journal* 46, 523–529.

Samson, J. and Holmes, J.C. (1985) The effect of temperature on rates of development of larval *Protostrongylus* spp. (Nematoda: Metastrongyloidea) from bighorn sheep, *Ovis canadensis canadensis* in the snail host *Vallonia pulchella. Canadian Journal of Zoology* 6, 1445–1448.

Sandground, J.H. (1929) *Ternidens deminutus* (Railliet and Henry) as a parasite of man in southern Rhodesia; together with observations and experimental infection studies on an unidentified nematode parasite of man from this region. *Annals of Tropical Medicine and Parasitology* 23, 23–32.

Sandground, J.H. (1931) Studies on the life-history of *Ternidens deminutus*, a nematode parasite of man, with observations on its incidence in certain regions of southern Africa. *Annals of Tropical Medicine and Parasitology* 25, 147–184.

Santos, F.T. dos, Pinto, V.M. and Graeff-Teixeira, C. (1996) Evidences against a significant role of *Mus musculus* as natural host for *Angiostrongylus costaricensis. Revista do Instituto Medicina Tropical de São Paulo* 38, 171–175.

Satrija, F. and Nansen, P. (1992a) Experimental concurrent infections with *Ostertagia ostertagi* and *Cooperia oncophora* in calves. *Bulletin of the Scandinavian Society for Parasitology* 2, 42.

Satrija, F. and Nansen, P. (1992b) Acquisition of inhibited early fourth stage *Ostertagia ostertagi* larvae in tracer calves grazed in late summer and early autumn. *Bulletin of the Scandinavian Society for Parasitology* 3, 20–22.

Sauerlander, R. (1976) Histological studies of the African giant snail (*Achatina fulica*) experimentally infected with *Angiostrongylus vasorum* or *Angiostrongylus cantonensis. Zeitschrift für Parasitenkunde* 49, 263–280.

Sauerlander, R. (1979) *Cepaea nemoralis* (Helicidae, Stylommatophora) als experimenteller Zwischenwirt für *Muellerius capillaris* (Protostrongylidae, Nematoda). *Zeitschrift für Parasitenkunde* 59, 53–66.

Sauerlander, R. and Eckert, J. (1974) Die Achatschnecke (*Achatina fulica*) als experimenteller Zwischenwirt für *Angiostrongylus vasorum* (Nematoda). *Zeitschrift für Parasitenkunde* 44, 59–72.

Sawabe, K. and Makiya, K. (1995) Comparative infectivity and survival of first-stage larvae of *Angiostrongylus cantonensis* and *Angiostrongylus malaysiensis. Journal of Parasitology* 81, 228–233.

Scarratt, W.K., Karzenski, S.S., Wallace, M.A., Chrisman, M.V., Saunders, G.K., Cordes, D.O. and Sponenberg, D.P. (1996) Suspected parelapostrongylosis in five llamas. *Progress in Veterinary Neurology* 7, 124–129.

Schad, G.A. (1956) Studies on the genus *Kalicephalus* (Nematoda: Diaphanocephalidae). I. On the life histories of the North American species *K. parvus*, *K. agkistrodontis* and *K. rectiphilus. Canadian Journal of Zoology* 34, 425–452.

Schad, G.A. (1962) Studies on the genus *Kalicephalus* (Nematoda: Diaphanocephalidae). II. A taxonomic revision of the genus *Kalicephalus* Molin, 1861. *Canadian Journal of Zoology* 40, 1035–1170.

Schad, G.A. (1977) The role of arrested development in the regulation of nematode populations. In: Esch, G.W. (ed.) *Regulation of Parasite Populations*. Academic Press, New York, pp. 111–167.

Schad, G.A. (1983) Arrested development of *Ancylostoma caninum* in dogs: influence of photoperiod and temperature on induction of a potential to arrest. In: Meerovitch, E. (ed.) *Aspects of Parasitology. Festschrift dedicated to the 50th Anniversary of the Institute of Parasitology of McGill University 1932–1982.* Institute of Parasitology, McGill University, Montreal, pp. 361–391.

Schad, G.A. and Kuntz, R.E. (1964) Speciation, zoogeography and host specificity in reptilian nematodes as illustrated by studies on the genus *Kalicephalus*. *Proceedings of the First International Congress of Parasitology Sept. 21–26, Rome, Italy*, pp. 514–515.

Schad, G.A. and Page, M.R. (1982) *Ancylostoma caninum*: adult worm removal, corticosteroid treatment, and resumed development of arrested larvae in dogs. *Experimental Parasitology* 54, 303–309.

Schad, G.A. and Warren, K.S. (1990) *Hookworm Disease: Current Status and New Directions*. Taylor and Francis, London.

Schad, G.A., Chowdhury, A.B, Dean, C.G., Kochar, V.K., Nawalinski, T.A., Thomas, J. and Tonascia, J.A. (1973) Arrested development in human hookworm infections: an adaptation to a seasonally unfavourable external environment. *Science* 180, 500–501.

Schlegel, M. (1933) Die Lungenwurmseuche beim Dachs. *Berlin Tierärztliche Wochenschrift* 49, 341–344.

Schlegel, M. (1934) Die Lungenwurmseuche beim Dachs. II. *Berlin Tierärztliche Wochenschrift* 50, 369–373.

Schneider, T., Epe, C. and Samson-Himmelstjerna, G. (1996) Species differentiation of lungworms (Dictyocaulidae) by polymerase chain reaction–restriction-fragment-length polymorphism of second internal transcribed spacers of ribosomal DNA. *Parasitology Research* 82, 392–394.

Schwartz, B. (1924) Preparasitic stages in the life history of the cattle hookworm (*Bunostomum phlebotomum*). *Journal of Agricultural Research* 29, 451–458.

Schwartz, B. and Price, E.W. (1929) The life history of the swine kidney worm. *Journal of Parasitology* 70, 613–614.

Schwartz, B. and Price, E.W. (1931) Infection of pigs through the skin with the larvae of the swine kidney worm, *Stephanurus dentatus*. *Journal of American Veterinary Medical Association* 79, 359–375.

Schwartz, B. and Price, E.W. (1932) Infection of pigs and other animals with kidney worms, *Stephanurus dentatus*, following ingestion of larvae. *Journal of the American Veterinary Medical Association* 81, 325–347.

Schwartz, B. and Alicata, J.E. (1934a) Life history of lungworms in swine. *United States Department of Agriculture Technical Bulletin* No. 456, pp. 1–42.

Schwartz, B. and Alicata, J.E. (1934b) The development of the trichostrongyle, *Nippostrongylus muris*, in rats following ingestion of larvae. *Journal of the Washington Academy of Sciences* 24, 334–338.

Schwartz, B. and Alicata, J.E. (1935) Life history of *Longistriata musculi*, a nematode parasitic in mice. *Journal of the Washington Academy of Sciences* 25, 128–146.

Schwartz, B. and Porter, D.A. (1938) The localization of swine lungworm larvae in the earthworm, *Helodrilus foetidus*. *Livro Jubilar Prof. Travassos, Rio de Janeiro, Brazil III*, pp. 429–440.

Scott, J.A. (1928) An experimental study of the development of *Ancylostoma caninum* in normal and abnormal hosts. *American Journal of Hygiene* 8, 158–204.

Scott, J.A. (1929) Experimental demonstration of a strain of the dog hookworm, *Ancylostoma caninum*, especially adapted to the cat. *Journal of Parasitology* 15, 209–215.

Scott, J.A. (1930) Further experiments with physiological strains of the dog hookworm, *Ancylostoma caninum*. *American Journal of Hygiene* 11, 149–158.

Seneviratna, P. (1959) Studies on *Anafilaroides rostratus* Gerichter, 1949 in cats. II. The life cycle. *Journal of Helminthology* 33, 109–122.

Seurat, L.G. (1916) Contributions à l'étude des formes larvaire des nématodes parasites hétéroxènes. *Bulletin Scientifique de la France et de la Belgique* 7, 297–377.

Seurat, L.G. (1920) Développment embryonnaire et evolution du *Strongylacantha glycirrhiza* Beneden (Trichostrongylidae). *Compte Rendu Société Biologie, Paris* 83, 1472–1474.

Sharma, R.L., Bhat, T.K. and Dhar, D.N. (1988) Control of sheep lungworm in India. *Parasitology Today* 4, 33–36.

Shaw, J.L. (1988) Arrested development of *Trichostrongylus tenuis* as third-stage larvae in red grouse. *Research in Veterinary Science* 45, 256–258.

Shaw, J.L. (1990) Effects of the caecal nematode *Trichostrongylus tenuis* on egg-laying by captive red grouse. *Research in Veterinary Science* 48, 59–63.

Shaw, J.L. and Moss, R. (1989a) The role of parasite fecundity and longevity in the success of *Trichostrongylus tenuis* in low density red grouse populations. *Parasitology* 99, 253–258.

Shaw, J.L. and Moss, R. (1989b) Factors affecting the establishment of the caecal threadworm *Trichostrongylus tenuis* in red grouse (*Lagopus lagopus scoticus*). *Parasitology* 99, 259–264.

Shaw, J.L., Moss, R. and Pike, A.W. (1989) Development and survival of the free-living stages of *Trichostrongylus tenuis*, a caecal parasite of red grouse *Lagopus lagopus scoticus*. *Parasitology* 99, 105–113.

Shaw, D.H., Conboy, G.A., Hogan, P.M. and Horney, B.S. (1996) Eosinophilic bronchitis caused by *Crenosoma vulpis* infection in dogs. *Canadian Veterinary Journal* 37, 361–363.

Shelton, G.C. and Griffiths, H.J. (1968a) Experimental host–parasite relationship studies with *Oesophagostomum columbianum* in sheep. I. Attempts to establish infection by different routes. *Research in Veterinary Science* 9, 354–357.

Shelton, G.C. and Griffiths, H.J. (1968b) Experimental host–parasite relationship studies with *Oesophagostomum columbianum* in sheep. II. Some effects of subcutaneous infections upon host immune and hypersensitive responses. *Research in Veterinary Science* 9, 358–365.

Shimshony, A. (1974) Observations on parasitic gastro-enteritis in goats in northern Israel. I. Clinical and helminthological findings. *Refuah Veterinarith* 31, 63–75.

Shoho, C. (1966) Observations on rats and rabbits infected with *Angiostrongylus cantonensis*. *British Veterinary Journal* 122, 251–258.

Shol, A.V. (1964) Diagnosis of *Setaria* infections in deer. In: Boev, S.N. (ed.) *Parasites of Farm Animals in Kazakhstan*, Vol. 3. Alma-Ata: Izdatel'stvo Akademii Nauk Kazakhstan SSR, 101–103. (In Russian.)

Shorb, D.A. (1939) Differentiation of eggs of various genera of nematodes parasitic in domestic ruminants in the USA. *USA Department of Agriculture Technical Bulletin* No. 694.

Shorb, D.A. (1940) A comparative study of the eggs of various species of nematode parasites in domestic ruminants. *Journal of Parasitology* 26, 223–231.

Shorb, D.A. (1942) Survival of sheep nematodes on pastures. *Journal of Agricultural Research* 65, 329–337.

Shostak, A.W. and Samuel, W.M. (1984) Moisture and temperature effects on survival and infectivity of first-stage larvae of *Parelaphostrongylus odocoilei* and *P. tenuis* (Nematoda: Metastrongyloidea). *Journal of Parasitology* 70, 261–269.

Silverman, P.H. and Campbell, J.A. (1958) Studies on parasitic worms of sheep in Scotland. I. Embryonic and larval development of *Haemonchus contortus* at constant conditions. *Parasitology* 49, 23–38.

Simaren, J.O. and Ogunkoya, O.A. (1970) A comparison of the development of *Nippostrongylus brasiliensis* larvae of various ages introduced orally and subcutaneously into laboratory rats. *Experientia* 26, 1023–1024.

Sinha, B.K. (1967) Earthworms, *Eutypheus waltoni* and *Pheretima* sp. as transport hosts of swine kidney worm, *Stephanurus dentatus* in Bihar, India (Preliminary observations). *Ceylon Veterinarian Journal* 15, 130–132.

Skalinskii, E.I. (1952) Migratory pathways of *Delafondia vulgaris* larvae in the body of the horse. *Trudi Vsesoyuznogo Instituta Eksperimental'noi Veterinarii* 19, 82–86. (In Russian.)

Skalinskii, E.I. (1954) The migration of larvae of *Delafondia vulgaris* and the pathological changes caused by them in the horse. *Trudi Gel'mintologicheskoi Laboratorii, Akademiya Nauk SSSR* 7, 392–393.

Skorping, A. (1982) *Elaphostrongylus rangiferi*: influence of temperature, substrate and larval age on the infection rate in the intermediate snail host, *Arianta arbustorum*. *Experimental Parasitology* 54, 222–228.

Skorping, A. (1985) *Lymnaea stagnalis* as experimental intermediate host of the protostrongylid nematode *Elaphostrongylus rangiferi*. *Zeitschrift für Parasitenkunde* 71, 265–270.

Skorping, A. and Halvorsen, O. (1980) The susceptibility of terrestrial gastropods of experimental infection with *Elaphostrongylus rangiferi* Mitskevich (Nematoda: Metastrongyloidea). *Zeitschrift für Parasitenkunde* 62, 7–14.

Slomke, A.M., Lankester, M.W. and Peterson, W.J. (1995) Infrapopulation dynamics of *Parelaphostrongylus tenuis* in white-tailed deer. *Journal of Wildlife Diseases* 31, 125–135.

Slonka, G.F. and Leland, S.E. (1970) *In vitro* cultivation of *Ancylostoma tubaeforme* from egg to fourth-stage larva. *American Journal of Veterinary Research* 31, 1901–1904.

Sly, D.L., Toft, J.D. II, Gardiner, C.H. and London, W.T. (1982) Spontaneous occurrence of *Angiostrongylus costaricensis* in marmosets (*Saguinus mystax*). *Laboratory Animal Science* 32, 286–288.

Smeal, M.G. and Donald, A.D. (1981) Effect on inhibition of development of the transfer of *Ostertagia ostertagi* between geographical regions of Australia. *Parasitology* 82, 389–399.

Smith, C.A. (1904) Uncinariasis in the South, with special reference to the mode of infection. *Journal of the American Medical Association* 43, 592–597.

Smith, G. (1989) Population biology of the parasitic phase of *Ostertagia circumcincta*. *International Journal for Parasitology* 19, 385–393.

Smith, G. and Grenfell, B.T. (1985) The population biology of *Ostertagia ostertagi*. *Parasitology Today* 1, 76–81.

Smith, H.J. (1974) Inhibited development of *Ostertagia ostertagi*, *Cooperia oncophora* and *Nematodirus helvetianus* in parasite-free calves grazing fall pastures. *American Journal of Veterinary Research* 35, 935–938.

Smith, H.J. and Archibald, R.M. (1965) The overwinter survival of ovine gastro-intestinal parasites in the Maritime Provinces. *Canadian Veterinary Journal* 6, 257–267.

Smith, W.D., Jackson, F., Jackson, E., Williams, J. and Miller, H.R.P. (1984) Manifestations of resistance to ovine ostertagiasis associated with immunological responses in the gastric lymph. *Journal of Comparative Pathology* 94, 591–601.

Snider, T.G., Williams, J.C., Sheehan, D.S. and Fuselier, R. H. (1981) Plasma pepsinogen, inhibited larvae development and abomasal lesions in experimental infections of calves with (*Ostertagia ostertagia*). *Veterinary Parasitology* 8, 173–184.

Soliman, K.N. (1952) Preparasitic stages of *Dictyocaulus viviparus* in southern England – II. *British Veterinary Journal* 108, 204–213.

Soliman, K.N. (1953a) Studies on the bionomics of the preparasitic stages of *Dictyocaulus viviparus* with reference to the same in the allied species in sheep '*D. filaria.*' *British Veterinary Journal* 109, 364–381.

Soliman, K.N. (1953b) Migration route of *Dictyocaulus viviparus* and *D. filaria* infective larvae to the lungs. *Journal of Comparative Pathology and Therapeutics* 63, 75–84.

Soliman, K.N. (1960) Morphological study on *Dicytocaulus arnfieldi* (Cobbold, 1884) Railliet and Henry, 1907, from a donkey in Egypt. *British Veterinary Journal* 116, 191–195.

Sollod, A.E., Hayes, T.J. and Soulsby, E.J.L. (1968) Parasitic development of *Obeliscoides cuniculi* in rabbits. *Journal of Parasitology* 54, 129–132.

Solomon, A., Paperna, I. and Markovics, A. (1996) The influence of aestivation in land snails on the larval development of *Muellerius cf. capillaris* (Metastrongyloidea: Protostrongylidae). *International Journal for Parasitology* 26, 363–367.

Soltys, A. (1964) Snails as intermediate hosts of nematodes of the family Protostrongylidae in sheep of the Lublin Palatinate. *Acta Parasitologica Polonica* 12, 233–237.

Sommerville, R.I. (1953) Development of *Ostertagia circumcincta* in the abomasal mucosa of sheep. *Nature, London* 171, 482–483.

Sommerville, R.I. (1954) The histotropic phase of the nematode parasite *Ostertagia circumcincta*. *Australian Journal of Agricultural Research* 5, 130–140.

Sommerville, R.I. (1957) The exsheathing mechanism of nematode infective larva. *Experimental Parasitology* 6, 18–30.

Sommerville, R.I. (1960) The growth of *Cooperia curticei* (Giles, 1892), a nematode parasite of sheep. *Parasitology* 50, 261–267.

Sonntag, D. (1991) Ecology of free-living stages of *Oesophagostomum quadrispinulatum* (Strongylida: Strongylidae) and migration of infective third-stage larvae. Inaugural Dissertation, Tierärzlichen Fakultät, Ludwig-Maximilians-Universität, Munich, Germany. 223 pp.

Sood, M.L. and Charanjit Kaur (1975) The effects of temperature on the survival and development of the infective larvae of twisted wireworm *Haemonchus contortus* (Rudolphi, 1803). *Indian Journal of Ecology* 2, 68–74.

Soulsby, E.J.L. (1965) *Textbook of Veterinary Clinical Parasitology*, Vol. I, *Helminths*. Blackwell Scientific Publications, Oxford, UK.

Soulsby, E.J.L. (1966) The mechanism of immunity to gastrointestinal nematodes. In: *Biology of Parasites. Proceedings of the Second International Conference of the World Association for the Advancement of Veterinary Parasitology*. Academic Press, New York, pp. 255–276.

Spindler, L.A. (1933) Development of the nodular worm, *Oesophagostomum longicaudum*, in the pig. *Journal of Agricultural Research* 46, 531–542.

Spindler, L.A. (1934) Field and laboratory studies on the behaviour of the larvae of the swine kidney worm, *Stephanurus dentatus*. *USA Department of Agriculture Technical Bulletin* No. 405, pp. 1–17.

Spindler, L.A. (1936) Effects of various physical factors on the survival of eggs and infective larvae of the swine nodular worm, *Oesophagostomum dentatum*. *Journal of Parasitology* (Suppl.) 22, 529.

Sprent, J.F.A. (1946a) Studies on the life history of *Bunostomum phlebotomum* (Railliet, 1900), a hookworm parasite of cattle. *Parasitology* 37, 192–201.

Sprent, J.F.A. (1946b) Some observations on the bionomics of *Bunostomum phlebotomum*, a hookworm of cattle. *Parasitology* 37, 202–210.

Spurlock, G.M. (1943) Observations on host–parasite relations between laboratory mice and *Nematospiroides dubius* Baylis. *Journal of Parasitology* 29, 303–311.

Stadelmann, H. (1892) Ueber den anatomischen Bau des *Strongylus convolutus* Ostertag nebst einigen Bemerkungen zu seiner Biologie. *Archiv für Naturgeschichte* 1, 149–176.

Stéen, M. (1991) Elaphostrongylosis, a clinical, pathological and taxonomic study with special emphasis on the infection in moose. Dissertation, Department of Veterinary Microbiology, Swedish University of Agricultural Sciences, Uppsala, Sweden.

Stéen, M., Chabaud, A.G. and Rehbinder, C. (1989) Species of the genus *Elaphostrongylus* parasite of Swedish Cervidae. A description of *E. alces* n.sp. *Annales de Parasitologie Humaine et Comparée* 64, 134–142.

Stéen, M., Blackmore, C.G.M. and Skorping, A. (1997) Cross-infection of moose (*Alces alces*) and reindeer (*Rangifer tarandus*) with *Elaphostrongylus alces* and *Elaphostrongylus rangiferi* (Nematoda, Protostrongylidae): effects on parasite morphology and prepatent period. *Veterinary Parasitology* 71, 27–38.

Stewart, T.B. (1954) The life history of *Cooperia punctata*, a nematode parasitic in cattle. *Journal of Parasitology* 40, 321–327.

Stewart, T.B. and Tromba, F.G. (1957) The control of the swine kidney worm *Stephanurus dentatus* through management. *Journal of Parasitology* (Suppl.) 43, 19–20

Stockdale, P.H.G. (1970a) Pulmonary lesions in mink with a mixed infection of *Filaroides martis* and *Perostrongylus pridhami*. *Canadian Journal of Zoology* 48, 757–759.

Stockdale, P.H.G. (1970b) The development, route of migration, and pathogenesis of *Perostrongylus pridhami* in mink. *Journal of Parasitology* 56, 559–566.

Stockdale, P.H.G. (1970c) The pathogenesis of the lesions elicited by *Aelurostrongylus abstrusus* during its prepatent period. *Pathologia Veterinaria* 7, 102–115.

Stockdale, P.H.G. and Anderson, R.C. (1970) The development, route of migration, and pathogenesis of *Filaroides martis* in mink. *Journal of Parasitology* 56, 550–558.

Stockdale, P.H.G., Fernando, M.A. and Lee, E.H. (1970) Age of infective larvae: a contributory factor in the inhibition of development of *Obeliscoides cuniculi* in rabbits. *Veterinary Record* 86, 176–177.

Stockdale, P.H.G. and Hulland, T. J. (1970) The pathogenesis, route of migration and development of *Crenosoma vulpis* in the dog. *Pathologia Veterinaria* 7, 28–42.

Stockdale, P.H.G., Fernando, M.A. and Craig, R. (1974) The development, route of migration and pathogenesis of *Crenosoma mephitidis* in the skunk (*Mephitis mephitis*). *Canadian Journal of Zoology* 52, 681–685.

Stödter, W. (1901) Die Strongyliden in dem Labmagen der gazähmten Wiederkäuer und die Magenwurmseuche. Dissertation, Hamburg, 108 pp.

Stoll, N.R. (1923) Investigations on the control of hookworm disease. XVIII. On the relation between the number of eggs found in human faeces and the number of worms in the host. *American Journal of Hygiene* 3, 156–179.

Stone, W.M. and Girardeau, M.H. (1966) *Ancylostoma caninum* larvae present in the colostrum of a bitch. *Veterinary Record* 79, 773–774.

Stone, W.M. and Peckham, J.C. (1970) Infectivity of *Ancylostoma caninum* larvae from canine milk. *American Journal of Veterinary Research* 31, 1693–1694.

Stoye, M. (1965) Untersuchungen über den Infektionsweg von *Bunostomum trigonocephalum* Rudolphi, 1808 (Ancylostomatidae) beim Schaf. *Zeitschrift für Parasitenkunde* 25, 526–537.

Stoye, M. (1973) Untersuchungen über die Möglichkeit pränataler und gälaktogener Infektionen mit *Ancylostoma caninum* Ercolani, 1859 (Ancylostomatidae) beim Hund. *Zentralblatt für Veterinarmedizin* 20, 1–39.

Stoye, M. and Schmelzle, H.M. (1986) Über den Ausbreitungsmodus der Larven von *Ancylostoma caninum* Ercolani, 1859 (Ancylostomatidae) im definitiven Wirt (Beagle). *Journal of Veterinarian Medicine* 33, 274–283.

Stradowski, M. (1971) The age of *Amidostomum anseris* (Zeder, 1800) Railliet and Henry, 1909 larvae and their invading activity. *Acta Parasitologica Polonica* 19, 63–68.

Stradowski, M. (1972) Attempts at experimental infection of the domestic duck *Anas platyrhynchos dom.* (L.) with *Amidostomum anseris* (Zeder, 1800) larvae (Nematoda). *Acta Parasitologica Polonica* 20, 179–187.

Stradowski, M. (1974) Development of eggs and larvae of *Amidostomum anseris* (Zeder, 1800) under laboratory conditions. *Acta Parasitologica Polonica* 22, 415–422.

Stradowski, M. (1977) Duration of prepatent, patent and postpatent periods of *Amidostomum anseris* (Zeder, 1800) infection in domestic geese. *Acta Parasitologica Polonica* 24, 249–258.

Stroh, G. and Schmid, F. (1938) *Protostrongylus capreoli* nov. sp., der häufigste Lungenwurm des Rehes. *Berliner Tierärztliche Wochenschrift* 9, 121–123.

Sukhdeo, M.V.K. and Mettrick, D.F. (1983) Site selection by *Heligmosomoides polygyrus* (Nematoda): effects of surgical alteration of the gastrointestinal tract. *International Journal for Parasitology* 13, 355–358.

Sultanov, M.A., Azimov, D.A. and Ubaidullaev, Y.U. (1975) Features of the biology of *Muellerius capillaris* (Mueller, 1889) in Uzbekistan. *Uzbekskii Biologicheskii Zhurnal* 19, 39–42. (In Russian.)

Supperer, R. (1958) Über die in der Aussenwelt ablaufende Entwicklungsphase von *Bunostomum phlebotomum* (Railliet, 1900), (Nematoda, Ancylostomidae). *Wiener Tierärztliche Monasschrift* 9, 553–560.

Svarc, R. (1968) Zur Frage der Pathologie und der Ocologie des Lungenwurmes *Muellerius capillaris*. *Studia Helminth* 2, 181–211.

Taffs, L.F. (1966) Helminths in the pig. *The Veterinary Record* 79, 671–692.

Taffs, L.F. (1969) Helminths of the pig: pathogenicity, diagnosis and control. *British Veterinarian Journal* 125, 304–310.

Taliaferro, W.H. and Sarles, M.F. (1939) The cellular reactions in the skin, lungs and intestine of normal and immune rats after infection with *Nippostrongylus muris*. *Journal of Infectious Diseases* 64, 157–192.

Talvik, H., Christensen, C.M., Joachim, A., Roepstorff, A., Bjørn, H. and Nausen, P. (1997) Prepatent periods of different *Oesophagostomum* spp. isolates in experimentally infected pigs. *Parasitology Research* 83, 563–568.

Taranik, K.T., Korzh, K.P. and Kolomatskaya, L.P. (1978) Occurrence of commensalism in the relationship between *Dictyocaulus viviparus* larvae and the fungi *Pilobolus* spp. (Phycomycetes, Mucoralis). I. *Vsesoyuznyi s'ezd Parazitotsenologov (Poltava, Sentyabr' 1978) Tezisy Dokladov Chast' 2 Kiev, USSR; 'Naukova Dumka'*, pp. 101–102. (In Russian.)

Taylor, E.L. (1935a) *Syngamus trachea*. The longevity of the infective larvae in the earthworms. Slugs and snails as intermediate hosts. *Journal of Comparative Pathology* 58, 149–165.

Taylor, E.L. (1935b) Seasonal fluctuation in the number of eggs of trichostrongylid worms in the faeces of ewes. *Journal of Parasitology* 21, 175–179.

Taylor, E.L. (1938) An extension to the known longevity of gapeworm infection in earthworms and snails. *Veterinary Journal* 94, 327–328.

Taylor, E.L. and Michel, J.F. (1952) Inhibited development of *Dictyocaulus* larvae in the lungs of cattle and sheep (Correspondence). *Nature, London* 169, 753.

Taylor, E.L. and Michel, J.F. (1953) The parasitological and pathologic significance of arrested development in nematodes. *Journal of Helminthology* 27, 199–205.

Tesh, R.B., Ackerman, L.J., Dietz, W.H. and Williams, J.A. (1973) *Angiostrongylus costaricensis* in Panama. Prevalence and pathologic findings in wild rodents infected with the parasite. *American Journal of Tropical Medicine and Hygiene* 22, 348–356.

Thatcher, V.E. and Scott, J.A. (1962) The life cycle of *Trichostrongylus sigmodontis* Baylis, 1945, and the susceptibility of various laboratory animals to this nematode. *Journal of Parasitology* 48, 558–561.

Theiler, A. and Robertson, W. (1915) Investigations into the life-history of the wire-worm in ostriches. *Third and Fourth Reports of the Director of Veterinary Research, Department of Agriculture, Union of South Africa.* pp. 292–345.

Theron, A. (1975) Recherches expérimentales sur l'évolution larvaire de *Skrjabingylus nasicola* (Nematoda: Metastrongyloidea) chez deux mollusques terrestres. *Vie et Milieu, C. (Biologie Terrestre)* 25, 49–54.

Thiengo, S.C., Vicente, J.J. and Pinto, R.M. (1997) Redescription of *Angiostrongylus (Parastrongylus) costaricensis* Morera and Cespedes (Nematoda, Metastrongyloidea) from a Brazilian strain. *Revista Brasiliera de Zoologie* 14, 839–844.

Thomas, R.J. (1957) A comparative study of the infective larvae of *Nematodirus* species parasitic in sheep. *Parasitology* 47, 60–65.

Thomas, R.J. (1959a) A comparative study of the life histories of *Nematodirus battus* and *N. filicollis*, nematode parasites of sheep. *Parasitology* 49, 374–386.

Thomas, R.J. (1959b) Field studies on the seasonal incidence of *Nematodirus battus* and *N. filicollis* in sheep. *Parasitology* 49, 387–410.

Thomas, R.J. and Stevens, A.J. (1960) Ecological studies on the development of the pasture stages of *Nematodirus battus* and *N. filicollis*, nematode parasites of sheep. *Parasitology* 50, 31–49.

Threlfall, W. (1965) Life-cycle of *Cyathostoma lari* Blanchard, 1849 (Nematoda, Strongyloidea). *Nature, London* 206, 1167–1168.

Threlkeld, W.L. (1934) The life history of *Ostertagia circumcincta. Technical Bulletin* No. 5, Virginia Polytechnic Institute, Virginia Agricultural Experimental Station, Blacksburg, Virginia, 24 pp.

Threlkeld, W.L. (1946) The life history of *Ostertagia ostertagi. Technical Bulletin* No. 100, Virginia Polytechnic Institute, Virginia Agricultural Experimental Station, Blacksburg, Virginia, 14 pp.

Threlkeld, W.L. (1947) Progress report on the parasitic stages of *Chabertia ovina. Journal of Parasitology* (Suppl.) 33, 12.

Threlkeld, W.L. (1948) The life history and pathogenicity of *Chabertia ovina. Virginia Agricultural Experiment Station, Technical Bulletin* No. 111, 27 pp.

Tiberio, S.R., Greiner, E.C. and Humphrey, P.P. (1983) A report of *Ollulanus tricuspis* and vomiting in cats from Florida. *Journal of American Animal Hospital Association* 19, 887–890.

Tindall, N.R. and Wilson, P.A.G. (1990a) A basis to extend the proof of migration routes of immature parasites inside hosts: estimated time of arrival of *Nippostrongylus brasiliensis* and *Strongyloides ratti* in the gut of the rat. *Parasitology* 100, 275–280.

Tindall, N.R. and Wilson, P.A.G. (1990b) An extended proof of migration routes of immature parasites inside hosts: pathways of *Nippostrongylus brasiliensis* and *Strongyloides ratti* in the rat are mutually exclusive. *Parasitology* 100, 281–288.

Tiunov, V.I. (1965) Duration of the prepatent period of development and life of two species of metastrongyles in the body of swine. *Materialy Nauchnoi Konferentsii Vsesoyuznogo Obshchestva Gel'mintologov, Moscow, USSR,* pp. 261–268. (In Russian.)

Todd, K.S., Levine, N.D. and Boatman, P.A. (1976) Effect of desiccation on the survival of infective *Haemonchus contortus* larvae under laboratory conditions. *Journal of Parasitology* 62, 247–249.

Trainer, D.O. (1973) Caribou mortality due to meningeal worm. *Journal of Wildlife Diseases* 9, 376–378.

Tromba, F.G. (1955) The role of the earthworm, *Eisenia foetida*, in the transmission of *Stephanurus dentatus. Journal of Parasitology* 41, 157–161.

Tromba, F.G. (1959) A technique for oral infection of earthworms. *Proceedings of the Helminthological Society of Washington* 26, 65–66.

Trushin, I.N. (1976) Freshwater snails in the life cycle of *Muellerius capillaris*. *Byulleten Vsesoyuznogo Instituta Gel'mintologii im. K.I. Skrjabina* 18, 90–95. (In Russian.)

Twohy, D.W. (1956) The early migration and growth of *Nippostrongylus muris* in the rat. *American Journal of Hygiene* 63, 165–185.

Tyler, G.V., Hibler, C.P. and Prestwood, A.K. (1980) Experimental infection of mule deer with *Parelaphostrongylus tenuis*. *Journal of Wildlife Diseases* 16, 533–540.

Udonsi, J.K. and Oranusi, N.A. (1991) Further observations on parastrongyliasis in Nigeria. I. Pattern of infection in the rat host. *Acta Parasitologica Polonica* 36, 83–86.

Upshall, S.M., Burt, M.D. and Dilworth, T.G. (1986) *Parelaphostrongylus tenuis* in New Brunswick: the parasite in terrestrial gastropods. *Journal of Wildlife Diseases* 22, 582–585.

Urquhart, G.M., Jarrett, W.R.H. and O'Sullivan, J.G. (1954) Canine tracheo-bronchitis due to infection with *Filaroides osleri*. *Veterinary Record* 66, 143–145.

Varughese, G. (1973) Studies on the life cycle and developmental morphology of *Cyclodontostomum purvisi* (Adams, 1933) a hookworm parasite of Malayan giant rats. *Southeast Asian Journal of Tropical Medicine and Public Health* 4, 78–95.

Vasilev, I. and Denev, I. (1972) Life-cycle and ecology of *Cyathostoma bronchialis*. *Izvestiya na Tsentralnata Khelmintologichna Laboratoriya* 15, 21–32. (In Russian.)

Veglia, F. (1916) The anatomy and life history of the *Haemonchus contortus* (Rud.). *The Third and Fourth Reports of the Director of Veterinary Research, Department of Agriculture, Union of South Africa*, pp. 349–500.

Veglia, F. (1924) Preliminary notes on the life history of *Oesophagostomum columbianum*. *9th and 10th Report of the Director of Veterinary Education and Research,* Department of Agriculture, Union of South Africa, pp. 811–823.

Veglia, F. (1928) Oesophagostomiasis in sheep (preliminary note). *13th and 14th Report of the Director of Veterinary Education and Research,* Department of Agriculture, Union of South Africa, pp. 755–797.

Verster, A., Collins, H.M. and Anderson, P.J.S. (1971) Studies on *Dictyocaulus filaria*. IV. The morphogenesis of the parasitic stages in lambs. *Onderstepoort Journal of Veterinary Research* 38, 199–206.

Volkov, F.A. (1983) *Ollulanus* infection in pigs in the Novosibirk region. *Profilaktika Nezaraznykh i Parazitarnykh Boleznei Zhivotnykh (Sbornik Nauchnykh Trudov),* pp. 148–150. (In Russian.)

Voronkova, Z.G. (1986) Distribution of *Ollulanus* infection in pigs in the northern Caucacus and lower Volga. *Byulleten' Vsesoyuznogo Instituta Gel'mintologii im. K.I. Skrjabina* No. 42, 39–43. (In Russian.)

Voronkova, Z.G., Alpatova, G.P. and Popov, M.A. (1985) *Ollulanus* infection in pigs (histological changes in the stomach). *Veterinariya Moscow, USSR* 1, 46–47. (In Russian.)

Waddell, A.H. (1969) The parasitic life cycle of the swine kidney worm *Stephanurus dentatus* Diesing. *Australian Journal of Zoology* 17, 607–618.

Waite, R.H. (1920) Earthworms – the important factor in the transmission of gapes in chickens. *Maryland Agricultural Experimental Station, Bulletin* No. 234, pp. 103–118.

Walker, H.D. (1886) The gapeworm of fowls (*Syngamus trachealis*). The earthworm (*Lumbricus terrestris*) its original host. Also, on the prevention of the disease called gapes, which is caused by this parasite. *Bulletin of the Buffalo Society of Natural Sciences* 5, 251–265.

Wallace, G.D. and Rosen, L. (1966) Studies on eosinophilic meningitis. 2. Experimental infection of shrimp and crabs with *Angiostrongylus cantonensis*. *American Journal of Epidemiology* 84, 120–131.

Wallace, G.D. and Rosen, L. (1967) Studies on eosinophilic meningitis. 4. Experimental infection of freshwater and marine fish with *Angiostrongylus cantonensis*. *American Journal of Epidemiology* 85, 395–402.

Wallace, G.D. and Rosen, L. (1969) Studies on eosinophilic meningitis. VI. Experimental infection of rats and other homiothermic vertebrates with *Angiostrongylus cantonensis*. *American Journal of Epidemiology* 89, 331–344.

Waller, P.J. and Donald, A.D. (1970) The response to desiccation of eggs of *Trichostrongylus colubriformis* and *Haemonchus contortus* (Nematoda: Trichostrongylidae). *Parasitology* 61, 195–204.

Waller, P.J. and Thomas, R.J. (1975) Field studies on inhibition of *Haemonchus contortus* in sheep. *Parasitology* 71, 285–291.

Wang, G.T. (1967) Effect of temperature and cultural methods on development of the free-living stages of *Trichostrongylus colubriformis*. *American Journal of Veterinary Research* 28, 1085–1090.

Wang, P.C. (1974) Studies on the development of *Metastrongylus apri* (Ebel, 1777). *Acta Zoological Sinica* 20, 364–377. (In Chinese.)

Wang, P.Q. (1983) Studies on the life-cycles of two species of duck gizzard worms, *Amidostomum anseris* (Zeder) and *Epomidostomum uncinatum* (Lundahl) (Nematoda: Trichostrongylidae). *Wuyi Science Journal* 3, 50–61.

Watkins, A.R.J. and Fernando, M.A. (1984) Arrested development of the rabbit stomach worm *Obeliscoides cuniculi*: manipulation of the ability to arrest through processes of selection. *International Journal for Parasitology* 6, 559–570.

Watkins, A.R.J. and Fernando, M.A. (1986a) Arrested development of the rabbit stomach worm *Obeliscoides cuniculi*: resumption of development of arrested larvae throughout the course of a single infection. *International Journal for Parasitology* 16, 47–54.

Watkins, A.R.J. and Fernando, M.A. (1986b) Arrested development of the rabbit stomach worm *Obeliscoides cuniculi*: varied responses to cold treatment by the offspring produced throughout the course of a single infection. *International Journal for Parasitology* 16, 55–61.

Watson, A. and Shaw, J.L. (1991) Parasites and Scottish ptarmigan numbers. *Oecologia* 88, 359–361.

Watson, H., Moss, R., Rothery, P. and Parr, R. (1984) Demographic causes and predictive models of population fluctuations in red grouse. *Journal of Animal Ecology* 53, 639–662.

Watson, H., Lee, D.L. and Hudson, P.J. (1987) The effect of *Trichostrongylus tenuis* on the caecal mucosa of young, old and anthelmintic-treated wild red grouse, *Lagopus lagopus scoticus*. *Parasitology* 94, 405–411.

Watson, H., Lee, D.L. and Hudson, P.J. (1988) Primary and secondary infections of the domestic chicken with *Trichostrongylus tenuis* (Nematoda), a parasite of red grouse, with observations on the effect of the caecal mucosa. *Parasitology* 97, 89–99.

Watson, T.G. (1983) Some clinical and parasitological features of *Elaphostrongylus cervi* infection in *Cervus elaphus*. *New Zealand Journal of Zoology* 10, 129.

Watson, T.G. (1984) Tissue worms in red deer. Symptoms and control. *AgLink Fpp 249* (lst Revision), Media Services, MAF, Wellington, New Zealand.

Watson, T.G. (1986) Efficacy of drenching red deer and wapiti with particular reference to *Elaphostrongylus cervi* and *Dictyocaulus viviparus*. *Deer Branch Course, No. 3*. Deer Branch of the New Zealand Veterinary Association, pp. 170–182.

Weber, J.M. (1986) Aspects quantitatifs du cycle de *Skrjabingylus nasicola* (Leuckart, 1842), nematode parasite des sinus frontaux des mustelids. Unpublished PhD thesis, University of Neuchâtel, Neuchâtel, Switzerland.

Webster, W.A. (1980) The direct transmission of *Andersonstrongylus captivensis* Webster 1978 (Metastrongyloidea: Angiostrongylidae) in captive skunks *Mephitis mephitis* (Schreber). *Canadian Journal of Zoology* 58, 1200–1203.

Webster, W.A. (1981) *Andersonstrongylus milksi* (Whitlock, 1956) n.comb. (Metastrongyloidea: Angiostrongylidae) with a discussion of related species in North American canids and mustelids. *Proceedings of the Helminthological Society of Washington* 48, 154–158.

Wehr, E.E. (1930) The occurrence of *Ornithostrongylus quadriradiatus* in the mourning dove. *Journal of Parasitology* 16, 167.

Wehr, E.E. (1937) Observations on the development of the poultry gapeworm *Syngamus trachea*. *Transactions of the American Microscopical Society* 56, 72–77.

Wehr, E.E. (1939) Domestic fowls as hosts of the poultry gapeworm. *Poultry Science* 18, 432–436.

Wehr, E.E. (1971) Nematodes. In: Davis, J.W., Anderson, R.C., Karstad, L. and Trainer, D.O. (eds) *Infections and Parasitic Diseases of Wild Birds*. Iowa State University Press, Ames, pp. 185–233.

Weinstein, P.P. (1996) Vitamin B_{12} changes in *Nippostrongylus brasiliensis* in its free-living and parasitic habitats with biochemical implications. *Journal of Parasitology* 82, 1–6.

Weinstein, P.P. and Jones, M.F. (1956) The *in vitro* cultivation of *Nippostronglys muris* to the adult stage. *Journal of Parasitology* 43, 215–231.

Weinstein, P.P., Rosen, L., Laqueur, G.L. and Sawyer, T.K. (1962) *Angiostrongylus cantonensis* infection in rats and rhesus monkeys and survival of the parasite *in vitro*. *Journal of Parasitology* 48, 51–52.

Weinstein, P.P., Rosen, L., Laqueur, G.L. and Sawyer, T.K. (1963) *Angiostrongylus cantonensis* infection in rats and rhesus monkeys, and observations on the survival of the parasite *in vitro*. *American Journal of Tropical Medicine and Hygiene* 12, 358–377.

Wells, H.S. (1931) Observations on the blood sucking activities of the hookworm *Ancylostoma caninum*. *Journal of Parasitology* 17, 167–182.

Wetzel, R. (1931) On the feeding habits and pathogenic action of *Chabertia ovina* (Fabricius, 1788). *North American Veterinarian* 12, 25–28.

Wetzel, R. (1937) Zur Entwicklung des Dachslungenwurmes *Filaroides falciformis* (Schlegel, 1933). *Sitzungsberichte der Gesellschaft naturforschender Freunde zu Berlin*, Part 1 (1–3), pp. 1–3.

Wetzel, R. (1938) Zur Biologie und systematischen Stellung des Dachslungenwurmes. *Livro Jub. Travassos*, pp. 531–536.

Wetzel, R. (1940) Zur Biologie des Fuchslungenwurmes *Crenosoma vulpis*, I. Mitteilung. *Archiv für Wissenschaftliche und Praktische Tierheilkunde* 75, 445–450.

Wetzel, R. (1941a) Zum Wirt-Parasiten-Verhältnis des Fuchslungenwurmes *Crenosoma vulpis*. *Deutsche Tierärztliche Wochenschrift* 49, 1–2, 28–30 3, 40–42.

Wetzel, R. (1941b) Zur Entwicklung des grossen Palisadenwurmes (*Strongylus equinus*) im Pferd. *Archiv für Wissenschaftliche und Praktische Tierheilkunde* 76, 81–118.

Wetzel, R. (1942) Über die Entwicklungsdauer der Palisadenwürmer im Körper des Pferdes und ihre praktische Auswertung. *Deutsche Tierärztliche Wochenschrift* 50, 443–444.

Wetzel, R. (1954) Helmintiasis intestinal del equino. *Revista de Medicina Veterinaria y Parasitologia. Caracas* 13, 17–25.

Wetzel, R. and Müller, F.R. (1935a) Die Lebensgeschichte des schachtelhalmförmigen Fuchslungenwurmes *Crenosoma vulpis* und seine Bekämpfung. *Deutsche Pelztierzüchter, München* 10, 361–365.

Wetzel, R. and Müller, F.R. (1935b) The life cycle of *Crenosoma vulpis*, the lungworm of foxes and ways and means of combat. *Fur Trade Journal of Canada* 13, 16–17.

Wetzel, R. and Enigk, K. (1937) Zur Biologie von *Graphidium strigosum*, dem Magenwurm der Hasen und Kaninchen. *Deutsch Tierärtzliche Wochenschrift* 45, 401–405.

Wetzel, R. and Enigk, K. (1938) Wandern der Larven des Palisadenwurmes (*Strongylus* spec.) der Pferde durch die Lungen? *Archiv für Wissenschaftliche und Praktische Tierheilkunde* 73, 83–93.

Wetzel, R. and Kersten, W. (1956) Die Leberphase der Entwicklung von *Strongylus edentatus*. *Wiener Tierärztliche Monatsschrift* 43, 664–673.

Whitlaw, H.A. and Lankester, M.W. (1994a) A retrospective evaluation of the effects of parelaphostrongylosis on moose populations. *Canadian Journal of Zoology* 72, 1–7.

Whitlaw, H.A. and Lankester, M.L. (1994b) The co-occurrence of moose, white-tailed deer and *Parelaphostrongylus tenuis* in Ontario. *Canadian Journal of Zoology* 72, 819–825.

Whitlaw, H.A., Lankester, M.W. and Ballard, W.B. (1996) *Parelaphostrongylus tenuis* in terrestrial gastropods from white-tailed deer winter and summer range in northern New Brunswick. *Alces* 32, 75–83.

Whitlock, J.M. and Le Jambre, L.F. (1981) On the taxonomic analysis of the genus *Haemonchus* Cobb, 1898. *Systematic Parasitology* 3, 7–12.

Wijers, D.J.B. and Smit, A.M. (1966) Early symptoms after infection of man with *Ancylostoma braziliense*. *Tropical and Geographical Medicine* 18, 48–52.

Williams, D.W. (1942) Studies on the biology of the nematode lungworms *Muellerius capillaris* in molluscs. *Journal of Animal Ecology* 11, 1–8.

Williams, J.C. and Bilkovich, F.R. (1971) Development and survival of infective larvae of the cattle nematode, *Ostertagia ostertagi*. *Journal of Parasitology* 57, 327–338.

Williams, J.C., Knox, J.W., Baumann, B.A., Snider, T.G., Kimball, M.D. and Hoerner, T.T. (1983) Seasonal changes of gastrointestinal nematode populations in yearling beef cattle in Louisiana, U.S.A. with emphasis on prevalence of inhibition in *Ostertagia ostertagi*. *International Journal for Parasitology* 13, 133–144.

Wilson, E.A., Leslie, A.S., Shipley, R.E. and Leiper, R.T. (1912) The grouse in disease. Part II, Chapters IV, V, VII, VIII. In: Leslie, A.S. and Shipley, A.E. (eds) *The Grouse in Health and Disease being the Popular Edition of the Report of the Committee of Inquiry on Grouse Disease*. Smith, Elder and Co., London, pp. 113–245.

Wilson, P.A.G., Gentle, M. and Scott, D.S. (1976a) Milk-borne infection of rats with *Strongyloides ratti* and *Nippostrongylus brasiliensis*. *Parasitology* 72, 355–360.

Wilson, P.A.G., Gentle, M. and Scott, D.S. (1976b) Dynamic determinants of the route of larval *Strongyloides ratti* in lactating rats and the control of experimental error in quantitative studies on milk transmission of skin-penetrating roundworms. *Parasitology* 73, 399–406.

Wissler, K. and Halvorsen, O. (1976) Infection of reindeer with *Elaphostrongylus rangiferi* (Nematoda: Metastrongyloidea) in relation to age and season. *Norwegian Journal of Zoology* 24, 462–463.

Wood, D.E. (1965) Experimental infection of a mongoose with *Angiostrongylus cantonensis* (Chen). *Journal of Parasitology* 51, 941.

Woodard, J.C., Zam, S.G., Caldwell, D.K. and Caldwell, M.C. (1969) Some parasitic diseases of dolphins. *Pathologia Veterinaria* 6, 257–272.

Woolf, A., Mason, C.A. and Kradel, D. (1977) Prevalence and effects of *Parelaphostrongylus tenuis* in a captive wapiti population. *Journal of Wildlife Diseases* 13, 149–154.

Worley, D.E. (1963) Experimental studies on *Obeliscoides cuniculi*, a trichostrongyloid stomach worm of rabbits. I. Host–parasite relationship and maintenance in laboratory rabbits. *Journal of Parasitology* 49, 46–50.

Xiao, B.N. and Kong, F.Y. (1987) A study of the life history of *Oesophagostomum kansuensis* Hsiung et Kung, 1955. *Acta Veterinaria et Zootechnica Sinica* 18, 179–183.

Yamini, B., Baker, J.C., Stroonberg, P.C. and Gardines, C.H. (1997) Cerebrospinal nematodiasis and vertebral chondrodysplasia in a calf. *Journal of Veterinary Diagnostic Investigations* 9, 451–454.

Yaron, V. (1968) The development of *Amidostomum anseris* (Zeder, 1800) in *Columbia livia dom.* *Helminthologia* 8–9, 195–199. (In Russian.)

Yokogawa, S. (1922) The development of *Heligmosomum muris* Yokogawa, a nematode from the intestine of the wild rat. *Parasitology* 14, 127–166.

Yoshida, Y. (1968) Pathobiologic studies on *Ancylostoma ceylanicum* infection. *International Congress of Tropical Medicine and Malaria (8th) Tehren, Iran*, pp. 170–171.

Yoshida, Y. (1971a) Comparative studies on *Ancylostoma braziliense* and *Ancylostoma ceylanicum*. I. The adult stage. *Journal of Parasitology* 57, 983–989.

Yoshida, Y. (1971b) Comparative studies on *Ancylostoma braziliense* and *Ancylostoma ceylanicum*. II. The infective larval stage. *Journal of Parasitology* 57, 990–992.

Yoshida, Y., Nakanishi, U. and Mitani, W. (1958) Experimental studies on the infection modes of *Ancylostoma duodenale* and *Necator americanus* in the definitive host (19 volunteers). *Japanese Journal of Parasitology* 7, 704–714. (In Japanese.)

Yoshida, Y., Okamota, K. and Chiu, J.K. (1972) Experimental infection of man with *Ancylostoma ceylanicum*. *Chinese Journal of Microbiology* 4, 157–167.

Yoshida, Y., Kondo, K., Kurimoto, H., Fukutome, S. and Shirasaka, S. (1974) Comparative studies on *Ancylostoma braziliense* and *Ancylostoma ceylanicum*. III. Life history in the definitive host. *Journal of Parasitology* 60, 636–641.

Yousif, F. and Ibrahim, A. (1978) The first record of *Angiostrongylus cantonensis* from Egypt. *Zeitschrift für Parasitenkunde* 56, 73–80.

Yousif, F. and Lammler, G. (1975) The suitability of several aquatic snails as intermediate hosts for *Angiostrongylus cantonensis*. *Zeitschrift für Parasitenkunde* 47, 203–210.

Zajicek, D. (1964) The embryonal and postembryonal development of *Amidostomum boschadis* Petrow and Fediuschin, 1949 (Nematoda). In: Ergens, R. and Ryšavý, B. (eds) *Parasitic Worms in Aquatic Conditions. Proceedings of Symposium, Prague, Oct. 29–Nov. 2, 1962.* Czeckoslovakia Academy of Sciences, Prague, pp. 137–141.

Zdarska, Z. (1960) Larvalni stadia cizopasnych cervu z nasich suchozemskych plzu. *Ceskoslovenska Parasitologie* 7, 355–379. (In Czechoslovakian.)

Zdzitowiecki, K. (1976) An experimental study on the infection of terrestrial and aquatic snails with *Muellerius capillaris* (Mueller, 1889) larvae (Nematoda, Protostrongylidae). *Acta Parasitologica Polonica* 24, 159–163.

Zurliiski, P. (1978) The effect of freezing, high temperature and desiccation on *Nematodirus spathiger* Railliet, 1896 ova and larvae. *Veterinarno-Meditsinski Nauki* 15, 107–116.

Zviaguintzev, S.N. (1934) Contribution to the life history of *Nematodirus helvetianus* May. *Trudi Dinamike. Razvit.* 8, 186–202.

Chapter 4
Order Oxyurida

Members of the order are unique microphagous nematodes which inhabit the posterior gut of various vertebrates and arthropods. Placed in their own order by Chabaud (1974), they constitute the only major group of nematodes with adult representatives in both vertebrates and invertebrates. There are two superfamilies. The Oxyuroidea contains numerous species in vertebrates. The Thelastomatoidea includes species found in invertebrates, especially herbivorous arthropods with a fermentation chamber as in cockroaches, diplopods and orthopterans (Adamson, 1989) and will not be discussed herein.

4.1

The Superfamily Oxyuroidea

The Oxyuroidea contains three families (Chabaud, 1974). The Pharyngodonidae includes parasites mainly of the posterior gut of herbivorous lower vertebrates, with a few species in mammals (Petter and Quentin, 1976). The Oxyuridae and the Heteroxynematidae contain many species in mammals and only a few species in birds; these parasites are especially common in the caeca of lizards, terrestrial tortoises, marsupials, rodents and primates. It has been suggested (Chitwood and Chitwood, 1950) that the forms in vertebrates are derived from those in arthropods. Anderson (1984) suggested that oxyuroids transferred from insects early in the evolution of tetrapods and gave rise to forms found today in modern vertebrates. Adapted mainly to terrestrial hosts, only a few oxyuroids were able to colonize certain fish and larval amphibians (tadpoles) with the necessary behavioural and ecological peculiarities to facilitate contaminative transmission (Moravec et al., 1992; Anderson and Lim, 1996). Morand et al. (1996) determined that body size of oxyuroids tended to increase with the body size of the host.

Members of the order Oxyurida are strictly monoxenous and transmission and development of species in both invertebrates and vertebrates are remarkably similar. Typically, female worms produce thick-shelled eggs which are elongated and flattened on one side. A subpolar operculum has been described on the eggs of most species. Eggs of some species may be deposited in an early stage of development and reach the infective stage only after they have been passed in faeces (e.g. *Passalurus ambiguus, Gyrinicola batrachiensis* and forms in insects). In many other members of the Oxyuroidea gravid females migrate to the anus of the host and deposit eggs in the perianal region, where they rapidly complete their development to the infective stage (e.g. *Enterobius vermicularis, Oxyuris equi, Skrjabinema ovis, Syphacia* spp.). Eggs are readily transferred from the perianal region to the host's mouth by grooming activities. Eggs dispersed into an environment with favourable conditions of humidity and temperature can become a continual source of eggs for oral infection. Eggs also adhere to faeces and are readily transmitted to coprophagic hosts such as lagomorphs and rodents. **Retrofection**, in which larvae hatched from eggs in the perianal region migrate into the rectum, has been proposed as an additional route of infection. Some authors question the evidence for retrofection, however.

Throughout the Oxyurida two moults occur during development of larvae in eggs. In *Hammerschmidtiella diesingi* and *Leidynema appendiculata* of cockroaches the fully formed larva is at the end of the second stage (Dobrovolny and Ackert, 1934; Todd, 1944). Kharichkova (1946) and Boecker (1953) reported two moults in the eggs of

Passalurus ambiguus of rabbits. Moults are often extremely difficult to prove because much of the cuticle may be resorbed during the moulting process, leaving only a delicate, easily overlooked membrane (Adamson, 1983). Also, the second moult may occur rapidly and only when the egg reaches the gut of the final host. In such cases the larva in the egg may be in the advanced second stage which, as Chabaud (1955) pointed out, is essentially the beginning of the third stage, generally the infective stage in secernentean nematodes. There are probably always two final moults in the definitive host.

Seurat (1912, 1913) and Petter (1969) reported the production of two kinds of eggs (**poecilogony**) in oxyuroids. In *Alaeuris caudatus*, *A. vogelsangi* and *Tachygonetria vivipara* of Testudinae and Iguanidae there are two types of females (**poecilogyny**). One female lays unembryonated, thick-shelled eggs which pass out of the host and are involved in transmission; the eggs embryonate to the third stage in the external environment. The other female produces thin-shelled eggs with fully developed larvae which, when deposited in the gut of the host, are autoinfective. In *Gyrinicola batrachiensis* of anuran tadpoles, individual female worms are didelphic and one branch of the genital system produces thin-shelled eggs which embryonate *in utero* into autoinfective larvae. The other branch produces unembryonated thick-shelled eggs which pass out of the host, embryonate and can then infect new hosts (Adamson, 1981a,b,c,d). Traumatic insemination, in which the male introduces sperm through the body wall of the fourth-stage female, is known in oxyuroids of the genera *Auchenacantha*, *Citellina*, *Passalurus* and probably *Austroxyuris* (Hugot *et al.*, 1982; Hugot, 1984; Adamson, 1989; Hugot and Bougnoux, 1987). Adamson (1989) suggested that severe competition to mate may have been the stimulus for the development of traumatic insemination of immature females.

In addition to the above-mentioned peculiarities of the Oxyurida, we now know that **haplodiploidy** is found in the order (Adamson, 1981c, 1989). In haplodiploidy, unfertilized eggs give rise to haploid males whereas fertilized eggs give rise to diploid females. First demonstrated in *Gyrinicola batrachiensis* of anuran tadpoles (Adamson, 1981c), the phenomenon has now been found in other Pharyngodonidae (*Mehdiella* spp., *Tachygonetria* spp., *Thelandros alatus*) as well as in Oxyuridae (e.g. *Passalurus ambiguus*, *Syphacia obvelata*) and Heteroxynematidae (*Aspiculuris tetraptera*) (see Adamson and Petter, 1982, 1983a,b; Adamson, 1984a, 1989). Haplodiploidy has also been reported in three species of the Thelastomatoidea of diplopods and insects (Adamson, 1984b; Adamson and Nasher, 1987) so in all probability it is a general characteristic of the order. For a detailed discussion of haplodiploidy and its implications refer to Adamson (1984a, 1989, 1990).

Ainsworth (1990) has reported male dimorphism in two species of Pharyngodonidae from lizards in New Zealand.

Family Pharyngodonidae

Gyrinicola

Gyrinicola was placed in its own family, the Gyrinicolidae, by Yamaguti (1938). Chabaud (1978) reduced the family to a subfamily of the Cosmocercidae

(Cosmocercoidea). Adamson (1981a) showed that the *Gyrinicola* belonged in the Pharyngodonidae.

G. batrachiensis (Walton, 1929)

Adamson (1981a) found this species commonly in the posterior end of the small intestine and in the large intestine of tadpoles of *Bufo americanus*, *Hyla versicolor*, *Pseudacris triseriata*, *Rana catesbeiana*, *R. clamitans*, *R. pipiens* and *R. sylvatica* in eastern Canada. The parasite occurred also in metamorphosing tadpoles but was totally absent from toads and frogs (i.e. adults) of the above species.

Female *G. batrachiensis* were didelphic and prodelphic (Fig. 4.1A). One uterus, coiled in the ventral half of the body of the worm, contained a single row of thin-shelled eggs (Fig. 4.1B) in various stages of development. Those in the vagina contained fully developed larvae (Adamson, 1981b). The 'shell' of these eggs was the vitelline membrane. There were two moults within these eggs, resulting in third-stage larvae (Adamson, 1983). The uterus coiled in the dorsal half of the body of the worm contained a single row of eggs with thick shells (Fig. 4.1C, D) and opercula. Eggs in the vagina were in the one- to eight-cell cleavage stage.

Larvae from thin-shelled eggs (Fig. 4.1F) were autoinfective and did not survive for more than 1 h in water or diluted buffer (Adamson, 1981b). Fully developed larvae were 366–469 µm in length. Tadpoles were successfully infected by oral inoculation of thin-shelled eggs.

Gravid females oviposited thick-shelled eggs in the eight-cell stage. These eggs were incubated in water at 20°C and were fully developed in 6 days, when larvae (presumably in the third stage) (Fig. 4.1E) were 232–238 µm in length (Adamson, 1981b). Tadpoles were infected by oral inoculation of fully developed thick-shelled eggs, which were regarded as the transmission and dispersal stage of the parasite. These eggs overwintered and were available to infect a new batch of tadpoles the following spring.

Growth of larvae in *R. clamitans* tadpoles at 20°C and 25°C was followed (Adamson, 1981b) (Fig. 4.1G, H). At 25°C and 20°C the third moult of males occurred in 5 and 9 days and in females in 6 and 9 days, respectively. At similar temperatures, the fourth moult in males (Fig. 4.1I) occurred in 8 and 16 days and in females (Fig. 4.1J) in 12 and 19 days, respectively. Females were gravid in 18 days at 25°C and in 33 days at 20°C. Males apparently died soon after inseminating females, since they were not found in experimentally infected tadpoles after 24 days at 25°C and 45 days at 20°C. Female worms, on the other hand, survived 40 days at 25°C and 60 days at 20°C.

Adamson (1981d) examined tadpoles of *R. clamitans* and *Bufo americanus* for *G. batrachiensis* in spring, summer and autumn. In *R. clamitans*, *G. batrachiensis* was lost during metamorphosis prior to the eruption of the forelimbs in June and July. Young-of-the-year tadpoles were apparently infected by ingesting embryonated thick-shelled eggs. In late autumn almost all tadpoles were infected and thin-shelled autoinfective eggs predominated in female worms. There was no evidence of transmission or autoinfection in winter but in early spring (April) female worms which had wintered in tadpoles deposited autoinfective thin-shelled eggs, resulting in a sharp increase in intensity. Female worms of the autoinfective generation matured in May and June and contained almost entirely thick-shelled eggs. Thus, in July, young-of-the-year tadpoles entered an environment recently contaminated with thick-shelled eggs.

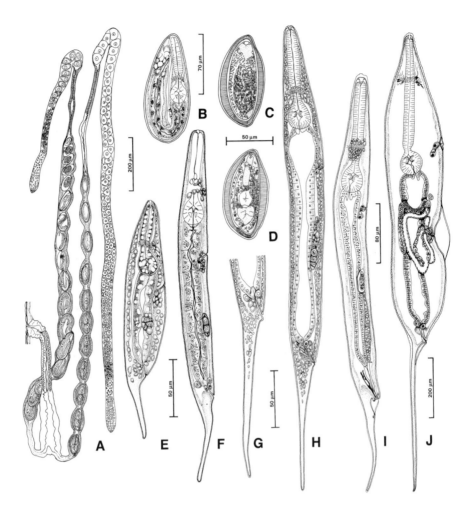

Fig. 4.1. Developmental stages of *Gyrinicola batrachiensis*: (A) female reproductive system with horns producing different eggs; (B) thin-shelled egg embryonated *in utero*; (C) thick-shelled egg, unembryonated; (D) thick-shelled egg, embryonated; (E) larva mechanically released from infective thick-shelled egg; (F) larva from infective thin-shelled egg; (G) third-stage female recovered from experimentally infected *Rana clamitans* held at 20°C, caudal end (12 days postinfection); (H) third-stage male recovered from experimentally infected *R. clamitans* held at 20°C (5 days postinfection); (I) moulting fourth-stage male recovered from experimentally infected *R. clamitans* held at 20°C (16 days postinfection); (J) moulting fourth-stage female recovered from experimentally infected *R. clamitans* held at 20°C (18 days postinfection). (After M.L. Adamson, 1981 – courtesy *Canadian Journal of Zoology*.)

G. batrachiensis underwent a single generation in *Bufo americanus*. Males were not found and females contained only thick-shelled eggs. Toads spent only a brief period as tadpoles and there was insufficient time for autoinfection to be effective.

Family Oxyuridae

Enterobius

Members of the genus are found in catarrhine primates and Ethiopian Sciuridae. The best known example is the pin or seat worm of humans.

E. vermicularis (Linnaeus, 1758)

This is a common parasite of the caecum and appendix of humans. It is cosmopolitan but occurs most commonly in temperate climates. Males are 2–5 mm and females 8–13 mm in length. Both sexes have long attenuated tails.

Transmission and development of *E. vermicularis* attracted the attention of several well-known figures in the last century, including Leuckart (1863), who infected himself and his students to study clinical effects and the strange behaviour of female worms.

Most development of eggs occurs *in utero*. Gravid female worms then move to the anus and deposit long trails of eggs in the perianal region. It has been estimated that each female worm can contain 4572–16,888 eggs (Reardon, 1938). A few hours (about 6) after being deposited in the perianal region, eggs reach the infective stage (Zawadowsky and Schalimov, 1929). There are presumably two moults in the egg.

The host becomes infected from ingesting eggs containing third-stage larvae. Eggs hatch in the duodenum. Larvae, 140–150 µm in length, develop in the small intestine and undergo two moults to the adult stage (for descriptions of larval stages see Hulinská, 1968). Young adults mate and migrate to the large intestine, where they attach to the mucosa. When females become gravid, they detach and move towards the anus. The time from infection to the release of eggs in the perianal region is 15–35 days (Cram, 1943; Kozlov, 1982). Cho *et al.* (1985) examined female worms from children. The worms were 4.10–9.90 mm in length and contained shelled eggs when they were about 5.5 mm in length. The number of eggs in the worms was positively correlated with length of the body and ranged from 19 ± 50 in worms 5.50–5.99 mm in length to 13,323 in worms 9.50–9.90 mm in length. Egg production began 28–32 days post infection.

Pinworms cause itching in the perianal region and infected individuals, especially children, scratch and transfer eggs on their hands to the mouth. Egg-laying activities take place mainly at night and soiled night clothes are a prime source of egg contamination (Schüffner, 1949). Eggs are tiny; they easily become airborne in dust and are available for inhalation (Nolan and Reardon, 1939). Because of the substantial output of eggs by female worms, the rapid maturation of eggs to the infective third stage and the ability of eggs to survive (Jones and Jacobs, 1941) and become disseminated in the environment, pinworm is ideally adapted for group transmission in families, schools, hospitals, mental institutions, etc.

Langhans (1926), Hamburger (1939), Schüffner (1949) and Schüffner and Swellengrebel (1949) believed that in certain conditions pinworm eggs may hatch on the anal mucosa and infective larvae migrate into the rectum and bowel, a type of transmission referred to as retrofection. Rarely, female worms will invade the reproductive tract of the female host and die there.

Oxyuris

O. equi (Schrank, 1788)

O. equi is a cosmopolitan parasite of the caecum and colon of horses. Females are 24–157 mm in length, depending on the length of the tail; there are apparently two types of females, one with a short tail and another with a long tail (Skrjabin *et al.*, 1960). Males are only 8–19 mm in length. Gravid females contain large numbers of eggs in the gastrula stage. They migrate to the anus and emerge head first, at the same time releasing a viscous grey fluid containing enormous numbers of eggs which adhere to the perianal region (Shul'ts and Krankrov, 1928). As she oviposits, the female shrinks and eventually passes out of the anus entirely and dies. Perroncito (1882) was apparently the first to appreciate that eggs of *O. equi* developed outside the host. Enigk (1949) reported that there were two moults in the egg and that about 3–5 days were required for development from the gastrula stage to the third and infective stage. Eggs adhering to the perianal region occasion a great deal of itching and many are rubbed off and dispersed in the environment. Fully developed eggs are fairly resistant and long lived. They contaminate the environment, especially stables, and are readily available for ingestion by horses.

According to Enigk (1949) larvae hatched in the small intestine of the final host and entered the mucosal crypts of the ventral colon and caecum, where they grew and moulted to the fourth stage in 3–11 days. The fourth stage had a modified buccal capsule with which to grasp mucosal tissue (Wetzel, 1930). These larvae apparently ingested mucosal tissue and perhaps some blood as they moved about from place to place. The fourth stage moulted to the subadult stage in about 45–60 days but females did not become gravid and lay eggs in the anal region until 139–156 days postinfection (Enigk, 1949).

Passalurus

P. ambiguus (Rudolphi, 1819)

P. ambiguus is a common, cosmopolitan parasite of the caecum and large intestine of rabbits and hares. Accounts of its development vary somewhat in detail from author to author and more research is needed. Several authors reported that female worms deposited eggs around the anus (Palimpsestov and Chebotarev, 1935; Kharichkova, 1946). According to Kharichkova (1946) eggs oviposited in the gastrula stage readily developed to the infective stage but, unlike most pinworms which develop rapidly, 7–8 days were required at 35–38°C to attain the infective stage. Two moults occurred in the eggs, one at 24 h and the other in 3 days. Resistance to desiccation increased with the development of the eggs.

Geller (1946a,b in Skrjabin *et al.*, 1960) reported that cleavage to the gastrula stage took place in the lower part of the large intestine and rectum and that females emerging from the anus of rabbits usually contained eggs in the gastrula stage capable of developing to the infective stage in the environment. Boecker (1953), on the other hand, claimed that females laid eggs in the gastrula stage in the lower intestine and that oviposition outside the body was unusual. Eggs passed out of the host in the mucus coating the faeces and developed outside the body. Boecker (1953) also reported two moults in the egg.

Kharichkova (1946) gave eggs to rabbits and recovered young adult worms 11 days later. According to Boecker (1953) eggs given orally to rabbits hatched in the caecum and moulted to fourth-stage larvae which were 700–800 μm in length. These larvae were found in the lumen, crypts and mucosa of the appendix and colon. Adults were always found in the lumen. In two rabbits the prepatent period was said to be 56 and 64 days. Worms apparently lived for 106 days in one rabbit.

Hugot *et al.* (1982) and Hugot (1984) reported that insemination takes place in a hypodermal pocket beneath the cuticle in the vulvar region. Also, a tubular egg receptacle formed from tissue of the ventral hypodermal cord in the vulvar region holds the total egg production of the female.

Skrjabinema

S. ovis (Skrjabin, 1915)

S. ovis is a cosmopolitan nematode of the colon of sheep and goats. Schad (1957) removed mature females from the rectum of infected hosts and allowed them to deposit eggs in tap water. Eggs were fully embryonated when laid. Eggs were also found on the perianal skin and in goats there was evidence that females migrated to this region at night. *In vitro* experiments indicated that eggs probably hatched in the small intestine of the host, since successive passages through gastric and intestinal fluids were necessary for hatching. Only fourth-stage larvae were recovered 12 and 17 days postinfection. In 17 days, fourth-stage larvae were up to 1.4 mm in length at the time of the fourth moult. Most worms were adults in 25 days. Sapozhnikov (1969) studied *S. ovis* in lambs in the CIS and reported two moults in the egg. Female worms moved from the intestine of the host and laid eggs in the perianal region. The third moult occurred 7 days and the fourth 18–19 days after sheep had ingested eggs. Worms were mature 24 days postinfection.

Syphacia

The parasites studied by Prince (1950) in rats and referred to as *S. obvelata* were *S. muris* (see Hussey, 1957). Prince questioned the results of Lawler (1939) who claimed to have infected mice by feeding them macerated, gravid female worms of *S. obvelata*.

S. muris Yamaguti, 1941

S. muris is the usual parasite of albino rats (Hussey, 1957). Prince's (1950) attempts to infect rats by feeding eggs from gravid female *S. muris* were unsuccessful. She found no evidence that eggs were ever deposited in the host. Larvated eggs, free larvae (140–220 μm in length) and egg shells were found on the perianal region of rats. She reported active larvae in the posterior region ('inch') of the large intestine and concluded that they were larvae which had hatched from eggs in the perianal region and entered the body by way of the anus. Retrofection and oral infection were regarded as the two modes of transmission.

Stahl (1963) infected rats 3–5 weeks of age with 50–100 eggs of *S. muris* given by stomach tube and examined them at periodic intervals. In 1 day, larvae were found in the caecum. Sheaths on some larvae indicated that a moult was taking place. On day 4, several females had been inseminated. In 5 days eggs appeared in some females; in 6 days worms were gravid; in 7 days females were packed with eggs and some were in the large intestine. Many worms ('migrators') were on, or in, mucus on faecal pellets. The activity of worms was 'feeble', suggesting that they were being carried along mainly by peristalsis. In 7.5 days 'larval worms were detected in the large intestine and caecum... indicating reinfection was occurring'. On day 8, eggs were detected in the anal region. Stahl (1963) noted that females were much more numerous than males (see Lewis and D'Silva, 1986, below).

Gulden (1967) reported that more eggs of *S. muris* were deposited during the day than at night. Mature females tended to emerge just before the rest period of the rat. It was not determined if females reached the anus by their own efforts or by faecal transport. Laying of eggs was stimulated by lower temperature and humidity and by higher oxygen reduction potentials. Eggs were infective immediately on laying and it was estimated that about 90% of them were ingested by the rat during the night following the day on which they were laid.

Lewis and D'Silva (1986) studied *S. muris* in rats and reported three moults during development. No moults were observed in the egg but they noted a moult as the larvae escaped from the egg in the rat (probably the second moult). A moult occurred in 40 h and another 66–72 h postinfection. In 5 days uteri contained eggs in the morula stage and in 7 days larvated eggs appeared in the perianal region. The number of eggs released by individual worms under experimental conditions was 421–542. Like Stahl (1963), Lewis and D'Silva (1986) reported that females were much more common than males in mature infections and they stated that most males were lost from the population in 4 days.

D'Silva (1992) reported that in depositing eggs on the perianal region of rats, *S. muris* followed a circadian rhythm. Most eggs were laid in daytime, with a peak around noon. Rats are nocturnal and normally feed and defaecate at night. Adults worms synchronize egg release during the day with the rest period of the rat, thus avoiding loss of eggs in the faeces at night. If rats are fed only during the daytime, the worms lay fewer eggs and switch egg-laying to the night time. The author failed to infect himself with *S. muris*.

S. obvelata (Rudolphi, 1802)

This is a common pinworm of the caecum of laboratory and wild mice (Hussey, 1957; Hugot, 1988). Philpot (1924) noted that eggs of *S. obvelata* were rare in the faeces of mice. He failed to infect mice with eggs or larvae that he had incubated. Lawler (1939) claimed to have infected mice by feeding them macerated, gravid female worms. According to Grice and Prociv (1993) eggs of *S. obvelata* died rapidly when dried or placed in water. The average number of eggs produced was 266–347 ($\bar{x}$ = 317), which was about 2–3% of the fecundity of *E. vermicularis*. Eggs kept at 30°C became infective between 6 and 42 h on floating cellophane membranes. The prepatent period in mice was 11–15 days.

Chan (1952) showed that the main mode of infection was by oral ingestion of larvated eggs contaminating the perianal region or the environment. He could not infect mice by feeding them newly hatched larvae or by inoculating eggs and free larvae into the anus. He did not believe that retrofection was a likely mode of transmission.

Chan (1952) reported that eggs, larvae and empty egg shells were present in the lower two-thirds of the intestine 1 h after mice were given larvated eggs. In 2 h, larvae were found in the caecum and small intestine. In 2 days, most larvae were found in the caecum and a moult was noted. In 5 days most females were inseminated and in 9–10 days most females were gravid. In 12 days eggs first appeared in the perianal region and some female worms were observed protruding from the anus. He noted that 'when disturbed, some [worms] disappeared, retreating inwards; others discharged eggs onto the perianal region . . . and subsequently dried up'. Gravid females continued to migrate out of the mice and by day 16 most mice were free of them.

Family Heteroxynematidae

Aspiculuris

A. tetraptera (Nitzsch, 1821)

A. tetraptera has been reported from the large and small intestine of mice (*Mus* spp.), rats (*Rattus* spp.) and various other rodents. It is a common parasite of laboratory mice. It is the only member of the Heteroxynematidae to have been studied but unfortunately important features of its development (e.g. moults) have not been elucidated. Female worms in the intestine produce eggs in an early stage of segmentation which embryonate in the external environment to the infective stage (presumably the third stage). Mice probably acquire their infections from ingesting eggs contaminating food and from coprophagy.

Philpot (1924) reported that eggs developed rapidly after leaving the host. He noted embryos moving in eggs after 68 h at 22°C and in 20 h at 27°C. Eggs incubated for 6–7 days at 25°C were apparently infective to mice, the hatched larvae passing to the caecum. In 90 h larvae left the caecum and all had reached the large intestine in 163 h. Males and females were recognized in 8–10 days and the adult characteristics were acquired between 14 and 18 days.

Hsü (1951) reported the results of placing eggs in a variety of media (horse serum, water, gastric juices, dilute HCl, sodium bicarbonate solution, etc.). He also reported the hatching of some larvae in water at 37°C in 48 h and under other conditions. He applied large numbers of larvated eggs to the anal region and vagina of 12 mice and found a few larvae and adults in their guts at various times thereafter. He regarded the results as evidence of retrofection.

Chan (1953) showed that eggs could withstand prolonged refrigeration and later (1955) followed development in mice given 200–700 eggs. Larvae were found through-out the large intestine of the mice in 24 h postinfection. However, in 72–144 h postinfection, larvae were found predominantly in the lower colon. Beginning on day 8, larvae reappeared in increasing numbers in the proximal part of the large intestine and their numbers decreased in the lower colon. On days 11–29, no worms were found in the first part of the colon. The results suggested that larvae initially moved to the lower part of the large intestine and then later recolonized the upper part of the intestine.

Mathies (1959) showed that *Mus musculus* was a much more suitable host for *A. tetraptera* than *Rattus norvegicus* and *Meriones unguiculatus*. He was not able to infect *Peromyscus maniculatus* and he regarded the gerbil and the rat as unsuitable hosts, since only a few retarded worms were found in them. Female mice became resistant to infection during the first oestrus whereas males gradually developed resistance as they advanced in age.

Anya (1966a) studied hatching and development of *A. tetraptera* at different temperatures and levels of oxygen. Anya (1966b) orally infected mice and concluded that eggs hatched in the lower intestine or in the caecum. After moulting, larvae migrated into the crypts of the colon. Larvae then entered the lumen of the colon, where they matured, the males in 20 days and the females in 23 days postinfection. Eggs were laid by females 24 days postinfection.

Derothe *et al.* (1997) showed that laboratory strains of mice were more resistant to infection than wild strains, probably because parasite pressure was severe in captivity.

References (Oxyurida)

Adamson, M.L. (1981a) *Gyrinicola batrachiensis* (Walton, 1929) n.comb. (Oxyuroidea; Nematoda) from tadpoles in eastern and central Canada. *Canadian Journal of Zoology* 59, 1344–1350.

Adamson, M.L. (1981b) Development and transmission of *Gyrinicola batrachiensis* (Walton, 1929) Adamson, 1981 (Pharyngodonidae: Oxyuroidea). *Canadian Journal of Zoology* 59, 1351–1367.

Adamson, M.L. (1981c) Studies on gametogenesis in *Gyrinicola batrachiensis* (Walton, 1929) (Oxyuroidea: Nematoda). *Canadian Journal of Zoology* 59, 1368–1376.

Adamson, M.L. (1981d) Seasonal changes in populations of *Gyrinicola batrachiensis* (Walton, 1929) in wild tadpoles. *Canadian Journal of Zoology* 59, 1377–1386.

Adamson, M.L. (1983) Ultrastructural observations on oogenesis and shell formation in *Gyrinicola batrachiensis* (Walton, 1929) (Nematoda: Oxyurida). *Parasitology* 86, 489–499.

Adamson, M.L. (1984a) Haplodiploidy in *Aspiculuris tetraptera* (Nitzch) (Heteroxynematidae) and *Syphacia obvelata* (Rudolphi) (Oxyuridae), nematode (Oxyurida) parasites of *Mus musculus*. *Canadian Journal of Zoology* 62, 804–807.

Adamson, M.L. (1984b) Anatomical adaptation to haplodiploidy in the oxyuroid (Nematoda) *Desmicola skrjabini* n.sp. from a diplopod in Gabon. *Annales de Parasitologie Humaine et Comparée* 59, 95–99.

Adamson, M.L. (1989) Evolutionary biology of the Oxyurida (Nematoda): biofacies of a haplodiploid taxon. *Advances in Parasitology* 28, 175–228.

Adamson, M.L. (1990) Haplodiploidy in the Oxyurida: decoupling the evolutionary processes of adaptation and speciation. *Annales de Parasitologie Humaine et Comparée* 65 (Suppl. 1), 31–35.

Adamson, M.L. and Nasher, A.K. (1987) *Hammerschmidtiella andersoni* sp.n. (Thelastomatidae: Oxyurida) from the diplopod *Archispirostreptus tumuliporus* in Saudi Arabia with comments on karyotype of *Hammerschmidtiella diesingi*. *Proceedings of the Helminthological Society of Washington* 54, 220–224.

Adamson, M.L. and Petter, A.J. (1982) Evidence of haplodiploidy in pharyngodonid (Nematoda: Oxyuroidea) parasites of *Testudo graeca*. *Annales de Parasitologie Humaine et Comparée* 57, 197–199.

Adamson, M.L. and Petter, A.J. (1983a) Haplodiploidy in pharyngodonid (Oxyuroidea: Nematoda) parasites of *Testudo graeca*. *Annales de Parasitologie Humaine et Comparée* 58, 267–273.

Adamson, M.L. and Petter, A.J. (1983b) Studies on gametogenesis in *Tachygonetria vivipara* Wedl, 1862 and *Thelandros alatus* Wedl, 1862 (Oxyuroidea: Nematoda) from *Uromastix acanthinurus* in Morocco. *Canadian Journal of Zoology* 61, 2357–2360.

Ainsworth, R. (1990) Male dimorphism in two new species of nematodes (Pharyngodonidae: Oxyurida) from New Zealand lizards. *Journal of Parasitology* 76, 812–822.

Anderson, R.C. (1984) The origins of zooparasitic nematodes. *Canadian Journal of Zoology* 62, 317–328.

Anderson, R.C. and Lim, L.H.S. (1996) *Synodontisia moraveci* n.sp. (Oxyuroidea: Pharyngodonidae) from *Osteochilus melanopleurus* (Cyprinidae) of Malaysia, with a review of pinworms in fish and a key to species. *Systematic Parasitology* 34, 157–162.

Anya, A.O. (1966a) Experimental studies on the hatching of the eggs of *Aspiculuris tetraptera* (Nematoda: Oxyuroidea). *Parasitology* 56, 733–744.

Anya, A.O. (1966b) Studies on the biology of some oxyurid nematodes. I. Factors in the development of eggs of *Aspiculuris tetraptera* Schulz. *Journal of Helminthology* 40, 253–260.

Boecker, H. (1953) Die Entwicklung des Kaninchenoxyuren *Passalurus ambiguus*. *Zeitschrift für Parasitenkunde* 15, 491–518.

Chabaud, A.G. (1955) Essai d'interprétation phylétique des cycles évolutifs chez les nématodes parasites de vértebres. *Annales de Parasitologie Humaine et Comparée* 30, 83–126.

Chabaud, A.G. (1974) No. 1. Keys to subclasses, orders and superfamilies. In: Anderson, R.C., Chabaud, A.G. and Willmott, S. (eds) *CIH Keys to the Nematode Parasites of Vertebrates*. Commonwealth Agricultural Bureaux, Farnham Royal, UK, pp. 1–17.

Chabaud, A.G. (1978) No. 6. Keys to the genera of the superfamilies Cosmocercoidea, Seuratoidea, Heterakoidea, and Subuluroidea. In: Anderson, R.C., Chabaud, A.G. and Willmott, S. (eds) *CIH Keys to the Nematode Parasites of Vertebrates*. Commonwealth Agricultural Bureaux, Farnham Royal, UK, pp. 1–71.

Chan, K.F. (1952) Life cycle studies on the nematode *Syphacia obvelata*. *American Journal of Hygiene* 56, 14–21.

Chan, K.F. (1953) The effect of storage at low temperature on the infectivity of *Aspiculuris tetraptera* eggs. *Journal of Parasitology* (Suppl.) 39, 42.

Chan, K.F. (1955) The distribution of larval stages of *Aspiculuris tetraptera* in the intestine of mice. *Journal of Parasitology* 41, 529–532.

Chitwood, B.G. and Chitwood, M.B. (1950) *Introduction to Nematology*. University Park Press, Baltimore, Maryland.

Cho, S.Y., Chang, J.W. and Jang, H.J. (1985) Number of intrauterine eggs in female *Enterobius vermicularis* by body length. *Korean Journal of Parasitology* 23, 253–259.

Cram, E.B. (1943) Studies on oxyuriasis. XXVIII. Summary and conclusions. *American Journal of Diseases of Children* 65, 46–59.

Derothe, J.M., Laubès, C., Orth, A., Renaud, F. and Moulia, C. (1997) Comparison between patterns of pinworm infection (*Aspiculuris tetraptera*) in wild and laboratory strains of mice, *Mus musculus*. *International Journal for Parasitology* 27, 645–651.

Dobrovolny, C.G. and Ackert, J.E. (1934) The life history of *Leidynema appendiculata* (Leidy), a nematode of cockroaches. *Parasitology* 26, 468–480.

D'Silva, J. (1992) *Syphacia muris* Yam (Nematoda: Oxyuroidea) oviposition and host behaviour. *Bangladesh Journal of Zoology* 20, 301–304.

Enigk, K. (1949) Zur Biologie und Bekämpfung von *Oxyuris equi*. *Zeitschrift für Tropenmedizin und Parasitologie* 1, 259–272.

Geller, E.R. (1946a) Analysis of populations of Oxyuridae in the different layers of the intestine and on autoinfection in oxyuriasis. *Meditsinskaya Parazitologiya i Paraziticheskie Bolezni* 5, 45–50. (In Russian.)

Geller, E.R. (1946b) On spontaneous recovery in oxyuriasis. *Gel'mintologicheskii sbornik, posvyashchennyi akademiku K.I. Skrjabin Izd. AN SSSR*, pp. 73–76. (In Russian.)

Grice, R.L. and Prociv, P. (1993) *In vitro* embryonation of *Syphacia obvelata* eggs. *International Journal for Parasitology* 23, 257–260.

Gulden, W.J.I. (1967) De rattemade *Syphacia muris* (Yamaguti, 1935). Thesis, Katholieke Universiteit, Nijmegen, The Netherlands.

Hamburger, F. (1939) Die Oxyurenneurose. *Medizinische Klinik* 35, 369–370.

Hsü, K.C. (1951) Experimental studies on egg development, hatching and retrofection in *Aspiculuris tetraptera*. *Journal of Helminthology* 25, 131–160.

Hugot, J.P. (1984) L'insémination traumatique chez les oxyures de dermpotères et de léporidés. Étude morphologique comparée. Considérations sur la phylogénèse. *Annales de Parasitologie Humaine et Comparée* 59, 379–385.

Hugot, J.P. (1988) Les nématodes Syphaciinae, parasites de rougeurs et de lagomorphes. *Mémoires du Muséum National d'Histoire Naturelle* 141, 13–149.

Hugot, J.P. and Bougnoux, M.E. (1987) Étude morphologique de *Austroxyuris finlaysoni* (Oxyuridae, Nematoda) parasite de *Petouroides volans* (Petauridae, Marsupialia). *Systematic Parasitology* 11, 113–122.

Hugot, J.P., Bain, O. and Cassone, J. (1982) Physiologie des invertébrés. Insémination traumatique et tube de ponte chez l'oxyure parasite du lapin domestique. *Compte Rendu Hebdomadaire des Séances de l'Académie des Sciences, Paris* 294, 707–710.

Hulinská, D. (1968) The development of the female *Enterobius vermicularis* and the morphogenesis of its sexual organ. *Folia Parasitologica* 15, 15–27.

Hussey, K.L. (1957) *Syphacia muris* v. *S. obvelata* in laboratory rats and mice. *Journal of Parasitology* 43, 555–559.

Jones, M.F. and Jacobs, L. (1941) Studies on oxyuriasis. XXIII. The survival of eggs of *Enterobius vermicularis* under known conditions of temperature and humidity. *American Journal of Hygiene* 33, 88–102.

Kharichkova, M.V. (1946) Biology of *Passalurus ambiguus* (Rud., 1819) in foxes in Russia. *Collected Papers on Helminthology Dedicated by his Pupils to K.I. Skrjabin in his 40th Year of Scientific, Educational and Administrative Achievement*, pp. 274–279. (In Russian.)

Kozlov, A.S. (1982) The course of enterobiasis in an adult (self-experiment). *Meditsinskaya Parazitologiya i Parazitarnye Bolezni* 60, 30–33. (In Russian.)

Langhans, G.L. (1926) Zur Biologie des *Oxyuris vermicularis*. *Archiv für Kinderheilkunde* 77, 27–37.

Lawler, H.J. (1939) Demonstration of the life history of the nematode *Syphacia obvelata* (Rudolphi, 1802). *Journal of Parasitology* 25, 442.

Leuckart, R. (1863) *Die menschlichen Parasiten und von die ihnen herrührenden Krankheiten. Ein Hand-und Lehrbuch für Naturforscher und Aerzte*. Leipzig and Heidelberg.

Lewis, J.W. and D'Silva, J. (1986) The life-cycle of *Syphacia muris* Yamaguti (Nematoda: Oxyuroidea) in the laboratory rat. *Journal of Helminthology* 60, 39–46.

Mathies, A.W. (1959) Certain aspects of the host–parasite relationship of *Aspiculuris tetraptera*, a mouse pinworm. I. Host specificity and age resistance. *Experimental Parasitology* 8, 31–38.

Morand, S., Legendre, P., Gardner, S.L. and Hugo, J.P. (1996) Body size evolution of oxyurid (Nematoda) parasites: the role of hosts. *Oecologia* 107, 274–282.

Moravec, F., Kohn, A. and Fernandes, B.M.M. (1992) Three new species of oxyuroid nematodes, including two new genera, from freshwater catfishes in Brazil. *Systematic Parasitology* 21, 189–201.

Nolan, M.O. and Reardon, L. (1939) Studies on oxyuriasis. XX. The distribution of the ova of *Enterobius vermicularis* in household dust. *Journal of Parasitology* 25, 173–177.

Palimpsestov, M.A. and Chebotarev, R.S. (1935) Zur Frage der Therapie bei Passalurose (*Passalurus ambiguus*) des Kaninchens. *Tierärztliche Rundschau* 41, 709–711.

Perroncito, E. (1882) I parassiti dell' uomo e degli animali utili. Delle più comuni malattie da essi prodotte, profilassi e cura relativa. Milano.

Petter, A.J. (1969) Deux cas de poecilogynie chez les oxyures parasites d'*Iguana iguana* (L.). *Bulletin du Muséum National d'Histoire Naturelle*, Serie 2, 41, 1252–1260.

Petter, A.J. and Quentin, J.C. (1976) No. 4. Keys to genera of the Oxyuroidea. In: Anderson, R.C., Chabaud, A.G. and Willmott, S. (eds) *CIH Keys to the Nematode Parasites of Vertebrates*. Commonwealth Agricultural Bureaux, Farnham Royal, UK, pp. 1–30.

Philpot, F. (1924) Notes on the eggs and early development of some species of Oxyuridae. *Journal of Helminthology* 2, 239–252.

Prince, M.J.R. (1950) Studies on the life cycle of *Syphacia obvelata*, a common nematode parasite of rats. *Science* 111, 66–68.

Reardon, L. (1938) Studies on oxyuriasis. XVI. The number of eggs produced by the pinworm, *Enterobius vermicularis*, and its bearing on infection. *Public Health Report* 53, 978–984.

Sapozhnikov, G.I. (1969) The life cycle of *Skrjabinema ovis* (Skrjabin, 1915). *Trudi Vsesoyuznogo Instituta Gel'mintologii* 15, 267–274. (In Russian.)

Schad, G.A. (1957) Preliminary observations of the sheep pinworm, *Skrjabinema ovis. Journal of Parasitology* (Suppl.) 43, 13.

Schüffner, W. (1949) Retrograde Oxyuren-Infektion, 'Retrofektion'. IV. Mitteilung. *Zeitschrift für Bakteriologie Abteilung* 154, 220–234.

Schüffner, W. and Swellengrebel, N.H. (1949) Retrofection in oxyuriasis. A newly discovered mode of infection with *Enterobius vermicularis. Journal of Parasitology* 35, 138–146.

Seurat, L.G. (1912) Sur les oxyures d'*Uromastix acanthinurus* Bell. *Compte Rendu des Séances de la Société de Biologie* 73, 223–226.

Seurat, L.G. (1913) Sur un cas de poecilogonie chez un oxyure. *Compte Rendu des Séances de la Société de Biologie* 74, 1089–1092.

Shul'ts, R.C. and Krankrov, A.A. (1928) Oxyuriasis of horses and methods for its diagnosis. *Vestnik Sovremennoi Veterinarii* 24, 720–723. (In Russian.)

Skrjabin, K.I., Shikhobalova, N.P. and Lagodovskaya, E.A. (1960) Oxyurata of animals and man. Part I. Oxyuroidea. In: Skrjabin, K.I. (ed.), *Essentials of Nematology*. Akademiya Nauk SSSR. Gel'mintologicheskaya Laboratoriya Osnovy Nematodologii. (Israel Program for Scientific Translations, Jerusalem, 1974.)

Stahl, W. (1963) Studies on the life cycle of *Syphacia muris*, the rat pinworm. *Keio Journal of Medicine* 12, 55–60.

Todd, A.C. (1944) On the development and hatching of the eggs of *Hammerschmidtiella diesingi* and *Leidynema appendiculata*, nematodes of roaches. *Transactions of the American Microscopical Society* 63, 54–67.

Wetzel, R. (1930) On the biology of the fourth-stage larva of *Oxyuris equi* (Schrank). *Journal of Parasitology* 17, 95–97.

Yamaguti, S. (1938) Studies on the helminth fauna of Japan. Part 23. Two new species of amphibian nematodes. *Japanese Journal of Zoology* 7, 603–607.

Zawadowsky, M.M. and Schalimov, L.G. (1929) Die Eier von *Oxyuris vermicularis* und ihre Entwicklungsbedingungen, sowie über die Bedingungen unter denen eine Autoinfektion bei Oxyuriasis unmöglich ist. *Zeitschrift für Parasitenkunde* 2, 12–42.

Chapter 5
Order Ascaridida

Except for the Ascaridoidea, our knowledge of the development and transmission of the Ascaridida is most incomplete. The order is diverse, however, and contains monoxenous superfamilies (e.g. Cosmocercoidea, Heterakoidea), autoinfective species (Atractidae) and heteroxenous groups such as the Ascaridoidea, Seuratoidea and Subuluroidea. Of the heteroxenous groups the subuluroids and some seuratoids (not the cucullanids) utilize arthropod intermediate hosts. Cucullanids and members of the Ascaridoidea basically utilize vertebrate intermediate hosts although transmission is frequently modified in the ascarioids by paratenesis and precocity.

5.1

The Superfamily Cosmocercoidea

The Cosmocercoidea is a rather heterogeneous group with the oviparous families Cosmocercidae and Kathlaniidae and the ovoviviparous Atractidae (see Chabaud, 1978). The development and transmission of the various members of the superfamily deserve more attention.

Family Cosmocercidae

Cosmocercids are parasites of the gut of amphibians and reptiles. The subfamily Gyrinicolinae of tadpoles has been transferred to the Oxyuroidea and placed in the family Pharygodonidae by Adamson (1981). The transmission of species of *Maxvachonia*, the only genus of the Maxvachoniinae (mainly of reptiles), has not been studied. The only other subfamily, the Cosmocercinae, is found mainly in amphibians and four species have been studied. Females produce thin-shelled eggs that larvate *in utero* (e.g. *Aplectana* spp.) or develop in the external environment into first-stage larvae (e.g. *Cosmocercoides variabilis*). Outside the host, eggs hatch and first-stage larvae develop and moult twice to the infective third stage. The final host becomes infected either orally (*Aplectana courdurieri*) or by skin penetration (*Cosmocerca commutata*, *Cosmocercoides variabilis*). In some species (e.g. *C. variabilis*) the parasites undergo a period of development in the lungs before establishing themselves in the intestine where they mature.

Aplectana

A. courdurieri Chabaud and Brygoo, 1958
This is a parasite of *Rana (Ptychadena) mascareniensis* in Madagascar (Chabaud and Brygoo, 1958a,b). The female contained a small number of large eggs, each containing a fully developed first-stage larva. Eggs usually hatched 3–4 h after being laid (they rarely hatched *in utero*) and developed to the third stage (about 760 μm in length) in water. At about 20°C the first moult occurred in 43 h and the second in 74 h. Development was slower at lower temperatures.

Third-stage larvae failed to penetrate when placed in water on the skin of frogs. However, larvae were ingested by, and developed in, tadpoles. At 25°C the third moult occurred in the intestine in about 8 days and the second in about 20 days. A few third-stage larvae left the intestine and were found in various tissues (liver, heart) of the tadpoles. The authors suggested that worms acquired by tadpoles could persist in the frog. Also, frogs may acquire infections from eating tadpoles.

A. macintoshii (Stewart, 1914)

This is a cosmopolitan parasite of the rectum of toads and frogs (Baker, 1987). Yuen (1965) found it in *Bufo melanostictus* in Singapore and Malaysia and studied its free-living development. Gravid females shed larvated eggs when placed in saline. Eggs hatched almost immediately and larvae developed rapidly at 26–29°C in a culture containing toad faeces. The first moult occurred 10–15 h and the third stage was attained 60–70 h after hatching. The cuticle of the second stage was retained as a sheath. Lateral alae were present and the cephalic structures weakly developed. The oesophagus was rhabditiform but slender and the intestine was packed with granules. The genital primordia were weakly developed. Larvae usually lay motionless unless disturbed, whereupon they moved rapidly.

Cosmocerca

C. commutata (Diesing, 1851)

This is a widely distributed parasite of toads and frogs. Fotedar and Tikoo (1968) studied material from *Bufo viridis* in Kashmir under the name of *C. kashmirensis* Fotedar, 1959 which Baker (1987) regarded as a synonym of *C. commutata*. Eggs hatched in 2–4 h at 22–32°C. The first moult occurred in 4–5 days. Sheathed larvae, regarded as second stage by the authors, penetrated the skin of *B. viridis*. Young worms were found in the lungs 3 days later and adult worms were recovered from the rectum 10–14 days postinfection. A lung migration is apparently a necessary part of the development.

Cosmocercoides

Vanderburgh and Anderson (1987a) have shown that the species of *Cosmocercoides* (i.e. *C. dukae*) living independently in terrestrial gastropods in North America (Ogren, 1953, 1959; Anderson, 1960; McGraw, 1968) is distinct from the species (*C. variabilis*) that reproduces in amphibians (Harwood, 1930; Baker, 1978; Vanderburgh and Anderson, 1987b). The confusion in the literature has arisen from the fact that amphibians occasionally feed on terrestrial gastropods: *C. dukae* infecting the latter can persist in the rectum of amphibians and be confused with *C. variabilis* (see also Bolek, 1997).

C. variabilis (Harwood, 1930)

This is a common parasite of the rectum mainly of Bufonidae but also of Hylidae and Miceohylidae (Vanderburgh and Anderson, 1986, 1987b; Baker, 1987; Joy and Bunten, 1997); reports in Ranidae need confirmation as they may refer to *C. dukae* acquired from gastropods. Eggs laid by gravid female worms were oval and thin-shelled (Fig. 5.1A). They developed rapidly in a culture containing toad faeces. The first-stage larva (Fig. 5.1B) was short, with a tapered tail and a rhabditiform oesophagus. First-stage larvae hatched from eggs and underwent two moults (Fig. 5.1C–E) to the infective third stage, which had broad lateral alae and a tapered tail ending in four small spines. The genital primordia were in the two-cell stage. Baker (1978) allowed third-stage larvae to penetrate the skin of parasite-free *Bufo americanus* at 14–18°C. Larvae were subsequently found in the body cavity, lungs and rectum. Entrance to the lungs was probably by direct penetration. Migration to the rectum was probably by the trachea and mouth. Moulting fourth-stage larvae occurred in the lungs and the rectum. Adults were found only in the rectum. Development of the female to the gravid stage

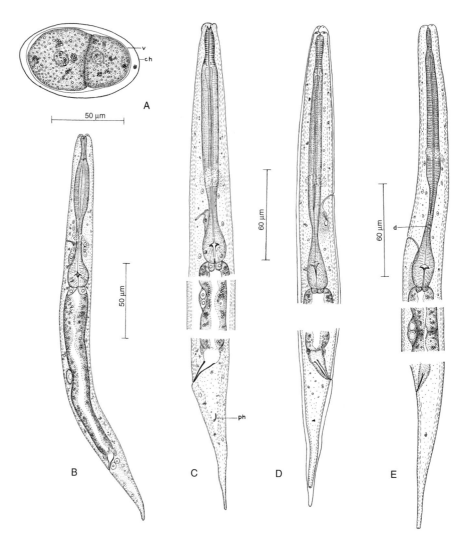

Fig. 5.1. Developmental stages of *Cosmocercoides dukae* (identical to those of *C. variabilis* of amphibians): (A) egg, unembryonated; (B) first-stage larva; (C) second-stage larva; (D) moulted late second-stage larva; (E) infective third-stage larva. (After R.C. Anderson, 1960 – courtesy *Canadian Journal of Zoology*.)

required more than 30 days at 14–18°C. However, worms with numerous eggs appeared 38 days postinfection.

Vanderburgh and Anderson (1987b) examined *Bufo americanus* from April to October in Ontario, Canada. Fourth-stage larvae were present in the lungs from late April to early May. There was evidence that larvae in lungs in spring migrated to the rectum and matured by the end of May. Transmission of *C. variabilis* apparently occurred throughout the season and larval and adult worms found in toads in early spring were probably acquired the previous year.

Family Kathlaniidae

Very little is known about the transmission of the Kathlaniidae (the Oxyascaridinae, with *Oxyascaris* and *Pteroxyascaris*, has been synonymized with the Cosmocercinae). There is an account in the literature of the development of *Cruzia americana* of the American opossum and a note on *Falcaustra wardi* of freshwater turtles. Also Bain and Philippon (1969) found kathlaniid-type larvae (possibly third stage) free in the Malpighian tubules of *Simulium damnosum* in Upper Volta. Possibly, kathlaniids in the lower vertebrates develop to the third stage outside the host and then invade various invertebrates which serve as paratenic hosts.

Cruzia

C. americana Maplestone, 1930

This is a parasite of the caecum and large intestine of American opossum (*Didelphis marsupialis virginiana*) in the USA. According to Crites (1956) transmission is direct. Eggs passed in faeces were in a morula or subvermiform stage. In favourable conditions of temperature, moisture and oxygen, a first-stage larva was formed in 7–9 days and the first moult took place by 10 days after the eggs had been passed in faeces. In the opossum, eggs with second-stage larvae hatched in the duodenum or upper ileum and moved to the caecum, where three moults to the adult stage occurred. The second moult took place 5 days, the third 10 days and the fourth 28–29 days postinfection. Fertilized eggs were passed by the host 46–48 days after infection.

Infection of the final host by second-stage larvae and the absence of an accompanying tissue phase is unusual and it would be useful if further studies were carried out on *C. americana*.

Falcaustra

Moravec *et al.* (1995) identified larvae as *Falcaustra* spp. in the following freshwater fishes from Texas: *Etheostoma fonticola*, *E. lepidotum*, *Cichlasoma cyanoguttatum*, *Lepomis auritus*, *Lepomis* sp. and *Gambusia affinsis*.

F. wardi (Mackin, 1936)

Bartlett and Anderson (1985) described third-stage larvae of *Falcaustra* in the tissues of the freshwater snail *Lymnaea stagnalis* in Ontario. A single male of *F. wardi* was found in the intestine of a laboratory-reared turtle (*Chelydra serpentina*) given ten larvae orally. It was suspected that the snails were serving as paratenic hosts of *F. wardi*.

Megalobatrachonema

M. terdentatum (Linstow, 1890)

This is a parasite of the stomach and intestine of salamanders (*Triturus cristatus, T. vulgaris, T. alpestris* and *T. palmatus*) in Europe. According to Barus and Groschaft (1962) eggs developed to the 30–40 cell stage *in utero* and were then laid and passed in faeces of the host. Eggs developed at 15–25°C into larvae in 48 h. Larvae moulted and hatched in 84 h. A second moult was not observed but the infective third stage was reached 72 h after hatching. These larvae survived for up to 16 days and then died. Neither the second nor third stages were resistant even to slight desiccation. Larvae apparently failed to penetrate the skin of salamanders but the latter could be infected *per os*. In 24 h larvae were found in the stomach of a *T. vulgaris* that had been given larvae. In 48 h two other infected salamanders had larvae in the stomach and intestine. Two salamanders were examined 16 days after infection and four fourth-stage larvae were found in one of them.

Petter and Chabaud (1971) cultured eggs of *M. terdentatum* from *T. vulgaris* at 19–20°C. Eggs contained motile larvae in a few days. In 8–10 days some larvae had hatched and some were surrounded by a sheath. In 11 days some larvae had a second loose cuticle and were regarded as third stage. They shed the cuticle in about 13 days and died after 17 days in culture. Third-stage larvae invaded and persisted in molluscs, oligochaetes and larval salamanders but did not grow significantly in the invertebrates and only slightly in the salamanders. Adult salamanders were infected with oligochaetes containing third-stage larvae but attempts to infect them directly with larvae were unsuccessful, indicating that the invertebrate host may be essential. In adult salamanders, third-stage larvae were first found in the body cavity, where they grew slowly for about 5.5 months. Worms in the third and fourth stages were found in the body cavity as well as in the intestine of wild-caught salamanders. The fourth moult took place in mid-March and the young adults appeared in the intestine. From April on, mature worms were found in the salamanders.

Family Atractidae

The Atractidae are found in amphibians, reptiles, mammals and fishes. They are unusual in that eggs hatch and larvae develop to the third stage *in utero*. Third-stage larvae autoinfect the host. Ivashkin and Babaeva (1973) noted that autoinfection is generally accompanied by high intensities, low pathogenicity and large larval forms. Furthermore, autoinfecting nematodes generally occupy capacious organs in the host. Their transmission from host to host is not understood. Petter (1966) claimed that tortoises became infected only after attaining sexual maturity and wondered if transmission occurred during mating.

Probstmayria

P. vivipara (Probstmayr, 1865)

This is a common parasite of the caecum and colon of equines (Jerke, 1902; Ransom, 1907). Female worms produce larvae which develop *in utero*. Adult worms are tiny (2.5–3.0 mm in length) with long filamentous tails. The female produces two to four large eggs at any one time which develop *in utero*. Larvae hatch and grow in the body of the female to the third stage when they are about 1.8 mm in length. Larvae are deposited in the lumen of the gut of the host and develop to maturity. As a result of autoinfection, extremely large numbers may build up in the host (Perroncito, 1882; Le Roux, 1924). It is not known precisely how the parasite transfers from host to host but it is presumed that various stages are passed in faeces and retain their infectivity for a period of time. Jerke (1902) noted that worms would remain alive in manure for 4–5 days.

Rondonia

Species of *Rondonia* are peculiar in that the vulva opens into the rectum. They are parasites of the intestine of fishes.

R. rondona Travassos, 1917

R. rondona has been found in various fishes (*Piaractus brachypomus, Myeletes torquatus, Doras granulosus, Myleus* sp.) in South America. The female is monodelphic and produces a small number of thin-shelled eggs which hatch and develop *in utero*. The uterus usually contains only a few larvae, which apparently become free only after bursting of the female body (i.e. **matricidal endotoky**) (Travassos, 1920; Da Costa, 1962) (cf. Baylis, 1936; Gallego, 1947). Da Costa (1962) believed that larvae which pass out of the host infect fish directly.

5.2

The Superfamily Seuratoidea

The seuratoids include a somewhat disparate group of genera believed, in some instances, to connect the Cosmocercoidea to the advanced Ascaridida or Spirurida (Chabaud, 1978). Unfortunately, our knowledge of the transmission and development of most of the species is very limited. Of the Seuratidae of reptiles and mammals, only three species have been studied. The Schneidernematidae of birds and mammals and the Chitwoodchabaudiidae of amphibians have not been investigated, although Puylaert (1970) claimed that third-stage larvae of *Chitwoodchabaudia skrjabini* were associated with chironomids found in the stomach of the host, *Xenopus laevis*. A few preliminary observations have been made on the Quimperiidae of fishes and amphibians. The Cucullanidae of fishes and turtles have been the subject of several recent studies.

Family Seuratidae

Seuratum

S. cadarachense Desportes, 1947

This is an intestinal parasite of the glirid rodent *Eliomys quercinus* in Europe. Quentin (1970a,b) and Quentin and Seureau (1975) took eggs from gravid females and infected adult *Locusta migratoria* kept at 27–28°C. The egg was thin-shelled and contained a fully developed first-stage larva with a prominent cephalic hook and a rounded, unarmed tail. After hatching in the midgut, larvae invaded the wall of the midgut (mesenteron) and some lodged at the base of the epithelium, where they became surrounded by haemocytes. Other larvae occurred in the haemocoel, where they became surrounded by capsules of haemocyte origin, often attached to muscles of the gut. The first moult occurred about 6–7 days postinfection. The second moult occurred about 14 days postinfection, when larvae were found in the third stage. The third stage was short and stout and had a pointed tail; the genital primordia were poorly developed. In the rodent host, Quentin (1970a,b) reported that the third moult occurred about 2 days postinfection; a fourth-stage larva was found in the epithelium of the small intestine at that time. Fourth-stage larvae were also found 4 and 6 days postinfection in experimentally infected *E. quercinus*.

S. nguyenvanaii Le-Van-Hoa, 1964

This is a parasite of the intestine of the house shrew *Suncus murinus* in southeast Asia (Vietnam). Le-Van-Hoa (1966) followed development in experimentally infected cockroaches (*Blatta orientalis* and *Periplaneta americana*). Larvae developed in the midgut for 6 days, then invaded the gut wall and became encapsulated in the haemocoel. The first moult occurred about 7 days and the second 8–9 days postinfection. Larvae developed to the fourth stage in 2 days in an experimentally infected house shrew.

Skrjabinelazia

S. galliardi Chabaud, 1973

This species was found in the lizard *Gonadotes humeralis* in Brazil. Chabaud *et al.* (1988) claimed that females produced two types of egg. One was thin-shelled and contained a third-stage larva, which was probably autoinfective. These eggs were produced by young females. Older worms produced red, thicker-shelled eggs with third-stage larvae which possibly pass out of the host. Eggs were given to unidentified crickets in which some third-stage larvae were subsequently found. The insect was regarded as a paratenic host.

Rhabbium

R. paradoxus Poiner, Chabaud and Bain, 1989

Adults of this species were found in the abdomen of worker ants (*Camponotus castaneus*) in Florida (Poinar *et al.*, 1989). This is probably an example of extreme precocity, as species of *Rhabbium* are normally parasites of reptiles; their transmission has not been investigated.

Family Quimperiidae

The biology of the quimperiids is little understood and only three preliminary observations have been made. Bain and Philippon (1969) found a type of larva in two of 1000 *Simulium damnosum* from Upper Volta which they thought belonged to the Quimperiidae. They felt that the insect was probably serving as a paratenic host. Moravec and Ergens (1970) found larval *Haplonema problematica* (syn. *Cottocomephoronema problematica*) encapsulated in the liver of some Mongolian fish (*Noemacheilus barbatulus*, *Phoxinus phoxinus*). Finally, Moravec (1974) made observations on the development of *Paraquimperia tenerrima* in eels (see below).

Paraquimperia

P. tenerrima (Linstow, 1878)

This common parasite of the intestine of European eels (*Anguilla anguilla*) was studied by Moravec (1974). Eggs laid by the female had thin, smooth shells and were unsegmented or in the first or second cleavage stages. Eggs embryonated at 20–25°C and active larvae were present by 3 days. No moulting was observed. In 5–6 days eggs hatched spontaneously. The free first-stage larva was long and the intestinal region was packed with granules; the tail was slightly tapered and had a rounded tip. These larvae moulted 3–4 days after hatching. The second-stage larvae were slender and had lateral alae and cervical papillae; the oesophagus was cylindrical and muscular and the genital primordia poorly developed. Six days after moulting, larvae died. Attempts to infect oligochaetes and snails with second-stage larvae were unsuccessful. A larva regarded as third stage was found in the intestine of a small eel given second-stage larvae 10 days earlier.

Family Cucullanidae

The cucullanids are peculiar intestinal parasites of fish and rarely turtles. The family is characterized by a highly developed buccal cavity formed from the oesophagus (i.e. oesophastome) (Berland, 1970). Details of the development and transmission of the cucullanids are still imperfectly known, especially as they concern species in marine hosts. There is some evidence that the group is primitively heteroxenous and uses vertebrates as intermediate hosts (e.g. *Dichelyne cotylophora, Truttaedacnitis truttae*) but that in some species the intermediate host has been replaced by a histotropic phase in the definitive host (e.g. *Cucullanus chabaudi, ?Dichelyne minutus, ?Cucullanus heterochrous*). More detailed experimental study is required on the transmission of the various species in the family.

Cucullanus

C. chabaudi Le-Van-Hoa and Pham-Ngoc-Khue, 1967
This is a common parasite of the fish *Pangasius pangasius* in Vietnam. According to Le-Van-Hoa and Pham-Ngoc-Khue (1967), eggs in gravid females were round, thin-shelled and embryonated. Eggs developed at room temperature in 2 days into first-stage larvae, which moulted in 4 days. Eggs hatched in 5–6 days. Larvae were said to develop in the swim bladder of the fish host for 3–4 weeks. In naturally infected fish, larvae moulting to the third stage were found in the liver. Larvae at the end of the third stage and of the fourth stage were found in the bile duct and gall bladder, where development to the adult stage occurred.

C. cirratus Muller, 1777
C. cirratus is a parasite of gadiform fishes. According to Valovaya (1978) eggs developed in water to the first stage and no moulting occurred. First-stage larvae hatched after 10–12 days of development of the eggs. Larvae remained alive for up to 2 weeks in water. Attempts to infect *Jaera albifrons* with first-stage larvae were unsuccessful. Second-, third- and fourth-stage larvae were found in *Gadus morhua* and *Eleginus navaga* by Valovaya (1977, 1979). Third-stage larvae were found in the gastric mucosa and free in the intestine; some were shedding the second-stage cuticle. The third moult occurred when larvae were 1.5–2.0 mm in length and began earlier in males than in females. Worms grew and developed rapidly after the fourth moult. It was suggested (Valovaya, 1979) that nematode eggs passed by infected fish were in the two- to four-cell stage. First-stage larvae hatch from eggs 10–14 days later, moult in 2–3 days and may survive in sea water as second-stage larvae for 1.5 months. If swallowed by fish, these

larvae penetrate the gastric mucosa, moult to the third stage and then migrate to the intestine, where third and fourth moults and maturation take place.

C. heterochrous Rudolphi, 1802

C. heterochrous is a parasite of plaice and flounders (Pleuronectiformes), in which they feed on the gut contents and the gut wall (Mackenzie and Gibson, 1970). Gibson (1972) studied the early development. Eggs developed and hatched in 7 days at 19°C. A moult was detected but it was not determined if it occurred in the egg or after hatching. Gibson (1972) reported on larvae in estuarine flounders (*Platichthys flesus*) in the River Ythan in Scotland and in marine flounders. In estuarine flounders, third-stage larvae appeared in January. They were encapsulated in the gut wall and numbers reached a maximum in April. They declined until by August none was found. Fourth-stage larvae first appeared in March; numbers reached a maximum in May and then declined during summer. They were found in the gut wall but were not encapsulated. Immature adults appeared in the gut lumen in March; they were most common in June and July and absent by January. Mature adults occurred in September; they reached a maximum in winter and then declined throughout the following spring.

In marine flounders, third-stage larvae appeared in November. Fourth-stage larvae increased in numbers during late autumn and were present in large numbers in winter and spring. Numbers of immature adults increased in spring and reached a peak in autumn. Numbers of mature worms reached a maximum in winter and declined in spring.

Dichelyne

Dichelyne spp. occur in fishes (especially perciforms) and turtles.

D. bullocki Stromberg and Crites, 1972

This is an intestinal parasite of the fish *Fundulus heteroclitus*. Its development and transmission were studied by Kuzia (1978). Eggs in gravid female worms were in the one- to eight-cell stage. Eggs developed rapidly at 23°C in tap water and in various salinities. First-stage larvae appeared in eggs in 65 h. The first moult was observed in the egg in 4 days. A second moult was reported in the egg in 7 days but no specimen was obtained with both the first and second cuticles. Most larvae hatched in 8 days and were assumed to be in the third stage. Larvae attached one end to the substrate and swayed from side to side; they remained active in cultures for up to 72 days. Eggs developed at 13°C and hatched in 21–33 days.

F. heteroclitus was placed with larvae that had hatched in culture and was examined later at various times. Three groups of fish were used. Group I fish were laboratory-reared (*n* = 9) and 5 months old: only third-stage larvae were found in these fish when they were examined up to 53 days postinfection. Group II fish were collected in an area where *D. bullocki* did not occur: only third-stage larvae were found in these fish even up to 168 days postinfection.

Group III fish (*n* = 70) had been collected at the same time and place as Group II fish and were examined 1–362 days after being exposed to eggs. Most fish contained only larvae identified as third stage (from 1 to 362 days). A few worms in the fourth stage were found in 14 fish examined 15–362 days postinfection. A single adult worm was found in the intestinal lumen of a fish examined 42 days postinfection.

In the first 4 days all larvae recovered were in the intestinal lumen. Five to 7 days after exposure, third-stage larvae were found in the intestinal wall and lumen. From 8 to 362 days, all third-stage larvae were in the wall of the intestine.

No nematode larvae were recovered from amphipods or copepods exposed to eggs and examined 5 days later.

D. cotylophora (Ward and Magath, 1917)

This is a common parasite of yellow perch (*Perca flavescens*) in Lake Erie, Ontario. Baker (1984) showed that the oval, thin-shelled eggs of the nematode embryonated in fresh water into first-stage larvae in 41 h at 23–25°C. At 62 h larvae pressed from eggs were moulting. A further moult apparently took place (71–87 h) and the third-stage larvae then hatched. The third-stage larva had a long club-shaped oesophagus and a long tapered tail ending in a rounded tip. Minnows (*Notropis cornutus*) were placed with third-stage larvae and examined at various times thereafter. By 8 days, moulting third-stage larvae were found in the intestine of the minnows and fourth-stage larvae were found 10–25 days postinfection, encapsulated in the liver. The worms were identical morphologically to larvae found in the intestine of naturally infected perch and it was concluded that prey fish such as minnows serve as intermediate hosts of *D. cotylophora* of perch. In perch, fourth-stage larvae apparently grew to adults in the gut without a histotropic stage. Larvae were acquired by perch in late summer, autumn and winter but they did not develop past the fourth larval stage until the following spring, at which time they developed rapidly into adults and produced eggs in early summer. Adult worms then disappeared in late summer.

D. minutus (Rudolphi, 1819)

D. minutus is an intestinal parasite of plaice and flounders (Pleuronectiformes). Janiszewska (1939) and Gibson (1972) described the larval stages. According to Gibson, eggs developed and hatched in 7 days at 19°C; a moult was not observed. Third- and fourth-stage larvae were found in the intestinal wall of the host. Adults occurred free in the gut lumen or attached to the gut wall (Mackenzie and Gibson, 1970).

In the River Ythan in Scotland, third-stage larvae were found in flounders (*Platichthys flesus*) throughout the year but their numbers increased in early spring, especially March to April (Gibson, 1972). Fourth-stage larvae appeared in March, reached maximum numbers in May and June and declined to low levels from August to November. Adults first appeared in April (immature) and May (mature) and reached a peak in numbers in June and July. Thereafter, they declined in numbers.

In marine flounders, third-stage larvae were first observed in January; their numbers increased by March and then declined by June. Fourth-stage larvae were first found in March. Immature adults appeared in June and mature adults in September and November.

Wulker (1930) suggested that decapods and cumaceans were intermediate hosts of *D. minutus* but Janizewska (1939) was not successful in infecting them. Markowski (1966) suggested that *Nereis diversicolor* was the intermediate host but Gibson (1972) failed to infect this polychaete as well as *Neomysis integer* and various amphipods.

Truttaedacnitis

Some authors (e.g. Moravec, 1979) regard *Truttaedacnitis* as a subgenus of *Cucullanus*. The nomenclature of Petter (1974) is accepted herein. Choudhury and Dick (1996) provided a useful cladistic analysis of the few species that they include in the genus, namely *T. heterodonti*, *T. sphaerocephala*, *T. truttae* and *T. pybusae*.

Moravec (1976) found what he regarded as third-stage larvae encapsulated in the gut wall of brook lamprey (*Lampetra planeri*) in Bohemia. Similar larvae, unidentified at the time, were noted by Moravec and Ergens (1970) in *Lampetra reissneri* in Mongolia. Moravec (1976) suggested that the larvae belonged to *Truttaedacnitis truttae* (syn. *Cucullanus truttae*) of salmonids and that the lamprey was either a paratenic ('reservoir') or intermediate host. Moravec and Malmquist (1977) found larval and adult worms in *L. planeri* collected in Sweden; they suggested that the adult nematodes in *L. planeri* in Europe were originally described as *Dacnitis stelmioides* Vessichelli, 1910 and that this name was a synonym of *T. truttae* (Fabricius, 1794), which is a parasite of the various salmonids in Europe. Moravec (1979) fed infected intestines of lamprey to *Oncorhynchus mykiss* and recovered adult worms in the trout several days later, confirming that specimens in lamprey and trout in Europe were conspecific and that the lamprey apparently served as an intermediate host.

Pybus *et al.* (1978a) identified nematodes found in the intestine of the brook lamprey (*Lampetra appendix*, syn. *L. lamotteni*) in Ontario, Canada, as *Truttaedacnitis stelmioides*. Third-stage larvae occurred in the liver and not the intestine, as in *T. truttae* in Europe. Larvae persisted in the liver of the ammocoete until the latter transformed into adults, whereupon the nematodes migrated to the gut and matured (see also Zekhnov, 1956; Shulman, 1957). Thus the ammocoete served as the intermediate host and the adult lamprey as the definitive host. There was no evidence that teleosts were involved in any part of the life cycle in the stream studied. Anderson and Bartlett (1993) felt that this was perhaps an example of extreme precocity in which the parasite began to mature in the intermediate host and eventually became independent of the definitive host (i.e. the phenomenon of **capture**). Anderson (1992) renamed the species in *L. appendix* in North America as *T. pybusae*, to distinguish it from *T. truttae* found in European salmonids (see also Choudhury and Dick, 1996).

T. pybusae Anderson 1992

Pybus *et al.* (1978b) found embryonating eggs of *T. pybusae* (see above) in water-filled glass funnels containing adult brook lampreys collected in spring. Eggs (63–87 × 51–68 μm) kept at 13°C embryonated and hatched in 18–20 days. Eggs kept at 4°C and 10°C did not embryonate beyond the tadpole stage. At 13°C larvae remained alive for up to 75 days. Newly hatched larvae (Fig. 5.2A) had lateral alae and were 380–447 μm in length. A moult was not reported. Some larvae found in the intestine of ammocoetes were identical to newly hatched larvae. Others were larger and one was moulting. The mean length of larvae increased with increasing length of the ammocoetes. Larvae were commonly found in the liver of ammocoetes and were presumably in the third stage (Fig. 5.2B). These larvae were 1.11–1.41 mm in length. Some resembled the largest larvae found in the intestine, which were in the fourth stage (Fig. 5.2C–G), and prevalence increased with size of the ammocoetes. Immature adult worms were found in the gut of transformers. Mature worms occurred in the gut of

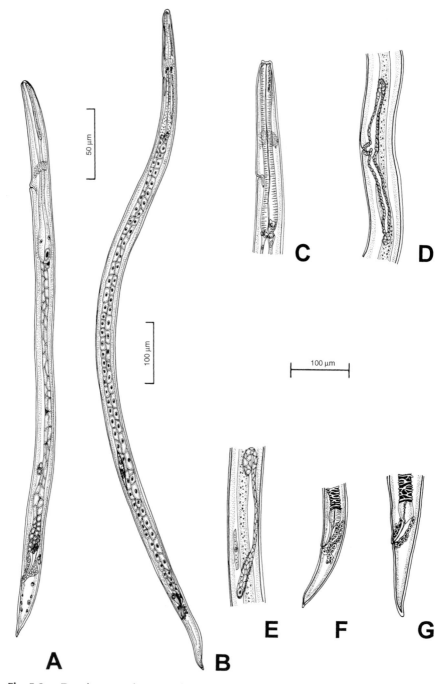

Fig. 5.2. Developmental stages of *Truttaedacnitis pybusae*: (A) newly hatched larva; (B) third-stage larva; (C) fourth-stage larva, anterior region; (D) female fourth-stage larva, genital primordium; (E) male fourth-stage larva, genital primordium; (F) male fourth-stage larva, tail end; (G) female fourth-stage larva, tail end. (After M. Pybus *et al.*, 1978 – courtesy *Canadian Journal of Zoology.*)

adult lamprey. It was hypothesized that eggs passed by adult lampreys hatch on the stream bed in spring and early summer. Newly hatched larvae are ingested by ammocoetes and remain in the intestine throughout summer. After attaining the third stage (Fig. 5.2B), larvae migrate to the liver and remain arrested for up to 4 years. During transformation of the host, larvae moult in the liver, re-enter the intestine and mature. Eggs would be passed from the lampreys and reinfect the stream bed inhabited by ammocoetes.

As indicated earlier, *T. pybusae* was not found in various fish (including salmonids) in the stream with infected lamprey and attempts to infect trout (*O. mykiss*) by feeding pieces of lamprey with larvae were unsuccessful.

Anderson (1996) restudied certain aspects of the transmission of *T. pybusae* in *L. appendix* in the same stream investigated by Pybus *et al.* (1978b). The gut in the spawning lampreys in early May degenerates and relatively few eggs of *T. pybusae* are passed into the spawning area. Eggs and worms tend to congregate in sac-like swellings in the intestine, filled with whitish mucus-like material. The ciliated lining was not present and the gut tended to constrict posteriorly, with the result that few eggs were able to pass out of the host. In addition, the few eggs passed would undoubtedly be dispersed in the rapidly flowing stream in May when spawning occurs, and it is difficult to imagine how they, and the larvae hatching from them, would become available to ammocoetes. An examination of prespawning adult lampreys, however, revealed that the gut containing adult nematodes was intact and evidently fully functional and eggs could readily pass out of the host. It is believed that this is the source of the eggs which contaminate lamprey beds where the ammocoetes occur. These eggs are passed into the lamprey beds weeks before spawning takes place and well before the gut of the lampreys degenerates during transformation into adulthood.

Attempts to infect *Oncorhynchus mykiss* with third-stage larvae from the liver of ammocoetes were unsuccessful. Some adult nematodes intubated into the stomach of *O. mykiss* attached to the gut wall and persisted in diminishing numbers for about 3 weeks. It was concluded, therefore, that *T. pybusae* is an independent parasite of the brook lamprey in North America and that teleosts which might consume an infected lamprey can serve for a brief period as postcyclic hosts.

T. sphaerocephalus (Rudolphi, 1809)

Khromova (1975) reported *T. sphaerocephalus* in Acipenseridae in the Azov and Caspian Seas and the lower reaches of the Volga. Fish became infected in summer. Eggs of the nematode developed and hatched in water and attempts to infect young sturgeon with hatched larvae were not successful. Many polychaetes, oligochaetes and crustaceans were examined for cucullanid larvae. *Cucullanus*-like larvae were said to occur in *Nereis diversicolor* in the Caspian Sea and it was suggested that this polychaete was an intermediate host.

T. truttae (Fabricius, 1794)

T. truttae (syn. *Dacnitis stelmioides*) is a common intestinal parasite of various salmonid fishes in the holarctic (Moravec, 1979, 1980; Butorina, 1988). The eggs were oval, thin-shelled and usually in the two-cell stage when laid (Moravec, 1979, 1980) and were 81–84 × 54–57 µm in size. In water at 22–24°C first-stage larvae appeared in the eggs in 3 days. The larvae moulted in 6–7 days and hatched (7–8 days). They did not develop further. These second-stage larvae were slender and had lateral alae and deirids. Attempts to infect oligochaetes, snails, copepods, larval chironomids, small fishes and trout (*Oncorhynchus mykiss*) with second-stage larvae were unsuccessful.

Encapsulated larvae on the intestine of *Lampetra planeri* had lateral alae and well-developed deirids and were 840–924 µm in length (Moravec, 1979). The undeveloped genital primordium was in the posterior third of the body. The oesophagus was club-shaped and a prominent nucleus was present near its junction with the intestine. Intestines with larvae were fed to trout (*O. mykiss*) which were examined at various times thereafter. In the gut of the trout, larvae attached to the mucosa of the anterior part of the intestine. In 20 days larvae moulted and fourth-stage larvae migrated to the pyloric caeca. The fourth moult was not observed but probably occurred 30–40 days postinfection. Gravid worms were found in 89 days in trout kept at 13–15°C.

5.3

The Superfamily Heterakoidea

Heterakoids are gut parasites with a prominent preanal sucker surrounded by a cuticularized ring (Chabaud, 1978). The superfamily consists of the Heterakidae and the Ascaridiidae. The former includes the Spinicaudinae of amphibians and reptiles and the Heterakinae, mainly of birds, including the well-known genus *Heterakis*. The Ascaridiidae includes only *Ascaridia*, mainly of birds, including poultry.

Heterakoids are monoxenous; eggs containing the infective stage are ingested by the host. Female worms produce unembryonated eggs which develop after being passed from the host. Most authors reported a single moult in the egg of those species that have been studied and concluded that the second-stage larva was infective. This conclusion conflicted with the fact that most authors detected only two moults in the avian host. However, Bain (1970) detected two moults in the egg of *Strongyluris brevicauda* and Araujo and Bressan (1977) found two moults in the egg of *Ascaridia galli* and it may be assumed that this is a general feature of the Heterakoidea.

Family Heterakidae

The spinicaudines (including *Spinicauda* and *Strongyluris*) seem to be the most primitive in the group and will be considered before the much better known *Heterakis* and *Ascaridia*.

Subfamily Spinicaudinae

Spinicauda

S. inglisi Chabaud and Brygoo, 1960
S. freitasi Chabaud and Brygoo, 1960

The two species occur together in the rectum of chameleons in Madagascar; adult males are readily separated but females are morphologically indistinguishable. Petter (1968a) made some tentative observations of mixed infections in chameleons given eggs derived from unidentified female worms. Eggs removed from the latter larvated in tap water in 14–15 days at 26°C. Larvae pressed from eggs were about 440 μm in length and had lateral alae. The tail was tapered and ended in a point. The oesophagus was rhabditiform but without a valve. About 100 eggs were given orally to each of 21 *Chamaeleo lateralis* collected in a region where *Spinicauda* spp. had not been reported. Chameleons were examined for worms 2–163 days later. The combined data suggested that worms developed in the intestine and adults matured in the rectum. Two types of larvae were noted but it was not possible to assign them to either *S. inglisi* or *S. freitasi*. Moulting larvae found at 20 days were 800–900 μm in length. Moulting worms were also found at 30 days and immature worms of both *S. inglisi* and *S. freitasi* appeared in the rectum as early as 29 days. At about 40 days worms were mature in the rectum.

More detailed study of these parasites is required, especially to determine if there are moults in the eggs and/or additional moults in the definitive host. It is most likely that two moults occur in the egg and what Petter (1968) regarded as the first-stage larva was, in fact, the third stage; the presence of alae in the larvae described by her, support this suggestion.

Strongyluris

S. brevicaudata Mueller, 1849

This is a parasite of the rectum of reptiles. Bain (1970) examined specimens from the rectum of *Agama agama* from Upper Volta, Africa. Eggs removed from female worms were placed in tap water at 23°C. In 12 days eggs contained larvae 250–270 μm in length. Bain (1970) noted that two moults took place in the egg and that by 21 days the third-stage larvae were fully developed and 250–315 μm in length. The head was provided with a ventral hook and the cephalic end was surrounded by about a dozen creases. The oesophagus was dilated posteriorly but lacked valves. The tail was pointed.

Bain (1970) allowed cockroaches (including *Blattella germanica*) and larval *Culex* sp. to ingest eggs with third-stage larvae. Eight days later she found some larvae in the body cavity of some of the cockroaches and in one of the mosquitoes. Five larvae were found encapsulated in the thorax of one adult *Culex* sp. 17 days after it was exposed to eggs.

Eggs with infective third-stage larvae were given to wild-caught *Lacerta muralis* and *Agama agama*, which were examined at various times (22–64 days) postinfection. Larvae were found in the lungs, body cavity and gut of the reptiles. Moulting worms were found on days 24, 53 and 64 but it was not certain if these worms were all the result of experimental infections, since control lizards also had worms. A study of the larval stages found indicated there were two moults in the definitive host but the precise timing of the moults was not determined.

Subfamily Heterakinae

Heterakis

H. gallinarum (Shrank, 1788)

H. gallinarum (syns *H. papillosa*, *H. vesicularis*, *H. gallinae*) is a common cosmopolitan caecal nematode of poultry, especially chickens and turkeys (Madsen, 1950). According to Graybill (1921) eggs are fully embryonated in 7–12 days at 18–29°C. Clapham (1933) reported that development to the infective stage occurred in 14–17 days and she regarded 26°C as optimal for development. Various authors have reported a single moult in the egg of *H. gallinarum*. However, as indicated above, it is likely that two moults occur in the egg. Clapham (1933, 1934) reported a moult in the egg and claimed that the cuticle was not shed until the larva reached the final host. Roberts (1937a,b) reported that 33°C was the optimum temperature for development of the egg and he claimed that the first larval moult occurred on day 4 at this temperature. At 27°C the first moult occurred on day 6 and larvae were infective 24 h after moulting. Numerous authors have reported that larvated eggs of *H. gallinarum* remain viable in the environment for long periods and are resistant to extremes of temperature and dry conditions (e.g. Osipov, 1957, 1958; Birova-Volosinovicova, 1965). Lund (1960) felt that eggs in soil a few inches below the surface were brought to the surface by earthworms, insect larvae and other agents.

Leuckart (1876) and Railliet and Lucet (1892) were first to show that chickens could be infected by orally inoculating them with embryonated eggs of the caecal worm. Numerous authors have confirmed these observations (Riley and James, 1921; Uribe, 1922; Dorman, 1928; Clapham, 1933; Roberts, 1937a,b; Lund, 1957; Vatne and Hansen, 1965; Lund and Chute, 1974).

Ackert (1917) showed that chickens could be infected by feeding them dung earthworms (*Helodrilus gieseleri*) exposed to larvated eggs of *H. gallinarum*. Frank (1953) showed that grasshoppers and flies could mechanically transfer eggs of *H. gallinarum*. Lund (1960, 1966) and Lund *et al.* (1966) followed up Ackert's preliminary observations and provided evidence that larvae of *H. gallinarum* could sequester themselves in tissues of *Allolobophora caliginosa*, *Eisenia foetida* and *Lumbricus terrestris* and that earthworms could serve as paratenic hosts ('vectors') of *H. gallinarum*. Spindler (1967) infected a few turkeys with sowbugs (*Porcellio scaber*) exposed to eggs of *H. gallinarum*; he concluded that transmission was 'mechanical' and resulted from unhatched eggs in the gut of the arthropods at the time they were fed to his experimental poults. Khaziev (1972) reported that guinea fowl in the CIS could be infected by feeding them *E. foetida* and *L. terrestris* exposed to eggs of *H. gallinarum*.

Eggs of *H. gallinarum* hatch in the small intestine of the definitive host and larvae migrate to the caeca, where they reach adulthood (Graybill, 1921; Uribe, 1922). Migration to the caeca was apparently completed in 17–48 h after the ingestion of eggs (Dorman, 1928; Clapham, 1933; Roberts, 1937a,b). Dorman (1928) and Clapham (1933) stated that all development took place in the lumen of the caeca. Uribe (1922), Baker (1933), Tyzzer (1934), Roberts (1937a,b) and Vatne and Hansen (1965) believed that larvae invaded the caecal wall and remained there for several days before re-entering the lumen. Itagaki (1930) suggested some larvae may invade caecal glands, where they are capable of maturing, but they do not as a rule invade the caecal wall. Madsen (1962b) felt that the differences reported by various authors might be the result of the 'interplay between resistance and the viability of larvae' and 'the conditions of the experiment'. Lund and Chute (1974) compared the reproductive potential of *H. gallinarum* in various galliforme species.

Graybill (1921) reported moults in the avian host at 9–10 and at 16 days. Clapham (1933) reported moults at 96 h and at 10 days. Roberts (1937a,b) reported three moults, namely at 4–6 days, 9–10 days and 14 days.

H. gallinarum is an important species in animal health because it transmits *Histomonas meleagridis*, the causal agent of blackhead (Graybill and Smith, 1920; Tyzzer, 1934; Niimi, 1937; Kendall, 1959; Gibbs, 1962; Lee, 1969, 1971) as well as

Histomonas wenrichi (see Lund, 1968). The histomonads feed and multiply in the germinal zone of the ovary of *H. gallinarum* (see Lee, 1969). They move to the oogonia and penetrate oocytes in the growth zone of the ovary. In the oocytes and newly formed eggs of *H. gallinarum* the parasitic protozoan feeds and divides; stages are similar to those found in the avian host but are smaller. The protozoan also invades the reproductive system of male *H. gallinarum* and Lee (1971) suggested that it could be transmitted from male to female *H. gallinarum* during copulation. In the avian host the protozoan escapes from the nematode and establishes itself in the caecum.

H. spumosa Schneider, 1866

H. spumosa is found in the upper colon and less commonly the caecum of rats and mice. Its transmission was studied experimentally in considerable detail by Winfield (1933) and Smith (1953). Winfield collected *H. spumosa* in rats (*Rattus norvegicus*) from Baltimore, Maryland, and established the parasite in laboratory-reared domestic rats and albino mice. At 30°C a morula appeared in eggs in 24 h. In 3 days eggs were in the tadpole stage and in 6 days most were in the vermiform stage. In 12–14 days almost all eggs were completely embryonated and infective. Embryonated eggs remained infective for at least 120 days at room temperature in moist conditions.

Eggs given to rats and mice apparently hatched in the upper small intestine and larvae passed into the colon, where they were found 72 h postinfection. Young worms seemed to localize in the colon and as they approached maturity on and about day 20 they tended to concentrate in the first 1–2 cm. There was no evidence that worms invaded the mucosa. The prepatent period in rats was 26–47 days. Winfield (1933) concluded that in rats female worms continued to grow for almost 100 days after starting to produce eggs. More than 50% of rats lost infections in about 150 days, although individuals retained infections for up to 312 days. The longest infections occurred in winter and the author wondered if seasonal changes in temperature were factors. Sex ratios favoured females (56.25% vs. 43.75%) and there was evidence that males died somewhat earlier than females. Egg production in young rats increased more rapidly and reached higher levels than in older rats. Also, there were indications that egg production did not diminish from the first few days until 230 days but that a peak was reached at about 150 days; female worms reached their maximum size in about 130 days. Nevertheless, egg production from a single infection in rats rose during the first 3 weeks and then declined gradually, presumably because worms were lost.

Smith (1953) described eggs of *H. spumosa* as oval to round and consisting of a single cell when oviposited. Eggs had a thick shell and a roughened mamillated surface. In fresh faeces, eggs were at the one-cell stage or just beginning the first cleavage. At 30°C the vermiform stage was reached in 6 days and by 8 days Smith observed a moult. The cuticle from this moult was shed prior to hatching, presumably because of the activity of the larva in the egg. The resulting larva had a 'boring tooth' and lateral alae.

Smith (1953) gave fully embryonated eggs to rats. Eggs hatched in the small intestine of rats in a few hours and larvae appeared in the colon by at least 48 h. In rats given large numbers of eggs, some larvae were found partially embedded in the mucosa for a brief time but none was found in this site after 4 days. A moult took place 6–9 days and another 15–19 days postinfection. Smith regarded the first moult in the rat as the second and the latter moult as the third, but it is clear from his descriptions of larvae that the first was the third moult and the second the fourth, which led directly to the adult stage. This agrees with Bain's (1970) conclusion that there must have been an additional moult in the egg and that the infective stage was in fact the third and not the second (see also *Ascaridia*). Fertilized eggs were noted by Smith in female worms 25 days after rats were infected, although the prepatent period was 30–46 days.

Larval nematodes found in the earthworm *Pheretima hilgendofi* in Japan proved to be *H. spumosa* when given to rats (Saitoh *et al.*, 1993). *Eisemia foetida* was allowed to ingest eggs of *H. spumosa*. Larvae invaded the epithelium of the gut and grew to 365.2–387 µm in length (cf. larvae in eggs 231.2–274.2 µm). Larvae from earthworms readily infected rats. In the Ogasawara islands, roof rats (*Rattus rattus*) flourish on both ground and trees. No earthworms occurred in their stomachs and prevalence of caecal worm was only 10%. On Torishima island, on the other hand, where rats must live on the ground, earthworms were found in the stomachs and caecal worm prevalence was 54%, indicating the importance of earthworms in the transmission of this nematode (cf. *Heterakis gallinarum*).

Family Ascaridiidae

Ascaridia

Species of this genus are common parasites of the intestine of gallinaceous birds. Eggs embryonate in the external environment under favourable conditions of temperature and moisture and are directly infective to the avian host. Authors have generally regarded the larva in the fully developed egg as the second stage. However, Araujo and Bressan (1977) showed that the infective stage in the egg of *A. galli* is the third stage (i.e. there were two moults in the egg) and there can be only two moults in the avian host.

A. columbae (Gmelin, 1790)
This is a parasite of pigeons (*Columba livia*). Its development and transmission are similar to those of *A. galli*. Unterberger (1868) gave embryonated eggs to pigeons and claimed to have found eggs in faeces of the birds 17–18 days later. Hwang and Wehr (1958) gave large numbers (500–1000) of eggs to pigeons and found some larvae in the liver and lungs. This led them to think that there might be a lung migration, as found in *Ascaris suum*. Later, however, they showed that larvae outside the intestine failed to develop (Hwang and Wehr, 1962; Wehr and Hwang, 1964). They reported that numerous larvae invaded the wall of the intestine before moving to the lumen, where they matured in 30–50 days. Melendez and Lindquist (1979) injected embryonated eggs into the wing vein of pigeons and later found larvae in granulomas in the lungs. Some larvae apparently found their way to the intestine, where they matured.

According to Vassilev (1993) *A. columbae* did not become established as adults in most gallinaceous birds or in domestic ducks given eggs. However, larvae grew for a time in the liver of ducks and partridges (*Alectoris chuka*) and these worms were infective to pigeons. Also adult worms appeared in the gut of some partridges given eggs.

A. compar (Schrank, 1790)
This is a parasite of the rock partridge (*Alectoris graeca*). According to Vassilev (1987) all development in the host takes place in the intestinal lumen. Gravid females were found 30 days postinfection and worms survived in the bird for up to 493 days.

A. dissimilis (Perez and Vigueras, 1931)

This is a parasite of the intestine of turkeys (Wehr, 1942). Pankavich *et al.* (1974) gave the prepatent period as 36–45 days, depending on the age of the birds. Horton-Smith *et al.* (1968) showed larvae invaded the wall of the intestine immediately after hatching and gave the prepatent period as 28 days. Norton *et al.* (1992) attributed high mortality in domestic turkeys to heavy infections with *A. dissimilis*.

A. galli (Schrank, 1788)

A. galli (syn. *A. lineata*, *A. perspicillus*) is a common intestinal parasite of the small intestine of chickens, turkeys, guinea fowl and game birds. By artificially hatching eggs, Araujo and Bressan (1977) discovered that there were two moults in the egg and that the infective larvae were in the third stage; they found two shed cuticles on larvae 11–14 days after eggs had been kept at 25°C. According to Ackert (1919) the infective stage is reached in eggs in 9 days at 28°C. The minimum time required for eggs to reach the infective stage was 5 days at 22–34°C, according to Reid (1959).

Eggs are resistant and may survive for several months in suitable moist conditions. Eggs ingested by the avian host hatch in the duodenum and jejunum within 24 h after being ingested (Ackert, 1923a,b; Guberlet, 1924; Tugwell and Ackert, 1952). It was noted by several authors that some larvae buried their heads in the intestinal mucosa 1–26 days postinfection and thereafter returned to the lumen and matured. Todd and Crowdus (1952) concluded that most *A. galli* mature without leaving the gut lumen. The significance of this tissue phase was discussed by Madsen (1962b) who did not regard it as a normal or necessary part of the biology of *A. galli* but a reflection of immune or other factors (see also Itagaki, 1927). Herd and McNaught (1975) concluded that the tissue phase was normal and its duration dose-dependent, i.e. it lasted from 3 to 16 days in a dose of 50 eggs and was abruptly terminated by a moult; at high doses (2000 eggs) it lasted for 54 days and the moult was delayed.

The prepatent period was given as 60 days by Ackert (1923a,b) and varied according to host age. It was 5–6 weeks in chickens under 3 months of age but up to 8 weeks in older birds (Ackert, 1931; Kerr, 1955).

Tugwell and Ackert (1952) and Deo and Srivastava (1955) claimed that there were three moults in the bird host, namely at 6–9 days, 14–15 days and 18–22 days postinfection. Khouri and Pande (1970) reported two moults at 6–9 days and 14–19 days.

Earthworms can harbour the eggs and larvae of *A. galli* (see Gurchenko, 1970a,b; Jacob *et al.*, 1970; Augustine and Lund, 1974) although there is apparently no evidence that larvae occur other than in the intestine of the annelids. Gurchenko (1970a,b) noted that *A. galli* can survive winter in *Eisenia foetida*.

Ikeme (1970) believed that some larvae of *A. galli* became inhibited in the third stage in the host and that this was dose-dependent. Inhibition may help to carry infection from one bird to another (see also Madsen, 1962a).

A. numidae (Leiper, 1908)

This is a parasite of the jejunum of guinea fowl (*Acryllium* spp., *Guttera* spp., *Numida* spp.). According to Macchioni (1971) larvae in the avian host are recovered mainly in the mucosa of the intestine and rarely in tissue. He preferred the use of 'mucus phase' rather than 'tissue phase' for this behaviour. The prepatent period was given as 27 days.

5.4

The Superfamily Ascaridoidea

The superfamily Ascaridoidea consists mainly of medium-sized to large nematodes with three lips sometimes separated by interlabia (Hartwich, 1974; Gibson, 1983). The worms usually inhabit the stomach and intestine of the definitive host and generally consume food ingested by the host.

The superfamily is divided into five families (Hartwich, 1974) (although Gibson (1983) included the heterocheilids in the Ascarididae). The Crossophoridae is restricted to two genera in hyracoids and the Heterocheilidae contains a single genus in sirenians. The Acanthocheilidae is restricted to elasmobranchs. The transmission of species in these three families has not been elucidated. The Anisakidae is a major group found in mammals, birds, reptiles and fishes. Transmission of a number of anisakids has been investigated, including such important species as *Pseudoterranova decipiens* of seals, *Anisakis simplex* of whales and members of the genus *Contracaecum* of fishes, birds and mammals. The Ascarididae is a major well-studied family with such significant genera as *Ascaris*, *Porrocaecum* and *Toxocara*. Nadler and Hudspeth (1998) have published a phylogeny of some ascaridoids based on ribosomal DNA and morphological characters.

Eggs of ascaridoids transmitted in terrestrial habitats are thick-shelled whereas those in species transmitted in aquatic habitats are generally thin-shelled. Eggs are unembryonated when laid and only develop when passed to the external environment in faeces of the definitive host. In the external environment eggs embryonate to first-stage larvae, which grow and moult to the second or third stage. In most strictly heteroxenous ascaridoids (e.g. those in marine mammals) there seems to be only one moult in the egg but more research is required to correlate moults in eggs with those in hosts. In the secondarily monoxenous forms (e.g. *Ascaris* spp. and *Toxocara canis*) there may be two moults in the egg. In species transmitted in aquatic habitats, larvae generally hatch from eggs in water and the sheathed second-stage larvae are free-living for variable periods of time. In species transmitted in terrestrial habitats, eggs do not hatch spontaneously and are markedly resistant and long-lived; eggs of these species hatch in suitable intermediate or definitive hosts, depending on the species.

A remarkable feature of the ascaridoids is that they are basically heteroxenous. Full appreciation of this fact was slow in coming. Cori, as early as 1898, correctly identified infective larvae of *Porrocaecum ensicaudatum* of passeriforme birds in earthworms. Fülleborn (1921a, 1929) suspected that the lung migration in *Ascaris* spp. represented a phase that once took place in an intermediate host. Thomas (1937a,b,c) reported that the ascaridoids of piscivorous birds (*Contracaecum rudolphi*) and predaceous fishes (*Raphidascaris acus*) were heteroxenous. Tiner (1949) and Sprent (1951) showed that ascaridoids of carnivores used small mammals as intermediate hosts and it finally became evident that heteroxeny was, in fact, a basic feature of ascaridoids which had to be taken into account in any consideration of their biology and evolution.

In most species, vertebrates serve as intermediate hosts in which development to the stage (generally the third, less commonly the fourth) infective to the definitive host takes place. The widespread use of vertebrates as intermediate hosts seems clearly to be a fundamental and primitive feature of the transmission of ascaridoids (Anderson, 1988). The use of invertebrate paratenic hosts to place second-stage larvae in the food chain of the vertebrate intermediate host is apparently also a basic and primitive feature of the transmission of ascaridoids (which probably originated in predators) (Anderson, 1988). In various species transmitted in aquatic habitats (e.g. *Contracaecum* spp., *Pseudoterranova decipiens*), paratenic hosts such as crustaceans (especially copepods), oligochaetes and larval insects ingest free larvae or eggs and transmit larvae to fish intermediate hosts. In species transmitted in terrestrial habitats (e.g. *Amplicaecum robertsi*), paratenic hosts probably are often earthworms, which ingest eggs passed in the faeces of the definitive host and transmit larvae to small mammal and marsupial intermediate hosts. Primitively, larvae in invertebrates remain undeveloped and the latter serve to place larvae in the food chain of the vertebrate intermediate host in whose tissues they will develop to the stage that is infective (primitively the beginning of the third stage) to the definitive host. The latter becomes infected by ingesting vertebrates containing infective third- or fourth-stage larvae, depending on the species (e.g. Angusticaecinae, most *Baylisascaris* spp., *Contracaecum* spp., *Raphidascaris acus*, *Toxascaris leonina*).

The above basic pattern of the transmission of ascaridoids has been modified in some species in that precocious development to the third or fourth stage occurs in the original invertebrate paratenic host (Anderson, 1988). Thus, invertebrates can serve, along with vertebrates, as intermediate hosts if the definitive host has broad feeding preferences (e.g. *Anisakis* spp.). In some species transmitted in aquatic habitats where development to the infective stage takes place in invertebrates, fish may function mainly as paratenic hosts (as in *Anisakis simplex*). Thus, the primitive or original invertebrate paratenic host and the original vertebrate intermediate host switch their functions in transmission. For example, *Hysterothylacium aduncum* of predaceous fishes develops to the infective stage in marine crustaceans and polychaetes but can persist unchanged in the tissues of prey fish which are eaten by the definitive host. *Porrocaecum angusticolle* of raptors is a comparable example of the same phenomenon in transmission under terrestrial conditions; in this species, development to the infective stage takes place in earthworms but insectivores serve as paratenic hosts.

If larvae acquire the ability to develop to the infective stage in invertebrates then the possibility exists for the elimination of the vertebrate intermediate host and to replace it with what was originally the invertebrate paratenic host. *Porrocaecum* spp. of birds and *Sulcascaris sulcata* of marine turtles, which use earthworms and molluscs, respectively, as intermediate hosts, may be examples of this phenomenon.

Precocial development in vertebrate intermediate hosts is also a feature of some ascaridoids. It is found in its simplest form in *Raphidascaris acus* of pike, in which the second-stage larvae develop to the fourth stage in the liver of perch. Precocity in the fish intermediate host is an important feature of the transmission of *Pseudoterranova decipiens*. Such precocity presumably accelerates the rate of development of the parasite to adulthood in the final host. Precocity may rarely lead to larvae reaching the fifth stage in the vertebrate intermediate host. This may explain the presence of adult *Hysterothylacium haze* in the body cavity of yellow-fin goby in Japan. Similar precocity is reported in *Hexametra angusticaecoides* in chameleons (intermediate host) and boas (final host) and in *Orneoascaris chrysanthemoides* of frogs (intermediate host) and night adders (final host). Anderson (1988) pointed out that such precocity would almost totally eliminate the time for worms to mature in the final host. In addition, it might have allowed some ascaridoids to adopt an independent existence in what was once an intermediate host (i.e. the parasite transfers from a predator host to a prey host).

In the most specialized life cycles of ascaridoids, one of the hosts has been eliminated and embryonated eggs in the second or early third stages are directly infective to the host. These represent **secondary monoxeny**, as they are presumably derived from an earlier heteroxeny. There are in the ascaridoids two routes towards secondary monoxeny.

1.   In the ascaridoids of predators, the vertebrate **intermediate** host is dropped and the definitive host serves as both intermediate and definitive host. A step in the loss of the intermediate host is revealed by *Toxascaris leonina* of canids and felids. The definitive host can be infected with eggs or mouse tissue containing larvae. If eggs are used to infect the host, there is a tissue phase in the gut wall before larvae enter the gut lumen and mature. If, on the other hand, larvae first develop in the tissue of mice and rabbits and are then given to cats or dogs, the tissue phase is eliminated and as a consequence the prepatent period is reduced by about 10–15 days.

In *Toxocara canis* and *T. cati*, eggs with third-stage larvae infect the host and there is the usual lung–tracheal or somatic migration. It is likely that dogs and cats usually get infected after ingesting eggs but larvated eggs will hatch in various vertebrates (from mice to humans) and wander as **visceral larval migrans** or become encapsulated. These larvae will also infect dogs and cats but their role in natural transmission is not clear. However, the presence of larvae which migrate in the body of the host, perhaps as a carry-over from an ancestral heteroxeny, led to **prenatal transmission** in *T. canis* and **transmammary transmission** in *T. cati*.

2.   The ascaridoids of non-predators such as humans, pigs, cows, rodents and horses (*Ascaris lumbricoides*, *A. suum*, *Parascaris equorum*, *Baylisascaris laevis*) have apparently achieved monoxeny by dropping the ancestral **predator** host (see also p. 7). These ascaridoids undertake the complex migration in the host that would be expected in an intermediate host. The novelty is the fact that they get to the gut and mature as they did in the ancestral predator host. The wandering of larvae in these ascaridoids gave rise to transmammary transmission (*Toxocara vitularum* in bovids and *T. pteropodis* of fruit bats).

Family Anisakidae

Anisakids are associated with aquatic organisms (fishes and marine mammals) and piscivorous birds. Transmission of species in the family is dependent upon water and usually involves aquatic invertebrates and fish intermediate or paratenic hosts. The family includes parasites of considerable economic and medical importance, e.g. *Anisakis simplex* and *Pseudoterranova decipiens*.

Subfamily Goeziinae

Members of the subfamily are stout stomach nematodes with cuticular rings with posteriorly directed spines. They occur in teleost fishes but have been reported sporadically in reptiles.

Goezia

G. ascaroides (Goeze, 1782)

G. ascaroides occurs in the stomach of the catfish (*Silurus glanis*) in Europe, including the Don and Dnieper rivers in Russia and brackish waters in the Azov and Caspian seas. It has been reported in *Oncorhynchus mykiss*, *Salmo dentex* and the marine fish *Trachinus vipera* (see Moravec, 1994). According to Mozgovoi *et al.* (1971) the tumours have openings into the stomach lumen and rounded, smooth-shelled, unsegmented eggs are released into the stomach and pass out in the faeces of the host. At 25–28°C eggs embryonated to first stage larvae in 2 days and the latter moulted. In an additional 3–4 days the larvae (170–180 μm in length) hatched and remained active in water for 2–3 weeks. The intermediate host is *Diaptomus castori*, in which larvae shed the first-stage cuticle and entered the haemocoel, where they grew and moulted into the third stage in 2–3 days. These early infective larvae were 430–460 μm in length and infective to fish. However, when they remained in the copepod for 15 days they grew to 1.65–1.71 mm in length and 59–73 μm in width. The fry of various forage fish (i.e. *Abramis brama*, *Blicca bjoerka*, *Alburnus alburnus*, *Scardinius erythrophthalmus*) as well as fry of *Siluris glanis* serve as paratenic hosts. The larvae become encapsulated in the gut wall of these fish. The capsules were 490–840 × 450–620 μm in size and contained one or two larvae 1.58–1.78 mm in length. The prepatent period in the catfish final host was estimated to be about 2 months.

Subfamily Anisakinae

The excretory pore is usually in the lip region. They are parasites mainly of marine mammals, turtles, fish-eating birds and elasmobranchs.

Anisakis

Adults of *Anisakis* spp. are parasites of the stomach and intestine of pinnipeds (elephant seal, fur seal, grey seal, leopard seal, monk seal, ringed seal, sea-lion, walrus) and cetaceans (dolphin, narwhal, porpoise, whales). Davey (1971), who revised the genus, recognized three valid species, namely: (i) *A. simplex* (Rudolphi, 1809) of both pinnipeds and cetaceans; (ii) *A. typica* (Diesing, 1860) of cetaceans; and (iii) *A. physeteris*

Baylis, 1923 of whales. Larvae of *Anisakis* spp. (Fig. 5.3D–F) have been reported in the body cavity or tissues of a great variety of marine and anadromous teleosts as well as squids and prawns. In recent years there have been numerous reports of enteritis caused by *Anisakis* spp. larvae in the stomach and intestine of humans who have consumed raw or poorly cooked marine fish or squid harbouring larvae (for reviews see Jackson, 1975; van Thiel, 1976; Smith and Wootten, 1978; Huang and Bussiéras 1988; Nagasawa, 1990a,b, 1993; Ishikura *et al.*, 1992).

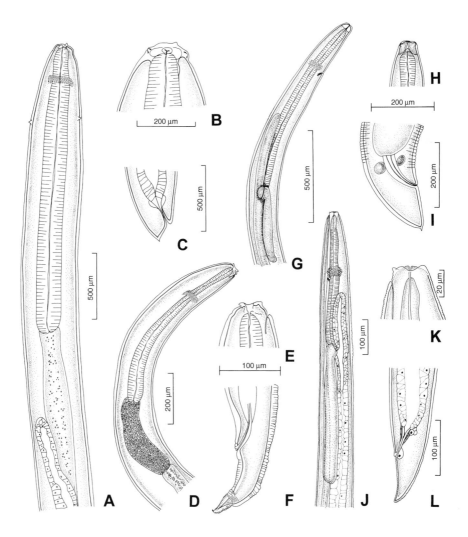

Fig. 5.3. Third-stage larvae of ascaridoids: (A)–(C) *Pseudoterranova decipiens*; (D)–(F) *Anisakis* sp. from *Latimeria chalumnae*; (G)–(I) *Hysterothylacium aduncum*; (J)–(L) *Contracaecum rudolphi*. (A–C, original R.C.A.; D–F, after R.C. Anderson, 1993 – courtesy *Environmental Biology of Fishes*; G–I, after F. Moravec *et al.*, 1985 – courtesy of *Folia Parasitologica*; J–L, after C.M. Bartlett – courtesy *Parasite*.)

Four larval types (I–IV) have been described (Shiraki, 1974). *Anisakis* Type I larvae from herring (*Clupea harengus*), mackerel (*Scomber scombrus*) and Norway haddock (*Sebastes marinus*) and identified as *A. simplex* (see Pippy and van Banning, 1975) is a parasite mainly of cetaceans and pinnipeds in cold temperate and polar waters (Davey, 1971). It is now recognized that *A. simplex* is a complex consisting of three sibling species. *A. simplex A* is mainly Mediterranean whereas *A. simplex B* is North Atlantic in distribution and the name *A. pegreffii* has been proposed for the former species (Nascetti *et al.*, 1986; Orecchia *et al.*, 1986; Siles *et al.*, 1997). Type II larvae have been assigned to *A. physeteris* (see Oshima, 1972; Suzuki and Ishida, 1979; Orecchia *et al.*, 1986). The taxonomic status of the other larval types is not clear. Mattiucci *et al.* (1997) found a Pacific and Austral population called *A. simplex C*. A new species (*A. ziphidarum*) has been described in beaked whales (*Mesoplodon layardii* and *Ziphius cavirostris*) in the Mediterranean and in South African waters (Paggi *et al.*, 1998). Smith and Wootten (1978) and Nagasawa (1990b) have published useful reviews of *Anisakis* and anisakiasis.

A. simplex (Rudolphi, 1809)

Adult *A. simplex* has been reported in numerous species of cetaceans (*Balaenoptera* spp., *Berardius bairdi*, *Delphinapterus leucas*, *Delphinus delphis*, *Globicephala scammoni*, *Hyperoodon ampullus*, *Kogio breviceps*, *Lagenorhynchus* spp., *Megaptera novaeangliae*, *Mesoplodon bidens*, *Monodon monoceros*, *Orcinus orca*, *Physeter catodon*, *Phocaena phocaena*, *Phocaenoides dalli*, *Pseudorca crassidens*, *Stenella caeruleoaeba*, *Steno bredanensis*) and pinnipeds (*Arctocephalus australis*, *Eumetopias jubatus*, *Halichoerus grypus*, *Hydrurga leptonyx*, *Monachus monachus*, *Mirounga* spp., *Odobenus rosmarus*, *Otaria byronia*, *Pusa hispida*, *Zalophus californianus*) (Davey, 1971).

Eggs of *A. simplex* were 40×50 μm in size with thin shells. Køie *et al.* (1995) claimed that a second moult occurred in the egg but this is unlikely, since studies indicate that the second moult occurs in euphausiid shrimps (see below). Eggs passed in faeces sank in sea water (Sluiters, 1974). Eggs developed and hatched in 4–8 days at 13–18°C and in 20–27 days at 5–7°C (van Banning, 1971). Second-stage larvae emerging from eggs were sheathed and 220–290 μm in length without the sheath (Smith and Wootten, 1978). Larvae survived ensheathed for 3–4 weeks in sea water at 13–18°C and for 6–7 weeks at 5–7°C (van Banning, 1971). Under natural conditions larvae are free to swim or drift in the water column and infect planktonic invertebrates (cf. *Pseudoterranova decipiens*).

Brattey and Clark (1992) studied eggs of *A. simplex* from a dolphin (*Lagenorhynchus albitustris*). Eggs hatched at all temperatures except –0.7°C. The number of days until hatching ranged from 3 at 24.3°C to 74–81 at 1.9°C. Survival times of larvae peaked at 75–105 days at 8.6°C, with a minimal survival time of 3–8 days at 24°C. This species is, therefore, adapted to cold temperatures. Harpacticoid copepods ingested newly hatched larvae but there was no evidence of growth or development up to 15 days.

Larvae identified as *A. simplex* have been found free in the haemocoel of euphausiid shrimps (Malacostraca, Eucarida, Euphausiacea) collected in the North Atlantic (Smith, 1971, 1983a,b). These larvae, found in the euphausiids *Nyctiphanes couchii*, *Thysanoessa inermis*, *T. longicauda* and *T. raschii*, were 4.2–23.6 mm in length. Smaller larvae (4.2–5.9 mm in length) seemed to be second stage in the process of moulting. Analyses of the morphology and dimensions of larvae from euphausiids and teleosts (including blue whiting, *Micromesisticus poutassou*) indicated that they were conspecific and in the third stage. In addition, Oshima (1969) and Oshima *et al.* (1968) infected *Euphausia pacifica* and *E. similis* with a species of *Anisakis* from dolphins and reported the second moult in these crustaceans.

Hays *et al.* (1998a) reported larval *A. simplex* in the euphausiid *Meganyctiphanes norvegica* and *Thysanoessa raschi* in the St Lawrence River estuary in Canada. Beluga, known to be a definitive host of *A. simplex*, is a year-round resident of the estuary. Larvae ranged in size from 10.0 to 39.30 ($\bar{x}$ = 26.74) mm in size and consisted of moulting second-stage larvae and fully grown third-stage larvae. Hays *et al.* (1998b) also reported on the presence of *A. simplex* in herring (95–99%) and capelin (5%) and noted no size or morphological differences in larvae from the two fish species or euphausiids. Thus, it is clear that euphausiids are intermediate hosts in which larvae develop from the second to the third stage and that fish are paratenic hosts.

In fish, coiled larvae were found encapsulated throughout the viscera. Larvae were found commonly on the surface of the liver and on the mesentery. Evidently encapsulation started shortly after larvae had invaded the body cavity of the fish (Prusevich, 1964; Smith, 1974). Smith (1974) experimentally transferred larvae from one fish to another. *A. simplex* larvae (Fig. 5.3B) have been reported in a wide range of marine and anadromous teleosts, including such commercially important species as herring, mackerel, halibut, cod and salmon in northern waters, where this infection is abundant in cetaceans and pinnipeds (Margolis and Arthur, 1979; McClelland *et al.*, 1990; Kino *et al.*, 1993; Aspholm, 1995; Brattey and Bishop, 1992). *A. simplex* is known as the 'herring worm' because of its high prevalence in herring.

Larval *A. simplex* have been found frequently in the tissues of myopsid and oegopsid cephalopods, some species of which are known to prey on euphausiids (Clarke, 1966) as well as herring and other fish (Smith, 1984). The role of squids in the transmission of *A. simplex* is unknown although they are important food items of some odontocete cetaceans (Slijper, 1962; Mackintosh, 1965; Nagasawa and Moravec, 1995) which are hosts of *A. simplex*.

In summary, eggs of *A. simplex* passed by marine mammals embryonate in sea water to second-stage larvae, which hatch (Fig. 5.4). The free-swimming sheathed larvae are ingested deliberately or accidentally by marine crustaceans (e.g. euphausiids). In the latter they invade the haemocoel and develop to the third stage. Growth in euphausiids is marked. Larvae ingested with euphausiids by teleosts penetrate the intestine and encapsulate in the tissues, especially on the mesentery and liver. Larvae also will invade squids that consume euphausiids. Marine mammals can acquire larvae from eating infected fish, crustaceans or squids. Thus, *A. simplex* is readily transmitted to pinnipeds as well as odontocete and baleen whales.

Contracaecum

The definitive hosts of members of the genus are piscivorous birds and mammals associated with fresh, brackish and sea water (e.g. cormorants, pelicans, seals). The parasites inhabit the stomach and consume food ingested by the host. When not feeding on ingesta they attach to the stomach wall. Round to oval, thin-shelled, unembryonated eggs, deposited by female worms in the stomach, pass out in the faeces of the host. In water, eggs embryonate into first-stage larvae. The latter develop further and moult into

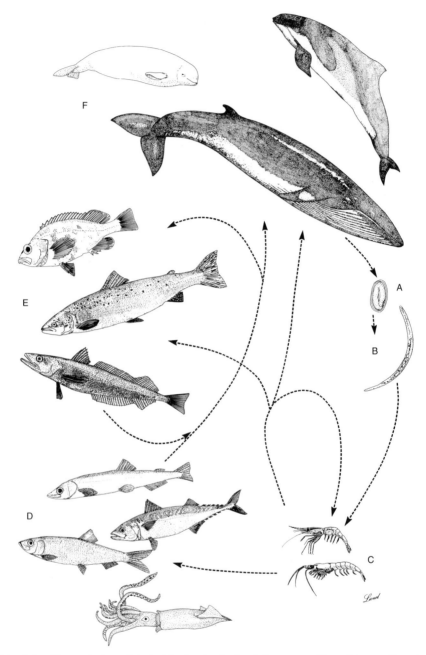

Fig. 5.4. Transmission of *Anisakis simplex* in the St Lawrence River Estuary, Canada: (A) embryonated egg; (B) free-living second-stage larva; (C) euphausiid intermediate hosts in which the third stage is attained with marked growth; (D,E) various fish and squid paratenic hosts in the food chain of (F) the three definitive cetacean hosts (beluga, minke and blue whales) in the estuary. (Prepared by Lionel Corriveau and Lena Measures – courtesy of the Institute Maurice-Lamontagne, Fisheries and Oceans, Mont-Joli, Quebec, Canada.)

the second stage, which has a cuticular tooth ventral to the oral opening, a ventriculus and a delicate ventricular appendix. The second-stage larva retains the cuticle of the first stage and is said to be sheathed. Larvae hatch in water and are ingested by invertebrate hosts, especially copepods, in the gut of which they exsheath and sequester themselves in the haemocoel, where they grow slightly or considerably, depending on the species; some authors report a moult in the invertebrates. Prey fish, which feed on copepods and other invertebrates, serve as vertebrate intermediate hosts in which larvae reach the third stage and usually grow substantially after becoming encapsulated on the mesentery. The third-stage larva has rudimentary lips but no interlabia. A tooth-like structure is present on the labium ventrolateral to the oral opening. The tail is short and curved ventrally and the ventricular appendix and caecum are well developed. Larvae may be found in a great variety of fishes in places where the latter are exposed to the eggs passed in the faeces of piscivorous birds and mammals. Also, larvae can probably pass from one fish intermediate host to another through predation and reinvasion of tissues of the new host.

Larvae of *Contracaecum* have been reported in a great variety of invertebrates, including coelenterates, ctenophores, gastropods, cephalopods, polychaetes, copepods, mysids, amphipods, euphausiids, decapods, echinoderms and chaetognaths (Norris and Overstreet, 1976).

Norris and Overstreet (1976) listed the invertebrates that have been reported to harbour larval stages of *Contracaecum* and *Thynnascaris*. Deardorff and Overstreet (1980a) placed *Thynnascaris* into synonymy with *Hysterothylacium* and pointed out that the definitive hosts of the latter genus are piscivorous fishes (see below) and not birds and mammals. Also, in *Contracaecum* the excretory pore is near the ventral interlabium whereas in *Hysterothylacium* it is in the region of the nerve ring. In the list provided by Norris and Overstreet (1976), therefore, '*C. aduncum*' and '*Thynnascaris*' refer to *Hysterothylacium* and not to *Contracaecum*.

The role played by various invertebrates in the natural transmission of *Contracaecum* spp. to fish intermediate hosts is not clear. Copepods are probably important as hosts which carry second-stage larvae to fish intermediate hosts (where development to the third stage occurs) or perhaps to certain invertebrates which consume them along with larvae.

C. microcephalum (Rudolphi, 1809)

According to Mozgovoi *et al.* (1968a) *C. microcephalum* is a parasite of piscivorous birds (Ciconiiformes, Anserinae). Copepods were regarded as first intermediate hosts and the second intermediate hosts were young freshwater fish, *Chironomus* spp. and larval Odonata of the genera *Agrion*, *Anax* and *Coenagrion*. The definitive hosts became infected from consuming fish containing larvae.

Semenova (1974) was unable to infect piscivorous birds and domestic ducklings with larvae (600–700 μm long) taken from experimentally infected copepods. However, infections could be established in birds with larvae more than 1.0 mm in length taken from naturally or experimentally infected fish and amphibians. In the bird, larvae grew to 2.2–2.5 mm before undergoing the third moult. The prepatent period was 35–40 days and the longevity of the worms was an additional 40–50 days. Juvenile worms were found mainly in the lower part of the oesophagus, while adults were found free or attached to the stomach wall, where an ulcer was created.

C. micropapillatum (Stossich, 1850)

C. micropapillatum was found in pelicans in Dagestan and Azerbaidzhan, CIS (Semenova, 1971, 1972, 1973, 1975). Eggs embryonated *in utero* to 8–16 cells. In water, eggs developed to first-stage larvae which moulted to second-stage larvae with a cephalic tooth, a well-developed excretory system and a gut with a ventriculus and intestinal caecum. Second-stage larvae hatched and were active for 1.5–2 days at 19–21°C and remained infective for 18–25 days.

According to Semenova (1979) the following copepods were infected experimentally with larvae of *C. micropapillatum*: *Acanthocyclops vernalis*, *A. viridis*, *Arctodiaptomus gracilis*, *Cyclops strenuus*, *Eucyclops leuckartii*, *E. macruroides*, *Macrocyclops albidus*, *M. fuscus*, *Mesocyclops crassus* and *M. leuckartii*. The second moult occurred in the haemocoel and muscles of the copepods in 4–6 days (Semenova, 1971) and larvae doubled in size in 15–20 days and became encapsulated in 110–140 days (Semenova, 1979). If swallowed by a frog or fish, larvae migrated to the body cavity of these hosts. Copepods were also observed to ingest third-stage larvae from dead infected copepods. Carp fry and amphipods (*Gammarus* sp.) also had a few larvae in their tissues after ingesting fully embryonated eggs or free larvae. Semenova (1973) reported that second-stage larvae invaded Odonata (especially *Agrion* sp. and *Coenagrion* sp.), frogs, tadpoles and fish. Larvae grew to 1.7–2.0 mm in 30 days in fish and 1.2–1.3 mm in Odonata (*Anax* sp. and *Aeschna* sp.). In young *Abramis brama* larvae developed faster and encapsulated earlier than in *Alburnus alburnus*, *Gambusia offinitis*, *Rutilus rutilus*, *Scardinius erythrophthalmus* and *Tinca tinca*.

Normally, third-stage larvae were encapsulated in fish at a length of 1.2–1.7 mm, when they were infective to pelicans. Third-stage larvae were also found in frogs and tadpoles, where they grew to 1.2–1.7 mm in 30–40 days.

C. multipapillatum (Drasche, 1882)

C. multipapillatum, a parasite of the proventriculus of water turkeys (*Anhinga anhinga*), cormorants, pelicans and herons in the USA (Deardorff and Overstreet, 1980b), has been studied by Huizinga (1967). Eggs were 65×58 μm in size with an outer sticky, slightly mamillated layer. In the uterus, eggs were unsegmented or in the first or second cleavage. At 21°C moulting larvae were present in 5–7 days in tap, lake or distilled water. The second-stage larva retained the cuticle of the first stage and hatched spontaneously in 5–7 days in lake water. Sheathed larvae remained alive and motile for 10–20 days in lake water, but died in about 7 days in sea water. Larvae were an average of 362 μm in length and the oesophagus was incompletely developed. The intestinal caecum was absent and the ventricular appendix was represented by a thin primordium (53 μm in length) extending parallel to the intestine.

The freshwater copepod *Cyclops vernalis* readily ingested second-stage larvae in Petri dishes and as many as 30 were removed from a single copepod placed with larvae. The sheath was shed immediately upon entry into the copepod. Seventeen days after ingestion by copepods, larvae were 485 μm in length. The ventricular appendix increased in length to 110 μm. The oesophagus and nerve ring were more clearly defined than previously.

Guppies (*Lebistes reticulatus*) did not become infected when placed with large numbers of free-swimming, second-stage larvae; exsheathed larvae were found in faecal material of the fish. Guppies readily became infected when allowed to ingest copepods containing larvae. In 10 days larvae moulted, when they were about 4–5 mm in length, but retained the cuticle of the second-stage. Larvae encapsulated on the mesentery and grew to a maximum of 22.8 mm in 30 days. The ventricular appendix and intestinal caecum were well developed. The large size of larvae (e.g. 15.0–21.6 mm at day 10) caused marked abdominal distension in the guppies.

Larvae similar to those found in experimentally infected guppies were found encapsulated in various freshwater fishes in Florida, including largemouth bass. Five parasite-free largemouth bass (*Micropterus salmoides*) were allowed to feed on infected guppies; at necropsy 7 days later, larvae were found encapsulated on the mesentery.

Deardorff and Overstreet (1980b) reported the presence of larvae of *C. multipapillatum* encapsulated in the liver, kidneys and mesentery of mullet (*Mugil cephalus*, *M. curema*) in the southern USA and stated that brown pelicans acquired infections by the time they were 2 weeks old.

Dick *et al.* (1987) reported heavy infections of *Contracaecum* spp. in stocked rainbow trout (*Oncorhynchus mykiss*) in a lake in Manitoba, Canada. Larvae were found in the intestine, body cavity and musculature of the fish. The presence of trout in the lake attracted various piscivorous birds, especially pelicans. Concurrent predation by trout on infected fathead minnows (*Pimephales promelas*), five-spined sticklebacks (*Culaea inconstans*) and nine-spined sticklebacks (*Pungitius pungitius*) increased the prevalence and intensity of the parasite in trout. When stocking of trout ceased, the use of the lake by piscivorous birds, especially pelicans, was reduced as was the intensity and prevalence of larvae in the indigenous minnows.

Larvae of *C. multipapillatum* were found in cichlids (*Cichlasoma urophthalmus*) (one to 18 per fish, $\bar{x} = 3.5$) collected in Mexico. Encapsulated larvae and adults were found in *Phalacrocorax olivaceus* (Vidal-Martinez *et al.*, 1994). Larvae were given orally to chickens, rats and ducks but none was found at necropsy. However, ten nematodes were found from the anterior intestine of a cat given larvae. The worms included four mature males, one gravid female, three non-gravid females and two larvae. Males were 19.8–29.8 mm in length and females 20.1–38.6 mm in length.

C. osculatum (Rudolphi, 1802)

C. osculatum is a parasite of seals. Nascetti *et al.* (1993), on the basis of multilocus electrophoresis study of adult worms from seals, recognized three species labelled *Contracaecum osculatum* A, B, and C. A occurred mainly in bearded seals (*Erignathus barbatus*) in the eastern and western North Atlantic. B occurred mainly in harp seals (*Pagophilus groenlandicus*) in the eastern and western North Atlantic and C occurred mainly in grey seals (*Halichoerus grypus*) in the eastern North Atlantic. Subsequently Orecchia *et al.* (1994) recognized two other species, called D and E, occurring mainly as adults in Weddell seals (*Leptonychotes weddelli*) and larvae in various fish in Ross and Weddell seals in the Antarctic.

Brattey (1995) was able to recognize *C. osculatum* A and B when electrophoretically studying larvae from various fish intermediate hosts around Newfoundland and Labrador. Larval *C. osculatum* B were relatively abundant in the fish, presumably because of the large populations of harp seals. Larval *C. osculatum* A were restricted to flatfish off Labrador, which accords with the distribution and benthic feeding habits of the bearded seal, the only definitive host of this species.

Valtonen *et al.* (1988) reported the species in seven species of fish and two species of seal from Bothnian Bay in the Baltic Sea. Highest prevalences were in salmon (*Salmo salar*) (20%), bull-rout (*Myxocephalus scorpius*) (20%), burbot (*Lota lota*) (16%) and cod (*Gadus morhua*) (15%). Prevalences were higher in larger fish but no seasonal variations were noted in prevalence and intensity. The parasite occurred sporadically in the resident ring seal (*Phoca hispida*) but a mean of 640 worms was found in grey seal which visited the area studied for a few weeks each spring. These animals probably maintain the infection in fish in the area. In grey seals, worms tended to occur in aggregations consisting of third- and fourth-stage larvae and adults.

McClelland and Ronald (1974a) found *C. osculatum* in the stomach of harbour (*Phoca vitulina*) and grey seals from Cape Breton Island, Canada, and in harp seals from the Gulf of Lawrence. Eggs in the vagina and uteri were multicellular and in sea water at 15°C developed and hatched to sheathed second-stage larvae which were 400 µm long, free-swimming and active. The sheath was expanded into fin-like folds. Larvae were artificially exsheathed and cultured to the early fifth stage.

Sudarikov and Rizhikov (1951) concluded that larvae (6.7–11.1 mm in length) in the endemic yellow-fin sculpin (*Cottocomephorus grewinki*) in Lake Baikal were *C. osculatum baicalensis*, which occurs in the Baikal seal (*Phoca siberica*). The crustacean *Macrohectopus branickii* is apparently the first intermediate host in Lake Baikal. It was concluded that two moults took place in the stomach of seals, a conclusion supported by *in vitro* experiments of McClelland and Ronald (1974a).

Davey (1969) reported that the optimum temperature for hatching of the eggs of *C. osculatum* was 16°C. Eggs required 2–6 months to develop at temperatures of 0–10°C and no development occurred at –20°C. The harpacticoid copepods *Tisbe furcata* and *Amphiascus similis* were infected with larvae but there was no development in them. Bier (1976) cultured eggs of *C. osculatum* and reported that larvae grew to an average of 6.50 mm in 32 weeks at 35°C.

Shiraki (1974) found larvae of *C. osculatum* in the brackish-water fish *Tribolodon hakonensis* in the northern sea of Japan.

Køie and Fagerholm (1993) suggested that two moults of *C. osculatum* (of grey seals) take place in the egg (but this requires confirmation – R.C.A.). Køie and Fagerholm (1995) concluded that calanoid copepods and, at certain times of the year, nauplius and cypris larvae of Curripedia may be important paratenic hosts in the Baltic Sea. Larvae which had increased in length from 300–320 µm to 400–420 µm in 2 weeks were found in liver sinusoids of nine-spined stickleback (*Pungitius pungitius*) which had been allowed to ingest ensheathed free larvae. Two cod (*Gadus morhua*) that had ingested eelpouts (*Zooarctes viviparus*) with larvae were examined a month later. A few dead encapsulated larvae were found in one cod. The other contained about 100 nodules (pyloric caecum and intestine) but about half contained no nematodes or their remains. Plaice (*Pleuronectes platessa*) were given larvae in the copepod *Acartia tonsa*. At necropsy larvae were found on the liver and mesentery 4 months later.

C. rudolphi Hartwich, 1964

C. rudolphi (syn. *C. spiculigerum* (Rudolphi, 1809)) is a common parasite of the proventriculus of cormorants (*Phalacrocorax* spp.) and pelicans (*Pelicanus occidentalis*). Thomas (1937a,b, 1940) reported that eggs developed in lake water and larvae hatched in 5 days at room temperature. When fed to guppies (*Lebistes reticulatus*) some larvae invaded the intestinal wall and became encapsulated on the mesentery and the viscera.

Huizinga (1966) collected adult worms from the proventriculus of *Phalacrocorax auritus*, *P. carbo* and *Pelicanus occidentalis* in Florida and Connecticut. Eggs, 64×58 µm in size, ovoid and smooth-shelled (cf. *C. multipapillatum*), sank in water. At 21°C first-stage larvae appeared active in eggs and in 5–7 days the first moult occurred; the first-stage cuticle was retained. At 13°C larvae reached the second stage in 25–30 days. At 21°C larvae hatched spontaneously in sea, tap or distilled water and retained the first-stage cuticle. Larvae (330 µm in length) remained active at 21°C for 15 days in sea water and then became less motile and died in 30 days. Larvae kept at 7–13°C remained viable for up to 6 months. The ventricular appendix was 65 µm long in second-stage larvae.

Cyclops vernalis, a freshwater copepod, readily ingested larvae and the latter migrated into the haemocoel of the crustacean. Growth in the copepod was slight (to 350 µm). The marine tide-pool copepod *Tigriopus californicus* also served as a suitable host and larvae were found in all regions of the haemocoel, including the caudal rami and antennae. Mortality in infected copepods was higher than in uninfected copepods.

Two hundred guppies (*Lebistes reticulatus*) and 20 killifish (*Fundulus heteroclitus*) were exposed to large numbers of second-stage larvae and examined immediately and up to 30 days postinfection. Most ingested larvae retained their sheaths but were passed, both dead and alive, in faeces of the guppies. Some motile exsheathed larvae, 500 µm in length, were noted in the mucosa of the gut of six guppies, 7 days postinfection. Three larvae digested from the mesentery of one guppy on day 16 were an average of 753 µm in length. Other guppies and all killifish were uninfected.

Twenty-five guppies were allowed to ingest infected *C. vernalis* and 25 were allowed to ingest infected *T. californicus*. Most larvae passed out in the faeces of the fish unsheathed but four guppies that ingested *C. vernalis* were infected. Two larvae found in guppies on day 9 were 0.65 and 1.09 mm long and one larva on day 37 was 1.49 mm long.

Five killifish were allowed to ingest infected *T. californicus*. At necropsy on day 18, three fish contained 30, four and two larvae encapsulated on the mesentery.

Huizinga (1966) reported encapsulated *Contracaecum* larvae in the mesentery of cunner (*Tautoglabrus adspersus*), winter flounder (*Pseudopleuronectes americanus*), killifish and sculpin (*Myoxocephalus octodecimspinosus*) indistinguishable from those found commonly in the stomach of cormorants. Huizinga (1971) noted that wild-caught nestling cormorants held in captivity until maturity lost their infections in 3 months.

According to Bartlett (1996) eggs of *C. rudolphi* from *Phalacrocorax auritus* in Nova Scotia hatched in 9–17 days in sea water at 15–20°C. Larvae were 223–255 µm in length when hatched. When ingested by copepods (*Tigriopus* sp.) larvae were 318–417 µm in 42 h and the ventricular appendix appeared, as well as a prominent excretory cell. Larvae from copepods readily invaded the tissues of guppies and mummichog (*Fundulus heteroclitus*). Amphipods served as a suitable paratenic host if they acquired larvae from copepods. Larvae given directly to fish and amphipods did not invade the tissues. Thus copepods are apparently significant and necessary precursor hosts. Larvae grew to third-stage larvae in guppies and reached a size of 3.1–3.9 mm (Fig. 5.3J–L) in the body cavity over a period of several days.

Mozgovoi *et al.* (1968b) reported that *C. rudolphi* in the CIS developed in the copepods *Diaptomus gracilis, Macrocyclops albidus, M. fuscus* and *Mesocyclops leuckartii*. Larval stages were also found in larval Odonata (*Coenagrion* and *Agrion*) and several cyprinids.

Hysterothylacium

Species of *Hysterothylacium* are found in the adult stage normally in the gut of fishes. Larvae have been reported widely in the tissues of a variety of marine fishes and invertebrates (including gastropods, squid, crabs, shrimps, starfish and arrow worms) serving as intermediate hosts (see, for example, Deardorff and Overstreet, 1981). In the literature, larvae and adults have frequently been referred to *Contracaecum* and *Thynnascaris* but Deardorff and Overstreet (1980a) have distinguished *Hysterothylacium* from *Contracaecum* (see comments under this genus above).

Knowledge of the development and transmission of *Hysterothylacium* species is still fragmentary. Eggs of the most typical species (*H. aduncum*) pass out in the faeces of the fish definitive host and embryonate to the second stage, which hatches and retains the first-stage cuticle as a sheath. Larvae are ingested by various invertebrates (e.g. mysids, copepods, isopods) in which they exsheath and invade the haemocoel. Development to the third stage takes place in invertebrates and the worms may then transfer to the definitive host, where they will mature, or they may transfer to fish paratenic hosts or other paratenic hosts in the tissues of which they will sequester themselves and be available through predation to the definitive host. The cephalic end of the third-stage larva (Fig. 5.3G–I) has four elevations and a tooth-like structure. The tail is conical and terminated by a sharp point. The intestinal caecum and ventricular appendix are well developed.

H. aduncum (Rudolphi, 1802)

H. aduncum (syn. *Contracaecum clavatum*) was first studied experimentally in fish in the Baltic, including the definitive host, the eelpout (*Zoarces viviparus*). Wülker (1929) proposed that there were three types of hosts in transmission, namely planktonic organisms (copepods, etc.), planktonivorous fish and piscivorous fish. Kahl (1936) suggested that second-stage larvae of *H. aduncum* would undergo partial development in a variety of intermediate hosts, including *Sagitta*, Colanidae, amphipods and medusae as well as prey fish such as *Ammodytes* spp. and *Merongus* spp. Kahl (1936) believed that second-stage larvae could directly infect fish intermediate hosts (e.g. *Merlangus merlangus*).

Markowski (1937) incubated eggs in sea water and fed them, when they were fully embryonated, to the marine copepods *Ascartia bifilosa* and *Eurytemora affinis*. Second-stage larvae exsheathed in the gut of copepods and migrated to the haemocoel. Markowski (1937) believed that several plankton-feeding fishes in the Baltic Sea with larvae encapsulated on the mesentery were the second intermediate hosts of the parasite which matured in eelpout.

Popova *et al.* (1964) reported that polychaetes (*Lepidonotus* sp.) were infected experimentally with larvae of *H. aduncum* and that natural infections occurred in species of *Gattiana, Harmothoe, Lepidonotus* and *Nereis* in the White Sea. The larval stages were infective to fish. The authors suggested that prey fish serve as paratenic hosts rather than true intermediate hosts.

Val'ter (1968, 1980) infected marine isopods (*Jaera albifrons*) with second-stage larvae and noted a natural infection in this crustacean in the White Sea. In the isopods the second-stage larvae exsheathed in the intestine and invaded the haemocoel. Larvae grew from 246–261 µm to 1.95–2.07 mm in length in 25 days. Larvae moulted 17–19 days postinfection. The resulting fully grown third-stage larvae were regarded as the stage infective to the definitive host.

Yoshinaga *et al.* (1987a) studied experimentally the transmission of *H. aduncum* in fresh water. The authors collected third-stage larvae from the mesentery and body cavity of Japanese smelt (*Hypomesus transpacificus*) and introduced them into the stomach of rainbow trout (*Salmo gairdneri*). Of 26 trout, 20 became infected. Third- and fourth-stage larvae were recovered 3–12 days postinfection. Adults were first recovered 11 days postinfection and gravid females in 59 days.

Eggs from female worms were incubated by Yoshinaga *et al.* (1987b). Sheathed second-stage larvae were present in 7 days (? at 15°C) and hatching occurred in 10 days. Mysids (*Neomysis intermedia*), exposed to larvated eggs and free larvae, became infected. Larvae were found in the haemocoel and the appendages of 13 of 15 mysids on day 5. One to eight third-stage larvae were found in the haemocoel of each of 17 of 19 mysids 30 days postinfection. Second-stage larvae from mysids early in the infection were 220–393 µm in length. Third-stage male larvae in mysids were 4.9–6.3 mm and females 3.9–6.8 mm in length.

According to Køie (1993), who claimed to have detected two moults in the egg, larvae were not infective to fish or non-crustacean invertebrates but developed in the copepod *Acartia tomsa*, harpacticoid copepods, various amphipods, isopods and mysids. Fish acquire larvae when they eat crustaceans with larvae longer than about 3 mm. She suggested that ctenophores, chaetognaths, polychaete crustaceans and ophiuroids may acquire larvae from eating infected crustaceans.

Marcogliese (1996) reported larvae of *H. aduncum* in eastern Canada in the hermit crab (*Pagurus acadianus*), the amphipods *Proboloides holmesi* and *Caprella linearis* and provided a table compiling new hosts reported since the paper of Norris and Overstreet (1976).

Yoshinaga *et al.* (1987b) pointed out that *H. aduncum* is of marine origin and reports of it in fresh water are probably the result of introductions by anadromous fish (Moravec *et al.*, 1985). Yoshinaga *et al.* (1987b) reported that in Lake Toro, Hokkaido, Japan, some fish (juvenile *Onchorynchus keta* and *Tridentiger obscurus*) which had never been in marine or brackish water became infected in the lake, probably through ingestion of *N. intermedia* infected by fish (*Platichtys stellatus*, *Salvelinus leucomaenis*) which entered the lake from the sea.

González (1998) described transmission of *H. aduncum* in Chilean marine farms.

H. analarum Rye and Baker, 1984

H. analarum is a parasite of the intestine of the centrachid fish *Lepomis gibbosus* in Ontario. Eggs were laid in the morula stage. At 20–23°C larvae in water were present in the eggs; the first-stage cuticle was loose by day 4 and larvae started to hatch in 5 days (Rye and Baker, 1992). Some larvae retained the loose first-stage cuticle where others shed it. Larvae, which were 190–249 µm in length, had a boring tooth, a distinct ventricular appendage, a small intestinal caecum and spiny protuberances on the tail extremity. Some copepods exposed to larvae ingested them and larvae remained alive in the haemocoel.

After examining *L. gibbosus* in spring, summer and autumn the authors concluded that there was one generation a year. Adult worms were common for a brief period in early summer and they released eggs which developed to the second stage in water. The fish acquired second-stage larvae probably from ingesting copepods in summer. Larvae invaded tissues of the fish and often became encapsulated on the intestinal serosa, where they developed to the third stage. In spring, third-stage larvae invaded the lumen of the intestine, where they gradually matured. Thus, the parasite uses the fish as both an intermediate and a final host.

H. bidentatum (Linstow, 1899)

H. bidentatum (syn. *Contracaecum bidentatum*) is a parasite of the stomach and oesophagus of the sterlet (*Acipenser ruthenus*) in the CIS. Geller (1957) reported that 94% of sterlets in the Volga River were infected. Geller and Babich (1953) reported that eggs were deposited in the 8- to 16-cell stage and were passed in the faeces of the fish host. Larvae moulted and hatched. The fully developed, sheathed, second-stage larvae were 250–300 μm in length. The third and infective stage was attained in *Gammarus* sp. (Amphipoda).

According to Geller (1957) infections in *A. ruthenus* reached a peak from September to October and most infected fish were 1–2 years old. In starving or moribund fish, adult worms (including gravid females) left the fish through the gill slits.

H. haze (Machida, Takahashi and Masuuchi, 1978)

H. haze (syn. *Thynnascaris haze*) is a parasite of the goby (*Acanthogobius flavimanus*) in the Bay of Tokyo, where it was associated with mass mortality (Yoshinaga *et al.*, 1988). When fish were 4–5 cm long in July, they became infected. Prevalence was as high as 70% by October. Most nematodes localized in the body cavity but some penetrated the liver, subcutaneous tissues and occasionally the orbit. They remained unencapsulated. Eggs in the uteri were spherical, thin-shelled and 46–59 μm in size. Embryonated eggs were noted in the body cavity and Machida *et al.* (1978) suggested that eggs or larvae were released into sea water upon the death of infected gobies.

Yoshinaga *et al.* (1988) noted that all stages could be found in infected goby. Eggs were 55–61 μm in size and embryonated to second-stage larvae (243–555 μm in length) which spontaneously hatched and then shed the first-stage cuticle. Third-stage larvae collected from goby were 0.57–3.37 mm in length. Fourth-stage and adult males were 1.7–17.4 mm; fourth-stage and adult females were 1.5–31.1 mm in length.

Yoshinaga *et al.* (1989) studied the development and transmission of *H. haze* in *A. flavimanus*. Second-stage larvae occurred in the gut wall. Third-stage larvae occurred in the gut wall, mesentery and body cavity. Fourth-stage larvae and adults were found in the body cavity. In heavily infected gobies, eggs and all stages from second-stage larvae to adults were found in the body cavity. Gobies were infected by feeding them viscera from heavily infected gobies or artificially incubated eggs containing second-stage larvae. Second- and third-stage larvae were found in the gut wall; fourth-stage larvae and adults were found in the body cavity of fish given second-stage larvae.

Various invertebrates (Polychaeta, Copepoda, Amphipoda, Mysidacea) were placed with the viscera of heavily infected gobies or artificially incubated eggs. Invertebrates were examined immediately or 7 days after the contact periods of 1–30 days. Second-stage larvae were found in *Amphithoe valida*, *Corophium uenoi*, *Grandidierella japonica* (Amphipoda), *Neomysis japonica* (Mysidacea) and *Nereis* sp. (Polychaeta). Larvae had grown to 450 μm in length.

Yoshinaga *et al.* (1989) concluded that *H. haze* has a direct life cycle in which fully developed eggs and second-stage larvae are directly infective to fish. Also, gobies may become infected by eating invertebrates containing second-stage larvae; the common goby feeds mainly on benthic invertebrates. The authors also suggested that eggs deposited in the body cavity are released into the water with the death of the host where they may infect other gobies directly, or indirectly through invertebrate hosts.

Anderson (1988) suggested that *H. haze* was an example of extreme precocity in which the fish intermediate host has also become a final host. The presence of adult worms in the body cavity, rather than the intestine, supports this suggestion.

Pseudoterranova

The genus contains three species: *P. kogiae* (Johnston and Mawson, 1939) from the pygmy sperm whale (*Kogia breviceps*), *P. ceticola* (Deardorff and Overstreet, 1981) from the dwarf sperm whale (*K. simus*) and *P. decipiens* (Krabbe, 1878) from the stomach of pinnipeds; the latter has now been divided into three sibling species (see below). Nothing is known about the transmission of the first two species but *P. decipiens*, the cod or sealworm, has been the subject of considerable research, because its larvae move through the food chain and come to reside and grow in the flesh of commercially important fish intermediate hosts, including Atlantic cod (*Gadus morhua*). A comprehensive multi-authored review of most aspects of *P. decipiens* in the North Atlantic has been published by Bowen (1990).

P. decipiens (Krabbe, 1878)

P. decipiens B (syns *Phocanema decipiens*, *Terranova decipiens*) is a common stomach parasite of Pinnipedia (seals, sea-lions, walruses), especially in temperate and polar regions of the world. The species has been intensively studied by Canadian scientists in the Gulf of St Lawrence and the Atlantic coast of Canada, where it occurs in grey seals (*Halichoerus grypus*), harbour seals (*Phoca vitulina*) and harp seals (*Pagophilus groenlandicus*) (Fig. 5.5). Grey seals are by far the most heavily infected species, followed by harbour seals. The parasite has also been reported in bearded seals (*Erignathus barbatus*). *P. decipiens* is relatively uncommon in harp seals and rarely noted in cetaceans (Templeman, 1990). Recent electrophoretic analyses of gene enzyme systems have revealed three sibling species: A, in grey seals in the northeast Atlantic; B, in harbour seals; C, in bearded seals (Mattiucci and Paggi, 1989; Paggi *et al.*, 1991). Di Deco *et al.* (1994) have used step-wise discriminant analysis to distinguish the three species (A, B, C) in North Atlantic seals. In addition a distinct species has been discovered in the sea-lion (*Otaria byronia*); 'hake' are probably the main intermediate host of this species (George-Nascimento and Llanos, 1995).

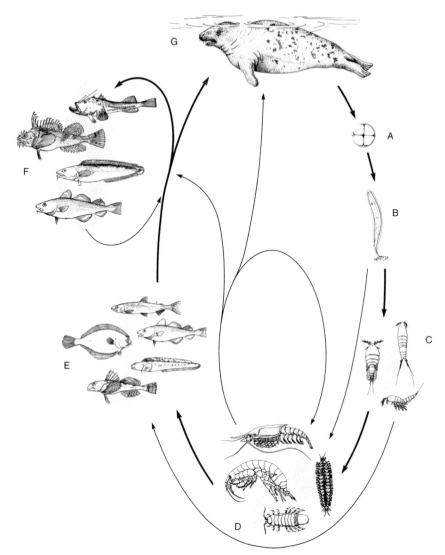

Fig. 5.5. Life cycle of *Pseudoterranova decipiens*: (A) partly embryonated ovum passed into sea water with seal faeces; (B) freshly hatched, ensheathed second-stage larva, adhered to substrate by caudal extremity; (C) early second-stage (third-stage?) larvae in haemocoel of benthic harpacticoid and cyclopoid copepods; (D) early to late second-stage larvae with possible moult (second moult) to third- (early third) stage larvae in haemocoel of benthic macroinvertebrates (mysids, amphipods, isopods and errant polychaetes); (E) early to late third-stage larvae in body cavities and musculature of benthophagous fish, e.g. (clockwise from top) smelt, juvenile cod, ocean pout, longhorn sculpin and American plaice; (F) late third-stage larvae in body cavities and musculature of demersal piscivorous fish including (top to bottom) monkfish, sea raven and mature cusk and cod; (G) third (moulting third-stage larva) and fourth (moulting fourth-stage larva) moults and development to adult in stomach of a pinniped host. (After G. McClelland *et al.,* 1990 – courtesy *Canadian Bulletin of Fisheries and Aquatic Sciences.*)

Brattey and Ni (1992) reported that *P. decipiens* B was rare in harp seals and Brattey and Stenson (1993) showed that the harbour seal and the grey seal were heavily infected with *P. decipiens* B in coastal Newfoundland and Labrador; smaller numbers of worms occur in the hooded seal (*Cystophora cristata*).

Thin-shelled eggs passed in faeces of infected seals were spherical, 40–50 µm in diameter, and in the 2- to 16-cell stage (McClelland *et al.*, 1990). Eggs settled in sea water and adhered to material in the substratum. They embryonated and hatched at a wide range of temperatures (1.7–26.0°C) in fresh, brackish and sea water but development time varied from 5 days at 22°C to 125 days at 1.7°C (Brattey, 1990; Burt *et al.*, 1990b; McClelland *et al.*, 1990). First-stage larvae in eggs moulted to the second stage, which retained the moulted cuticle. Measures and Hong (1995) used transmission electron microscopy to show that only a single moult occurs in the egg (cf. Køie *et al.*, 1995). At 5°C most (85%) eggs hatched in 5–8 days and at 12°C most (85%) hatched in 21–24 days (Burt *et al.*, 1990b). According to Measures (1996) eggs kept at 0°C, 5°C, 10°C and 15°C in fresh, brackish water and sea water developed except at 0°C. Larvae hatched in 57 days at 5°C, 21–24 days at 10°C and 10–11 days at 15°C. The average hatch was 95–99%. Few eggs kept at 0°C for 1 year hatched when placed at 18–20°C. McConnell *et al.* (1997) reported that fresh eggs settled at a rate of 1.01×10^{-4} m per second and could be transported 50 km in 100 m of water in 12 days before landing on the bottom sediment. Sheathed larvae hatching from eggs were 200–215 µm in length. They adhered to the substrata by their tails and continually arched and extended their bodies (McClelland, 1982). Larvae survived for only a few hours at 20–25°C but at 4°C they remained viable for 120–140 days.

In experiments, ensheathed larvae were readily predated by copepods, including marine benthic harpacticoids and cyclopoids, marine pelagic calanoids and freshwater cyclopoids (McClelland, 1982, 1990). Larvae exsheathed in the gut of copepods and penetrated the midgut wall into the haemocoel, where they grew rapidly. Growth was six times faster at 15°C than at 5°C and larvae reached a length of 2 mm in the calanoid *Tortanus discaudata* (see Jarecka *et al.*, 1988), at which time they might have been large enough to infect fish intermediate hosts. However, in most copepods they grew only to 300–500 µm and were probably not infective to fish. McClelland (1982) pointed out that copepods are too short-lived to sustain prolonged development of larval *P. decipiens*.

Amphipods (*Gammarus lawrencianus*, *G. oceanica*, *Unciola irrorata*) as well as zoea larvae of crabs (*Cancer* spp. and *Hyas* spp.) ingested larvae of *P. decipiens* but few exsheathed and invaded the haemocoel (Jarecka *et al.*, 1988; McClelland, 1990). However, numerous amphipods became infected when they were allowed to feed on copepods containing larvae. According to McClelland (1995) larvae in *G. lawrencianus* and *U. irrorata* were 0.82–7.08 mm in length but only those equal or greater than 1.41 mm in size were infective to fish. The larvae reached 23.6–30.7 mm in length in 56 days in smelt (*Osmerus mordox*) (Fig. 5.3A–C). Polychaetes, mysids, isopods, cumaceans, decapods and gastropods also acquired larvae from ingesting copepods (McClelland, 1990). Larvae in amphipods grew to 2–3 mm in length in 4 weeks at 15°C and up to 10 mm in 12 weeks. According to Jarecka *et al.* (1988) *P. decipiens* will also grow in the haemocoel of larval crabs, although these pelagic larvae are unlikely important hosts; larvae of *P. decipiens* are relatively rare in pelagic consumers of plankton such as mackerel and herring (McClelland, 1990). Natural infections of

P. decipiens have been reported in polychaetes, mysids, amphipods, isopods and decapods but intensity is usually 1% and prevalence less than 1% (Marcogliese, 1996; Marcogliese and Burt, 1993). There is probably a moult to the third stage in macroinvertebrates but more evidence is required. McClelland and Ronald (1974b) reported that larvae which emerged from eggs in culture developed directly to the infective stage without evidence of a moult.

Although larval *P. decipiens* will invade copepods and can readily transfer to larger long-lived invertebrates, where substantial growth occurs, it is unlikely that seals become infected from consuming invertebrates. The main source of infection is undoubtedly fish which have acquired larvae from eating infected macroinvertebrates. Larvae of sealworm infect a wide range of fish hosts in the North Atlantic and adjacent waters. Abundance and intensities of larvae are usually greatest in piscivorous demersal fish (i.e. those living near the bottom), such as goosefish (*Lophius americanus*), Atlantic cod (*Gadus morhua*) and sea raven (*Hemitripterus americanus*). However, the greatest densities (numbers of larvae per unit weight of the host) are found in small benthophagous fishes such as ocean pout (*Macrozoarces americanus*), sculpins (*Myoxocephalus octodecemspinosus*, *M. scorpius*), American plaice (*Hippoglossoides platessoides*) and, in coastal waters, pelagic feeders such as smelt (*Osmerus eperpanus*, *O. mordax*) (Jensen *et al.*, 1994). Also, larval *P. decipiens* have been experimentally transferred by means of amphipods to a variety of fish species (McClelland *et al.*, 1990). It has been reported that larvae 2–5 mm long reached a length of 23–31 mm in smelt held at 15°C and 10–20 mm in mummichog (*Fundulus heteroclitus*) and winter flounder (*Pseudopleuronectes americanus*).

It is well known that the parasite can transfer from one fish to another (Scott, 1953; Burt *et al.*, 1990a; Jensen, 1997) and larger piscivorous fishes (e.g. Atlantic cod, goosefish and sea raven) may accumulate large numbers of larvae. However, seals feed mainly on smaller fish, which are probably the main source of infection (McClelland *et al.*, 1990).

Scott (1953) gave larvae 30–50 mm in length to harbour seals, which passed eggs 17–22 days later. Worms were lost from the seals soon after patency. McClelland and Ronald (1974b) gave larvae 40–60 mm in length to a harp seal. In 15 days all males were mature and two of 25 females were gravid. McClelland (1980a) reported that in the stomach of seals the third moult occurred 2–5 days and the fourth 5–15 days postinfection. McClelland (1980b) compared infections in harbour and grey seals. Eggs were found in faeces of harbour seals 21 (17–30) days postinfection and the patent period was 15–45 days. In grey seals, the prepatent period was 19 (16–23) days and the patent period was 20–60 days. In 6 weeks females and males in harbour seals were 40.8–76.2 mm and 45.5–60.8 mm in length, respectively, and it was estimated that the mean number of eggs produced by a female was 156,000. In grey seals comparable data for females and males were 69.7–104.3 mm and 53.8–72.7 mm and the mean number of eggs produced per female was 366,000. Marcogliese (1997) counted eggs in adult female *P. decipiens* from grey seals collected on Antocosta Island, in the Gulf of St Lawrence, in 1992. Uterine egg counts were 54,916.9 ± 51,866.4. The relationship of eggs found in uteri and eggs passed in faeces of the host has not been determined.

McClelland (1980a) noted that seals with existing or recent infections resisted challenge infections although surviving worms grew normally. Resistance to reinfection subsided when seals were maintained free of *P. decipiens* for 2–6 months.

It is evident that most growth of *P. decipiens* takes place in the third larval stage in fish and because of this remarkable precocity, maturation in the seal host is extremely rapid (a matter of 2–3 weeks). At the same time the patent period is short (usually about 2–3 weeks, with a maximum of 6 weeks) (McClelland, 1980b). Such marked precocity in the infective stage in fish, along with a short prepatent period in the definitive host, is probably advantageous to the parasite. Although *P. decipiens* can be transmitted year round to seals through the ingestion of fish, the breeding periods might be an important time for transmission because of the addition to the seal population of large numbers of pups. The short prepatent period would enable pups to make a major contribution to the parasite population in the region of the natal area before they disperse. On Sable Island several thousand grey seal pups are born from late December until late January and they begin to feed independently 17 days after birth (Stobo and Zwanenburg, 1990; Thompson *et al.*, 1990). Pups feed in the region of the natal area throughout February and March and perhaps well into June (Thompson *et al.*, 1990) before they disperse. There is, therefore, ample time for pups to acquire infections and to make a major contribution to the parasite populations in crustaceans and fish surrounding the natal area. There is evidence that infections in pups rapidly increase until March (Stobo and Zwanenburg, 1990). Young naïve pups are probably more susceptible to infection than juveniles or adults which have, or had, substantial infections in earlier years and have become resistant. Also, the percentage of worms maturing in naïve pups and their egg output might be expected to be greater than in older animals. It is perhaps worth noting that ascaridoids of terrestrial mammals are parasites overwhelmingly of young animals (e.g. *Ascaris* spp., *Toxocara* spp.).

McClelland *et al.* (1990) reported 'dramatic increases' in recent years in the abundance of *P. decipiens* in fish from waters surrounding Sable Island where grey seals have increased and a substantial source of these infections could be from pups. The parasite was most numerous in large demersal piscivorous fishes (*Lophius americanus*, *Brosma brosme*, *Gadus morhua* and *Hemitripterus americanus*) and in small benthophagous fishes (*Hippoglossoides platessoides*, *Marozoarces americanus*, *Myxocephalus octodecemspinosus* and *Scophthalmus aquosus*).

P. decipiens C of the bearded seal in northern waters in Norway develops in the liver of the rough dab (*Hippoglossoides platessoides*) and Greenland halibut (*Reinhardtius hippoglossoides*) (see Paggi *et al.*, 1991; Bristow and Berland, 1992).

Jensen *et al.* (1994) found larval *P. decipiens* in demersal and benthic fishes in two areas in Norway, with intensities of infection being higher in shallower waters. Intensities were highest in bullrout (*Myxocephalus scorpius*), cust (*Brome broma*) and Atlantic cod. Both grey and common seals are common in one of the two areas studied and in one area grey seals apparently acquire infections from eating sculpins. In the other area common seals seem to acquire infections from eating cod which have eaten infected sculpins.

Although the seal stocks were decimated by phocine distemper, larvae remained prevalent in the various fish intermediate hosts in the North Sea (Aspholm *et al.*, 1995). Aspholm *et al.* (1995) reported that prevalence and abundance of larvae in cod decreased with increasing distance from a haul-out island in the outer Oslofjord in the North Sea. Eleven fish species examined were uninfected. In addition to cod, 93% of bullrout (*M. scorpius*) were infected. It was suggested that common seals in the region acquire infection from eating cod, which acquired their infection from eating bullrout. However, of 45 seals examined only two harboured sexually mature worms.

Sulcascaris

S. sulcata (Rudolphi, 1819)

S. sulcata is a parasite of the stomach of marine turtles (*Caretta caretta, Chelonia mydas*) (Sprent, 1977). Larval stages have also been found in the adductor muscle of sea scallops (*Amusium balloti* and *Chlamys* sp.) and other marine molluscs (Cannon, 1978; Lichtenfels *et al.*, 1978; Lester *et al.*, 1981). According to Berry and Cannon (1981) eggs were thin-shelled with a rugose surface and 86–100 × 61–75 μm in size. They were deposited in the 8- to 16-cell stage and developed in sea water at 25°C. Two moults occurred in the eggs; in 5 days third-stage larvae were 300–440 μm in length and had a small boring tooth and a caudal mucron. Eggs hatched in 7 days and third-stage larvae survived in sea water for 3–4 days.

Third-stage larvae were drawn into the siphon of gastropods (*Polinices sordidus*) and bivalves (*Melina ephippium, Pinctada* spp.); larvae were found 11–210 days later in the liver and gonads of the molluscs and in the adductor muscles of the bivalves. Larvae grew in the molluscs to a length of about 5 mm and moulted to the fourth stage in about 115 days postinfection. The latter grew to a maximum of about 26.7 mm in 177 days and the genital primordium had developed substantially.

Third- and fourth-stage larvae similar to those found in experimentally infected molluscs were found in naturally infected sea scallops (*A. balloti*). Larvae were confined to a small sector of the adductor muscle adjacent to the loop of the gut. Fourth-stage larvae from naturally-infected scallops were given to 6-month-old loggerhead turtles (*C. caretta*). Within 2 days fourth-stage larvae attached to the base of the oesophagus and stomach. In 7–35 days worms were found in the stomach and some had moulted to the adult stage. It was estimated that 5 months were required for worms to become gravid but this conclusion is not supported by experimental evidence.

In south central Queensland, loggerhead turtles congregate regularly to breed in major scallop beds, where they feed on scallops and other molluscs. It is in such areas that *S. sulcata* is transmitted.

Subfamily Raphidascaridinae

The excretory pore is usually near the nerve ring in members of the subfamily. Species occur in marine and freshwater teleosts (eight genera) and elasmobranchs (one genus). Some species of *Heterotyphlum* have been reported as sporadic accidental parasites of piscivorous birds.

Raphidascaris

The genus contains five species, two of which occur in freshwater fishes (Smith, 1984a). Of these, *R. acus* has been studied in some detail. Transmission involves, firstly, an aquatic invertebrate paratenic host which acquires larvae from ingesting eggs of the parasite and, secondly, a fish intermediate host which consumes paratenic hosts and in which worms develop in the liver to the stage infective to the predaceous definitive host.

R. acus (Bloch, 1779)

R. acus is an holarctic parasite of the intestine, pylorus and, less commonly, the stomach of predatory fishes mainly of the families Esocidae and Salmonidae, although it has been reported in the gut of fish belonging to other families (Mozgovoi, 1953; Shulman, 1958; Moravec, 1971a; Smith, 1984a). The parasite is common in *Esox* spp. in North America (Smith, 1984a).

Thomas (1937c, 1940) fed embryonated eggs of *R. acus* (syn. *R. canadensis*) to fish and dragonfly nymphs (*Tetragoneuria* spp.) and subsequently recovered larvae from their tissues. Engashev (1965), Kosinova (1965) and Supriaga and Mozgovoi (1974) concluded that various invertebrates (mainly arthropods) were intermediate or paratenic hosts of *R. acus*. Moravec (1970a,b), Alvarez-Pellitero (1979) and Torres and Alvarez-Pellitero (1988) successfully infected chironomids, amphipods and oligochaetes by allowing them to ingest eggs. Engashev (1965) reported that predatory larval insects could acquire larvae of the parasite by predating infected invertebrates. Moravec (1996) reported a single third-stage larva of *R. acus* in one of 775 *Gammarus fossarum* in the Czech Republic, showing that the larvae may develop in invertebrates, but the significance of this observation is still not clear (see below).

It was earlier reported by Thomas (1940), Engashev (1965) and Supriaga and Mozgovoi (1974) that larvae from eggs invaded and developed in the liver of fish and it was eventually realized that fish were intermediate hosts. Detailed studies of the development and transmission of *R. acus* were conducted by Moravec (1970a,b) and Smith (1984a,b, 1986).

Eggs were 90–111 × 83–101 μm in size and, *in utero*, contained one to four cells. The egg envelop was 25 μm thick and hyaline. According to Smith (1984b) eggs embryonated to second-stage larvae in 5 days at 22°C. By day 30 about 10% of the eggs in culture had spontaneously hatched. Embryonated eggs were placed with amphipods (*Hyalella azteca*) and chironomid larvae (*Tendipes* sp.). Second-stage larvae were recovered later from these invertebrates, which could serve as paratenic hosts. Darters (*Etheostoma caeruleum*, *E. nigrum*), yellow perch (*Perca flavescens*) and rainbow trout (*Oncorhynchus mykiss*) were orally infected with embryonated eggs. Second- and third-stage larvae were found in the wall of the gut, on the mesentery and in the liver of the darters and perch. The second and third moults occurred in these fish. Moulting second-stage larvae were 1.2–1.3 mm in length. Moulting third-stage larvae were 2.2–2.5 mm in length in wild-caught perch and 2.4–3.9 mm in experimentally infected perch. Pike (*Esox lucius*), rainbow trout and brook trout (*Salvelinus fontinalis*) were given fourth-stage larvae. The fourth moult occurred in about 1 week, when nematodes were about 6 mm in length. Adult *R. acus* were found in trout 2 weeks postinfection. Mature males and gravid females were found 5 weeks postinfection in pike at 12–14°C.

Smith (1986) studied the seasonal transmission of *R. acus* in two small lakes on Manitoulin Island, Ontario. Second-stage larvae were found in a few dragonfly nymphs and caddisfly larvae collected in the lakes where pike were heavily infected with *R. acus*. Second-, third- and fourth-stage larvae were found in the liver of yellow perch, Iowa darter (*Etheostoma exile*), bluntnose minnow (*Pimephales notatus*), fathead minnow

(*P. promelas*) and blacknose shiner (*Notropis heterolepis*). Yellow perch was the dominant intermediate host, in which prevalence was 100% and intensity as high as 928. Both prevalence and intensity increased with length and age of perch. Analysis of field data indicated that second-stage larvae, acquired by perch from invertebrates in summer, developed in the liver to the fourth stage by November. By the following summer larvae were encapsulated but remained alive for at least another year; since larvae lived for at least 2 years, they tended to accumulate in larger fish. Valtonen *et al.* (1994) described lesions in the liver and pancreas of roach (*Rutilus rutilus*) infected with larvae of *R. acus* in Finland.

Pike fed regularly on perch and the parasite was acquired in late autumn; in the following winter and spring, prevalence in pike was 100% and intensity over 200. In summer, however, prevalence declined and intensity dropped to 15. J.D. Smith (1986) concluded that seasonality of infection in pike was probably controlled by the timing of predation on perch and the rate of development and longevity of the parasite, and not by any seasonal variability in the availability of larvae in perch, which did not vary significantly. Transmission to pike apparently continued in summer but the low intensity may have resulted from low recruitment into and rapid turnover of the parasite population.

Moravec (1970a) studied the seasonal dynamics of *R. acus* in the River Bystřice in the Czech Republic and reported an annual cycle in which egg-producing females were present in *Salmo trutta* only from May to the beginning of July. Seasonal changes were probably due to changes in food consumption of the definitive host. The main intermediate hosts were *Cottus gobio*, *C. poecilopus* and *Neomacheilus barbatulus*. Larval *Prodiamesa olivacea* (Chironomidae) were paratenic hosts for second-stage larvae.

Family Ascarididae

Ascaridids are mainly parasites of terrestrial hosts and their transmission commonly involves terrestrial invertebrates and small mammal paratenic or intermediate hosts. The eggs of some species are directly infective to the definitive host, however.

Subfamily Ascaridinae

Members of the subfamily are parasites of terrestrial mammals.

Ascaris

A. lumbricoides (Linnaeus, 1758)
A. suum Goeze, 1782
A. lumbricoides of humans and *A. suum* of swine are the largest and best known round-worms inhabiting the intestines of their hosts. The two species are morphologically indistinguishable but can only be cross-transmitted with difficulty between humans and pigs (de Boer, 1935; Takata, 1951; Soulsby, 1965; Crompton, 1989). Epidemiological studies have generally failed to reveal a clear relationship between the presence of

A. suum in swine and human infection with *A. lumbricoides* (see Payne *et al.*, 1925; Caldwell and Caldwell, 1926; Roberts, 1934). Anderson *et al.* (1993) and Anderson and Jaenike (1997) decided after study using enzyme electrophoresis and mitochondrial DNA sequencing and other molecular methods that *Ascaris* from humans and pigs were 'involved in separate transmission cycles in Guatemala'. Anderson (1995) reported, however, that molecular data indicated that sporadic human cases in the USA may have been of porcine origin. Nevertheless, the bulk of the evidence seems to lead to the conclusion that the swine and human ascarids are sibling species that probably do not cross-breed or commonly infect the alternative host. Larvae of both *Ascaris* spp. will, however, migrate to the lungs of humans, sheep, cattle, rodents and lagomorphs, where they may be associated with respiratory distress referred to in cattle as atypical interstitial pneumonia (Kennedy, 1954; Fitzgerald, 1962; Johnson, 1963; McDonald and Chevis, 1965; Greenway and McCraw, 1970 a,b; McCraw and Greenway, 1970; Maruyama *et al.*, 1996).

The discovery of the transmission of *Ascaris* spp. was not straight forward. Grassi (1888), Lutz (1888), Epstein (1892), Wharton (1915) and Lane (1917) were certain that transmission was direct through the ingestion of eggs; Epstein infected children and found eggs in their faeces 80 days later. Stewart (1916a,b,c, 1917) confirmed Davaine's (1863) observation that when embryonated eggs were fed to rodents, they hatched in the small intestine and larvae could be found in the faeces a few hours later. He also observed that not all larvae passed out of the body; he detected larvae in the liver and lungs 4–10 days postinfection and reported that some larvae migrated from the lungs to the trachea and then to the stomach and intestine. Stewart was unsuccessful in infecting pigs with eggs and concluded, erroneously, that rodents served as intermediate hosts; the latter were believed to pass altered larvae in their faeces which contaminated the environment (Ransom, 1922).

Ransom and Foster (1917, 1919, 1920) and Ransom and Cram (1921) infected guinea pigs, rabbits, goats and young pigs with eggs and suggested that early failures to infect swine were the result of age resistance in the host. They discovered a moult in the egg and confirmed that there was indeed a lung–tracheal migration. Martin (1926) and Roberts (1934) confirmed these findings and completed the life cycle in detail. Finally, Koino (1922) infected himself with *A. lumbricoides* and proved that the larvae underwent a lung–tracheal migration, as had been reported in *A. suum*. For a comprehensive review of the discovery of the life cycle of *Ascaris* spp., refer to Schwartz (1959), Taffs (1961) and Grove (1990). Kean *et al.* (1978) reproduced a number of classic papers on ascariasis.

Eggs of *A. lumbricoides* are 65–75×35–$50\,\mu$m and those of *A. suum* 50–70×40–$60\,\mu$m in size (Soulsby, 1965; Faust *et al.*, 1975). The outer surface of the eggs is coarsely mamillated and sticky. Eggs are passed from the host in the one-cell stage. The optimum temperature for development was 30–$33°$C but eggs developed satisfactorily at lower temperatures as well (Ransom and Foster, 1920). The latter authors observed a moult in the first-stage larva in the egg at 13–18 days at 30–$33°$C and Alicata (1934) reported that eggs were not infective before this moult was completed. More recently, a second moult has been reported in larvae in the eggs of *A. lumbricoides* and *A. suum* (Henner, 1959; Thust, 1966; Araujo, 1972a; Maung, 1978; Artigas and Ueta, 1989). Maung (1978) reported considerable variability in the timing of the moults but at $28°$C the first occurred in 15–24 days and the second after

17 days. According to Artigas and Veta (1989) eggs were fully embryonated in 12.7 days at 27°C. The first moult occurred in 15.7 days and the second in 18.3 days. Only eggs cultured for 18 days were infective to mice, in which larvae wandered in the liver and the lungs. If two moults do occur in the eggs then Maung's (1978) suggestion that the second moult is initiated in eggs but the final shedding of the cuticle (ecdysis) takes place in the liver seems reasonable (see below). Fully embryonated eggs of *Ascaris* spp. can remain viable for up to 5 years under favourable conditions (Roberts, 1934; Kalbe, 1955).

Humans and pigs become infected when they ingest embryonated eggs. Eggs hatched in the intestine and the released larvae were about 224–263 μm in length, according to Douvres *et al.* (1969). Larvae were rather stout with conical tails. They invaded the gut wall and entered the hepatic portal system, which carried them to the liver in 1–2 days. Moulting larvae in the liver 28 h postinfection were 172–252 μm in length and in 4 days third-stage larvae were 533–619 μm in length (Douvres *et al.*, 1969). According to Rhodes *et al.* (1977) and Murrell *et al.* (1997) larvae invade the caecum and colon and not the intestine. Murrell *et al.* (1997) found larvae as early as 3 h postinfection in the mucosa and in the liver in 6 h. By 24 h most larvae had disappeared from the mucosa. According to Kelley *et al.* (1957) and Douvres *et al.* (1969) most third-stage larvae had migrated to the lungs by 5–6 days and were 0.98–1.68 mm in length in 6–7 days. Larvae then migrated up the trachea, were swallowed and reached the intestine in 8–9 days, where at 10 days they moulted for the third time when they were 1.57–1.87 mm in length. According to Kelley *et al.* (1957) migration to the gut continued over a period of 15 days when almost all larvae had left the lungs. Larvae at 15 days were on average 2.73 mm in length. According to Roberts (1934) the final moult took place in the intestine 21–29 days postinfection, when worms were 17.5–22.5 mm in length. Worms were mature in 50–55 days and eggs appeared in faeces in 60–62 days. Olsen *et al.* (1958) reported that some adults could live for 55 weeks in swine but most were gradually lost by 23 weeks. Olsen *et al.* (1958) estimated that a female worm could produce 2 million eggs a day.

A. lumbricoides seems to behave in humans essentially as *A. suum* does in pigs, although it is usually reported that the second moult takes place in the lungs rather than in the liver. The evidence for this conclusion is not clear. Since there is considerable variability in the size of moulting larvae it is possible that moulting forms could be found in both the liver and lungs of the host and this could result in some confusion.

Jungerson *et al.* (1997) transferred gravid female worms orally into pigs and noted that the production of eggs ceased in 2–3 weeks but resumed in 3–6 days after the oral inoculation of male worms into the same pigs. Roepstorff *et al.* (1997) infected pigs with 100 to 10,000 eggs of *A. suum*. Larvae were found in the liver in 3 days and in the lungs on day 7. However, most larvae were eliminated from the intestine by 17–21 days postinfection. Regardless of the number of eggs used to infect the pigs, the results were always small aggregated populations in the pigs by day 28. Boes *et al.* (1998) showed that infections of *A. suum* in a pig population were highly overdispersed, like *A. lumbricoides* in humans.

Baylisascaris

Members of the genus are parasites of the intestine of Carnivora and Rodentia. They possess cervical alae, an area rugosa and pitted eggs (Sprent, 1968). Fully embryonated eggs contain second-stage larvae which develop to the third and infective stage in vertebrate intermediate hosts. Two species in mustelids (*B. columnaris*, *B. devosi*), one in procyonids (*B. procyonis*) and one in ursids (*B. transfuga*) are known to develop to the infective stage in small mammals, which can serve as intermediate hosts. A fifth species (*B. tasmaniensis*) occurs in carnivorous marsupials and develops in small prey marsupials, which serve as intermediate hosts. Embryonated eggs of a sixth species in large rodents (*B. laevis*) as well as *B. procyonis* of raccoons can directly infect the definitive host; second-stage larvae undertake a liver–lung migration and develop to the third or fourth stage before migrating via the trachea to the intestine, where they grow into adults (cf. *Ascaris lumbricoides* and *A. suum*). Morphologically, *B. laevis* is similar to forms in carnivores and is probably a capture derived from heteroxenous forms in carnivores. Anderson (1999) reported larvae of *Baylisascaris* sp. in the brain of a 3-day-old moribund lamb, suggesting prenatal transmission.

B. columnaris (Leidy, 1856)

This is a common parasite of skunks (*Mephitis chenga*, *M. mephitis*, *M. mesomelas*, *M. nigra*, *Spilogale putorius*) in North America. Tiner (1952, 1953a) and Sprent (1952a) embryonated eggs taken from worms from skunk, orally inoculated mice (white mice and *Peromyscus leucopus*) and subsequently found encapsulated third-stage infective larvae, mainly in the wall of the lower ileum, caecum and rectum. Tiner (1953a) noted that infected mice remained healthy and did not exhibit neurological signs (cf. *B. procyonis*). Sprent (1952a) gave embryonated eggs to mice and noted that larvae reached the lungs by the first day and were rapidly distributed in the body. In about 10 days encapsulation took place in the heart, kidneys and liver. A few infected mice displayed neurological signs. Tiner (1953a) gave a skunk a mouse with numerous encapsulated larvae. The skunk died 44 days later and 485 worms (some mature) were found in the intestine.

Berry (1985) restudied *B. columnaris*. Eggs were 60–88 × 58–72 μm in size with thin, mamillated and sticky shells. Eggs reached the second stage in 11–16 days at 25°C. Some larvae escaped from the first cuticle while in the egg; infective larvae were 310–380 μm in length. Meadow voles (*Microtus pennsylvanicus*) were given embryonated eggs. Most larvae were found in the intestinal wall and on the mesentery 1–2 days later but some were found in the liver, lungs, kidneys, diaphragm, heart, brain and carcass. Third-stage larvae were 1.32–1.74 mm in length and encapsulated 7 days postinfection, mainly in the wall of the caecum and in the mesentery.

B. devosi Sprent, 1952

B. devosi (syn. *Ascaris mustelarum* in Sprent, 1952b) is known from the marten (*Martes americanus*) and fisher (*M. pennanti*) in North America (Sprent, 1968). It is morphologically similar to *B. columnaris* and has been transmitted to skunk and ferrets (Sprent, 1952b). According to Sprent (1953a) eggs were 58–77 × 51–61 μm in size. Eggs were incubated on wet charcoal at 20–27°C; second-stage larvae were infective and 200–308 μm in length after 12 days. Eggs hatched in white mice; second-stage larvae penetrated the intestinal wall and migrated to the liver and mesenteric tissues. In 3 days larvae were found in the heart, lungs, brain, kidneys and especially subcutaneous tissues of the neck region and the back over the intercostal spaces. Larvae grew in the tissues of the mice; in 8–12 days they moulted to the third stage and became encapsulated. Encapsulated larvae were 1.12–1.91 mm in length and remained alive for at least 6 months. Young ferrets were infected by feeding them mice containing larvae. In ferrets, the third moult occurred in 3–4 days and the fourth 2–3 weeks postinfection. Fifth-stage worms were mature in 56 days but continued to grow for up to 6 months.

B. laevis (Leidy, 1856)

B. laevis is a parasite of the intestine mainly of marmots (*Marmota broweri*, *M. caligata*, *M. flaviventris*, *M. monax*, *M. olympus*) and ground squirrels (*Citellus beecheyi*, *C. parryi*, *C. richardsoni* and *C. undulatus*) in North America. It has been reported in *M. menzbieri* in Siberia (Sprent, 1968) and is morphologically similar to *B. columnaris* (see Tiner, 1951; Sprent, 1968). *B. laevis* is probably a capture from carnivores.

Babero (1957, 1959, 1960a,b) studied *B. laevis* from *C. franklini*, *C. tridecemlineatus* and *M. monax*. Eggs, 60–80 × 50–70 µm in size, reached the infective stage in distilled water in 17 days at 25–30°C. Second-stage larvae were 140–320 µm in length. In the definitive host, larvae from eggs migrated through the gut wall into the portal vein and the liver, where the second moult occurred in 10–12 days, after which they grew slightly to 1.3 mm in length. Some early fourth-stage larvae also appeared in the liver. From the liver, larvae entered the postcaval circulation and passed to the heart and lungs. This migration occurred over several days and was carried out by larvae of varying sizes. The minimum time for larvae to reach the lungs was 34 days and the minimum time at which fourth-stage larvae were found in the intestine of the natural host was 42 days (Babero, 1960a,b).

B. procyonis (Stefanski and Zarnowski, 1951)

B. procyonis is a common parasite of raccoons (*Procyon lotor*) in North America. There are various reports of *B. columnaris* in raccoons (Tiner, 1949, 1951; Sprent, 1951) that probably refer to *B. procyonis*. Tiner (1949) first reported that larvae of the species from raccoons caused fatal neurological disease in grey squirrels (*Sciurus carolinensis*), white-footed mice (*Peromyscus leucopus*), house mice (*Mus musculus*), cotton rats (*Sigmodon hispidus*), hamsters and guinea pigs. At necropsy he found encapsulated third-stage larvae in the myocardium, pericardium, walls of the caval veins, lungs and under the pleura of squirrels. He also observed that two wild-caught fox squirrels (*Sciurus niger*) from a woodlot where an infected raccoon was taken had similar larvae in their tissues. Infections were established in both raccoons and skunks by feeding them cotton rats with 15- to 20-day-old capsules containing larvae apparently derived from raccoons. One skunk was given an infected *P. leucopus* and 44 days later 485 mature adults were found in its intestine. Tiner (1951) reported that a single larva of *B. procyonis* in the medulla or spinal cord of a mouse was fatal. He reported that 'the raccoon ascarid grows more rapidly than the skunk ascarid and to a larger size'.

Tiner (1949) and Berry (1985) showed that *B. procyonis* and *B. columnaris* could be cross-transmitted between skunks and raccoons. There is no evidence, however, that cross-infection occurs in nature and hybridization experiments would be a useful endeavour. Berry (1985) reported that the oesophagus of female *B. procyonis* was slightly shorter than in *B. columnaris*. The second- and third-stage larvae of *B. procyonis* were slightly shorter than those of *B. columnaris* (second-stage 310–380 µm; third-stage 1.32–1.74 mm). Lateral alae in third-stage *B. procyonis* were wider than those in *B. columnaris*. According to Tiner (1953b) third-stage larvae of *B. procyonis* were encapsulated in the thoracic viscera of mice and there was a tendency for some larvae to reach the central nervous system early in infections, where they elicited neurological signs (Tiner, 1953b), whereas larvae of *B. columnaris* rarely reached the CNS before 30 days and tended to be overcome there. Berry (1985) provided some electrophoretic evidence that *B. procyonis* and *B. columnaris* have different alleles in the 6-phosphogluconate dehydrogenase system.

Experimental evidence suggests that larvae of *B. procyonis* are more likely to cause neurological disease in intermediate hosts than larvae of *B. columnaris*. Ascaridoid larvae presumed to belong to either *B. columnaris* or *B. procyonis* have been observed frequently in the CNS of animals exhibiting neurological signs, including groundhogs (Richter and Kradel, 1964; Swerczek and Helmboldt, 1970; Fleming and Caslick, 1978; Fleming *et al.*, 1979; Kazacos *et al.*, 1981a), rabbits (Jacobson *et al.*, 1976), squirrels (Fritz *et al.*, 1968; Schueler, 1973; Coates *et al.*, 1995), nutria (Dade *et al.*, 1977), birds (Sass and Gorgacz, 1978; Winterfield and Thacker, 1978; Richardson *et al.*, 1980; Reed *et al.*, 1981; Kazacos, 1982; Kazacos *et al.*, 1982a,b; Armstrong *et al.*, 1989; Kwiecien *et al.*, 1993; Suedmeyer *et al.*, 1996; Williams *et al.*, 1997), chinchillas (Sanford, 1989), guinea pigs (Andel, 1995), dogs (Rudmann *et al.*, 1996) and primates (Campbell *et al.*, 1997; Ball *et al.*, 1998). In some cases neurological signs and mortality in animals have been traced back convincingly to raccoons (Jacobson *et al.*, 1976; Richardson *et al.*, 1980; Reed *et al.*, 1981; Kazacos *et al.*, 1982a,b, 1984).

The effects of *B. procyonis* on experimentally infected squirrels, monkeys, swine and chickens have been reported by Kazacos *et al.* (1981b, 1984), Kazacos and Wirtz (1983) and Kazacos and Kazacos (1984). Human cases have been reported by Fox *et al.* (1985) and Cunningham *et al.* (1994). For useful summaries of the biology of *B. procyonis* the reader is referred to Kazacos (1983, 1997).

Raccoons are also readily infected by the ingestion of embryonated eggs of *B. procyonis* and it is probably by this route that most young raccoons acquire infections.

B. schroederi McIntosh, 1939

Specimens were found in a giant panda (*Ailuropoda melanocleuca*) brought to the New York Zoo from China 2 months earlier. According to Li (1989) larvae hatched and penetrated the intestinal wall and spread by the circulatory system to the liver, lungs, brain, kidney and spleen of white mice. The second moult occurred in the liver on day 18. The third-stage larvae remained mainly in the liver for at least 1 year.

B. tasmaniensis Sprent, 1970

B. tasmaniensis was found in the Tasmanian devil (*Sarcophilus harrisi*), the quoll (*Dasyurus viverrinus*) and the spotted tiger cat (*Dasyurops maculatus*) in Tasmania and is the only member of the genus known in marsupials. In devils, worms were found in the stomach, usually on the outside of a mass of hair, bone or wool. In quolls, worms were found mainly in the anterior part of the small intestine (Sprent, 1970a).

Eggs were cultured on moist charcoal at 22°C. Larvae moulted by 20 days and were 192–256 µm in length (Sprent *et al.*, 1973). Eggs given to mice hatched and second-stage larvae migrated to the liver, lungs, brain, eyes and kidneys but mainly to the mesentery and intestinal wall, where they became encapsulated and grew to a length of 0.53–1.60 mm in 2–6 weeks.

Third-stage larvae in mouse tissues were given to several wild-caught marsupials presumed to be uninfected. The results were difficult to interpret. Fourth-stage larvae and adults were collected from *Dasyuroides byrnei* and fourth-stage larvae and adults were collected from *S. harrisi*. In the devil the third moult apparently occurred when larvae were 2.0–2.5 mm in length. Fourth-stage larvae found in the animals were 2.7–10.7 mm and the fourth moult occurred when worms were about 11.6 mm. The smallest female with eggs was 25 mm in length.

Munday and Gregory (1974) found larval stages in granulomas on the viscera of the wombat (*Vombatus wisinus*) and used them to infect devils, indicating that in natural conditions transmission of *B. tasmaniensis* is dependent on marsupial intermediate hosts in the same way that other species in placental animals depend on small mammals for their transmission.

B. transfuga (Rudolphi, 1819)

B. transfuga is a parasite of the intestine of bears and pandas (*Ailurus fulgens, Melursus ursinus, Thalarctos maritimus, Ursus* spp.). Sprent (1952a) gave embryonated eggs to white mice. Second-stage larvae apparently reached the lungs on the first day and encapsulated larvae were noted in the intestinal wall in 8 days. Thereafter, encapsulated larvae were observed on the rectum, caecum and subcutaneous tissues of the neck, back and intercostal regions.

Papini *et al.* (1993) gave embryonated eggs of *B. transfuga* from a polar bear to 3-day-old chicks (*Gallus domesticus*). Numerous larvae were noted in the liver of the chicks for 30 days but a few larvae reached the brain. Papini and Casarosa (1993) and Papini *et al.* (1994, 1996) gave larvated eggs to mice and rabbits, in which they subsequently found larvae in various tissues but not in the brain. Eggs inoculated into the peritoneal cavity of mice hatched and larvae migrated to the intestine, liver, lungs, brain and carcass (Papini *et al.*, 1995). Some eggs hatched in subcutaneous sites after inoculation under the skin and larvae were found in various sites in the body, including the brain.

Lagochilascaris

Species of the genus are small to moderately sized ascaridoids reported in lions (*L. major*), cougars (*L. buckleyi*), opossums (*L. turgida, L. sprenti*) and humans (*L. minor*). In unusual hosts, and occasionally in the usual host, members of the genus have a tendency to leave the gut and mature in abscesses and granulomas in various regions of the body (for reviews see Sprent, 1971; Bowman *et al.*, 1983; Smith *et al.*, 1983).

L. minor (Leiper, 1909)

In cats and dogs, adult worms occurred in inflamed lesions in the larynx, pharynx and rhinopharynx (Volcán *et al.*, 1992). Eggs passed in faeces of the host were unsegmented and 66 and 56 µm in size, with thick, sculptured shells. Infected larvae appeared in eggs in 2 weeks at room temperature when incubated in water (Volcán *et al.*, 1982). Cats given eggs orally did not become infected (Campos *et al.*, 1992).

Volcán and Medrano (1990) gave embryonated eggs of *L. minor* from a human infection to agoutis (*Dasyprocta leporina*) and examined them 14–46 days later. Third-stage larvae were found in granulomas in the skeletal muscles; in 46 days three larvae were an average of 7.02 mm in length.

Volcán *et al.* (1992) and Freire Filha and Campos (1992) gave eggs to white mice and reported that larva were found in the lungs and later in the superficial adipose tissue associated with muscles, where they became encapsulated by 14 days. In 40 days larvae were in advanced third stage and 4.4–7.6 mm in length. Cats fed mice infected 40 days earlier passed eggs in 17–20 days. At necropsy, adult worms were found on the surface of the larynx, pharynx and rhinopharynx. One cat developed a fistula in the posterior wall of the pharynx. Also, destructive lesions at the base of the tongue consisted of sacs with inflamed walls containing mucus and eggs as well as adult worms moving in and out of the sac. Cats given embryonated eggs or mice inoculated with eggs 3 days earlier remained uninfected.

L. minor has been reported sporadically in humans in South America and Mexico, where it is associated generally with purulent abscesses in the region of the ear, neck, jaw, orbit, mastoid process and retropharyngeal tissues (Vargas-Ocampo and Alvarado-Aleman, 1997; Moraes *et al.*, 1983, 1985; Botero and Little, 1984; Osterburg, 1992; Aguilar-Nascimento *et al.*, 1993; Ollé-Goig *et al.*, 1996; Calvopiña *et al.*, 1998). These abscesses, containing mature worms and advanced larvae (often in large numbers) and eggs, may break through the skin. Worms have also been reported in the lungs and brain (Rosemberg *et al.*, 1986). Adult female worms were 6–20 mm and males 5.0–17.0 mm in length (Sprent, 1971).

L. minor is apparently a heteroxenous ascaridoid of carnivores which will reach the infective third stage or early fourth stage in rodents. Humans can also serve as a definitive host but the route of infection is still unknown. The animal research reported above suggests that humans might acquire this strange nematode from eating uncooked or poorly cooked flesh of some rodents serving as intermediate hosts. There is, however, no explanation for the large number of worms at various stages of development reported in human infections. Wild hosts of *L. minor* may include ocelots (Brenes-Madrigal *et al.*, 1972) and the wild dog (*Speothos venaticus*) (Volcán and Medrano, 1991).

L. sprenti Bowman, Smith and Little, 1983

This species was found in the stomach of opossums (*Didelphis virginiana*) in Louisiana. Smith *et al.* (1983) cultured eggs from worms found in opossums for 40 days at room temperature and used them to infect mice; the eggs contained larvae (presumably second-stage) 410–460 µm in length. Larvae were found in the mucosa of the small intestine of the mice in the first 4 h postinfection. In 4.5 h they were recovered from the liver and in 24 h from the lungs. In 2–4 days larvae were found in the lungs, subcutaneous connective tissues and the skeletal musculature. After 6 days larvae were found almost exclusively encapsulated in the musculature. By 75 days larvae were 3.7–4.7 mm in length. Abscesses with adult worms were found in the head, neck and forelegs of a few of the infected mice. Viable larvae were also found in the skeletal musculature of hamsters, jirds, rats, rabbits, rhesus monkeys and stumptail macaques given larvated eggs.

No larvae were found in six opossums given embryonated eggs and examined 1–128 days later. Developing larvae and adult worms were found in nine of 13 adult opossums fed carcasses of infected mice. One opossum became patent (apparently at 21 days postinfection) and continued to pass eggs for at least 131 days. The authors reported adult worms, embryonated eggs and free larvae in a tumour-like mass on the mesentery of a naturally infected opossum.

Parascaris

P. equorum (Goeze, 1782)

P. equorum is an important cosmopolitan parasite of Equidae. Excellent reviews have been published by Clayton (1978, 1986) and Drudge and Lyons (1983). Infections occur mainly in suckling and weanling foals; prevalence and intensity decrease markedly with host age.

Eggs are spherical with roughened shells and have a sticky surface, which causes them to adhere firmly to each other and objects in the environment. Eggs embryonate to sheathed second-stage larvae in 10 days at 25–35°C. Embryonated eggs can remain viable for years and thus pastures or stables have only to be contaminated with eggs once every few years for the perpetuation of the parasite. Foals become infected by ingesting eggs contaminating their food and water and the environment generally. Lyons *et al.* (1996) demonstrated the transmission of *P. equorum* to a foal confined to a parasite-free stall; they wondered if 'recycling' of eggs during coprophagia might have occurred and resulted in exposure of the foal to embryonated eggs. Attempts to demonstrate transmammary (lactogenic) or prenatal transmission have apparently not been successful. Eggs hatched in the gut of foals and larvae invaded the wall of the small intestine and migrated by the portal system to the liver, which they reached in 2 days. In 7–14 days larvae migrated to the lungs. From 15 to 23 days larvae invaded the alveoli, migrated up the respiratory system to the pharynx and were swallowed. In 10–12 weeks postinfection worms matured mainly in the duodenum and proximal jejunum. Males were about 10 cm and females about 15 cm in length at this time. It has been estimated that one mature female can produce 100,000 eggs daily. The prepatent period was from 72 to 115 days. Foals are normally infected soon after birth and maintain infections for 6–12 months, after which they are lost because worms become senescent and immunity develops.

Pilitt *et al.* (1979) described the early stages in horses.

Toxascaris

Members of the genus have narrow cervical alae and lack both an *area rugosa* and an oesophageal ventriculus. Eggs are smooth-shelled (cf. *Toxocara* and *Baylisascaris*). Members of the genus occur in canids and felids.

T. leonina (Linstow, 1902)

This is a cosmopolitan parasite of the gut of Canidae and Felidae, including cats, dogs, foxes, wolves, lynx, leopards, jaguars, tigers and lions (Sprent, 1959a). Fülleborn (1922) fed embryonated eggs to a dog and 13 days later found partially encapsulated larvae (some moulting) in the liver, lungs and gut wall; larvae were 210–700 μm in length. Wright (1935) reported that eggs hatched in the duodenum and larvae completed the second moult and penetrated the small intestine. Larvae grew and then emerged into the lumen in 9–10 days where the third moult occurred in 18 days. Matoff (1949, cited in Mozgovoi and Shakhmatova, 1973) gave eggs to mice and demonstrated that larvae established themselves in the muscles, grew and became encapsulated, and remained alive for more than 5 months. Sprent (1952a) reported that larvae invaded the intestine of mice, and migrated into somatic tissues in a week. Cats could be infected by feeding them mice with larvae in the tissues (Sprent, 1953b).

Matoff and Vasilev (1958b) also noted that larvae developed in tissues of rabbits, mice and rats. These larvae matured in the intestinal lumen of dogs without a preliminary migration in the intestinal wall, which occurred when second-stage larvae from eggs were used for infection.

Sprent (1959a) compared *T. leonina* from cats and dogs. The parasite from dogs could be transmitted to cats and vice versa. Dogs and cats could be infected by: (i) oral inoculation of embryonated eggs containing second-stage larvae, and by (ii) feeding mouse tissues containing third-stage larvae. Second-stage larvae hatched from eggs in the gut of cats and invaded the intestine, where they moulted to the third stage when they were 420–560 μm in length. The third moult took place in the gut, when worms were 560–740 μm in length, and the final moult occurred 6 weeks postinfection, when worms were 7.7 mm in length. Eggs appeared (presumably in faeces) in 74 days. Attempts to infect cats with larvae from tissues of mice were not highly successful and only fourth-stage larvae were attained. Six dogs were orally inoculated with eggs; fourth-stage larvae and adults were found in three dogs examined 38–49 days postinfection. A dog given larvae in mouse tissue passed eggs in its faeces 8 weeks postinfection and continued to do so for 36 weeks. Okoshi and Usui (1968) claimed that dogs were easily infected with the canine strain of *T. leonina*, but not cats; whereas a feline strain developed readily in both dogs and cats with a prepatent period of 50–63 days. Also, eggs from cheetahs, lions and tigers infected dogs and cats.

Dubey (1967) examined mice infected with eggs of *T. leonina* 1–20 days postinfection. Larvae were confined to the gut and mesentery for the first 6 days and were present in the musculature from 7 to 29 days. Third-stage larvae were found in the muscles as early as 9 days. Steffe (1983) reported transmammary transmission of larvae to young mice.

Kudryavtsev (1971, cited in Mozgovoi and Shakhmatova, 1973) infected dogs with eggs and with larvae from mice. In dogs infected directly with eggs the prepatent period was 42–49 days. In dogs fed mouse tissues with third-stage larvae the prepatent period was 11–15 days earlier.

According to Kudryavtsev (1974) larvae reached the infective stage in 8–9 days in eggs at 27°C. In rodents, larvae moulted a second time 7–8 days postinfection. Larvae (third-stage) from rodents given to dogs and foxes matured in 18 days. In contrast, larvae given in eggs required 14–15 additional days to mature.

Prokopic and Figallová (1982a,b) reported experiments comparing the behaviour of *T. leonina*, *Ascaris suum*, *Toxocara canis* and *T. cati* in white mice.

Subfamily Toxocarinae

The subfamily consists of *Paradujardinia* of sirenians, *Porrocaecum* of terrestrial birds and *Toxocara* of terrestrial mammals.

Porrocaecum

Members of the genus are widely distributed parasites of the intestine of birds. The most primitive mode of transmission occurs in species such as *P. angusticolle* and *P. talpae* of raptors. Eggs passed in faeces of the raptor definitive host embryonate to first-stage larvae, which moult to the second stage. Eggs must be ingested by earthworms, in which they hatch. Second-stage larvae invade the blood vessels of the annelid and develop to the third stage. Shrews, and perhaps other small mammals that consume earthworms, serve as paratenic hosts which transfer the parasite to the carnivorous definitive hosts. Species in ducks, passerines and other birds which do not normally consume small mammals have eliminated the vertebrate paratenic host in their transmission and the definitive host becomes infected directly from ingesting earthworms. Osche (1959) noted the morphological similarities within this latter group of worms, consisting of *P. crassum*, *P. ensicaudatum* and *P. semiteres*, as well as the fact that they were transmitted directly through the agency of earthworms; he referred to them as the '*ensicaudatum* group'.

P. angusticolle (Molin, 1860)

This species occurs in the intestine of hawks (e.g. *Accipiter* spp., *Buteo* spp., *Milvus* spp.) throughout the world. Mozgovoi and Shakhmatova (1979) studied the transmission to the kite *Milvus korschun* in the Altai Territory of the CIH. Eggs developed on damp filter paper at 18–20°C and moulted in 5–7 days. Earthworms were allowed to ingest eggs. The latter hatched in the intestine and larvae migrated to the blood vessels, where the second moult occurred. Third-stage larvae persisted for long periods in the earthworms. Fifteen to 20% of earthworms taken from soil under trees frequented by raptorial birds contained one or two larvae of *Porrocaecum*. Attempts to infect white mice, guinea pigs and wild shrews were not successful. However, shrews and moles in the locality where hawks roosted had larvae in the intestine and, in the case of shrews, also under the skin. A kite chick was given 18 larvae from shrews; 6 days later immature *Porrocaecum* sp. were found.

P. crassum (Deslongchamps, 1824)

P. crassum is a fairly common parasite of waterfowl, especially ducks (Anatidae). According to Mozgovoi (1952) eggs kept at 22.0–32.5°C on moist filter paper developed into larvae in a few days and larvae then moulted. He fed eggs to ducks and noted that they passed unchanged through the gut of the birds. Attempts to infect insects, fish, snails and crustaceans were not successful. However, eggs given to earthworms hatched and larvae migrated to the blood vessels, where they grew, moulted and attained the infective stage. Earthworms were given to ducks and the larvae freed from the earthworms entered the submucosa of the gizzard. They remained in the gizzard for 7 days before entering the intestine, where they matured in 3 weeks. According to

Mozgovoi (1954) *P. crassum* is most frequently found in ducks in southern Russia, where infections are highest in May and June, especially in birds 1–3 months old. Worms survived the winter in earthworm intermediate hosts.

Supriaga (1972) reported that the following species of earthworms could serve as intermediate hosts: *Allolobophora caliginosa, A. dubiosa, A. jassyensis, A. terrestris, Criodrilus lacuum, Dendrobaena mariupoliensis, D. schmidti, D. subrubicunda, Eisenia fetida, E. rosea, E. ukrainae, Eiseniella teraedra, Eophila montana* and *Octolasium complanatum*. Eggs which had embryonated contained second-stage larvae. Eggs hatched in the intestine of earthworms and larvae migrated to the blood vessels, where they grew and moulted to the third stage in 2.5 months at 23–30°C.

P. ensicaudatum (Zeder, 1800)

P. ensicaudatum is a common, cosmopolitan parasite of the intestine of passerine birds, including thrushes, starlings and grackles. Cori (1898) identified larvae found in the blood vessels of *Lumbricus terrestris* as *Spiroptera turdi*, now known as a synonym of *P. ensicaudatum*, a common parasite of *Turdus* spp. in Europe (Hartwich, 1959). Since then *P. ensicaudatum* larvae have been reported to develop to the third and infective stage in numerous species of earthworms, including *Allolobophora caliginosa, A. chlorotica, Eisenia fetida, Eudrilus eugeniae* and *L. terrestris* (see Osche, 1955; Bohm and Supperer, 1958; Baer, 1961; Levin, 1961; Demshin, 1975; McNeill and Anderson, 1990a).

Several authors carried out preliminary experimental studies of *P. ensicaudatum*, notably Levin (1956, 1961), Rysavy (1958), Baer (1961) and Jogis (1967). The most complete study, using substantial numbers of susceptible laboratory-reared avian hosts, was carried out by McNeill (1988) and McNeill and Anderson (1990a,b) (Fig. 5.6).

Eggs of *P. ensicaudatum* were 57–80 × 92–104 μm in size, with a thick shell and a surface pattern of interconnecting ridges, giving rise to a pitted appearance. Cytoplasm completely filled the interior of the eggs (McNeill and Anderson, 1990a). Eggs at room temperature developed into second-stage larvae in 25 days; larvae retained the cuticle of the first moult. Larvated eggs remained viable at 5°C for at least 3 years. Second-stage larvae were long (690–735 μm) and slender, with lateral alae and a developing oblong ventriculus. The genital primordium was not detected.

The earthworms *E. fetida* and *E. eugeniae* proved to be suitable intermediate hosts when allowed to ingest larvated eggs with their food (McNeill and Anderson, 1990a). In 21 days second-stage larvae were found mainly in the ventral blood vessel. Larvae moulted to the third stage in 23 days; they grew rapidly during the first 4 weeks of infection and more slowly thereafter until the maximum size (3.50–4.04 mm) was reached in 8–10 weeks.

Attempts to infect starlings (*Sturnus vulgaris*) with third-stage larvae after 4–6 weeks of development in earthworms were unsuccessful. However, starlings given larvae after development for 7–8 or more weeks in earthworms were consistently infected. Third-stage larvae had a rounded anterior end and a tapered tail. Lateral alae were present and lips were absent. The ventriculus was well developed and the genital primordia consisted of only two to four cells.

In starlings, third-stage larvae invaded the koilin matrix of the gizzard within 2 h and moulted there in 36 h, thus avoiding a host response by remaining in a relatively inert environment (NcNeill and Anderson, 1990b). Fourth-stage larvae moved directly

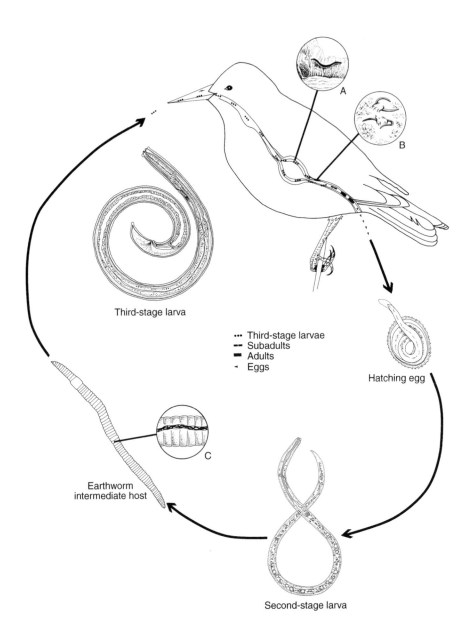

Fig. 5.6. Development and transmission of *Porrocaecum ensicaudatum*: (A) larva in koilin matrix of gizzard; (B) fourth-stage larvae embedded in intestinal mucosa; (C) third-stage larvae in ventral blood vessel of earthworm. (Original, U.R. Strelive.)

from the gizzard into the mucosa of the duodenum 48 h postinfection. The final moult occurred in the intestinal mucosa 11 days postinfection and adult worms entered the intestinal lumen in 28 days. Eggs were first detected in faeces of starlings 33 days postinfection and the longevity of the infection was approximately 70 days. Starlings were susceptible to repeated infections and the number of adult worms that matured in the intestine was independent of the number of larvae ingested.

All passerine birds given third-stage larvae from earthworms became infected, including starlings, robins (*Turdus migratorius*), red-winged blackbirds (*Agelaius phoenicius*), house sparrows (*Passer domesticus*), zebra finches (*Poephila guttata*) and brown-headed cowbirds (*Molothus ater*). However, attempts to infect galliformes (*Gallus gallus, Colinus virginianus*), anseriformes (*Anas platyrhynchos*) and falconiformes (*Falco sparvarius*) were not successful. It was concluded that *P. ensicaudatum* is specific to passerine birds. However, although red-winged blackbirds, cowbirds and house sparrows could be infected experimentally, they were not found infected naturally, presumably because they do not ingest earthworms. The findings support those of Jogis (1970) who reported that *P. ensicaudatum* only matured in Sturnidae and Turdidae.

Starlings, robins and grackles (*Quiscalus versicolor*) are commonly infected in southern Ontario. Examination of large numbers of starlings and grackles revealed a seasonal pattern of parasite acquisition and loss which was common to both the migratory grackles and non-migratory starlings. Individuals of both species were uninfected in early spring but became infected as soon as the ground thawed and earthworms became available. Prevalence was similar in adult and young-of-the-year birds but intensity was significantly higher in the latter, presumably because of their higher food intake. Prevalence and intensity declined throughout late summer and fall and dropped to zero in winter as the population of adult worms in the birds became senescent and died.

P. semiteres (Zeder, 1800)

This species (syn. *P. heteroura*) has been reported in various birds of the genera *Calidris, Charadrius, Lanius, Philomachus, Turdus* and *Vanellus* in Europe. Mozgovoi and Bishaeva (1959) and Jogis (1967) reported that it developed to the infective stage in *Eisenia fetida* and *Lumbricus rubellus*. Moravec (1971b) found larvae (4.38–4.84 mm in length) in these earthworm species living in river banks in the Czech Republic.

P. talpae (Schrank, 1788)

Osche (1955) found larvae (7.94–8.54 mm in length) of *P. talpae* encapsulated on the intestine and mesentery of the shrews *Neomys fodiens, Sorex araneus* and *Talpa europaea* in Germany. Osche (1955) gave larvae from *N. fodiens* to a 'Mausebussard' (presumably *Buteo buteo*) believed to be free of nematodes. Eggs were found in the buzzard's faeces 21 days later and five adult worms were found in the small intestine at necropsy 24 days postinfection. Worms were said to resemble *P. angusticolle*, which was then placed into the synonymy of *P. talpae* – a decision not accepted by Hartwich (1959).

Toxocara

As indicated in the introduction to the ascaridoids, members of the genus *Toxocara* have abandoned heteroxeny but the larval stages will wander in the tissues of invertebrates and vertebrates which may serve as paratenic hosts. The tendency of larvae to wander in tissues has given rise in the genus to prenatal and transmammary transmission.

T. canis (Werner, 1782)

T. canis is a common parasite of the intestine of dogs and, less commonly, cats. Many aspects of *Toxocara canis* are reviewed in Lewis and Maizals (1993). Fully developed larvae appeared in eggs in 9 days when incubated at 26–30°C (Schacher, 1957) and in 11–18 days when kept at 30°C (Noda, 1961). Larvae were 348–532 μm in length according to Schacher (1957) and 360–434 μm according to Nichols (1956). However, Brunaská *et al.* (1995), using ultrastructural methods, reported two moults in the eggs and Bowman *et al.* (1993) published a detailed ultrastructural study of infective larvae. According to Feney-Rodriguez *et al.* (1988) *T. canis* eggs, like those of *A. suum*, will not embryonate in darkness, although those of *Toxascaris leonina* will do so. Eggs hatched in the intestine of mice and larvae underwent a somatic migration and eventually became encapsulated in 12 days, mainly in the subcutaneous tissues of the back, legs and chest (Sprent, 1952a). Larvae did not grow or develop in mice (Nichols, 1956). Lee (1960), Higashikawa (1961), Burren (1971) and Dunsmore *et al.* (1983) also reported larvae in the liver, brain (especially the cerebellum), kidneys and heart of infected mice. Larvae are known to invade the tissues of many animals, including birds and humans. Recently, Pahari and Sasmal (1990) showed that larvae persisted for long periods mainly in the liver of Japanese quail (*Coturnix japonica*) given embryonated eggs. Infection of mice with larvae of *T. canus* reduced aggressive behaviour and increased the level of flight and defensive behaviour (Cox and Holland, 1998). Apparently the behaviour of mice is influenced by the number of larvae in the brain. The tendency of larval *T. canis* to invade tissues accounts for its importance as an actual or potential pathogen of humans, resulting in ocular and other disturbances, depending on the number of worms present and their location in the body (Wilder, 1950; Smith and Beaver, 1955; Beaver, 1956; Sprent and Jones, 1977).

Dogs became infected by ingesting embryonated eggs or animal tissues containing second-stage larvae (Noda, 1957; Sprent, 1957). However, the behaviour of larvae differed, depending on the age and sex of the dog (Fig. 5.7). In young dogs (less than about 3 months) eggs hatched in the duodenum; larvae entered the lymphatics and venous capillaries, which carried them to the liver in 1–3 days (Webster, 1956, 1958). Larvae then passed to the heart via the hepatic vein and vena cava. They passed to the lungs, where they grew and moulted. When they were about 0.80–0.95 mm in length, the larvae (some of which were passing up the trachea and into the oesophagus) moulted to the third stage. In the gut the third moult occurred in 13 days and the fourth in 19–27 days. Eggs first appeared in the faeces 4–5 weeks postinfection.

In older dogs, however, larvae rarely passed from the lungs to the trachea (**lung–tracheal migration**) (Noda, 1957, 1959). Most entered the pulmonary veins and were carried back to the heart and pumped into the tissues of the dog. They remained in the same stage even after arriving in the tissues. The importance of this **somatic migration** (as distinct from the lung–tracheal migration) is that it facilitates prenatal transmission, i.e. larvae pass from the pregnant bitch to the developing fetuses (Fülleborn, 1921a,b; Shillinger and Cram, 1922; Augustine, 1927; Yutuc, 1949; Stone and Girardeau, 1968; Burke and Roberson, 1985). The bitch appears resistant to the parasite and larval development to the adult stage is inhibited (Scothorn *et al.*, 1965). Mozgovoi and Nosik (1960) concluded that larvae remained in the fetal liver until birth, when they continued their migration to the lungs, trachea and gut, growing and

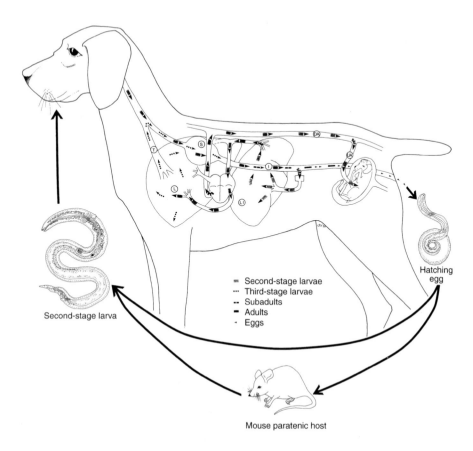

Fig. 5.7. Development and transmission of *Toxocara canis*. DA = dorsal aorta; I = intestine; L = lung; LI = liver; S = stomach; T = trachea; UA = umbilical artery. Transmammary transmission not included. (Original, U.R. Strelive.)

moulting *en route*. In addition to prenatal transmission, transmammary transmission may sometimes occur but it is less important than prenatal transmission (Burke and Roberson, 1985).

Dubinsky *et al.* (1995) found anti-*Toxocara* antibodies in small mammals, mainly in urban and rural areas within the Slovak Republic. The most significant hosts were *Mus musculus*, *Apodemus agrarius* and *Micromys minutus*. *Toxocara* larvae were found in 6.3% of 476 small mammals examined, indicating that they may play a significant role in transmission.

T. cati (Shrank, 1788)

T. cati (syn. *T. mystax*) is a cosmopolitan parasite of felines, including domestic cats and larger cats such as cougars and tigers. According to Nichols (1956) second-stage larvae in eggs were morphologically indistinguishable from those of *T. canis* except that those of *T. cati* were slightly wider (see also Sprent and Barrett, 1964). Sprent (1956) found larvae in tissues of earthworms and cockroaches exposed to larvated eggs. In mice, larvae were found in the liver, lungs and muscles. Also, larvae were found in tissues of dogs, lambs and chickens given eggs.

Eggs given to cats hatched and second-stage larvae entered the stomach and intestinal wall (Sprent, 1956). Development was extremely variable and difficult to interpret. In 3 days larvae appeared in the liver and lungs. In 5 days some larvae were found in the trachea. In 10–13 days numerous larvae appeared in the lungs and some moulting larvae were found in the stomach wall. By 15 days larvae were found in the stomach contents. By 19 days fourth-stage larvae first appeared. In 42 days small larvae were present in muscles and lungs but fourth-stage larvae and adults were present in the stomach contents. In cats given mice containing second-stage larvae, larvae were found in stomach wall in 2–3 days. In 6 days moulting larvae were in the stomach wall. In 10 days larvae were 0.50–1.24 mm long and some were moulting. Fourth-stage larvae and young adults appeared in the intestine in 21 days (cf. *Toxascaris leonina*). In 35–42 days adult worms were present.

Dubey (1968) showed that larvae of *T. cati* from eggs migrated to the liver, lungs, kidneys, and brains of mice and finally settled in musculature within 24 h. By day 5, larvae had left the liver and lungs and by the third week almost all larvae were in the musculature. Dubey (1967) estimated that an experimentally infected cat passed an average of 19,000 eggs per day and this output continued for at least 3 months.

Swerczek *et al.* (1971) experimented with kittens delivered by Caesarean section and raised colostrum free. Kittens allowed to nurse on infected queens became infected. Kittens not allowed to nurse did not become infected. Larvae were recovered from mammary glands and milk of infected queens. The results demonstrated both transmammary transmission and also lack of prenatal transmission in cats.

T. pteropodis (Baylis, 1936)

T. pteropodis is a parasite of fruit bats (*Pteropus alecto, P. geddiei* and *P. poliocephalus*) in Queensland, Australia and Vanuatu. Prociv (1983) found larvae in the liver of adult bats and in the mammary glands of lactating females, as well as developing larvae in the intestine of young bats as early as 2 days. Juvenile bats commenced to pass eggs when they were 2 months old and expelled the worms after weaning in about 5 months. Bats maintained on milk and sugar usually retained their worms for several weeks but tended to expel them when transferred to a fruit diet.

Eggs were unsegmented when passed in faeces of juvenile bats. They embryonated in 10 days at 20–26°C on leaves and even on glass slides. Two moults apparently occurred in the eggs. Larvae from eggs were 220–450 μm in length. Larvae from eggs given to fruit bats migrated to, and remained in, the liver, where they grew to a length of 550–730 μm in 6 months. Worms in the mammary glands of parturient females were 705–955 μm in length. It was concluded that fruits on which bats feed became contaminated with eggs, which were ingested along with the fruit. Larvae migrated to the liver of the bat and those in females migrated to the mammary glands about the time of parturition. Ingested in milk by the juvenile bats, larvae grew rapidly in the intestine of the latter and produced eggs in about 2 months.

In subsequent experiments Prociv (1989a) reported that four pups of an experimentally infected female *Pteropus poliocephalus* became infected after suckling. In two of the pups, eggs first appeared in the faeces 35–48 days postpartum. In the other two pups, infertile eggs were passed in 37 and 42 days; in one pup, fertile eggs started to be passed after an additional 23 days while the other expelled a single female after passing unfertilized eggs for about 4 weeks. The precise duration of patency is unknown but under natural conditions juvenile bats are free of nematodes when about 5 months old (Prociv, 1989b,c).

It is postulated that juvenile bats defaecate while attached to their mothers, especially when nursing. The mothers groom the young and ingest faeces contaminated with eggs of *T. pteropodis*. The parents and juveniles pass these unembryonated eggs into the environment; after developing, the eggs are infective and may reach the bats through accidental coprophagy.

Prociv (1985) and Prociv and Brindley (1986) reported that larvae given to white mice migrated to the liver. The possibility that hepatitis and other clinical signs in humans on Palm Island, Queensland, were caused by larval *T. pteropodis* has been explored (Moorhouse, 1982; Prociv *et al.*, 1986).

T. vitulorum (Goeze, 1782)

T. vitulorum (syn. *Neoascaris vitulorum*) is mainly a parasite of species of Asian *Bos* spp. and *Bubalus bubalis* but it has also been reported in goats and sheep (Warren, 1970; Roberts, 1993b). Adult parasites are found in the intestine of calves but rarely in adult animals. Warren (1971) noted that most early parasitologists believed *T. vitulorum* was transmitted prenatally, since mature worms were reported in calves about 2 weeks after birth (Boulenger, 1922; Griffiths, 1922; Schwartz, 1925; Vaidyanathan, 1949). Also, a number of workers reported infections in young animals (calves, lambs and kids) following experimental oral infections of their mothers and assumed that prenatal transmission had occurred (Herlich and Porter, 1953; Srivastava and Mehra, 1955; Matoff and Vasilev, 1958a, 1959; Vasilev, 1959; Cvetković and Nevenić, 1960). In addition, most attempts to infect adult cattle and older calves, as well as sheep and goats, by oral inoculation of larvated eggs were unsuccessful.

According to Warren (1971) development to the second stage in the eggs took about 50 days at 22°C. Larvae expressed from eggs were 345–398 μm in length with a well-defined ventriculus. The optimum temperature for eggs to develop in water was 10–12 days at 28–30°C (Chauhan and Pande, 1981). A second moult may occur in the egg but this has not yet been confirmed by ultrastructural studies. Under suitable conditions embryonated eggs can persist for several months and perhaps years (Roberts, 1989). In one of several experiments Warren (1971) gave larvated eggs to pregnant cows. The dams gave birth and calves were allowed to suckle. Eggs of *T. vitulorum* appeared in faeces of calves 26–33 days after birth and adults were found at necropsy. He did not find larvae in colostrum collected from infected cows before the birth of calves but they were found consistently in milk collected 2–18 days after birth. Larvae in milk were 0.76–1.47 mm in length and considered to be third stage. Calves withdrawn from dams at birth did not become infected but foster calves allowed to suckle infected cows did so. The results, confirmed by Mia *et al.* (1975), showed that *T. vitulorum* is transmitted through the milk of the host.

Cows acquire infection by ingesting eggs contaminating the environment. Eggs hatched in the small intestine in 2–8 h; the larvae invaded the mucosa, entered capillaries and were carried in the portal vessels to the liver (Roberts, 1990a). Some larvae reached the lungs, muscles, kidneys and brain. Larvae grew about 10% but apparently did not moult. These larvae may persist in the tissue of a cow for at least 5 months after an initial infection but the precise time is unknown (Roberts, 1993a).

In the pregnant cow, larvae grew to about 500–600 µm in length in the liver and lungs 1–8 days before parturition and some migrated to the mammary glands, where they grew to about 0.75–1.47 µm in length and then passed into the milk some 7 days after parturition, when they were available to the suckling calf (Roberts, 1990a,b) (cf. the hookworm *Uncinaria lucasi*). Larvae persisted in milk for about 8 days after parturition (Roberts, 1993a).

Worms established themselves in the duodenum of calves and apparently moulted on day 12, when the calf was 10–12 days old (Warren, 1971; Mia *et al.*, 1975). The prepatent period in calves was 22.3 ± 1.6 days according to Roberts (1989); egg output reached a peak in about 5 weeks and declined to zero in about 3 months. Peak egg production was about 100,000 eggs per mature female per day. Adult worms eventually became senescent in 35 ± 12 days and were passed in faeces of the host (Roberts, 1990a). Excellent reviews of *T. vitularum* have been published by Roberts (1990b, 1993b).

Subfamily Angusticaecinae

The subfamily is confined to terrestrial reptiles, including tortoises, lizards and snakes, but most are parasites of the latter two. A few species, which are probably captures from reptiles, have been reported in amphibians. Eggs are spherical with pitted surfaces and they generally embryonate to the second stage, although in some species (e.g. *Hexametra boddaertii*) there are said to be two moults in the egg. There is limited experimental evidence that eggs will hatch in certain invertebrates (e.g. earthworms, snails, mosquito larvae) and persist in the tissues of these organisms, which could then serve as paratenic hosts; however, there is no field evidence that paratenesis of this type is significant in transmission. Eggs of the various species will hatch in a variety of potential or actual vertebrate intermediate hosts including frogs, rodents and, in Australia, small marsupials (see, for example, *Amplicaecum robertsi* and *Ophidascaris labiatopapillosa*). Second-stage larvae invade the tissues of the vertebrate intermediate host and grow to the stage infective to the definitive host, usually the third. Growth of larvae in the intermediate host can be spectacular and exceed that which eventually will take place in the gut of the definitive host as they reach adulthood (see, for example, *Hexametra boddaertii* and *Amplicaecum robertsi*). The marked precocity in vertebrate intermediate hosts of many ascaridoids of snakes may be related to the intermittent feeding of the final host. Since the worms feed on food consumed by the snake, there are probably advantages in maturing early to take full advantage of the food provided by the ingested intermediate host.

Some interpretations about the growth, moulting patterns and movements of larval and subadult ascaridoids in definitive hosts seem to be based, unfortunately, on limited data complicated by the use in feeding experiments of wild-caught animals of uncertain age and parasitological history.

Amplicaecum

The 11 species reported in the genus are found mainly in reptiles. *A. robertsi* is the only species known in pythons (Sprent and Mines, 1960). Two species have been reported in amphibians: *A. numidicum* has been reported only in frogs and may be a capture from reptiles; *A. involutum* is reported from frogs and toads as well as reptiles, raising the possibility that it is a parasite essentially of snakes and reptiles which develops precociously in amphibian intermediate hosts.

A. robertsi Sprent and Mines, 1960
A. robertsi occurs in the oesophagus and stomach of carpet pythons (*Morelia spilotes variegatus*) in Queensland, Australia (Sprent, 1963a,b). Worms were clustered in the host and the anterior fifth of the body of each worm was inserted into a common opening in a nodule in the wall of the gut; worms were often associated with *Ophidascaris moreliae* and they fed on ingesta in the stomach.

Eggs cultured on moist charcoal were fully embryonated in 10–12 days at 21°C. Eggs were 79–105 × 66–95 μm in size with a pitted surface. Fully developed second-stage larvae were 380–580 μm in length, with tapered, pointed tails and an obvious genital primordium in the posterior fifth of the body.

Eggs with second-stage larvae hatched in mice, rats and guinea pigs and larvae invaded the liver, where they developed to the third stage; the second moult occurred in 5–6 days, when larvae were about 520–690 μm in length. Third-stage larvae grew markedly after the second moult and measured 56–78 mm in 274 days. The excretory system and the gut were highly developed in large third-stage larvae. The caecum appeared when larvae were 3–5 mm in length and gradually increased in length as the latter grew. The poorly developed genital primordium was elongate and about 200 μm in length. Growth of larvae was most marked and rapid in indigenous hosts (*Melomys cervinipes*, *M. lutillus* and *Rattus assimilis*).

Large third-stage larvae (> 26 mm) from the liver of mice and rats infected carpet pythons. The third moult occurred over a period of about 6 weeks and the fourth in 56–112 days. Lengths of the various stages were highly variable and there was considerable overlap in the lengths of larger third-stage larvae from the liver of mice and rats (56–78 mm) and adult worms found in experimentally infected carpet pythons (combined ranges of males and females 34–181 mm), indicating that the most important growth of *A. robertsi* took place precociously in the liver of the intermediate host. Thus, most of the energy utilized by fourth-stage larvae and adult worms was expended in reproduction and not body growth. This is in contrast to *Ophidascaris moreliae* (see below) in which most growth took place in the reptile definitive host.

Third-stage larvae of *A. robertsi* have been found in the liver and peritoneal cavity of naturally infected mammals in Queensland, namely tree kangaroos (*Dendrolagus lumholtzi*), bandicoots (*Isoodon macrourus*, *Perameles nasuta*), phascogales (*Phascogale tapoatafa*), wallabies (species not indicated), bats (*Rhinolophus* sp.), rats (*Rattus assimilis*,

Uromys caudimaculatus) and opossums (*Trichosurus caninus*) (Sprent, 1963b). Larvae were 5–100 mm in length.

Attempts to infect mealworms and cockroaches with eggs containing second-stage larvae were unsuccessful although eggs hatched in the crop. Some second-stage larvae were found in snails (*Physastra* sp.) and earthworms allowed to ingest eggs. Second-stage larvae failed to develop in cold-blooded vertebrates (fish, frogs, lizards and snakes) although they persisted in the tissues for several days. Larvae were found in the liver of finches, fowl, budgerigars, canaries, pigeons and a black duck given larvated eggs. These larvae were two to three times longer than second-stage larvae from eggs.

Hexametra

Species of *Hexametra* occur in the oesophagus, stomach, intestine and cloaca of lizards (four species) and snakes (? eight species) (Mozgovoi, 1953; Sprent, 1978). The genus is recognized, in part, by the presence of six uterine branches.

H. angusticaecoides Chabaud and Brygoo, 1960

This is a parasite of the gut of snakes (*Acrantophis dumerili*, *Ithycyphus miniatus*, *Lioheterodon madagascariensis*, *Madagascarophis columbrina* and *Sanzinia madagascariensis*) in Madagascar (Ghadirian, 1968). *H. angusticaecoides* was originally described on the basis of adult worms from the intestine and immature stages from the body cavity and intestine of chameleons (*Chamaeleon oustaleti*, *C. verrucosus*) (Chabaud and Brygoo, 1960; Chabaud *et al.*, 1962). Eggs embryonated in 15 days at 35°C and the second-stage larvae hatched spontaneously. Larvae ingested by larval *Culex* and *Chironomus*, as well as *Periplaneta orientalis*, entered the Malpighian tubules, where they persisted with little growth. Larvae ingested by tadpoles became encapsulated.

In mice and chameleons, second-stage larvae apparently passed into the body cavity and subcutaneous tissues and grew. Attempts were made to infect chameleons with larvae hatched from eggs, with larvae found in experimentally infected *Culex* and mice, and with larvae taken from other chameleons. Unfortunately, the chameleons used were wild-caught and their histories unknown; thus, the data are difficult to interpret. Nevertheless, the authors concluded that in chameleons the second moult occurred in about 30 days when larvae were 5.0 mm in length, the third in 60 days when larvae were 25 mm in length and the fourth in 120 days when larvae were 3–4 cm in length. Development took place in the body cavity or subcutaneous tissues. There was no evidence to indicate how some adults attained the gut but it was suggested that it was by way of the lungs.

These data were interpreted by Chabaud *et al.* (1962) as evidence of the transformation of a primitively heteroxenous life cycle to a monoxenous life cycle. Sprent (1982) thought it was evidence that 'survival and continuance of egg-laying can occur in both prey and predator' and evidence that 'this evolutionary "shortcut" from a prey group to predator group was probably in some instances followed by suppression of the egg-laying stage in the prey host so that the prey became an intermediate host'. Sprent (1978) even postulated that *Hexametra* spp. 'originated in lizards, especially chameleons in Africa by this process'. Anderson (1988), on the other hand, interpreted the presence of worms in the body cavity and subcutaneous tissues of chameleons as evidence of extreme precocity of larvae of an ascaridoid of boas in a reptilian intermediate host.

Ghadirian (1968) suggested that transmission might be monoxenous in snakes in warm humid regions or involve only an invertebrate paratenic host, whereas in dry areas a vertebrate intermediate host was more likely to be utilized.

H. boddaertii (Baird, 1860)

H. boddaertii (syn. *Hexametra quadricornis* of Araujo, 1970) is a parasite of snakes (*Bothrops* spp., *Crotalus* spp., *Mastigodryas boddaertii*, *Philodryas scholti* and *Pseudoboa trigemina*) in North and South America and the West Indies (Sprent, 1978).

Araujo (1970) and Saliba and Araujo (1972) reported two moults in the eggs. Eggs hatched in mice; larvae migrated to the liver and, in 16 days, entered the body cavity and remained there for more than a year, growing to about 8 cm in length without moulting. Infective larvae given to rattlesnakes migrated to the body cavity and were found coiled under the serosa of the stomach and anterior part of the intestine. Fourth-stage larvae were present in 23 days and adult worms were found in the gut of the snakes in 120 days.

H. quadricornis (Wedl, 1861)

This is a parasite of snakes (Elapidae–*Bungarus fasciatus*, *Naja* spp.; Viperidae–*Bitis* spp., *Cerastes vivera*, *Echis carinatus*, *Vipera* spp., *Trimeresurus* spp., *Zamenis gemonensis*; Colubridae–*Boaedon olivaceus*, *Boiga* spp., *Coluber* spp., *Elaphe quadrivirgata*, *Leptodira duchesnii*, *Malpolon monspessulanus*, *Natrix* spp., *Psammodynastes pulverentulus* and *Pseudaspis cana*). The species has been found in Africa, southern Europe, the eastern Mediterranean region, India, Malaysia, Taiwan and Indonesia (Sprent, 1978).

Hörchner (1962) reported that eggs from worms found in the African puff adder (*Bitis arietans*) reached the first stage in 16 days at 17–19°C and second-stage larvae were present 12 days later. Araujo (1971) believed that two moults occurred in the eggs. Hörchner (1962) gave large numbers of infective eggs (50,000 each) to four white mice. Larvae invaded the liver and moulted to the third stage in 5–6 days. Third-stage larvae, 3.2–4.5 mm in length, were found in the peritoneal cavity of the other mice 9, 10 and 16 days postinfection. He gave larvae found in the body cavity of one mouse to another mouse. In 5 days larvae found in the body cavity were 6–10 mm in length. Encapsulated larvae 12–22 mm in length were found on the mesentery of other mice infected 35–45 days previously. Mice with larvae were fed to two *Coluber jugularis casprus*, which were examined 56 and 67 days later. Fourth-stage larvae and immature adults (16–43 mm long) were found in the body cavity of the vipers. The author suggested that worms were not found in the intestine because *C. j. casprus* is not a suitable host. Saliba and Araujo (1972) concluded that larval *H. quadricornis* in mice reached the body cavity by invading the intestinal villous capillaries and the liver, from which they emerged into the body cavity. Larvae were seen in the intestinal capillaries in 90 min and in the liver in 6 h.

Petter (1968b) studied *H. quadricornis* from *Bitis arietans* and *Naja haje* from Africa and *Vipera aspis* from France. She concluded, from various feeding experiments, that second-stage larvae could persist in the liver, intestinal wall and peritoneum of reptiles, birds and mammals but growth and the second moult occurred only in small mammals like mice, rats and insectivores. Second-stage larvae given to snakes crossed the intestinal wall and entered the body cavity, where growth and the third and fourth moults occurred. Growth and the times of moults were highly variable.

According to Araujo (1970) eggs embryonated to the infective stage in 17 days at 25°C. Larvae were found in mouse liver 6 h after oral inoculation. Larvae moulted in about 6 days. In 16 days larvae migrated into the body cavity, where they remained free for at least a year. Growth in the mice was marked and inversely proportional to the number of larvae present but they apparently remained in the third stage. In *Crotalus durissus* third-stage larvae appeared under the serosa in the body cavity in 15 days. Fourth-stage larvae appeared in the peritoneum and gut lining in 22–23 days and adults in the intestine in 110–117 days.

Ophidascaris

Members of the genus are parasites of the oesophagus and stomach of snakes and lizards (Ash and Beaver, 1963). Species have well-developed interlabia and more or less square lips. Females have two uteri and the vulva is behind the middle of the body. Three species infect pythons in India, southeast Asia, Africa and Australia (Sprent, 1969). The parasites feed on ingesta in the gut of the definitive host and remain attached to the gut wall at other times, sometimes clustered together with their heads buried in the centre of an elevated nodule, or looped through the stomach wall with the extremities extending into the lumen (Ash and Beaver, 1963; Sprent, 1970b). In pythons, more than one species of ascaridoid may share the same attachment site. Eggs are spherical or subspherical, with large or small pits in the shell. The genus has about 25 known species but the biology of only two has been investigated in any detail. In mice, Sprent (1969) reported that growth of larvae of four species from pythons was as follows after 3 months: *O. filaria* (Dujardin, 1845) and *O. moreliae* Sprent, 1969 – 6 mm; *O. baylisi* Robinson, 1934 and *O. infundiculum* (Linstow, 1903) – 40 mm.

O. labiatopapillosa Walton, 1927

This is a parasite of the snakes *Heterodon cyclopion*, *H. platyrhinos*, *Lampropeltis getulus* and *Natrix sipedon* in the USA (Ash and Beaver, 1963). Walton (1937) found larvae (up to 12 cm in length) encapsulated in the stomach wall, the mesentery and body muscles of amphibians (*Amphiuma tridactylum*, *Rana aesopus*, *R. catesbiana* and *R. sphenocephala*) collected in Florida. Snakes (*Natrix rhombifera* and *N. sipedon*) given larvae from *Rana* spp. contained immature and mature *O. labiatopapillosa* when examined 4 weeks postinfection. Infective eggs were given to young *Rana pipiens*; larvae were found in the stomach wall of the frogs 4 weeks later.

O. moreliae Sprent, 1969

O. moreliae (syn. '*O. filaria*' of Sprent, 1969) is a parasite of the stomach and oesophagus of the carpet python (*Aspidites melanocephalus* and *Morelia spilotes variegatus*) in Australia and is often found attached to the same site as *Amplicaecum robertsi* (Sprent, 1970b).

Eggs, 78–85 × 68–80 μm in size were embryonated in 14–21 days and larvae moulted in 12 days (a temperature of '70°C' given by Sprent, 1970b, is obviously an error). Eggs containing second-stage larvae remained viable and infective for 7 years in moist charcoal. Eggs hatched in mammals (laboratory mice and rats, *Rattus fuscipes*, guinea pigs, an opossum and a bandicoot); second-stage larvae migrated to the liver and lungs and appeared encapsulated in subcutaneous tissues, especially of the neck, where growth to about 8 mm occurred in 5 weeks. The second moult was not observed but larvae were infective to pythons.

In pythons, third-stage larvae migrated to the lungs, where they remained for at least 3 months. They grew to a maximum length of about 45 mm before the third moult in 117 days. During the third moult, larvae moved up the trachea and attached to the oesophagus as fourth-stage larvae. They eventually moved to the stomach, where the fourth moult occurred in 318 days, when they were 37–55 mm long. Thus, the main growth of the nematode occurred in the lungs and stomach of the definitive host and not in the mammalian intermediate host (cf. *Amplicaecum robertsi*).

Polydelphis

According to Sprent (1969, 1978) the genus contains the single species *P. anoura*. Other nominal species (*P. attenuata*, *P. bicornuta*, *P. oculata*, *P. mucronata*), with the possible exception of *P. brachycheilos*, were regarded as synonyms.

P. anoura Dujardin, 1845

P. anoura has, apparently, a wide distribution in pythons (*Aspidites ramsayi*, *Chondropython viridis*, *Liasis* spp., *Morelia spilotes* and *Python* spp.) in Africa, India, Thailand, Malaysia, Papua New Guinea and Australia. Sprent (1959b) reported that second-stage larvae of *P. anoura* from *M. spilotes* invaded the intestinal wall of rodents and grew to 2.1 mm in length. Kutzer and Grünberg (1965) and Kutzer and Lamina (1965) reported that second-stage larvae of *P. anoura* (syn. *P. attenuata*) from *Python reticulatus* reached the liver and lungs of rodents and moulted 4–10 days postinfection. The infective stage (5.12 mm maximum in length) was reached in 40 days. Kutzer and Grünberg (1965) reported that infective larvae from mice reached adulthood in the stomach and intestine of snakes after two moults.

Sprent (1970c) studied *P. anoura* from *M. spilotes* from Queensland, Australia. Larvae in eggs moulted in 10 days and were infective in 14 days. Second-stage larvae were 354–487 μm in length. In mice and rats, second-stage larvae hatched and migrated mainly to the liver, lungs and intestinal wall, where they became encapsulated. Larvae were infective to pythons in 39–42 days, when they were 5–6 mm in length. Moults were not noted but it was believed that larvae at this time were in the third stage. The genital primordium was an oval body on the ventral side of the intestine about half-way along the body. *Liasis childreni*, *L. emethistinus*, *L. fuscus* and *M. spilotes* were given nodules containing larvae from the small intestine of experimentally infected rodents. The third moult took place in the stomach of the snakes by 8–14 days, when larvae were 3.5–7.0 mm in length. Growth and development of the fourth stage commenced in the oesophagus and stomach but growth was slow and adults were not found until 183 days.

Travassosascaris

Members of the genus occur in rattlesnakes and have four uteri and interlabia.

T. araujoi Sprent, 1978

T. araujoi (syn. *Polydelphis quadrangularis* Araujo, 1969 not Schneider, 1866) is a common parasite of *Crotalus durissus terrificus* in Brazil. According to Araujo (1971, 1972b) larvae in eggs moulted twice and attained the third stage (444–492 μm in length) in about 7 days at 20°C. In mice, larvae migrated to the liver in 3 days. They became encapsulated in 20 days when they were about 2.5–5.7 mm in length and the sexes could be identified. Larvae continued to grow in the liver and by 367 days males were an average of 29.9 mm and females 32.0 mm in length. The genital primordia were well developed by 173 days and the four rudiments of the uteri were well defined.

Larvae were infective to snakes after about 28 days of growth in the liver of mice and were about 4.9 mm in length at this time. These small larvae given to snakes invaded the stomach wall or body cavity and developed to the fourth stage, before returning to the gut lumen to moult and reach adulthood. Larger larvae (e.g. those 19.8 mm in length after 173 days) from mice, remained in the lumen of the gut of snakes and matured in about 366 days.

Subfamily Heterocheilidae

Members of this subfamily are found mainly as parasites of aquatic reptiles (especially crocodilians) and Sirenian mammals (Gibson, 1983). Little is known about the biology of these nematodes.

Brevimulticaecum

B. tenuicolle (Rudolphi, 1819)

This is a parasite of the stomach of the alligator (*Alligator mississippiensis*) (Sprent, 1979). Larvae were found encapsulated in the stomach muscles of *Rana catebeina*, *R. sphenocephala* and *Siren lacertina* in Florida by Walton (1937), who gave 50 larvae to a young alligator (*A. mississippiensis*). The alligator died 3 weeks later and Walton found several immature males and females of *M. tenuicolle* in this host. Moravec and Kaiser (1994) reported similar larvae (1.96–3.77 mm in length) in abdominal capsules in *Hyla minuta* in Trinidad.

5.5

The Superfamily Subuluroidea

Subuluroids are lipless nematodes in which the anterior lobes of the oesophagus form a varied and complex pharyngeal region important in the systematics of the group (Chabaud, 1978). A prominent preanal sucker is present. Eggs are thick-shelled and contain a fully developed first-stage larva *in utero*. This larva is characterized by an oral opening slightly ventral in position and a three-pronged hook on the anterior margin of the oral opening. The tail is short and blunt. Two prominent glands are present in the pseudocoelom in the oesophageal region. These glands have long ducts which pass to the cephalic region.

The superfamily, which is biologically very homogeneous, is divided into the Maupasinidae, with a single species found in African elephant shrews (Macroscelididae), and the Subuluridae, found in various orders of birds (passeriformes, galliformes, gruiformes), marsupials, rodents, lemurs, tarsiers and monkeys. Alicata (1939) and Cuckler and Alicata (1944) were first to show the importance of insects as intermediate hosts of the subuluroids and their relationship to the Spirurida. Current evidence suggests that species studied are not highly specific in their use of intermediate hosts.

Eggs hatch in the gut of the insect (usually orthopterans, dermapterans or coleopterans) and larvae penetrate the gut wall and attain the body cavity. Shortly after the first moult, larvae start to become encapsulated. The thin, transparent capsules, containing one or two larvae, are generally attached to the outer wall of the intestine. Soon after the second moult, the early third-stage larva undergoes a peculiar transformation within its capsule. The body shortens and thickens markedly and the extremities bend dorsally. The larva eventually becomes a rounded mass in the capsule with its body in a complete coil, the head and tail adjacent to each other. The intestine is cramped and often doubled back on itself. The oesophagus has a large valved bulb. Lateral alae are present. Such larvae are instantly recognizable as subuluroids and have been reported a number of times as such in insects (Chabaud, 1954; Petter, 1960; Nassi and Dupouy, 1977).

Once in the intestine or caecum of the final host, larvae escape the capsules and within a few hours they stretch and regain an elongated shape. Development occurs in the lumen of the gut, usually the caecum, with the usual two moults to the adult stage. During development from the third through the fourth stage, the complex head structures of the adult make their appearance as documented by Quentin (1969) in *Subulura williaminglisi*.

Family Maupasinidae

Maupasina

M. weissi Seurat, 1913

This species was first found in the caecum of *Elephantulus rozeti* in Tunisia. Seurat (1917) found larvae in viscera of elephant shrews and concluded that transmission was direct. However, Quentin and Verdier (1979) have shown that *M. weissi* is heteroxenous. Eggs deposited by female worms were round and relatively thin-shelled. The eggs underwent a period of 'maturation' in the caecum of the host, during which the shell became thickened and brown. These latter eggs were given in food to four *Locusta migratoria* dissected 30 and 45 days later. Three of the four locusts contained infective larvae, apparently encapsulated in fat tissue at the level of ovaries and oviducts. Head structures of the encapsulated third stage were complex, with pharyngeal teeth. Genital primordia were minute. Quentin and Verdier (1979) suggested that the larvae found in Algeria by Nassi and Dupouy (1977) – encapsulated in a tenebrionid beetle, *Pimelia* (*Homalopus*) *arenacea*, and identified as '*Subulura* sp.' – were probably infective larvae of *M. weissi*.

Family Subuluridae

Allodapa

A. suctoria (Molin, 1860)

A. suctoria (syns *Subulura brumpti*, *Subulura suctoria*) is a common parasite of the caecum of poultry in many parts of the world. Alicata (1939), Cuckler and Alicata (1944), Abdou and Selim (1957, 1963) and Barus *et al.* (1967) reported its development in coleopterans (*Alphitobius diaperinus*, *Ammophorus insularis*, *Blaps polycresta*, *Dactylosternum* sp., *Dermestes vulpinus*, *Gonocephalum seriatum*, *Ocnera hispida* and *Tribolium castaneum*), dermapterans (*Euborellia annulipes*) and orthopterans (*Conocephalus saltator*, *Oxya chinensis*). Cuckler and Alicata (1944) followed development in grasshoppers (*C. saltator* and *O. chinensis*) (temperature not given) and chickens. Eggs given to grasshoppers hatched in the gut in 4–5 h. Some larvae had invaded the body cavity in 24 h. The first moult took place 4–5 days postinfection. The oral opening of the second-stage larva was terminal. The oesophagus was expanded posteriorly but lacked a bulb. The glands noted in the first stage were absent. By 7–8 days the host started to encapsulate the larva. The capsules were usually located on the wall of the intestine and some capsules contained two larvae. The second moult occurred 13–15 days postinfection. Within a day of moulting, larvae began to contract; the anterior and posterior extremities bent dorsally and eventually formed the characteristic rounded body in the capsule (Fig. 5.8). In chickens, larvae appeared in the caeca in 24 h where growth and development took place. There was no evidence that larvae invaded the caecal wall. The third moult was not recorded but the fourth was noted 18 days postinfection; about 24 days later the parasites were mature and eggs appeared in the faeces (at 42 days).

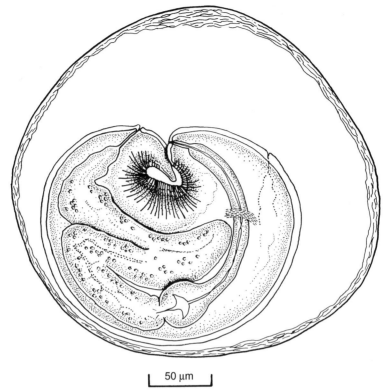

50 μm

Fig. 5.8.   Third-stage larva of *Allodapa suctoria* in capsules (After J. Alicata, 1939 – courtesy *Journal of Parasitology.*)

Karunamoorthy *et al.* (1994) found one to five larvae in *Alphitobius diaperinus* inhabiting chicken houses in India. Infective larvae appeared in beetles 12–16 days postinfection and the prepatent period in chickens was 30–35 days.

Quentin and Poinar (1973) studied the early development of *A. suctoria* in grasshoppers.

Subulura

S. bolivari (Lopez-Neyra, 1922)

Chabaud (1954) collected specimens of *S. bolivari* from a little owl (*Athene noctua*) from Morocco. He fed eggs of female specimens to several species of insects and found third-stage larvae in the coleopterans *Blaps pinguis, Gonocephalum rusticum* and *Pimelia rugosa* 17 days later. Larvae were in large capsules attached to the intestinal wall.

S. jacchi (Marcel, 1857)

Chabaud and Lariviere (1955) collected specimens from a Brazilian marmoset (*Hapale jacchus*) which died in the Jardin des Plantes in Paris. They successfully infected the cockroach, *Blabera fusca*, held at 25°C. Development in the roach was asynchronous and the timing of the moults was not determined. Larvae were in delicate capsules or free in the body cavity; the authors felt that free larvae grew faster than encapsulated larvae.

S. otolicni (Van Beneden, 1890)

This is a parasite of the caecum of the lemurs *Galago alleni* and *Galagoides demidovii*. Quentin and Tcheprakoff (1969) successfully infected the dermapterans *Anisolabis annulipes*, *Diaperasticus erythrocephalus* and *Labidura riparia* by allowing them to consume crushed female worms; the insects were kept at 22°C. Two days after exposure, first-stage larvae were found in the haemocoel. They were encapsulated at the end of the first stage and thereafter on the intestinal wall anterior to the Malpighian tubules. The first moult occurred 6–7 days postinfection. In 22 days most larvae were in the third stage. The authors gave two white mice some third-stage larvae: two fourth-stage females were found in one of them 4 days later.

S. williaminglisi Quentin, 1965

This is a parasite of the caecum of various African murid and cricetid rodents, including *Cricetomys gambianus*, *Hybomys univittatus* and *Thamnomys rutilans* in North Africa. Quentin (1969) took eggs from worms found in rodents and gave them to the dermapterans *Anisolabis annulipes*, *Diaperasticus erythrocephalus* and *Labidura riparia*, which he maintained at 22°C. Larvae were found apparently in all three species when they were dissected at various times postinfection. Larvae hatched in the gut of the insects within 14 h. The first moult took place in the body cavity 11–14 days and the second 21–23 days postinfection. Third-stage larvae then underwent the characteristic contraction. Larvae were said to be encapsulated in the intestinal wall anterior to the Malpighian tubules.

Infective larvae were given orally to unspecified rodents (apparently African Muridae). In the duodenum and caecum the third-stage larva lost its spherical shape within 48 h. By 4 days larvae were in the fourth stage and males and females could be differentiated by size and tail length. By 14 days larvae were triple the length at 4 days. By 20 days worms were moulting to the adult stage.

Tarsubulura

T. perarmata (Ratzel, 1868)

T. perarmata is a small species described from tarsiers (*Tarsius* spp.) and tupaids (*Tupaia glis* and *T. minor*) in southeast Asia (e.g. Malaysia and Sri Lanka). Quentin *et al.* (1977) infected crickets (*Oxya* sp. and *Valanga* sp.) with eggs from worms found in *Tarsius spectrum* collected near Kuala Lumpur. At 28°C the first moult occurred in 14 days; some larvae moulted the second time in 3 weeks. By 28 days all larvae were in the third stage. The second and third stages were in capsules attached to the wall of the midgut in front of the Malpighian tubules.

References (Ascaridida)

Abdou, A.H. and Selim, M.K. (1957) On the life cycle of *Subulura suctoria*, a caecal nematode of poultry in Egypt. *Zeitschrift für Parasitenkunde* 18, 20–23.

Abdou, A.H. and Selim, M.K. (1963) The life cycle of *Subulura suctoria* (Mol. 1860) Railliet and Henry, 1919 in the domestic fowl. *Zeitschrift für Parasitenkunde* 23, 45–49.

Ackert, J.E. (1917) A means of transmitting the fowl nematode *Heterakis papillosa* Bloch. *Science* 66, 394.

Ackert, J.E. (1919) Studies on the development of *Ascaridia perspicillum*, parasitic in fowls. *Proceedings of the American Society of Zoology, 17th Annual Meeting, St Louis, December*, pp. 331–332.

Ackert, J.E. (1923a) On the habitat of *Ascaridia perspicillum* (Rud.). *Journal of Parasitology* 10, 101–103.

Ackert, J.E. (1923b) On the life history of the fowl nematode, *Ascaridia perspicillum* (Rud.). *Anatomical Records* 26, 356.

Ackert, J.E. (1931) The morphology and life history of the fowl nematode *Ascaridia lineata* (Schneider). *Parasitology* 23, 360–379.

Adamson, M.L. (1981) *Gyrinicola batrachiensis* (Walton, 1929) n.comb. (Oxyuroidea; Nematoda) from tadpoles in eastern and central Canada. *Canadian Journal of Zoology* 59, 1344–1350.

Aguilar-Nascimento, J.E., Silva, G.M., Tadano, T., Filho, M.V., Akiyama, A.M.P. and Castelo, A. (1993) Infection of the soft tissue of the neck due to *Lagochilascaris minor*. *Transactions of the Royal Society of Tropical Medicine and Hygiene* 87, 198.

Alicata, J.C. (1934) Observations on the period required for *Ascaris* eggs to become infective. *Proceedings of the Helminthological Society of Washington* 1, 12.

Alicata, J.E. (1939) Preliminary note on the life history of *Subulura brumpti*, a common cecal nematode of poultry in Hawaii. *Journal of Parasitology* 25, 179–180.

Alvarez-Pellitero, M.P. (1979) Observaciones sobre el ciclo vital de *Raphidascaris acus* en los ambientes naturales de los ríos de León. *Anales de la Facultad de Veterinaria de León, Universidad de Oviedo* 25, 129–154.

Andel, R.A. van., Franklin, C.L., Besch-Williford, C., Riley, L.K., Hook, R.R. Jr and Kazacos, K.R. (1995) Cerebrospinal larva migrans due to *Baylisascaris procyonis* in a guinea pig colony. *Laboratory Animal Science* 45, 27–30.

Anderson, B.C. (1999) Congenital *Baylisascaris* sp. larval migrans in a newborn lamb. *Journal of Parasitology* 85, 128–129.

Anderson, R.C. (1960) On the development and transmission of *Cosmocercoides dukae* of terrestrial molluscs in Ontario. *Canadian Journal of Zoology* 38, 801–825.

Anderson, R.C. (1988) Nematode transmission patterns. *Journal of Parasitology* 74, 30–45.

Anderson, R.C. (1992) *Nematode Parasites of Vertebrates: their Development and Transmission*, 1st edn. CAB International, Wallingford, UK.

Anderson, R.C. (1996) Additional observations on the development and transmission of *Truttaedacnitis pybusae* Anderson, 1992 (Seuratoidea: Cucullanidae) of the brook lamprey, *Lampetra appendix* (DeKay, 1842). *Parasite* 3, 33–37.

Anderson, R.C. and Bartlett, C.M. (1993) The significance of precocity in the transmission of the nematode parasites of vertebrates. *Canadian Journal of Zoology* 71, 1917–1922.

Anderson, T.J.C. (1995) *Ascaris* infections in humans from North America: molecular evidence for cross infection. *Parasitology* 110, 215–219.

Anderson, T.J.C. and Jaenike, J. (1997) Host specificity, evolutionary relationships and macrogeographic differentiation among *Ascaris* populations from humans and pigs. *Parasitology* 115, 325–343.

Anderson, T.J.C., Romero-Abal, M.E. and Jaenike, J. (1993) Genetic structure and epidemiology of *Ascaris* populations: pattern of host affiliation in Guatemala. *Parasitology* 107, 319–334.

Araujo, P. (1970) Considerações sôbre a evolução de *Hexametra quadricornis* (Wedl, 1861) (Nematoda), parasita de ofídios. *Revista da Faculdade de Farmácia e Bioquimica, São Paulo Universidade* 8, 193–242.

Araujo, P. (1971) Considérations sur la deuxième mue des larves d'Ascarides parasites de serpents. *Annales de Parasitologie Humaine et Comparée* 46, 605–612.

Araujo, P. (1972a) Observações pertinentes às primeiras ecdises de larvas de *Ascaris lumbricoides, A. suum* e *Toxocara canis*. *Revista do Instituto de Medicina Tropical de São Paulo* 14, 83–90.

Araujo, P. (1972b) Observations sur le cycle biologique de l'ascaride *Polydelphis quadrangularis* (Schneider, 1866) parasite du serpent crotale. *Annales de Parasitologie Humaine et Comparée* 47, 91–120.

Araujo, P. and Bressan, M.C.R.V. (1977) Considerations sur la deuxième mue des larves d'*Ascaridia galli*. *Annales de Parasitologie Humaine et Comparée* 52, 531–537.

Armstrong, D.L., Montani, R.J., Doster, A.R. and Kazacos, K.R. (1989) Cerebrospinal nematodiasis in macaws due to *Baylisascaris procyonis*. *Journal of Zoo and Wildlife Medicine* 20, 354–359.

Artigas, P. de T. and Ueta, M.T. (1989) Sobre a evolucão de *Ascaris lumbricoides* Linnaeus, 1758, na fase larvar endovular. *Memórias do Instituto Butantan* 51, 15–24.

Ash, L.R. and Beaver, P.C. (1963) Redescription of *Ophidascaris labiatopapillosa* Walton, 1927, an ascarid parasite of North American snakes. *Journal of Parasitology* 49, 765–770.

Aspholm, P.E. (1995) Short communication *Anisakis simplex* Rudolphi, 1809, infection in fillets of Barents Sea cod *Gadus morhua* L. *Fisheries Research* 23, 375–379.

Aspholm, P.E., Ugland, K.I., Jodestøl, K.A. and Berland, B. (1995) Sealworm (*Pseudoterranova decipiens*) infection in common seals (*Phoca vitulina*) and potential intermediate fish hosts from outer Oslo fjord. *International Journal for Parasitology* 25, 367–373.

Augustine, D.L. (1927) Development in prenatal infestation of *Belascaris*. *Journal of Parasitology* 13, 256–259.

Augustine, P.C. and Lund, E.E. (1974) The fate of eggs and larvae of *Ascaridia galli* in earthworms. *Avian Diseases* 18, 394–398.

Babero, B.B. (1957) *Ascaris laevis* migration in experimental hosts. *Journal of Parasitology* (Suppl.) 43, 121.

Babero, B.B. (1959) Pathology resulting from experimental infections by *Ascaris laevis* Leidy. *Transactions of the American Microscopical Society* 78, 330–335.

Babero, B.B. (1960a) Studies on the larval morphology of *Ascaris laevis* Leidy, 1856. *American Midland Naturalist* 64, 349–361.

Babero, B.B. (1960b) On the migration of *Ascaris laevis* Leidy, 1856 in some experimentally infected hosts. *Transactions of the American Microscopical Society* 79, 439–442.

Baer, J.G. (1961) Host reactions in young birds to naturally occurring superinfestations with *Porrocaecum ensicaudatum*. *Journal of Helminthology* (Suppl.) 35, 1–4.

Bain, O. (1970) Cycle évolutif de l'Heterakidae *Strongyluris brevicaudata* (Nematoda). Mise en évidence de deux mues dans l'oeuf. *Annales de Parasitologie Humaine et Comparée* 45, 637–653.

Bain, O. and Philippon, B. (1969) Recherche sur des larves de nématodes Ascaridida trouvées chez *Simulium damnosum*. *Annales de Parasitologie Humaine et Comparée* 44, 147–156.

Baker, A.D. (1933) Some observations on the development of the caecal worm, *Heterakis gallinae* (Gmelin, 1790) Freeborn, 1923, in the domestic fowl. *Scientific Agriculture* 13, 356–363.

Baker, M.R. (1978) Transmission of *Cosmocercoides dukae* (Nematoda: Cosmocercoidea) to amphibians. *Journal of Parasitology* 64, 765–766.

Baker, M.R. (1984) On the biology of *Dichelyne (Cucullanellus) cotylophora* (Ward and Magath, 1917) (Nematoda, Cucullanidae) in perch (*Perca flavescens*) from Lake Erie, Ontario. *Canadian Journal of Zoology* 62, 2062–2073.

Baker, M.R. (1987) *Synopsis of the Nematoda Parasitic in Amphibians and Reptiles.* Memorial University of Newfoundland Occasional Papers in Biology, No. 11 325 pp.

Balbuena, J.A., Karlsbakk, E., Saksvik, M., Kvenseth, A.M. and Nyland, A. (1998) Research notes. New data on the early development of *Hysterothylacium aduncum* (Nematoda, Anisakidae). *Journal of Parasitology* 84, 615–617.

Ball, R.L., Dryden, M., Wilson, S. and Veatch, I. (1998) Cerebrospinal nematodiasis in a white-handed gibbon (*Hylobates lar*) due to *Baylisascaris* sp. *Journal of Zoo and Wildlife Medicine* 29, 221–224.

Banning, P. van (1971) Some notes on a successful rearing of the herring-worm, *Anisakis marina* L. (Nematoda: Heterocheilidae). *Journal du Conseil* 34, 84–88.

Bartlett, C.M. (1996) Morphogenesis of *Contracaecum rudolphii* (Nematoda: Ascaridoidea), a parasite of fish-eating birds, in its copepod precursor and fish intermediate hosts. *Parasite* 4, 367–376.

Bartlett, C.M. and Anderson, R.C. (1985) Larval nematodes (Ascaridida and Spirurida) in the aquatic snail, *Lymnaea stagnalis. Journal of Invertebrate Pathology* 46, 153–159.

Barus, V. and Groschaft, J. (1962) *Megalobatrachonema terdentatum* (Linstow, 1890) Hartwich, 1960 (Nematoda, Subulascarididae) in Czechoslovakia, and its development. *Helminthologia* 4, 67–78.

Barus, V., Busa, V., Rysavy, B. and Lorenzo-Hernandez, N. (1967) Distribucion del nematodo *Subulura suctoria* (Molin, 1860) en Cuba y observaciones de su ciclo evolutivo. *Poeyana Instituto de Biologia, Serie A* 48, 1–11.

Baylis, H.A. (1936) The nematode genus *Rondonia* Travassos, 1920. *Annals and Magazine of Natural History* 17, 606–610.

Beaver, P.C. (1956) Parasitological reviews. Larva migrans. *Experimental Parasitology* 5, 587–621.

Berland, B. (1970) On the morphology of the head in four species of Cucullanidae. *Sarsia* 43, 15–64.

Berry, G.N. and Cannon, L.R.G. (1981) The life history of *Sulcascaris sulcata* (Nematoda: Ascaridoidea), a parasite of marine molluscs and turtles. *International Journal for Parasitology* 11, 43–54.

Berry, J.F. (1985) Phylogenetic relationship between *Baylisascaris* spp. Sprent, 1968 (Nematoda: Ascarididae) from skunks, raccoons, and groundhogs in south Ontario. Unpublished MS thesis, University of Guelph, Guelph, Ontario.

Bier, J.W. (1976) Experimental anisakiasis: cultivation and temperature tolerance determination. *Journal of Milk and Food Technology* 39, 132–137.

Birova-Volosinovicova, V. (1965) Studium vyvinu vajicok *Heterakis gallinarum* v laboratornych podmienkach. *Biologia, Bratislava* 20, 122–126. (In Czechoslovakian.)

Boer, E. de (1935) Experimenteel onderzoek betreffende *Ascaris lumbricoides* von mensch en varken. *Tijdschrift voor Diergeneeskunde, Utrecht* 62, 665–673.

Boes, J., Medley, G.F., Eriksen, L., Roepstorff, A. and Nansen, P. (1998) Distribution of *Ascaris suum* in experimentally and naturally infected pigs and comparison with *Ascaris lumbricoides* infection in humans. *Parasitology* 117, 589–596.

Bohm, L.K. and Supperer, R. (1958) Beiträge zur Kenntnis tierischer Parasiten III. *Zentralblatt für Bakteriologie, Parasitenkunde, Infecktionskrankheiten und Hygiene. Abteilung 1 Originale* 172, 298–309.

Bolek, M.G. (1997) Seasonal occurrence of *Cosmocercoides dukae* and prey analysis in the blue-spotted salamander, *Ambystoma laterale*, in southeastern Wisconsin. *Journal of the Helminthological Society of Washington* 64, 292–295.

Botero, D. and Little, M.D.C. (1984) Two cases of human *Lagochilascaris* infection in Colombia. *American Journal of Tropical Medicine and Hygiene* 33, 381–386.

Boulenger, C.L. (1922) On *Ascaris vitulorum* Goeze. *Parasitology* 14, 87–92.

Bowen, W.D. (ed.) (1990) Population biology of sealworm (*Pseudoterranova decipiens*) in relation to its intermediate and seal hosts. *Canadian Bulletin of Fisheries and Aquatic Sciences* 222, 306 pp.

Bowman, D.D., Smith, J.L. and Little, M.D. (1983) *Lagochilascaris sprenti* sp.n. (Nematoda: Ascarididae) from the opossum, *Didelphis virginiana* (Marsupialia: Didelphidae). *Journal of Parasitology* 69, 754–760.

Bowman, D.D., Oaks, J.A. and Grieve, R.B. (1993) Ultrastructure of the infective stage larva of *Toxocara canis* (Nematoda: Ascaridoidea). *Journal of the Helminthological Society of Washington* 60, 183–204.

Brattey, J. (1990) Effect of temperature on egg hatching in three ascaridoid nematode species from seals. In: Bowen, W.D. (ed.) *Population Biology of Sealworm* (Pseudoterranova decipiens) *in Relation to its Intermediate and Seal Hosts. Canadian Bulletin of Fisheries and Aquatic Sciences* 222, 27–39.

Brattey, J. (1995) Identification of larval *Contracaecum osculatum* s.l. and *Phocascaris* sp. (Nematoda: Ascaridoidea) from marine fishes by allozyme electrophoresis and discriminant function analysis of morphometric data. *Canadian Journal of Fisheries and Aquatic Sciences* 52 (Suppl. 1) 116–128.

Brattey, J. and Bishop, C.A. (1992) Larval *Anisakis simplex* (Nematoda: Ascaridoidea) infection in the musculature of Atlantic cod, *Gadus morhua*, from Newfoundland and Labrador. *Canadian Journal of Fisheries and Aquatic Sciences* 49, 2635–2647.

Brattey, J. and Clark, K.J. (1992) Effect of temperature on egg hatching and survival of larvae of *Anisakis simplex* B (Nematoda: Ascaridoidea). *Canadian Journal of Zoology* 70, 274–279.

Brattey, J. and Ni, I.H. (1992) Ascaridoid nematodes from the stomach of harp seals, *Phoca groenlandica*, from Newfoundland and Labrador. *Canadian Journal of Fisheries and Aquatic Sciences* 49, 956–966.

Brattey, J. and Stenson, G.B. (1993) Host specificity and abundance of parasitic nematodes (Ascaridoidea) from the stomachs of five phocid species from Newfoundland and Labrador. *Canadian Journal of Zoology* 71, 2156–2166.

Brenes-Madrigal, R.R., Ruiz, A. and Frenkel, J.K. (1972) Discovery of *Lagochilascaris* sp. in the larynx of a Costa Rica ocelet (*Felis pardalis mearns*). *Journal of Parasitology* 58, 978.

Bristow, G.A. and Berland, B. (1992) On the ecology and distribution of *Pseudoterranova decipiens* C (Nematoda: Anisakidae) in an intermediate host, *Hippoglossoides platessoides*, in northern Norwegian waters. *International Journal for Parasitology* 22, 203–208.

Brunaská, M., Dubinsky, P. and Reiterova, K. (1995) *Toxocara canis*: ultrastructural aspects of larval moulting in the maturing eggs. *International Journal for Parasitology* 25, 683–690.

Burke, T.M. and Roberson, E.L. (1985) Prenatal and lactational transmission of *Toxocara canis* and *Ancylostoma caninum*: experimental infection of the bitch before pregnancy. *International Journal for Parasitology* 15, 71–75.

Burren, C.H. (1971) The distribution of *Toxocara* larvae in the central nervous system of the mouse. *Transactions of the Royal Society of Tropical Medicine and Hygiene* 65, 450–453.

Burt, M.D.B., Campbell, J.D., Likely, C.G. and Smith, J.W. (1990a) Serial passage of larval *Pseudoterranova decipiens* (Nematoda: Ascaridoidea) in fish. *Canadian Journal of Fisheries and Aquatic Sciences* 47, 693–695.

Burt, M.D.B., Smith, J.W., Jarecka, A., Pike, A.W., Wootten, R. and McClelland, G. (1990b) *Pseudoterranova decipiens* (Nematoda: Ascaridoidea): time of development to hatching of larvae at different temperatures and salinities. In: Bowen, W.D. (ed.) *Population Biology of*

Sealworm (Pseudoterranova decipiens) *in Relation to its Intermediate and Seal Hosts. Canadian Bulletin of Fisheries and Aquatic Sciences* 222, 41–45.

Butorina, T.E. (1988) *On the Role of Lampreys in the Life Cycle of Nematodes in Infecting Salmonids on Kamchatka.* Publication of the Laboratory of Population Biology, Institute for Marine Biology, Far East Branch, No. 4C, pp. 66–67. Academy of Sciences of the USSR, Vladivostok.

Caldwell, F.C. and Caldwell, E.L. (1926) Are *Ascaris lumbricoides* and *Ascaris suilla* identical? *Journal of Parasitology* 13, 141–145.

Calvopiña, M., Guevara, A.G., Herrera, M., Serrano, M. and Guiderian, R.H. (1998) Treatment of human *Lagochilascaris* with ivermectin. *Transaction of the Royal Society of Tropical Medicine and Hygiene* 92, 223–224.

Campbell, G.A., Hoover, J.P., Russel, W.C. and Breazile, J.E. (1997) Naturally occurring cerebral nematodiasis due to *Baylisascaris* larval migration in two black and white ruffed lemurs (*Varecia variegata variegata*) and suspected cases in three emus (*Dromaius novaehollandiae*). *Journal of Zoo and Wildlife Medicine* 28, 204–207.

Campos, D.M.B., Komma, M.D., Barbosa, W., Santos, M.A., Souza, L.C.S., Pinto, R.N.L., Barcelos, M., Carneiro, J.R. and Evangilista, A. (1987) Notas parasitológicas sobre lagochilascarise humana em Goias. *Revista de Patologia Tropical* 16, 129–142.

Campos, D.M.B., Freire Filha, L.G., Vieira, M.A., Pacô, J.M. and Maia, M.A. (1992) Experimental life cycle of *Lagochilascaris minor* Leiper, 1909. *Revista do Instituto de Medicina Tropical de São Paulo* 34, 277–287.

Cannon, L.R.G. (1978) A larval ascaridoid nematode from Queensland scallops. *International Journal for Parasitology* 8, 75–80.

Chabaud, A.G. (1954) Sur le cycle évolutif des Spirurides et de nématodes ayant une biologie comparable. Valeur systématique des caracteres biologique. *Annales de Parasitologie Humaine et Comparée* 29, 42–88.

Chabaud, A.G. (1978) No. 6. Keys to genera of the Superfamilies Cosmocercoidea, Seuratoidea, Heterakoidea and Subuluroidea. In: Anderson, R.C., Chabaud, A.G. and Willmott, S. (eds) *CIH Keys to the Nematode Parasites of Vertebrates.* Commonwealth Agricultural Bureaux, Farnham Royal, UK, pp. 1–71.

Chabaud, A.G. and Brygoo, E.R. (1958a) Cycle évolutif dun nématode cosmocercide, parasite de grenouilles malgaches. *Comptes Rendus de l'Academie des Sciences* 246, 1771–1773.

Chabaud, A.G. and Brygoo, E.R. (1958b) Description et cycle évolutif d'*Aplectana courdurieri* n.sp (Nematoda, Cosmocercidae). *Mémoires de l'Institut Scientifique de Madagascar. Série A. Biologie Animale* 12, 159–176.

Chabaud, A.G. and Brygoo, E.R. (1960) Nématodes parasites de caméléons malgaches. *Mémoires de l'Institut Scientifique de Madagascar* 14, 125–159.

Chabaud, A.G. and Lariviere, M. (1955) Cycle évolutif d'un ascaride: *Subulura jacchi* (Marcel, 1857) parasite des primates, chez la blatte Blabera fusca. *Comptes Rendus des Séances de la Société de Biologie* 149, 1416–1419.

Chabaud, A.G., Brygoo, E.R. and Petter, A.J. (1962) Cycle évolutif de l'ascaride des caméléons malgaches. *Bulletin de la Société Zoologique de France* 87, 515–532.

Chabaud, A.G., Bain, O. and Poinar, G.O. Jr (1988) *Skrjabinelazia galliardi* (Nematoda, Seuratoidea): complements morphologiques et cycle biologique. *Annales de Parasitologie Humaine et Comparée* 63, 278–284.

Chauhan, P.P.S. and Pande, B.P. (1981) Observations on the embryonic development of bubaline and bovine strains of *Neoascaris vitulorum* (Goeze, 1782) Travassos, 1927 eggs. *Indian Journal of Animal Science* 51, 439–455.

Choudhury, A. and Dick, T.A. (1996) Observations on the morphology, systematics, and biogeography of the genus *Truttaedacnitis* (Nematoda: Cucullanidae). *Journal of Parasitology* 82, 977–987.

Clapham, P.A. (1933) On the life-history of *Heterakis gallinae*. *Journal of Helminthology* 11, 67–86.

Clapham, P.A. (1934) Some observations on the response of chickens to infestation with *Heterakis gallinae*. *Journal of Helminthology* 12, 71–78.

Clarke, M.R. (1966) A review of the systematics and ecology of oceanic squids. *Advances in Marine Biology* 4, 91–300.

Clayton, H.M. (1978) Ascariasis in foals. *Veterinary Record* 102, 553–556.

Clayton, H.M. (1986) Ascarids. Recent advances. Life cycle. *Veterinary Clinics of North America: Equine Practice* 2, 313–328.

Coates, J.W., Siegert, J., Bowes, V.A. and Steer, D.G. (1995) Encephalitic nematodiasis in a Douglas squirrel and a rockdove ascribed to *Baylisascaris procyonis*. *Canadian Veterinary Journal* 36, 566–569.

Cori, C.J. (1898) Beiträge zur Biologie von *Spiroptera turdi* Molin. Ueber das Vorkommen der Jungendstadien dieses Nematoden im Bauschgefässe des Regenwormes. *Lotos* 18, 23–30.

Cox, D.M. and Holland, C.V. (1998) The relationship between numbers of larvae recovered from the brain of *Toxocara canis*-infected mice and social behaviour and anxiety in the host. *Parasitology* 116, 599–594.

Crites, J.L. (1956) Studies on the morphology, taxonomy and life history of *Cruzia americana* Maplestone, 1930 a parasitic nematode of *Didelphis marsupialis virginiana*. Unpublished PhD thesis, Ohio State University, Columbus, Ohio.

Crompton, D.W.T. (1989) Biology of *Ascaris lumbricoides*. In: Crompton, P.W.T., Nesheim, M.C. and Pawlowski, Z.S. (eds) *Ascariasis and its Prevention and Control*. Taylor and Francis, London, pp. 9–44.

Cuckler, A.C. and Alicata, J.E. (1944) The life history of *Subulura brumpti*, a cecal nematode of poultry in Hawaii. *Transactions of the American Microscopical Society* 63, 345–357.

Cunningham, C.K., Kazacos, K.R., McMillan, J.A., Lucas, J.A., McAuley, J.B., Wozniak, E.J. and Weiner, L.B. (1994) Diagnosis and management of *Baylisascaris procyonis* infection in an infant with nonfatal meningoencephalitis. *Clinical Infectious Diseases* 18, 868–892.

Cvetković, L. and Nevenić, V. (1960) Epizootiology of *Neoascaris vitulorum* in calves. *Acta Veterinaria* 10, 49–59.

Da Costa, S.C.G. (1962) Aspectos biologicos do genero *Rondonia* Travassos, 1920 (Nematoda, Atractidae). *Arquivos do Museu Nacional Rio de Janeiro* 52, 75–78.

Dade, A.W., Williams, J.F., Trapp, A.L. and Ball, W.H. (1977) Cerebral nematodiasis in captive nutria. *Journal of Veterinary Medical Association* 171, 885–886.

Davaine, C.J. (1863) Nouvelle recherches sur le développement et la propagation de l'ascaride lumbricoide et du trichocéphale de l'homme. *Compte Rendus de la Société de Biologie* 4, 261–265.

Davey, J.T. (1969) The early development of *Contracaecum osculatum*. *Journal of Helminthology* 43, 293–298.

Davey, J.T. (1971) A revision of the genus *Anisakis* Dujardin, 1845 (Nematoda: Ascaridata). *Journal of Helminthology* 45, 51–72.

Deardorff, T.L. and Overstreet, R.M. (1980a) Review of *Hysterothylacium* and *Iheringascaris* (both previously – *Thynnascaris*) (Nematoda: Anisakidae) from the northern Gulf of Mexico. *Proceedings of the Biological Society of Washington* 93, 1035–1079.

Deardorff, T.L. and Overstreet, R.M. (1980b) *Contracaecum multipapillatum* (= *C. robustum*) from fishes and birds in the northern Gulf of Mexico. *Journal of Parasitology* 66, 853–856.

Deardorff, T.L. and Overstreet, R.M. (1981) Larval *Hysterothylacium* (= *Thynnascaris*) (Nematoda: Anisakidae) from fishes and invertebrates in the Gulf of Mexico. *Proceedings of the Helminthological Society of Washington* 48, 113–126.

Demshin, N.I. (1975) Phylogenetic and ecological factors in the formation of parasitic relationships in helminths developing with oligochaetes and leeches. *Trudi Biologo-Pochvennovo Instituta* 26, 21–32.

Deo, P.G. and Srivastava, H.D. (1955) Studies on the biology and life-history of *Ascaridia galli*, Schrank, 1788 (Abstract). *Proceedings of the 42nd Indian Science Congress*, Part III, pp. 282–283.

Di Deco, M.A., Orecchia, P., Paggi, L. and Petrarca, V. (1994) Morphometric stepwise discriminant analysis of three genetically identified species within *Pseudoterranova decipiens* (Krabbe, 1878) (Nematoda: Ascaridida). *Systematic Parasitology* 29, 81–88.

Dick, T.A., Papst, M.H. and Paul, H.C. (1987) Rainbow trout (*Salmo gairdneri*) stocking and *Contracaecum* spp. *Journal of Wildlife Diseases* 23, 242–247.

Dorman, H.P. (1928) Studies on the life cycle of *Heterakis papillosa* Bloch. *Transactions of the American Microscopical Society* 47, 379–413.

Douvres, F.W., Tromba, F.G. and Malakatis, G.M. (1969) Morphogenesis and migration of *Ascaris suum* larvae developing to the fourth stage in swine. *Journal of Parasitology* 55, 689–712.

Drudge, J.H. and Lyons, E.T. (1983) Ascariasis. In: Robinson, N.E. (ed.) *Current Therapy in Equine Medicine*. W.B. Saunders Co., Philadelphia, Pennsylvania, pp. 262–267.

Dubey, J.P. (1967) Egg production of *Toxocara cati* (Correspondence). *Veterinary Record* 81, 671–672.

Dubey, J.P. (1968) Migration of *Toxocara cati* larvae in mice. *Tropical and Geographical Medicine* 20, 172–176.

Dubinsky, P., Havasiorá-Reiterová, K., Petko, B., Hovorka, I. and Tomasovicova, O. (1995) Role of small mammals in the epidemiology of toxocariasis. *Parasitology* 110, 187–193.

Dunsmore, J.D., Thompson, R.C.A. and Bates, I.A. (1983) The accumulation of *Toxocara canis* larvae in the brains of mice. *International Journal for Parasitology* 13, 517–521.

Engashev, V.G. (1965) Study of the life cycle of *Raphidascaris acus* (Bloch, 1779). *Materialy Nauchnoi Konferentsii Vsesoyuznogo Obshchestva Gel'mintologov* Part 2, 89–94. (In Russian.)

Epstein, A. (1892) Ueber die Uebertrangund der menschlichen Spulwurms (*Ascaris lumbricoides*). *Verhandlungen Gesellschaft Kinderheilk, Wiesbaden (1891)* 9, 1–16.

Fagerholm, H.P., Nausen, P., Roepstorff, A., Frandsen, F. and Eriksen, L. (1998) Growth and structural features of the adult stage of *Ascaris suum* (Nematoda, Ascaridoidea) from experimentally infected domestic pigs. *Journal of Parasitology* 84, 269–277.

Faust, E.C., Beaver, P.C. and Jung, R.C. (1975) *Animal Agents and Vectors of Human Diseases*. Lea and Febiger, Philadelphia, Pennsylvania.

Feney-Rodriguez, S., Cuéllar Del Hoyo, C. and Guillén-Llera, J.L. (1988) Comparative study of the influence of light on embryonization of *Toxocara canis*, *Toxascaris leonina* and *Ascaris suum*. *Revista Ibérica de Parasitologia* 48, 395–401.

Fitzgerald, P.R. (1962) The pathogenesis of *Ascaris lumbricoides* var. *suum* in lambs. *American Journal of Veterinary Research* 23, 731–736.

Fleming, W.J. and Caslick, J.W. (1978) Rabies and cerebrospinal nematodiasis in woodchucks (*Marmota monax*) from New York. *Cornell Veterinarian* 68, 391–395.

Fleming, W.J., Georgi, J.R. and Caslick, J.W. (1979) Parasites of the woodchuck (*Marmota monax*) in central New York State. *Proceedings of the Helminthological Society of Washington* 46, 115–127.

Fotedar, D.N. an4d Tikoo, R. (1968) Studies on the life cycle of *Cosmocerca kashmirensis* Fotedar, 1959, a common oxyurid nematode parasite of *Bufo viridis* in Kashmir. *Indian Science Congress Association Proceedings* 55 (III), 460.

Fox, A.S., Kazacos, K.R., Gould, N.S., Heydemann, P.T., Thomas, C. and Boyer, K.M. (1985) Fatal eosinophilic meningoencephalitis and visceral larva migrans caused by the raccoon ascarid *Baylisascaris procyonis*. *New England Journal of Medicine* 312, 1619–1623.

Frank, J.F. (1953) A note on the experimental transmission of enterohepatitis of turkeys by arthropods. *Canadian Journal of Comparative Medicine and Veterinary Science* 17, 230–231.

Freire Filha, L.G. and Campos, D.M.B. (1992) Development of *Lagochilascaris minor* Leiper, 1909, in inbred C57B1/6 mice. *Revista de Patologia Tropical* 21, 219–233.

Fritz, T.E., Smith, D.C. and Flynn, R.J. (1968) A central nervous system disorder in ground squirrels (*Citellus tridecemlineatus*) associated with visceral larva migrans. *Journal of American Veterinary Medical Association* 153, 841–844.

Fülleborn, F. (1921a) Über die Wanderung von Askaris und anderen Nematodenlarven im Körper und intrauterine Askarisinfektion. *Archiv für Schiffs- und Tropen-Hygiene, Pathologie und Therapie Exotischer Krankheiten* 25, 146–149.

Fülleborn, F. (1921b) Askarisinfektion durch Verzehren eingekapselter Larven und über gelungene intrauterine Askarisinfektion. (Vorläufige Mitteilung). *Archiv für Schiffs- und Tropen-Hygiene, Pathologie und Therapie Exotischer Krankheiten* 25, 367–375.

Fülleborn, F. (1922) Über Infektionsversuche mit *Toxascaris* (Vorläufige Mitteilung). *Archiv für Schiffs- und Tropen-Hygiene, Pathologie und Therapie Exotischer Krankheiten* 26, 59–60.

Fülleborn, F. (1929) On the larval migration of some parasitic nematodes in the body of the host and its biological significance. *Journal of Helminthology* 7, 15–26.

Gallego, J.B. (1947) Revision de la familia Atractidae Travassos, 1920 con descripcion de dos nuevas especies. *Revista Iberica de Parasitologia* 7, 3–90.

Geller, E.R. (1957) Epizootiology of *Contracaecum* infection of sterlet (*Acipenser ruthenus*). *Zoologicheski Zhurnal* 36, 1441–1447. (In Russian.)

Geller, E.R. and Babich, L.A. (1953) The biology of *Contracaecum bidentatum* (Linstow, 1855). *Papers on Helminthology presented to Academician K.I. Skrjabin on his 75th birthday*. Izatelstvo Akademii Nauk SSSR, Moscow, pp. 133–138. (In Russian.)

George-Nascimento, M. and Llanos, A. (1995) Micro-evolutionary implications of allozymic and morphometric variations in sealworms *Pseudoterranova* sp. (Ascaridoidea: Anisakidae) among sympatric hosts from the south eastern Pacific Ocean. *International Journal for Parasitology* 25, 1163–1171.

Ghadirian, E. (1968) Nématodes parasites d'ophidiens malgaches. *Mémoires du Muséum National d'Histoire Naturelle Zoologie Série A* 54, 1–54.

Gibbs, B.J. (1962) The occurrence of the protozoan parasite *Histomonas meleagridis* in the adults and eggs of the cecal worm *Heterakis gallinae*. *Journal of Protozoology* 9, 288–293.

Gibson, D.I. (1972) Contributions to the life histories and development of *Cucullanus minutus* Rudolphi, 1819 and *C. heterochrous* Rudolphi, 1802 (Nematoda: Ascaridida). *Bulletin of the British Museum (Natural History) Zoology*, 22, 153–170.

Gibson, D.I. (1983) The systematics of ascaridoid nematodes – a current assessment. In: Stone, A.R., Platt, H.M. and Khalil, L.F. (eds) *Concepts in Nematode Systematics*. The Systematics Association Special Volume No. 2, Academic Press, New York, pp. 321–338.

González, L. (1998) The life cycle of *Hysterothylacium aduncum* (Nematoda: Anisakidae) in Chilean marine farms. *Aquaculture* 162, 173–186.

Grassi, B. (1888) Weiteres zur Frage der Ascarisentwicklung. *Zentralblatt für Bakteriologie, Parasitenkunde und Infektionskrankheiten* 3, 748–749.

Graybill, H.W. (1921) Data on the development of *Heterakis papillosa* in the fowl. *Journal of Experimental Medicine* 34, 259–270.

Graybill, H.W. and Smith, T. (1920) Production of fatal blackhead in turkeys by feeding embryonated eggs of *Heterakis papillosa*. *Journal of Experimental Medicine* 31, 647–655.

Greenway, J.A. and McGraw, B.M. (1970a) *Ascaris suum* infection in calves. I. Clinical signs. *Canadian Journal of Comparative Medicine* 34, 227–237.

Greenway, J.A. and McGraw, B.M. (1970b) *Ascaris suum* infection in calves. II. Circling and marrow eosinophil responses. *Canadian Journal of Comparative Medicine* 34, 238–246.

Griffiths, J.A. (1922) Prenatal infection with parasitic worms. *Veterinary Journal* 78, 478–481.

Grove, D.I. (1990) *A History of Human Helminthology*. CAB International, Wallingford, UK.

Guberlet, J.E. (1924) Notes on the life history of *Ascaridia perspicillum* (Rud.). *Transactions of the American Microscopical Society* 44, 152–156.

Gurchenko, R.N. (1970a) Development of *Ascaridia galli* in earthworms. *Veterinariya, Moscow* 47, 72–73. (In Russian.)

Gurchenko, R.N. (1970b) The longevity of *Ascaridia galli* larvae in earthworms. *Byulleten Vsesoyuznogo Instituta Gel'mintologii im. K.I. Skrjabin* 4, 33–34. (In Russian.)

Hartwich, G. (1959) Revision der parasitischen Vogel Nematoden Mittleuropas. 1. Die Gattung *Porrocaecum* Railliet et Henry, 1912 (Ascaridoidea). *Mitteilungen aus dem Zoologischen Museum* 35, 107–147.

Hartwich, G. (1974) No. 2. Keys to genera of the Ascaridoidea. In: Anderson, R.C., Chabaud, A.G. and Willmott, S. (eds) *CIH Keys to the Nematode Parasites of Vertebrates*. Commonwealth Agricultural Bureaux, Farnham Royal, UK, pp. 1–15.

Harwood, P.D. (1930) A new species of *Oxysomatium* (Nematoda) with some remarks on the genera *Oxysomatium* and *Aplectana* and observations on the life history. *Journal of Parasitology* 17, 61–73.

Hays, R., Measures, L.N. and Huot, J. (1998a) Euphausiids as intermediate hosts of *Anisakis simplex* in the St. Lawrence estuary. *Canadian Journal of Zoology* 76, 1226–1235.

Hays, R., Measures, L.N. and Huot, J. (1998b) Capelin (*Mallotus villosus*) and herring (*Clupea harengus*) as paratenic hosts of *Anisakis simplex*, a parasite of beluga (*Delphinapterua leucas*) in the St. Lawrence estuary. *Canadian Journal of Zoology* 76, 1–7.

Henner, S. (1959) Untersuchungen über Häutungen von Larven verschiedener Ascaridenarten während ihrer präparasitischen Phase. Inaugural – Dissertation zur Erlangung der veterinärmedizinischen Doktorwürde der Tierärztlichen Fakultät der Ludwig-Maximilians-Universität zu München, 38 pp.

Herd, R.P. and McNaught, D.J. (1975) Arrested development and the histiotropic phase of *Ascaridia galli* in the chicken. *International Journal for Parasitology* 5, 401–406.

Herlich, H. and Porter, D.A. (1953) Prenatal infection of a calf with the nematode *Neoascaris vitulorum*. *Journal of Parasitology* (Suppl.) 39, 33–34.

Higashikawa, H. (1961) Experimental studies on visceral larva migrans. *Shikoku Acta Medica* 17, 1–20. (In Japanese.)

Hörchner, F. (1962) Ein Beitrag zur Kenntnis des Entwicklungszyklus von *Hexametra quadricornis* Wedl, 1862, Mozgovoy, 1951 (Nematoda: Ascaridae). *Zeitschrift für Parasitenkunde* 21, 187–194.

Horton-Smith, C., Long, P.L. and Lee, D.L. (1968) Observations on the tissue and post-tissue larval development of *Ascaridia dissimilis* Vigueras 1931, and a comparison with that of *Ascaridia galli*, in turkeys. *Parasitology* 58, 709–714.

Huang, W. and Bussiéras, J. (1988) Anisakidae and human anisakiasis. Part One: literature data. *Annales de Parasitologie Humaine et Comparée* 63, 119–132.

Huizinga, H.W. (1966) Studies on the life cycle and development of *Contracaecum spiculigerum* (Rudolphi, 1809) (Ascaridoidea: Heterocheilidae) from marine piscivorous birds. *Journal of the Elisha Mitchell Scientific Society* 82, 181–195.

Huizinga, H.W. (1967) The life cycle of *Contracaecum multipapillatum* (von Drasche, 1882) Lucker, 1941 (Nematoda: Heterocheilidae). *Journal of Parasitology* 53, 368–375.

Huizinga, H.W. (1971) Contracaeciasis in pelicaniform birds. *Journal of Wildlife Diseases* 7, 198–204.

Hwang, J.C. and Wehr, E.E. (1958) Observations on early development of *Ascaridia columbae* in pigeons (Abstract). *Journal of Parasitology* 44, 26–27.

Hwang, J.C. and Wehr, E.E. (1962) Observations on the life history of *Ascaridia columbae* (Abstract). *Journal of Parasitology* 48, 40.

Ikeme, M.M. (1970) Retarded metamorphosis in larvae of *Ascaridia galli* following repeated challenge of poultry with infective eggs. *The Veterinary Record* 87, 725–726.

Ishikura, H., Kikuchi, K., Nagasawa, K., Ooiwa, T., Takamiya, H., Sato, N. and Sugane, K. (1992) Anisakidae and anisakidosis. *Progress in Clinical Parasitology* 3, 43–102.

Itagaki, S. (1927) On the life history of the chicken nematode, *Ascaridia perspicillum*. *Proceedings of the World's Poultry Congress (Ottawa, Canada)*, pp. 339–344.

Itagaki, S. (1930) The nature of the parasitic nodules in the caecal wall of fowls and the development of *Heterakis vesicularis*. *Report of the Proceedings of the World's Poultry Congress (London, England)*, pp. 517–520.

Ivashkin, M.V. and Babaeva, M.B. (1973) The 'internal cycle' of viviparous nematodes from the gastro-intestinal tract of animals. In: Gagarin, V.G. (ed.) *Problemy Obshchei i Prikladnoi Gelmintologii*. Izdatel'stvo, Moscow, pp. 61–68. (In Russian.)

Jackson, G.J. (1975) The 'new disease' status of human anisakiasis and North American cases: a review. *Journal of Milk and Food Technology* 38, 769–773.

Jackson, G.J. (1995) Role of hyperbenthic crustaceans in the transmission of marine helminth parasites. *Canadian Journal of Fisheries and Aquatic Sciences* 54, 815–820.

Jacob, P.D., Varghese, C.G., Georgekutty, P.T. and Peter, C.T. (1970) A preliminary study on the role of grasshoppers (*Oedaleus abruptus and Spathosternum parasiniferum*) in the transmission of *Ascaridia galli* (Schrank, 1788) Freeborn, 1923 in poultry. *Kerala Journal of Veterinary Science* 1, 65–70.

Jacobson, H.A., Scanlon, P.R. and Nettles, V.F. (1976) Epizootiology of an outbreak of cerebrospinal nematodiasis in cottontail rabbits and woodchucks. *Journal of Wildlife Diseases* 12, 357–360.

Janiszewska, J. (1939) Studien über die Entwicklung und die Lebensweise der parasitischen Würmer in der Flunder (*Pleuronectes flesus* L.). *Memoires de l'Academie Polonais du Science serie B* 14, 1–68.

Jarecka, L., Choudhury, O. and Burt, M.D.B. (1988) On the life cycle of *Pseudoterranova decipiens*: experimental infections of micro- and macroinvertebrates (Abstract). *Bulletin of the Canadian Society of Zoologists* 19, 32.

Jensen, T. (1997) Experimental infection/transmission of sculpins (*Myoxocephalus scorpius*) and cod (*Gadhus morhua*) by sealworm (*Pseudoterranova decipiens*) larvae. *Parasitology Research* 83, 380–382.

Jensen, T., Anderson, K. and Des Clers, S. (1994) Sealworm (*Pseudoterranova decipiens*) infections in demersal fish from two areas in Norway. *Canadian Journal of Zoology* 72, 598–608.

Jerke, H.W.M. (1902) Eine parasitische Anguillula des Pferdes. *Archiv für Wissenschaftliche und Praktische Tierheilkunde* 29, 113–127.

Jogis, V.A. (1967) Life-cycle of *Porrocaecum semiteres* (Zeder, 1800) (Nematoda: Ascaridata). *Parazitologiya* 1, 213–218. (In Russian.)

Jogis, V.A. (1970) Experimental study of the specificity of *Porrocaecum ensicaudatum* (Zeder, 1800) (Ascaridata) *Parazitologiya* 4, 563–568. (In Russian.)

Johnson, A.A. (1963) Ascarids in sheep. *New Zealand Veterinary Journal* 11, 69–70.

Joy, E.J. and Bunten, C.A. (1997) *Cosmocercoides variabilis* (Nematoda, Cosmocercoidea) populations in the eastern American toad *Bufo americanus* (Salientas: Bufonidae) from western Virginia. *Journal of the Helminthological Society of Washington*, 64, 102–105.

Jungerson, G., Eriksen, L., Nansen, P. and Fagerholm, H.P. (1997) Sex-manipulated *Ascaris suum* infections in pigs: implications for reproduction. *Parasitology* 115, 439–442.

Kahl, W. (1936) Beitrag zur Kenntnis des Nematoden *Contracaecum clavatum* Rud. *Zeitschrift für Parasitenkunde* 8, 509–520.

Kalbe, I. (1955) Untersuchungen über die Entwicklungsfähigkeit der Eier von *Ascaris lumbricoides* L. und *Parascaris equorum* Goeze nach Abwasseraufenthalt und unter anderen biologischen Bedingungen. *Archiv für Experimentelle Veterinärmedizin* 9, 557–568.

Karunamoorthy, G., Chellappa, D.J. and Anandan, R. (1994) The life history of *Subulura brumpti* in the beetle *Alphitobius diaperinus*. *Indian Veterinary Journal* 71, 12–15.

Kazacos, K.R. (1982) Contaminative ability of *Baylisascaris procyonis* infected raccoons in an outbreak of cerebrospinal nematodiasis. *Proceedings of the Helminthological Society of Washington* 49, 155–157.

Kazacos, K.R. (1983) *Raccoon Roundworms* (Baylisascaris procyonis) – *a Cause of Animal and Human Disease*. Station Bulletin No. 422, Department of Veterinary Microbiology, Pathology and Public Health, Agriculture Experimental Station, Purdue University, West Lafayette, Indiana, 22 pp.

Kazacos, K.R. (1997) Visceral, ocular, and neural larva migrans. In: Connor, D.H., Chandler, F.W., Schwartz, D.A., Manz, H.J. and Lack, E.E. (eds.) *Pathology of Infectious Diseases*. Vol. II. Appleton and Lange, Stamford, Connecticut, pp. 1459–1473.

Kazacos, K.R. and Kazacos, E.A. (1984) Experimental infection of domestic swine with *Baylisascaris procyonis* from raccoons. *American Journal of Veterinary Research* 45, 1114–1121.

Kazacos, K.R. and Wirtz, W.L. (1983) Experimental cerebrospinal nematodiasis due to *Baylisascaris procyonis* in chickens. *Avian Diseases* 27, 55–65.

Kazacos, K.R., Appel, G.O. and Thacker, H.L. (1981a) Cerebrospinal nematodiasis in a woodchuck suspected of having rabies. *Journal of American Veterinary Medical Association* 179, 1102–1104.

Kazacos, K.R., Wirtz, W.L., Burger, P.P. and Christmas, C.S. (1981b) Raccoon ascarid larvae as a cause of fatal central nervous system disease in subhuman primates. *Journal of the American Veterinary Medical Association* 179, 1089–1094.

Kazacos, K.R., Kazacos, E.A., Render, J.A. and Thacker, H.L. (1982a) Cerebrospinal nematodiasis and visceral larva migrans in an Australian (Latham's) brush turkey. *Journal of American Veterinary Medical Association* 181, 1295–1298.

Kazacos, K.R., Winterfield, R.W. and Thacker, H.L. (1982b) Etiology and epidemiology of verminous encephalitis in an emu. *Avian Diseases* 26, 389–391.

Kazacos, K.R., Vestra, W.A. and Kazacos, E.A. (1984) Raccoon ascarid larvae (*Baylisascaris procyonis*) as a cause of ocular larva migrans. *Investigative Ophthalmology and Visual Science* October 1984, 1177–1183.

Kean, B.H., Mott, K. and Russell, A.J. (eds) (1978) *Tropical Medicine and Parasitology*. Classic Investigations, Vol. 2. Cornell University Press, Ithaca, and London.

Kelley, G.W., Olsen, L.S. and Hoerlein, A.B. (1957) Rate of migration and growth of larval *Ascaris suum* in baby pigs. *Proceedings of the Helminthological Society of Washington* 24, 133–136.

Kendall, S.B. (1959) The occurrence of *Histomonas meleagridis* in *Heterakis gallinae*. *Parasitology* 49, 169–172.

Kennedy, P.C. (1954) The migration of the larvae of *Ascaris lumbricoides* in cattle and their relation to eosinophilic granulomas. *Cornell Veterinarian* 44, 531–565.

Kerr, K.B. (1955) Age of chickens and the rate of maturation of *Ascaridia galli*. *Journal of Parasitology* 41, 233–235.

Khaziev, G.Z. (1972) The role of earthworms in the epizootiology of *Ascaridia* and *Heterakis* of chickens and guinea fowl. *Trudi Bashkirskogo Sel'skokhozyaistvennogo Instituta* 17, 97–112. (In Russian.)

Khouri, S.R. and Pande, B.P. (1970) Observations on post-embryonic development in relation to tissue-phase of *Ascaridia galli* in laboratory-raised chicks. *Indian Journal of Animal Science* 40, 61–72.

Khromova, L.A. (1975) Development of *Dacnitis sphaerocephalus caspicus* (Nematoda: Cucullanidae). *Zoologicheskii Zhurnal*, 54, 449–452. (In Russian.)

Kino, H., Watanabe, K., Matsumoto, K., Veda, M., Sigiura, M., Suzuki, H., Takai, T., Tsuboi, H., Sano, M., Fujiv, Y. and Kagei, N. (1993) Occurrence of anisakiasis in the western part of Shizuoka Prefecture, with special reference to the prevalence of anisakid infections in sardine, *Engraulis japonica. Japanese Journal of Parasitology* 42, 308–312.

Køie, M. (1993) Aspects of the life cycle and morphology of *Hysterothylacium aduncum* (Rudolphi, 1802) (Nematoda, Ascaridoidea, Anisakidae). *Canadian Journal of Zoology* 71, 1289–1296.

Køie, M. and Fagerholm, H.P. (1993) Third-stage larvae emerge from eggs of *Contracaecum osculatum* (Nematoda, Anisakidae). *Journal of Parasitology* 79, 777–780.

Køie, M. and Fagerholm, H.P. (1995) The life cycle of *Contracaecum osculatum* (Rudolphi, 1802) *sensu stricto* (Nematoda, Ascaridoidea) in view of experimental infections. *Parasitology Research* 81, 481–489.

Køie, M., Berland, B. and Burt, M.D.B. (1995) Development to third-stage larvae occurs in the eggs of *Anisakis simplex* and *Pseudoterranova decipiens* (Nematoda, Ascaridoidea, Anisakidae). *Canadian Journal of Fisheries and Aquatic Sciences* 52, 134–139.

Koino, S. (1922) Experimental infection of the human body with ascarids. *Japan Medical World* 2, 317–320.

Kosinova, V.G. (1965) Life cycle of the nematode *Raphidascaris acus* (Bloch, 1779) Railliet and Henry, 1915 – a parasite of fish. *Materialy Nauchnoi Konferentsii Vsesoyuznogo Obshchestva Gel'mintologov* Part 2, 128–131. (In Russian.)

Kudryavtsev, A.A. (1971) Toxocariasis of polar fox (history of the problem, biology of the pathogen, prophylaxis, and therapeutics). *Kand. Dissertatsiya, Moscow.* (Cited in Mozgovoi and Shakhmatova, 1973, *Osnovy Nematodologii*, Vol. 23.) (In Russian.)

Kudryavtsev, A.A. (1974) A study of the biology of *Toxascaris leonina* (Linstow, 1902). *Gel'minty Zhivotnykh, Cheloveka i Rastenii na Yuzhnom Urale, Vypusk 1. Ufa, USSR.* Akademiya Nauk SSSR, Bashkir Filial Instituta Biologii, pp. 124–131. (In Russian.)

Kutzer, E. and Grünberg, W. (1965) Parasitologie und Pathologie der Spulwurmkrankheit der Schlangen. *Zentralblatt für Veterinarische Medizin* 12, 155–175.

Kutzer, E. and Lamina, J. (1965) Zur Biologie einiger Schlangen Ascariden. *Zeitschrift für Parasitenkunde* 25, 211–230.

Kuzia, E.J. (1978) Studies on the life history, seasonal periodicity and histopathology of *Dichelyne bullocki* Stromberg and Crites (Nematoda: Cucullanidae) a parasite of *Fundulus heteroclitus* (L.). Unpublished PhD thesis, University of New Hampshire, Durham, New Hampshire.

Kwiecien, J.M., Smith, D.A., Key, D.W., Swinton, J. and Smith-Maxie, L. (1993) Encephalitis attributed to larval migration of *Baylisascaris* species in emus. *Canadian Veterinary Journal* 34, 176–178.

Lane, C.L. (1917) Major Stewart on *Ascaris* infection. *Indian Medical Gazette* 52, 301.

Le Roux, P.L. (1924) Helminths collected from equines in Edinburgh and in London. *Journal of Helminthology* 2, 111–134.

Lee, D.L. (1969) The structure and development of *Histomonas meleagridis* (Mastigamoebidae: Protozoa) in the female reproductive tract of its intermediate host, *Heterakis gallinarum. Parasitology* 59, 877–884.

Lee, D.L. (1971) The structure and development of *Histomonas meleagridis* in the male reproductive tract of its intermediate host, *Heterakis gallinarum* (Nematoda). *Parasitology* 63, 439–445.

Lee, H.F. (1960) Effects of superinfection on the behaviour of *Toxocara canis* larvae in mice. *Journal of Parasitology* 46, 583–588.

Lester, R.J.G., Blair, D. and Heald, D. (1981) Nematodes from scallops and turtles from Shark Bay, Western Australia. *Australian Journal of Marine and Freshwater Research* 31, 713–717.

Leuckart, R. (1876) *Die Menschlichen Parasiten und die von ihnen Herrührenden Krankheiten*, Vols 2 and 3. Leipzig, pp. 513–882

Le-Van-Hoa (1966) Cycle évolutif de *Seuratum nguyenvanaii* Le-Van-Hoa, 1964, parasite de la musaraigne, *Suncus murinus* (L.) au Viet-Nam. *Bulletin de la Société Pathologie Exotique* 57, 124–128.

Le-Van-Hoa and Pham-Ngoc-Khue (1967) Morphologie et cycle évolutif de *Cucullanus chabaudi* n.sp., parasite des poissons, *Pangasius pangasius* H.B. (*P. buchanani*) du Sud-Viet-Nam. Note Preliminaire. *Bulletin de la Société Pathologie Exotique* 60, 315–318.

Levin, N.L. (1956) Life history studies on *Porrocaecum ensicaudatum*, an avian nematode. Unpublished PhD thesis, University of Illinois, Normal, Illinois.

Levin, N.L. (1961) Life history studies on *Porrocaecum ensicaudatum*, an avian nematode. I. Experimental observations in the chicken. *Journal of Parasitology* 47, 38–46.

Lewis, J.W. and Maizals, R.M. (eds) (1993) *Clinical, epidemiological and molecular perspectives. Ascaridoidea*. Toxocara and *Toxocariasis*. British Society for Parasitology with the Institute of Biology, 169 pp.

Li, J.H. (1989) Development of *Baylisascaris schroederi* in mouse. *Chinese Journal of Veterinary Science and Technology* 8, 24–25.

Lichtenfels, J.R., Bier, J.W. and Madden, P.A. (1978) Larval anisakid (*Sulcascaris*) nematodes from Atlantic molluscs with marine turtles as definitive hosts. *Transactions of the American Microscopical Society* 97, 199–207.

Lund, E.E. (1957) Growth and development of *Heterakis gallinae* in turkeys and chickens infected with *Histomonas meleagridis*. *Journal of Parasitology* 43, 297–301.

Lund, E.E. (1960) Factors influencing the survival of *Heterakis* and *Histomonas* on soil (Abstract). *Journal of Parasitology* 46, 38.

Lund, E.E. (1966) The significance of earthworm transmission of *Heterakis* and *Histomonas*. *Proceedings of the First International Congress of Parasitology, Rome, 1964* 1, 371–372. Pergamon Press, Oxford, UK.

Lund, E.E. (1968) Acquisition and liberation of *Histomonas wenrichi* by *Heterakis gallinarum*. *Experimental Parasitology* 22, 62–67.

Lund, E.E. and Chute, A.M. (1974) The reproductive potential of *Heterakis gallinarum* in various species of galliform birds; implications for survival of *H. gallinarum* and *Histomonas meleagridis* in recent times. *International Journal for Parasitology* 4, 455–461.

Lund, E.E., Wehr, E.E. and Ellis, D.J. (1966) Earthworm transmission of *Heterakis* and *Histomonas* to turkeys and chickens. *Journal of Parasitology* 52, 899–902.

Lutz, A. (1888) Zur Frage der Uebertragung des menschlichen Spulwurms. Weitere Mitteilungen. *Zentralblatt für Bakteriologie, Parasitenkunde und Infektionskrankheiten* 3, 425–428.

Lyons, E.T., Swerczek, T.W., Tolliver, S.C. and Drudge, J.H. (1996) Natural superinfection of *Parascaris equorum* in a stall confined orphan horse foal. *Veterinary Parasitology* 66, 119–123.

Macchioni, G. (1971) The life cycle of *Ascaridia numidae* (Leiper, 1908) in the guinea fowl (*Numida meleagris*). *Veterinaria Italiana* 22, 488–493.

Machida, M. Takahashi, K. and Masuuchi, S. (1978) *Thynnascaris haze* n.sp. (Nematoda, Anisakidae) from goby in the Bay of Tokyo. *Bulletin of the National Science Museum, Series A (Zool.)* 4, 241–244.

Mackenzie, K. and Gibson, D.I. (1970) Ecological studies of some parasites of plaice *Pleuronectes platessa* L. and flounder *Platichthys flesus*. *Symposium of the British Society of Parasitology* 8, 1–42.

MacKinnon, B.M. and Burt, M.D.B. (1992) Functional morphology of the female reproductive tract of *Pseudoterranova decipiens* (Nematoda) raised *in vivo* and *in vitro*. *Zoomorphology* 112, 237–245.

Mackintosh, N.A. (1965) *The Stocks of Whales*. Fishing News (Books), London.

Madsen, H. (1950) Studies on species of *Heterakis* (nematodes) in birds. *Danish Review of Game Biology* 1, 1–43.

Madsen, H. (1962a) On the interaction between *Heterakis gallinarum*, *Ascaridia galli*, 'Blackhead' and the chicken. *Journal of Helminthology* 36, 107–142.

Madsen, H. (1962b) The so-called tissue phase in nematodes. *Journal of Helminthology* 36, 143–148.

Marcogliese, D.J. (1996) Larval parasitic nematodes infecting marine crustaceans in eastern Canada. 3. *Hysterothylacium aduncum*. *Journal of the Helminthological Society of Washington* 63, 12–18.

Marcogliese, D.J. (1997) Fecundity of sealworm (*Pseudoterranova decipiens*) infecting grey seals (*Halichoerus grypus*) in the Gulf of St. Lawrence, Canada: lack of density-dependent effects. *International Journal of Parasitology* 27, 1401–1409.

Marcogliese, D.J. and Burt, M.D.B. (1993) Larval parasitic nematodes infecting marine crustaceans in eastern Canada, Passamaquoddy Bay, New Brunswick. *Journal of the Helminthological Society of Washington* 60, 100–104.

Margolis, L. and Arthur, J.R. (1979) Synopsis of the parasites of fishes of Canada. *Bulletin of the Fisheries Research Board of Canada* 199, 1–270.

Markowski, S. (1937) Über die Entwicklungsgeschichte und Biologie des Nematoden *Contracaecum aduncum* (Rudolphi, 1802). *Bulletin de l'Académie Polonaise des Sciences et Lettres, Série B* 2, 227–247.

Markowski, S. (1966) The diet and infection of fishes in Cavendish Dock, Barrow-in-Furness. *Journal of Zoology London* 150, 183–197.

Martin, H.M. (1926) Studies on *Ascaris lumbricoides*. *Research Bulletin of the Nebraska Agricultural Experimental Station* No. 37, 78 pp.

Maruyama, H., Nawa, Y., Noda, S., Mimori, T. and Choi, W.Y. (1996) An outbreak of visceral larva migrans due to *Ascaris suum* in Kyushu, Japan. *Lancet* 347, 1766–1767.

Matoff, K. and Vasilev, I. (1958a) The lamb as the host of *Neoascaris vitulorum* (Goeze, 1782) Travassos, 1927. *Izvestiya na Instituta po Sravnitelna Patologiya na domashnite Zhivotni, Sofia* 7, 281–297. (In Russian.)

Matoff, K. and Vasilev, I. (1958b) Uber die Biologie von *Taxasaris leonina* (Linstow, 1902, Leiper, 1907). *Zeitschrift für Parasitenkunde* 19, 111–135.

Mattiucci, S. and Paggi, L. (1989) Multilocus electrophoresis for the identification of larval *Anisakis simplex* A and B and *Pseudoterranova decipiens* A, B, and C from fish. In: Möller (ed.) *Nematode Problems in North Atlantic Fish*. Report from a workshop in Kiel 3–4 April, 1989. ICES Mariculture Communication, C.M. 1989/F, 6.

Mattiucci, S., Nascetti, G., Bullini, L., Orecchia, P. and Paggi, L. (1986) Genetic structure of *Anisakis physeteris* and differentiation from the *Anisakis simplex* complex (Ascaridida: Anisakidae). *Parasitology* 93, 383–387.

Mattiucci, S., Nascetti, G., Cianchi, R., Paggi, L., Arduin, P., Margolis, L., Brattey, J., Webb, S., D'Amelio, S., Crecchia, P. and Bullini, L. (1997) Genetic and ecological data on the *Anisakis simplex* complex, with evidence for a new species (Nematoda, Ascaridoidea, Anisakidae). *Journal of Parasitology* 83, 401–416.

Mattiucci, S., Paggi, L., Nascetti, G., Ishikura, H., Kikuchi, K., Sato, N., Cianchi, R. and Bullini, L. (1998) Allozyme and morphological identification of *Anisakis*, *Contracaecum* and *Pseudoterranova* from Japanese waters (Nematoda, Ascaridoidea). *Systematic Parasitology* 40, 81–92.

Maung, M. (1978) The occurrence of the second moult of *Ascaris lumbricoides* and *Ascaris suum*. *International Journal for Parasitology* 8, 371–378.

McClelland, G. (1980a) *Phocanema decipiens*: growth, reproduction, and survival in seals. *Experimental Parasitology* 49, 175–187.

McClelland, G. (1980b) *Phocanema decipiens*: molting in seals. *Experimental Parasitology* 49, 128–136.

McClelland, G. (1982) *Phocanema decipiens* (Nematoda: Anisakinae): experimental infections in marine copepods. *Canadian Journal of Zoology* 60, 502–509.

McClelland, G. (1990) Larval sealworm (*Pseudoterranova decipiens*) infections in benthic macrofauna. In: Bowen, W.D. (ed.) *Population Biology of Sealworm* (Pseudoterranova decipiens) *in Relation to its Intermediate and Seal Hosts. Canadian Bulletin of Fisheries and Aquatic Sciences* 222, 47–65.

McClelland, G. (1995) Experimental infection of fish with larval sealworm, *Pseudoterranova decipiens* (Nematoda, Anisakidae), transmitted by amphipods. *Canadian Journal of Fisheries and Aquatic Sciences* 52 (Suppl. 1), 140–155.

McClelland, G. and Ronald, K. (1974a) *In vitro* development of the nematode *Contracaecum osculatum* Rudolphi, 1802 (Nematoda: Anisakinae). *Canadian Journal of Zoology* 52, 847–855.

McClelland, G. and Ronald, K. (1974b) *In vitro* development of *Terranova decipiens* (Nematoda) (Krabbe, 1878). *Canadian Journal of Zoology* 52, 471–479.

McClelland, G., Misra, R.K. and Martell, D.J. (1990) Larval anisakine nematodes in various fish species from Sable Island Bank and vicinity. In: Bowen, W.D. (ed.) *Population Biology of Sealworm* (Pseudoterranova decipiens) *in Relation to its Intermediate and Seal Hosts. Canadian Bulletin of Fisheries and Aquatic Sciences* 222, 83–113.

McConnell, C.J., Marcogliese, D.J. and Stacey, M.W. (1997) Settling rate and dispersal of seal worm eggs (Nematoda) determined by using a revised protocol for myxozoan spores. *Journal of Parasitology* 83, 203–206.

McCraw, B.M. and Greenway, J.A. (1970) *Ascaris suum* infection in calves. III. Pathology. *Canadian Journal of Comparative Medicine* 34, 247–255.

McDonald, F.E. and Chevis, R.A.F. (1965) *Ascaris lumbricoides* in lambs. *New Zealand Veterinary Journal* 13, 41.

McGraw, J.C. (1968) A study of *Cosmocercoides dukae* (Holl, 1928) Wilkie, 1930 (Nematoda: Cosmocercidae) from amphibians in Ohio. Unpublished PhD thesis, Ohio State University, Columbus, Ohio.

McNeill, M.A. (1988) Development, pathogenicity and epizootiology of *Porrocaecum ensicaudatum* (Nematoda: Ascaridoidea) in birds. Unpublished PhD thesis, University of Guelph, Guelph, Ontario.

McNeill, M.A. and Anderson, R.C. (1990a) Development of *Porrocaecum ensicaudatum* (Nematoda: Ascaridoidea) in terrestrial oligochaetes. *Canadian Journal of Zoology* 68, 1476–1483.

McNeill, M.A. and Anderson, R.C. (1990b) Development of *Porrocaecum ensicaudatum* (Nematoda: Ascaridoidea) in starlings (*Sturnus vulgaris*). *Canadian Journal of Zoology* 68, 1484–1493.

Measures, L.N. (1996) Effect of temperature and salinity on development and survival of eggs and free-living larvae of sealworm (*Pseudoterranova decipiens*). *Canadian Journal of Fisheries and Aquatic Sciences* 53, 2804–2807.

Measures, L.N. and Hong, H. (1995) The number of moults in the egg of seal worm *Pseudoterranova decipiens* (Nematoda: Ascaridoidea): an ultrastructural study. *Canadian Journal of Fisheries and Aquatic Sciences* 52, 156–160.

Melendez, R.D. and Lindquist, W.D. (1979) Experimental life cycle of *Ascaridia columbae* in intravenously infected pigeons, *Columba livia*. *Journal of Parasitology* 65, 85–88.

Meuwissen, M. (1997) Investigation on the transmission of *Toxocara canis* Werner 1782 (Anisakidae) from the red fox (*Vulpes vulpes*) to dog (beagle). Dissertation, Aus dem Institut für Parasitologie der Tierärztlichen Hoch Schule Hannover, 181 pp.

Mia, S., Dewan, M.L., Uddin, M. and Chowdhury, M.U.A. (1975) The route of infection of buffalo calves by *Toxocara* (*Neoascaris*) *vitulorum*. *Tropical Animal Health Production* 7, 153–156.

Migajska, H. (1994) The distribution and survival of eggs of *Ascaris suum* in six different natural soil profiles. *Acta Parasitologica* 38, 170–174.

Moorhouse, D.E. (1982) Toxocariasis: a possible cause of Palm Island mystery disease. *Medical Journal of Australia* 1, 172–173.

Moraes, M.A.P., Arnaud, M.V.C. and DeLima, P.E. (1983) Novos casos de infecção humana por *Lagochilascaris minor* Leiper, 1909, encon trados no estado do Pará, Brasil. *Revista de Instituto de Medicina Tropical do São Paulo* 25, 139–146.

Moraes, M.A.P., Arnaud, M.V.C., de Macedo, R.C. and Anglada, A.E. (1985) Infecção pulmonar fatal por *Lagochilascaris* sp., provavelmente *Lagochilascaris minor* Leiper, 1909. *Revista de Instituto de Medicina Tropical do São Paulo* 27, 46–52.

Moravec, F. (1970a) On the life history of the nematode *Raphidascaris acus* (Bloch, 1779) in the natural environment of the river Bystrice, *Czechoslovakia. Journal of Fish Biology* 2, 313–322.

Moravec, F. (1970b) Study of the development of *Raphidascaris acus* (Bloch, 1779) (Nematoda: Heterocheilidae). *Věstník Ceskoslovenské Spolećnosti Zoologické* 34, 33–49.

Moravec, F. (1971a) Nematodes of fishes in Czechoslovakia. *Acta Scientiarum Naturalium Academiae Bohemoslovacae, Brno* 5, 1–49.

Moravec, F. (1971b) A new natural intermediate host of the nematode *Porrocaecum semiteres* (Zeder, 1800). *Folia Parasitologica* 18, 26.

Moravec, F. (1974) Some remarks on the development of *Paraquimperia tenerrima* Linstow, 1878 (Nematoda: Quimperiidae). *Scripta Facultatis Scientiarum Naturalism Universitatis Purkynianae Brunensis Biologia* 4, 135–142.

Moravec, F. (1976) Occurrence of the encysted larvae of *Cucullanus truttae* (Fabricius, 1794) in the brook lamprey, *Lampetra planeri* (Bl). *Scripta Facultatis Scientiatum Naturalism Universitatis Purkynianae Brunensis Biologia* 6, 17–20.

Moravec, F. (1979) Observations on the development of *Cucullanus* (*Truttaedacnitis*) *truttae* (Fabricius, 1794) (Nematoda: Cucullanidae). *Folia Parasitologica* 26, 295–307.

Moravec, F. (1980) Biology of *Cucullanus truttae* (Nematoda) in a trout stream. *Folia Parasitologica* 27, 217–226.

Moravec, F. (1994) *Parasitic Nematodes of Freshwater Fish in Europe*. Academia, Praha and Kluwer, The Netherlands, 473 pp.

Moravec, F. (1996) The amphipod *Gammarus fossarum* as a natural true intermediate host of the nematode *Raphidascaris acus. Journal of Parasitology* 82, 668–669.

Moravec, F. and Ergens, R. (1970) Nematodes from fishes and cyclostomes of Mongolia. *Folia Parasitiologica* 17, 217–232.

Moravec, F. and Kaiser, H. (1994) *Brevimulticaecum* sp. larvae (Nematoda: Anisakidae) from the frog *Hyla minuta* Peters in Trinidad. *Journal of Parasitology* 80, 154–156.

Moravec, F. and Malmquist, B. (1977) Records of *Cucullanus truttae* (Fabricius, 1794) (Nematoda: Cucullanidae) from Swedish brook lampreys, *Lampetra planeri* (Bloch). *Folia Parasitiologica* 24, 323–329.

Moravec, F., Nagasawa, K. and Urawa, S. (1985) Some fish nematodes from freshwater in Hokkaido, Japan. *Folia Parasitologica* 32, 305–316.

Moravec, F., Huffman, D.G. and Swim, D.J. (1995) The first record of fish as paratenic hosts of *Falcaustra* spp. (Nematoda: Kathlaniidae). *Journal of Parasitology* 81, 809–812.

Morrow, D.A. (1968) Pneumonia in cattle due to migratory *Ascaris lumbricoides* larvae. *Journal of the Veterinary Medical Association* 15, 184–189.

Mozgovoi, A.A. (1952) The life cycle of *Porrocaecum crassum*, a nematode of aquatic birds. *Doklady Akademii Nauk SSSR* 83, 335–336. (In Russian.)

Mozgovoi, A.A. (1953) Ascaridata of animals and man and the diseases caused by them. In: Skrjabin, K.I. (ed.) *Essentials of Nematology*, Vol. II, Part I. (Israel Program for Scientific Translations, Jerusalem, 1968.)

Mozgovoi, A.A. (1954) On the study of the epizootiology of *Porrocaecum* in aquatic birds. *Trudi Gel'mintologicheskoi Laboratorii Akademiya Nauk SSSR* 7, 196–199. (In Russian.)

Mozgovoi, A.A. and Bishaeva, L. (1959) On the evolution cycle of *Porrocaecum heteroura* (Ascaridata: Anisakidae). *Helminthologia* 1, 195–197. (In Russian.)

Mozgovoi, A.A. and Nosik, A.F. (1960) A study of intrauterine infection of animals with helminths. *Trudi Gel'mintologicheskoi Laboratorii Akademiya Nauk SSSR* 10, 143–148. (In Russian.)

Mozgovoi, A.A. and Shakhmatova, V.I. (1973) Ascaridata of animals and man and diseases caused by them. In: Skrjabin, K.I. (ed.) *Fundamentals of Nematology*, Vol. 23, Part III. (Translated by Amerind Publishing, India, 1985.)

Mozgovoi, A.A. and Shakhmatova, V.I. (1979) Study of the life cycle of anisakids from predatory birds on Lake Utkul' in the Altai Territory. *Sibirskogo Otdeleniyi Akademii Nauk* (*Ekologiya i morfologiya gel'mintov zapadnoi Sibiri*) 38, 94–103. (In Russian.)

Mozgovoi, A.A., Semenova, M.K. and Shakhmatova, V.I. (1968a) Life cycle of *Contracaecum microcephalum* (Ascaridata: Anisakidae) a parasite of fish-eating birds. *Papers in Helminthology presented to Academician K.I. Skrjabin on his 90th birthday*. Iztatelstro Akademii Nauk SSSR, Moscow, pp. 262–272. (In Russian.)

Mozgovoi, A.A., Shakhmatova, V.I. and Semenova, M.K. (1968b) Life cycle of *Contracaecum spiculigerum* (Ascaridata: Anisakidae), a parasite of domestic economically important birds. *Trudi Gel'mintologicheskoi Laboratorii Akademiya Nauk SSR* 19, 129–136. (In Russian.)

Mozgovoi, A.A. and Shakhtmatova, V.I. and Semanova, M.K. (1971) On the life cycle of *Goezia ascaroides* (Ascaridata: Goeziidae), a nematode of freshwater fish. *Sbornik Rabot po Gelmintologi*. Publishing House Kolos, Moscow, pp. 259–265. (In Russian.)

Munday, B.L. and Gregory, G.G. (1974) Demonstration of larval forms of *Baylisascaris tasmaniensis* in the wombat (Vombatus urisinus). *Journal of Wildlife Diseases* 10, 241–242.

Murrell, K.D., Eriksen, L., Nansen, P., Slotved, H.C. and Rasmussen, T. (1997) *Ascaris suum*: a revision of its early migratory path and implications for human ascariasis. *Journal of Parasitology* 83, 255–260.

Nadler, S.A. and Hudspeth, D.S.S. (1998) Ribosomal DNA and phylogeny of the Ascaridoidea (Nematoda: Secernentea): implications for morphological evolution and classification. *Molecular Genetics and Evolution* 10, 221–236.

Nagasawa, K. (1990a) The life-cycle of *Anisakis simplex*: a review. In: Ishikuva, K. and Kikuchi, K. (eds) *Intestinal Anisakiasis in Japan*. Springer-Verlag, Tokyo, pp. 31–40.

Nagasawa, K. (1990b) *Anisakis* larvae in intermediate and paratenic hosts in Japan. In: Ishikuva, K. and Kikuchi, K. (eds) *Infected Fish, Sero-immunological Diagnosis and Prevention*. Springer-Verlag, Tokyo, pp. 23–39.

Nagasawa, K. (1993) Review of human pathogenic parasites in the Japanese common squid (*Todarodes pacificus*) from the sea of Japan. In: Okutami, I., O'Dor, R.K. and Kubodera, I. (eds) *Recent Advances in Fisheries Biology*. Tokai University Press, Tokyo, pp. 293–312.

Nagasawa, K. and Moravec, F. (1995) Larval anisakid nematodes of Japanese common squid (*Todarodes pacificus*) from the sea of Japan. *Journal of Parasitology* 81, 69–75.

Nascetti, G., Ciauchi, R., Mattiucci, S., D'Amelio, S., Orecchia, P., Paggi, L., Brattey, J., Berland, B., Smith, J.W. and Bullini, L. (1993) Three sibling species within *Contracaecum osculatum* (Nematoda, Ascaridida, Ascaridoidea) from the Atlantic Arctic-Boreal region: productive isolation and host preferences. *International Journal for Parasitology* 23, 105–120.

Nascetti, G., Paggi, L., Orecchia, P., Smith, J.W., Mattiucci, S. and Bullini, L. (1986) Electrophoretic studies on the *Anisakis simplex* complex (Ascaridida: Anisakidae) from the Mediterranean and North-east Atlantic. *International Journal for Parasitology* 16, 633–640.

Nassi, H. and Dupouy, J. (1977) L'encapsulement de la larve de troisième stade de *Subulura* sp. (Nematoda, Ascaridida) chez *Pimelia (Homalopus) arenacea* Solier 1836 (Coleopt. Tenebr.); le rôle de la crypte de régénération. *Société Zoologique de France* 102, 243–249.

Nichols, R.L. (1956) The etiology of visceral larva migrans. I. Diagnostic morphology of infective second-stage *Toxocara* larvae. *Journal of Parasitology* 42, 349–362.

Niimi, D. (1937) Studies on blackhead. II. Mode of infection. *Journal of the Japanese Society of Veterinary Science* 16, 23–26.

Noda, R. (1957) Experimental studies on *Toxocara canis* infections in puppies. *Bulletin of the University of Osaka Prefecture, Series B, Agriculture and Biology* 7, 47–55. (In Japanese.)

Noda, R. (1959) A report on the experimental prenatal infection with *Toxocara canis* (Werner, 1782) in the dog. *Bulletin of the University of Osaka Prefecture, Series B, Agriculture and Biology* 9, 77–82. (In Japanese.)

Noda, R. (1961) Studies on the development of eggs of the dog ascarid, *Toxocara canis* (Werner, 1782), with an observation on its infection in mice. *Bulletin of the University of Osaka Prefecture, Series B, Agriculture and Biology* 11, 65–75. (In Japanese.)

Norris, D.E. and Overstreet, R.M. (1976) The public health implications of larval *Thynnascaris* nematodes from shellfish. *Journal of Milk and Food Technology* 39, 47–54.

Norton, R.A., Hopkins, B.A., Skeeles, J.K., Beasley, J.N. and Kreeger, J.M. (1992) High mortality of domestic turkeys associated with *Ascaridia dissimilis. Avian Diseases* 36, 469–473.

Ogren, R.E. (1953) A contribution to the life cycle of *Cosmocercoides* in snails (Nematoda: Cosmocercidae). *Transactions of the American Microscopical Society* 72, 87–91.

Ogren, R.E. (1959) The nematode *Cosmocercoides dukae* as a parasite of the slug. *Proceedings of the Pennsylvania Academy of Science* 33, 236–241.

Okoshi, S. and Usui, M. (1968) Experimental studies on *Toxascaris leonina*. V. Experimental infection of dogs and cats with eggs of canine, and feline strains. *Japanese Journal of Veterinary Science* 30, 81–91. (In Japanese.)

Ollé-Goig, J.E., Recacoechea, M. and Feeley, T. (1996) First case of *Lagochilascaris minor* infection in Bolivia. *Tropical Medicine and International Health* 1, 851–853.

Olsen, L.S., Kelley, G.W. and Sen, H.G. (1958) Longevity and egg production of *Ascaris suum. Transactions of the American Microscopical Society* 77, 380–383.

Orecchia, P., Paggi, L, Mattiucci, S., Smith, J.W., Nascetti, G. and Bullini, L. (1986) Electrophoretic identification of larvae and adults of *Anisakis* (Ascaridida: Anisakidae). *Journal of Helminthology* 60, 331–339.

Orecchia, P., Mattiucci, S., D'Amelio, S., Paggi, L., Plötz, J., Cianchi, R., Nascetti, G., Ardvino, P. and Bullini, L. (1994) Two new members in the *Contracaecum osculatum* complex (Nematoda, Ascaridoidea) from the Antarctic. *International Journal for Parasitology* 24, 367–377.

Osche, G. (1955) Über Entwicklung, Zwischenwirt und Bau von *Porrocaecum talpae, Porrocaecum ensicaudatum und Habronema mansioni* (Nematoda). *Zeitschrift für Parasitenkunde* 17, 144–164.

Osche, G. (1959) Über Zwischenwirte, Fehlwirte und die Morphogenese der Lippenregion bei *Porrocaecum* – und *Contracaecum* – Arten (Ascaridoidea: Nematoda). *Zeitschrift für Parasitenkunde* 19, 458–484.

Oshima, T. (1969) On the first intermediate host of *Anisakis. Saishin Igaku* 24, 401–404. (In Japanese.)

Oshima, T. (1972) Anikakis and anisakiasis in Japan and adjacent area. *Progress in Medical Parasitology in Japan*, Vol. 4. Meguro Parasitological Museum, Tokyo, pp. 305–393.

Oshima, T. (1979) In vitro development of Anisakis larvae collected from fishes caught in Juroshio current and their identification by the adults (Abstract). XIV Pacific Science Congress, August, 1979, Khabarovsk U.S.S.R. Committee F. Marine Sciences. Section F111, Biological Productivity of the Pacific Ocean. Moscow, U.S.S.R. p. 291.

Oshima, T., Kobayashi, A., Kumada, T., Koyama, T., Kagei, N. and Nemoto, T. (1968) Experimental infection with second-stage larvae of *Anisakis* sp. of *Euphausia similis* and *Euphausia pacifica*. *Japanese Journal of Parasitology* 17, 585. (In Japanese.)

Osipov, A.N. (1957) The ability of *Heterakis gallinarum* eggs to survive in cold conditions. *Trudi Moskovskoi Veterinarnoi Akademii* 19, 350–355. (In Russian.)

Osipov, A.N. (1958) Epizootiology of *Heterakis* in chickens. *Trudi Moskovskoi Veterinarnoi Akademii* 27, 196–218. (In Russian.)

Osterburg, B.F.J. (1992) *Lagochilascaris minor* in Surinam. *Tropical and Geographical Medicine* 44, 154–159.

Paggi, L., Nascetti, G., Webb, S.C., Mattiucci, S., Cianchi, R. and Bullini, L. (1998) A new species of *Anisakis* Dujardin, 1845 (Nematoda, Anisakidae) from beaked whales: allozyme and morphologic evidence. *Systematic Parasitology* 40, 161–174.

Paggi, L., Nascetti, G., Cianchi, R, Orecchia, P, Mattiucci, S., D'Amelio, S., Berland, B., Brattey, J., Smith, J.W. and Bullini, L. (1991) Genetic evidence for three species within *Pseudoterranova decipiens* (Nematoda, Ascaridida; Ascaridoidea) in the North Atlantic and Norwegian and Barents Seas. *International Journal for Parasitology* 21, 195–212.

Pahari, T.K. and Sasmal, N.K. (1990) Infection if Japanese quail with *Toxocara canis* larvae and establishment of patent infection in pups. *Veterinary Parasitology* 35, 357–364.

Pankavich, J.A., Emro, J.E., Poeschel, G.P. and Richard, G.A. (1974) Observations on the life history of *Ascaridia dissimilis* (Perez Vigueras, 1931) and its relationship to *Ascaridia galli* (Schrank, 1788). *Journal of Parasitology* 60, 963–971.

Papini, R. (1997) Lack of vertical transmission with *Baylisascaris transfuga* larvae in mice. *Revue de Médecine Vétérinaire* 148, 27–28.

Papini, R. and Casarosa, L. (1994) Observations on the infectivity of *Baylisascaris transfuga* eggs for mice. *Veterinary Parasitology* 51, 283–288.

Papini, R., Cavicchio, P. and Casarosa, L. (1993) Experimental infection in chickens with larvae of *Baylisascaris transfuga* (Nematoda: Ascaridoidea). *Folia Parasitologica* 40, 141–143.

Papini, R., Renzoni, G., Malloggi, M. and Casarosa, L. (1994) Visceral larval migrans in mice experimentally infected with *Baylisascaris transfuga* (Ascaridae: Nematoda). *Parasitologica* 36, 321–329.

Papini, R., Demi, S. and Croce, G.D. (1996) Observations on the migratory behaviour of *Baylisascaris transfuga* larvae in rabbits. *Revue de Médecine Vétérinaire* 147, 893–896.

Payne, F.K., Ackert, J.E. and Hartman, E. (1925) The question of the human and pig *Ascaris*. *American Journal of Hygiene* 5, 90–101.

Perroncito, E. (1882) *I Parassiti dell'Uomo e degli Animali Utili. Delle Più Comuni Malattie da essi Prodotte Profilassi e Cura Relativa*. Milano xii, 506 pp.

Petter, A.J. (1960) Sur une larve de subuluride, parasite de la blatte germanique (*Blatella germanica* L.). *Comptes Rendus des Séances de la Société de Biologie* 154, 300–301.

Petter, A.J. (1966) Equilibre des éspèces dans les populations de nématodes parasites du colon des tortues terrestres. *Mémoires du Muséum National d'Histoire Naturelle* 39, 1–252.

Petter, A.J. (1968a) Cycle évolutif de 2 espèces d'Heterakidae parasites de caméléons malgaches. *Annales de Parasitologie Humaine et Comparée* 43, 693–704.

Petter, A.J. (1968b) Observations sur la systématique et le cycle de l'ascaride *Hexametra quadricornis* (Wedl, 1862). *Annales de Parasitologie Humaine et Comparée* 43, 655–691.

Petter, A.J. (1974) Essai de classification de la famille des Cucullanidae. *Bulletin du Muséum National d'Histoire Naturelle, Paris, 3e ser.*, no. 255, *Zoologie* 177, 1469–1490.

Petter, A.J. and Chabaud, A.G. (1971) Cycle évolutif de *Megalobatrachonema terdentatum* (Linstow) en France. *Annales de Parasitologie Humaine et Comparée* 46, 463–477.

Pilitt, P.A., Lichtenfels, J.R. and Madden, P.A. (1979) Differentiation of fourth and early fifth stages of Parascaris equorum (Goeze, 1782) Nematoda: Ascaridoidea. *Proceedings of the Helminthological Society of Washington* 46, 15–20.

Pippy, J.H. and Banning, P. van (1975) Identification of *Anisakis* larva (I) as *Anisakis simplex* (Rudolphi, 1809, det. Krabbe, 1878) (Nematoda: Ascaridata). *Journal of The Fisheries Research Board of Canada* 32, 29–32.

Poinar, G.O. Jr, Chabaud, A.G. and Bain, O. (1989) *Rabbium paradoxus* sp.n. (Seuratidae: Skrjabinelaziinae) maturing in *Camponotus castaneus* (Hymenoptera: Formicidae). *Proceedings of the Helminthological Society of Washington* 56, 120–124.

Popova, T.I., Mozgovoi, A.A. and Dmitrenko, M.A. (1964) Biology of Ascaridata of animals from the White Sea. *Trudi Gel'mintologicheskoi Laboratorii Akademiya Nauk SSR* 14, 163–169. (In Russian.)

Prociv, P. (1983) Observations on the transmission and development of *Toxocara pteropodis* (Ascaridoidea: Nematoda) in the Australian Grey-Headed Flying-Fox, *Pteropus poliocephalus* (Pteropodidae: Megachiroptera). *Zeitschrift für Parasitenkunde* 69, 773–781.

Prociv, P. (1985) Observations on *Toxocara pteropodis* infections in mice. *Journal of Helminthology* 59, 267–275.

Prociv, P. (1989a) Observations on egg production by *Toxocara pteropodis*. *International Journal for Parasitology* 19, 441–443.

Prociv, P. (1989b) Larval migration in oral and parenteral *Toxocara pteropodis* infections and a comparison with *T. canis* dispersal in the flying fox, *Pteropus poliocephalus*. *International Journal for Parasitology* 19, 891–196.

Prociv, P. (1989c) Intraovular development and moulting of *Toxocara pteropodis*. *International Journal for Parasitology* 19, 749–755.

Prociv, P. and Brindley, P.J. (1986) Oral, parenteral and paratenic infections of mice with *Toxocara pteropodis*. *International Journal for Parasitology* 16, 471–474.

Prociv, P., Moorhouse, D.E. and Mak, J.W (1986) Toxocariasis – an unlikely cause of Palm Island mystery disease. *Medical Journal of Australia* 145, 14–15.

Prokopic, J. and Figallová, V. (1982a) The migration of larvae of *Toxascaris leonina* (Linstow, 1902) in experimentally infected white mice. *Folia Parasitologica* 29, 233–238.

Prokopic, J. and Figallová, V. (1982b) Migration of some roundworm species in experimentally infected white mice. *Folia Parasitologica* 29, 309–313.

Prusevich, T.O. (1964) On the formation of capsules around larvae of *Anisakis* sp. in the tissues of the shorthorn sculpin *Myxocephalus scorpius*. *Trudi Murmanskogo Biologicheskogo Instituta* 5, 265–273. (In Russian.)

Puylaert, F.A. (1970) Description de *Chitwoodchabaudia skryabini* g.n., sp.n. (Chitwood-chabaudiidae fam. nov.), parasite de *Xenopus laevis victorianus* Ahl. (Cosmocercoidea – Nematoda – Vermes). *Revue de Zoologie et de Botanique Africaines* 81, 369–382.

Pybus, M.J., Anderson, R.C. and Uhazy, L.S. (1978a) Redescription of *Truttaedacnitis stelmioides* (Vessichelli, 1910) (Nematoda: Cucullanidae) from *Lampetra lamottenii* (Lesueur, 1827). *Proceedings of the Helminthological Society of Washington* 45, 238–245.

Pybus, M.J., Uhazy, L.S. and Anderson, R.C. (1978b) Life cycle of *Truttaedacnitis stelmioides* (Vessichelli, 1910) (Nematoda: Cucullanidae) in America brook lamprey (*Lampetra lamottenii*). *Canadian Journal of Zoology* 56, 1420–1429.

Quentin, J.C. (1969) Cycle biologique de *Subulura williaminglisi* Quentin, 1965. Ontogénèse des structures céphaliques. Valeur phylogénétique de ce caractère dans la classification des nématodes Subuluridae. *Annales de Parasitologie Humaine et Comparée* 44, 451–484.

Quentin, J.C. (1970a) Sur le cycle évolutif de *Seuratum cadarachense* Desportes, 1947 et ses affinités avec ceux des némátodes subulures (Ascaridida) et rictulaires (Spirurida). *Annales de Parasitologie Humaine et Comparée* 45, 605–628.

Quentin, J.C. (1970b) Cycle du némátode *Seuratum cadarachense* Desportes, 1947. Affinités entre les cycle évolutif des Seuratidae et des Rictulariidae. *Comptes Rendus de l'Academie de Science, Paris* 270, 2311–2314.

Quentin, J.C. and Poinar, G.O. Jr (1973) Comparative study of the larval development of some heteroxenous subulurid and spirurid nematodes. *International Journal for Parasitology* 3, 809–827.

Quentin, J.C. and Seureau, C. (1975) Sur l'organogenése de *Seuratum cadarchense* Desportes, 1947 (Nematoda, Seuratoidea) et les réactions cellulaires de l'insecte *Locusta migratoria*, hôte intermédiaire. *Zeitschrift für ParasitenKunde* 47, 55–68.

Quentin, J.C. and Tcheprakoff, R. (1969) Cycle biologique de *Subulura otolicni* (Van Beneden, 1890). *Bulletin du Muséum National d'Histoire Naturelle, Série 2e*, 41, 571–578.

Quentin, J.C. and Verdier, J.M. (1979) Cycle biologique de *Maupasina weissi* Seurat, 1913 (Nematode Subuluroidea), parasite du Macroscélide. Ontogénèse des structures cephaliques. *Annales de Parasitologie Humaine et Comparée* 54, 621–635.

Quentin, J.C., Krishnasamy, M. and Tcheprakoff, R. (1977) Cycle biologique de *Tarsubulura perarmata* (Ratzel, 1868). *Annales de Parasitologie Humaine et Comparée* 52, 159–170.

Railliet, A. and Lucet, A. (1892) Observations et expériences sur quelques helminthes du genre *Heterakis* Dujardin. *Bulletin de la Société Zoologique de France* 17–19, 117–120.

Ransom, B.H. (1907) *Probstmayria vivipara* (Probstmayr, 1865) Ransom, 1907, a nematode of horses heretofore unreported from the USA. *Transactions of the American Microscopical Society* 27, 33–40.

Ransom, B.H. (1922) Some recent additions to the knowledge of ascariasis. *Journal of the American Medical Association* 79, 1094–1097.

Ransom, B.H. and Cram, E.B. (1921) The course of migration of *Ascaris* larvae. *American Journal of Tropical Medicine* 1, 129–160.

Ransom, B.H. and Foster, W.D. (1917) Life history of *Ascaris lumbricoides* and related forms. (Preliminary note). *Journal of Agricultural Research* 11, 395–398.

Ransom, B.H. and Foster, W.D. (1919) Recent discoveries concerning the life history of *Ascaris lumbricoides. Journal of Parasitology* 5, 93–99.

Ransom, B.H. and Foster, W.D. (1920) Observations on the life history of *Ascaris lumbricoides. United States Department of Agriculture Technical Bulletin* No. 817, 47 pp.

Reed, W.M., Kazacos, K.R., Dhillon, A.S., Winterfield, R.W. and Thacker, H.L. (1981) Cerebrospinal nematodiasis in bobwhite quail. *Avian Diseases* 25, 1039–1046.

Reid, W.M. (1959) Effects of temperature on the development of the eggs of *Ascaridia galli. Journal of Parasitology* 46, 63–67.

Rhodes, M.B., McCullough, R.A., Mebus, C.A., Klucas, C.A., Ferguson, D.L. and Twiehaus, M.J. (1977) *Ascaris suum*: Hatching of embryonated eggs in swine. *Experimental Parasitology* 42, 356–362.

Richardson, J.A., Kazacos, K.R. and Thacker, H.L. (1980) Verminous encephalitis in commercial chickens. *Avian Diseases* 24, 498–503.

Richter, C.B. and Kradel, D.C. (1964) Cerebrospinal nematodiasis in Pennsylvania groundhogs (*Marmota monax*). *American Journal of Veterinary Research* 25, 1230–1235.

Riley, W.A. and James, L.G. (1921) Studies on the chicken nematode, *Heterakis papillosa* Bloch. *Journal of the American Veterinary Medical Association* 59, 208–217.

Roberts, F.H.S. (1934) The large roundworm of pigs, *Ascaris lumbricoides* L. 1758. Its life history in Queensland, economic importance and control. *Bulletin of the Animal Health Station Yeerongpilly, Queensland, Australia*, No. 1, 81 pp.

Roberts, F.H.S. (1937a) Studies on the life history and economic importance of *Heterakis gallinae* (Gmelin, 1790) Freeborn, 1923, the caecum worm of fowls. *Australian Journal of Experimental Biology and Medicine* 15, 429–439.

Roberts, F.H.S. (1937b) Studies on the biology and control of the large roundworm of fowls *Ascaridia galli* (Schrank 1788) Freeborn 1923. *Queensland Department of Agriculture and Stock Bulletin* No. 2, pp. 1–106.

Roberts, J.A. (1989) The extraparasitic life cycle of *Toxocara vitulorum* in the village environment of Sri Lanka. *Veterinary Research Communications* 13, 377–388.

Roberts, J.A. (1990a) The egg production of *Toxocara vitulorum* in Asian buffalo (*Bubalus bubalis*). *Veterinary Parasitology* 37, 113–120.

Roberts, J.A. (1990b) The life cycle of *Toxocara vitulorum* in Asian buffalo (*Bubalus bubalis*). *International Journal for Parasitology* 20, 833–840.

Roberts, J.A. (1993a) The persistence of larvae of *Toxocara vitulorum* in Asian buffalo cows. *Buffalo Journal* 9, 247–251.

Roberts, J.A. (1993b) *Toxocara vitulorum* in ruminants. *Helminthological Abstracts* 62, 151–174.

Roepstorff, A., Eriksen, L., Slotred, H.C. and Nansen, P. (1997) Experimental *Ascaris suum* infection in the pig: worm population kinetics following single inoculations with three doses of infective eggs. *Parasitology* 115, 443–445.

Rosemberg, S., Lopes, M.B.S., Masuda, Z., Campos, R. and Vieira Bressan, M.C.R. (1986) Fatal encephalopathy due to *Lagochilascaris minor* infection. *American Journal of Tropical Medicine and Hygiene* 35, 575–578.

Rudmann, D.G., Kazacos, K.R., Storandt, S.T., Harris, D.L. and Janovitz, E.B. (1996) *Baylisascaris procyonis* larva migrans in a puppy: a case report and update for the veterinarian. *Journal of the American Animal Hospital Association* 32, 73–76.

Rye, L.A. and Baker, M.R. (1992) The life history of *Hysterothylacium analarum* Rye and Baker, 1984 (Nematoda: Anisakidae) in *Lepomis gibbosus* (Pisces: Centrarchidae) in southern Ontario, Canada. *Canadian Journal of Zoology* 70, 1576–1584.

Rysavy, B. (1958) Der Entwicklungszyklus von *Porrocaecum ensicaudatum* Zeder, 1800 (Nematoda: Anisakidae). *Acta Veterinaria Academiae Scientiarum Hungaricae* 9, 317–323.

Saitoh, Y., Yokokura, Y. and Itagaki, H. (1993) Earthworms as a transport host of the rat cecal worm, *Heterakis spumosa*. *Japanese Journal of Parasitology* 42, 392–397.

Saliba, A.M. and Araujo, P. (1972) Pathological changes experimentally induced by the larvae of *Hexametra quadricornis* (Wedl, 1861) in mice. *Folia Clinica et Biologica* 1, 3–7.

Sanford, S.E. (1989) Cerebrospinal nematodiasis caused by *Baylisascaris procyonis* in chinchillas. *Journal of Veterinary Diagnostic Investigation* 3, 77–79.

Santos, M.A.Q. dos, Campose, D.M.B., Komma, M.D. and Barnaké, W. (1987) *Lagochilascaris minor* (Leiper, 1909) em abscessodentario em Goiânia. *Revista de Patologia Tropical* 16, 129–142.

Sass, B. and Gorgacz, E.J. (1978) Cerebral nematodiasis in a chukar partridge. *Journal of the Veterinary Medical Association* 173, 1248–1249.

Schacher, J.F. (1957) A contribution to the life history and larval morphology of *Toxocara canis*. *Journal of Parasitology* 43, 599–612.

Schueler, R.L. (1973) Cerebral nematodiasis in a red squirrel. *Journal of Wildlife Diseases* 9, 58–60.

Schwartz, B. (1925) Occurrence of *Ascaris* in cattle in the USA. *North American Veterinarian* 6, 24–30.

Schwartz, B. (1959) Evolution of knowledge concerning the roundworm *Ascaris lumbricoides*. *Annual Report of the Smithsonian Institution, Washington, DC*, pp. 465–481.

Scothorn, M.W., Koutz, F.R. and Groves, H.F. (1965) Prenatal *Toxocara canis* infection in pups. *Journal of the American Veterinary Medical Association* 146, 45–48.

Scott, D.M. (1953) Experiments with the harbour seal, *Phoca vitulina*, a definitive host of a marine nematode, *Porrocaecum decipiens. Journal of the Fisheries Research Board of Canada* 10, 539–547.

Semenova, M.K. (1971) Life-cycle of *Contracaecum micropapillatum* (Ascaridata, Anisakidae). *Trudi Gel'mintologicheskoi Laboratorii (Teoreticheskie Voprosy Obshchei Gel'mintologii)* 22, 148–152. (In Russian.)

Semenova, M.K. (1972) Study of the biology of *Contracaecum micropapillatum* (Ascaridata: Anisakidae). *Problemy Parasiztologii. Trudi VII Nauchnoi Konferentsii Parazitologov USSR, Pt. II, Izdatel'stvo Nauk Dumka*, pp. 238–240. (In Russian.)

Semenova, M.K. (1973) Reservoir hosts of *Contracaecum micropapillatum* (Ascaridata: Anisakidae). *Trudi Gel'mintologicheskoi Laboratorii (Ekologiya i taksonomiya gel'mintov)* 23, 136–140. (In Russian.)

Semenova, M.K. (1974) The development of *Contracaecum microcephalum* (Anisakidae) in the definitive host. *Trudi Gel'mintologicheskoi Laboratorii (Ekologiya i geografiya gel'mintov)* 24, 153–160. (In Russian.)

Semenova, M.K. (1975) The morphology of *Contracaecum micropapillatum* (Stossich, 1890) Baylis, 1920 (Ascaridata: Anisakidae) in ontogenesis. *Trudi Gel'mintologicheskoi Laboratorii (Issledovanie po sistematike, zhiznennym tsiklam i biolhimii gel'mintov)* 25, 145–156. (In Russian.)

Semenova, M.K. (1979) The role of copepods in the life-cycle of *Contracaecum micropapillatum* (Ascaridata, Anisakidae). *Trudi Gel'mintologicheskoi Laboratorii (Gel'minty zhivotnykh i rastenii)* 29, 126–129. (In Russian.)

Seurat, L.G. (1917) Sur l'évolution du *Maupasina weissi* Seurat (Heterakidae). *Comptes Rendus des Séances de l'Academie des Sciences* 164, 1017–1018.

Shillinger, J.E. and Cram, E.B. (1922) Parasitic infestation of dogs before birth. *Journal of the American Veterinary Medical Association* 16, 63 and 200–203.

Shiraki, T. (1974) Larval nematodes of family Anisakidae (Nematoda) in the northern Sea of Japan – as a causative agent of eosinophilic phlegmon or granuloma in the human gastro-intestinal tract. *Acta Medica et Biologica* 22, 57–98.

Shulman, S.S. (1957) Material on the parasite fauna of lampreys from the basin of the Baltic and White Seas. In: G.K. Petrashevskii (ed.) *Parasites and Diseases of Fish. Izvestiya Gosudarstvennogo Nauchno-issle-skogo Instituta Ozernogo i Rechnogo Rybnogo Khozyaistva* 42, 282–299. (In Russian.)

Shulman, S.S. (1958) Zoogeography of parasites of USSR freshwater fishes. In: Dogiel, V.A., Petrushevski, G.K. and Polyanski, Yu.I. (eds) *Parasitology of Fishes*. TFH (Great Britain), UK, pp. 180–229. (Translated from the Russian by Z. Kabata.)

Siles, M., Cuéllar, C. and Perteguer, M.J. (1997) Genomic identification of *Anisakis simplex* isolates. *Journal of Helminthology* 71, 73–75.

Slijper, E.J. (1962) *Whales*. Hutchinson & Co., London.

Sluiters, J.F. (1974) *Anisakis* sp. larvae in the stomach of herring (*Clupea herengus* L.). *Zeitschrift für Parasitenkunde* 44, 279–288.

Smith, J.D. (1984a) Taxonomy of *Raphidascaris* spp. (Nematoda, Anisakidae) of fishes, with a redescription of *R. acus. Canadian Journal of Zoology* 62, 685–694.

Smith, J.D. (1984b) Development of *Raphidascaris acus* (Nematoda, Anisakidae) in paratenic, intermediate, and definitive hosts. *Canadian Journal of Zoology* 62, 1378–1386.

Smith, J.D. (1986) Seasonal transmission of *Raphidascaris acus* (Nematoda), a parasite of freshwater fishes, in definitive and intermediate hosts. *Environmental Biology of Fishes* 16, 295–308.

Smith, J.L., Bowman, D.D. and Little, M.D. (1983) Life cycle and development of *Lagochilascaris sprenti* (Nematoda: Ascarididae) from opossums (Marsupialia: Didelphidae) in Louisiana. *Journal of Parasitology* 69, 736–745.

Smith, J.W. (1971) *Thysanoessa inermis* and *T. longicaudata* (Euphausiidae) as first intermediate hosts of *Anisakis* sp. (Nematoda: Ascaridata) in the northern North Sea, to the north of Scotland and at Faroe. *Nature* 234, 478.

Smith, J.W. (1974) Experimental transfer of *Anisakis* sp. larvae (Nematoda: Ascaridida) from one fish host to another. *Journal of Helminthology* 48, 229–234.

Smith, J.W. (1983a) Larval *Anisakis simplex* (Rudolphi, 1809, det. Krabbe, 1878) and larval *Hysterothylacium* sp. (Nematoda: Ascaridoidea) in euphausiids (Crustacea: Malacostraca) in the North-East Atlantic and northern North Sea. *Journal of Helminthology* 57, 167–177.

Smith, J.W. (1983b) *Anisakis simplex* (Rudolphi, 1809, det. Krabbe, 1878) (Nematoda: Ascaridoidea): morphology and morphometry of larvae from euphausiids and fish, and a review of the life-history and ecology. *Journal of Helminthology* 57, 205–224.

Smith, J.W. (1984) Larval ascaridoid nematodes in myopsid and oegopsid cephalopods from around Scotland and in the northern North Sea. *Journal of Marine Biology Association* 64, 563–572.

Smith, J.W. and Wootten, R. (1978) *Anisakis* and anisakiasis. In: Lumsden, W.H.R, Muller, R. and Baker, J.R. (eds) *Advances in Parasitology*, Vol. 16. Academic Press, New York, pp. 93–163.

Smith, M.H.D. and Beaver, P. (1955) Visceral larva migrans due to infection with dog and cat ascarids. *Pediatric Clinics of North America* 2, 163–168.

Smith, P.E. (1953) Life history and host–parasite relations of *Heterakis spumosa*, a nematode parasite in the colon of the rat. *American Journal of Hygiene* 57, 194–221.

Soulsby, E.J.L. (1965) *Textbook of Veterinary Clinical Parasitology*, Vol. 1. *Helminths*. Blackwell Scientific Publications, Oxford, UK.

Spindler, L.A. (1967) Experimental transmission of *Histomonas meleagridis* and *Heterakis gallinarum* by the sow-bug *Porcellio scaber*, and its implications for further research. *Proceedings of the Helminthological Society of Washington* 34, 26–29.

Sprent, J.F.A. (1951) On the migratory behaviour of the larvae of various *Ascaris* species in mice. *Journal of Parasitology* (Suppl.) 37, 21.

Sprent, J.F.A. (1952a) On the migratory behaviour of the larvae of various *Ascaris* species in white mice. I. Distribution of larvae in tissues. *Journal of Infectious Diseases* 90, 165–176.

Sprent, J.F.A. (1952b) On an *Ascaris* parasite of the fisher and marten, *Ascaris devosi* sp.nov. *Proceedings of the Helminthological Society of Washington* 19, 27–37.

Sprent, J.F.A. (1953a) On the life history of *Ascaris devosi* and its development in the white mouse and the domestic ferret. *Parasitology* 42, 244–258.

Sprent, J.F.A. (1953b) Intermediate hosts in *Ascaris* infection. *Journal of Parasitology* (Suppl.) 39, 38.

Sprent, J.F.A. (1956) The life history and development of *Toxocara cati* (Schrank, 1788) in the domestic cat. *Parasitology* 46, 54–78.

Sprent, J.F.A. (1957) The development of *Toxocara canis* (Werner, 1782) in the dog. *Journal of Parasitology* (Suppl.) 43, 45.

Sprent, J.F.A. (1959a) The life history and development of *Toxascaris leonina* (von Linstow, 1902) in the dog and cat. *Parasitology* 49, 330–371.

Sprent, J.F.A. (1959b) Observations on the development of ascaridoid nematodes of the carpet snake. *Journal of Parasitology* (Suppl.) 45, 35.

Sprent, J.F.A. (1963a) The life history and development of *Amplicaecum robertsi*, an ascaridoid nematode of the carpet python (*Morelia spilotes variegatus*). I. Morphology and functional significance of larval stages. *Parasitology* 53, 7–38.

Sprent, J.F.A. (1963b) The life history and development of *Amplicaecum robertsi*, an ascaridoid nematode of the carpet python (*Morelia spilotes variegatus*). II. Growth and host specificity of larval stages in relation to the food chain. *Parasitology* 53, 321–337.

Sprent, J.F.A. (1968) Notes on *Ascaris* and *Toxascaris*, with a definition of *Baylisascaris* gen.nov. *Parasitology* 58, 185–198.

Sprent, J.F.A. (1969) Studies on ascaridoid nematodes in pythons: speciation of *Ophidascaris* in the Oriental and Australian regions. *Parasitology* 59, 937–959.

Sprent, J.F.A. (1970a) *Baylisascaris tasmaniensis* sp.nov. in marsupial carnivores: heirloom or souvenir? *Parasitology* 61, 75–86.

Sprent, J.F.A. (1970b) Studies on ascarioid nematodes in pythons; the life-history and development of *Ophidascaris moreliae* in Australian pythons. *Parasitology* 60, 97–122.

Sprent, J.F.A. (1970c) Studies on ascaridoid nematodes in pythons: the life history and development of *Polydelphis anoura* in Australian pythons. *Parasitology* 60, 375–397.

Sprent, J.F.A. (1971) Speciation and development in the genus *Lagochilascaris*. *Parasitology* 62, 71–112.

Sprent, J.F.A. (1977) Ascaridoid nematodes of amphibians and reptiles: *Sulcascaris*. *Journal of Helminthology* 51, 379–387.

Sprent, J.F.A. (1978) Ascaridoid nematodes of amphibians and reptiles: *Polydelphis, Travassosascaris* n.g. and *Hexametra*. *Journal of Helminthology* 52, 355–384.

Sprent, J.F.A. (1979) Ascaridoid nematodes of amphibians and reptiles: *Multicaecum* and *Brevimulticaecum*. *Journal of Helminthology* 53, 91–116.

Sprent, J.F.A. (1982) Host–parasite relationships of ascaridoid nematodes and their vertebrate hosts in time and space. *Deuxième Symposium sur la Spécificité Parasitaire des Parasites des Vertébrés. Mémoires du Muséum National D'Histoire Naturelle Serie A, Zoologie*, pp. 255–264.

Sprent, J.F.A. and Barrett, M.G. (1964) Large roundworms of dogs and cats: differentiation of *Toxocara canis* and *Toxascaris leonina*. *Australian Veterinary Journal* 40, 166–171.

Sprent, J.F.A. and Jones, H.I. (1977) Toxocariasis. *Australian Family Physician* 6, 6 pp.

Sprent, J.F.A. and Mines, J.J. (1960) A new species of *Amplicaecum* (Nematoda) from the carpet snake (*Morelia argus variegatus*): with a re-definition and a key for the genus. *Parasitology* 50, 183–198.

Sprent, J.F.A., Lamina, J. and McKeown, A. (1973) Observations on migratory behaviour and development of *Baylisascaris tasmaniensis*. *Parasitology* 67, 67–83.

Srivastava, H.D. and Mehra, K.N. (1955) Studies on the life history of *Ascaris vitulorum* (Goeze, 1782), the large intestinal roundworm of bovines. *Proceedings of 42nd Indian Science Congress, Part III*, p. 354.

Steffe, G.A. (1983) Zum Verhaltenalaktogen übertragener Larven von *Ancylostoma caninum* Ercolani, 1859 (Ancylostomatidae), *Toxocara canis* Werner, 1782 (Anisakidae) und *Toxascaris leonina* Linstow, 1902 (Ascaridae) in der Maus. *Inaugural – Dissertation, Tierärztliche Hochschule, Hannover*, GFR 60 pp.

Stewart, F.H. (1916a) On the life history of *Ascaris lumbricoides*. *British Medical Journal* 2, 5–7.

Stewart, F.H. (1916b) The life history of *Ascaris lumbricoides*. *British Medical Journal* 2, 474.

Stewart, F.H. (1916c) Further experiment on *Ascaris* infection. *British Medical Journal* 2, 486–488.

Stewart, F.H. (1917) On the life history of *Ascaris lumbricoides*. *British Medical Journal* 2, 753–754.

Stobo, W.T. and Zwanenburg, K.C.T. (1990) Grey seal (*Halichoerus grypus*) pup production on Sable Island and estimates of recent production in the Northwest Atlantic. In: Bowen, W.D. (ed.) *Population Biology of Sealworm* (Pseudoterranova decipiens) *in Relation to its Intermediate and Seal Hosts. Canadian Bulletin of Fisheries and Aquatic Sciences* 222, 171–184.

Stone, W.M. and Girardeau, M. (1968) Transmammary passage of *Ancylostoma caninum* larvae in dogs. *Journal of Parasitology* 54, 426–429.

Sudarikov, V.E. and Rizhikov, K.M. (1951) Notes on the bionomics of *Contracaecum osculatum baicalensis*, a nematode of the Baikal seal. *Trudi Gel'mintogicheskoi Laboratorii Akademiya Nauk SSR* 5, 59–66. (In Russian.)

Suedmeyer, W.K., Bermudez, A. and Kazacos, K.F. (1996) Cerebellar nematodiasis in an emu (*Dromaius novaehollandiae*). *Journal of Zoo and Wildlife Medicine* 27, 544–549.

Supriaga, V.G. (1972) Development of a nematode, *Porrocaecum crassum* in the intermediate host, earthworms. *Zoologicheskii Zhurnal* 51, 1123–1128. (In Russian.)

Supriaga, V.G. and Mozgovoi, A.A. (1974) Biological peculiarities of *Raphidascaris acus* (Anisakidae: Ascaridata), a parasite of freshwater fish. *Parazitologiya* 8, 494–503. (In Russian.)

Suzuki, T. and Ishida, K. (1979) *Anisakis simplex* and *Anisakis physeteris*: physiochemical properties of larval and adult hemoglobins. *Experimental Parasitology* 48, 225–234.

Swerczek, T.W. and Helmboldt, C.F. (1970) Cerebrospinal nematodiasis in groundhogs (*Marmota monax*). *Journal of American Veterinary Medical Association* 157, 671–674.

Swerczek, T.W., Nielson, S.W. and Helmboldt, C.F. (1971) Transmammary passage of *Toxocara cati* in the cat. *American Journal of Veterinary Research* 32, 89–92.

Taffs, L.F. (1961) Immunological studies on experimental infections of pigs with *Ascaris suum* Goeze, 1782. I. An introduction with a review of the literature and the demonstration of complement-fixing antibodies in the serum. *Journal of Helminthology* 35, 319–344.

Takata, I. (1951) Experimental infection of man with *Ascaris* of man and the pig. *Kitasato Archives* 23, 49–59.

Templeman, W. (1990) Historical background to the sealworm problem in eastern Canadian waters. In: Bowen, W.D. (ed.) *Population Biology of Sealworm* (Pseudoterranova decipiens) *in Relation to its Intermediate and Seal Hosts. Canadian Bulletin of Fisheries and Aquatic Sciences* 222, 1–16.

Thiel, P.H. van (1976) The present state of anisakiasis and its causative worms. *Tropical and Geographical Medicine* 28, 75–85.

Thomas, L.J. (1937a) On the life cycle of *Contracaecum spiculigerum* (Rud.). *Journal of Parasitology* 23, 429–431.

Thomas, L.J. (1937b) Further studies on the life cycle of *Contracaecum spiculigerum*. *Journal of Parasitology* 23, 572.

Thomas, L.J. (1937c) Life cycle of *Raphidascaris canadensis* Smedley, 1933, a nematode parasite from pike (*Esox lucius*). *Journal of Parasitology* 23, 572.

Thomas, L.J. (1940) Life cycle studies on *Contracaecum spiculigerum*, a nematode from the cormorant, *Phalacrocorax auritus* and other fish-eating birds (Abstract). *International Congress (3rd) for Microbiology, New York. Report of Proceedings* 11, 883.

Thompson, D., Mansfield, A.W., Beck, B., Bjørge, A., Bowen, D., Hammill, M., Hauksson, E., Myers, R., Ni, I-Hsun and Zwanenburg, K. (1990) Seal Ecology. Group Report 3. In: Bowen, W.D. (ed.) *Population Biology of Sealworm* (Pseudoterranova decipiens) *in Relation to its Intermediate and Seal Hosts. Canadian Bulletin of Fisheries and Aquatic Sciences* 222, 163–170.

Thust, R. (1966) Elektronenmikroskopische Untersuchungen über den Bau des larvalen Integumentes und zur Häutungsmorphologie von *Ascaris lumbricoides*. *Zoologischer Anzeiger* 177, 411–417.

Tiner, J.D. (1949) Preliminary observations on the life history of *Ascaris columnaris*. *Journal of Parasitology* (Suppl.) 35, 13.

Tiner, J.D. (1951) The morphology of *Ascaris laevis* Leidy, 1856, and notes on ascarids in rodents. *Proceedings of the Helminthological Society of Washington* 18, 126–131.

Tiner, J.D. (1952) Speciation in the genus *Ascaris*: additional experimental and morphological data. *Journal of Parasitology* (Suppl.) 38, 27.

Tiner, J.D. (1953a) Fatalities in rodents caused by larval *Ascaris* in the central nervous system. *Journal of Mammalogy* 34, 153–167.

Tiner, J.D. (1953b) The migration, distribution in the brain and growth of ascarid larvae in rodents. *Journal of Infectious Diseases* 92, 105–113.

Todd, A.C. and Crowdus, D.H. (1952) On the life history of *Ascaridia galli*. *Transactions of the American Microscopical Society* 71, 282–287.

Torres, P. and Alvarez-Pellitero, M.P. (1988) Observaciones sobre el ciclo vital de *Rhabdidascaris acus* (Nematoda, Anisakidae) *in vitro* de los estadios iniciales e infestaciones experimentales de macroivertebrados y *Salmo gairdneri*. *Revista Ibérica de Parasitologica* 48, 41–50.

Travassos, L. (1920) Esboço de uma chave geral dos nematodeos parasitos. *Revista de Veterinaria e Zootechnia* 10, 59–70.

Tugwell, R.L. and Ackert, J.E. (1952) On the tissue phase of the life cycle of the fowl nematode *Ascaridia galli* (Schrank). *Journal of Parasitology* 38, 277–288.

Tyzzer, E.E. (1934) Studies on histomoniasis or 'blackhead' infections in the chicken and the turkey. *Proceedings of the American Academy of Arts and Science* 69, 189–264.

Unterberger, A. (1868) Ueber das Vorkommen und die Enticklungsgeschichte von *Ascaris maculosa*. *Oesterreichische Vierteljahresschrift für Wissenschaftliche Veterinärkunde* 30 (1), 38–42.

Uribe, C. (1922) Observations on the development of *Heterakis papillosa* Bloch in the chicken. *Journal of Parasitology* 8, 167–176.

Vaidyanathan, S.N. (1949) *Ascaris vitulorum* – prenatal infection in calves. *Indian Veterinary Journal* 26, 228–230.

Val'ter, E.D. (1968) On the participation of isopods in the life cycle of *Contracaecum aduncum* (Ascaridata, Anisakoidea). *Parazitologiya* 2, 521–527. (In Russian.)

Val'ter, E.D. (1980) Observations on the development of *Contracaecum aduncum* (Ascaridata) in *Jaera albifrons* (Crustacea). *Trudi Belomorskoi Biologicheskoi Stantsii Moskovskogo Gosudarstvennogo Universiteta* (*Biologiya Belogo morya*) 5, 155–164. (In Russian.)

Valovaya, M.A. (1977) The morphology of *Cucullanus cirratus* Mueller, 1777 (Nematoda: Cucullanidae). *Parazitologiya* 11, 424–430. (In Russian.)

Valovaya, M.A. (1978) Embryonic and postembryonic development of *Cucullanus cirratus* Muller, 1777 (Nematoda, Cucullanidae). *Parazitologiya* 12, 426–433. (In Russian.)

Valovaya, M.A. (1979) The biology of *Cucullanus cirratus* Muller, 1777 (Nematoda, Cucullanidae). *Parazitologiya* 13, 540–544. (In Russian.)

Valtonen, E.T., Fagerholm, H.P. and Helle, E. (1988) *Contracaecum osculatum* (Nematoda: Anisakidae) in fish and seals in Bothnian Bay (Northeastern Baltic Sea). *International Journal for Parasitology* 18, 365–370.

Valtonen, E.T., Haaparanta, A. and Hoffman, R.W. (1994) Occurrence and histological response of *Raphidascaris acus* (Nematoda: Ascaridoidea) in roach from four lakes differing in water quality. *International Journal for Parasitology* 24, 197–206.

Vanderburgh, D.J. and Anderson, R.C. (1986) The relationship between nematodes of the genus *Cosmocercoides* Wilkie, 1930 (Nematoda: Cosmocercoidea) in toads (Bufo americanus) and slugs (Deroceras laeve). *Canadian Journal of Zoology* 65, 1650–1661.

Vanderburgh, D.J. and Anderson, R.C. (1987a) Seasonal changes in prevalence and intensity of *Cosmocercoides dukae* (Nematoda: Cosmocercoidea) in Deroceras laeve (Mollusca). *Canadian Journal of Zoology* 65, 1662–1665.

Vanderburgh, D.J. and Anderson, R.C. (1987b) Preliminary observations on seasonal changes in prevalence and intensity of *Cosmocercoides variabilis* (Nematoda: Cosmocercoidea) in *Bufo americanus* (Amphibia). *Canadian Journal of Zoology* 65, 1666–1667.

Vargas-Ocampo, F. and Alvarado-Aleman, F.J. (1997) Infestation from *Lagochilascaris minor* in Mexico. *International Journal for Dermatology* 36, 56–58.

Vassilev, I. (1959) The goat (*Capra hircus*) as host of *Neoascaris vitulorum* (Goeze, 1782) Travassos, 1927. *Compte Rendu de l'Académie Bulgare des Sciences* 12, 597–600.

Vassilev, I. (1987) On the development of *Ascaridia compar* (Schrank, 1790) in the host. *Khelmintologiya* 23, pp. 30–35.

Vassilev, I. (1993) On the ecology of *Ascaridia columbae* (Gmelin, 1790). *Helminthologia* 30, 135–138.

Vatne, R.D. and Hansen, M.F. (1965) Larval development of caecal worm (*Heterakis gallinarum*) in chickens. *Poultry Science* 44, 1079–1085.

Vessichelli, N. (1910) Di un nuovo *Dacnitis* parassita del *Petromyzon planeri*. *Monitore Zoologica Italiano* 21, 304–307.

Vidal-Martinez, V.M., Osorio-Sarabia, D. and Overstreet, R.M. (1994) Experimental infection of *Contracaecum multipapillatum* (Nematoda: Anisakinae) from Mexico in the domestic cat. *Journal of Parasitology* 80, 576–579.

Volcán, G.S. and Medrano, P.C.E. (1990) Induced infection in the wild rodent *Dasyprocta leporina* (Rodentia: Dasyproctidae) with larval eggs of *Lagochilascaris minor* (Nematoda: Ascarididae). *Revista do Instituto de Medicina Tropical de São Paulo* 32, 395–402.

Volcán, G.S. and Medrano, P.C.E. (1991) Infección natural de *Speothos venatricus* (Carnivora: Canidae) por estadios adultos de *Lagochilascaris* sp. *Revista de Patologia Tropical* 16, 1–6.

Volcán, G.S., Ochoa, F.R., Medrando, C.E. and Valera, Y. (1982) *Lagochilascaris minor* infection in Venezuela. *American Journal of Tropical Medicine and Hygiene* 31, 1111–1113.

Volcán, G.S., Medrano, P.C.E. and Payares, G. (1992) Experimental heteroxenous cycle of *Lagochilascaris minor* Leiper, 1909 (Nematoda: Ascarididae) in white mice and in cats. *Memórias do Instituto Oswaldo Cruz* 87, 525–532.

Walton, A.C. (1937) The Nematoda as parasites of amphibians. III. Studies on life histories. *Journal of Parasitology* 23, 299–300.

Warren, E. G. (1970) Studies on the morphology and taxonomy of the genera *Toxocara* Stiles, 1905 and *Neoascaris* Travassos, 1927. *Zoologischer Anzeiger* 185, 393–442.

Warren, E.G. (1971) Observations on the migration and development of *Toxocara vitulorum* in natural and experimental hosts. *International Journal for Parasitology* 1, 85–99.

Webster, G.A. (1956) A preliminary report on the biology of *Toxocara canis* (Werner, 1782). *Canadian Journal of Zoology* 34, 725–726.

Webster, G.A. (1958) On prenatal infection and the migration of *Toxocara canis* Werner, 1782 in dogs. *Canadian Journal of Zoology* 36, 435–440.

Wehr, E.E. (1942) The occurrence in the USA of the turkey ascarid *Ascaris dissimilis* and observations on its life history. *Proceedings of the Helminthological Society of Washington* 9, 73–74.

Wehr, E.E. and Hwang, J.C. (1959) Further observations on the life history and development of *Ascaridia columbae* (Gmelin, 1790) Travassos, 1913 in the pigeon (Abstract). *Journal of Parasitology* 45, 43.

Wehr, E.E. Hwang, J.C. (1964) The life cycle and morphology of *Ascaridia columbae* (Gmelin, 1790) Travassos, 1913 (Nematoda: Ascarididae) in the domestic pigeon (*Columba livia domestica*). *Journal of Parasitology* 50, 131–137.

Wharton, L.D. (1915) The development of the eggs of *Ascaris lumbricoides*. *Philippine Journal of Science, Section B, Tropical Medicine* 10, 19–23.

Wilder, H.C. (1950) Nematode endophthalmitis. *Transactions of the American Academy of Ophthalmology Nov.–Dec.*, pp. 99–109.

Williams, C.K., McKown, R.D., Veatch, J.K. and Applegate, R.D. (1997) *Baylisascaris* sp. found in a wild northern bobwhite (*Colinus virginianus*). *Journal of Wildlife Diseases* 33, 158–160.

Winfield, G.F. (1933) Quantitative experimental studies on the rat nematode *Heterakis spumosa* Schneider, 1866. *The American Journal of Hygiene* 17, 168–228.

Winterfield, R.W. and Thacker, H.L. (1978) Verminous encephalitis in the emu. *Avian Diseases* 22, 336–339.

Wright, W.H. (1935) Observations on the life history of *Toxascaris leonina* (Nematoda: Ascaridae). *Proceedings of the Helminthological Society of Washington* 2, 56.

Wülker, G. (1929) Der Wirtwechsel der parasitischen Nematoden. *Verhandlungen der Deutschen Zoologischen Gesellschaft* 33, 147–157.

Wulker, G. (1930) Über Nematoden aus Nordseetieren I. *Zoologischer Anzeiger* 87, 293–302.

Yoshinaga, T., Ogawa, K. and Wakabayashi, H. (1987a) Experimental life cycle of *Hysterothylacium aduncum* (Nematoda: Anisakidae) in freshwater. *Fish Pathology* 22, 243–251.

Yoshinaga, T., Ogawa, K. and Wakabayashi, H. (1987b) New record of third-stage larvae of *Hysterothylacium aduncum* (Nematoda: Anisakidae) from *Neomysis intermedia* (Crustacea: Mysidae) in a freshwater lake in Hokkaido, Japan. *Nippon Suisan Gakkaishi* 53, 63–65.

Yoshinaga, T., Ogawa, K. and Wakabayashi, H. (1988) Developmental morphology of *Hysterothylacium haze* (Nematoda: Anisakidae). *Fish Pathology* 23, 19–28.

Yoshinaga, T., Ogawa, K. and Wakabayashi, H. (1989) Life cycle of *Hysterothylacium haze* (Nematoda: Anisakidae: Raphidascaridinae). *Journal of Parasitology* 75, 756–763.

Yuen, P.H. (1965) Some studies on the taxonomy and development of some rhabditoid and cosmocercoid nematodes from Malayan amphibians. *Zoologischer Anzeiger* 174, 275–298.

Yutuc, L.M. (1949) Prenatal infection of dogs with ascarids, *Toxocara canis* and hookworms, *Ancylostoma caninum*. *Journal of Parasitology* 35, 358–360.

Zekhnov, M.I. (1956) The parasite fauna of *Lampetra*. *Uchenye Zapiski Vitebskogo Veterinarnogo Instituta* 14, 187–191. (In Russian.)

Chapter 6

Order Spirurida – Suborder Camallanina

Females of most Spirurida produce eggs containing fully developed first-stage larvae (larvae are considerably specialized in the Filarioidea) which only develop to the third and infective stage in the tissues of arthropod intermediate hosts. Members of the Gnathostomatoidea are the exception since eggs, deposited in an undeveloped state, embryonate to second-stage larvae and hatch in water. Transmission in aquatic habitats depends upon crustaceans and larval insects as intermediate hosts. Transmission in terrestrial habitats depends upon terrestrial insect and crustacean intermediate hosts.

The Camallanina contains the Camallanoidea of the gut and the Dracunculoidea mainly of the deeper tissues and cavities. Intermediate hosts are copepods.

6.1

The Superfamily Camallanoidea

Camallanoids are parasites of the stomach and intestines of lower predaceous vertebrates (Chabaud, 1975). The superfamily consists of eight clearly related genera. Of the some 150 species described throughout the world, about 40 occur in amphibians and reptiles (especially turtles) (Baker 1987), the remainder in marine, estuarine and freshwater fishes (Ivashkin *et al.*, 1971). The transmission and development of the former group have been little studied but a number of species in fish have been investigated. Metchnikoff (1866) and Leuckart (1876) reported the development in copepods of *Camallanus lacustris* of European fishes. This was one of the first demonstrations of heteroxeny in the Nematoda (cf. *Mastophorus muris*) and it led directly to Fedchenko's (1871a) great work on the human guinea worm.

Certain features of development are similar in all the species studied. First-stage larvae hatch *in utero* and are either passed into water with faeces of the host or gravid females protrude from the anus and rupture on contact with water. The first-stage larvae are slender, with attenuated tails, and their vigorous activity attracts copepods and perhaps other crustaceans which devour them. In the crustacean intermediate host, the larvae enter the haemocoel and develop to the third-infective stage. This larva is generally rather short and stout and readily recognized as a camallanoid, because it has already acquired the large buccal cavity peculiar to the camallanoids. The tail is always armed with a few terminal spines. The predator definitive host can become infected by ingesting copepods with larvae. However, the larvae will persist in the gut or become encapsulated in the tissues of planktonivorous fishes which consume copepods. These paratenic hosts move the larvae in the food chain to the piscivorous definitive hosts. Larvae may grow to the fourth stage in the tissues of the paratenic hosts. Jackson and Tinsley (1998) found larvae in aquatic toads (Pipidae) in Africa.

Practically all studies of camallanoids have implicated copepods as intermediate hosts. It is worth mentioning that Linstow (1909) claimed to have found larvae of *Camallanus lacustris* in the isopod *Asellus aquaticus* and Fusco (1980) reported that some larvae of *Spirocamallanus cricotus* developed successfully in white shrimp (*Penaeus setiferus*).

Family Camallanidae

Camallanus

C. anabantis Pearse 1937

This species is common in various fishes in India, including the type host *Anabas testudineus* from freshwater swamps (Soota, 1984; De, 1993). Invasion of fish occurs in spring and summer and they grow in the fish during the monsoon and autumn. The proportion of males in the worm population increases in the early autumnal period and then decreases rapidly after they have fertilized the females. Larvae are released from females in late winter and early spring.

C. cotti Fujita 1927

Campana-Rouget *et al.* (1976) reported that *C. cotti* (syn. *C. fotedari* Raina and Dhar, 1972) of Asian origin was established in aquarium guppies (*Lebistes reticulatus* and *Danio rero*) in France. The parasite developed in *Cyclops* sp. Guppies fed infected copepods had adult worms 10 weeks later.

C. lacustris (Zoega, 1776)

Since the early observations of Metchnikoff (1866) and Leuckart (1876) there have been additional studies of *C. lacustris*, notably by Kupriyanova (1954), Campana-Rouget (1961) and Moravec (1969, 1971a,b). This parasite is common in predatory European fishes of the Percidae, Salmonidae, Gadidae and perhaps Esocidae and Siluridae (Moravec, 1971a). First-stage larvae are 440–580 μm in length and develop to the infective third stage (660–880 μm) in the cyclopoid copepods *Megacyclops viridis*, *Macrocyclops albidus*, *Acanthocyclops vernalis*, *Mesocyclops leuckarti*, *Eucyclops serrulatus* and *Cyclops strenuus* (see Moravec, 1969). Larvae from the copepods reached the adult stage in 91 days in *Perca fluviatilis*, the fourth moult of the males occurring 35 days and of the females 67 days postinfection.

Chandra and Chubb (1992) reported that first-stage larvae survival was related to temperature. Larvae survived 69 days at 5°C and 28 days at 20°C. The ability to invade copepod was lost after 7 days at 5°C, but was retained for 23 days at 20°C.

Moravec (1971a) noted that larvae from copepods developed to the adult stage in various predatory fishes whereas larvae remained in the third stage in non-predatory Cyprinidae and Cobitidae, which evidently serve as paratenic hosts. In *Leuciscus cephalus*, a predatory cyprinid, the parasite developed to the adult stage but more slowly than in other predatory fish such as *P. fluviatilis*.

C. oxycephalus Ward and Magath, 1916

C. oxycephalus has been studied in considerable detail in Lake Erie (Stromberg and Crites, 1974a,b, 1975a,b; Crites, 1976). The nematode is a common parasite of a variety of piscivorous and planktonivorous fishes in the Great Lakes and elsewhere in North America (Fig. 6.1). In summer the gravid females protruded from the anus of the host and ruptured, releasing larvae. Larvae (629–645 μm in length) that were passed into water attached by their tails to objects in the environment and wriggled vigorously. They penetrated the intestine of copepod predators (*Cyclops bicuspidatus*, *C. vernalis*) and developed in the haemocoel to the infective stage. The first moult occurred 3 days

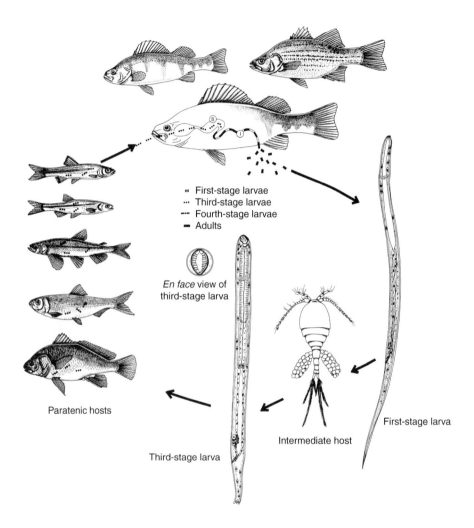

First-stage larvae
Third-stage larvae
Fourth-stage larvae
Adults

En face view of
third-stage larva

Paratenic hosts

Intermediate host

First-stage larva

Third-stage larva

Fig. 6.1.   Development and transmission of *Camallanus oxycephalus* in Lake Erie.
I = intestine; S = stomach. (Illustrations of fish adapted from *Freshwater Fishes of
Canada* by W.B. Scott and E.J. Crossman, 1998, Fisheries Research Board of Canada;
original, U.R. Strelive.)

postinfection at 25°C and the second on day 6. The stout infective larva (450–671 μm
in size) was orange in colour and had a well-developed buccal capsule and three
well-defined mucrons on the tip of the tail; it lay coiled and inactive in the haemocoel.
In the fish host, the third moult occurred 9–10 days postinfection. The males
underwent the fourth moult 17–18 days and the females 24 days postinfection.
Mucrons, found in both the third and fourth stages, were lost at the final moult.

   Small forage fishes like spot-fin shiner (*Notropis spilopterus*), emerald shiner
(*N. atherinoides*), smelt (*Osmerus mordax*), freshwater drum (*Aplodinotus grunniens*) and

alewife (*Alosa pseudoharengus*) harboured larvae and even adult worms, which could be transferred to piscivorous fish such as white bass (*Morone chrysops*), yellow perch (*Perca flavescens*) and smallmouth bass (*Micropterus dolomieui*). In some forage fishes, the worms occurred in the intestine but encapsulated fourth-stage larvae were also found on the mesentery of freshwater drum and were infective to bass.

C. *oxycephalus* occurred in all size groups of white bass in Lake Erie but prevalence and intensity were higher in larger fish (91%) than in smaller fish (72–86%). During July and August most worms in white bass were in the third and fourth stages. This indicated that adults die in summer during or after the release of larvae and fish acquire new infections during the same general period. This new population of worms remained arrested in fish over winter and only matured in spring, when cyclopoid copepods were most abundant.

C. *truncatus* (Rudolphi, 1814)

This parasite, found in numerous fish species in western Europe and the CIS, developed in *Mesocyclops leuckarti* and *Acanthocyclops viridis* (see Kupriyanova, 1954) and *Macrocyclops albidus*, *Cyclops strenuus* and *Megacyclops viridis* (see Moravec, 1971b). First-stage larvae were 440–560 µm in length. At 19–21°C the first moult took place 1–2 days and the second 5–6 days postinfection; third-stage larvae (790–910 µm) were infective to perch (*Perca fluviatilis*), in which the worms were fully developed about 25–32 days postinfection (Kupriyanova, 1954).

Evlanov (1995) related the presence of C. *truncatus* in *Perca fluviatilis* in Russia to host size and season. Large fish ensured a high intensity of larvae in copepods whereas fish less than 150 mm in length rejected the parasites. A decrease in intensity in larger fish occurred during winter and spring.

C. *zacconis* Li, 1941

This parasite from *Zacco temmincki* in China reached the infective stage in 7 days in *Mesocyclops leuckarti* at 20–25°C (Wang and Ling, 1975). The first-stage larvae were 440–460 µm in length and the infective larvae 420–460 µm in length. The tail of the latter was tricuspid.

Paracamallanus

P. *cyathopharynx* (Baylis, 1923)

This is a common intestinal parasite of catfish of the family Clariidae in Africa (e.g. *Clarias lazera*). It developed in *Mesocyclops leuckarti*; the first moult took place 24 h and the second 8–9 days postinfection at 23–24°C (Moravec, 1974). First-stage larvae were 287–315 µm and third-stage larvae 508–574 µm in length.

P. *singhi* (Ali, 1957)

P. *singhi* of the fish *Channa striata* in India developed in *Pachycyclops bistriatus* (see De *et al*. 1984). First-stage larvae were 224–250 µm in length. The first moult occurred 3–6 days postinfection (at 29–30°C) and the second 8–11 days postinfection (at 20–22°C). Third-stage larvae were 452–708 µm in length. Larvae grew and moulted the third time 10 days postinfection in C. *striata*.

P. sweeti Moorthy, 1937

P. sweeti, a parasite of the piscivorous freshwater fish *Ophicephalus gachua* in India, developed to the third stage in *Mesocyclops leuckarti* and *M. hyalinus,* which are common in ponds (Moorthy, 1938a). First-stage larvae were 240 μm in length. The first moult occurred 24–36 h and the second 5–12 days postinfection, depending on the temperature. Third-stage larvae were 560 μm in length. Suitable paratenic hosts were the smaller forage fish *Lepidocephalichthys thermalis, Barbus puckelli, B. (Puntius) ticto* and *Gambusia* sp.; the latter was introduced to India from Europe. Third-stage larvae were found in, and encapsulated on, the intestine of the paratenic hosts.

Procamallanus

P. fukiensis Wang and Ling, 1975

This nematode from *Parasilurus asotus* in China developed in *Thermocyclops oithonoides, Eucyclops serrulatus* and *Mesocyclops leuckarti.* Development took 21 days at 9–21°C and 7–11 days at 25–32°C. Third-stage larvae reached the adult stage in 15 days in the final host (Wang and Ling, 1975). The tail of the infective larva ends in two points. Infective larvae given to a catfish matured in 15 days.

P. laeviconchus (Wedl, 1862)

This nematode is a common parasite of African freshwater fishes, especially siluroids, and less frequently of species of Mormyridae, Characidae, Tetraodontidae and Cichlidae. It developed in *Mesocyclops leuckarti,* the first moult occurring 1–2 days and the second 5–6 days postinfection at 23–24°C (Moravec, 1975). Third-stage larvae were found attached to the intestinal mucosa of *Gambusia affinis* allowed to ingest infected copepods. This fish probably serves as a paratenic host predated by siluroids, in which the nematode matures.

P. neocaballeroi (Caballero-Deloya, 1977)

P. neocaballeroi is a parasite of the intestine of *Astyanax fasciatus* in Mexico (Moravec and Vargas-Vazquez, 1996). In *Mesocyclops* sp. at 21–22°C the first moult occurred on day 3 and the second moult 4–5 days postinfection. First-stage larvae were 593–613 μm in length and third-stage larvae were 645–685 μm in length. Larvae grew only slightly during development from the first to the third stage. In the fish host, larvae moulted in 10 and in 14–15 days at 25–32°C. The prepatent period was about 2 months.

P. rebecae Andrade-Salas, Pineda-Lopez and Garcia-Magaña, 1994

This is a parasite of the intestine of cichlid fishes in Mexico (Moravec *et al.,* 1995). First-stage larvae were 488–508 μm in length. *P. rebecae* developed to the infective stage in *Mesocyclops* sp. At 21–22°C the two moults occurred on day 3 and on days 5–6 and third-stage larvae were 870–1105 μm in length. In the definitive host (*Cichlasoma urophthalmus*) moults occurred 13–14 days and 42 days postinfection. The prepatent period was probably about 2–3 months but was not definitely established. Guppies (*Poecilia reticulata*) can serve as paratenic hosts.

P. spiculogubernaculus Agarwal, 1958

De *et al.* (1986) infected *Mesocyclops obsolatus* and *M. oithonoides* with *P. spiculogubernaculus* from naturally infected fish (*Heteropneustes fossilis*). First-stage larvae were 458–573 μm in length. At 20–28°C the first moult occurred 3–8 days and the second 4–12 days postinfection. Infective larvae were 807–1159 μm in length.

Sinha (1988) reported that the species developed in *Cyclops vicinus*, *Mesocyclops leukarti* and *M. hyalinus*. In the usual host (*H. fossilis*) the worms matured in 138–164 days. In three other fish species (*Clarias batrachus*, *Lepidocephalichthyes guntea* and *Puntius conchonius*) larvae remain alive in the stomach but failed to develop.

Neocamallanus

N. ophiocephali (Pearse, 1933)

N. ophiocephali (syn. *C. adamsi* Bushirullah, 1974) from *Channa striatus* (syn. *Ophiocephalus striatus*) in Bangladesh developed to the third stage in *Mesocyclops leuckarti* and *Thermocyclops crassus* (see Bashirullah and Ahmed, 1976a). First-stage larvae were 264–286 μm in length. The first moult occurred 5 days and the second 10 days postinfection at 24°C. Third-stage larvae were 541–606 μm in length.

Spirocamallanus

Moravec and Amin (1978) regard *Spirocamallanus* as a subgenus of *Procamallanus*.

S. cearensis (Pereira, Vianna, Diaz and Azevedo, 1936)

The parasite has been reported in the small intestine of *Astyanax bimaculatus* in Brazil. It developed in *Diaptomus cearensis* and *D. azevedoi* (see Pereira *et al.*, 1936). *Curimatus elegans* was regarded as a paratenic host. First-stage larvae were 460–560 μm in length and third-stage larvae 660–760 μm. This species is probably a synonym of *S. hilarii* (Vaz and Pereira, 1934).

S. cricotus Fusco and Overstreet, 1978

S. cricotus has been reported in the gut of at least 13 species of marine and estuarine fishes in the northern Gulf of Mexico, including the Atlantic croaker (*Micropogonias undulatus*). The parasite developed in the harpacticoid copepod *Tigriopus californicus* (see Fusco, 1980). First-stage larvae were 444–493 μm in length. The first moult occurred 6 days and the second 10–11 days postinfection at 23–26°C. Third-stage larvae were 548–1000 μm in length. Some larvae apparently developed in the white shrimp (*Penaeus setiferus*) but the role of these crustaceans in transmission is not known. Overstreet (1973) found two larvae of *S. cricotus* (syn. *S. pereirae*) in the intestine of a naturally infected *P. setiferus* and (Overstreet, 1978) similar larvae in the squid, *Loliguncula brevis*.

S. fulvidraconis (Li, 1935)

S. fulvidraconis is found in the oesophagus and stomach of numerous fish species (including the type host *Pseudopaguis fulvidraco*) in China and the CIS. Li (1935) described its development in *Cyclops albidus, C. bicusbidatus, C. phaleratus, C. serratus, C. vicinus* and *C. magnus*. First-stage larvae were 455 μm and third-stage larvae 1015 μm in length.

S. mysti (Karve, 1952)

S. mysti (syn. *S. intestinecolas* (Bashirulla, 1973)) of the freshwater fish *Mystus vittatus* developed in *Mesocyclops crassus* and *M. leuckarti*. Larvae (329–486 μm) invaded the haemocoel of the copepods by 3 h at 16–20°C (De, 1995). The first moult occurred 4–6 days after infection and the second in 8 days. Third-stage larvae (616–860 μm) were found mainly in the cephalothorax. In fish the third moult took place 15–16 days postinfection and the fourth in 37 days in males and 67 days in females. Bashirulla and Ahmed (1976b) reported that the first moult occurred in 39 h and the second in 72 h at 25–26°C. First-stage larvae were 316–334 μm and third-stage larvae 616–726 μm in length.

Species in amphibians and turtles

Thurston (1970) reported that *Spirocamallanus xenopodis* (Baylis, 1929) and *Camallanus johni* Yeh, 1960 of the stomach of African clawed toads (*Xenopus* spp.) developed to the third stage in *Thermocyclops infrequens* and *Mesocyclops leuckarti* in about 15 days at 22–25°C.

Bartlett and Anderson (1985) found third-stage larvae, presumably of *Serpinema trispinosum* (Leidy, 1852) of turtles, in the tissues of the freshwater snail *Lymnaea stagnalis* in a pond in Ontario and suggested that the snail was serving as a paratenic host. Moravec *et al.* (1998) found similar larvae in fish serving apparently as paratenic hosts.

6.2

The Superfamily Dracunculoidea

Dracunculoids occur in tissues and serous cavities mainly of fishes, reptiles, birds and mammals (Chabaud, 1975) and rarely amphibians (Petter and Planelles, 1986). After insemination, the female grows markedly and fills with huge numbers of first-stage larvae, which must be dispersed into the environment where they are available to copepod intermediate hosts. In many species which have been studied, the fully gravid female must be in contact with fresh water which causes her to burst, thus releasing the larvae into the environment. The female may elicit a skin lesion to gain access to fresh water or she may migrate into the rectum and protrude from the body of the host to achieve the same result. In some highly specialized species, larvae released within the host make their way to the tissues, including the blood, where they are available to ectoparasitic crustaceans which serve, or probably serve, as intermediate hosts. Most dracunculoids that have been studied occur in hosts which have contact with fresh water during some part of their lives. Unfortunately, species in marine fishes have not been studied.

Family Dracunculidae

The family consists of a small number of recognized species in two genera: *Dracunculus* occurs in mammals and reptiles (snakes and turtles) and *Avioserpens* in aquatic birds. In contrast to the philometrids of fishes, males are relatively large and easily observed. The fully developed adults of both sexes generally occur under the skin of the host. First-stage larvae, which are extremely numerous in the gravid female, are released into the water through a skin lesion, initially in the form of a blister containing myriads of larvae which breaks when immersed in water. Thereafter, larvae can be released from the female worm repeatedly when the ulcer over the worm is bathed in water. Larvae wriggle vigorously and tend to remain suspended in the water column, where they attract copepods which ingest them. Larvae develop to the infective third stage in the haemocoel of the copepod. Infection of the final host is through the ingestion of copepods containing infective larvae or, especially in carnivorous hosts, through the agency of fish and frog paratenic hosts. One species, *Dracunculus medinensis*, is a celebrated parasite of humans in parts of Africa and southeast Asia. For a comprehensive review of this important human pathogen the reader is referred to Muller (1971).

Dracunculus

Ivashkin *et al.* (1971) followed Mirza (1957) in recognizing only two species of *Dracunculus*, namely *D. medinensis* (L., 1758) in mammals and *D. oesophagea* (Polonio, 1859) in reptiles. We follow Vaucher and Bain (1973), Chabaud (1975) and Baker (1987), who preferred to recognize several distinct species in reptiles and mammals (see also Muller, 1971). The biology of two species in mammals, namely *D. insignis* and *D. medinensis*, has been well studied. Those in reptiles have been very incompletely studied.

D. medinensis (Linnaeus, 1758)

The development and transmission of the human guinea worm were first outlined by Fedchenko in a justly celebrated paper published in 1871 when he was only 27 years of age; he died 2 years later in a climbing accident on Mont Blanc. A reproduction of his paper appeared in 1971 in the *American Journal of Tropical Medicine and Hygiene* (Fedchenko, 1871b) on the occasion of the centenary of its publication. Several authors have made major contributions to our knowledge of the general biology of *D. medinensis*, namely Moorthy and Sweet (1936, 1938), Moorthy (1938a,b), Onabamiro (1954, 1956,) and Muller (1968, 1971). *D. medinensis* is readily transmissible to dogs, cats and a number of non-human primates (Muller, 1971). For recent reviews of dracunculiasis in Ghana, see Hunter (1996, 1997a,b).

The gravid female, which occurs under the skin, is one of the largest nematodes known, measuring up to 80 cm in length in humans. The typical blister and ulcer occurs most commonly on the lower extremities, especially on the upper foot. First-stage larvae (608 μm in length) released into water have long tails, move actively in the water column and are predated readily by copepods. Larvae invade the haemocoel and develop to the third stage in a variety of cyclopoid copepods, including species such as *Cyclops vernalis americana*, which is unlikely to be a natural intermediate host. Muller (1971) listed numerous species which were shown to be naturally infected in endemic areas, or likely to be suitable hosts, namely: *Cyclops bisetosus*, *C. decipiens*, *C. fimbriatus*, *C. hyalinus*, *C. inopinus*, *C. iranicus*, *C. karvei*, *C. leuckarti*, *C. microspinulosus*, *C. nigerianus*, *C. oithonoides*, *C. rylovi*, *C. tinctus*, *C. varicans* and *C. vermifes*. According to Moorthy (1938b) the first moult occurred 5–7 days and the second 8–12 days postinfection in copepods kept at 32–39°C. At lower temperatures (13–21°C) the first moult occurred 8–10 and the second 13–16 days postinfection. The third stage (451 μm in length) was fully formed in 14–20 and 17–24 days, respectively, at temperatures of 32–39°C and 13–21°C. Other authors have reported similar results (Muller, 1971).

The development of *D. medinensis* has been followed in dogs, cats and rhesus monkeys (Moorthy and Sweet, 1938; Onabamiro, 1956; Muller, 1968). Muller (1968) found larvae in the duodenal wall 13 h after infection, on the mesentery for up to 12 days and in the thoracic and abdominal muscles in 15 days where they reached adulthood. Moorthy and Sweet (1938) found fertilized females in a dog (infected twice) 103 and 132 days postinfection. Females were fertilized by at least 3½ months, according to Onabamiro (1956). Females moved to the extremities 8–10 months postinfection and contained larvae at 10 months. Males died between 3 and 7 months postinfection and became encapsulated (Muller, 1971).

In humans, gravid females emerged 10–14 months postinfection, based on observations of infections in travellers who had spent short periods in endemic areas. The prepatent period reported in dogs and monkeys is similar to that in humans.

Humans acquire the parasite from drinking water contaminated with infected copepods. Ponds, cisterns and step wells are important sources of infection in many endemic areas. Muller (1971) stated that 'dracunculiasis is primarily a disease of poverty, particularly of remote rural communities without adequate water supplies'. For example, Yelifari *et al.* (1997) found seven species and two subspecies of copepods in man-made water reservoirs in northern Ghana. Prevalence of infection in the copepods was 0.3%. The distribution of the intermediate hosts is restricted to ponds and wells and the disease can be prevented by simple public health measures, i.e. by the provision of drinking water free of infected copepods.

D. insignis (Leidy, 1858)

The biology of this species has been studied by Crichton and Beverley-Burton (1974, 1975, 1977). It is a common parasite of raccoons (*Procyon lotor*) in Ontario, Canada. It is an annual species and gravid females emerged in legs in spring and early summer (April to June). First-stage larvae were 596–857 μm in length. Larvae infected *Cyclops bicuspidatus thomasi* and *C. vernalis*. At 24°C the first moult occurred 8–9 days and the second 13–16 days postinfection. Third-stage larvae were 595–703 μm in length. In experimentally infected raccoons, third-stage larvae were found in the gut wall and mesentery 1 day postinfection. Larvae were found in intercostal muscles by day 5 and in subcutaneous tissue of the thorax and abdomen by day 7. Worms were in the fourth stage by 19 days and were sexually distinct by 34 days, when they were present in subcutaneous tissues of the thorax, abdomen and inguinal region. Male worms were mature at 60 days and females at 65–70 days. Females with larvae were found in the extremities as early as 120 days postinfection. The prepatent period was 354 (309–410) days. Development was similar in mink (*Mustela vison*) although, in Ontario, mink were only found infected in regions where there were raccoons (Crichton and Beverley-Burton, 1974). Ferrets were infected with material from raccoons but attempts to infect a dog were unsuccessful. Eberhard *et al.* (1988) studied development of *D. insignis* in ferrets. Gravid females were found as early as 128 days postinfection and first-stage larvae, present in female worms as early as 200 days, were infective to copepods. Male worms were found up to 200 days postinfection.

Tadpoles and adult *Rana pipiens* and *R. clamitans* were suitable paratenic hosts of *D. insignis* (see Crichton and Beverley-Burton, 1977). Larvae moved about freely in the tissues of frogs and were found in the wall of the gut and musculature of the front legs and pelvic girdle. Larvae from experimentally infected frogs had grown to 570–698 μm in length, in contrast to larvae from copepods, which were 434–605 μm in length. Larvae from frogs readily infected a raccoon. Eberhard and Brandt (1995) showed that tadpoles of *Xenopus laevis* and bullfrogs (*Rana* sp.) would ingest copepods infected with larvae of *D. insignis* and that in *X. laevis* larvae persisted through metamorphosis of the frogs to adults, although some larvae were lost in discarded caudal tissue.

D. doi Chabaud, 1960

Vaucher and Bain (1973) found *D. doi* in the snake *Sanzinia madagascariensis*. First-stage larvae developed in the haemocoel of *Cyclops strenuus* collected in France.

Two male worms were found under the skin of *Python rejus* given 19 infective larvae 4 months earlier.

D. ophidensis Brackett, 1934

Fully developed adult males and females were usually found in subcutaneous tissues of garter snakes (*Thamnophis sirtalis*) in Michigan, although gravid females were once found in the body cavity (Brackett, 1938). Gravid females were associated with swelling of the epidermis but Brackett (1938) did not report the presence of characteristic blisters (the present author has seen typical blisters on the skin of a large naturally infected anaconda; the blisters contained a milky fluid with large numbers of larvae, which readily developed in the haemocoel of *Cyclops vernalis*). Larvae of *D. ophidensis* developed in the haemocoel of *C. viridis*. Brackett (1938) reported one moult, presumably the second, 12–15 days after copepods had ingested larvae. Infective larvae were found unchanged in the body cavity of tadpoles allowed to ingest infected copepods. Snakes fed tadpoles containing larvae became infected. Adult worms occurred around the heart, along the tongue and oesophagus, in membranes lining the body cavity and in subcutaneous tissues of snakes given infected copepods. The results were inconclusive, however, because the snakes may have been infected naturally and no controls were used.

Avioserpens

A. mosgovoyi Supryaga, 1965

This parasite was studied by Supryaga (1965, 1967, 1969, 1971). Adults were found in subcutaneous tissues, mainly in the submaxillary areas, of coots (*Fulica atra*) and grebes (*Podiceps cristatus*, *P. grisegena* and *P. ruficollis*) in the CIS. It also infected mallards (*Anas platyrhynchos*). The presence of numerous worms resulted in the formation of large tumours, sometimes as large as the head of the bird. Larvae were passed into the water from lesions produced by gravid worms. They developed into third-stage larvae in several species of cyclopoid copepods as well as *Diaptomus gracilis*. At 22–29°C the first moult took place at 6–8 days postinfection and the second 2–3 days later. Moravec and Scholz (1990) found third-stage larvae, probably of this species, in *Cyclops strenuus* in the Czech Republic.

Various species of fish (roach, gobies and sticklebacks), frogs and larvae of dragonflies were suitable paratenic hosts, the larvae persisting in the tissues for 2–2.5 months.

In the final host, larvae migrated to the serosa and mesentery, where the third moult took place 4–5 days postinfection. Larvae then migrated via the air sacs to subcutaneous tissues, where males moulted on day 12 and females on days 13–14 postinfection. Females began to discharge larvae 4 weeks postinfection and continued to do so for about 13 days. Males apparently persisted in the tissues for up to 260 days and may be capable of fertilizing females of subsequent infections.

Transmission is most likely in stagnant water bodies with high concentrations of copepods and paratenic hosts. Thus, prevalences and intensities in birds in the CIS were highest from July to September, when these conditions prevail.

A. taiwana (Sugimoto, 1919)

The parasite occurs in domestic ducks in China. Suitable intermediate hosts were *Cyclops strenuus*, *Eucyclops serrulatus*, *Mesocyclops leuckarti* and *Thermocyclops hyalinus* (see Wang *et al.*, 1983). The first moult occurred 3–4 days and the second 7 days postinfection. Males were found on the mesentery and females in subcutaneous tissue of ducklings 18 and 20 days postinfection, respectively. According to Truong-Tan-Ngog (1937) the parasite occurred in ducks in the dry season and formed tumours, mainly in the inferior mandibles and chin and less commonly on the shoulders and shanks.

Family Philometridae

The philometrids are common and exclusively parasites of fish. They occupy the same niche in fish as the filarioids do in terrestrial vertebrates and, like the filarioids, they have diversified extensively in their hosts. The gravid female is generally well known because it is large, packed with larvae and frequently occurs in readily observed parts of the host's body – and even, at certain times of the year, protrudes from the anus. In the more primitive species, the gravid female occurs under the skin of the body (*Philometroides cyprini*), the fins (*Philometroides huronensis*, *P. sanquinea*, *Philonema fujimotoi*), the cheek pouches (*Philometroides nodulosa*) and, rarely, the gill arteries (*Philometra obturans*). These kinds of females elicit a break in the tissues which exposes them to fresh water, whereupon they burst and release their numerous larvae. In more specialized species, females become gravid in the body cavity and migrate to the anal region, expose themselves to water and burst (*Philometra abdominalis*, *P. ovata*). In some species (*Philonema oncorhynchi*, *P. agubernaculum*), gravid females and larvae pass out with the reproductive products of the host during spawning, a process probably under the control of host hormones. Species of *Ichthyofilaria* produce microfilariae-like larvae which can be found in the blood or skin of the host, where they are available to blood-sucking crustaceans which may serve as intermediate hosts (Appy *et al.*, 1985).

With a few exceptions, philometrids utilize copepods as intermediate hosts and development in the haemocoel to the infective third stage is temperature dependent. Many species of fish, especially those which are planktonivorous, acquire their infections from ingesting copepods. However, transmission of some species in predaceous fishes (e.g. *P. obturans* of pike) is probably largely dependent upon paratenesis.

Once the definitive host acquires infective larvae, the latter generally migrate to the serosa of the swimbladder or kidneys, where they grow into fifth-stage males and females. The early adults are extremely small (2–4 mm in length). Once the female is inseminated she migrates to the definitive site, grows markedly and becomes gravid. She may undertake this migration in steps (e.g. *P. huronensis*). There is evidence in some species that the tiny males die and are resorbed after they have inseminated the females. In other species, males seem to persist for prolonged periods after inseminating the females, although their subsequent role in reproduction is uncertain.

Most species studied are strictly annual. Infective larvae are generally acquired in summer and early fall, with the ingestion of infected copepods, but the female worms do not become gravid until spring. The period when the gravid females appear generally coincides with the times when copepods are abundant and also, frequently, when fish are spawning (for an extreme example of this phenomenon see *P. oncorhynchi* of anadromous sockeye salmon).

Philometra

P. abdominalis Nybelin, 1928
This is a common parasite of the body cavity of Cyprinidae in Europe (i.e. *Gobio gobio, Leuciscus cephalus, L. leuciscus* and *Phoxinus phoxinus*). Its development and transmission were studied by Molnar (1967) and Moravec (1977a,b). Female worms became gravid mainly in July and August. They migrated to the anus and pushed part of their bodies into the water, ruptured and released larvae (480–543 μm in length); larvae may also be shed with reproductive products of the host. Larvae developed in *Acanthocyclops vernalis, Diacyclops bisetosus, Macrocyclops albidus, M. fuscus* and *Megacyclops viridis*. At 20–24°C, the first moult occurred 5–6 days and the second 7–9 days postinfection. Third-stage larvae were 459–531 μm in length. Early stages in the fish were found under the serosa of the swimbladder, where they matured and mated. Fertilized females moved into the body cavity and became gravid from late June to September of the following year.

P. cylindracea (Ward and Magath, 1917)
Larvae of *P. cylindracea* from *Perca flavescens* developed in *Cyclops vernalis* to the infective stage in 7–10 days (Molnar and Fernando, 1975). Ten days after copepods ingested larvae, they were placed in an aquarium with 2-year-old perch. Fourth-stage larvae were found in the body cavity and under the serosa of perch a week later. By 4 weeks both sexes were 2.1–2.8 mm in length, at 2 months the females were 4.6–5.6 mm in length. Most worms were in the body cavity but a few were in the swimbladder.

P. fujimotoi Furuyama, 1932
This species occurs in the fins and body cavity of *Ophiocephalus argus* in Japan. Furuyama (1934) suggested that *Cyclops leuckarti, C. serrulatus, C. signatus* and *C. strenuus* were suitable intermediate hosts. He fed larvae from copepods to *O. argus* and found female worms in the fins 4 months later.

P. kotlani (Molnar, 1969)
This is a parasite of the body cavity of *Aspius aspius* in Hungary, where it was studied by Molnar (1969a,b). Fish acquired infections in summer. Larvae remained in the region of the swimbladder but fertilized females entered the body cavity and became gravid there in May.

P. lateolabracis Yamaguti, 1935
This is a parasite of the body cavity and gonads of free-living and cultured red sea bream (*Pagrus major*). Sakaguchi *et al.* (1987) believed that the worms took 2 years to mature in the host and that after shedding eggs they died and became encapsulated.

P. obturans (Prenant, 1886)

P. obturans, a common parasite of pike (*Esox lucius*) in Europe, has been studied by Molnar (1976, 1980), Moravec (1978a) and Moravec and Dykova (1978). It is an unusual philometrid because the gravid and subgravid females occur in the vascular system of pike, especially the ventral aorta. The extremities of the parasites often extended into gill arteries. The gravid worms penetrated the wall of the gill arteries and apparently exposed themselves to water, whereupon they burst, releasing the larvae (460–500 µm in length). Larvae developed into the infective stage (708–789 µm in length) in about 10 days at 20–22°C in *Acanthocyclops vernalis, A. viridis, Cyclops strenuus, Eucyclops serrulatus, Macrocyclops albidus* and *M. fuscus.* In pike, third-stage larvae migrated and matured in the abdominal cavity and also the vitreous humour of the eyes. After fertilization, females invaded gill arteries and became gravid. Perch (*Perca fluviatilis*) and rudd (*Scardinius erythrophthalmus*) served as paratenic hosts, the infective larvae localizing in the vitreous humour of the eyes. Gravid females have been found throughout the year in pike, perhaps because transmission is not dependent on the ingestion by pike of copepods, which have annual life cycles, but on the ingestion of infected paratenic hosts, which are available throughout the year.

P. ovata (Zeder, 1803)

This is a common parasite of the abdominal cavity of numerous species of Cyprinidae (e.g. *Abramis ballerus, A. brama, Rutilus rutilus*) in Europe and Asia. Its biology has been investigated by Molnar (1966) and Moravec (1980). Gravid females penetrated the rectal region of the host; upon being exposed to water they ruptured and released larvae. Larvae developed to the third stage in *Acanthocyclops vernalis, A. viridis, Cyclops strenuus, Macrocyclops albidus* and *Megacyclops gigas.* At 25°C the first moult occurred 3–4 days and the second 5–7 days postinfection. In cyprinids, larvae developed in the serosa of the swimbladder and the body cavity. Mating took place in July and early August and females were then found in the body cavity. Females became gravid and released larvae in May and early June of the year following infection of the final host. According to Molnar (1966) most gravid female worms were found in fish that were also infected with *Ligula* sp. (Cestoda).

P. sibirica (Bauer, 1946)

P. siberica is a parasite of the body cavity of *Coregonus alba* in Finland, the Baltic Sea basin, northern Russia (Karelia) and Siberia (Moravec, 1994). According to Korenchenko (1994) larvae were found in *Acanthocyclops* sp., *Cyclops gracilis scutifer* and *Heterocope borealis* in Lake Gekovo in the Russian Far East.

Philometroides

P. cyprini (Ishii, 1931)

Vasilkov (1968) and Vismanis (1970) studied this species (syn. *P. lusiana*) in the carp *Cyprinus carpio* in the CIS. It is an annual species which developed in *Acanthocyclops vernalis, A. viridis, Cyclops strenuus, Eucyclops serrulatus, E. macruroides, Macrocyclops albidus* and *Mesocyclops leuckarti.* Carp acquired infections in June in Europe. Larvae ingested by carp congregated near the swimbladder, gonads and kidneys and developed to adulthood. In one month females were fertilized and then migrated to the scale pouches and grew rapidly. Larvae appeared in female worms in the middle of May and were released at the end of May.

P. huronensis Uhazy, 1977

P. huronensis is a common parasite of white sucker (*Catostomus commersoni*) in Lake Huron, Ontario, where it has been studied in detail by Uhazy (1977a,b, 1978). The prominent, red, gravid females appeared subcutaneously in the fins of suckers from April to June. Coagulative necrosis of the fish's epidermis over a part of the gravid female exposed the latter to water, which resulted in bursting and release of first-stage larvae (Uhazy, 1978). Remains of spent females were resorbed by the host. First-stage larvae (362–406 µm in length) were ingested by, and developed to the infective stage in, the haemocoel of *Cyclops bicuspidatus thomasi* and *C. vernalis* (Fig. 6.2). In *C. bicuspidatus* held at 10°C the first moult occurred 14–18 days and the second 30 days postinfection; in *C. vernalis* kept at 20–23°C comparable figures were 6–9 and 14–20 days. Infected copepods became lethargic and settled to the bottom of the dish containing them. Infective third-stage larvae were 389–414 µm in length.

Seven laboratory-reared 13-month-old white suckers were given *C. bicuspidatus* containing larvae and kept at various temperatures for prolonged periods (up to 218 days). As early as 63 days postinfection, minute adult worms were found in the peritoneum around the swimbladder. Mature females remained uninseminated and were 2.5–3.5 mm in length. Mature males were 2.4–2.6 mm in length. Both were found near the swimbladder up to 218 days postinfection.

Seasonal examination of white suckers in Lake Huron revealed the pattern of transmission and development of *P. huronensis* (Fig. 6.3). Fourth-stage larvae and uninseminated but mature females were found in the peritoneum around the swimbladder of suckers in July, showing that transmission started before July. This agreed with experimental evidence that copepods infected in early spring (April–June) would have infective larvae in the latter part of June. Increased intensity of fourth-stage larvae from early summer to early autumn showed that transmission continued during summer. By September almost all females found were inseminated and the numbers increased near the bases of the fins at the same time as the numbers near the swimbladder decreased from September to December. Eggs first appeared in females as early as September and there was a marked increase in the size of worms from autumn to spring, when females began to appear in fins and were 35–101 mm in length and fully gravid. The fate of males after they had inseminated the females is unknown but presumably they died and were resorbed. *P. huronensis* is, therefore, an annual species. In spring and early summer, as fish are losing the previous year's infection, they are acquiring a new one which indicates, given the high prevalence of infection in all age classes of white suckers in Lake Huron, that the fish do not develop a protective immunity to subsequent infection. Uhazy (1977b) noted that, although transmission of *P. huronensis* coincided with spawning, it was probably independent of associated physiological changes in the fish since gravid female worms occurred in the fins of sexually immature suckers.

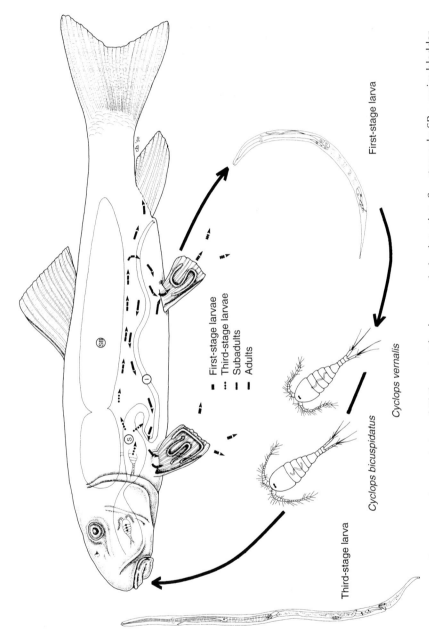

Fig. 6.2. Development and transmission of *Philometroides huronensis*. I = intestine; S = stomach; SB = swim bladder. (Original, U.R. Strelive.)

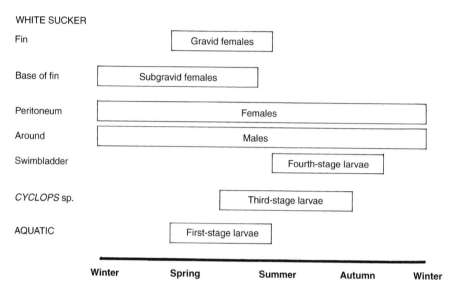

Fig. 6.3.   Annual life cycle of *Philometroides huronensis* in white suckers (*Catostomus commersoni*) in southern Lake Huron, Ontario, Canada. (After L. Uhazy, 1977 – courtesy *Canadian Journal of Zoology.*)

P. nodulosa (Thomas, 1929)

P. nodulosa occurs in the body cavity of white sucker (*Catostomus commersoni*) and other fish species in North America. Dailey (1966) infected *Cyclops thomasi* and *C. vernalis*, in which the third stage was attained in 20 days. Larvae given to white suckers were found in the swimbladder and kidneys 9 days postinfection. Worms migrated to the subcutaneous tissues of the head (cheek galleries), where the females were inseminated and became gravid in the spring; males died and were resorbed.

P. parasiluri (Yamaguti, 1935)

This species occurs in *Percottus glehni* in the CIS. According to Ermolenko (1984) the intermediate host is *Acanthocyclops viridis*.

P. sanguinea (Rudolphi, 1819)

According to Wierzbicki (1960) this is an annual species in *Carassius carassius* in Poland. Gravid females first appeared in the fins in September and their numbers increased in autumn and winter; worms were absent by June. He successfully infected *Cyclops kolensis* with first-stage larvae.

Yashchuk (1970, 1971, 1974) reported studies of this species in reservoirs in the CIS. Suitable intermediate hosts were *Acanthocyclops nanus*, *A. languidoides*, *Cyclops strenuus* and *Diaptomus gracilis*. Larvae reached the infective stage in 10–11 days at 18–19°C and in 26–27 days at 10°C. Ouk and Chun (1973) and Nakajima and Egusa (1977) reported studies of the species in *Carassius auratus gibalio* in Korea and Japan under the name *Philometroides crassii*.

Philonema

P. agubernaculum Simon and Simon, 1936

This species has been found commonly in the body cavity of landlocked salmon (*Salmo salar*) and brook trout (*Salvelinus fontinalis*) in Maine. Meyer (1960) claimed to have infected copepods but it is not clear which species served as intermediate hosts and he failed to infect fish given copepods with larvae. Vik (1964) found nine larvae in 4500 digested smelt (*Osmerus mordax*), an important prey of salmon and trout in the study area in Maine. Subadult *P. agubernaculum* were found in a hatchery trout given larvae by stomach tube 2 months earlier. First-stage larvae occurred in roe and milt of salmon stripped at spawning time.

Chacko (1976) infected *Cyclops bicuspidatus*. Subadult worms were found in the body cavity of *Oncorhynchus mykiss* 128 days after being given copepods with third-stage larvae and adult worms were found 205 days postinfection.

P. oncorhynchi Kuitunen-Ekbaum, 1933

This nematode is found in the body cavity of sockeye salmon (*Onchorhynchus nerki*) and other salmon in various parts of the world, including the Pacific coast of Canada, where it has been studied in *O. nerki* in Cultus Lake, British Columbia, by Platzer and Adams (1967), Adams (1969), Ko and Adams (1969), Lewis *et al.* (1974) and Bashirullah (1967, 1983). After emerging as fry, *O. nerki* spend 1–2 years in Cultus Lake (the nursery lake). They then migrate to sea and return some 2 years later to the natal streams, where they spawn and die. Gravid female worms were found only in spawning salmon. These worms were readily expelled when roe was stripped and it is believed that they normally pass out of the fish with the reproductive products during spawning from mid-November to mid-December. Gravid female worms rapidly burst in fresh water, releasing myriads of first-stage larvae (416–527 μm in length) which survived for up to 25 days and remained suspended in the water column. Larvae developed to the infective stage in *Cyclops bicuspidatus*. Development to the infective third stage (788–1038 μm in length) in the copepod was temperature dependent (17–19 days at 15°C; 30–31 days at 10°C; 23–25 days at 4°C). Based on these data it was suggested that some copepods with infective larvae would be in Cultus Lake as early as mid-January and probably as late as May, since *C. bicuspidatus* is long-lived. These infected copepods provided the source of larvae for the previous year's young fish which would migrate seaward from mid-April to the first week in May. Some infected copepods might even be available to the new group of fish that become free-swimming in May. Thus, young salmon migrating to sea were heavily infected with third-stage larvae, which were found mainly in the tunica adventitia of the swimbladder and pneumatic duct and less commonly in the parietal peritoneum. Larvae provided the source of worms found in the body cavity of fish returning to Cultus Lake to spawn years later.

Experiments in which naturally infected fish were held in fresh water for 2 years indicated that worms failed to develop for the first 6 months but grew rapidly for the next 8–16 months, had attained the fourth stage by 15 months and were subadult in the body cavity in 20 months. Worms containing eggs were found in salmon migrating up the Fraser River in September. Worms with eggs containing mainly morulae were found in salmon arriving in Cultus Lake in September. By November, fully gravid female worms were present in spawning fish. Thus, both fish and parasite became gravid at the same time. It has been suggested that the final maturation of the worms is controlled by hormones of the host (Adams, 1969).

Family Anguillicolidae

The family includes *Molnaria* and *Skrjabillanus* (syn. *Agrachanus*) of teleost fishes and *Anguillicola* of eels.

Molnaria and *Skrjabillanus*

Molnaria intestinalis (syn. *M. erythrophthalmi*) and *Skrjabillanus scardinii* occur on the serosa of the swimbladder, kidneys and intestine as well as on the mesentery of *Scardinus erythrophthalmus* in the CIS. According to Tikhomirova (1971, 1975, 1980) first-stage larvae of both species were carried in blood to the skin, where small accumulations could be found. The larva is short and has a hook-like tail. She reported that the ectoparasitic branchiurans, *Argulus coregoni* and *A. foliaceus*, ingested larvae while feeding. Larvae invaded the haemocoel of the crustaceans and moved to the legs, where development to the third stage took place. At 26°C, the first moult took place in 4 days and the second 8–9 days postinfection. The third-stage larva was long and had a rounded tail. Third-stage larvae were said to leave the mouth parts of the branchiurans while the latter were feeding. Larvae then invaded the fish, migrated to the serous membranes and reached adulthood. Mating occurred in 34–36 days and larvae were shed by female worms 46–48 days postinfection.

Moravec (1978b) found a larva in *Argulus foliaceus* which he first identified as *M. erythrophthalmi* but later (1985) regarded as *Skrjabillanus tincae*. Moravec (1985) found *S. tincae* in 15% of tench (*Tinca tinca*) in the Czech Republic, mainly under the serosa of the posterior part of the swimbladder and in the urinary system. He concluded that tench acquired new infections mainly in late summer and autumn (August–November) and from February to March. Females became gravid from October to May and died after releasing larvae. Only tench 2 years of age and older were infected, and prevalence and intensity increased with fish age.

Anguillicola

Members of the genus *Anguillicola* occur in the swimbladders of eels. There are five species, once restricted to eels in the Pacific region (Moravec and Taraschewski, 1988):

1. *A. globiceps* Yamaguti, 1935 in *Anguilla japonica* in Japan and China.
2. *A. australiensis* Johnston and Mawson, 1940 in *Anguilla reinhardtii* in South Australia.
3. *A. crassus* Kuwa, Nimi and Itaki, 1974 in *Anguilla japonica* in Japan and China.
4. *A. novaezelandae* Moravec and Taraschewski, 1988 in *Anguilla australis* and *A. dieffenbachii* in New Zealand.

5. *A. papernai* Moravec and Taraschewski, 1988 in *Anguilla moszambica* in the Republic of South Africa.

Until recently no member of the genus *Anguillicola* occurred in European or North American eels. Paggi *et al.* (1982) reported a species of *Anguillicola*, now identified as *A. novaezelandae* (see Moravec and Taraschewski, 1988), in eels (*Anguilla anguilla*) in Lake Bracciano in Italy. Apparently, this species has not been found elsewhere in Europe.

Neumann (1985) reported swimbladder nematodes in eels in the Weser-Ems River region of northern Germany. Subsequently this nematode, *A. crassus*, was reported in many regions of Europe, including the British Isles (Pilcher and Moore, 1993), and is now considered a significant pathogen of the European eel (Tarachewski *et al.*, 1987).

Fries *et al.* (1996) reported *A. crassus* in eels (*Anguilla rostratus*) in a south Texas aquaculture facility but there is no evidence that it has spread from the facility. It has also been reported in eels in Chesapeake Bay and in Hudson River localities (Barse and Secor, 1999). It was not found in eels in Lake Ontario and the St Lawrence River in Ontario, Canada (R.C. Anderson, 1998–1999, unpublished data).

European eels spawn from May to June (Vladykov and March, 1975) in the Sargasso Sea (Schmidt, 1922; Harden-Jones, 1968; Tesch, 1977). The larva that hatches from the eggs is a transparent, ribbon-like creature with the shape of a willow leaf. Referred to as **leptocephali**, the larvae drift in the North Atlantic current, probably feeding on plankton, and taking about 3 years to reach coastal Europe. In coastal waters leptocephali transform into miniature transparent eels referred to as **glass eels**. The latter grow somewhat and become pigmented, and are referred to as **elvers** (Tesch, 1977). The glass eels and elvers invade freshwater streams, rivers and lakes. They are carnivorous and feed on crustaceans and insects. When larger, the eels are markedly piscivorous. When they reach maximum size, in up to 10 years, they move down streams and enter the sea to begin their prolonged journey (6000–7000 km) to the Sargasso Sea, which they reach in an estimated 5 months (Tesch, 1977).

Pacific eel hosts of *Anguillicola* spp. behave like European eels in that they migrate as immature adults to regions of the south Pacific, where they spawn, and the leptocephali are distributed by ocean currents to suitable coastal areas (Tsukamato, 1992).

The eel fishing industry is based on: (i) the capture of fully grown eels before or during their migration to sea; (ii) the capture of glass eels and elvers for consumption during their migration upstream; (iii) the capture of migrating elvers, which are then raised in cage culture or (iv) placed in lakes and ponds for future harvest (Tesch, 1977). The ill-advised transfer of elvers of Pacific eels to European waters inevitably resulted in the introduction of swimbladder nematodes to European eels.

A. crassus Kuwahara, Niimi, and Itagaki, 1974

Nagasawa *et al.* (1994) have provided a helpful review of early Asian work on *A. crassus*. Hirose *et al.* (1976) concluded that *A. crassus* adults sucked blood. The female ruptured and released eggs which were 80–92 × 62–71 µm in size, each containing a second-stage larva retaining the first-stage cuticle. Eggs hatched in the swimbladder lumen (Hirose *et al.*, 1976; Egusa and Hirose, 1983) and larvae made their way to the gut by way of the pneumatic duct. Eggs taken from female worms hatched in fresh water at various temperatures (Kim *et al.*, 1989). The newly hatched sheathed larvae were 240–270 μm in length (Hirose *et al.*, 1976); they attached to the substratum by the caudal extremities and wriggled intensively (Kim *et al.*, 1989). Sheathed larvae taken from the swimbladder survived in fresh water for 5 months at 24°C and for 8 months at 7°C (Hirose *et al.*, 1976). Larvae ingested by *Eucyclops serrulatus* penetrated into the haemocoel, where they exsheathed and grew. Kim *et al.* (1989) collected crustaceans in water near an eel farm. The only crustacean that contained larvae was *Thermocyclops hyalinus*, which they also infected experimentally. Intensity in experimental infections was 2.5 in *E. serrulatus* (Hirose *et al.*, 1976) and 2.9 in *T. hyalinus* (Kim *et al.*, 1989).

According to Kim *et al.* (1989) the second moult in *T. hyalinus* took place in 4 days at 25.0–26.7°C. Larvae were 920–950 μm at 18 days. According to Hirose *et al.* (1976) larvae were infective to eels in 1 week. Egusa (1978) and Kim *et al.* (1989) claimed that larvae were infective to eels in 4–7 days.

Egusa (1978) reported that infected larvae penetrated the gut and moved into the body cavity of the eel. They then moved to the region of the liver and invaded the wall of the swimbladder from 1 week to 1 month; the smallest larva was only 2 mm in length.

Kim *et al.* (1989) fed infected *T. hyalinus* to eels. Larvae found in the swimbladder in 7 days were 920 μm in length and by 9 days were 950 μm in length. Sexes were indistinguishable in 38 days (males 1.68 mm and females 3.75 mm in length). In 50 days eggs were seen in females which were 8.7 mm in length; males were 6.6 mm in length.

Anguillicola crassus has been studied extensively in Europe where, as indicated above, it now occurs commonly in the European eel. According to De Charleroy *et al.* (1990) eggs as well as larvae reach the gut of the eels. It was estimated that a single female contains 100,000–150,000 eggs (Thomas and Ollevier, 1993a) or 500,000 eggs (Kennedy and Fitch, 1990), although in neither case is it clear if these figures include incompletely embryonated eggs. Petter *et al.* (1989, 1990) reported that the eggs were 90–100 × 75–90 μm in size and readily hatched at 20–22°C. The sheathed second-stage larvae were 230–290 μm in length with a conical tail. Larvae remained viable in water for 15 days at 18–29°C and for 40 days at 12°C. Thomas and Ollevier (1993a) noted that hatching of eggs taken from female worms occurred between 10 and 30°C and was inversely related to temperature. De Charleroy *et al.* (1989) reported that sheathed larvae survived for 15 days at 30°C, 23 days at 21°C and 42 days at 4°C. Kennedy and Fitch (1990) suggested that larvae survived longer in 20% sea water than in fresh water and higher levels of salinity. Reimer *et al.* (1994) suggested that transmission can occur in brackish water.

In Europe a number of copepod species have been shown experimentally to be suitable for the development of *A. crassus*. These include *Acanthocyclops rubustus* (see Petter *et al.*, 1989), *Paracyclops fimbriatus* (see Haenen *et al.*, 1989; De Charleroy *et al.*, 1990; Thomas and Ollevier, 1993b), *Cyclops vicinus* and *Macrocyclops albidus* (see Kennedy and Fitch, 1990; Moravec and Konecny, 1994).

Petter *et al.* (1990) and Bonneau *et al.* (1991) showed that the ostracod *Cypria opthalmica* was a suitable intermediate host. Moravec and Konecny (1994) showed that larvae would develop in the ostracod *Notodromus monacha*. Kennedy and Fitch (1990) suggested, but apparently did not prove, that juvenile *Gammarus* sp., *Diaptomus gracilis* and the brackish water species *Eurytemora affinis* were suitable intermediate hosts.

In the copepod intermediate host, the second moult occurred in 12 days at 20–22°C and the larvae remained coiled in the haemocoel (Petter *et al.*, 1989). The infective larvae had narrow lateral alae and a conical unarmed tail. The cephalic extremity had a cuticular thickening in the form of an inverted V. The larvae were 760 (530–950) μm in length.

Petter *et al.* (1989) showed that the third-stage larvae of *A. crassus* became encapsulated in the body cavity of experimentally infected guppies (*Lebistes reticulatus*), raising the likelihood that the larger piscivorous eels would become infected from eating fish paratenic hosts. Subsequently Haenen and Van Banning (1990) and Haenen *et al.* (1994) in The Netherlands found larvae in the swimbladder of smelt (*Osmerus epertanus*), ruffe (*Gymnocephalus cernuus*), perch (*Perca fluviatilis*), zander (*Stizostedion lucioperca*) and three-spined stickleback (*Gasterosteus aculeatus*). Höglund and Thomas (1992) found larvae in the wall of the swimbladders and in the gut of black goby (*Gobius niger*) and ruffe in the Baltic in Sweden. De Charleroy *et al.* (1990) recovered third-stage larvae from the body cavity of carp (*Cyprinus carpio*) and ide (*Leuciscus idus*) given copepods containing infective larvae of *A. crassus*.

Thomas and Ollevier (1992a) found larvae in the following fishes in Belgium: ruffe, pumpkinseed (*Lepomis gibbosus*), brown bullhead (*Ictalurus nebulosus*), zander, stickleback, tilapia (*Oreochromis niloticus*), perch, gungeon (*Gobio gobio*), chub (*Leuciscus cephalis*), nose-carp (*Chondrostoma nasus*), ide, rudd (*Scardinius erythrophthalmus*), roach (*Rutilus rutilus*) and tench (*Tinca tinca*). Prevalences ranged from 95.7% in ruffe to 1.5% in tench. Larvae were found in or around the swimbladder.

Pazooki and Székely (1994) found larvae in various fishes in Lake Velence in Hungary, i.e. ruffe, zander, pumpkinseed, Chinese rasbora (*Pseudorasbora parva*), roach, common carp (*Cyprinus carpio*), bream (*Abramis brama*), bleak (*Alburnus alburnus*). Prevalences (with large sample sizes) were 95–100% in bleak and ruffe and the mean intensity was 184.

Reimer *et al.* (1994) reported a few *A. crassus* (third- and fourth-stage larvae) in the swimbladders of deep-snouted pipefish (*Syngnathus typhie*) and black goby (*Gobius niger*) in the Baltic Sea, where the parasite occurs in eels in brackish water (Reimer, 1987).

Finally Székely (1994, 1995) examined fish in Lake Balaton, Hungary. Prevalences were high in bream, bleak, asp (*Aspius aspius*), white bream (*Blicca bjoerkna*), gibal carp (*Carasius auratus gibalio*), Chinese rasbora, betterlung (*Rhodeus amarus*), roach, rudd, tench, pumpkinseed, perch, ruffe, pike perch (*Stizostedion lucioperca*), pike (*Esox lucius*), catfish (*Silurus glanis*) and river goby (*Neogobius fluviatilis*). Where samples were more than ten the prevalences were 36–100%. Székely *et al.* (1996) studied larvae in various fish paratenic hosts. They noted that some larvae moved freely in the body cavity and could be found in the gonads, the intestinal wall and the swimbladder. Other larvae became more or less encapsulated by an inflammatory response. The river goby proved to be the most suitable host among the 22 fish species studied and few if any larvae elicited host reaction in this host.

In *Perca fluviatilis* and *Lepomis gibbosus* fourth-stage *A. crassus* were found in the wall of the swimbladder (De Charleroy *et al.*, 1990; Thomas and Ollevier, 1992b;

Moravec, 1996). Moravec (1996) found larvae in the aquatic snail *Galba corvus* exposed to infected copepods.

A number of authors have infected eels with larvae from various paratenic hosts (Haenen and van Banning, 1991; Höglund and Thomas, 1992; Székely, 1996). Haenen *et al.* (1989, 1991) fed infected *Paracyclops fimbriatus* to elvers of European eels. Larvae migrated through the intestinal wall into the peritoneal cavity and reached the swimbladder lumen in 170 h. De Charleroy *et al.* (1990) allowed glass eels to feed on copepods with larvae and noted that the larvae remained viable in the body cavity. Larvae found in the swimbladder of smelt and ruffe paratenic hosts produced infection in eels.

Moravec *et al.* (1994) gave larvae from *Cyclops strenuus* to eels (placed in captivity as elvers and kept for 1 year in the laboratory). At 20–22°C development from the third to the fourth stage took 3 weeks. Young adults appeared in about 1 month and non-embryonated eggs first appeared in female worms in 6–7 weeks. The prepatent period was about 3 months and the patent period about 1 month.

Molnár (1993) and Würtz *et al.* (1996) reported marked reduction in oxygen in the swimbladder of infected eels. Sprengel and Lüchtenberg (1991) suggested that swimming speed was reduced in infected eels. Molnár (1993) believed that the eels natural resistance was adversely affected by damage in the wall of the swimbladder. Haenen *et al.* (1996) noted that the severity of lesions was related to the level of infection, especially in eels that had been reinfected. Boon *et al.* (1990) asserted that change in the amount of plasma protein and haematocrit were due to the migration of larval stages in the wall of the swimbladder and the blood-sucking activities of fourth stage and adult worms. According to Knopf *et al.* (1998) larvae of *A. crassus* failed to invade the swimbladder of eels at low temperatures (e.g. 4°C) and adult worms in eels were damaged by low temperatures. The results suggest that spread of *A. crassus* in northern regions may be restricted by low ambient temperatures.

Molnár *et al.* (1991, 1994), Molnár (1993, 1994), Csaba and Láng (1992) and Csaba *et al.* (1993) reported in detail the pathology in heavily infected eels from Lake Balaton, a shallow, warm body of water in Hungary where a massive die-off of eels had occurred. They described chronic degenerative, inflammatory and proliferative changes in the swimbladder wall associated with numerous larvae, as well as acute hyperplasia and hyperaemia. The result of these changes was a marked reduction in the size of the swimbladder lumen. The pathological changes often resulted in mortality in the eel population (cf. Lake Balaton in Hungary). Molnár *et al.* (1994) noted that by 1991 the entire eel population in Lake Balaton was infected with *A. crassus*. Marked differences in the prevalences and intensities occurred in different regions of the lake and by 1993 a host–parasite equilibrium had become established.

A. globiceps Yamaguti, 1935

According to Wang and Zhao (1980) and Huang (1981), *A. globiceps*, a common swimbladder parasite of eels (*Anguilla japonica*), developed in the haemocoel of *Eucyclops serratus* and *Thermocyclops taikoknensis*.

Family Micropleuridae

Micropleura

Members of this genus are found in the body cavity of crocodiles and turtles.

M. indica Khera, 1951

Siddiqi and Jairajpuri (1963) claimed that *M. indica* Khera, 1951 from the freshwater turtle (*Lissemys punctata*) in India developed to the infective stage in *Cyclops* spp. The first moult occurred 36–48 h postinfection and larvae were typically dracunculoid.

Family Daniconematidae

The family includes a single genus and species, found under the serosa of the swimbladder and intestine of the European eel (*Anguilla anguilla*).

Daniconema

D. anguillae Moravec and Køie, 1987

These fine thread-like nematodes with dome-shaped dorsoventral cephalic papillae occur in the serosa of the swimbladder and the intestine of *Anguilla anguilla* in Europe. It is suggested by Moravec (1994) that the first-stage larvae are found in the blood and the intermediate host may be blood-sucking branchiurids (*Argulus*). Larvae regarded as 'microfilaria of *Philometra* sp.' by Elkan and Reichenbach-Klinke (1974) probably belong to *D. anguillae*.

References (Camallanina)

Adams, J.R. (1969) Migration route of invasive juvenile *Philonema oncorhynchi* (Nematoda: Philometridae) in young salmon. *Journal of the Fisheries Research Board of Canada* 26, 941–946.

Appy, R.G., Anderson, R.C. and Khan, R.A. (1985) *Ichthyofilaria canadensis* n.sp. (Nematoda: Dracunculoidea) from eelpouts (*Lycodes* spp.). *Canadian Journal of Zoology* 63, 1590–1592.

Baker, M.R. (1987) *Synopsis of the Nematoda Parasitic in Amphibians and Reptiles*. Memorial University of Newfoundland Occasional Papers in Biology, No. 11, 325 pp.

Barse, A.M. and Secor, D.H. (1999) An exotic nematode parasite of the American eel. *Fisheries* 24, 6–24.

Bartlett, C.M. and Anderson, R.C. (1985) Larval nematodes (Ascaridida and Spirurida) in the aquatic snail, *Lymnaea stagnalis*. *Journal of Invertebrate Pathology* 46, 153–159.

Bashirullah, A.K.M. (1967) The development and maturation of *Philonema* species (Nematoda: Philometridae) in salmonid hosts with different life histories. *Dissertation Abstracts* 27 (9) 33–32.

Bashirullah, A.K.M. (1983) Survival and development of *Philonema oncorhynchi* (Nematoda: Philometridae) in different recipient fish hosts. *Rivista di Parassitologia* 44, 489–496.

Bashirullah, A.K.M. and Ahmed, B. (1976a) Development of *Camallanus adamsi* Bashirullah, 1974 (Nematoda: Camallanidae) in cyclopoid copepods. *Canadian Journal of Zoology* 54, 2055–2060.

Bashirullah, A.K.M. and Ahmed, B. (1976b) Larval development of *Spirocamallanus intestinecolas* (Bashirullah, 1973) Bashirullah, 1974 in copepods. *Rivista Parassitologia* 37, 303–311.

Bonneau, S., Blanc, G. and Petter, A. (1991) Studies on the biology of the first larval stages of *Anguillicola crassus* (Nematoda, Dracunculoidea): specificity of the intermediate host and influence of temperature on the rate of development. *Bulletin Français de la Pêche et de la Pisciculture* 320, 1–6.

Boon, J.H., Cannaerts, V.H.H., Augustijn, H., Machiels, M.A.M., De Charleroy, D. and Ollevier, F. (1990) The effect of different infection levels with infective larvae of *Anguillicola crassus* on haematological parameters of European eel (*Anguilla anguilla*). *Aquaculture* 87, 243–253.

Brackett, S. (1938) Description and life history of the nematode *Dracunculus ophidensis* n.sp., with a redescription of the genus. *Journal of Parasitology* 24, 353–361.

Campana-Rouget, Y. (1961) Remarques sur le cycle évolutif de *Camallanus lacustris* (Zoega, 1776) et la phylogenie des Camallanidae. *Annales de Parasitologie Humaine et Comparée* 36, 425–434.

Campana-Rouget, Y., Petter, A.J., Kremer, M., Molet, B. and Miltgen, F. (1976) Présence du nématode *Camallanus fotedari* dans le tube digestif de poissons d'aquarium de diverses provenances. *Bulletin de l'Academie Vétérinaire de France* 49, 205–210.

Chabaud, A.G. (1975) No. 3. Keys to genera of the order Spirurida. Part 1. Camallanoidea, Dracunculoidea, Gnathostomatoidea, Physalopteroidea, Rictularioidea and Thelazioidea. In: Anderson, R.C., Chabaud, A.G. and Willmott, S. (eds) *CIH Keys to the Nematode Parasites of Vertebrates*. Commonwealth Agricultural Bureaux, Farnham Royal, UK, pp. 1–27.

Chacko, A.J. (1976) Life history and control of *Philonema agubernaculum* Simon and Simon (Nematoda: Philometridae) from Palisades Reservoir, Idaho. *Dissertation Abstract International* 36B, 3172.

Chandra, K.J. and Chubb, J.C. (1992) Survival, activity and penetration of the first-stage larva of *Camallanus lacustris* (Zoega, 1776) (Nematoda: Camallanidae). *Research and Review in Parasitology* 52, 95–98.

Crichton, V.F.J. and Beverley-Burton, M. (1974) Distribution and prevalence of *Dracunculus* spp. (Nematoda: Dracunculoidea) in mammals in Ontario. *Canadian Journal of Zoology* 52, 163–167.

Crichton, V.F.J. and Beverley-Burton, M. (1975) Migration, growth and morphogenesis of *Dracunculus insignis* (Nematoda: Dracunculoidea). *Canadian Journal of Zoology* 53, 105–113.

Crichton, V.F.J. and Beverley-Burton, M. (1977) Observations on the seasonal prevalence, pathology and transmission of *Dracunculus insignis* (Nematoda: Dracunculoidea) in the raccoon (*Procyon lotor* (L.)) in Ontario. *Journal of Wildlife Diseases* 13, 273–280.

Crites, J.L. (1976) An alternative pathway in the life cycle of *Camallanus oxycephalus* Ward and Magath 1916 (Nematoda: Camallanidae). *Journal of Parasitology* 62, 166.

Csaba, G. and Láng, M. (1992) Examination on the cause of the mass eel death in Lake Balaton in 1991. *Halászat* 85, 14–17.

Csaba, G., Láng, M., Sálti, G., Ramotsa, J., Glavis, R. and Rátz, F. (1993) The nematode *Anguillicola crassus* (Nematoda, Anguillicolidae), and its role in the death of eels in Lake Balaton (Hungary) in 1991. *Magyar Állatorvosok Lapja* 1993/1, 11–21.

Dailey, M.D. (1966) Biology and morphology of *Philometroides nodulosa* (Thomas, 1929) n.comb. (Philometridae: Nematoda) in the western white sucker (*Catostomus commersoni*). Unpublished PhD thesis, Colorado State University, Fort Collins, Colorado.

De, N.C. (1993) Seasonal dynamics of *Camallanus anabantis* in the climbing perch, *Anabas testudineus*, from the freshwater swamps near Kalyani, West Bengal. *Folia Parasitologica* 40, 49–52.

De N.C. (1995) On the development of *Spirocamallanus mysti* (nematoda: Camallanidae). *Folia Parasitologica* 42, 135–142.

De, N.C., Samanta, P. and Majumdar, G. (1984) Aspects of the developmental cycle of *Neocamallanus singhi* Ali, 1957 (Nematoda: Camallanidae). *Folia Parasitologica* 31, 303–310.

De, N.C., Sinha, R.K. and Majumdar, G. (1986) Larval development of *Procamallanus spiculogubernaculus* Agarwal, 1958 (Nematoda: Camallanidae) in copepods. *Folia Parasitologica* 33, 51–60.

De Charleroy, D., Thomas, K., Belpaire, C. and Ollevier, F. (1989) The viability of the free-living larvae of *Anguillicola crassus*. *Journal of Applied Ichthyology* 5, 154–156.

De Charleroy, D., Grisez, L., Thomas, K., Belpaire, C. and Ollevier, F. (1990) The life cycle of *Anguillicola crassus*. *Diseases of Aquatic Organisms* 8, 77–84.

Eberhard, M.L. and Brandt, F.H. (1995) The role of tadpoles and frogs as paratenic hosts in the life cycle of *Dracunculus insignis* (Nematoda: Dracunculoidea). *Journal of Parasitology* 81, 792–793.

Eberhard, M.L., Ruiz-Tiben, E. and Wallace, S.V. (1988) *Dracunculus insignis*: experimental infection in the ferret, *Mustela putorius furo*. *Journal of Helminthology* 62, 265–270.

Egusa, S. (1978) Anguillicolosis of eels. In: Egusa, S. (ed.) *Infectious Diseases of Fish*. Koseisha Koseikaku, Tokyo, pp. 501–506. (In Japanese.)

Egusa, S. and Hirose, H. (1983) Anguillicolosis of eels. In: Egusa, S. (ed.) *Fish Pathology* (*Infectious and Parasitic Diseases*). Koseisha Koseikaku, Tokyo, pp. 308–310. (In Japanese.)

Elkan, E. and Reichenback-Klinke, H.H. (1974) *Color Atlas of the Diseases of Fishes, Amphibians and Reptiles*. T.F.H. Publications, Hong Kong, 256 pp.

Ermolenko, A.V. (1984) The systematic position and biology of *Philometra parasiluri* Yamaguti, 1935 (Nematoda, Philometridae). In: *Parazity zhivotnykh i rastenii*. Vladivostok U.S.S.R. Akademii Nauk SSR. Dal'nevostochnyi Nauchnyi Tsentr, Biologii-Pochvennyi Instituta, pp. 78–81. (In Russian.)

Evlanov, L.A. (1995) The reproductive structure of the parasitic nematode *Camallanus truncatus* and the factors determining its changes. *Parazitologiya* 29, 417–423. (In Russian.)

Fedchenko, A.P. (1871a) Concerning the structure and reproduction of the guinea worm (*Filaria medinensis* L.). *Izvestiia Imperatorskago Obshchestva Liubitelei Estestvoznaniia Antropologii i Ethnografii, Moskova* 8, 71–81. (In Russian.)

Fedchenko, A.P. (1871b) Concerning the structure and reproduction of the guinea worm (*Filaria medinensis* L.). Reprinted (1971) in: *American Journal of Tropical Medicine and Hygiene* 20, 511–523.

Fries, L.T., Williams, D.J. and Johnson, S.K. (1996) Occurrence of *Anguillicola crassus*, an exotic parasitic swim bladder nematode of eels in the southwestern USA. *Transactions of the American Fisheries Society* 125, 794–797.

Furuyama, T. (1934) On the morphology and life history of *Philometra fujimotoi* Furuyama, 1932. *Keijo Journal of Medicine* 5, 165–177.

Fusco, A.C. (1980) Larval development of *Spirocamallanus cricotus* (Nematoda: Camallanidae). *Proceedings of the Helminthological Society of Washington* 47, 63–71.

Haenen, O.L.M. and Van Banning, P. (1990) Detection of larvae of *Anguillicola crassus* (an eel swimbladder nematode) in freshwater fish species. *Aquaculture* 87, 103–109.

Haenen, O.L.M. and Van Banning, P. (1991) Experimental transmission of *Anguillicola crassus* (Nematoda: Dracunculoidea) larvae from infected prey fish to the eel *Anguilla anguilla*. *Aquaculture* 92, 115–119.

Haenen, O.L.M., Grisez, L., De Charleroy, D., Belpaire, C. and Ollevier, F. (1989) Experimentally induced infections of European eel *Anguilla anguilla* with *Anguillicola crassus* (Nematoda, Dracunculoidea) and subsequent migration of larvae. *Diseases of Aquatic Organisms* 7, 97–101.

Haenen, O.L.M., Grisez, L., De Charleroy, D., Belpaire, C. and Ollevier, F. (1991) Artificial infection of the European eel with third-stage larvae of the nematode *Anguillicola crassus*. *Journal of Aquatic Animal Health* 3, 263–265.

Haenen, O.L.M., Van Banning, P. and Dekker, W. (1994) Infection of eel *Anguilla anguilla* (L.) and smelt *Osmerus eperlanus* (L.) with *Anguillicola crassus* (Nematoda, Dracunculoidea) in the Netherlands from 1986 to 1992. *Aquaculture* 126, 219–229.

Haenen, O.L.M., van Wijngaarden, T.A.M., van der Heijden, M.H.T., Höglund, J., Cornelissen, J.B.J.W., van Leengoed, L.A.M.G., Borgsteede, F.H.M. and van Muiswinkel, W.B. (1996) Effects of experimental infections with different doses of *Anguillicola crassus* (Nematoda, Dracunculoidea) on European eel (*Anguilla anguilla*). *Aquaculture* 141, 41–57.

Harden-Jones, F.R. (1968) *Fish Migration*. E. Arnold Publications, London, 325 pp.

Hirose, K., Sekino, T. and Egusa, S. (1976) Notes on the egg deposition, larval migration and intermediate host of the nematode *Anguillicola crassa* parasitic in the swimbladder of eels. *Fish Pathology* 11, 27–31.

Höglund, J. and Thomas, K. (1992) The black goby *Gobius niger* as a potential paratenic host for the parasitic nematode *Anguillicola crassus* in a thermal effluent of the Baltic. *Diseases of Aquatic Organisms* 13, 175–180.

Huang, L.F. (1981) Studies on the life cycle of *Anguillicola globiceps*. *Zoology Magazine* 1, 24–25.

Hunter, J.M. (1996) An introduction to guinea worm on the eve of its departure: dracunculiasis transmission, health effects, ecology and control. *Social Science and Medicine* 43, 1399–1425.

Hunter, J.M. (1997a) Geographic patterns of guinea worm infestation in Ghana: an historical contribution. *Social Sciences and Medicine* 43, 1399–1425.

Hunter, J.M. (1997b) Bora holes and the vanishing of guinea worm disease in Ghana's upper region. *Social Science and Medicine* 45, 71–89.

Ivashkin, V.M., Sobolev, A.A. and Khromova, L.A. (1971) *Essentials of Nematology. Vol. 22. Camallanata of Animals and Man and Diseases Caused by Them*. Academy of Sciences of USSR, Helminthological Laboratory, Moscow. (Israel Program for Scientific Translations, Jerusalem, 1977.)

Jackson, J.A. and Tinsley, R.C. (1998) Hymenochirine anurans (Pipidae) as transport hosts in camallanid nematode life-cycles. *Systematic Parasitology* 39, 141–151.

Kennedy, C.R. and Fitch, D.J. (1990) Colonization, larval survival and epidemiology of the nematode *Anguillicola crassus*, parasitic in the eel, *Anguilla anguilla*, in Britain. *Journal of Fish Biology* 36, 117–131.

Kim, G., Kim, E.B., Kim, J.Y. and Chun, S.K. (1989) Studies on the nematode *Anguillicola crassus* parasitic in the air bladder of the eel. *Journal of Fish Pathology* 2, 1–118. (In Korean with English abstract.)

Knopf, K., Würtz, J., Sures, B. and Taraschewski, H. (1998) Impact of low water temperatures on the development of *Anguillicola crassus* in the final host *Anguilla anguilla*. *Diseases of Aquatic Organisms* 33, 143–149.

Ko, R.C. and Adams, J.R. (1969) The development of *Philonema oncorhynchi* (Nematoda: Philometridae) in *Cyclops bicuspidatus* in relation to temperature. *Canadian Journal of Zoology* 47, 307–312.

Korenchenko, E.A. (1994) First data on the biology of *Philonema sibirica*. *Parazitologiya* 27, 385–390. (In Russian.)

Kupriyanova, R.A. (1954) Contribution to the biology of the fish nematodes, *Camallanus lacustris* and *C. truncatus* (Nematoda: Spirurida). *Doklady Akademii Nauki SSSR* 97, 373–376. (In Russian.)

Kuwahara, A., Niimi, A. and Itagaki, H. (1974) Studies on a nematode parasitic in the air bladder of the eel. I. Description of *Anguillicola crassa* n.sp. (Philometridea, Anguillicolidae). *Japanese Journal of Parasitology* 23, 275–279.

Leuckart, R. (1876) *Die Menschlichen Parasiten und die von Ihnen Herrührenden Krankheiten.* Vols 2,3. Lief., Leipzig, pp. 513–882.

Lewis, J.W., Jones, D.R. and Adams, J.R. (1974) Functional bursting by the dracunculoid nematode *Philonema oncorhynchi. Parasitology* 69, 417–427.

Li, H.C. (1935) The taxonomy and early development of *Procamallanus fulvidraconis* n.sp. *Journal of Parasitology* 21, 103–113.

Linstow, O. (1909) *Parasitische Nematoden.* Süsswasserfauna Deutschlands (Brauer), Heft, 15 pp. 47–83.

Metchnikoff, I. (1866) *Entgegnung auf die Erwiderung des Herrn Prof. Leuckart in Giessen, in Betreff der Frage ueber die Nematodenentwicklung.* Göttingen, 23 pp.

Meyer, M.C. (1960) *Philonema agubernaculum* and other related dracunculoids infecting salmonids. *Libro Homenaje ae Dr. Eduardo Caballero y Caballero Jubileo 1930–1960,* Mexico, pp. 487–492.

Mirza, M.B. (1957) On *Dracunculus* Reichard, 1759 and its species. *Zeitschrift für Parasitenkunde* 18, 44–47.

Molnár, K. (1966) Life history of *Philometra ovata* (Zeder, 1803) and *Ph. rischta* Skrjabin, 1917. *Acta Veterinaria Academiae Scientarium Hungaricae* 16, 227–241.

Molnár, K. (1967) Morphology and development of *Philometra abdominalis* Nybelin, 1928. *Acta Veterinaria Academiae Scientarium Hungaricae* 17, 293–300.

Molnár, K. (1969a) Host–parasite relationship between fish nematodes (Genus: *Thwaitia*) and their hosts. *Acta Veterinaria Academiae Scientarium Hungaricae* 19, 427–433.

Molnár, K. (1969b) Morphology and development of *Thwaitia kotlani* sp.n. (Philometridae, Nematoda). *Acta Veterinaria Academiae Scientarium Hungaricae* 19, 137–143.

Molnár, K. (1976) Data on the developmental cycle of *Philometra obturans* (Prenant, 1886) (Nematoda: Philometridae). *Acta Veterinaria Academiae Scientiarum Hungaricae* 26, 183–188.

Molnár, K. (1980) Recent observations on the developmental cycle of *Philometra obturans* (Prenant, 1886) (Nematoda: Philometridae). *Parasitologia Hungarica* 13, 65–66.

Molnár, K. (1993) Effect of decreased oxygen content on eels (*Anguilla anguilla*) infected by *Anguillicola crassus* (Nematoda: Dracunculoidea). *Acta Veterinaria Hungarica* 41, 349–360.

Molnár, K. (1994) Formation of parasitic nodules in the swim bladder and intestinal walls of the eel *Anguilla anguilla* due to infections with larval stages of *Anguillicola crassus. Diseases of Aquatic Organisms* 20, 163–170.

Molnár, K. and Fernando, C.H. (1975) Morphology and development of *Philometra cylindracea* (Ward and Magath, 1916) (Nematoda: Philometridae). *Journal of Helminthology* 49, 19–24.

Molnár, K., Székely, C. and Baska, F. (1991) Mass mortality of eels in Lake Balaton due to *Anguillicola crassus* infection. *Bulletin of the European Association of Fish Pathologists* 11, 211–212.

Molnár, K., Székely, C. and Perényi, M. (1994) Dynamics of *Anguillicola crassus* (Nematoda: Dracunculoidea) infection in eels of Lake Balaton, Hungary. *Folia Parasitologica* 41, 193–202.

Moorthy, V.N. (1938a) Observations on the life history of *Camallanus sweeti. Journal of Parasitology* 24, 323–342.

Moorthy, V.N. (1938b) Observations on the development of *Dracunculus medinensis* larvae in cyclops. *American Journal of Hygiene* 27, 437–460.

Moorthy, V.N. and Sweet, W.C. (1936) Guinea-worm infection of *Cyclops* in nature. *Indian Medical Gazette* 71, 568–570.

Moorthy, V.N. and Sweet, W.C. (1938) Further notes on the experimental infection of dogs with dracontiasis. *American Journal of Hygiene* 27, 301–310.

Moravec, F. (1969) Observations on the development of *Camallanus lacustris* (Zoega, 1776) (Nematoda: Camallanidae). *Véstnik Československe Spolećnosti Zoologické* 33, 15–33.

Moravec, F. (1971a) On the problem of host specificity, reservoir parasitism and secondary invasions of *Camallanus lacustris* (Nematoda; Camallanidae). *Helminthologia* 10, 107–114.

Moravec, F. (1971b) Some notes on the larval stages of *Camallanus truncatus* (Rudolphi, 1814) and *Camallanus lacustris* (Zoega, 1776) (Nematoda; Camallanidae). *Helminthologia* 10, 129–135.

Moravec, F. (1974) The development of *Paracamallanus cyathopharynx* (Baylis, 1923) (Nematoda: Camallanidae). *Folia Parasitologica* 21, 333–343.

Moravec, F. (1975) The development of *Procamallanus laeviconchus* (Wedl, 1862) (Nematoda: Camallanidae). *Véstnik Československe Spolećnosti Zoologické* 39, 23–38.

Moravec, F. (1977a) The life history of the nematode *Philometra abdominalis* in the Rokytka Brook, Czechoslovakia. *Véstnik Československe Spolećnosti Zoologické* 41, 114–120.

Moravec, F. (1977b) The development of the nematode *Philometra abdominalis* Nybelin, 1928 in the intermediate host. *Folia Parasitologica* 24, 237–245.

Moravec, F. (1978a) The development of the nematode *Philometra obturans* (Prenant, 1886) in the intermediate host. *Folia Parasitologica* 25, 303–315.

Moravec, F. (1978b) First record of *Molnaria erythrophthalmi* larvae in the intermediate host in Czechoslovakia. *Folia Parasitologica* 25, 141–142.

Moravec, F. (1980) Development of the nematode *Philometra ovata* (Zeder, 1803) in the copepod intermediate host. *Folia Parasitologica* 27, 29–37.

Moravec, F. (1985) Occurrence of the endoparasitic helminths in tench (*Tinca tinca*) from the Macha Lake fishpond system. *Véstnik Československe Spolećnosti Zoologické* 49, 32–50.

Moravec, F. (1994) *Parasitic Nematodes of Freshwater Fishes in Europe.* Academia, Prague, and Kluwer, Dordrecht, The Netherlands.

Moravec, F. (1996) Aquatic invertebrates (snails) as new paratenic hosts of *Anguillicola crassus* (Nematoda: Dracunculoidea) and the role of paratenic hosts in the life cycle of this parasite. *Diseases of Aquatic Organisms* 27, 237–239.

Moravec, F. and Vargas-Vázquez, J. (1996) The development of *Procamallanus* (*Spirocamallanus*) *neocaballeroi* (Nematoda: Camallanidae), a parasite of *Astyanax fasciatus* (Pisces) in Mexico. *Folia Parasitologica* 43, 61–70.

Moravec, F. and Amin, A. (1978) Some helminth parasites, excluding Monogenea, from fishes of Afghanistan. *Acta Scientiarum Naturalium Academiae Scientiarum Bohemicae Brno* 12, 1–45.

Moravec, F. and Dykova, I. (1978) On the biology of the nematode *Philometra obturans* (Prenant, 1886) in the fishpond system of Macha Lake, Czechoslovakia. *Folia Parasitologica* 25, 231–240.

Moravec, F., Mendoza-Franco, E. and Vivas-Rodriguez, C. (1998) Fish as paratenic hosts of *Serpinema trispinosum* (Leidy, 1852) (Nematoda: Camallanidae). *Journal of Parasitology* 84, 454–456.

Moravec, F. and Konecny, R. (1994) Some new data on the intermediate and paratenic hosts of the nematode *Anguillicola crassus* Kuwahara, Niimi et Itagaki, 1974 (Dracunculoidea), a swim bladder parasite of eels. *Folia Parasitologica* 41, 65–70.

Moravec, F. and Scholz, T. (1990) First record of *Avioserpens* larvae (Nematoda) from the naturally infected intermediate host. *Folia Parasitologica* 37, 93–94.

Moravec, F. and Taraschewski, H. (1988) Revision of the genus *Anguillicola* Yamaguti, 1935 (Nematoda: Anguillicolidae) of the swimbladder of eels, including descriptions of two new species, *A. novaezelandiae* sp.n. and *A. papernai* sp.n. *Folia Parasitologica* 35, 125–146.

Moravec, F., Dicave, D., Orecchia, P. and Paggi, L. (1994) Experimental observations on the development of *Anguillicola crassus* (Nematoda: Dracunculoidea) in its definitive host, *Anguilla* (Pisces). *Folia Parasitologica* 41, 138–148.

Moravec, F., Mendoza-Franco, E., Vargas-Vazquez, J. and Vivas-Rodriguez, C. (1995) Studies on the development of *Procamallanus* (*Spirocanallanus*) *rebecae* (Nematoda: Camallanidae) a parasite of cichlid fishes in Mexico. *Folia Parasitologica* 42, 281–292.

Moravec, F. and Vargas-Vázquez, J. (1996) The development of *Procamallamus* (*Spirocamallamus*) *neocaballeroi* (Nematoda: Camallanidae), a parasite of *Astyanax fasciatus* (Pisces) in Mexico. *Folia Parasitologica* 43, 61–70.

Moravec, F., Mendoza-Franco, E. and Vivas-Rodriguez, C. (1998) Fish as paratenic hosts of *Serpinema trispinosum* (Leidy, 1852) (Nematoda: Camallanidae). *Journal of Parasitology* 84, 454–456.

Muller, R. (1968) Studies on *Dracunculus medinensis* (Linnaeus). I. The early migration route in experimentally infected dogs. *Journal of Helminthology* 42, 331–338.

Muller, R. (1971) *Dracunculus* and dracunculiasis. *Advances in Parasitology* 9, 73–153.

Nagasawa, K., Kim, Y.G. and Hirose, H. (1994) *Anguillicola crassus* and *A. globiceps* (Nematoda: Dracunculoidea) parasitic in the swimbladder of eels (*Anguilla japonica* and *A. anguilla*) in East Asia: a review. *Folia Parasitologica* 41, 127–137.

Nakajima, K. and Egusa, S. (1977) Studies on the philometrosis of crucian carp. IV. Invasion and growth of larvae in *Cyclops*. *Fish Pathology, Japan* 12, 191–197.

Neumann, W. (1985) Schwimmblasenparasit *Anguillicola* bei Aalen. *Fischer und Teichwirt* 36, 322.

Onabamiro, S.D. (1954) The diurnal migration of *Cyclops* infected with the larvae of *Dracunculus medinensis* (Linnaeus) with some observations on the development of the larval worms. *West African Medical Journal* 3, 180–194.

Onabamiro, S.D. (1956) The early stages of the development of *Dracunculus medinensis* (Linnaeus) in the mammalian host. *Annals of Tropical Medicine and Parasitology* 50, 157–166.

Ouk, D.H. and Chun, S.K. (1973) Life cycle and chemotherapeutic control of a filarial worm, *Philometroides carassii* parasitic in *Carassius auratus*. *Bulletin of the Korean Fisheries Society* 6, 112–122. (In Korean.)

Overstreet, R.M. (1973) Parasites of some penaeid shrimps with emphasis on reared hosts. *Aquaculture* 2, 105–140.

Overstreet, R.M. (1978) *Marine maladies? Worms, Germs and Other Symbionts from the Northern Gulf of Mexico*. Mississippi-Alabama Sea Grant Consortium, 140 pp.

Paggi, L., Orecchia, P., Minervini, R. and Mattiucci, S. (1982) Sulla comparsa di *Anguillicola australienis* Johnson and Mawson, 1940 (Dracunculoidea: Anguillicolidae) in *Anguilla anguilla* del Lago de Bracciano. *Parassitolgia* 24, 139–144.

Pazooki, J. and Székely, C. (1994) Survey of the paratenic hosts of *Anguillicola crassus* in Lake Velence, Hungary. *Acta Veterinaria Hungarica* 42, 87–97.

Pereira, C., Vianna, M., Dias, E. and Azevedo, P. (1936) Biologia do nematoide *Procamallanus cearensis* n.sp. *Archivos do Instituto Biologico* 7, 209–226.

Petter, A.J. and Planelles, G. (1986) Un nouveau genre de Dracunculidae (Nematoda) parasite d'amphibien. *Bulletin du Muséum National d'Histoire Naturelle, A (Zoologie, Biologie et Ecologie Animales)* 8, 123–132.

Petter, A.J., Fontaine, Y.A. and Le Belle, N. (1989) Etude du développement larvaire de *Anguillicola crassus* (Dracunculoidea, Nematoda) chez un Cyclopidae de la region parisienne. *Annales de Parasitologie Humaine et Comparée* 64, 347–355.

Petter, A.J., Cassone, J. and Le Belle, N. (1990) Observations sur la biologie des premiers stades larvaires d'*Anguillicola crassus*, nématode parasite d'anguille. *Annales de Parasitologie Humaine et Comparée* 65, 28–31.

Pilcher, M.W. and Moore, J.F. (1993) Distribution and prevalence of *Anguillicola crassus* in eels from the tidal Thames catchment. *Journal of Fish Biology* 43, 339–344.

Platzer, E.G. and Adams, J.R. (1967) The life history of a dracunculoid, *Philonema oncorhynchi*, in *Oncorhynchus nerka*. *Canadian Journal of Zoology* 45, 31–43.

Reimer, L.W. (1987) Bewegungen der Aale der Ostsee vor der Laichwanderung auf Grund ihrer Parasitierung. *Wissenschaftliche Zeitschrift Pädagogische Hochschule Güstrow, Mathematik-nat. Facultat.* 2, 157–166.

Reimer, L.W., Hildebrand, A., Scharberth, D. and Walter, U. (1994) *Anguillicola crassus* in the Baltic Sea: field data supporting transmission in brackish waters. *Diseases of Aquatic Organisms* 18, 77–79.

Sakaguchi, S., Shibahara, T. and Yamagata, Y. (1987) Parasitic ecology of a *Philometra lateolabracis* parasite of the red sea bream. *Bulletin of the National Research Institute of Aquaculture* 12, 73–78. (In Japanese.)

Schmidt, J. (1922) The breeding places of the eel. *Philosophical Transactions of the Royal Society, London Series B* 211, 179–208.

Siddiqi, A.H. and Jairajpuri, M.S. (1963) On *Micropleura indica* Khera, 1951 (Nematoda: Dracunculidae) from a new host *Lissemys punctata*, with studies on its life history. *Zeitschrift für Parasitenkunde* 23, 99–105.

Sinha, A.K. (1988) On the life cycle of *Procamallamus spiculogubernaculus* (Camallanidae) (Agrarwal, 1958) a new nematode of fishes. *Revista di Parasitologia* 5, 111–116.

Soota, T.D. (1984) Studies on the nematode parasites of vertebrates. I. Fishes. *Records of the Zoological Surveys of India. Miscellaneous Publication Occasional Paper* No. 54, 1–352.

Sprengel, G. and Lüchtenberg, H. (1991) Infection by endoparasites reduces maximum swimming speed of European smelt, *Osmerus eperlanus*, and the European eel *Anguilla anguilla*. *Diseases of Aquatic Organisms* 11, 31–35.

Stromberg, P.C. and Crites, J.L. (1974a) The life cycle and development of *Camallanus oxycephalus* Ward and Magath, 1916 (Nematoda: Camallanidae). *Journal of Parasitology* 60, 117–124.

Stromberg, P.C. and Crites, J.L. (1974b) Survival, activity and penetration of the first-stage larvae of *Camallanus oxycephalus* Ward and Magath, 1916. *International Journal for Parasitology* 4, 417–421.

Stromberg, P.C. and Crites, J.L. (1975a) Population biology of *Camallanus oxycephalus* Ward and Magath, 1916 (Nematoda: Camallanidae) in white bass in western Lake Erie. *Journal of Parasitology* 61, 123–132.

Stromberg, P.C. and Crites, J.L. (1975b) An analysis of the changes in the prevalence of *Camallanus oxycephalus* (Nematoda: Camallanidae) in western Lake Erie. *Ohio Journal of Science* 75, 1–6.

Supryaga, A.M. (1965) The life cycle of *Avioserpens mosgovoyi* (Camallanata: Dracunculidae), a nematode of birds (Preliminary report). *Materialy Nauchno Konferentsii Vsesoyuznogo Obschchestva Gelmintologii*, Part IV, pp. 275–277. (In Russian.)

Supryaga, A.M. (1967) Life cycle of *Avioserpens mosgovoyi* (Camallanata: Dracunculidae), a parasite of aquatic and marsh birds. *Problemy Parazitologii*, pp. 199–202. (In Russian.)

Supryaga, A.M. (1969) The life-span of *Avioserpens mosgovoyi* in final hosts. *Problemy Parazitologii*, Part I, pp. 245–246. (In Russian.)

Supryaga, A.M. (1971) Life cycle of *Avioserpens mosgovoyi* (Camallanata: Dracunculidae), nematodes of aquatic birds. *Sbornik Rabot po Gel'mintologii pos vyashchen 90-letiyu so dnya rozhdeniya Akademika K.I. Skrjabina. Moscow.* Izdatel'stvo 'KOLOS'. pp. 374–383. (In Russian.)

Székely, C. (1994) Paratenic hosts for the parastic nematode *Anguillicola crassus* in Lake Balaton, Hungary. *Diseases of Aquatic Organisms* 18, 11–20.

Székely, C. (1995) Dynamics of *Anguillicola crassus* (Nematoda: Dracunculoidea) larval infections in paratenic host fishes of Lake Balaton, Hungary. *Acta Veterinaria Hungarica* 43, 401–422.

Székely, C. (1996) Experimental studies on the infectivity of *Anguillicola crassus* third-stage larvae (Nematoda) from paratenic hosts. *Folia Parasitologica* 43, 305–311.

Székely, C., Pazooki, J. and Molnar, K. (1996) Host reaction in paratenic fish hosts against third-stage larvae of *Anguillicola crassus*. *Diseases of Aquatic Organisms* 26, 173–180.

Taraschewski, H., Moravec, F., Lamah, T. and Anders, K. (1987) Distribution and morphology of two helminths recently introduced into European eel populations: *Anguillicola crassus* (Nematoda, Dracunculoidea) and *Paratenuisentis ambiguus* (Acanthocephala, Tenuisentidae). *Diseases of Aquatic Organisms* 3, 167–176.

Tesch, F.W. (1977) *The Eel: Biology and Management of Anguillid Eels.* Chapman & Hall, London, 434 pp.

Thomas, K. and Ollevier, F. (1992a) Paratenic hosts of the swimbladder nematode *Anguillicola crassus*. *Diseases of Aquatic Organisms* 13, 165–174.

Thomas, K. and Ollevier, F. (1992b) Population biology of *Anguillicola crassus* in the final host *Anguilla anguilla*. *Diseases of Aquatic Organisms* 14, 163–170.

Thomas, K. and Ollevier, F. (1993a) First estimate of the egg production of *Anguillicola crassus* (Nematoda: Dracunculoidea). *Folia Parasitologica* 40, 104.

Thomas, K. and Ollevier, F. (1993b) Hatching, survival, activity and penetration efficiency of second-stage larvae of *Anguillicola crassus* (Nematoda). *Parasitology* 107, 211–217.

Thurston, J.P. (1970) Studies on some protozoa and helminth parasites of *Xenopus*, the African clawed toad. *Revue Zoologie et de Botanique Africaines* 82, 349–364.

Tikhomrova, V.A. (1971) Biology of Skrjabillanidae (Camallanata). *Trudi Gelmintologische Laboratorii* 22, 208–211. (In Russian.)

Tikhomirova, V.A. (1975) The life cycle of nematodes of the family Skrjabillanidae. *Voprosy ekologii zhivotnykh Vypusk 2, Kalinin, USSR*, pp. 100–113. (In Russian.)

Tikhomirova, V.A. (1980) Nematodes of the family Skrjabillanidae (Nematoda: Camallanata). *Parazitologiya* 14, 258–262. (In Russian.)

Truong-Tan-Ngog (1937) Filariose du canard domestique en Conchinchine due à *Oshimaia taiwana* (Sugimoto, 1919). *Bulletin de la Société Pathologie Exotique* 30, 775.

Tsukamoto, K. (1992) Discovery of the spawning area for Japanese eel. *Nature* 356, 789–791.

Uhazy, L.S. (1977a) Development of *Philometroides huronensis* (Nematoda: Dracunculoidea) in the intermediate and definitive hosts. *Canadian Journal of Zoology* 55, 265–273.

Uhazy, L.S. (1977b) Biology of *Philometroides huronensis* (Nematoda: Dracunculoidea) in the white sucker (*Catostomus commersoni*). *Canadian Journal of Zoology* 55, 1430–1441.

Uhazy, L.S. (1978) Lesions associated with *Philometroides huronensis* (Nematoda: Philometridae) in the white sucker (*Catostomus commersoni*). *Journal of Wildlife Diseases* 14, 401–408.

Vasilkov, G.V. (1968) The life cycle of *Philometra lusiana* (Nematoda: Dracunculidae). *Trudi Vsesoyuznogo Instituta Gelmintologii* 14, 156–160. (In Russian.)

Vaucher, C. and Bain, O. (1973) Développment larvaire de *Dracunculus doi* (Nematoda), parasite d'un serpent Malgache, et description de la femelle. *Annales de Parasitologie Humaine et Comparée* 48, 91–104.

Vik, R. (1964) Notes on the life history of *Philonema agubernaculum* Simon and Simon, 1936 (Nematoda). *Canadian Journal of Zoology* 42, 511–512.

Vismanis, K.O. (1970) The life cycles of the causative agent of philometrosis in carp. *Trudi Baltiiskogo Nauchno-Issledovatel Ped. Rybnogo Khozyaistra*, 4, 403–414.

Vladykov, V.O. and March, H. (1975) Distribution of leptocephali of the two species of *Anguilla* in the western North Atlantic based on collections made between 1968–1993. *Syllogeus* 6, 38 pp.

Wang, P. and Ling, X. (1975) Some nematodes of the suborder Camallanata from Fujian Province, with notes on their life histories. *Acta Zoologica Sinica* 21, 350–358. (In Chinese with English summary.)

Wang, P. and Zhao, Y. (1980) Observations on the life history of *Anguillicola globiceps* (Nematoda: Anguillicolidae). *Acta Zoologica Sinica* 26, 243–249. (In Chinese.)

Wang, P., Sun, Y. and Zhao, Y. (1983) Studies on the life history and epidemiology of *Avioserpens taiwana* (Sugimoto, 1919) of the domestic duck in Fujian. *Acta Zoologica Sinica* 29, 350–357. (In Chinese.)

Wierzbicki, K. (1960) Philometrosis of crucian carp. *Acta Parasitologica Polonica* 8, 181–196.

Würtz, J., Taraschewski, H. and Pelster, B. (1996) Changes in gas composition in the swimbladder of the European eel (*Anguilla anguilla*) infected with *Anguillicola crassus* (Nematoda). *Parasitology* 112, 233–238.

Yashchuk, V.D. (1970) Experimental infection of invertebrates with *Philometra sanguinea* larvae. *Byulleten Vsesoyuznogo Instituta Gelmintologi im K.I. Skrjabina* 4, 183–187.

Yashchuk, V.D. (1971) Life cycle of *Philometra sanguinea* of carp. *Veterinariya, Moscow* 48, 73–75. (In Russian.)

Yashchuk, V.D. (1974) Intermediate hosts of *Philometroides sanguinea*. *Veterinariya, Moscow* 7, 75–76. (In Russian.)

Yelifari, L., Frempong, E. and Olsen, A. (1997) The intermediate hosts of *Dracunculus medinensis* in northern region, Ghana. *Annals of Tropical Medicine and Parasitology* 91, 403–409.

Chapter 7

Order Spirurida – Suborder Spirurina

The Spirurina includes a diverse group of superfamilies which use arthropod intermediate hosts. Species of Gnathostomatoidea use copepod intermediate hosts but in other superfamilies intermediate hosts are insects or various crustaceans other than copepods. First-stage larvae are generally provided with cephalic hooks and spines.

7.1

The Superfamily Gnathostomatoidea

Gnathostomatoids constitute a small group of genera characterized by massive, complex pseudolabia, and often spinous cephalic inflations (Chabaud, 1975a). Four of the five genera occur in the lower vertebrates and one (*Gnathostoma*) in mammals. Some species of *Gnathostoma* have been fairly well studied because of their significance to human and animal health. Other members have been little studied.

Family Gnathostomatidae

Subfamily Spiroxyinae

Spiroxys

S. contortus (Rudolphi, 1819)

S. contortus has been reported in western Europe and the CIS as well as in Tunisia and Cuba (Baker, 1987). It is, however, most widely recognized as a common parasite of the gastric mucosa of turtles in eastern North America (Hedrick, 1935). North American hosts include *Chelydra serpentina*, *Chrysemys picta*, *Deirochalys reticularia*, *Emydoidea blandingii*, *Graptemys geographica*, *Kinosternum subrubrum*, *Pseudemys scripta*, *Sternotherus odoratus*, *Terrapene carolina* and *Trionyx spiniferus* (see Baker, 1987).

Hedrick (1935) studied development and transmission of *S. contortus* in North America. Eggs were deposited by females in the one- to two-cell stage. When passed in faeces of the turtle they were in the 4- to 16-cell stage. In water at 23–25°C, eggs were fully developed in 84 h and hatched as second-stage larvae (175–294 μm in length) within the cuticle of the first-stage larvae (Bartlett and Anderson, 1985a). The caudal end of the sheath attached the larvae in some way to the substrate and the larvae thrashed from side to side. Larvae were readily ingested by *Cyclops albidus*, *C. brevispinosus*, *C. serrulatus* and *Mesocyclops leuckarti*, in the haemocoel of which they developed to the infective third stage. Larvae moulted in 10–14 days according to Hedrick (1935), and 9–12 days according to Bartlett and Anderson (1985a). Larvae continued to grow slightly (from 1.05–1.95 to 2.08–2.20 mm in length) after the moult to the third stage in the copepod.

Hedrick (1935) found third-stage larvae of *S. contortus* encapsulated on the mesentery of various potential paratenic hosts in Michigan, namely mud minnows (*Umbra lima*), bullheads (*Ameiurus nebulosus*), tadpoles and frogs (*Rana clamitans*), larval and adult newts (*Triturus viridescens*) and dragonfly nymphs (probably *Sympetrum* sp.). Khromova (1969, 1971) and Bartlett and Anderson (1985a) found infective larvae of *S. contortus* in the snails *Lymnaea ovata* in the CIS and *L. stagnalis* in Ontario, respectively. McAllister *et al.* (1993a) found encapsulated larvae of *S. contortus* in the North American spiny softshell turtle (*Apalone spinifera*) in Texas.

Attempts to transmit *S. contortus* experimentally to turtles have been inconclusive (Hedrick, 1935; Bartlett and Anderson, 1985a). Usually larvae invaded the gastric mucosa and even the serosal part of the stomach and failed to develop. Hedrick (1935) noted that larvae in the gastric mucosa were usually found in wild turtles during the spring, and adult worms rarely before 1 June. He also mentioned that adult worms in turtles collected in late summer and fall appeared 'old' and were often detached in the gut lumen. These observations suggest that *S. contortus* may be an annual species which is acquired in summer and which overwinters as larvae. The latter only mature the following spring and summer and after reproduction the adults die and are replaced by newly acquired larvae (cf. *Camallanus oxycephalus* and *Philometroides huronensis*). Some physiological factors in the host may inhibit development of the larvae until the following spring and summer, when copepods are blooming. Such factors, if they are not taken into account, might make it difficult to infect turtles experimentally. Possibly a sojourn in a vertebrate paratenic host is essential, as in *Gnathostoma spinigerum* and *G. doloresi* (see below).

S. japonica Morishita, 1926

S. japonica, a parasite of *Rana nigromaculata*, developed in *Mesocyclops leuckarti* (see Hasegawa and Otsuru, 1978). Paratenic hosts were *Misgurnus anguillicaudatus* and tadpoles of *Rana rugosa*.

Subfamily Gnathostomatinae

Gnathostoma

Members of the genus generally occur in tumours in the stomach wall of carnivorous mammals but two species occur in the kidneys of otters (*Lutra* spp.) and two species occur in the oesophagus of mink and weasels. Posterior extremities of worms generally protrude through orifices in the tumour leading to the lumen of the stomach, to aid in fertilization and release of eggs. In some species only anterior extremities occur in the tumour.

All *Gnathostoma* spp. produce eggs with a prominent swelling at one or both ends. Eggs passed from the host are either unsegmented or in the two- to four-cell stage. Eggs develop in water into first-stage larvae which moult in about 8–10 days and then hatch as second-stage larvae in 12–14 days. Free second-stage larvae retain the cuticle of the first stage and are highly active in water. The ensheathed first-stage larva (about 250–300 μm in size) has an attenuated tail and a rounded anterior end with a tooth-like structure. Prommas and Daengsvang (1933) and Yoshida (1934) first showed that larvae of *Gnathostoma spinigerum* were readily ingested by, and developed in, cyclopoid copepods. Larvae rapidly shed the first-stage cuticle in the intestine of the copepod, penetrated the intestine and settled in the haemocoel. They developed rapidly in the copepod and underwent the second moult a few days later.

The third-stage larvae have the spinose head bulb characteristic of adults and are about 500 µm in length. These early third-stage larvae were generally not infective to the definitive hosts (cf. *G. hispidum*). Prommas and Daengsvang (1936, 1937) discovered in Thailand that larvae of *G. spinigerum* would invade tissues of fish (*Clarias batrachus* and *Ophicephalus striatus*) where they generally became encapsulated in muscles and grew markedly into advanced third-stage larvae (about 3–4 mm in length), which were then infective to dogs and cats. These advanced third-stage larvae can pass unchanged from one paratenic host to another during predation and scavenging. Thus, infective larvae transfer readily from strictly aquatic paratenic hosts (i.e. fish) to a great variety of terrestrial or semi-terrestrial hosts like frogs, snakes, lizards, turtles, birds and mammals (including humans). The definitive host acquires infection from eating the flesh of paratenic hosts harbouring advanced third-stage larvae. In some species, larvae may migrate extensively in the definitive host before establishing themselves in the stomach. Moults in the final host are unreported.

Miyazaki (1952a, 1954, 1960, 1966) published an extensive review of Japanese work on *Gnathostoma* and gnathostomiasis and Daengsvang (1980) reviewed studies carried out in Thailand.

G. doloresi Tubangui, 1925

This is a parasite of the stomach mainly of wild boar and less commonly of domestic pigs in the Philippines, Japan and southeast Asia (Miyazaki, 1954, 1960). Eggs developed and hatched in water in 7 days at 27°C. First-stage larvae were 180–237 µm in length. The first intermediate hosts were *Cyclops strenuus*, *C. vicinus*, *Eucyclops serrulatus* and *Mesocyclops leuckarti*, in which development was completed in 7 days at 27°C. Known naturally infected paratenic hosts in Japan are salamanders (*Hynobius naevius* and *H. stejnegeri*) (Miyazaki and Ishii, 1952; Miyazaki, 1954) and snakes (*Trimeresurus elegans*, see Miyazaki and Kawashima, 1962; *T. f. flavoviridis*, see Tada *et al.*, 1969). Frogs, toads, mice and rats have been shown experimentally to serve as paratenic hosts (Daengsvang, 1980). Nawa *et al.* (1993) found larvae of *G. doloresi* in bluegill (*Lepomis macrochirus*) in Japan. Imai *et al.* (1989) gave larvae from the snake *Aghistrodon halys* to a pig. Eggs appeared in the faeces in 87 days and two adult worms were found at necropsy. Larvae must spend some time in paratenic hosts before they will develop to adulthood in swine (cf. *G. hispidum*).

G. hispidum Fedchenko, 1872

This gnathostome occurs in the stomach of wild and domestic pigs. Golovin (1956) described its development in cyclopoid copepods and in fish, frogs and reptiles. He infected pigs by feeding them copepods or flesh of paratenic hosts. Daiya (1969) reported the presence of larvae in birds (*Mergus albellus*, *Falco peregrinus* and *Accipiter gentilis*). Dissamarn *et al.* (1966) described development in *Mesocyclops leuckarti* and infected swine with larvae directly from copepods. Fish (*Anabas* spp. and *Ophicephalus striatus*) as well as white rats were suitable experimental paratenic hosts. Daengsvang (1971, 1980) reported that *Cyclops varicans* and *M. leuckarti* were suitable intermediate hosts and that larvae readily encapsulated in tissues of experimentally infected fish, amphibians and rodents (rats and mice) (see also Koga and Ishii, 1990). According to Wang *et al.* (1976), eggs of *G. hispidum* developed at 28°C into first-stage larvae in 7–8 days. Larvae then moulted and hatched 9–10 days later. Larvae developed in

Acanthocyclops virides, Cyclops strenuus, C. vicinus, Eucyclops serrulatus, Macrocyclops albidis, Mesocyclops leuckarti, Thermocyclops hyalinus oithoides and *T. oithoides*. Various species of fish harboured larvae after being given infected copepods. Third-stage larvae invaded rats and the rodent *Ochotona daurica* and grew in tissues for about 10 days. The authors also reported that pigs could become infected from ingesting copepods with larvae. Akahane *et al.* (1982, 1983) identified larval *G. hispidum* from loaches (*Misgurnus anguillicaudatus*) from China. The larvae grew and became encapsulated in rats. Larvae from rats were given to a pig which passed eggs 5 months later. Eggs developed into sheathed larvae which hatched in 16 days at 20–21°C and readily infected *E. serrulatus* and *C. vicinus*. Third-stage larvae were also successfully transferred to goldfish (larvae were found in the liver), frogs and mice (larvae were found in muscles and liver).

According to Koga and Ishii (1990) humans in Japan can acquire *G. hispidum* from eating raw loaches imported from southeast Asia, especially mainland China. They showed that larvae persisted in a monkey and in rats for prolonged periods. They suggest that larvae may persist in human tissues for the lifetime of the individual.

G. nipponicum Yamaguti, 1941

This is a parasite of the oesophagus of weasels (*Mustela sibirica itatsi*) in Japan (Yoshida, 1934; Miyazaki, 1952b, 1954; Arita, 1953; Yamaguchi *et al.*, 1956; Koga and Ishii, 1981). It develops in *Cyclops vicinus, Eucyclops serrulatus* and *Mesocyclops leuckarti*. Experimentally, frogs were infected. Ando *et al.* (1992, 1994) infected *C. vicinus* and *Thermocyclops hyalinus* and the resulting third-stage larvae were 520 μm in length in the haemocoel of the copepods. Advanced third-stage larvae (1.0–1.1 mm in length) were present in the muscles of loaches given larvae from copepods. Larvae also appeared in the tissues of frogs, quail and snakes given larvae. Weasels given larvae passed eggs 65–90 days postinfection. In the Ueno district of Japan, loaches (*Misgurnus anguillicaudatus*), catfish (*Silurus asotus*) and snakes (*Elaphe quadrivirgata*) contained larvae. In weasels larvae moved from the stomach and invaded muscles, where they developed into young adults in 40 days. These young adults invaded the oesophageal wall and became encapsulated in a tumour within 60 days. Ferrets and mink could also be infected. In rats, advanced third-stage larvae remained in the stomach for 12 h postinfection and then moved into muscles within 48 h.

According to Oyamada *et al.* (1996a) larvae developed in 12 days at 25°C in *Acanthocyclops vernalis, E. serrulatus* and *Macrocyclops fuscus*. The larvae developed into advanced third-stage larvae in loaches (*M. anguillicaudatus*). Oyamada *et al.* (1995a,b, 1996a,b,c,d, 1997, 1998) reported advanced larvae in the fishes *Silurus asotus, Misgurnus anguillicaudatus, Chaenogobius urotainia, Oncorhynchus masou* and *Triloboladus hakonenis*. In addition, larvae were found in the tissues of rats (*Rattus norvegicus*) and the water shrew *Chimarrogale himalayica* (Okyama *et al.*, 1996). Sohn *et al.* (1993, 1996) reported larvae in imported loaches in Korea.

G. procyonis Chandler, 1942

G. procyonis is a common stomach parasite of raccoons (*Procyon lotor*) in the southern USA (Fig. 7.1). At room temperature, eggs developed into first-stage larvae which moulted in 8–10 days and hatched between 12 and 14 days (Ash, 1962a). Larvae developed in *Cyclops bicuspidatus, C. vernalis* and *Macrocyclops albidis* in 7–8 days to the early third stage. Advanced third-stage larvae were found in tissues of

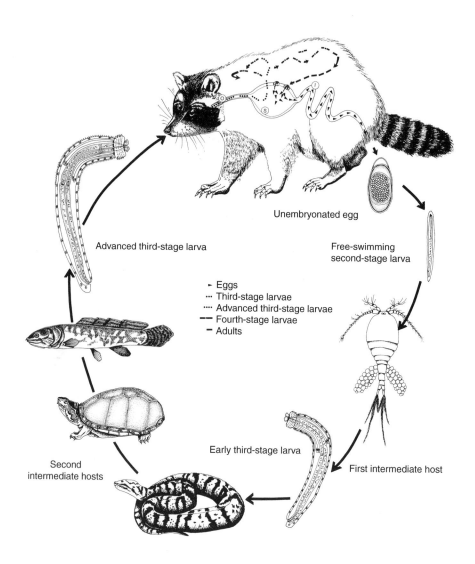

Fig. 7.1. Development and transmission of *Gnathostoma procyonis.* I = intestine; O = oesophagus; S = stomach. (Original, U.R. Strelive.)

naturally infected snakes (*Agkistrodon piscivorus, Lampropeltis getulus* and *Natrix sipedon*), turtles (*Graptemys pseudogeographica, Kinosternon subrubrum, Pseudemys scripta* and *Terrapene carolina*), the alligator (*Alligator mississippiensis*) and the bowfin (*Amia calva*). In raccoons, advanced third-stage larvae penetrated the stomach or duodenum and underwent a prolonged (2–4 months) tissue migration before returning to the stomach, where they matured 3–6 months postinfection (Ash, 1962b).

G. spinigerum Owen, 1836

This is a common stomach parasite of cats and dogs, especially in Asia. It is the main cause of human gnathostomiasis and has, therefore, been the subject of a number of investigations. Heydon (1929) studied the hatching of eggs in water. Prommas and Daengsvang (1933, 1936, 1937), Yoshida (1934), Daengsvang and Tansurat (1938), Africa *et al.* (1936a,b) and Refuerzo and Garcia (1938) revealed the role of copepods and various paratenic hosts in its transmission.

Eggs developed to the second stage and started to hatch in about 9 days at 25–31°C. First-stage larvae were about 300 μm in length. The nematode reached the early third stage in *Cyclops strenuus, C. varicans, C. vicinus, E. agilis, Eucyclops serrulatus, Mesocyclops leuckarti* and *Thermocyclops* sp. in about 7–8 days at 29–31°C. Third-stage larvae are about 500 μm in length. In fish, early third-stage larvae increased in size in the viscera and muscles and became encapsulated as early as 6 days postinfection. Advanced third-stage larvae were found in various animals given infected copepods, including fish, frogs, toads, reptiles, chickens, various rodents, tree shrews, pigs and primates. Larvae also were found in experimentally infected crabs (*Paratelphusa sexpunctatum* and *Potamon smithenus*). Advanced third-stage larvae were readily transferred experimentally from one type of animal to another (Daengsvang 1968, 1971; Daengsvang *et al.*, 1971).

Chandler (1925) found larvae of *G. spinigerum* in *Python reticulatus* and cobras (*Naja* spp.) in India. Since that time, larvae have been found in a great variety of naturally infected animals in enzootic areas (Africa *et al.* 1936a,b; Daengsvang and Tansurat, 1938; Miyazaki, 1960; Daengsvang, 1980), including at least 15 species of fish, eels, several species of snakes and frogs, a variety of predominantly piscivorous birds (e.g. herons, egrets, grebes, mergansers, ducks, rails, kingfishers) as well as crows, hawks, owls and weasels. These findings indicate the remarkable ability of advanced third-stage larvae to transfer from one host to another under natural conditions.

Cats and dogs acquire infections from eating paratenic hosts containing advanced third-stage larvae. In cats and dogs, larvae penetrate the stomach or intestine and enter the liver. They then wander in the muscles or connective tissues and grow before reinvading the wall of the stomach, where they mature and elicit the tumours associated with adults (Ueki, 1957). Prommas and Daengsvang (1937) gave the prepatent period as 198 and 223 days, Miyazaki (1960) as 100 days, and Nishikubo (1963) as 3–5 months. Daengsvang *et al.* (1971) claimed that larvae from *Ophicephalus striatus* penetrated the skin of dogs and cats; four cats passed eggs 2–7 months postinfection and eggs appeared in faeces of dogs 3–4 months postinfection.

G. vietnamicum Le Van Hoa, 1965

Like *G. miyazakii* Anderson, 1964 of *Lutra canadensis*, this gnathostome occurs in the kidney of its host (*Lutra elioti*). It developed to the second stage in *Mesocyclops leuckarti* (see Daengsvang, 1980).

Echinocephalus

E. pseudouncinatus Millemann, 1951

Millemann (1951, 1963) found larval and adult *E. pseudouncinatus* in the spiral valve of the elasmobranchs *Heterodontus francisci* and *Myliobatis californicus* in the Gulf of

California. He also found larvae encapsulated in the foot of the abalone (*Haliotis corrugata*) and concluded that they were the intermediate host. Pearse and Timm (1971) found larvae of *E. pseudouncinatus* in gonads of sea urchins (*Centrostephanus coronatus*) in southern California; prevalences were 40–78% in three localities. Larvae were said to suppress gametogenesis.

E. overstreeti Deardorff and Ko, 1983

Beveridge (1987) reviewed the various hosts of *E. overstreeti* in South Australia. Nematodes undergoing the final moult in elasmobranchs indicated that scallops (*Pecten albus* and *Chlamy bifrons*) are possible intermediate hosts.

E. sinensis Ko, 1975

Adult and larval *E. sinensis* were found in the spiral valve of eagle rays (*Aetabatus flagellum*) in Deep Bay, Hong Kong (Ko, 1975, 1976, 1977; Ko *et al.*, 1975). Two size groups of larval *E. sinensis* were found in the genital ducts and Leydig tissues of oysters (*Crassostrea gigas*). The smaller larvae were regarded as second stage and the larger as third stage. It was noted that rays preyed on oysters and it was concluded that oysters served as intermediate hosts of the parasite.

Ko (1976) reported that the highest prevalence and intensity of *E. sinensis* in oysters in Hong Kong was in August. Larvae recovered at various times in the year were given to kittens, monkeys and puppies. The larvae penetrated the gastrointestinal tract and some wandered in the tissues of these animals. Infections, however, were only achieved with larvae collected from August to October, suggesting that temperature might be a factor in infectivity.

Ko (1977) acclimated oysters with larvae of *E. sinensis* to various temperatures from 5 to 33°C. Larvae acclimated to 33°C or 28°C for 2 days readily invaded tissues of kittens. Larvae acclimated to 20–24, 15 and 5°C rarely invaded tissues of kittens.

Given the systematic affinities of the genus *Echinocephalus*, it is possible that the intermediate hosts of both *E. sinensis* and *E. pseudouncinatus* are arthropods, probably marine crustaceans such as copepods, and that molluscs, echinoderms and other marine organisms serve as paratenic or second intermediate hosts in which growth occurs. The smaller larvae of *E. sinensis* found in the oysters may be younger than the larger larvae (cf. *Spiroxys contortus*).

7.2

The Superfamily Physalopteroidea

Family Physalopteridae

Among the three subfamilies of the Physalopteridae (there is one family in the superfamily), only species in the Physalopterinae have been investigated. This subfamily consists of seven or eight closely related genera (Chabaud, 1975a) which occur mainly in the stomach of reptiles, birds, mammals and, rarely, amphibians and fish. Transmission and development of some species in mammals have been rather intensively studied. There is only limited information on species in reptiles and nothing on species in birds.

The physalopterines are usually found firmly attached to the gastric mucosa with the aid of large dentate pseudolabia and a collarette which presses into the mucosa. Studies of two species (*Physaloptera maxillaris* and *Turgida turgida*) using marked food indicate that the parasites do not normally feed on the gastric mucosa but detach and feed on food in the stomach. Studies of *P. maxillaris* showed that development from the third to later stages is dependent on the presence of ample food in the stomach. In the unfed host, the nematode remains in the third stage attached to the gastric mucosa. Despite the fact that worms do not normally feed on host tissues, there is evidence from experiments (in which various animals on similar diets were given larvae) that some species (e.g. *P. maxillaris*, *T. turgida*) are markedly host specific.

Eggs deposited by females are oval, with thick, smooth shells, and each contains a fully developed first-stage larva. Eggs can survive for long periods in the environment, given suitable moisture conditions. Alicata (1937) was the first to show that eggs of a physalopterine (*T. turgida*) would hatch in cockroaches (*Blattella germanica*) and larvae would develop in capsules on the gut of the insect. Since that time, a number of species have been studied in more detail and various insects have been implicated in the transmission of different species; cockroaches are, incidentally, among the least suitable of hosts. The first-stage larva invades the wall of the gut, usually the ileum or rectum but sometimes other sites as well, and in suitable sites becomes surrounded by a syncytium composed of hypertrophied epithelial cell nuclei and cytoplasm (Poinar and Quentin, 1972; Poinar and Hess, 1974; Cawthorn and Anderson, 1977; Gray and Anderson, 1982c). The syncytium eventually becomes surrounded by a capsule consisting of an inner layer of loose fibrous tissue and an outer layer of muscles and peritoneum of the gut wall. This latter layer is often infiltrated with haemocytes. The capsule containing the larva bulges from the gut into the haemocoel and is often pedunculate. The syncytium gradually disappears as the larva grows and there is much evidence that it is used as a source of nutrients by the latter. The larva undergoes the usual two moults to the third stage, which has the cephalic characteristics of the adult form and is, therefore, readily recognizable as a physalopteroid. In unsuitable insect hosts, the capsule frequently becomes heavily infiltrated by haemocytes and melanization may take place. Cawthorn and Anderson (1977) noted that *P. maxillaris* developed in the ileum of crickets and the colon of cockroaches. In the latter, haemocytes frequently passed through capsules which were thinner than those in crickets and larvae became melanized.

The final host can acquire infections from ingesting insects containing infective larvae. The latter attach to the stomach wall and, depending on the presence of food in the stomach, grow to adulthood. There is considerable indirect evidence from experiments and field observations that paratenesis is widely used in transmission throughout the group. Larvae ingested by possible paratenic hosts generally attach to the gastric mucosa and can persist in this site for varying periods of time and be available to predaceous final hosts. Infective larvae of *Physaloptera rara* of canids and felids have been found encapsulated in the wall of the gut and on the mesentery of rattlesnakes, which are considered important paratenic hosts in the western USA. A number of authors (Cram, 1932; Boughton, 1935; Campbell and Lee, 1953; Dixon and Roberson, 1967) have reported unidentified larval physalopteroids in the muscles of grouse and quail in North America. One wonders if these might be larvae of a species found in raptorial birds, since attempts to induce infective larvae of certain mammal physalopterines to invade birds have been unsuccessful.

Subfamily Physalopterinae

Heliconema

Members of the genus are mainly parasites of anguilliform fishes but they have also been reported from teleosts, skates, rays and snakes.

H. brooksi Crites and Overstreet, 1991
Crites and Overstreet (1991) found a larval nematode moulting its second cuticle under the rostral exoskeleton of a white shrimp, *Penaeus setiferus* (Penaeidae). The larva had features of physalopteroids of fish as well as those of *H. brooksi* of the shrimp eel (*Ophichthus gomesi*). Larvae were 1.52 mm long.

Physaloptera

P. hispida Schell, 1950
P. hispida occurs in the stomach of the cotton rat (*Sigmodon hispidus littoralis*). Adults are attached to the gastric mucosa and females shed eggs which are passed with faeces of the host. Schell (1952a,b) reported that the nematode developed to the infective stage in ground beetles (*Harpalus* sp.), European earwigs (*Forficula auricularia*) and German cockroaches (*Blattella germanica*). Larvae hatched in the mesenteron and invaded the wall of the colon. Larvae became surrounded by thin-walled capsules (containing one or more larvae) which often protruded from the surface of the colon and were pedunculate.

The infective third stage was attained in 30–35 days; the sex of larvae could not be determined at this stage. Larvae from various intermediate hosts readily infected cotton rats. Worms developed in the pyloric region of the stomach and were found separately for the first 30–40 days. Later, they occurred in compact masses. Between 60 and 65 days postinfection, copulating pairs were noted. Fully developed eggs were present in female worms 73–90 days after rats were infected. Cotton rats could be superinfected. Also, albino rats and Norway rats were infected experimentally.

Schell (1952a) claimed that rats could be infected with second-stage larvae taken 19–20 days postinfection from intermediate hosts. Second-stage larvae attached to the stomach wall and developed into third-stage larvae in 17 days.

P. maxillaris Molin, 1860

P. maxillaris is a common stomach parasite of skunk (*Mephitis mephitis*) in North America. Hobmaier (1941) claimed to have infected cockroaches with eggs of this species but no mention was made of the final host from which the parasite was collected, or of the locality. It is not known, therefore, if *P. maxillaris* from skunk was indeed the source of the material studied. Subsequently, the biology of the species from skunk was studied in detail in Ontario by Lincoln and Anderson (1973, 1975) and by Cawthorn and Anderson (1976a,b,c, 1977). The parasite was usually found firmly attached to the gastric mucosa but worms detached to feed on stomach contents; there was no evidence that they fed on host tissues. Eggs (Fig. 7.2A) fed to crickets (*Acheta pennsylvanicus*, syn. *Gryllus pennsylvanicus*) hatched and most larvae (Fig. 7.2B) invaded the ileum, where they developed and became encapsulated (Fig. 7.2C–F). Both nymphal and adult crickets were infected and it was possible to superinfect them. Larvae 9 days after infection (approximately 1.7–2.0 mm shortly after the second moult) were infective to skunk even though they were much shorter than larvae which had remained in crickets for 25 days (3.8–4.2 mm in length).

Skunks were readily infected with larvae from experimentally infected crickets. In well-fed skunks, third-stage larvae attached to the gastric mucosa and developed. The third moult occurred 5 days and the fourth 10 days postinfection, and females deposited eggs 41–45 days postinfection. Development was consistently asynchronous.

Few worms developed beyond the third stage in skunk maintained on subsistence diets and none developed in skunk deprived of food for 70 days, although third-stage larvae were apparently normal and attached to the stomach wall. The significance of these findings was revealed by examinations of wild skunk. Adult worms were most abundant in skunk in summer and rare in winter. Third-stage larvae were most abundant in late autumn and mid-winter and least abundant in spring and mid-summer. Developing fourth-stage larvae were most abundant in spring and least abundant in early autumn. These data suggested that the third-stage larvae acquired in late autumn, when food became scarce, failed to develop and were the arrested overwintering stage, which initiated the spring cycle when skunk began to feed again. The adult stage is relatively short-lived and by autumn most have reproduced and passed out of the host.

Larvae given to frogs (*Rana pipiens*) and garter snakes (*Thamnophis sirtalis*) attached to the stomach and persisted for prolonged periods without development. They may serve as paratenic hosts although the ingestion of crickets is likely to be the most important source of infection in skunk.

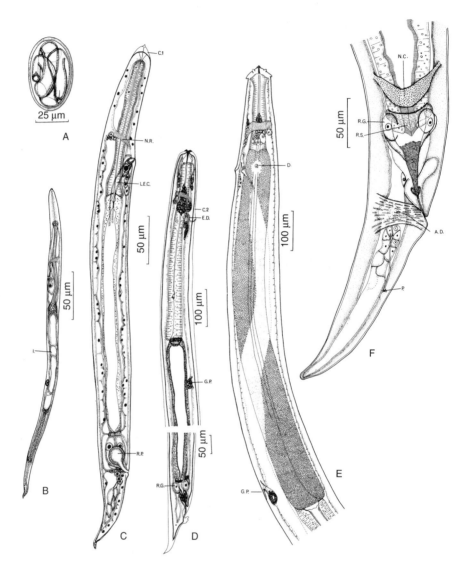

Fig. 7.2. Developmental stages in the development and transmission of
Physaloptera maxillaris: (A) egg; (B) first-stage larva; (C) larva undergoing the first
moult; (D) larva undergoing the second moult; (E) third-stage larva, anterior end;
(F) third-stage larva, posterior end. C.1 = first-stage cuticle; C.2 = second-stage
cuticle; D = derid; E.D. = terminal excretory duct; E.C. = excretory canal; E.P.
= excretory pore; G.P. = genital primordium; I. = intestine; L.E.C. = lateral
excretory canal; L.G. = lateral ganglion; N.S. = nerve commissure; N.R. = nerve
ring; P. = phasmid; R.G. = rectal gland; R.P. = rectal plug; R.S. = rectal sphincter;
V.G. = ventral ganglion. (After R.C. Lincoln and R.C. Anderson, 1975 – courtesy
Canadian Journal of Zoology.)

P. maxillaris is apparently highly specific to skunk. Attempts to infect dogs, cats, mink, ferrets, rats and raccoons maintained on diets similar to those fed to skunks were unsuccessful.

P. praeputialis Linstow, 1889

This species is found in the stomach of cats. Petri and Ameel (1950) infected German cockroaches, camel crickets (*Ceutophilus* spp.) and field crickets with eggs. Zago Filho (1957, 1962a) followed development of larvae in experimentally infected crickets (*Acheta assimilis*). First-stage larvae (approximately 340 µm in length) invaded the rectum of *A. assimilis* and elicited typical capsules, which were described and illustrated. The first moult occurred 12 days and the second 23 days postinfection at 22–24°C. Third-stage larvae were 2.0–2.7 mm in length. Of *A. assimilis* collected in the field, 8.37% contained infective larvae of *P. praeputialis*, suggesting that this insect is an important natural intermediate host. Other insect species were allowed to ingest eggs (i.e. *Apotetamenus clypeatus*, *Aracnominus speluncae*, *Gryllulus sigilatus*, *Miogryllus verticalis*, *Periplaneta americana* and *P. australasiae*) but, compared with *A. assimilis*, relatively few were infected when examined several days later and few larvae were found in them.

In cats, larvae established themselves in the stomach and the third moult took place 45 days and the fourth 95 days postinfection. The prepatent period in cats was 131–156 days. Velikanov and Sharpilo (1984) claimed to have infected cats with larvae from *Varanus griseus* and *Hemiechinus auratus* from Turkmenia and recovered adult *P. praeputialis* (parasites were adult 50–60 days postinfection). Similar larvae were found in other reptiles and mammals, indicating the probable importance of paratenesis in transmission.

P. rara Hall and Wigdor, 1918

P. rara is a common stomach parasite of coyotes (*Canis latrans*) and less commonly of dogs, foxes and cats in central and western USA. Reports in certain other hosts, including badgers and raccoons, need confirmation. Petri (1950) apparently infected German cockroaches and beetles (*Tribolium confusum*) and gave third-stage larvae from cockroaches to cats and dogs, from which he recovered immature *P. rara* a month later. Petri and Ameel (1950) infected ground beetles (*Harpalus* sp.) and crickets (*Acheta assimilis*). Widmer (1967) found larvae attached to the gastric mucosa of rattlesnakes (*Crotalus viridis*) in Colorado and fed them (Widmer 1970) to cats, from which he recovered adult *P. rara*. Olsen (1971) found larvae encapsulated in the wall of the oesophagus, stomach, intestine and mesentery of rattlesnakes. He infected cats with larvae from snakes and recovered adult worms, from which he collected eggs with which to infect *A. assimilis* and *Melanoplus femurrubrum*. In *A. assimilis* (at 23–27°C), the first moult took place 11–12 days and the second 13–14 days postinfection. Third-stage larvae continued to increase in length for up to 152 days when they reached the size of those found in snakes. Cats given larvae from rattlesnakes passed eggs 75–79 days later. Infective larvae given to *Rana pipiens* were passed in faeces for up to 21 days. Larvae given to chickens were passed in faeces within 36 h of infection. A few larvae persisted in mice for several days.

Thubunaea

T. baylisi Akhtar, 1939

Velikanov (1992) gave eggs of *T. baylisi* from lizards in Russia to the beetle *Trigonoscalis gigas* and described the resultant encapsulated third-stage larvae found in this insect 35 days post-infection.

Turgida

T. turgida (Rudolphi, 1819)

T. turgida is a common stomach parasite of opossums (*Didelphis virginiana, D. paraguayensis*) in North and South America. As indicated earlier, it was the first member of the Physalopteroidea to be studied when Alicata (1937) discovered that it would develop to the infective stage in the German cockroach (*Blattella germanica*). Since that time detailed studies have been carried out by Zago Filho (1958, 1962b) and Gray and Anderson (1982a,c). Most worms were found attached to the greater curvature of the corpus of the stomach immediately posterior to the fundus. Larvae tended to be scattered but adults occurred mainly in one to three groups. Attachment sites were usually pinpoint ulcers but numerous large adult worms were rarely associated with large ulcers. Nevertheless, experiments with marked food indicated that *T. turgida* consumed food in the stomach of the opossum (Gray and Anderson, 1982b).

Zago Filho (1958, 1962b) followed development of *T. turgida* in *Acheta assimilis* and *Miogryllus verticalis* which he regarded as the best intermediate hosts of the various species of crickets given eggs. At 24–26°C the first moult occurred 10 days and the second 16 days postinfection. Gray and Anderson (1982c) reported that most eggs hatched in the ileum of *Acheta pennsylvanicus*. The first-stage larvae (179 ± 17.6 µm in length) invaded the gut wall in 4 h. By 108 h a capsule had formed around larvae in the epithelial syncytium. The number of larvae in capsules in crickets varied from 1 to 35 and it was believed that smaller capsules frequently merged into larger capsules during their development. At 30°C the first moult occurred 5 days and the second 10 days postinfection. The third stage reached its maximum length (approximately 1.9 mm) in 14 days. Gray and Anderson (1982c) also infected grasshoppers (*Melanoplus sanguinipes* and *Schistocerca gregaria*) and found the typical capsules mainly in the ileum and rarely in the colon. The beetles *Tribollium confusum* and *Tenebrio molaris* were unsuitable as intermediate hosts. According to Zago Filho (1958) cockroaches and beetles were not suitable intermediate hosts.

T. turgida is evidently specific to certain opossums. Attempts to infect cats, dogs, raccoons and the little water opossum (*Lutreolina crassicaudata*) were unsuccessful (Zago Filho, 1958; Gray and Anderson, 1982b). In opossums, third-stage larvae attached to the gastric mucosa and developed asynchronously. The third moult took place as early as 15 days and the fourth as early as 35 days postinfection. The prepatent period was 83–105 days. Second-stage larvae and larvae undergoing the second moult were not infective to opossums. It was estimated that, because of asynchronous development, an opossum infected only once could retain a patent infection for 360–400 days. Also, infections were successfully challenged, indicating that the final host does not develop a protective immunity to the presence of the parasite. Third- and fourth-stage larvae were found in opossums in Florida mainly from May to August, suggesting that this is a major period of transmission in that region.

Alicata (1937) gave larvae to one each of a dog, cat, rabbit, guinea pig, rat and chick. Later, larvae were found in the washings of the stomach of the cat and the rabbit and encapsulated larvae were found in the stomach of the rat. Zago Filho (1958) gave infective larvae to various animals and noted that undeveloped larvae persisted in the stomach of rats, guinea pigs and mice for 40–80 days. Gray (1981) gave infective larvae to raccoons (*Procyon lotor*), cats (*Felis catus*), groundhogs (*Marmota monax*), guinea pigs (*Cavia porcellus*), rats (*Rattus norvegicus*), mice (*Mus musculus*), white-footed mice (*Peromyscus maniculatus*), rabbits (*Oryctolagus cuniculus*), ruffed grouse (*Bonasa umbellus*), garter snakes (*Thamnophis sirtalis*) and frogs (*Rana pipiens*). Numerous undeveloped third-stage larvae were found attached to the stomach of all these hosts when they were examined 5–55 days later. Only a single larva was found in a grouse, however, and the site of the parasite was not recorded. Some of these kinds of animals could serve as paratenic hosts but their role in field transmission is unknown. Gray (1981) took larvae from experimentally infected frogs and snakes and successfully infected opossums with them.

Skrjabinoptera

S. phrynosoma (Ortlepp, 1922)

S. phrynosoma is a common stomach worm of Texas horned toads (*Phrynosoma cornutum*). Its transmission reveals unusual adaptations for survival in arid conditions. According to Lee (1955) eggs in the uteri became enclosed in thick-walled capsules, each containing five to 69 eggs; the capsules were thought to be secreted by the uterine epithelium. Gravid female worms passed out in host faeces; the eggs within their capsules, as well as in the body of the dead female, survived for prolonged periods under dry conditions. According to Lee (1955, 1957) workers of the Texas agricultural ant (*Pogonomyrmex barbatus* var. *molefaciens*) collected the dead females with their egg capsules and transferred them to their nests, where they were eaten by larval ants, in which development to the third stage took place. Lee was not successful in infecting workers experimentally but he found larvae of *S. phrynosoma* in naturally infected callous, pupal and larval ants. Each larva probably invaded a fat cell, where initial development took place. In ants, the fully formed third-stage larvae (3.5–7.3 mm in length) were found in membranous capsules of host origin in the body cavity and each capsule contained two or three larvae. Heavy infections caused distension and loss of colour of the stomach of infected ants. Encapsulated larvae were given to 27 small horned toads (less than 50 mm snout to vent) and dissected at intervals thereafter. All toads were infected and all except five had the same number of worms as larvae given.

Abbreviata

Capsules containing physalopterid larvae are commonly found in the gastric tissue of snakes (Elapidae) and agamid, gekkonid, pygopodid, scindid and varonid lizards in western Australia (Jones, 1995). The larvae are probably from species of *Abbreviata*, which are common in the stomach of large Australian reptiles that feed on smaller reptiles.

A. caucasica (Linstow, 1902)

Larvae found in *Blattella germanica* in the zoo in Paris were identified by Petter (1960) as *A. caucasica*. Numerous eggs being passed by gorillas may have been attributable to this species, which could readily be transmitted in zoo situations by means of cockroaches. Poinar and Quentin (1972) collected specimens of *A. caucasica* from *Cercocebus aterrimus* and infected *B. germanica* and grasshoppers (*Schistocerca gregaria*). Eggs hatched in the midgut 1.5 h after being ingested. First-stage larvae (240 µm in length) entered the wall of the colon, became encapsulated and developed to the third stage (1.7 mm in length).

A. kazachstanica Markov and Paraskiv, 1956

This is a common parasite of reptiles in eastern CIS (Baker, 1987). Gafurov and Lunkina (1970) found larvae of the parasite in six species of tenebrionid beetles in Tadzhikistan. Kabilov and Siddikov (1978) experimentally infected *Colliptamus barbarus, C. intalicus, Gryllus bimaculatus, Opiolopus oxyanus, Pimelia verrucosa, Prosodes pygmae, Semenovia timberlana, Stalagmoptera confusa* and *Tentiria gigas. G. bimaculatus, P. verrucosa* and *S. confusa* were also found naturally infected. According to Kabilov (1980a) various tenebrionids (*Dila laevicollis, Prosodes* sp., *P. nitida, P. baeri, Pimelia verrucosa, Stalagmoptera confusa*), gryllids (*G. bimaculatus*) and mantids (*Hierodula kazachstanica*) in the Uzbak were suitable intermediate hosts of this parasite found in *Ophisaurus aphodus*. Various other insects (acrids, tettigonids and additional tenebrionids) were infected in the laboratory. Frogs and geckoes can serve as paratenic hosts.

A. turkomanica Andrushko and Markov, 1956

According to Annaev and Mushkambarova (1975) larvae tentatively identified as *A. turkomanica* occurred in *Pisterotarsa gigantea zoubkoffi* in Turkmania. Larvae were given to the reptiles *Phrynocephalus helioscopus* and *Eremias velox* but failed to develop in these hosts. They developed to the fourth stage in 73 days in *Agama sanguinolenta* and were mature in 101 days in *Phrynocephalus mystaceus. Varanus griseus* was also infected with larvae that had been 15 days in *Phrynocephalus* spp.

Pseudophysaloptera

P. vincenti Quentin, 1969

Quentin (1969b) found a physalopteroid third-stage larva in the dermapteran *Labidura riparia* in Africa. He gave the larva to a young rodent (*Thamnomys surdaster*) from which he later recovered a male worm identified as *P. vincenti*, which is a parasite of the lemur *Galago demidovii* in Africa.

Basir (1948) found physalopteroid larvae in the same species of earwig in northern India.

7.3

The Superfamily Rictularioidea

Family Rictulariidae

Rictularioids constitute a homogeneous group of about 50 species divided into two genera and several subgenera (Quentin, 1969a,b; Chabaud, 1975a). The group lacks pseudolabia and has instead a denticulate, hexagonal oral opening and a sizeable buccal cavity with teeth. The oral opening and the buccal cavity are generally displaced to the dorsal side of the cephalic extremity. The presence of numerous large body spines is also diagnostic. Worms are found free in the lumen or firmly attached to the mucosa of the intestine. Quentin (1969d) recognized five subgenera in *Pterygodermatites.*

Witenberg (1928) fed young dogs the viscera of reptiles in which he had found rictularioid larvae. He recovered adult *Pterygodermatites affinis* (syn. *Rictularia cahirensis*) in the dog and correctly hypothesized that the worms underwent a development in some coprophagous insect. Oswald (1958a) showed that *P. coloradensis* (syn. *Rictularia coloradensis*) of white-footed mice would attain the infective stage in various insects. Since that time, the development of a few other species has been investigated, mainly by French workers.

Eggs are oval, with smooth, thick shells and each contains a fully developed first-stage larva. Eggs hatch in the gut of the insect intermediate host and larvae invade the wall of the ileum and localize in the region immediately posterior to its junction with the Malpighian tubules (Seureau, 1973). The larva elicits the formation of a syncytium of epithelial cells which becomes surrounded by a fibrous capsule, which lies between the circular muscles and the epithelium of the ileum (Seureau, 1973). Larvae undergo two moults and attain the infective stage inside the capsules. The time for development to the third stage is dependent on temperature. In three species (*P. hispanica, P. affinis* and *R. proni*) the first moult occurred 2–3 days and the second 6–11 days postinfection at 28°C. The third-stage larva is characteristic of the group and readily recognizable. It is unusually short and stout and has a well-developed buccal capsule with denticles. It has prominent cross-striated lateral alae which begin at the level of the buccal cavity and terminate near the anus. The genital primordia are poorly differentiated although the sexes can be recognized. The tail is sharply pointed.

Development in the final host has not been followed in as much detail as that in the intermediate host. The third moult takes place early (2–3 days) and the fourth moult as early as 4 days (e.g. *R. proni*) but the female worms are not gravid until about 30–40 days. The early fourth stage is similar to the third stage but the characteristic head and cuticular body structures of the rictularioids appear during its development to the adult stage (Quentin, 1970a). Paratenesis occurs in species in carnivores.

Rictularia

R. cristata Froelich, 1802

R. cristata (syn. *R. amurensis*), from rodents in eastern Europe and the CIS, was studied by Morozov (1960). He experimentally infected the diplopod *Chromatoiulus projectus kochi* and obtained infective larvae 33–40 days later. He found larvae in three of the 514 diplopods collected in the field and concluded that this arthropod and related species were natural intermediate hosts. He obtained fourth-stage larvae 5–7 days postinfection in white mice given infective larvae.

R. proni Seurat, 1915

This parasite, found in the intestine of the rodent *Apodemus sylvaticus* in France, was studied by Quentin (1970b) and Quentin and Poinar (1973). Larvae developed in orthopterans (*Locusta migratoria*, *Oedipoda germanica*, *Omocestus raymondi*), dermapterans (*Forficula auricularia*) and coleopterans (*Tenebrio molitor*). Larvae were in capsules in the ileum behind its junction with the Malpighian tubules (Seureau, 1973). At 28°C the first moult took place 3 days and the second about 11 days postinfection. Larvae were fully developed (496 μm in length) by 15 days.

In *A. sylvaticus*, larvae moulted 48–72 h and 4 days postinfection. Larvated eggs were noted in females 30 days postinfection.

Pterygodermatites

P. affinis (Jagerskiold, 1904)

P. affinis (syn. *Rictularia cahirensis*) is a widespread parasite of the intestine of carnivores (i.e. Canidae, Felidae, Mustelidae, Viverridae). Witenberg (1928) and Gupta and Pande (1970a) found encapsulated larvae of *P. affinis* in the wall of the intestine and on the mesentery of reptiles (i.e. *Hemidactylus flaviridis*). As indicated above, Witenberg (1928) infected dogs with larvae from reptiles. Gupta and Pande (1970a) successfully infected a cat with similar larvae; larvae given to a guinea pig were later found encapsulated on the stomach.

Quentin *et al.* (1976b) found larvae encapsulated in the insect *Tachyderma hispida* in Algeria. Larvae, which were in the wall of the ileum behind its junction with the Malpighian tubules, were given to a young cat, which passed eggs 38 days later. In *Locusta migratoria* at 28°C the first moult took place in the ileum in 2–3 days and the worms were in the third stage (725–745 μm in length) by 5–7 days. According to Quentin *et al.* (1976b), unidentified larvae found by Chabaud (1954b) in *Akis elegans* in Morocco belonged to *P. affinis*.

P. coloradensis (Hall, 1916)

This is a parasite of rodents in North America. Oswald (1958a) experimentally infected roaches (*Blatta orientalis*, *Blattella germanica*, *Parcoblatta pennsylvanica*, *P. virginica*, *Periplaneta americana*, *Supella supellectilium*), field crickets (*Acheta assimilis*), camel crickets (*Ceuthophilus* sp.) and beetles (*Chlaenius* sp., *Dicaelus sculptulis*, *Tenebrio molitor*) with eggs from worms found in white-footed mice (*Peromyscus leucopus*). He found larvae in naturally infected field crickets (*Ceuthophilus gracilipes*, *Ceuthophilus* sp.) and rarely in wood roaches (*Parcoblatta pennsylvanica* and *P. virginica*).

At 22–26°C, eggs (Fig. 7.3A) hatched as early as 3 h in the midgut of German cockroaches. Larvae (Fig. 7.3B–I) were found 24 h postinfection in the epithelium of the hindgut, where they elicited typical epithelial syncytia surrounded by capsules. The first moult occurred 7–8 days (Fig. 7.3) and the second 12–13 days later. Each capsule (Fig. 7.3H) contained one to 12 larvae.

Oswald (1958b) gave larvae to mice (*Mus musculus*). In 3 h larvae were found, usually in the upper third and occasionally in the middle third of the small intestine. A moult occurred 6–8 days postinfection which he interpreted as the third moult. Since the adult characteristics of the worm appeared at this time, it is clear that what he regarded as the third moult was in fact the fourth moult. He noted sperm in males 10 days postinfection and gave evidence that females were inseminated at about that time. Eggs were present in females by about 30 days and eggs appeared in faeces of mice as early as 33 days.

Oswald (1958b) reported a sharp drop in the number of worms in his infected mice between days 6 and 10 and suggested that the moulting period (to the adult stage) was a critical period. He recorded another substantial decline in the numbers of worms after 60 days and by 80 days few worms were present. This indicated that worms became senescent after 60 days, though he suggested that immunity could also be involved in the loss of worms.

P. desportesi (Chabaud and Rousselot, 1956)

This is a parasite of the murid *Lophuromys sikapusi* in Africa. The larvae reached the infective stage in delicate capsules in the intestinal wall of the dermapterans *Anisolabis annulipes* and *Diaperasticus erythrocephalus* maintained at 22°C (Quentin, 1969c). The first moult was observed 11 days and the second 17–24 days postinfection. Larvae were given to a wild-caught *Praomys morio* (Muridae) and 108 hours later a fourth-stage female ready to moult and a young male were found in the intestine of this rodent.

P. hispanica Quentin, 1973

This is a parasite of the murid *Apodemus sylvaticus* in Europe and has been studied by Quentin and Seureau (1974). The parasite developed in locusts (*Locusta migratoria*) maintained at 28°C. Larvae localized in the ileum near the proctodeal valve. Development to the third stage was rapid; the first moult occurred 3 days and the second 5 days postinfection. Third-stage larvae (450–530 µm in length) were fully developed in 6 days. Larvae provoked certain cellular reactions in the gut wall which resulted in capsules being expelled into the haemocoel; capsules were commonly found in this location 10 or more days postinfection. Quentin and Seureau (1974) noted that larvae in eggs (354 µm in length) were unusually large and that growth in the insect host was correspondingly reduced.

In experimentally infected *A. sylvaticus*, the fourth stage was attained in 3 days and subadult worms appeared 1 week postinfection. Females were gravid in 38 days.

P. nycticebi (Monnig, 1920)

This is a parasite of New World monkeys (*Leontopithecus rosalia*, *Pithecia pithecia*, *P. monachus*). Montali *et al.* (1983) reported high intensity of infections in two *L. rosalia*

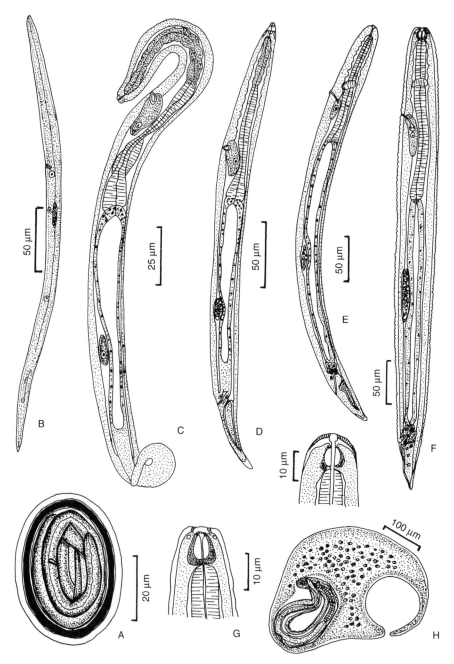

Fig. 7.3. Developmental stages of *Pterygodermatitis coloradensis*: (A) egg; (B) first-stage larva (day 1); (C) first-stage larva (day 3); (D) moulting first-stage larva; (E) second-stage larva; (F) third-stage larva; (G) third-stage larva, anterior end; (H) encapsulated larva from body cavity of intermediate host (*Parcoblatta virginica*); (I) second-stage larva, beginning of second moult at 9 days. (After V.H. Owsald, 1958 – courtesy *Transactions of the American Microscopical Society*.)

in a zoo in Washington, DC. Cockroaches (*Blattella germanica*) exposed to eggs contained infective larvae 19 days later. Naturally infected cockroaches were also discovered in the zoo and were presumably the source of infection in monkeys.

Eggs obtained from faeces of *L. rosalia* and *P. pithecia* in the Oklahoma City Zoo were given in ground meat to *Blattella germanica* by Yue and Jordan (1986). Infective larvae were found in the cockroaches 20 days later. Larvae were given to four golden hamsters (*Mesocricetus auratus*); two gravid female *P. nycticebi* were found in one of the hamsters 28 days postinfection. Some *L. rosalia* in the Oklahoma City Zoo were known to have been infected in the zoo, perhaps through the agency of cockroaches.

P. taterilli (Baylis, 1928)

P. taterilli, collected from the duodenum of the *Taterillus gracilis* (Cricetidae) from Africa, was studied by Quentin (1970c). Development occurred in *Locusta migratoria* and *Shistocerca gregaria* kept at 27–28°C. Larvae invaded the ileum by 3 h and the first moult occurred in 4 days. Larvae were still in the second stage (455 µm in length) 8 days postinfection. Fourth-stage larvae were found in a *T. gracilis* 96 h after being given third-stage larvae.

P. zygodontomis (Quentin, 1967)

This is a parasite of *Zygodontomys lasiurus* (Cricetidae) in Brazil. Quentin (1969b) studied its development in the dermapteran *Doru lineare* and the cockroach *Periplaneta americana*. Larvae invaded the intestinal wall immediately posterior to its junction with Malpighian tubules. At 25°C the first moult occurred 6–7 days and the second about 11 days postinfection.

The Superfamily Thelazioidea

The superfamily consists of three families, united mainly on the basis of cephalic structures but otherwise highly diverse in their biology (Chabaud, 1975a). The transmission and development of the ovoviviparous and oviparous eyeworms (*Thelazia* and *Oxyspirura*) of birds and mammals and the rhabdochonids (*Rhabdochona*) of fishes and primates have been studied. The transmission of members of the family Pneumospiruridae (*Pneumospirura*, *Metathelazia* and *Vogeloides*), mainly of carnivorous mammals, has not been studied.

Family Thelaziidae

Subfamily Thelaziinae

Thelazia

Members of the genus are parasites of the orbits (under the lids, conjunctiva and nictitating membrane, and in the lachrymal glands and ducts) of birds and mammals. Transmission and development of some species in mammals but none in birds have been elucidated. Female worms produce large numbers of thin-shelled eggs which embryonate *in utero* into active, sheathed, fully differentiated first-stage larvae, which are deposited into the lachrymal secretions of the host. Klesov (1949) and Krastin (1949a) first showed how transmission occurred in species found in mammals. Larvae are ingested by muscid flies which feed about the eyes. In the fly, larvae lose their sheaths (i.e. hatch), penetrate the gut and enter the haemocoel. In some species, larvae are said to invade testes or, more commonly, egg follicles (*T. rhodesi*, *T. skrjabini*), where they become encapsulated and grow into large third-stage larvae (approximately 2–5 mm in length). Other species are said to develop in the fat body (*T. callipaeda*). Some authors claimed that larvae of *T. gulosa* developed in egg follicles but others reported that they developed in capsules attached to the body wall or the fat body. Development in the fly is temperature dependent. The infective stage is reached in about 2 weeks at 25°C. The third-stage larva is long with prominent transverse striations, a short buccal cavity and muscular oesophagus, and a short, bluntly tapered or rounded tail, depending on the species. These larvae break out of their capsules and migrate to the head of the fly, which they leave when the latter feeds about the eyes of the definitive host. In the orbit the worms reach adulthood in about 1 month. Kennedy (1993) showed that the prevalences of two eyeworm species (*T. gulosa* and *T. skrjabini*) in cattle in Alberta were related to pasture type that determined the abundance of its vector (*Musca autumnalis*).

It has been suggested that the filarioids of the Onchocercidae may have originated from ancestors with life cycles similar to those of *Thelazia* spp. (Anderson, 1957b).

T. californiensis Price, 1930

This is mainly a parasite of dogs (*Canis familiaris*) but it has also been reported in sheep (*Ovis aries*), deer (*Odocoileus hemionus*), coyotes (*Canis latrans*), cats (*Felis catus*) and bears (*Ursus americanus*) and, rarely, humans in western North America. Burnett *et al.* (1957) allowed laboratory-reared flies to feed on the sheathed larvae and discovered developmental stages in *Fannia canicularis*. Similar larvae were found in wild-caught *F. benjamini* collected in an enzootic area in California.

T. callipaeda Railliet and Henry, 1910

This is a parasite of canids (*Canis familiaris*, *Nyctereutes procyonoides*, *Vulpes fulva*), cats (*Felis catus*), rabbits (*Oryctolagus cuniculus*) and, rarely, humans in the CIS, China, Japan, India, Burma and Korea. Larvae ingested by *Phortica variegata* penetrated the gut wall in a few hours. They remained in the abdominal haemocoel for about 2 days. On the third day larvae invaded the fat body (of the female) and the testes, where they subsequently became encapsulated, grew and moulted twice to the third stage (Kozlov, 1962, 1963). The first moult took place in 7–8 days and the second 19–21 days postinfection. Infective larvae were about 2.46–3.08 mm in length and the genital primordia were precociously developed (i.e. the testes extended to the cephalic end and then bent and extended posteriorly; two uteri extended well behind the vagina and vulva primordia).

In the orbit of the definitive host, Kozlov (1962) noted a moult 7–9 days postinfection (since he reported the presence of spicules and preanal papillae, this moult was probably the fourth and not the third). He claimed that males were mature 13–14 days postinfection. Females were fully grown in 19 days and eggs were present in the uteri. By 22–25 days, uteri were distended with eggs containing larvae. He found motile larvae in the vagina of a female 55 days postinfection. According to Kozlov (1962) mature worms occurred in dogs throughout the year. Thus, the minimum duration of *T. callipaeda* in the final host was probably about 9 months and generations may overlap.

T. gulosa (Railliet and Henry, 1910)

This is a parasite mainly of cattle (*Bos grunniens* and *B. taurus*) in the CIS and Europe, where its transmission has been studied. Like *T. lacrymalis* of horses, it is now widely distributed in North America where it is transmitted by the face fly, *Musca autumnalis*, which was introduced to the continent in the early 1950s and has spread widely in the northern USA and the Canadian provinces (Greenberg, 1971). In the Ukraine the intermediate host is reported to be *Musca larvipara* (see Klesov, 1950) but this identification was questioned by Skrjabin *et al.* (1967). In the Far East of the CIS, *T. gulosa* is transmitted by *Musca amica*. In the Crimea, *M. vitripennis* is a vector (Krastin, 1950a,b; Skrjabin *et al.*, 1967). In Europe and North America *M. autumnalis* is the major vector (Vilagiova, 1967; Branch and Stoffolano, 1974; Moolenbeek and Surgeoner, 1980; Geden and Stoffolano, 1981, 1982). According to Krastin (1950a,b) larvae developed in egg follicles of the female fly. Geden and Stoffolano (1982) reported that larvae rapidly invaded the midgut of the fly (in 1–4 h). Larvae developed in capsules attached to the body wall or the fat body in experimentally infected flies kept at 28–30°C (see also Moolenbeek and Surgeoner, 1980). The third stage was reached in 9 days.

Kennedy and MacKinnon (1994) noted that the distribution of *T. gulosa* differed from that of *T. skrjabini* in the orbits of cattle. The two major lachrymal ducts associated with the Harderian gland contained 58% of the *T. skrjabini* whereas the large ventral duct of the orbital glands contained 58% of the *T. gulosa*. The eyeworms occurred more often (90%) in a duct as a single species than as a mixed infection of the two species (10%).

T. lacrymalis (Gurlt, 1831)

T. lacrymalis is a parasite of equines, especially horses (*Equus caballus*) in Europe, Asia and South America and more recently in North America. Skrjabin *et al.* (1967) reported developmental stages in *Musca osiris* in the CIS. Ivashkin *et al.* (1979) reported larvae in *M. autumnalis*. Lyons *et al.* (1980) experimentally infected *M. autumnalis* and recovered third-stage larvae (1.8–2.9 mm in length) 12–15 days postinfection (the temperature was not indicated). Its spread in recent years in horses in North America is probably the result of the introduction of *M. autumnalis*.

T. lessei Railliet and Henry, 1910

T. lessei, a parasite of the dromedary (*Camelus dromedarius*) and the Bactrian camel (*C. bactrianus*) in the CIS and India, is transmitted by *Musca lucidula* in Turkmenia (Dobrynin, 1972, 1974). Infected flies were found around the eyes of dromedaries from the end of May to the beginning of October.

T. rhodesi (Desmarest, 1822)

This is a parasite of cattle (*Bos taurus*), buffalo (*Bubalus buballis*), zebu (*Bos indicus*), bison (*Bison bonasus*) and less commonly of horses (*Equus caballus*), sheep (*Ovis aries*) and goats (*Capra hircus*). In the Ukraine, Klesov (1949) reported that *Musca autumnalis* and *M. larvipara* were vectors. Klesov (1950) and Krastin (1949a,b, 1950a) concluded that the intermediate hosts were *Musca convexifrons* and *M. larvipara*.

According to Klesov (1950) and Vilagiova (1967) first-stage larvae invaded and developed in ovarian follicles of *M. larvipara* and *M. autumnalis* (only the female flies of these species attack the definitive host). Upon reaching the infective third stage (in 15–30 days), larvae migrated to the mouthparts of the fly. Adults appeared in the orbits of an experimentally infected host in 20–25 days. Krastin (1958) found a gravid female in a calf infected 45 days earlier.

Grétillat and Touré (1970) reported that *Musca sorbens* was a suitable intermediate host in Senegal.

Keiserovskaya (1975) confirmed that *M. autumnalis* and *M. larvipara* were suitable intermediate hosts and that the infective stage was reached in 18–25 days. In experimentally infected cattle, *T. rhodesi* developed to adulthood in 13–14 days and motile larvae were observed in female worms in 25 days. *T. rhodesi* survived in the final host for over 9 months and flies transmitted infections mainly during July and August in Azerbaijzhan. Miyamoto *et al.* (1981) experimentally infected *Musca hervei*. The parasite developed equally in both sexes of the fly. At 20, 25 and 30°C the infective stage was active in 60, 15 and 12 days, respectively.

T. skrjabini Erschow, 1928

This is a parasite of cattle (*Bos grunniens, B. taurus*) in the CIS and Poland. According to Krastin (1952) and Skrjabin *et al.* (1967) the intermediate hosts are *Musca amica* and *M. vitripennis*. It now occurs commonly in cattle in North America. The parasite is reported to develop in the egg follicles. The third-stage larva was long with an unusually short oesophagus and tail (the anus was almost subterminal).

O'Hara and Kennedy (1991) orally inoculated *Musca autumnalis* with first-stage larvae of *T. skrjabini*, which has recently been introduced to North America. They found 315 capsules containing larvae in 95 flies – 20 in the head and 295 in the abdomen. Of the latter, two were among Malpighian tubules, 20 were unattached in the body cavity and the remainder were attached to fat body tissues lining the abdominal wall. First-stage larvae were 221–284 μm in length. Second-stage larvae were found in capsules as early as 3 days but most were 6–10 days postinfection. Moulting from the second to the third stage occurred between 9 and 15 days. Third-stage larvae, present 9–49 days postinfection, were 2.10–2.86 mm in length. The cuticle had annular-like transverse grooves extending along half to two-thirds of the body. Six inner papillae and four submedian cephalic papillae were present. Larvae emerged from the labellum of the flies.

Subfamily Oxyspirurinae

Oxyspirura

Some 84 species have been reported in the orbits of birds in at least 43 families worldwide (Addison and Anderson, 1969). Species of *Oxyspirura* produce small, oval, smooth-shelled eggs; these are deposited into lachrymal secretions which carry them via the lachrymal ducts to the buccal cavity, where they are swallowed and passed in faeces of the host. The transmission of only one species has been elucidated.

O. mansoni (Cobbold, 1879)

O. mansoni (syn. *O. parvorum*) is well known as a parasite of chickens and domesticated ducks (*Anas platyrhynchos*) but it has been reported in a wide variety of hosts, including numerous passeriformes (Addison and Anderson, 1969). The parasite is usually found in the inner corner of the orbit and under the nictitating membrane. It is also found in the nasal sinuses and the lachrymal glands. Fielding (1926, 1927), Kobayashi (1927) and Sanders (1929), apparently working independently, discovered that the intermediate host of *O. mansoni* was the cockroach *Pycnoscalus surinamensis*, commonly found in places where chickens and ducks are raised. This cockroach has a circumtropical distribution and has spread from its origin in the East Indies throughout the tropical and subtropical regions of the world as a result of human trade. The world distribution of *O. mansoni* in domesticated birds closely parallels that of *P. surinamensis*. Fielding (1928) noted that larvae were free in the haemocoel of the cockroach from 10 to 14 days and became encapsulated 17–18 days postinfection. He observed a moult on the 25th day and by 52 days the infective stage was attained.

Fielding (1926, 1927) infected ducks and chickens by allowing them to feed on cockroaches with infective larvae and concluded that larvae left the insect in the crop and migrated up the oesophagus to the mouth and from there to the orbits by way of the nasolachrymal ducts. Kobayashi (1927) infected chickens by feeding them cockroaches and also by placing larvae in the eyes. He observed two moults in the final host, one 3–5 days and the other 15 or 19 days postinfection, and stated that worms were mature in 3–4 weeks.

Sanders (1929) transmitted *O. mansoni* to chickens by feeding them infected cockroaches and by placing larvae in the orbits. He confirmed that larvae left the cockroach in the crop of the final host and migrated rapidly to the eyes (within 20 min) by way of the oesophagus, mouth and lachrymal ducts (Fig. 7.4). Larvae passing into the gizzard were destroyed. He noted that some worms matured in the tear-sac and released their eggs in the lachrymal fluid. Larvae reached the infective stage in cockroaches in 50–100 days. In chickens the worms matured in 48 days. Sanders also successfully infected various passeriforme birds (*Agelaius phoeniceus*, *Aphelocoma cyanea*, *Dolichonyx oryzivorus*, *Lanius ludovicianus*) and a pigeon (*Columba livia*).

Brenes *et al.* (1962) successfully infected chickens, pigeons and the whistling duck (*Dendrocygna a. autumnalis*). Schwabe (1950, 1951) studied the effects of *O. mansoni* on chickens and noted that the eyes of a single bird could accommodate 200 worms.

Family Rhabdochonidae

Rhabdochona

Members of the genus occur in the intestine of freshwater fishes. The genus is divided into subgenera on the basis of whether eggs are more or less smooth-shelled or have floats and filaments (Moravec, 1975). Current understanding of transmission has had a somewhat circuitous history. Weller (1938) found a few first-stage larvae in the intestine of an amphipod allowed to ingest eggs of *Rhabdochona ovifilamenta*. Gustafson (1939) claimed to have infected mayfly nymphs of the genus *Hexagenia* with eggs of a *Rhabdochona* sp. from *Aplodinotus grunniens* from Illinois and that young adult worms were recovered from *Ameiurus melas* 3 days after giving the fish 20-day-old larvae. Later, Gustafson (1942) said that 'several species of *Rhabdochona*' developed in 'various mayfly nymphs'. He claimed that a nymph of *Hexagenia* contained encapsulated male worms 7 months after it was given eggs and that natural infections of adult *R. cascadilla* occurred in *Hexagenia*. Gustafson (1942) reported that mayflies were intermediate hosts of *R. cascadilla*, *R. cotti* and *R. decaturensis* and that stonefly nymphs were also intermediate hosts of *R. cotti*, but again he provided no details to substantiate his claims. Schtein (1959) described larvae of *R. denudata* of Cyprinidae in mayfly nymphs in the CIS. Vojtkova (1971) reported similar larvae in caddisfly nymphs in the Czech Republic. Poinar and Kannangara (1972) found mature worms identified as a new species (*R. praecox*) in the hepatopancreas of freshwater crabs (*Paratelphus rugosa*) from mountain streams in Sri Lanka.

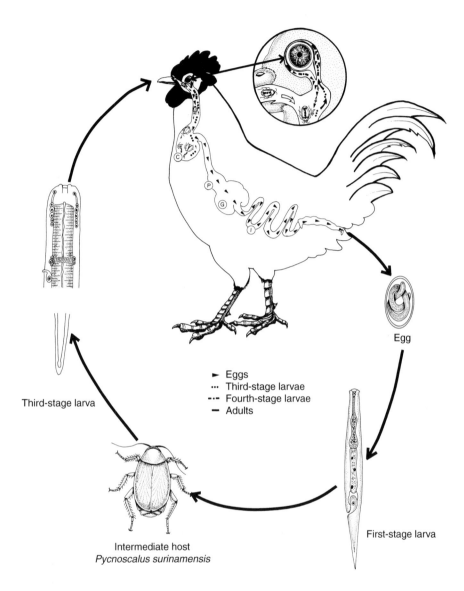

Fig. 7.4. Development and transmission of *Oxyspirura mansoni* of chickens. C = crop; G = gizzard; I = intestine; P = proventriculus. (Original, U.R. Strelive.)

Moravec (1972, 1976, 1977) and Byrne (1992a,b) finally provided more convincing evidence of the importance of mayflies in the transmission of species of *Rhabdochona* in Europe and Canada. Also, Anderson (1988) concluded that the presence of adult *Rhabdochona* spp. in mayflies and crabs was an example of extreme precocity, which markedly reduces the time for the parasite to produce gametes and fertilized eggs in the final host, and that the phenomenon was possibly related to the limited time available for transmission because of the behaviour of the definitive host, which may remain in the transmission site for a brief period of time.

R. canadensis Moravec and Drai 1971

Barger and Janovy (1994) reported this species in *Cyprinella lutrensis* in the Platt River in Nebraska. The ephemeropterans *Trichorythodes* sp. and *Caenis* sp. proved to be suitable experimental intermediate hosts. Attempts to infect a number of fish other than *C. lutrenis* were not successful.

R. cascadilla Wigdor, 1918

Byrne (1992b) noted that the prevalence and intensity of *R. cascadilla* in the common shiner, *Notropis cornutus*, were greatest from early May to early August in a stream in Ontario. Thereafter, the parasite largely disappeared. Gravid females were present only in June and July. Larvae were most prevalent from mid-July to early September. The intermediate host was not discovered although it was believed to be a mayfly. As mentioned earlier, Gustafson (1942) claimed to have found adult *R. cascadilla* in *Hexagenia* sp.

R. ergensi Moravec, 1968

R. ergensi is a parasite of *Noemacheilus barbatulus* in Europe. Its development was studied by Moravec (1972). Eggs were oval and thick-shelled and had several long filaments attached to each pole. Eggs hatched and first-stage larvae invaded the body cavity of the thorax and abdomen of nymphs of *Habroleptoides modesta* kept at 13–15°C. Developing larvae moulted 13 and 16 days postinfection, by which time they were in thin-walled capsules in muscles and fat body tissues mainly of the dorsal regions of the abdomen and thorax. Fully formed third-stage larvae were present 22 days postinfection. The third-stage larva had a long unarmed buccal cavity and the genital primordia were precociously developed. The tail was tapered and sharply pointed. According to Moravec (1972) some larvae moulted to the fourth stage. Mayfly nymphs with larvae were pipetted into the stomach of *N. barbatulus* kept at 18°C. Larvae, released in the stomach of the fish, attached to the mucosa. Development was markedly asynchronous. Only one moult was observed, in males 6–20 days and in females 33–43 days postinfection. Larvated eggs were produced by females in 43 days.

R. hellichi (Šrámek, 1901)

Found in *Barbus barbus* in the Jihava River in the Czech Republic, this species reached the infective stage in two species of trichopterans (caddis flies), namely *Hydropsyche augustipennis* and *H. pellucidula* (see Moravec, 1995). The larvae found in the caddis flies were regarded as fourth stage.

R. phoxini Moravec, 1968

R. phoxini, a parasite of the cyprinid *Phoxinus phoxinus* in Europe, was studied by Moravec (1976, 1977). Eggs with larvae were smooth-shelled or covered partially with proteinaceous material. Moravec gave eggs to the isopod *Asellus aquaticus* and to nymphs of the mayflies *Cloeon* sp., *Ecdyonurus* sp., *Habrophlebia fusca* and *H. lauta*. Eggs hatched; larvae (210–219 μm in length) invaded and developed in the body cavity (mainly of the abdomen of *Habrophlebia* spp.). Moults occurred 12–16 days and 20–36 days postinfection. After about 30 days, third-stage larvae were surrounded by delicate capsules. The third-stage larva (1.29–1.40 mm in length) had a long, tubular, unarmed buccal cavity. The tail was tapered and sharply pointed and the genital primordia were precociously developed. After 30 days a third moult took place 'during the following weeks.' By 172 days postinfection, fourth-stage larvae as well as third-stage larvae were present in the mayflies.

Mayfly nymphs with larvae were given to young *Phoxinus phoxinus* and a *Rutilus rutilus*. After being released in the intestine from the intermediate host, larvae attached to the mucosa and developed. Third-stage larvae apparently moulted 3 days and fourth-stage larvae 22–38 days postinfection. It was estimated that females started to produce eggs in about 2 months postinfection.

Moravec (1977) studied the transmission of *R. phoxini* in *P. phoxinus* in the Rokytka Brook in the Czech Republic, where the intermediate hosts were *Ecdyonurus dispar*, *Ephemera danica*, *Habrophlebia fusca* and *H. lauta*. Mature female worms were present in the fish host almost exclusively from May to July. During this period gravid females moved to the posterior region of the intestine and oviposited, after which adult worms of both sexes largely disappeared from the fish and were replaced by larvae acquired during the ingestion of nymphs and imagoes of 'summer species' of mayflies (*E. dispar* and *Habrophlebia* spp.).

R. rotundicaudatum Byrne, 1992

Byrne (1992a,b) reported on the epizootiology of *R. rotundicaudatum* found in various fishes in a stream in Ontario. Prevalence and intensity of this species were followed in the common shiner (*Notropis cornutus*). Adult worms appeared in large numbers in spring. Female worms rapidly matured and were gravid in June and July but disappeared from the fish thereafter. Larvae were most common from mid-July to early September. The intermediate host was the mayfly *Ephemera simulans*, in which some worms developed to the adult stage. *E. simulans* was first seen on the wing on May 27–28. Prior to emergence the advanced nymphs were readily recovered from shallow stream riffle areas where various fish spawn. Thus, the emergence of the mayflies and the concentration of the definitive host in the same region coincided and the parasite was transmitted, frequently in the adult stage, to the fish, in which it rapidly became gravid.

R. zacconis Yamaguti, 1935

Moravec *et al.* (1998) studied *R. zacconis* in *Tribolodon hakonensis* in the Okitsu River, Japan. Gravid females were present from April to June and in August to September. They were absent in July. New infections were acquired mainly (but not only) during May to July. The seasonal changes are probably related to water temperature and the availability of infective larvae in intermediate hosts.

Trichospirura

T. leptostoma Smith and Chitwood, 1967

T. leptostoma is a parasite of the pancreatic ducts of primates (*Callithrix jacchus*, *Sanguinus oedipus*, *Callicebus moloch*, *Saimiri sciureus*, *Aotus trivirgatus*, *Callimaco goeldi*) in South America (Cosgrove *et al.*, 1970; Orihel and Seibold, 1971; Beglinger *et al.*, 1988). The presence of the worms has been associated with wasting disease in marmosets (Pfister *et al.*, 1990).

Beglinger *et al.* (1988) and Pfister *et al.* (1990) reported larvae in cockroaches in rooms housing marmosets. Illgen-Wilcke *et al.* (1992) gave eggs (15–20 × 40–50 µm in size and larvated) to cockroaches (*Blatella germanica* and *Supella longipalpa*). Eggs hatched in the gut of insects and first-stage larvae, which were 185 µm in length, invaded the haemocoel and grew in muscles of the thorax and legs. The first moult occurred in 25–30 days and the second in 37–40 days at 23–25°C. Third-stage larvae were 980 µm in length. Larvae were given to marmosets, which passed eggs 7–9 weeks later. Infections remained patent for 2 years.

7.5

The Superfamily Spiruroidea

This superfamily once included thelazioids, gnathostomatoids, habronematoids, rictularioids and physalopteroids (Chitwood and Chitwood, 1950). The removal and elevation to superfamily status of the latter groups (Chabaud, 1975b) have reduced the once great Spiruroidea to four small families (Gongylonematidae, Spiruridae, Spirocercidae and Hartertiidae) with 21 genera found predominantly in birds and mammals. The transmission and development of some species in all four families have been studied.

Species of the Spiruroidea are mainly stomach parasites. Transmission and development are fairly uniform in the superfamily. Apparently all species produce thick-shelled resistant eggs containing a fully differentiated first-stage larva which has a cephalic hook and rows of minute spines around the rather blunt anterior end. The tail of the first-stage larva is often blunt and surrounded by a circlet of minute spines. Eggs hatch in the gut of insects and larvae invade the haemocoel. Larvae develop in the haemocoel or in other tissues such as muscles and fat body and undergo two moults. Second and third larval stages eventually become encapsulated. Third-stage larvae are generally large and possess some of the cephalic characteristics of adults. The caudal extremity of the third-stage larva of many species has terminal spines or tubercles; in other species the terminal end is rounded and unornamented. Paratenesis is a common phenomenon in the transmission of spiruroids and the third-stage larvae of several species have been found in tissues of a variety of vertebrates which ingest infected insects such as dung beetles.

Family Gongylonematidae

Gongylonema

Members of the genus occur embedded in the mucosa and submucosa of the anterior region of the gut of birds and mammals. The genus is easily recognized because the cuticle in adults is covered by large verruciform thickenings, especially prominent on the anterior part of the body. The oval, thick-shelled egg contains a first-stage larva when released by the female worm into the lumen of the gut. The larva has a cephalic hook and numerous rows of minute spines encircle the anterior region of the body (Alicata, 1935). The tip of the tail is rounded and surrounded by a circle of small spines. Ransom and Hall (1915) were apparently first to show that a species of *Gongylonema* (i.e. *G. pulchrum*) developed in the haemocoel of dung beetles (Coleoptera) and cockroaches

(Dictyoptera), the former being of prime importance in transmission to ruminants. Larvae of species of *Gongylonema* develop initially in the haemocoel of the insect intermediate host but eventually most third-stage larvae become encapsulated in muscles (Quentin *et al.*, 1986b).

Desportes *et al.* (1949), Chabaud (1954b), Quentin (1969d) and Quentin *et al.* (1986b) called attention to the fact that the morphology of infective larvae can be used to distinguish species. Morphological characters in the larvae which can be used include the length of the body (e.g. 1.05–1.10 mm in *G. neoplasticum* to 3.2–4.2 mm in *G. problematicum*), length of the buccal cavity (e.g. 50–55 μm in *G. neoplasticum* to 25–35 μm in *G. problematicum*) and the structure of the caudal extremity (e.g. bifurcated and ornamented with denticles in *G. neoplasticum* and round and smooth in *G. problematicum*). The length of the oesophagus in relation to body length also helps to distinguish species in the larval stage.

G. congolense Fain, 1955

This is a common parasite of ducks (Anatidae) and gallinaceous birds (including chickens and wild fowl) in Zaire and Rwanda (Quentin *et al.*, 1986b). It developed to the infective stage in the muscles of *Locusta migratoria* (Orthoptera). The infective larva was 3.5 mm in length, the oesophagus and buccal cavity were short, and the tail tip had two lateral groups of five tiny elevations. Larvae provoked hypertrophy of the muscle fibres and fragmentation of myofibrils.

G. dupuisi Quentin, 1965

G. dupuisi was found in three *Mastomys* sp. in North Africa. According to Quentin (1969d) eggs hatched and larvae developed in the dermapteran *Anisolabis annulipes*. At 20°C larvae were found in the body cavity 6 days after the insect was given eggs. In 9 days larvae were found free in the haemocoel and in 19 days larvae were found in flat capsules in the fat body. Third-stage larvae (Fig. 7.5A) found in 33 days were 1.04 mm in length and encapsulated.

G. ingluvicola Ransom, 1904

G. ingluvicola was found originally in chickens (Ransom, 1904). Cram (1935) found this parasite in the oesophagus of mountain quail (*Oreortyx picta*) in western North America. Larvae reached the infective stage in *Blattella germanica*. The larva was 1.7 mm in length and 'inconspicuous conical papillae' were present on the tip of the tail.

G. ivaschkini Minailova, 1977

Third-stage larvae of this species, found in *Blaps fausti* in Turkmenia, were given to rabbits (Minailova, 1977). Larvae were found in the oesophageal wall of rabbits by 16 days. Adult worms were found 40 days postinfection. In 85 and 153 days, female worms contained ova without larvae, presumably because the host was unsuitable; the natural definitive host for the parasite is unknown.

G. mucronatum Seurat, 1916

Quentin and Seguignes (1979) found this species in the oesophageal mucosa of Tunisian hedgehogs (*Erinaceus algirus*). *Locusta migratoria* proved to be a suitable experimental intermediate host. At 30°C, third-stage larvae first appeared in 30 days.

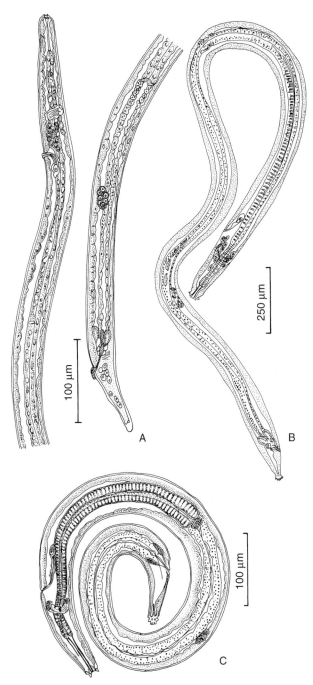

Fig. 7.5. Infective third-stage larva: (A) *Gongylonema dupuisi*; (B) *Protospirura numidica*; (C) *Mastophorus muris*. (A, after J.C. Quentin, 1969 – courtesy *Annales de Parasitologie Humaine et Comparée*; B, after J.C. Quentin *et al.*, 1968 – courtesy *Annales de Parasitologie Humaine et Comparée*; C, after J.C. Quentin, 1970 – courtesy *Annales de Parasitologie Humaine et Comparée*.)

Third-stage larvae from capsules in the abdominal muscles of the locust were 2.1–2.3 mm in length and the caudal end was smooth and rounded. The oesophagus was long.

G. neoplasticum (Fibiger and Ditlevsen, 1914)

This parasite of the stomach of rats developed in cockroaches and *Tenebrio molitor* (see Fibiger and Ditlevsen, 1914). Dittrich (1963) found infective larvae in *Periplaneta americana* and adults in *Rattus norvegicus* in the Leipzig Zoo. Fibiger (1913, 1919) reported cancers in the stomach of infected rats but the lesions he observed are now believed to be non-malignant and associated with a diet deficient in vitamin A (Hitchcock and Bell, 1952).

G. problematicum Schulz, 1924

Chabaud (1954) collected this species from the stomach of gerbils (*Meriones libycus* and *M. persicus*) in Iran. Eggs hatched and larvae developed in the beetle *Blaps* sp. In 46 and 73 days, numerous capsules containing larvae were found in the beetles. An infective larva was 4.2 mm in length and the tail was rounded and unornamented.

G. pulchrum Molin, 1857

G. pulchrum (syn. *G. scutata*) is a common cosmopolitan parasite of cattle, sheep, goats, camels, pigs, equines and cervids, where it occurs in the mucosae of the mouth, oesophagus, tongue and, rarely, the rumen. The parasite is transmissible to rabbits. Third-stage larvae have been found in a great variety of beetles of the Scarabaeidae, Tenebrionidae, Hydrophilidae and Histeridae and it will also develop in cockroaches (*Blattella germanica*) (Ransom and Hall, 1915; Lucker, 1932; Alicata, 1935; Cebotarev and Poliscuk, 1959, 1961; Popova, 1960; Ivashkin and Khromova, 1961b; Stewart and Kent, 1963; Fincher *et al.*, 1969; Sultanov and Kabilov, 1969; Oguz, 1970; Ramishvili, 1973; Kudo *et al.*, 1996). Beetles of the following genera have been shown to be suitable intermediate hosts: *Amphimallon, Aphodius, Blaps, Caccobius, Chironitis, Copris, Dermestes, Geotrupes, Gymnopleurus, Liatongus, Oniticella, Onitis, Onthophagus, Pentodon, Scarabaeus* and *Sisyphus*. Infective larvae can overwinter in dung beetles (Popova, 1960).

Alicata (1935) studied development of *G. pulchrum* in *Blattella germanica*. Larvae invaded the crop and entered the haemocoel; numerous first-stage larvae were found in the haemocoel 48 h postinfection. The first moult occurred 19 days postinfection. Second-stage larvae then invaded the body musculature (especially the ventral part of the abdomen), where they moulted 29–32 days postinfection. Third-stage larvae were eventually encapsulated. Oguz (1970) reported that the first moult occurred 18 days and the second 30 days postinfection in *B. germanica*. Third-stage larvae were 1.9–2.4 mm in length and were encapsulated in 32 days. The caudal end had five projections. The oesophagus extended about two-thirds the length of the body.

Popova (1960) gave infective larvae of *G. pulchrum* from dung beetles to rabbits and obtained adult worms 78–85 days later. Popova (1960) reported that the first moult in rabbits occurred 11 days and the second 36 days postinfection. The prepatent period was 56 days. Observations of Popova on infections in rabbits were confirmed by Gupta (1970) and Oguz (1970). Lucker (1932) transmitted *G. pulchrum* from ruminants to pigs and Alicata (1935) transmitted it to guinea pigs.

G. soricis Fain, 1955

Quentin and Gunn (1981) found this species commonly in the oesophagus of the insectivores *Crocidura flavescens* and *C. poensis* in Togo. Numerous infective larvae were encapsulated in the abdominal muscles of *Periplaneta orientalis* given eggs of the parasite 22 days earlier and kept at 29°C. The larvae were 1.5–1.6 mm in length. The oesophagus of the larva was long and the caudal end had numerous (12) projections.

Family Spiruridae

Protospirura

P. muricola Gedoelst, 1916

P. muricola (syn. *P. bonnei*) is a common stomach parasite of rats. Brumpt (1931) found this parasite commonly in rats (*Rattus norvegicus*) in Venezuela. The parasite developed in experimentally infected cockroaches (*Blattella germanica, Periplaneta orientalis* and *Rhyparobia maderae*). First-stage larvae entered the haemocoel of the insects and eventually became encapsulated. The larvae were 1.4 mm long by 15 days and 2.7 mm by 52 days. Over 1000 capsules were found in the abdominal cavity and smaller numbers in the thoracic cavity and muscles of one large *R. maderae*. The caudal end of the infective larva had 15 small points.

Quentin (1969a) found *P. muricola* in various African murid rodents (*Hybomys univittatus, Mastomys* sp. and *Rattus alexandrina*). He successfully infected the dermapterans *Anisolabis annulipes, Diaperasticus erythrocephalus* and *Labidura riparia* kept at 22°C. The first-stage larva was 280 µm in length and had a cephalic hook, rows of spines near the anterior end and lateral alae. The tail ended in three points. Larvae became encapsulated in the haemocoel of the insect host 8–10 days postinfection. The first moult took place 11 days and the second 20 days postinfection. In 23 days the third-stage larva was 1.9 mm in length and had cephalic characters similar to those in the adult. Genital primordia were poorly developed. The tail ended in nine small points.

Quentin (1969a) transmitted *P. muricola* to the murids *H. univittatus, Praomys jacksoni* and *Thamnomys rutilans*. Third-stage larvae grew between days six and ten postinfection, followed by the third moult. The fourth moult was not observed but was believed to have taken place 15–18 days postinfection. Eggs were embryonated in females in 40 days; worms were fully grown at this time.

Baylis (1928) noted what appeared to be *P. muricola* in the lorisid, *Perodicticus potto*, in Africa. Foster and Johnson (1939) reported *P. muricola* in captive primates (*Aotus zonalis, Ateles dariensis* and *Cebus capucinus*) in Panama. The cockroach *Leucophaea maderae* proved to be the intermediate host. Fatalities, associated with obstruction, tissues invasion, pressure and secondary infections, were especially common in infected *C. capucinus*.

P. numidica Seurat, 1914

This parasite has been found in the lower oesophagus and stomach of palaearctic and nearctic rodents and carnivores (palaearctic: *Arvicanthis barbarus, Felis ocreata, Genetta afra*; nearctic: *Calomys callosus, Canis latrans, Dipodomys ordii, Eutamius minimus, Onychamys leucogaster, Peromyscus crinitus, P. maniculatus, P. truei, Rheithrodontomys megalotis* and *Zygodontomys lasiurus*) (Seurat, 1914; Butler and Grundmann, 1954; Quentin *et al.*, 1968). Quentin *et al.* (1968) regarded the New World form as a subspecies (*P. n. criceticola*) of the Old World form (*P. n. numidica*).

Crook and Grundmann (1964) showed that the widely distributed beetle *Eleodes tuberculata* was a suitable intermediate host in the USA. Quentin *et al.* (1968) showed that the earwig *Doru lineare* and the cockroach *Periplaneta americana* were suitable intermediate hosts in Brazil. Both insects were abundant in habitats occupied by infected rodents (e.g. *Zygodontomys lasiurus*). Healey and Grundmann (1974) infected grasshoppers (*Melanoplus atlantis*, *M. femurrubrum*) and crickets (*Gryllus pennsylvanicus*) in the USA.

Eggs of *P. numidica* were oval and thick-shelled (Quentin *et al.*, 1968). In the insect intermediate host, first-stage larvae entered the haemocoel, where development took place and they became encapsulated. Dyer and Olsen (1967) reported that the first moult occurred 7–9 days and the second 22 days postinfection (temperature not given). Quentin *et al.* (1968) stated that the first moult in the insect intermediate host occurred 15 days and the second 26–27 days postinfection. Healey and Grundmann (1974) reported that, in grasshoppers and crickets, capsules were found 20–30 days postinfection in the muscles and Malpighian tubules but after 40 days most capsules were free in the haemocoel.

Infective third-stage larvae (Fig. 7.5B) were 3.0–3.8 mm in length according to Crook and Grundmann (1964), 3.3 mm according to Quentin *et al.* (1968) and 2.6–2.7 mm according to Dyer and Olsen (1967). The buccal cavity was well developed and the tail terminated in a cluster of 12–15 points. The genital primordia were poorly developed.

Dyer and Olsen (1967) gave encapsulated larvae to *Peromyscus maniculatus* and reported that the first moult in the stomach took place 11 days and the second 15 days postinfection. By 17 days all worms were in the fifth stage. The prepatent period was 35–42 days.

Spirura

Species of *Spirura* are found in the wall of the oesophagus and stomach of mammals. The genus is easily recognized by the presence in adults of one or two ventral cuticular bosses in the cervical region. Eggs are oval and smooth-shelled and each contains a fully developed first-stage larva with a prominent cephalic hook. Near the blunt anterior end there are several rows of tiny spines. The tail ends in a sharp point but slightly anterior to the point there is a circle of small spines. Eggs hatch and larvae invade and develop in the haemocoel of insects. The third stage is encapsulated by the host. Highly developed genital primordia seem to be characteristic of infective third-stage larvae of all species of *Spirura*.

S. guianensis (Ortlepp, 1924)

S. guianensis occurs in neotropical primates and marsupials. Quentin (1973) infected *Locusta migratoria* with eggs from parasites found in *Caluromys philander*. Second-stage larvae were observed in the haemocoel 10 days postinfection. By 12 days all larvae found were encapsulated in fat tissue. Larvae had reached the third stage by 24 days. There was slight growth after the second moult and by 32 days larvae were 3.3 mm in length. The tail end of the third-stage larva had several small processes. Genital primordia were well developed.

Quentin (1973) distinguished head structures of various stages in detail and noted that the ventral cervical boss characteristic of the genus first appeared during development of the third stage in the definitive host. The fourth stage was recognized by its unique cephalic structures.

S. infundibuliformis McLeod, 1933

S. infundibuliformis is a parasite of the stomach of ground squirrels (*Spermophilus* spp.) in western North America. The parasite developed to the infective stage in grasshoppers collected in places where *S. richardsoni* was prevalent, i.e. *Melanoplus infantilis, Aeropidellus clavatus, Aulocara elliottii* and *Camnula pellucida* (Anderson *et al.*, 1993). The parasite also developed in laboratory-reared crickets (*Acheta pennsylvanicus*). In the insect the eggs hatched and larvae invaded the haemocoel within 2 days. At ambient temperatures of 20–30°C, the first moult occurred in 6–7 days and the second in 11–12 days at which time larvae occurred in capsules attached to the body wall. Infective larvae were 1.64–1.86 μm in length and the tip of the tail had a terminal spike surrounded by about four shorter spikes.

Third-stage larvae were given to rats, gerbils (*Meriones* sp.), hamsters (*Mesocricetus auratus*) and a single white-footed mouse (*Peromyscus leucopus*). At necropsy no worms were found in rats or the mouse. A third-stage larva was found in a gerbil 7 days postinfection and two third-stage larvae were found in a gerbil examined at 18 days. Two to eight third-stage larvae were found in each of six hampsters 8–37 days postinfection. Three contained one to three fourth-stage larvae and two each a single immature male worm.

S. malayensis Quentin and Krishnasamy, 1975

This species from slow lorises (*Nycticebus caucang*) and tree shrews (*Tupaia minor*) in Malaysia developed in *Blattella germanica* at 25°C (Quentin, 1975). The first-stage larva was 244 μm in length. Infective larvae were 4.1–4.2 mm in length. The genital primordium was 410 μm in length. The tail terminated in a group of spines.

S. rytipleurites (Deslongchamps, 1824)

This species was divided by Chabaud (1954b) into two subspecies. *S. r. rytipleurites* is a parasite of the cat and was studied by Galeb (1878) and Stefanski (1934). *S. r. seurati* Chabaud, 1954b is a parasite of hedgehogs (*Arinaceus algirus*) and was studied by Chabaud (1954b).

S. r. rytipleurites reached the infective stage in oriental cockroaches. Larvae were found in capsules mainly in the abdomen. Larvae were 13 mm in length and had prominent deirids and lateral alae (Stefanski, 1934).

S. r. seurati reached the infective stage in the beetles *Blaps maroccana, B. pinguis, Morica favieri, M. planata* and *Pimelia rugosa* (see Chabaud, 1954b). Attempts to infect *Akis elegans* and *Periplaneta americana* were unsuccessful. First-stage larvae were 300 μm in length. The first moult in the insect host occurred 10 days postinfection. The second moult took place 20 days postinfection, when larvae were 5–6 mm in length. After the second moult, larvae continued to grow and in 56 days were 12 mm in length. The genital primordia also continued to grow markedly after the second moult. Cervical bosses appeared and lateral alae and deirids were present. The terminal end of the tail was variable; in some larvae it was smooth and in others it had several tiny spines.

Seurat (1911, 1913a) had also observed similar unusually large larvae in insects (*Ontophagus* sp.) and concluded that they were in the fourth stage, because of the advanced state of the reproductive system. Chabaud (1954b) concluded, somewhat reluctantly, that *S. rytipleurites* developed directly from the third stage to the sexually active form, that there were only three moults and that it was the fourth stage which matured. However, precocity in the development of the reproductive system in the third-stage larva is now well documented in other *Spirura* spp. as well as in the Acuarioidea of birds (see section 7.7). It is likely that further studies of the development of *S. rytipleurites* in the definitive host will reveal the missing third moult.

S. talpae (Gmelin, 1790)

S. talpae is a common stomach parasite of *Talpa europaea* (Insectivora) in France. Chabaud and Mahon (1958) gave eggs of the nematode to larvae of species in one of three insect genera (*Cetonia*, *Orythyrea* or *Tropinota*) although the authors stated that the insects probably belonged to *Cetonia aurata*. The insects were kept at approximately 18°C and examined 25–51 days and 8 months later. Larvae of *S. talpae* were found in capsules with double walls in the body cavity of the insects. Larvae (3.4–4.1 mm in length) were in the third stage in 37 days. Genital primordia were well developed. Larvae continued to grow slightly and on day 51 postinfection were 4.3–4.7 mm in length, with genital primordia 180–240 µm in length. In 8 months larvae were 5 mm and the genital primordium was 250 µm in length. Third-stage larvae had alae and deirids and the tail tip was rounded and unornamented. The authors contrasted the level of development of the reproductive system and the length of the body in the third stage with that noted earlier by Chabaud (1954b) in *S. r. seurati*.

Family Spirocercidae

The Spirocercidae consists of three subfamilies. The Spirocercinae and the Ascaropsinae have weakly developed lips. In Ascaropsinae the buccal cavity or pharynx has distinctive rugous or annular thickenings. The Mastophorinae have highly developed pseudolabia divided into six denticulate lobes. With the exception of *Spiralatus* spp. of birds, all species in the family occur in mammals.

Subfamily Spirocercinae

Cyathospirura

C. chabaudi Gupta and Pande, 1981

Gupta and Pande (1981) found encapsulated larvae in the stomach and intestine of wall lizards (*Hemidactylus flaviviridis*) in India. Infective larvae were 1.4–1.5 mm in length. The caudal end terminated in a group of eight spines. Genital primordia were weakly developed. Larvae developed to the adult stage in the stomach of dogs. The third moult apparently occurred 15 days postinfection. Adults were found 24 days postinfection. Worms were not found in rabbits, guinea pigs, piglets, pigeons or chicks given infective larvae.

Spirocerca

S. lupi (Rudolphi, 1809)

S. lupi (syn. *S. sanguinolenta*) is a large (males 3–4 cm, females 6–7 cm), reddish parasite found commonly in tumours in the wall of the oesophagus, stomach and dorsal aorta mainly of dogs, foxes, jackals and wolves in warm latitudes of the world (Fitzsimmons, 1960). According to Faust (1929) tumours opened by small, smooth, round apertures into the gut lumen and a milky purulent fluid carried eggs of the parasite through the apertures into the lumen of the gut. Eggs (22–27 μm × 8–12 μm in size) passed in faeces of the host were elongated, with parallel sides and thick shells. According to Faust (1929) and Chhabra and Singh (1972a) the first-stage larvae were short, tapered posteriorly and 110–124 μm in length. The cephalic end was surrounded by seven rows of minute spines and the tail ended in three points; the middle point was longer than the other two.

Grassi (1888) thought that *Blatta orientalis* was a suitable intermediate host but this has been disproved by subsequent study. Seurat (1916) found larvae in beetles (*Akis goryi, Copris hispana, Geotrupes douei, Gymnopleurus sturmi, Scarabaeus sacer* and *S. variolosus*) in North Africa. Faust (1928) reported that *Canthon* sp. was an intermediate host in China but Théodoridès (1952) believed that the beetle mentioned was *Paragymnopleurus* sp. Ono (1929, 1932, 1933) found infective larvae in the dung beetles *Gymnopleurus mopsus, G. sinnatus* and *Scarabaeus sacer* in Manchuria. He also reported larvae in the dragonfly *Anax parthenope*. Bailey (1963) and Bailey *et al.* (1963) in the USA experimentally infected *Canthon depressipennis, C. pilarius, Geotrupes blackburnii, Orthophagus hecate, O. pennsylvanicus, O. sylvanicus* and *Phanaeus vindex*. Indian authors have reported the following as intermediate hosts: *Calotes versicolor, Catharsius pithecius, Euoniticellus pallipes, Gymnopleurus koenigi, Histor maindronii, Hybosorus lutarius, H. orientalis, Oniticellus pallens, O. pallipes, Onitis philemon, Onthophagus bonasus, O. dama, O. deflexicollis, O. gazelle, O. mopsus* and *O. quadridentatus* (Anantaraman and Jayalakshmi, 1963; Chowdhury and Pande, 1969; Sen and Anantaraman, 1971; Chhabra and Singh, 1973, 1977). According to Chhabra and Singh (1972a) infective third-stage male larvae were 1.7 mm in length. Female larvae were 2.3 mm in length. Larvae were found generally in delicate capsules in the haemocoel. The buccal cavity had thick walls and the tail was short. The posterior end of the tail had five spinous processes.

Canids acquire *S. lupi* by ingesting infected dung beetles (Fitzsimmons, 1960) or vertebrate paratenic hosts. Seurat (1916) reported that infective third-stage larvae were very common in animals in the high plateau region of Algeria. He reported larvae in such animals as hedgehogs (*Erinaceus algirus*), chickens, mice (*Mus musculus, M. norvegicus*) and even equines (*Equus asinus*). Larvae were encapsulated in the external tunic of the stomach and in the mesentery; larvae were also found on the surface of the liver. Faust (1927) reported infective larvae encapsulated on the mesentery of *Erinaceus dealbatus* in China. Ono (1933) found larvae in hedgehogs, chickens and ducks in

Manchuria. Ryzihkov and Nazarova (1959) reported larvae in several species of birds and the polecat (*Putorius eversmani*) in the CIS. Fitzsimmons (1960) found larvae in lizards, insectivores and, especially, in the crop of poultry and francolins in Africa. He regarded poultry and francolins (*Francolinus bicalcaratus*) as prime sources of infection, since dogs were fed entrails of the birds. In India, lizards (*Calotes versicolor*) and shrews (*Suncus suncus*) have been reported to harbour larvae (Sen and Anantaraman, 1971) and Chhabra and Singh (1972b) reported that the larvae became encapsulated in experimentally infected cats, primates (*Macaca mulatta*), rabbits, guinea pigs, rats, mice, chickens, lizards (*Hemidactylus* sp.) and toads (*Bufo melanosticus*).

Bailey (1963) and Bailey *et al.* (1964) believed that chickens were the main paratenic hosts of *S. lupi* in rural districts near Auburn, Georgia (USA) where dogs were frequently fed entrails of chickens and game birds.

Faust (1927) and Hiyeda and Faust (1929) collected encapsulated infective larvae from hedgehogs in China and fed them to dogs and cats to study the behaviour of the parasite in the definitive host. The host invariably vomited 30–60 mins after being infected. They concluded that larvae penetrated the stomach wall, passed through the gastroepiploic and portal veins, and then through the liver and lung capillaries to the left heart and the arterial system. Larvae attached themselves to the intima of the abdominal and lower thoracic aorta, migrated within the aortic wall upward and produced extensive lesions. Hu and Hoeppli (1936, 1937) repeated the above experiments in dogs, however, and reported that larvae first penetrated into the stomach wall and migrated towards the gastric serosa. They then travelled by way of the coronary and the gastroepiploic and coeliac arteries towards the upper abdominal and lower thoracic portions of the aorta. Worms travelled mainly in the wall of the blood vessels and reached the oesophagus by migrating in connective tissue between the upper thoracic aorta and the oesophagus.

According to Nazarova (1964) third-stage larvae penetrated the stomach of dogs and accumulated around small arteries by the end of the second day. By day 4 they were in the arterial wall migrating towards the aorta, where they reached the fifth stage in 3 months. Migration to the oesophagus was noted 102 days postinfection.

Chhabra and Singh (1972c) claimed that in dogs the fourth stage was reached in 4–5 weeks and the adult stage in 10 weeks. At 20 weeks worms were in tumours in the oesophagus but there were no openings to allow eggs to enter the oesophagus. Chhabra and Singh (1972d) reported the prepatent period in three dogs as 149–170 days. Sen and Anantaraman (1971) found larvae in the stomach of dogs in 3–11 days, in the aorta in 7–109 days and in the oesophagus in 93–227 days postinfection. The prepatent period was 121–124 days.

S. lupi has been associated with aortic and oesophageal sarcoma in dogs (Seibold *et al.*, 1955; Ribelin and Bailey, 1958; Bailey, 1963).

Vigisospirura

V. potekhina (Petrov and Potekhina, 1953)

This is a parasite of the stomach of carnivores (*Lynx rufus*, *Meles meles*) in North America and the CIS (Wong *et al.*, 1980). It is transmissible to dogs (Gafurov, 1969).

Gafurov (1969) infected beetles and reported that first-stage larvae became encapsulated in 8–9 days and that the two moults occurred in 15–16 and 28–32 days postinfection. He found infective larvae in various beetles in the CIS (*Blaps fausti*, *Cyphogenia gibba*, *Pachyscelis banghaasi*, *P. laevicollis*, *Pisterotarsa kiritshenkoi*, *Scarabaeus sacer*, *Stalagmoptera incostata*, *Trigonoscelis ceromatica* and *T. gemmulata*). In dogs the third and fourth moults were observed and eggs were passed 45–50 days postinfection.

Subfamily Ascaropsinae

Ascarops

A. strongylina (Rudolphi, 1819)

A. strongylina is a common parasite of the stomach mainly of wild and domesticated Suidae. It is, however, transmissible to lagomorphs, guinea pigs and cattle. The egg was elliptical and slightly flattened at the poles, and had a smooth, thick shell with numerous tiny punctuations (Alicata, 1935). The first-stage larva was about 150 µm in length. It had a prominent cephalic hook and the anterior end had several rows of delicate spines. The tail ended in a short conical structure.

Numerous dung beetles and a few other types of beetles have been shown to serve as intermediate hosts of *A. strongylina*. Dung beetles of the following genera are considered important: *Aphodius*, *Caccobius*, *Canthon*, *Ceratophius*, *Copris*, *Geotrupes*, *Gymnopleurus*, *Oniticellus*, *Onthophagus*, *Phanaeus* and *Trox* (Ono, 1932; Alicata, 1935; Porter, 1939; Stewart and Kent, 1963; Shmitova, 1964; Dimitrova, 1966; Chowdhury and Pande, 1969; Fincher *et al.*, 1969; Slepnev, 1971; Zajicek and Pav, 1972). Beetles associated with plant roots have also been found infected; these belonged to the genera *Cotinis*, *Dyscinetus*, *Hybosorus*, *Phyllophaga* and *Phyllophagus* (Fincher *et al.*, 1969).

Shmitova (1963) and Varma *et al.* (1976) reported that infective larvae of *A. strongylina* became encapsulated in the tissues (especially the stomach) of mice. Varma *et al.* (1976) reported that larvae recovered from mice grew to adulthood in rabbits. Shmitova (1963) reported that larvae became encapsulated in lizards and frogs and Dimitrova (1966) found larvae in wild vertebrates (i.e. *Apodemus sylvaticus*, *Crocidura suaveolens*, *Ophisaurus apodus* and *Talpa europaea*) in Bulgaria. Gupta (1969) found larvae in *Natrix piscator* which he successfully transmitted to rabbits, which passed eggs 41 days postinfection. McAllister *et al.* (1993b) found larvae in the Mediterranean geckos (*Hemidactylus turcicus*) in Texas. This reptile was introduced into the New World around the turn of the century and is now found in various places around the Gulf Coast.

According to Shmitova (1964) larvae hatched in the gut of beetle intermediate hosts within 3 days and migrated into the haemocoel, where development took place. The first and second moults occurred 16–18 and 25–27 days postinfection, respectively. Larvae became encapsulated by the host. Slepnev (1971) reported that development to the third stage in *Geotrupes stercorarius* required 28–30 days at 17–21°C. Moults took place 16–19 and 28–31 days postinfection. According to Alicata (1935), third-stage larvae were 1.9–2.3 mm in length and the tip of the tail possessed a smooth knob-like process.

Shmitova (1964) reported that infective larvae invaded the mucosa of the fundus of the stomach of pigs and within 5–9 days migrated to the pyloric region. The third moult occurred 4–5 days and the fourth 20 days postinfection. Worms were mature in 25–31 days and eggs were laid 46–50 days postinfection. The life span of the parasite was about 10.5 months.

Physocephalus

P. sexalatus (Molin, 1860)

This species occurs in the stomach and, rarely, the small intestine of wild and domestic Suidae, peccary and, less commonly, tapirs, equines, cattle and lagomorphs. Its transmission is similar to that of *Ascarops strongylina* and infective larvae of both species are found in similar intermediate hosts. The first-stage larva is similar to that of *A. strongylina* (see Alicata, 1935). Alicata (1935) found first-stage larvae in the haemocoel of beetles 24 h postinfection. By 16 days larvae were encapsulated in Malpighian tubules and were moulting for the first time. The second moult occurred 34 days postinfection and capsules enclosing larvae increased in size as larvae grew. Fully formed capsules were attached to Malpighian tubules or free in the haemocoel. Third-stage larvae were about 1.6 mm in length according to Alicata (1935) and could readily be distinguished from those of *A. strongylina* by the presence on the tip of the tail of about 20–23 digitiform processes; the tail of *A. strongylina* lacks terminal processes.

Infective larvae of *P. sexalatus* have been found encapsulated (often along with *A. strongylina*) in tissues of numerous species of dung beetles of the genera *Aphodius, Ataenius, Ateuchus, Boreocanthon, Canthon, Ceratophius, Copris, Deltochilum, Dichotomius, Geotrupes, Glomeris, Gymnopleurus, Melanocanthon, Oeceoptoma, Onthophagus, Passalus, Phanaeus, Pinotus, Scarabaeus* and *Trox* (Seurat, 1913b; Cram, 1928; Alicata, 1935; Porter, 1939; Stewart and Kent, 1963; Dimitrova, 1966; Fincher *et al.*, 1969; Baruš *et al.*, 1970).

Encapsulated infective larvae of *P. sexalatus* occur commonly in the tissues (especially on the gut and mesentery) of amphibians, reptiles, birds and mammals which have ingested infected beetles (Cram, 1930; Alicata, 1935; Chandler, 1946; Allen and Spindler, 1949; Tromba, 1952; Dimitrova, 1966; Krahwinkel and McCue, 1967). Krahwinkel and McCue (1967) reported that 49% of 72 wild birds of 14 passeriforme species in the southeastern USA had larval *P. sexalatus* encapsulated in their digestive tracts. Tromba (1952) reported larvae in bats (*Myotis lucifugus*), Chandler (1946) reported them in armadillos (*Dasypus novenicinctus*) and Bartlett *et al.* (1987) reported *P. sexalatus*-like larvae in shorebirds (*Bartramia longicauda* and *Numenius americanus*).

Simondsia

S. paradoxa Cobbold, 1864

This is a parasite of the stomach of swine and occurs along with *Physocephalus sexalatus* and *Ascarops strongylina*. Trifonov (1962) gave a pig several samples of wild-caught beetles (*Caccobius schreberi*) containing infective larvae. He found adult *S. paradoxa* in the stomach of the pig 44 days after the last sample was fed. Dimitrova (1965a) found larvae in dung beetles in Bulgaria. Eighteen of 31 species were infected. The most heavily infected species were *Caccobius schreberi, Copris hispanus, Geotrupes mulator, G. spiniger, Onthophagus taurus* and *Sisyphus schaefferi*. Pigs were successfully infected with larvae from the latter two species. Dimitrova (1965b) found infective larvae in *Apodemus sylvaticus, Crocidura suaveolens, Erinaceus rumenicus, Rana ridibunda* and *Talpa europaea* in areas where pigs were abundant. *Cavia cobaya, Mesocricetus auratus, Mus musculus, Oryctolagus cuniculus, Passer domesticus* and *Triturus triturus* were infected experimentally indicating the potential range of paratenic hosts. Gupta and Pande (1970b) in India found larval *S. paradoxa* in beetles (*Onthophagus* sp., *Oniticellus pallens*) as well as in insectivores (*Crocidura coerulea*) and lizards (*Hemidactylus flaviviridis*). Larvae were given to piglets and adult worms were recovered from them.

Streptopharagus

S. kutassi Schulz, 1927
Chabaud (1954) found infective third-stage larvae free in the body cavity of *Pimelia angustata* from Mauritania. Larvae were given to a young white rat which was examined for worms 10 days later; late third-stage larvae were found in the stomach. Subsequently, Chabaud (1954) found adult *S. kutassi* and larvae identical to those found earlier in *P. angustata* in rodents (*Citellus fulves, Meriones libycus* and *M. persicus*) from Iran. Infective larvae were 2.9–3.2 mm in length; genital primordia were poorly developed and the tail ended in a cluster of digitiform spines.

S. pigmentatus (Linstow, 1897)
S. pigmentatus is a parasite of the stomach and small intestine of Old World monkeys of the genera *Cercopithecus, Colobus, Erythrocebus, Hylobates, Macaca* and *Papio*. It is common in the Japanese monkey (*Macaca fuscata*) according to Machida *et al.* (1978).

Machida *et al.* (1978) found larval worms encapsulated in the haemocoel of beetles (*Geotrupes laevistriatus, Onthophagus atripennis* and *O. ater*) which had fed on the faeces of *M. fuscata* infected with *S. pigmentatus*. Male larvae were 2.2–2.45 mm and female larvae were 2.37–2.52 mm in length. Larvae were given to a crab-eating monkey (*Macaca fascicularis*) and 69 days later two preadult *S. pigmentatus* were found in its intestine.

Subfamily Mastophorinae

Mastophorus

Members of the genus are parasites mainly of the stomach, and less commonly of the oesophagus and duodenum, of rodents of the families Microtidae and Muridae. Worms are medium to large in size. The oral opening is surrounded by two large, lateral, trilobed lips with dentate borders.

M. muris (Gmelin, 1790)

This is a common parasite of rats and other rodents (e.g. *Apodemus sylvaticus, Calomys callosus, Clethrionomys glareolus, Mus musculus, Rattus alexandrinus, R. assimilis* and *R. norvegicus*). The species has a number of synonyms listed by Wertheim (1962), including: *Protospirura ascaroidea* Hall, 1916; *P. gracilis* Cram, 1924; *P. columbiana* Cram, 1926; *P. marsupialis* Baylis, 1934; *P. glareoli* Soltys, 1949; and *P. bestiarum* Kreis, 1953.

Eggs of *M. muris* are oval with thick, smooth shells (Quentin, 1970a). The first-stage larva (222–285 µm in length) had a cephalic hook and rows of tiny spines in the cephalic region. The tail had a circle of minute spines a short distance from the apex, which also had a few spines. First-stage larvae hatched in various insects and invaded the tissues, where they developed to encapsulated third-stage larvae. Leuckart (1865) and Marchi (1871) fed eggs to flour beetles (*Tenebrio molitor*) and recovered encapsulated larvae with which they infected mice. This was probably the first demonstration of heteroxeny in the Nematoda. Cram (1924, 1926) infected *Phyllodromia germanica*. Miyata (1939) reported that fleas (*Ceratophyllus anisus, Ctenopsyllus segnis, Nosopsyllus fasciatus* and *Xenopsylla cheopis*) were suitable intermediate hosts in Japan. Beaucournu and Chabaud (1963) found larvae encapsulated in *Ctenophthalmus arvernus* collected on *Apodemus sylvaticus* and in *C. agyrtes* from *Clethrionomys glareolus*. Golvan and Chabaud (1963) found encapsulated larvae in the abdomen of the dipterans *Phlebotomus ariasi* and *P. perniciosus* in France. Shogaki *et al.* (1972) showed that *Blattella germanica* was an important intermediate host in Nagoya City, Japan.

Quentin (1970a) studied the development of *M. muris* in *Locusta migratoria*, the dermapteran *Labidura riparia* and the cockroach *Periplaneta americana*, which he maintained at 27–28°C. Larvae became encapsulated in the haemocoel and fat tissue by 15 days. The third-stage larva (Fig. 7.5C) was about 1.0 mm in length and had lateral alae and a long buccal cavity. The caudal end terminated in a group of tubercles. Genital primordia were poorly developed. In rats the third moult occurred in 8 days and adult worms were present 28 days postinfection. Quentin (1970a) described and compared head structures of the various stages in detail.

Family Hartertiidae

Hartertia

Members of the genus are sizable worms with large pseudolabia, each slightly divided into three lobes. They are parasites mainly of the intestine, rarely the gizzard, of birds and have been little studied.

H. gallinarum (Theiler, 1919)

This is a parasite found in the intestine of fowl in Africa. Superficially it resembles *Ascaridia galli*. Theiler (1919) showed in South Africa that the thick-shelled eggs passed by the definitive host were ingested by workers of the termite *Hodotermes pretoriensis*. First-stage larvae penetrated the gut of the termite and developed in the haemocoel into unusually large infective larvae, which varied from 20 to 30 mm in length and produced obvious swelling of the abdomen of infected termites. No soldier termites were infected.

Chickens are fond of termites and it was the custom to dig up nests to expose the insects and allow chickens to feed on them. The worms reached adulthood in chickens in 3 weeks.

H. obesa Seurat, 1915

Annaev and Meredov (1990) found larvae (8.32–12.88 mm in length) of *H. obesi* in termites (*Anacanthotermes turbestanicus*) in Turkmenia. Some birds in the area were the presumed hosts.

7.6

The Superfamily Habronematoidea

Head structures characterize habronematoids; the pseudolabia are not large and median lips are present (Chabaud, 1975b). Nevertheless, the superfamily is biologically diverse. It includes economically important and well-studied groups such as the tetramerids (with *Tetrameres*) of the proventriculus of birds and noted for their peculiar sexual dimorphism, as well as the habronematids (with *Habronema*, *Draschia* and *Parabronema*) which are transmitted by adult muscid dipterans to horses and certain ruminants. Because several species in the above-mentioned groups are pathogens of poultry and draught animals, they were subject to investigation early in this century. The superfamily includes the fish-dwelling cystidicolids, a number of which have been studied rather intensively in recent years. The superfamily also includes aberrant genera such as *Hedruris* of lower vertebrates and the enigmatic crassicaudines of cetaceans. The superfamily is uniformly heteroxenous and paratenesis occurs in some groups.

Family Habronematidae

Draschia, Habronema, Parabronema

Members of these genera occur in the stomach of horses and certain ruminants, including camels and elephants. Females, which occur in small tumours in the stomach wall, deposit oval thin-shelled eggs. The latter usually hatch in the stomach releasing small, poorly differentiated larvae with an anterior spine-like tooth. Larvae pass out with faeces of the host. Ransom (1911, 1913) in the USA first discovered that the larvae of *Habronema muscae* of horses developed in larvae of muscid flies inhabiting dung. First-stage larvae of various species in the three genera are ingested by fly larvae. They penetrate the gut of the latter and enter the haemocoel, where they remain for a short time before invading tissues, usually cells of the Malpighian tubules or fat body, where development takes place to the second stage. Second-stage larvae then enter the haemocoel and develop further until the second moult. This phase of development takes place in the pupa of the host. Development of the nematode and the fly is synchronized. The third larval stage is only reached about the time the adult fly emerges from the pupa. The third stage then migrates to the head of the fly, where it continues to grow for a few days and becomes infective. When infected flies feed on a moist warm surface, such as around the nares or mouth of the definitive host, the larvae are stimulated to break out of the mouth parts of the fly. They make their way to the mouth of the host and are swallowed. Transmission is, therefore, by the oral route, as one would expect in a parasite of the stomach. Larvae, however, frequently leave flies feeding about the eyes or on skin lesions (Bull, 1919). These larvae do not develop in these locations but may provoke an inflammatory reaction which eventually results in their death. The condition is known as **cutaneous habronemiasis**. Larvae may enter damaged blood vessels and be carried to the lungs, where they elicit a granulomatous condition and are overcome.

Some authors have suggested that development and transmission of these kinds of habronematids is a step towards those of filarioids (Chandler *et al.*, 1941; Bain, 1981c) but Anderson (1957b) regarded the few similarities that seem to exist as superficial and the result of convergence.

D. megastoma (Rudolphi, 1819)

This common parasite of the stomach of horses was studied by Roubaud and Descazeaux (1921). First-stage larvae in dung invaded larvae of *Fannia* spp., *Musca domestica*, *M. humilis*, *M. sorbens*, *Muscina stabulans* and *M. terraereginae*. The larva (110 μm in length) left the alimentary canal of the muscid larva and entered the body cavity. It remained in the haemocoel for about 3 days and then invaded the Malpighian tubules. Development of the insect and nematode was synchronized and if that of the former was retarded (e.g. by temperature), then development of the latter was also retarded. The first moult took place in 3–4 days and at this time larvae had a sausage shape. The second stage grew (990 μm) and the Malpighian tubules containing larvae degenerated. This occurred at the time the insect was in the form of a pupa. In 8 days larvae entered the body cavity and were surrounded by a sheath derived from the degenerate Malpighian tubules. Advanced second-stage larvae then moulted to the third stage about the time the adult fly emerged from the pupa (about 9 days postinfection). Third-stage larvae migrated to the head of the fly and were found in the labium 13 days postinfection. They continued to grow until they reached the infective stage in 15–16 days and were 2.5–3.0 mm in length.

When an infected fly was feeding around the nostrils or mouth of horses, infective larvae were stimulated to migrate into the labellum when it rested on a warm surface. Larvae broke out of the labium and eventually reached the buccal cavity of the horse and were swallowed. Roubaud and Descazeaux (1921) regarded *Musca domestica* as the most efficient vector of *D. megastoma*. Larvae were apparently unable to escape from the labium of *Muscina stabulans*. Horses may also get infected from ingesting flies. In the stomach of the horse the larvae invaded the mucosa and reached the adult stage in about 2 months.

According to Skvortsov (1937), eggs of *D. megastoma* hatched in manure or water in 48 h; after invading larval flies, they reached the Malpighian tubules in 3 days and were encapsulated during the pupal stage. Gorshkov (1948) experimentally infected larvae of *Musca domestica*. At 20–22°C he reported the first moult at 4–5 days and the second moult in the pupa in 8–11 days. Third-stage larvae were fully developed in 13–15 days. Larvae appeared in the mouthparts of the fly 2–3 days after its emergence from the pupa. Kawamata (1961) found larvae in *Musca tempestiva*, *M. vicinina* and *Stomoxys calcitrans* in Japan. He claimed that larvae were encapsulated in muscles 96 h after they had penetrated the fly larva (other authors insist, however, that the Malpighian tubules are the site of development).

H. microstoma (Schneider, 1866)

H. microstoma of horses developed to the infective stage in the fat body of larvae and pupae of the blood-sucking stable fly *Stomoxys calcitrans*. Third-stage larvae, which were 1.6 mm in length, appeared in the mouthparts of the fly in about 20 days postinfection. Roubaud and Descazeaux (1922b) discovered the intriguing fact that infective larvae of *H. microstoma* interfered with the ability of the fly to rasp a hole in the skin of the host in order to get blood. Rather, the infected fly reverted to its ancestral feeding behaviour and fed on moist places, including those about the nares and mouths of horses. Ivashkin and Khromova (1961) reported that *Lyperosia viritans* was a common vector in Turkmenia.

H. muscae (Carter, 1861)

H. muscae of horses developed to the infective stage in larvae and pupae of *Musca domestica*, *M. humilis*, *M. lusoria*, *M. terraereginae*, *M. ventrosa*, *Pseudopyrellia* spp. and *Sarcophaga* spp. (Ransom, 1911, 1913; Hill, 1918, 1920; Roubaud and Descazeaux, 1922a; Mello and Cuocolo, 1943). Larvae invaded and developed in a fat body cell. Development was similar to that of *D. megastoma*. Third-stage larvae in the labium of the fly were 2.4–2.8 mm in length.

P. skrjabini Rasowska, 1924

This is an abomasal worm of camels (*Camelus dromedarius*). Ivashkin (1956, 1958, 1959) studied its transmission in Mongolia and Kazakhstan. Elongated, thin-shelled eggs with larvae were passed in faeces of camels. Larvae of the stomoxyid fly *Lyperosia titillans* living in camel dung ingested the eggs. First-stage larvae hatched in the insect, invaded the haemocoel and became encapsulated in tissues. Second-stage larvae appeared in pupae. Third-stage larvae appeared in the metamorphosed fly in 15–17 days at 20–28°C and were 1.78–1.85 mm in length, with a rounded unarmed tail. Ivashkin believed that under natural conditions of temperature the full development in the insect would be about 28–32 days. Infected flies were collected from the middle of June until their demise at the end of September but the most intense transmission occurred from mid-July to mid-August (Ivashkin, 1959). The final host was infected *per os* in the usual way and it was estimated the worms could live for 15–19 months in camels.

Cyrnea

Members of the genus occur in the proventriculus of birds. Only three species have been studied. Eggs are oval, with smooth, thick shells (Fig. 7.6B). The first-stage larva is small. The cephalic end has a dorsally placed hook and several rows of tiny posteriorly directed spines which occur on the body immediately behind the cephalic extremity. The tail ends bluntly but has a tiny, terminal spine and a circlet of about seven tiny spines. Larvae develop in the haemocoel and tissues of orthopteran insects. Some larvae become encapsulated but others do not. There are the usual two moults during development. The third-stage larva has complex cephalic structures and a long divided oesophagus. The genital primordium is a small oval group of cells. The tail is short and ends in a rounded swelling covered with minute spines.

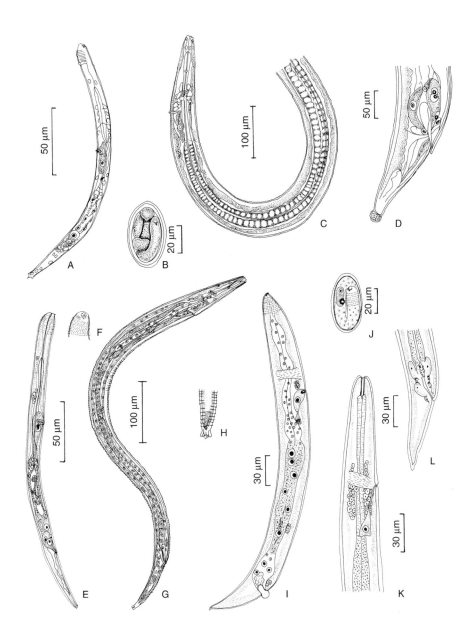

Fig. 7.6. *Cyrnea (Procyrnea) mansioni*: (A) first-stage larva; (B) egg; (C) infective third-stage larva, anterior end; (D) infective third-stage larva, posterior end. *Tetrameres (Tetrameres) cardinalis*: (E) first-stage larva; (F) first-stage larva, anterior end; (G) infective third-stage larva; (H) third-stage larva, tail end. *Geopetitia aspiculata*: (I) first-stage larva; (J) egg; (K) infective third-stage larva, anterior end; (L) infective third-stage larva, posterior end. (A–D, after J.C. Quentin *et al.*, 1983 – courtesy *Annales de Parasitologie Humaine et Comparée*; E–H, after J.C. Quentin and H. Barre, 1976 – courtesy *Annales de Parasitologie Humaine et Comparée*; I–L, after C.M. Bartlett *et al.*, 1984 – courtesy *Journal of Wildlife Diseases*.)

C. colini Cram, 1927

C. colini (syn. *Seurocyrnea colini*) occurs in the wall of the proventriculus (near the gizzard) of quail (*Colinus virginianus*), prairie chicken (*Tympanuchus americanus*), sharp-tailed grouse (*Pedioecetes phasianellus*) and turkey (*Meleagris gallopava*) in the USA. Cram (1931, 1933a) successfully infected *Blattella germanica* and nymphs (but not adults) of *Melanoplus femurrubrus*. Larvae developed among tissues of the insect and were never encapsulated. Third-stage larvae were fully developed and 3.3–3.7 mm in length in 18 days; they did not grow between then and 45 days. Larvae were given to quail and chickens, in which the nematodes developed. Fourth-stage worms and immature adults (males 5.5 mm and females 6.3 mm in length) were found 13 days postinfection. Mature worms were found in quail 41 days postinfection. Turkeys given larvae did not become infected.

C. eurycerca Seurat, 1914

Seureau and Quentin (1983) studied development of this species from the francolin (*Francolinus bicalcaratus*) in Togo. Eggs were given to the orthopteran *Tylotropidius patagiatus* (Acrididae). In 24 h at 30°C, larvae were found in the haemocoel. Larvae grew slowly in the haemocoel for about 5 days and became encapsulated. Some larvae then left their capsules and invaded the fat body, where they developed. The first moult took place in 6–10 days. The second-stage larva was at first encapsulated but left the capsule when it was about 1.2–1.3 mm in length. By 10 days, larvae (2.0 mm in length) were found in the haemocoel, where the second moult occurred in 14 days. Third-stage larvae were 3.0–3.2 mm in length and unencapsulated.

C. mansioni (Seurat, 1914)

Quentin *et al.* (1983) collected this species (syn. *Habronema mansioni*) from the falconiformes *Milvus migrans* and *Accipites badius* in Togo and studied its development in the orthopteran *Tylotropidius patagiatus* (Acrididae) at 30°C (Fig. 7.6A–D). Growth was rapid after the third day. The first moult occurred in 3–6 days and the second 8–10 days postinfection. There was no growth after the second moult. In 10 days all larvae were uniformly at the infective third stage. Infective larvae were 1.7–1.8 mm in length and found usually free in the haemocoel during dissection of the insects. However, in 16 days a larva was found encapsulated in the fat body.

Hsu and Chow (1938) found third-stage larvae encapsulated in the stomach wall of toads (*Bufo bufo asiaticus*) in China which they fed to hawks (*Cerchneis tinnunculus japonicus, Falco subbuteo jakutensis, Accipiter nisus nisosimilis*). Adult *C. mansioni* were recovered from the proventriculus ('stomach') of the birds when they were examined 18–44 days later.

Hadjelia

H. truncata (Creplin, 1825)

H. truncata has been found in the gizzard of a number of bird species in Europe and Asia (Skrjabin and Sobolev, 1963) including the hoopoe (*Upupa epops*) in France. Chabaud (1951, 1954c) fed eggs to various insects and reported development in the haemocoel of the beetles *Asida jurinei, A. sericea* and *Phylan abbreviatus*. The terminal end of the first-stage larva had seven to eight spines. There was a ventral cephalic spine and apparently rows of delicate spines some distance from the cephalic extremity. At about 18–20°C the first moult took place in the insect in 25 days and the second in 50 days. Growth was most marked after 35 days (from 1.8 to 4.0 mm). Third-stage larvae had large pseudolabia and a divided oesophagus, and the genital primordia were not highly developed. The tail ended in a peculiar unarmed button-like appendage.

Appleby *et al.* (1995) reviewed reports in 'lofted' pigeons (*Columbia livia domestica*) and described serious lesions in the gizzard in birds from Cyprus.

Procyrnea

P. zorillae (Seurat, 1919)

Velikanov (1988) found infective third-stage larvae in the tissues of reptiles (*Agama caucacisa, Coluber karelina, Echis multisquamatus, Elaphe dione, Psammophis lineatum, Spalerosophis diadema* and *Varanus griseus*) and hedgehogs (*Hemiechinus auritus*) in Turkmenia. He gave larvae from *H. auritus* to young owls (*Athene noctua*) and recovered adult worms identified as belonging to *P. zorillae*. Worms were apparently mature in the proventriculus and gizzard by 17 days postinfection, although larvae were also present on days 24 and 26. Tenebrionid beetles (*Ocnera* sp., *Trigonoscetis gigas*) were given eggs and third-stage larvae were recovered from them 32 days later. Larvae from beetles were 2.9–3.6 mm in length whereas those in a reptile paratenic host were 8.6–15.0 mm in length, indicating substantial growth in the latter. Velikanov (1988) suggested that *P. zorillae* was specific to owls (Strigidae).

Sicarius

S. dipterum (Popowa, 1927)

S. dipterum is a parasite of the gizzard mucosa of the hoopoe (*Upupa epops*) in France. Chabaud (1951, 1954c) fed eggs to various insects and reported development in the body cavity of the beetles *Asida jurinei, A. sericea, Phylan abbreviatus* and *Tentyria mucronatus*. The first-stage larva had a cephalic hook and a rounded tail surrounded by a circle of spines. At about 20°C the first moult occurred in 20 days, followed by rapid growth. The second moult occurred in 30 days and the growth ceased. Larvae were generally free in the haemocoel but towards the end of the second stage some larvae were in delicate capsules. The third-stage larva was about 1.5 mm in length. The cephalic end was complex and similar to that in the adult. The tail end was pointed and covered with numerous posteriorly directed spines.

Family Hedruridae

Hedruris

Hedruris, the only member of the family, contains parasites of the intestine of fishes, amphibians and reptiles. The genus is peculiar because of its four complex lips, an undivided oesophagus and the fact that the terminal end of the female is modified into an attachment or prehensile organ. Development of only two species has been investigated. Both species develop into immature adults in isopod intermediate hosts and this probably accelerates gamete production in the final host.

H. androphora Nitzsch, 1821

This is a parasite of salamanders (Urodela) in Europe. Leuckart (1876) noted that *H. androphora* reached the adult stage in the body cavity of isopods (*Asellus*). Petter (1971) found specimens in the smooth newt (*Triturus vulgaris*) in France and studied its development in *Asellus aquaticus* allowed to ingest eggs and kept at 20°C. The larvated egg was oval, thick-shelled and operculate (both poles). The sides of the egg have two large masses with a central nipple-like elevation. First-stage larvae invaded the haemocoel of *A. aquaticus* and eventually lodged in capsules composed of large cells mainly in the head and first thoracic segments. The larva was short with a cephalic hook and a short pointed tail. In 12 days larvae were in the second stage and in 14 days third-stage larvae were found. In 16 days fourth-stage larvae were found, some surrounded by the cuticle of previous moults. In 31 days subadults were found (females 5.7 – 5.8 mm and males 5.1 mm in length). It was believed that worms only reached maturity in the salamander final host. Petter (1971) found subadult worms in delicate capsules in the body cavity of isopods in a pond with infected salamanders.

The presence of extreme precocity in *H. androphora* in its aquatic intermediate host is evidence that transmission takes place within confined limits of time and space, probably during the brief period when adult salamanders come together in ponds to spawn. Hagstrom (1979) noted that, in Sweden, newts spend only short periods of time in water (in spring and early summer).

H. ijimai Morishita, 1926

Hasegawa and Otsuru (1979) studied this common parasite of *Rana ornativentris* and *R. rugosa* in Japan. Immature adult worms were found in *Asellus hilgendorfi* inhabiting the same region as the frogs. They infected isopods experimentally. First-stage larvae invaded the haemocoel, first in the head region and then more posteriorly in the body of the isopod. At 20°C there were four moults: at 7–15 days, at 20–23 days, at 27–36 days and at 36–45 days postinfection. Infected isopods were given to three *R. brevipoda porosa*. One of two frogs examined 5 days later and one examined in 10 days had female worms containing eggs. Thus, this is also a parasite which exhibits extreme precocity. Although development in the intermediate host is prolonged, the production of larvated eggs in the frog host is extremely rapid.

Family Tetrameridae

The family includes the well-studied genera *Tetrameres* and *Microtetrameres* as well as *Geopetitia* of the proventriculus of birds. The transmission of the Crassicaudinae (with *Crassicauda* and *Placentonema*) of cetaceans has not been adequately studied.

Subfamily Crassicaudinae

Crassicauda

Species of the genus are long worms found in the urogenital system of whales.

C. boopis Baylis, 1920

C. boopis (syn. *C. pacifica*) has been found commonly in the kidneys, renal veins and intrarenal ureters of fin whales (*Balaenoptera physalus*) in the North Atlantic (Lambertsen, 1985). A female was about 157 cm in length and the vulva opened a short distance from the caudal extremity, which lay free in the renal ducts of the host. Eggs were oval, with thick, smooth shells, and on average 55.0 × 37.1 µm in size. Eggs contained fully developed larvae.

Lambertsen (1986) reviewed early reports of *Crassicauda* spp. and described the prevalence and lesions associated with *C. boopis* in fin whales. Eggs of *C. boopis* were found in urine. Larvae found in urine were also assigned to *C. boopis* although no comparative descriptions of free larvae and larvae from eggs were provided by the author, who concluded that transmission occurred when the suckling calf ingested larvae shed in the urine of the cow. Such an interpretation is in conflict with the obvious thick shells of the eggs of *C. boopis*, and the systematic position of *Crassicauda* spp. as members of the heteroxenous habronematoids, which have been shown consistently to use arthropod intermediate hosts in which development to the third and infective stage takes place.

Later Lambertsen (1992) reported that *C. boopis* in renal veins can provoke a chronic occlusive phlebitis, even extending into the vena cava, and can compromise the health of the infected whale. The parasite has also been found commonly in humpback (*Megaptera novaeangliae*) and blue whales (*Balaenoptera musculus*) and in Icelandic waters the prevalence in fin whale was 95%. Lambertsen (1992) reported worms in the renal veins of a newborn fin whale calf, which indicates that transplacental transmission may occur (one assumes the female parent would acquire larvae of the parasite by feeding on crustacean intermediate hosts and/or paratenic hosts – R.C.A.).

Subfamily Tetramerinae

Tetrameres and *Microtetrameres*

Adults of species of *Tetrameres* and *Microtetrameres* are parasites of the proventriculus of birds. *Microtetrameres* is sometimes regarded as a subgenus of *Tetrameres* but differences in adult and larval morphology justify its recognition as a distinct genus. Female worms are deep red and typically embedded in the gastric glands. The body of the female is fusiform or spherical in shape but the cephalic and caudal extremities have the usual nematode shape and structure. In *Tetrameres* the central part of the body is globular or spindle-shaped and divided into four equal sectors (i.e. *tetra-meres*) by longitudinal grooves in the cuticle. In *Microtetrameres* the central part of the body is twisted into two to three tight, thick coils. In mature females of both genera the central part of the body is occupied largely by expanded uteri coiled about the intestine and containing enormous numbers of small eggs. Male worms, which are tiny and have the usual nematode shape, are generally found on the mucosa or in the lumen of the proventriculus. They have sometimes been reported from the mucosa and in the crypts associated (or not) with females. The evidence suggests that males move about freely, the better to reach and fertilize the stationary females. Tails of the latter are directed towards the lumen of the proventriculus and, since the vulva is near the anus, eggs are passed out directly into the lumen of the proventriculus or carried there by secretions of the glands. The small, oval eggs have thick, smooth shells, the poles of which appear swollen and set apart from the rest of the egg by delicate circular grooves forming opercula. The poles in some species have tufts of short filaments. Each egg contains a fully developed first-stage larva when it is deposited by the female. The first-stage larva is short and rather stout (Fig. 7.6E,F). The cephalic end has a hook placed to the ventral side of the oral opening. Several circles of tiny posteriorly directed spines surround the head a short distance from the oral opening. In *Tetrameres* the rounded tail of the first-stage larva is encircled by minute spines (e.g. *T. cardinalis*, *T. confusa*, *T. crami*). In *Microtetrameres* (e.g. *M. centuri*, *M. corax*, *M. inermis*) the tail is sharply pointed and unarmed (cf. also the third-stage larva below).

The earliest observations on transmission were made by Rust (1908) in Germany who reported in an abstract that *Tetrameres fissispina* (syn. *Tropidocerca fissispina*) of Anatidae developed in *Daphnia pulex*. Somewhat earlier, Linstow (1894) had described *Spiroptera pulicis* (syn. *Filaria pulicis*) in *Gammarus pulex* in Germany which Seurat (1919) subsequently recognized as a larval *T. fissispina*. Since these early tentative observations, the development and transmission of a number of species have been investigated, some more thoroughly than others.

Transmission of species of *Tetrameres* takes place either in water (e.g. species in waterfowl) or on land (e.g. species in terrestrial birds). In the former, intermediate hosts are generally aquatic crustaceans (amphipods, cladocerans) and in the latter, mainly terrestrial insects and isopods. The various species do not seem highly specific in their use of intermediate hosts. In the latter, eggs hatch and first-stage larvae penetrate the gut wall and enter the haemocoel, where development to the third and infective stage takes place. The developing larvae are generally free in the haemocoel for several days and then become encapsulated. Larvae moult twice and after a few additional days are infective to the definitive host. Infective larvae are about 1.5–2.0 mm in length. The tail of the third stage of species of *Tetrameres* is rounded and armed with a group of bristles or blunt appendages (Fig. 7.6G, H), whereas the tail of the larvae of species of *Microtetrameres* is tapered and ends in a sharp point or, more commonly, a tiny, rounded mucron. In the final host, female third-stage larvae rapidly invade the gastric glands and in a day or two moult to the fourth stage. At this time the developing genital tract begins to grow in a spiral fashion around the intestine. In about 1–2 weeks the female moults and begins to acquire the characteristic shape and colour of the adult, a process which is completed in another 3 weeks when she is gravid. The female ingests blood and the gland containing her degenerates and transforms into a fibrous capsule open to the lumen of the proventriculus.

It is beyond the scope of this review to pass judgement on the validity of the various species mentioned in the following summary, although *T. crami* may be a synonym of *T. fissispina*. Likewise *T. americana* and *T. mohtedai* may be synonyms of *T. confusa*; Zago Filho and Pereiro Barretto (1962) regarded *T. americana* as a synonym of *T. confusa*.

Tetrameres

T. americana Cram, 1927

This is a common parasite of chickens and bobwhite quail (*Colinus virginianus*) in North America. Cram (1931, 1937) reported that suitable intermediate hosts were the grasshoppers *Scyllina cyanipes* in Puerto Rico and *Melanoplus femurrubrum* and *M. differentialis* in the USA. Larvae in the tissues of grasshoppers were active for about 10 days before they invaded various tissues, chiefly the muscles, and became encapsulated. Third-stage larvae were 1.8–1.9 mm in length. Cram (1937) experimentally infected chickens, a pigeon (*Columba livia*), two bobwhite quail and a ruffed grouse (*Bonasa umbellus*). She was essentially unsuccessful in infecting domestic ducks. In chickens, female worms were in the gastric glands in 14 days and had the morphological features characteristic of the adult. In 16–19 days fully developed males were found in the glands and by the end of this period the females were fully grown. In 29 days most males were found in the lumen of the proventriculus and females for the first time were red. In 35 days females were gravid but the eggs unembryonated. In 45 days eggs were embryonated.

T. cardinalis Quentin and Barre, 1976

This species was found in the northern cardinal (*Cardinalis cardinalis* (syn. *Richmondia cardinalis*)) in Mexico and its development in *Locusta migratoria* was studied by Quentin and Barre (1976). Each operculum on the eggs had a tuft of short filaments. Eggs hatched in *L. migratoria* and at 28°C first-stage larvae (Fig. 7.6E,F) appeared in the haemocoel within 24 h. The first moult started in 4 days and was completed in 5 days. The second moult started around day 7. At this time some larvae were in the fat body. In 11 days numerous larvae were in thin capsules in the fat body. The cuticle of third-stage larvae (Fig. 7.6G) had numerous deep transverse striations, imparting a crenulated appearance to the surface of the body. The long tapered tail ended in five to eight rounded protuberances (Fig. 7.6H). Quentin and Barre (1976) gave larvae to a cardinal and 11 days later obtained and described fourth-stage males and females.

T. confusa Travassos, 1917

Zago Filho and Pereira Barretto (1962) studied the development of this species, which was found in chickens in Brazil. They successfully infected the grasshoppers *Eutryxalis filata filata* and *Orphulella punctata*. The third stage (1.6–1.7 mm in length) was reached in 9–10 days at 25.6–27.7°C. Larvae were found encapsulated in the abdominal muscles. Eggs failed to hatch in various beetles, cockroaches and crickets.

T. crami Swales, 1933

This is a common parasite of wild and domesticated ducks in North America. According to Swales (1936) eggs remained viable for 10 months when stored at 4–5°C. He allowed *Daphnia pulex* to ingest eggs but noted that they passed through this cladoceran unhatched. Eggs hatched; larvae invaded and developed in the haemocoel of the amphipods *Hyalella azteca* (syn. *H. knickerbockeri*) and *Gammarus fasciatus*. The times of the moults in the amphipods were not determined precisely and the temperature was not stated. Second-stage larvae appeared 7–11 days postinfection and then became encapsulated in various parts of the body, especially in the coxal plates on the inner surface of the dorsal shield. A second moult occurred in the capsules. Third-stage larvae required a period of growth and development before they were infective to ducks. Worms were sexually mature in ducks in 33 days.

T. fissispina (Diesing, 1861)

This is a parasite of Anatidae. Rust (1908) noted that it developed in the cladoceran *Daphnia pulex*. Garkavi (1949a) reported that *Gammarus lacustris* was a suitable intermediate host, in which the parasite reached the infective stage in 8–18 days, depending on the temperature. There were two moults in the intermediate hosts. According to Petrov (1970), *Gammarus lacustris* was infected throughout the year with *T. fissispina* and the peak prevalence occurred in June.

The parasite matured in 18 days in ducklings; there were two moults during development in this host. Cvetaeva (1960) noted that males and females occurred together in glands 10 days postinfection but by 12 days males were scarce in the gastric glands and by 18 days none was found. She concluded that development of the males and insemination of the females occurred early in the infections and in the gastric glands.

Kovalenko (1960) showed that *Gammarus maeoticus* and *G. locusta* were suitable intermediate hosts and claimed that infective larvae occurred in various fish paratenic hosts along the coast of the Sea of Azov (i.e. *Caspialosa brashnikovi maeotica, Lucioperca lucioperca, Neogobius fluviatilis, N. melanostomum, Rutilus rutilus* and *Scardinius erythrophthalmus*).

T. mohtedai Bhale Rao and Rao, 1944

This is a parasite of chickens in various parts of the world, including India, where it was first discovered. Sundaram *et al.* (1963) followed development in grasshoppers (the genus and species were not specified). Numerous first-stage larvae were found in the haemocoel of infected grasshoppers within 24 h; the first moult occurred in 2–3 days and the second in about 11–12 days postinfection. Larvae became encapsulated 1–2 days after the second moult and were found in the head, on membranes suspending the ovaries and in the fat body. Infective larvae (1.7 mm in length) were given to chickens, which passed eggs in their faeces 44 days later. Attempts to infect chickens with encapsulated third-stage larvae which had developed in grasshoppers for 12 days were not successful, indicating that larvae required a period of development after the second moult in the grasshoppers. The authors believed that 17 days of development was necessary for larvae to become infective but they failed to give the temperature at which the grasshoppers were kept.

According to Ramaswamy and Sundaram (1979) the coleopterans *Llatongus cinctus* and *Sphaeridium quinquemaculatus* have also been reported as suitable intermediate hosts. Lim (1975) noted that infections of *T. mohtedai* took place in chicks less than 2 weeks old in brooders in Singapore and discovered that the intermediate host was the moth *Setomorpha rutella*, the larvae of which live in stored grain. Larval insects ingested eggs of *T. mohtedai* and infective larvae occurred in the adult moth, which was eaten by the chickens. Lim (1975) reported the prepatent period in chickens as 36–43 days. Finally, Ramaswamy and Sundaram (1979) successfully infected the isopod *Porcellio laevis*, in which the moults occurred on days 4 and 8 and third-stage larvae became encapsulated by 13 days postinfection. The prepatent period was 46–55 days.

T. pattersoni Cram, 1933

Cram (1933b) infected the grasshoppers *Chortophaga viridifasciata* and *Melanoplus femurrubrum* with this species from bobwhite quail (*Colinus virginianus*). Third-stage larvae (1.4 mm in length) were encapsulated in the muscles and in the 'mesenteries' of the body cavity 24 days after the grasshoppers were given eggs. She was not successful in infecting two adult turkeys, a pigeon and a duck.

Microtetrameres

M. centuri Barus, 1966

Ellis (1969a) collected this species from meadowlarks (*Sturnella magna, S. neglecta*) in the USA and studied development in grasshoppers (*Melanoplus* spp.) at an unspecified temperature. The first moult was said to have occurred 8–16 days and the second 9 days postinfection. Third-stage larvae (1.1–2.0 mm in length) were encapsulated in the haemocoel (up to eight larvae were in each capsule but usually one). Ellis (1969b) recovered adult male worms from a canary (*Serinus canarius*) experimentally infected with third-stage larvae from grasshoppers.

M. corax Schell, 1953

Bethel (1973) collected specimens from magpies (*Pica pica hudsonia*) in Colorado, USA. Eggs hatched in the intestine of grasshoppers (*Melanoplus* spp.) and appeared in the thoracic haemocoel in 24 and 36 h. Larvae in the haemocoel moulted to the second stage 10–14 days postinfection. Third-stage larvae (1.9–2.3 mm in length) were recovered 27–56 days postinfection, mainly in the thoracic haemocoel and less frequently in the fat body of the abdomen. At 56 days larvae were in thin-walled capsules loosely connected to surrounding tissues.

M. helix Cram, 1927

Cram (1934b) gave eggs of this species, which is normally found in Corvidae, to grasshoppers (*Melanoplus bivittatus* and *M. femurrubrum*) and a cockroach (*Blattella germanica*). Numerous third-stage larvae (2.3–2.6 mm in length) were found in the grasshoppers but only a single larva in the cockroach. Larvae were in thick-walled, semi-transparent capsules in the body cavity and muscles of the legs and head. Larvae were given to an adult pigeon (*Columba livia*), which became infected.

M. inermis (Linstow, 1879)

Quentin *et al.* (1986a) collected this species from the African weaver finch, *Ploceus aurantius*, from Togo. The orthopterans *Tylotropidius patagiatus* and *Locusta migratoria* held at 28–30°C were suitable intermediate hosts, the larvae developing in the fat body.

A moult occurred 7–8 days postinfection and by 18 days the infective stage (1.5–1.6 mm in length) was attained.

Geopetitia

G. aspiculata Webster, 1971

Bartlett *et al.* (1984) reported this species in capsules attached to the serosa of the posterior end of the oesophagus, the proventriculus and the anterior part of the gizzard of 12 passeriforme, coraciiforme and charadriiforme species of birds in a zoo in Canada, where it was associated with considerable mortality. The parasite has never been found outside zoos. Worms were extensively coiled within the capsules; the posterior end of both sexes extended through the wall of the capsule into the lumen of the proventriculus. Some large capsules contained as many as 50 worms and some occupied much of the body cavity. Eggs in gravid females were oval, with smooth, thick shells and 46–50 × 24–28 µm in size (Fig. 7.6J). The first-stage larva (340 µm) in the egg was fully developed and had a delicate hook near the oral opening and circles of fine, posteriorly directed spines a short distance behind the cephalic extremity. Eggs hatched in, and first-stage larvae (Fig. 7.6I) invaded, the fat body of crickets (*Acheta domesticus*), where they developed to the infective stage at 22°C. Larvae developed asynchronously in the crickets and moulting second-stage larvae were found as late as 42 days. Third-stage larvae were in amber-coloured capsules, indicating that the cricket is not an entirely suitable host. The third-stage larva (Fig. 7.6K,L) were 1.4–1.5 mm in length and had a long unarmed buccal cavity, a clearly divided oesophagus and weakly developed, double lateral alae. Sexes could be recognized by the position of the genital primordium (a short distance anterior to the anus in females). The tail ended in a small mucron. Infective larvae were given to a cutthroat finch (*Amadina fasciata*) which was examined 35 days later. One male and one female adult were found in a membranous capsule on the proventricular serosa, with their extremities protruding into the lumen of the proventriculus.

French *et al.* (1994) reviewed all the many reports of *G. aspiculata* associated with mortality in insectivorous birds in zoological gardens in North America and reported new cases from Lincoln Park Zoological gardens. The nematodes developed in *Blattella germanica, Supella sepellectilium* and *Acheta domestica*. The infective stage was attained in 35 days and was 1.30–1.83 µm in length. Zebra finches (*Taeniopygia guttata*) were given larvae. In 24–48 h postinfection, larvae were found by sectioning at the junction of the proventriculus and the ventriculus. Gross lesions were observed 2 weeks postinfection as raised nipple-like nodules on the serosal surface of the greater curvature of the proventriculus. In 14 weeks the periventricular nodule enlarged and the mass of worms in the nodule matured.

Family Cystidicolidae

Members of the family are found in marine and freshwater fishes. They occur mainly in the intestine but some are found in the stomach and pyloric caeca. Several species are adapted to the swimbladder of physostomous fishes. In the past few years knowledge of the transmission and development of the cystidicolids has expanded considerably. They all produce thick-shelled eggs which contain fully developed first-stage larvae. The eggs may have polar plug-like structures, lateral floats or filaments. Eggs are passed into water with faeces of the host and are ingested by various aquatic organisms which may serve as intermediate hosts. The latter are frequently nymphal stages of aquatic insects (e.g. Ephemeraptera) or crustaceans (e.g. Amphipoda, Decapoda). Eggs hatch in the intermediate host; the first-stage larva penetrates the gut and invades the haemocoel where development commences. Larvae of some species remain free in the haemocoel until development is completed. Others may invade the tissues (e.g. muscles) and eventually become encapsulated. There are always two moults in the intermediate host. In the cystidicolids there is generally marked growth and development after the second moult before the larvae are infective to the final host. Thus, third-stage larvae as long as 10 mm may be found in some intermediate hosts. During development after the second moult, the oesophagus differentiates and lengthens markedly. At the same time the genital primordia grow considerably and the sexes are generally easily recognized. Fish acquire their infections by ingesting arthropods containing third-stage larvae. There is limited evidence that prey fish may serve as paratenic hosts of some species. There are undoubtedly two moults in the final host during development to adulthood although this has not always been documented. There is little evidence that a tissue phase occurs and the various species seem to develop in the site which they will occupy as adults. Infective larvae of swimbladder nematodes reach their final site by migrating up the duct connecting the swimbladder to the intestine.

Ascarophis

Members of the genus are thread-like nematodes found in the digestive tract of marine and estuarine fishes (Ko, 1986). Eggs are oval and thick-shelled with filaments on one or both poles. A number of authors have assigned large third-stage larvae found in marine decapods to species of *Ascarophis*. Uspenskaya (1953, 1954) found larvae assigned to *A. filiformes* and *A. morrhuae* in decapods (*Enalus gaimardi*, *Eupagurus pubescens*, *Hetairus polaris*, *Pagurus pubescens*, *Pandalus borealis* and *Spirontocaris spinus*) in the Bering Sea. Uzmann (1967) found an *Ascarophis* larva in a lobster (*Homarus americana*) in North America. Petter (1970) assigned larvae found in the crab, *Carcinus maenas*, off the Brittany coast in France to *A. morrhuae*; an unidentified larva of *Ascarophis* was also found in an unidentified crab found in the stomach of *Trigla* sp.

Tsimbalyuk *et al.* (1970) reported *A. pacificus* larvae in the body cavity of crustaceans (*Anisogammarus kygi*, *A. ochotensis*, *A. tiuschovi*, *Idothea ochotensis* and *Pagurus middendorffii*) from the littoral zone of Big Shantar Island in the Okhotsk Sea; one to four larvae were found per host. Poinar and Kuris (1975) found third-stage larvae of *Ascarophis* sp. in shore crabs (*Hemigrapsus oregonensis*) and porcelain crabs (*Pachycheles rudis*) in California. Larvae were in capsules in various regions of the crustaceans. Poinar and Thomas (1976) distinguished, but did not assign to species, four kinds of *Ascarophis* third-stage larvae in marine decapods in California. Two types occurred in *Callianassa californiensis*, one type in *Pagurus samuelis* and *P. granosimanus*, and one type in *Pachycheles pubescens* and *Pugettia producta*. All larvae occurred in capsules in the haemocoel.

Appy and Butterworth (1983) found adult *Ascarophis* sp. in estuarine *Gammarus tigrinus* in Canada. Fagerholm and Butterworth (1988) reported mature male and female *Ascarophis* sp. in *Gammarus oceanicus* in the northern Baltic Sea and also in *Gammarus* sp. in estuarine localities in the New Brunswick region of the northwestern Atlantic. Jackson *et al.* (1997) found adult *Ascarophis* spp. in *Gammarus* spp. in Quebec. These reports are regarded herein as examples of extreme precocity of a nematode of fish in its intermediate host. There is no evidence that the parasites pass from one intermediate host to another.

Capillospirura

C. pseudoargumentosa (Appy and Dadswell, 1978)

This (syn. *Caballeronema pseudoargumentosa*) is a common parasite of the mucosal lining of the stomach of the shortnose sturgeon (*Acipenser brevirostrum*) in New Brunswick, Canada. Appy and Dadswell (1983) have elucidated its development. The oval egg was fully larvated *in utero*. It had a polar swelling to which were attached numerous filaments. The first-stage larva was fully developed, with a pointed tail and a long thread-like buccal cavity. Eggs hatched and larvae developed to the third stage in the amphipods *Gammarus fasciatus* and *G. tigrinus* but not in *Crangon septemspinosa*, *G. duebeni*, *G. oceanicus* and *Hyalella azteca*. It also did not develop in isopods, decapods and mysids. At 10–14°C development to the third stage in *G. tigrinus* took 28–40 days and at 21–25°C it took 10–15 days. There was considerable growth of larvae after the second moult (i.e. from 1.5 to 2.3–2.6 mm in length). The third-stage larva had pseudolabia with a conical process and well-defined submedian labia. The buccal cavity was elongate and the oesophagus was long and divided. The tail ended in a rounded drop-like mucron. Sexes were recognizable and the genital primordia well developed. Infective larvae fed to shortnose sturgeon moulted from the third to the fourth stage within 15 days. Appy and Dadwell (1978) found and described moulting fourth-stage larvae in wild-caught sturgeons. Third-stage larvae given to rainbow trout (*Oncorhynchus mykiss*) were found undeveloped in the stomach of one of these fish 2 days later. None was found in a trout examined 18 days postinfection. Because of these observations the authors doubted if fish serve as paratenic hosts of *C. pseudoargumentosa* and they noted that shortnose sturgeon are rarely piscivorous.

Cystidicola

Members of the genus occur, often in large numbers, in the swimbladder of physostomous fishes. The literature is somewhat confusing because until recently the name of one species restricted to Salmonidae in North America was known as *C. cristivomeri* and the name of another species in both Coregonidae and Salmonidae was known as *C. stigmatura*. A third species from Europe was known as *C. farionis*. It is now recognized, however, that the species found exclusively in Salmonidae in North America and which has an egg with lateral swellings is correctly called *C. stigmatura* (syn. *C. cristivomeri*) (see Black, 1983a). The only other species found in both Salmonidae and Coregonidae in Europe and North America and which has an egg with polar filaments is known as *C. farionis*.

Baylis (1931) and Awachie (1973) found larvae in *Gammarus pulex* in Britain which they identified as *C. farionis*, a common parasite of *Salmo trutta*. Mamaev (1971) reported larval *C. farionis* in *Anisogammarus* sp. in the CIS. Smith and Lankester (1979) questioned the validity of the report of Bauer and Nikolskaya (1952) of a larval *C. farionis* in *Pontoporeia affinis* in Lake Ladgoda in the CIS. Smith and Lankester (1979) and Black and Lankester (1980) firmly established the importance of amphipods and mysids in the transmission of *Cystidicola* spp.

C. farionis Fischer, 1798

Lankester and Smith (1980) reported *C. farionis* in North America in *Coregonus clupeaformis*, *C. artedii*, *C. hoyi*, *C. nigripinnus*, *Oncorhynchus gorbuscha*, *O. kisutch*, *O. mykiss*, *O. tshawytscha*, *Osmerus mordax*, *Prosopium cylindraceum*, *S. trutta*, *Salvelinus fontinalis*, *S. namaycush* and *S. namaycush* × *fontinalis*. The parasite rarely, if ever, reaches maturity in the last five hosts, however.

Smith and Lankester (1979) followed the development of *C. farionis* in the amphipods *Gammarus fasciatus*, *Hyalella azteca* and *Pontoporeia affinis*. *Gammarus pseudolimnaeus* and *Mysis relicta* were apparently unsuitable hosts. The egg was oval with polar filaments and contained a short first-stage larva with a tapered but blunt tail. First-stage larvae (Fig. 7.7B) were found in the haemocoel of *H. azteca* at 19°C 24 h after eggs were ingested. In *P. affinis* maintained at 5–7°C, larvae were not found in the haemocoel until 10–20 days. Development in the haemocoel was asynchronous and temperature dependent. In *G. fasciatus* held at 16–18°C and 12–14°C second-stage larvae first appeared 2.5 and three weeks postinfection, and third-stage larvae 3 and 5 weeks postinfection, respectively. In *H. azteca* held at 12–14°C and 5–7°C, second-stage larvae were first seen 5 and 7 weeks and third-stage larvae 6 and 8 weeks postinfection, respectively. In *P. affinis* at 5–7°C second-stage larvae were first seen 6 weeks and third-stage larvae 7 weeks postinfection. Larvae were always free in the haemocoel, generally in the thorax and less commonly in the abdomen, head and pereiopods. The third-stage larva at the time of the second moult was only 450 µm in length but larvae grew substantially after this moult and eventually attained a maximum length of 3.0–4.5 mm. The larva (Fig. 7.7A,C) had an elongate buccal cavity and a long divided oesophagus. The genital primordia grew markedly after the second moult (when it was a small cluster of cells) until it was an elongate structure beside the intestine. The tail was cone-shaped and blunt.

Black and Lankester (1980) gave infective third-stage larvae from whitefish (*Coregonus clupeaformis*) to *Oncorhynchus mykiss* and found a third-stage larva in the stomach 6–8 h later. Other third-stage larvae were found in the swimbladder in 16–17 h postinfection. A larva was found in the pneumatic duct 18 h postinfection. In the swimbladder, male worms moulted the third time 1–19 (the first figure may be an error) days and the females 12–74 days postinfection. Males moulted the fourth time in 74–92 days and females in 111–112 days. Mature males were found 112 days and a mature female 235 days postinfection.

Black (1983c) noted that, although *C. farionis* has been reported widely in *Salvelinus namaycush* in various parts of Canada, mature worms only occur in this fish in the northwest. He suggested the northwestern population of *C. farionis* is a strain which

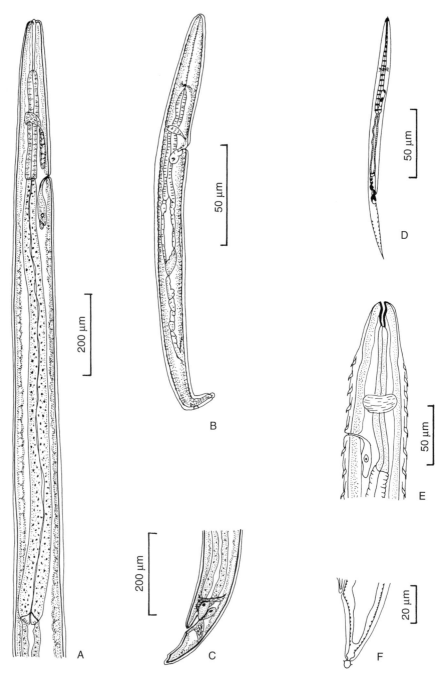

Fig. 7.7. *Cystidicola farionis*: (A) infective third-stage larva; (B) first-stage larva; (C) infective third-stage larva, caudal end. *Spinitectus micracanthus*; (D) first-stage larva; (E) infective third-stage larva, anterior end; (F) infective third-stage larva, caudal end. (A–C, after J.D. Smith and M.L. Lankester, 1979 – courtesy *Canadian Journal of Zoology*; D–F, after E.J. Keppner, 1975 – courtesy *The American Midland Naturalist*.)

survived, along with its host, in a Bering refugium during the last glaciation and spread from there to restricted waters in the northwest with the retreat of the glaciers.

C. stigmatura (Leidy, 1886)

C. stigmatura (syn. *C. cristivomeri*) is restricted to species of the genus *Salvelinus* in North America, namely lake trout (*S. namaycush*), arctic char (*S. alpinus*) and brook trout (*S. fontinalis*) (see Lankester and Smith, 1980). This species was originally described from the Great Lakes but with the decline of lake trout it has apparently become extinct in its type locality (Black, 1983b).

The egg of *C. stigmatura* is thick-shelled and oval with a plug-like structure on each pole. The sides of the eggs have large float-like structures which readily distinguish this species from *C. farionis* (see Ko and Anderson, 1969, under *C. cristivomeri*). Smith and Lankester (1979) studied development of *C. stigmatura* in the opossum shrimp (*Mysis relicta*). Eggs hatched in the shrimp and first- and second-stage larvae were found free in the haemocoel inside pereiopods or near thoracic muscles at the bases of appendages. Third-stage larvae were found coiled within thin-walled capsules in muscles of pereiopods. At 4–5°C the second stage was attained in 10 weeks and the third in 17 weeks. At 9–11°C the second stage was present in 3 weeks and the third in 8 weeks. Morphogenesis was similar to that of *C. farionis*. The second moult occurred when the larva was about 720 μm in length. Thereafter, the larva grew to 5.7 mm in length. The genital primordium was precociously developed (up to 1.3 mm in length) in the most advanced third-stage larvae. Naturally infected *M. relicta* found in Burchell Lake, Northwest Territories, were 5.9–10.8 mm in length and two types of gonads were recognizable. One was convoluted and lay behind the midbody; the other was straight and longer and extended equally anterior and posterior to the midbody. Smith and Lankester (1979) felt that the first was probably the male gonad and the second the female gonad. Amphipods were not suitable intermediate hosts.

Black and Lankester (1980) gave larvae to lake trout held at 4–10°C. The third moult of both males and females took place in 20–21 days and the fourth moult took place in males 84–104 days and in females 85–203 days postinfection.

Black and Lankester (1981) and Black (1985) estimated that *C. stigmatura* can live for at least 10 years in lake trout and arctic char and probably grow continuously. In the swim bladder, lengths of female worms and the proportion of females reaching sexual maturity were inversely related to intensity. Also, the lengths of mature females were positively correlated with the rate at which they produced eggs. Egg output was density dependent. Also, reproductive output of the parasites in individual fish, as estimated by the number of free eggs in the swimbladder, was overdispersed in the host population and concentrated in the oldest age classes of char. Because of the latter, the parasite is probably highly susceptible to any reduction in the numbers of larger fish. Black (1984) estimated that in Gavia Lake, Northwest Territories, exploitation of larger fish decreased nematode egg production by about one half during a 7-year period. It was suggested that decline of the definitive host by overfishing or other factors has probably resulted in the extinction of *C. stigmatura* in some lakes, including the Great Lakes where it once existed (Black 1983b, 1985).

Black (1983c) analysed the current host and geographic distribution of *C. stigmatura* in North America and concluded that this species arose and survived in the Upper Mississippi River Refugium during the Wisconsin glaciation. Char, the intermediate host (*M. relicta*) and *C. stigmatura* dispersed in lakes along the route of the retreating glacier in a northwesterly direction. This explanation accounted for the known past and present distribution of *C. stigmatura* and *M. relicta* in North America.

Cystidicoloides

C. tenuissima (Zeder, 1800)

This is a common parasite of the stomach mainly of salmonid fishes in Europe, Asia and North America. Choquette (1955) found adult worms (called *Metabronema salvelini*) in speckled trout (*Salvelinus fontinalis*) in Quebec and larvae in the haemocoel of mayfly nymphs (*Hexagenia recurvata* and *Polymitarcys* sp.). He fed infected mayflies to trout and recovered adult worms 60–70 days later.

Moravec (1971a) infected mayfly nymphs (*Ephemera danica*, *Habrophlebia lauta* and *Habroleptoides modesta*) with eggs from worms found in *Salmo trutta* in the Czech Republic. The egg was oval with a thick, smooth shell and contained a first-stage larva with an attenuated tail. Eggs hatched in the gut of the mayfly and first-stage larvae invaded the abdominal cavity and thorax. At 13–15°C the first moult occurred 5–18 days and the second 23–26 days postinfection. Third-stage larvae were never encapsulated and moved freely in the body cavity, in the thoracic muscles and less commonly in the legs and head of the mayfly. During the second moult, larvae were 1.6–3.4 mm in length. Larvae continued to grow and in 38 days were 2.8–4.1 mm in length. Those in naturally infected mayflies were 4.1–7.9 mm in length. In addition to the body, the genital primordia had grown considerably in the larger larvae. The third-stage larva had an elongated buccal cavity and clearly divided oesophagus. The tail was short, rounded and usually ended in a small protuberance. Moravec (1971b) determined that in the River Bystrice in the Czech Republic *Ephemera* sp. and *H. modesta* were natural intermediate hosts of *C. tenuissima*.

Moravec (1971a) infected *S. trutta* maintained at 18°C. Male worms were found in the stomach 12 days postinfection and females had completed the final (fourth) moult. It was suggested that eggs are produced in a month or more after fish are infected. De and Moravec (1979) infected rainbow trout (*Oncorhynchus mykiss*) and determined that third-stage larvae attached to the stomach mucosa and developed to adulthood. At 11–16°C the third moult occurred 10–14 days postinfection. The final moult in the males occurred in 28 days; the most advanced females were found in 34 days but they were still in the fourth stage. In fish kept at 7–11°C, the third moult of male and female larvae occurred in 15 days and in the females the fourth moult occurred in 32 days. Larvae were also given to two non-salmonid species: loach (*Neomacheilus barbatulus* – Cobidae) and minnows (*Phoxinus phoxinus* – Cyprinidae). Larvae in the third and fourth stages were found in the loach, which may serve as a paratenic host.

Moravec (1971b) concluded that *C. tenuissima* had two generations a year in the River Bystrice, since gravid female worms were found only from May to July and from September to October. In the Cernovicky Brook in the Czech Republic the only intermediate host was *Ephemera danica* and the only definitive host was *S. trutta* (see Moravec and De, 1982). Third-stage larvae occurred throughout the year in trout; gravid females with larvated eggs were present mainly from May to December and occasionally from February to March. Infective larvae were found attached to the gastric mucosa in two of 26 bullhead (*Cottus gobio*) and it was suggested this fish, a prey of trout, may serve as a paratenic host.

Moravec and De (1982) felt that water temperature was a key factor in determining the rate of development of *C. tenuissima* in its various hosts. Also, seasonal changes in the population dynamics of the different species of mayflies are probably important in the epizootiology of the nematode.

Salvelinema

Members of the genus are known from the swimbladder of salmonid fishes in the Pacific. Only two species (*S. salmonicola* and *S. walkeri*) are probably valid.

Koshida (1905, 1910) identified the amphipod *Gammarus* sp. as the intermediate host of *S. salmonicola* in Japan. Margolis and Moravec (1982) believed that large larvae (4.5–9.2 mm) found free in haemocoel of the amphipod *Ramellogammarus vancouverensis* from a creek on Vancouver Island, Canada, belonged to *S. walkeri*, which was found in coho salmon (*Oncorhynchus kisutch*) smolts and in *Salmo clarki* collected in the same stream. Moravec and Nagasawa (1986) described *Salvelinema* larvae from the haemocoel of the amphipod *Paramoera japonica* from the Rokumaibashi River, Aomori Prefecture, Japan. They believed that larvae probably belonged to *S. salmonicola* known from Japanese salmonids (e.g. *Salvelinus leucomaenis*).

Spinitectus

Members of the genus are widely distributed parasites of the intestine and pyloric caeca of fishes and amphibians. They have characteristic transverse, denticulate rings on the body. These denticulate rings also appear in the third-stage larva. Thus, adults and infective larvae are easily recognized to genus. Gustafson (1939) reported that *S. gracilis* attained the infective stage in mayflies but provided few details. Yamaguti and Nishimura (1944) and Johnson (1966) recognized worms found in shrimp (*Caridina weberi sumatrensis* and *C. denticulata*) in Japan and India, respectively, as larval *Spinitectus*. Since these early tentative observations, four species (one in amphibians and three in fishes) have been investigated. The egg, which is oval with a thick, smooth shell, contains a fully developed first-stage larva. The first-stage larva has not been adequately studied but it has an anterior pointed tooth and an elongate (e.g. *S. micracanthus*) or short tail (e.g. *S. gracilis*). Eggs hatch in the gut of larval stages of aquatic insects (e.g. mayflies) and crustaceans. First-stage larvae invade the haemocoel of the arthropod and enter muscle cells, where they grow for a time and eventually become encapsulated. There are two moults in the intermediate host and the infective third-stage larva has the characteristic body ornamentation found in the adults.

S. carolini Holl, 1929

This is a common parasite of freshwater fishes. Its development and transmission were studied by Jilek and Crites (1980, 1981, 1982a). Mayfly nymphs (*Ephemerella doris*, *Heptagenia marginalis* and *Hexagenia limbata*), larval plecopterans (*Neoperla clymene*) and larval dipterans (*Chironomus tentans*) were found naturally infected in the USA. In addition, mayflies (*Baetis carolina, Caenis simulans, E. doris, H. marginalis, H. limbata* and *Stenonema frontale*) as well as odonatans (*Gomphus quadricolor, Ischnura verticalis* and *Pachydiplax longipennis*) and plecopterans (*Acroneuria brevicauda*) were infected experimentally by allowing them to ingest eggs.

Eggs hatched in the midgut of the insect and were found in the haemocoel in 1–6 h. At 20°C the first moult commenced 18 h postinfection in the haemocoel and was completed by 36 h. The second-stage larva grew for about 5 days and then invaded abdominal muscles and became encapsulated. The second moult took place 8 days postinfection but larvae were not infective until 14 days postinfection. Jilek and Crites (1980) suggested that heavy infections of *S. carolini* in mayfly nymphs may inhibit metamorphosis.

In the fish host, infective-stage larvae remained in the lumen of the intestine, invaded the submucosa or penetrated the gut and became encapsulated on the mesentery. The third moult apparently occurred 2 days after the fish was infected and the fourth moult in 14 days. Gravid females were present in 21 days.

More than 80% of third, fourth and adult stages of *S. carolini* given to fish were recovered 14 days later at necropsy. This suggested that transmission from fish to fish through predation may occur.

S. gracilis Ward and Magath, 1917

S. gracilis is a common parasite of freshwater fishes in North America. It has also been reported in amphibians. As indicated earlier, Gustafson (1939) reported *S. gracilis* in mayflies. Keppner (1975) recovered larvae from naturally infected *Hexagenia* sp. nymphs. Jilek and Crites (1981) found (in Ohio) natural infections in nymphs of mayflies (*Caenis simulans, Heptagenia marginalis* and *Hexagenia limbata*) and plecopterans (*Neoperla clymene*). They experimentally infected mayflies (*Baetis carolina, C. simulans, Ephemerella doris, H. marginalis, H. limbata, Stenonema frontale*), odonatans (*Gomphus quadricolor, Ischnura verticalis* and *Pachydiplax longipennis*), a plectopteran (*Acroneuria brevicauda*) and a collembolan (*Podura aquatica*).

According to Jilek and Crites (1982b) eggs hatched in the gut of larval insects and entered the haemocoel within 6 h. The first-stage larva was short and had a pointed cephalic tooth and a rounded tail. The first moult began in 18 h and was completed by 36 h at 20°C. Second-stage larvae grew within the haemocoel for 6 days and then invaded abdominal muscles, where they became encapsulated. The second moult occurred 8 days postinfection. Larvae were not infective until about 15 days post-infection, i.e. 6–7 days after the second moult in the insect. Larvae were 1.7–1.9 mm in length and the genital primordia were well developed.

In experimentally infected *Lepomis cyanellus*, infective larvae were present 2 days postinfection in the intestinal lumen, in the submucosa and encapsulated on the mesentery. The third moult was said to have occurred in 3 days. The final moult occurred in 15 days and gravid female worms were present in 24 days. Apparently, worms will reach the adult stage in the tissues of the host as well as in the lumen of the intestine.

All third, fourth and adult stages of *S. gracilis* given orally to fish were recovered 14 days later, suggesting that the parasite could be transferred from fish to fish by predation.

S. micracanthus Christian, 1972

This is a parasite of bluegill (*Lepomis macrochirus*) in the USA. Keppner (1975) fed eggs to nymphs of mayflies (*Hexagenia* sp.) and examined them at various times thereafter. Eggs hatched within 6 h in the posterior part of the midgut and the anterior part of the hindgut. Within 12 h larvae (Fig. 7.7D) were found in the haemocoel and some had entered individual longitudinal muscle cells in the abdomen. The muscle cells eventually disintegrated as the larvae grew and a delicate capsule was formed around the larva or larvae (one to three larvae were in each capsule). The first moult occurred 10 days and the second 19–20 days postinfection, when larvae were 1.3–1.4 mm in length. There was rapid growth after the second moult and in 25 days the cuticle had 16–17 circles of cuticular spines in larvae that were 2.2–2.4 mm in length (Fig. 7.7E–F). There were four combs per circle. The first three circles had five to seven spines per comb whereas the remainder had four to six spines per comb. The primordium of the vulva was visible and the male genital primordium was thick posteriorly and thin anteriorly. The tail of the larva had a small spike covered with minute spines. Keppner (1975) gave infective larvae to *L. macrochirus*, which were examined at various times thereafter. Worms were always found with their heads embedded in the villi of the intestine a short distance behind the pyloric caeca. Larvae grew slowly and he recognized only one moult, around 16 days postinfection. In 26 days adults were present and sperm appeared in males. In 36 days, developing eggs were present in females.

S. ranae Morishita, 1926

Hasegawa and Otsuru (1977) in Japan recovered larval *S. ranae* in freshwater shrimp (*Paratya compressa improvisa*) as well as adults in *Cynops pyrrhogaster*, *Rana ornativentris* and *R. rugosa*. Larvae from shrimp were given orally to *R. brevipoda porosa*, *R. rugosa* and *R. tagoi* kept at room temperature in spring to autumn and at 22–26°C in winter. In *R. brevipoda*, in spring and summer, larvae moulted 6–7 days postinfection and again between 25 and 45 days postinfection. Larvae failed to develop in frogs in late autumn or winter even at warm temperatures. In *R. tagoi* the fourth moult occurred 25 days postinfection. Larvae grew rapidly in *R. rugosa* and mature worms were found in 15 days. Larvae persisted in *C. pyrrhogaster* without development.

7.7

The Superfamily Acuarioidea

Members of the Acuarioidea are small to medium-sized nematodes inhabiting the upper alimentary tract mainly of birds. Four genera and one species of another genus are found in the stomach of mammals. The parasites in mammals are presumably derived from those in birds. Only one family (Acuariidae) is recognized. The species are characterized by the presence of peculiar cephalic structures in the form of cordons extending from the cephalic extremity (Acuariinae except for *Paracuaria*, in which they are absent in all stages), cordons restricted to the anterior extremity only (Seuratiinae) or **ptilina** (a term proposed by Wong and Lankester (1984) for variously modified cephalic ornaments which extend posteriorly from the oral opening) in the form of cylindrical horns, blades or shields (Schistorophinae) (Chabaud, 1975b); however, in *Schistogendra* typical ptilina are absent. The function of these cephalic structures is unknown but it has been suggested that in species which attach to the oesophagus and proventriculus they may aid in feeding. Bartlett (1991) has described a gelatinous cap-like structure termed a **pileus** around the anterior end of two species of acuarioids and suggested that it is by means of this structure that these nematodes remain attached and not the cordons – which may, however, aid in directing food from the gut of the host to the oral opening of the worm. The function of cordons in the many species which occur under the koilin or in the muscles of the gizzard (including all the Schistorophinae and the Seuratiinae and most of the Acuariinae) is problematic.

Although most adult acuarioids occur in the gizzard of birds, a few species are found in the oesophagus (e.g. *Cosmocephalus obvelatus*, *Skrjabinocerca americana*). Adult *Skrjabinoclava inornatae* are found in both the proventriculus and the posterior half of the oesophagus. Other acuarioids are found exclusively in the proventriculus (e.g. *Echinuria uncinata*, *Skrjabinoclava* spp. except *S. inornatae*). In the oesophagus and proventriculus, worms are found with their heads embedded in the mucosa, where they elicit pin-point lesions; the worms may move from one attachment site to another, leaving behind lesions devoid of worms. Worms in the gizzard are usually found free under the koilin but some species occur in muscles (e.g. *Acuaria hamulosa*, *Syncuaria squamata*). Adult *Paracuaria adunca* and *Chordatortilis crassicauda* occur under the koilin at the junction of the proventriculus and gizzard. Acuarioids occurring in the gizzard must have some way to discharge their eggs into the lumen of the gut. This has not been sufficiently investigated. Cram (1931) reported openings through the gizzard lining associated with *A. hamulosa*. Alicata (1938) found adult *A. hamulosa* mainly in tissues of the gizzard at its junction with the intestine and perhaps it may be relatively easy for worms to oviposit into the gut lumen in such sites. Adult *S. squamata* are embedded in the musculature of the gizzard but their caudal ends extend through the koilin into the lumen; in piscivorous birds the koilin is thin.

Most acuarioids occur in birds living in aquatic habitats (i.e. Procellariiformes, Pelecaniformes, Ciconiiformes, Anseriformes, Charadriiformes and Coraciiformes) and relatively few in birds associated with terrestrial habitats (i.e. Falconiformes, Galliformes and Passeriformes).

All acuarioids produce oval, smooth and thick-shelled eggs, each of which contains a small first-stage larva. The cephalic end is simple and apparently unarmed. The tail is cone-shaped and devoid of spines. The German parasitologist Hamann (1893) provided the first clues to the transmission of acuarioids when he reported larvae of *E. uncinata*, the proventricular worm of ducks and geese, in the cladoceran *Daphnia pulex* in ponds inhabited by ducks and geese near Berlin. Hamann gave a correct account of how transmission probably occurs. A few years later Piana (1897) in Italy showed that *Dispharynx nasuta* of the proventriculus of gallinaceous and passerine birds developed in terrestrial isopods. Thus, the two general patterns of transmission, one in aquatic and the other in terrestrial environments, were established late in the 19th century. Despite these early beginnings, many years were to elapse before Cram (1931, 1934a), Cuvillier (1934) and Alicata (1938) investigated experimentally the transmission and development of some acuarioids in terrestrial hosts and showed the importance of various insects and isopods as intermediate hosts. Garkavi (1949b, 1950, 1953, 1956) investigated the development and transmission of *Streptocara crassicauda* of Anatidae and revealed the importance of aquatic crustaceans and the possible role of fish paratenic hosts. In the past 20 years there has been renewed interest in the acuarioids and a number of more detailed studies have been carried out on species in waterfowl and shorebirds. Except for *S. crassicauda* of the Seuratiinae, all species studied belong in the Acuariinae. Little is known about the transmission of the Schistorophinae.

Species of acuarioids in terrestrial hosts develop successfully in a great variety of arthropod intermediate hosts, including isopods, grasshoppers, beetles and even diplopods. A particular species is often capable of developing in a wide variety of intermediate hosts. In *Acuaria* spp., development of larvae generally commences in the haemocoel but eventually larvae invade muscles and attain the infective third stage within muscle fibres. There are the expected two moults in the insect.

Acuarioids which parasitize aquatic hosts develop to the third stage in the haemocoel of aquatic crustaceans (e.g. *E. uncinata* in cladocerans; *C. obvelatus*, *P. adunca*, *S. americana* and *Skrjabinoclava morrisoni* in amphipods; *Desportesius invaginata* and *S. squamata* in ostracods; and *S. inornatae* in fiddler crabs). The same species may utilize several different kinds of crustaceans. For example, *E. uncinata* is said to develop in amphipods, ostracods and conchostracans as well as cladocerans although the most important hosts in transmission may be cladocerans. Other acuarioid species may be much more specific in their ability to develop in crustaceans (e.g. *Skrjabinoclava* spp.).

Larvae normally remain unencapsulated in crustaceans but capsules have been noted rarely around larvae in long-standing infections. Rates of development of larvae in intermediate hosts are difficult to compare because development is temperature dependent and experimentally infected invertebrates have been maintained at various temperatures. At approximately 20–25°C, the maximum time for development in acuarioids which have been studied was about 40 days in *S. prima* and the minimum time 11 days in *E. uncinata*.

Third-stage larvae vary considerably in morphology in the Acuarioidea. In species of *Acuaria*, *Cheilospirura* and *Skrjabinoclava*, the posterior quarter or fifth of the body is bent dorsally and the tail is always armed with spines or tubercles. In other species the tail is unarmed and generally conical.

Most hosts of acuarioids probably acquire their infections from ingesting arthropod intermediate hosts containing infective third-stage larvae. In piscivorous birds such as cormorants, herons, gulls and bitterns, however, transmission depends on frog and fish paratenic hosts which consume the crustacean intermediate hosts. Infective larvae become encapsulated in the intestine or on the mesentery of the paratenic host and retain their infectivity for the final hosts. There is evidence that larvae can readily transfer from one paratenic host to another (e.g. from a forage fish to a piscivorous fish).

Some gizzard worms undergo their entire development within the gizzard of the final host where the adults are found (e.g. *A. hamulosa*). Other gizzard worms first develop to the subadult stage in the proventriculus before invading the gizzard (e.g. *P. adunca*, *S. squamata*). Of the oesophageal worms, *S. americana* develops entirely in the oesophagus whereas *S. inornatae* and *C. obvelatus* reach the subadult stage in the proventriculus before colonizing the oesophagus. Proventricular worms seem to develop entirely in the proventriculus; larval *Skrjabinoclava* spp. are often found in the proventriculus of shorebirds (Wong and Anderson, 1987b, 1990a).

In all species thoroughly studied there are the usual two moults leading to the adult stage in the final host. The appearance of cordons seems to be variable depending on the species. Cordons are sometimes present on the third stage (e.g. *D. nasuta*, *S. americana*) but in most species studied cordons first appear during development to the fourth stage in the definitive host. Cordons in the third and fourth stages are often simpler than those of the adult which replace them after the fourth moult.

The rate of development to mature adults in the final host varies in those acuarioids which have been studied, from a maximum of 76 days in *A. hamulosa* to only 9 days in *S. crassicauda*. Wong *et al.* (1989) and Bartlett *et al.* (1989) suggested that there is a relationship between the level of development of genital primordia in third-stage larvae and the rapidity with which larvae attain adulthood in the definitive host. For example, in *S. americana*, *P. adunca* and *S. inornatae* genital primordia in the third stage are precociously developed and all three species mature rapidly in the definitive host (less than 18 days).

Anderson *et al.* (1996) contrasted the acuarioid fauna in African–European waders with those in North America and concluded that waders (Charadriidae and Scolopacidae) in various distinct flyways throughout the world may have more or less distinct acuarioid species acquired from the ingestion of crustacean intermediate hosts in staging and wintering areas.

Family Acuariidae

Subfamily Acuariinae

Acuaria

Members of *Acuaria* occur under the koilin of the gizzard mainly of terrestrial birds. Eggs ingested by various insects (especially Orthoptera and Coleoptera) and diplopods hatch and first-stage larvae invade the haemocoel and start to develop. Larvae eventually invade muscle cells, where they become encapsulated as third-stage infective larvae. These larvae have well-developed pseudolabia and the cephalic region is slightly inflated behind the oral region and separated from the main body by grooves. Lateral alae are present. The posterior part of the body is generally bent dorsally and the terminal end of the tail is ornamented with numerous pointed or rounded protuberances.

A. anthuris (Rudolphi, 1819)

A. anthuris is a gizzard parasite of species of the corvid genera *Pica* and *Garrulus* (cf. *A. depressa*). Its development was studied in *Locusta migratoria* by Quentin *et al.* (1972) in France; the source of the parasite was *Pica pica*. Development in locusts was followed at three different temperatures. The first-stage larva (250–260 µm in length) was slender, with an attenuated tail terminated by a delicate mucron. At 22–24°C the first moult occurred on day 7 and larvae were in the third stage by day 13. At 25–28°C the first moult occurred on day 5 and the second on day 9. At 34–37°C the first moult occurred on day 3. At 22–28°C larvae were found free in the mesenteron 3 h post-infection. At 48 h larvae were in the anterior region of the abdomen. In 3 days larvae were active in the abdominal haemocoel. In 7 days larvae were found in muscles of the first abdominal segment and by day 13 larvae were in the third stage, coiled within muscle fibres. Seureau (1973) described changes in muscle fibres associated with larvae. The third-stage larvae were 800–860 µm in length. The posterior quarter of the body was curved dorsally. The tail ended in a digitiform process which was accompanied by about six additional ventroterminal and two dorsoventral spines. Grooves setting off pseudolabia were interpreted by Quentin *et al.* (1972) as cordons.

Rietschel (1973) collected *A. anthuris* from the gizzard and successfully infected the grasshoppers *Chorthippus longicornis* and *Locusta migratoria*, in which development was completed in 21 days at 35°C.

A. depressa (Schneider, 1866)

A. depressa is a gizzard parasite of crows (Corvidae spp.) (see Quentin *et al.*, 1972). Cram (1934) and Rietschel (1973) studied the biology of this species under the name of *A. anthuris*, according to Quentin *et al.* (1972). Cram (1934) found that *A. depressa* from *Corvus brachyrhynchos* and *C. ossifragus* of the USA developed to the infective stage in *Melanoplus femurrubrum* and in unidentified crickets. The larvae were fully developed 28–30 days postinfection. Rietschel (1973) collected worms from *Corvus corone* in Europe and successfully infected *Locusta migratoria* and *Chorthippus longicornis*. Quentin and Seureau (1983) described the morphology of the third-stage larva which had lateral alae; the tail was ornamented with numerous tubercles and the caudal end was bent dorsally. The genital primordium was weakly developed.

A. gruveli (Gendre, 1913)

Quentin and Seureau (1983) found this parasite in the mucosa of the gizzard of the gallinaceous *Francolinus bicalcaratus* in Togo. They gave eggs to *Tylotropidus patagiatus* (Acrididae) and followed development in this insect kept at 30°C. In 24 h active first-stage larvae were found in the haemocoel. The first-stage larva was about 200 µm long. After 3 days larvae were found mainly in the dorsal muscles of the abdomen but a few were already in the thoracic muscles. Larvae were inside muscle fibres; sometimes several larvae were in a single fibre. The first moult occurred 5 days and the second 8 days postinfection. Third-stage larvae were 540 and 610 µm in length and had lateral alae. The posterior part of the body was bent dorsally and the tail ended in a digitiform point dorsal to which were three additional but smaller processes (Fig. 7.8A).

A. hamulosa (Diesing, 1851)

A. hamulosa (syn. *Cheilospirura hamulosa*) is a common cosmopolitan parasite found under the koilin or in the muscular wall of the gizzard of chickens and turkeys. Cram (1931) and Cuvillier (1934) showed that grasshoppers (*Melanoplus differentialis*, *M. femurrubrum*, *Paroxya clavuliger*) were suitable intermediate hosts in continental USA. Alicata (1938) in Hawaii successfully infected grasshoppers (*Atraciomorpha ambigua*, *Conocephalus sallator*, *Oxya chinensis*), a sandhopper (*Orchestia platensis*), beetles (*Carpophilus dimidiatus*, *Dactylosternum abdominale*, *Epitragus diremptus*, *Euxestus* sp., *Litargus baleatus*, *Palorus ratzeburgi*, *Tenebroides nana*, *Tribolium castaneum*, *Typhaea stercorea*) and weevils (*Oxydema fusiforme*, *Sitophilus oryzae*). Alicata also found natural infections in *E. diremptus*, *O. platensis* and *T. nana* and in unidentified grasshoppers. In Cuba, Refuerzo (1940) infected the grasshoppers *Aeolopus tamulus* and *Oxya sinensis*.

In the CIS, Dotsenko (1953) infected orthopterans (*Aeolopus* sp., *Primnoa ussriensis* and *Setrix japonica*) and beetles (*Decticus verrucivorus*, *Gampsocleis sadahovi* and *Phaenorepterus falcata*) and found the orthopteran *Odealius infermalis* naturally infected.

Several other authors working in Cuba have reported that various insects and diplopods were suitable intermediate hosts of *A. hamulosa* (see Jurášek *et al.*, 1970; Moya and Ovies, 1982). The arthropods infected experimentally or found naturally infected were: Insecta – *Carcinophora americana*, *Doru taeniata*, *Dirmestes ater*, *Euborellia annulipes*, *Labidura bidens*, *L. reporia*, *Marava unidentata*; Diplopoda – *Microspirobolus* sp. and *Stilpnochlora caouloniana*.

In India, Kalyanasundarum (1977) found larvae of *A. hamulosa* in *Oxya nitidula* and *Spathosthernum prasinifierum*.

First-stage larvae of *A. hamulosa* migrated into the haemocoel of insects and developed mainly in muscles, where they became encapsulated, according to Cram (1931). Alicata (1938) reported, however, that in grasshoppers first-stage larvae were found free in the body cavity for about 11 days, at which time they began to moult. Second-stage larvae were also found free in the body cavity and they moulted 17 days postinfection. After the third stage was attained, the larvae began to become encapsulated in musculature and became tightly coiled. The caudal part of the body was bent dorsally. The tail of the third-stage larva terminated in a finger-like projection with three additional, but shorter, finger-like projections ventral to the longer terminal end of the tail. Cordons have not been reported in the third stage. According to Cram (1931) the infective larva was 700 µm in length.

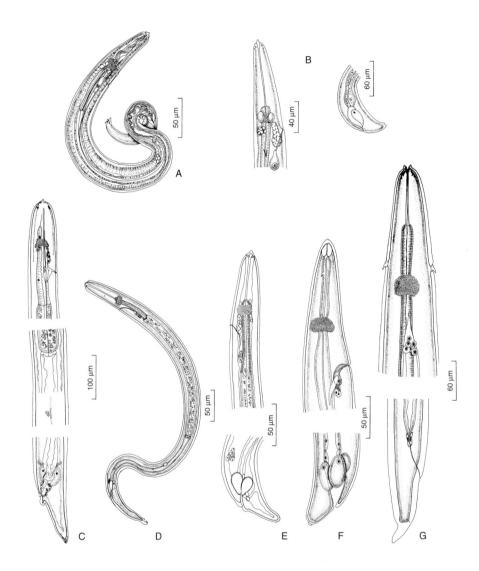

Fig. 7.8. Infective third-stage larvae of Acuarioidea in birds: (A) *Acuaria gruveli*; (B) *Desportesius invaginatus*; (C) *Skrjabinocerca americana*; (D) *Skrjabinoclava morrisoni*; (E) *Syncuaria squamata*; (F) *Streptocara crassicauda*; (G) *Ancyracanthopsis schikhobalova*. (A, after J.C. Quentin and C. Seureau, 1983 – courtesy *Annales de Parasitologie Humaine et Comparée*; B, after A. G. Chabaud, 1950 – courtesy *Annales de Parasitologie Humaine et Comparée*; C, after C.M. Bartlett, R.C. Anderson and P.L. Wong, 1989 – courtesy *Canadian Journal of Zoology*; D, after P.L. Wong and R.C. Anderson, 1987 – courtesy *Canadian Journal of Zoology*; E, after P.L. Wong and R.C. Anderson, 1987 – courtesy *Canadian Journal of Zoology*; F, after R.J. Laberge and J.D. McLaughlin, 1989 – courtesy *Canadian Journal of Zoology*; G, after P.L. Wong and R.C. Anderson, 1994 – courtesy *Systematic Parasitology*.)

Cram (1931) experimentally infected young chickens but attempts to infect a pigeon (*Columba livia*), a ruffed grouse (*Bonasa umbellus*), a bobwhite quail (*Colinus virginianus*) and a turkey (*Meleagris gallopavo*) were not successful; the latter four species were adults. Infective larvae in chickens apparently burrowed under the gizzard lining at its junction with the proventriculus and were found in this site 11 days postinfection. Later they were found well distributed on the surface of the muscle wall of the gizzard and at 25 days were penetrating the musculi laterali and the musculi intermedii. Development was completed and worms matured in about 76 days postinfection. The final location of adult worms was always in the musculi intermedii. She reported that there were openings through the gizzard lining for the passage of eggs to the gut lumen.

Alicata (1938) followed development in chickens. Third-stage larvae penetrated the gizzard lining within 24 h. By 12 days they were penetrating into the musculature and were moulting. In the same site worms moulted again 16–21 (or more) days postinfection. By 42 days the young adults were entirely in the gizzard musculature, mainly near the intestinal opening. He found females with embryonated eggs in the uteri 90 days postinfection.

Cheilospirura

C. spinosa (Cram, 1927)

This species occurs in the gizzard of ruffed grouse (*Bonasa umbellus*), bobwhite quail (*Colinus virginianus*) and sharp-tailed grouse (*Tympanuchus phasianellus*) in the USA. Cram (1931) fed eggs to various insects and found development in the muscles of grasshoppers (*Melanoplus femurrubrum, M. differentialis*). In 25 days third-stage larvae were encapsulated. The posterior quarter of the infective larva was curved dorsally and the tail ended in a digitiform process accompanied by four spines (cf. *A. hamulosa*). The larva was 850 µm long.

Cram (1931) successfully infected quail and ruffed grouse but not chickens. Worms originating from ruffed grouse were transferred successfully to quail. In 14 days, worms were in the fourth stage or were immature adults beneath the koilin. Thirty-two days postinfection unembryonated eggs were present in the uteri of female worms and in 45 days females were ovipositing larvated eggs.

Cosmocephalus

C. obvelatus (Creplin, 1825)

C. obvelatus has been reported from numerous species of piscivorous birds but is probably mainly a parasite of the oesophagus of gulls (Laridae) spending time in fresh-water habitats. Its development and transmission were investigated by Wong and Anderson (1982) in a breeding colony of ring-billed gulls (*Larus delawarensis*) in Lake Ontario, Canada.

Eggs were oval (Fig. 7.9A) and contained a first-stage larva (Fig. 7.9B) 189–205 µm in length. The buccal cavity was represented by a long filament. The tail was long and tapered to a fine point. Eggs hatched and larvae developed (Fig. 7.9C) in the haemocoel of the amphipods *Crangonyx laurentianus*, *Gammarus fasciatus* and

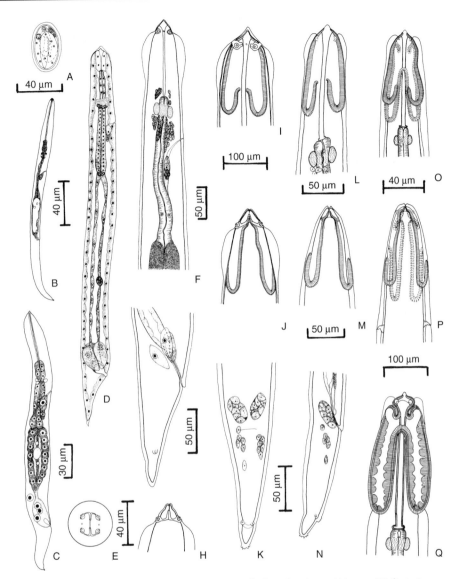

Fig. 7.9. Developmental stages of *Cosmocephalus obvelatus*: (A) egg; (B) first-stage larva (1 day postinfection); (C) first-stage larva (5 days postinfection); (D) moulting first-stage larva; (E) third-stage larva, *en face* view; (F) third-stage larva, anterior end; (G) third-stage larva, posterior end; (H) third-stage larva, dorsoventral view; (I) late third-stage larva, head region, lateral view (2 days postinfection); (J) late third-stage larva, head region, dorsoventral view (2 days postinfection); (K) male fourth-stage larva, caudal end, ventral view (5 days postinfection); (L) fourth-stage larva, anterior end, lateral view (5 days postinfection); (M) fourth-stage larva, anterior end, dorsoventral view (5 days postinfection); (N) male fourth-stage larva, posterior end, ventral view (5 days postinfection); (O) late fourth-stage larva, anterior end, lateral view (11 days postinfection); (P) late fourth-stage larva, anterior end, dorsoventral view (11 days postinfection); (Q) adult, anterior end, lateral view. (After P.L. Wong and R.C. Anderson, 1982 – courtesy *Canadian Journal of Zoology*.)

Hyalella azteca. Most larvae were found in the head region. At 20°C the first moult (Fig. 7.9D) occurred 10 days and the second 18 days postinfection. Third-stage larvae (Fig. 7.9E–H) were 1.6–2.4 mm in length 31 days postinfection. Prominent triangular pseudolabia were present. Sexes could be identified in the third stage by the moderately developed genital primordia. The tail was unornamented and ended in a blunt point.

In experimentally infected young gulls, infective larvae invaded the proventriculus, developed (Fig. 7.9I,J) and moulted 3.5 days postinfection. In 5 days fourth-stage larvae (Fig. 7.9K–N) were found attached to the posterior third of the oesophagus. The fourth moult (Fig. 7.9O,P) occurred 11 days postinfection, when worms were found through-out the oesophagus. Female worms were gravid in 27 days. Cordons appeared during development of the third stage in the final host. The resulting fourth stage had cordons which were recurved but not anastomosing. The complex adult cordons (Fig. 7.9Q) appeared during development of the fourth stage.

On April 2, when adult gulls had recently returned to the nesting site, prevalence was 74%. It rose to 100% as the season progressed. Intensity in birds also rose from April 2 to May 7 and remained unchanged to the end of June. During this period larval stages were commonly found in the birds. In birds hatched in the colony, prevalence reached 100% in 28-day-old birds and intensity increased up to the time of fledging. There was evidence from the population structure of worms in the young birds that fourth-stage and adult worms were sometimes regurgitated along with food during feeding by parents.

Infective larvae of *C. obvelatus* were found in various fishes (i.e. *Gasterosteus ocullatus*, *Cottus* sp., *Notropis hudsonius*, *Osmerus mordax*, *Semotilus atromaculatus*) collected in waters adjacent to the gull colony. These and other species of fish, which had died naturally in the lake, were presumably the source of *C. obvelatus* in gulls.

Desportesius

D. brevicaudatus (Dujardin, 1845)

This species has been found in the gizzard of *Botaurus stellaris* and *Ixobrynchus minuta* (syn. *Ardetta minuta*) in the CIS. According to Mozgovoi *et al.* (1965), an infective larva was found in a dragonfly larva (*Anax* sp.). They also found larvae in various fish species, including sticklebacks, catfish, pike and tench. They found juvenile worms in a stork and a heron fed infected sticklebacks and showed that larvae could transfer from one fish to another.

D. invaginatus (Linstow, 1901)

D. invaginatus (syn. *D. spinulatus*) from under the gizzard lining of cattle egrets (*Bubulus ibis*) was studied by Chabaud (1950). Eggs hatched and larvae developed to the infective stage in the haemocoel of the ostracods *Cyprinotus salinus* and *Pionocypris v. vidua*. At 25°C the first moult occurred in 5 days and the second 9–10 days postinfection. Larvae (Fig. 7.8B) were fully developed 12–13 days postinfection. Larvae given to tadpoles of *Discoglossus plexus* were found encapsulated in fatty tissue on the peritoneum. Larvae given to cyprinids (close to *Barbus*) were found encapsulated on the intestine.

Development in the gizzard of the cattle egret was completed in about 14 days. Cordons made their appearance during the development of the fourth stage and, unlike those in the adults, were anastomosing but non-recurrent.

Dispharynx

D. nasuta (Rudolphi, 1819)

D. nasuta (syn. *D. spiralis*) is a cosmopolitan parasite of the proventriculus of gallinaceous and passerine birds. Piana (1897) and Cram (1931) first showed that *D. nasuta* would develop in Isopoda. Cram (1931) infected *Armadillidium vulgare* and *Porcellio scaber* in the USA and Birová *et al.* (1974a,b) infected *Cubaris* sp. and *Porcellionides pruinosus* in Cuba. According to Cram (1931) larvae developed in unspecified tissues of the isopod and were unencapsulated. She observed a moult 14 days postinfection; larvae were fully developed in 26 days and were 2.9–3.2 mm in length. The tail tapered to a rounded tip slightly set off from the rest of the tail. Birová *et al.* (1974a,b) found moulting larvae 10–15 days postinfection. By 15–16 days larvae were fully developed with delicate, short, simple, non-anastomosing and non-recurrent cordons. They believed that there was only one moult in the isopod and suggested that a moult occurred in the egg (this is most unlikely – R.C.A.). Larvae were 2.7–3.6 mm in length in 20 days.

Cram (1931) acquired eggs of *D. nasuta* from worms in ruffed grouse (*Bonasa umbellus*) and infected isopods. Infective larvae from isopods were given to ruffed grouse, bobwhite quail (*Colinus virginianus*), pigeons (*Columba livia*) and chickens. Worms were recovered from all these hosts except chickens. Females were gravid 27 days postinfection. Cuvillier (1937) pointed out that *D. nasuta* from wild birds (grouse and quail) was not highly infective to chickens and turkeys even though it could readily be transmitted to pigeons. He suggested that physiological strains of *D. nasuta* existed.

Birova *et al.* (1974b) infected chickens with larvae from isopods and examined them for worms at intervals thereafter. They claimed to have recognized moults on days 4, 8 and 10 (since there was variation in the times of moulting they may have been misled into believing that three moults occurred rather than two – R.C.A.). Cordons in fourth-stage larvae were long and only slightly recurrent. The female worms had larvated eggs 19–21 days postinfection.

Echinuria

E. uncinata (Rudolphi, 1819)

E. uncinata is a common cosmopolitan proventricular nematode of waterfowl (Anatidae). It is regarded as an important pathogen of waterfowl in places where there is intense transmission. Hamann (1893) reported larvae (of '*Filaria uncinata*') in naturally infected *Daphnia pulex* in Germany. Romanova (1947) reported that infective larvae developed in the haemocoel of *D. magna* and *D. pulex*. The first moult occurred 6 days postinfection. At 17–23°C the infective stage was attained in 14 days and at 25–29°C in 12 days. Garkavi (1960) reported that development in cladocerans (*Daphnia* spp.) took 5 days at 26–30°C and 6 days at 16–20°C. Larvae were not encapsulated. In ducks there were two moults and by 34 days females with unembryonated eggs were found in the proventriculus.

Kotelnikov (1961) successfully infected the cladoceran *Ceriodaphnia* sp., the amphipod *Gammarus* sp., the isopod *Asellus aquaticus* and an unidentified ostracod. *Ceriodaphnia* sp. was most susceptible and larvae completed their development as rapidly in this species as in species of *Daphnia*, i.e. 11 days at 18–20°C. The parasite matured in 40 days in ducks. Misiura (1970) infected the ostracod *Heterocypris incongruens* as well as *D. magna* and *D. pulex*; development was similar in the three species. At 19–25°C, the first moult occurred 9–14 days and the second 14–21 days postinfection. First-stage larvae were 103–104 μm in length. Infective larvae were 0.6–1.5 mm in length; the buccal cavity was long and the tail conical and unarmed.

Potekhina (1968) studied *E. uncinata* in *D. magna* in a pond in the Moscow Zoo. The parasite appeared in *D. magna* from the end of May to the beginning of June and infective larvae appeared from the middle of June. Mortality in waterbirds began in July and was greatest in August.

Guilhon *et al.* (1971) studied experimentally *E. uncinata* from waterfowl in a zoo park in France. After hatching, larvae rapidly invaded the haemocoel of *D. pulex*. At 18°C the second stage was reached in 8–9 days and by 14–15 days infective larvae were present. At 12°C, 19 days were required for development to the third stage. Larvae were not encapsulated and occurred in various parts of the body, even the antennae. In experimentally infected ducks, infective larvae invaded the proventriculus, especially near the gizzard. The third moult took place 48 h postinfection and cordons (non-anastomosing) first appeared in the fourth stage. The final moult occurred six days postinfection. The prepatent period was 30–50 days in six experimentally infected ducks. Infections often caused a fatal proventriculitis with oedema and nodule formation which caused occlusion and malfunction of the proventriculus.

Austin and Welch (1972) studied *E. uncinata* in Manitoba. Larvae were found commonly in naturally infected *D. magna*, *D. pulex* and *Simocephalus vetulus* in a pond frequented by numerous waterfowl. Numbers of larvae reached a peak in August and most larvae were found in *D. magna*. In addition to the three cladocerans mentioned, larvae also reached the third stage in the conchostracan *Lynceus brachyurus* but not in several other crustaceans or aquatic insects allowed to ingest eggs. At 20–24°C the third stage was reached in 10 days in *Daphnia* spp. Larvae were 1.0–1.9 mm in length and possessed clearly defined, chevron-shaped cordons.

Development of *E. uncinata* in ducks was similar to that reported by Guilhon *et al.* (1971) although the fourth moult was said to occur 20 days postinfection and the time of the final moult was not determined accurately. In experimental studies, mallards (*Anas platyrhynchos*), domestic ducks (*A. p. domestica*), gadwalls (*A. strepera*) and pintails (*A. acuta*) were most susceptible to infection. No worms were recovered from shovellers (*A. clypeata*), blue-winged teal (*A. discors*) and coots (*Fulica americana*). Infections lasted for at least 11 months in mallards.

Paracuaria

P. adunca (Creplin, 1846)

Paracuaria is an unusual acuarioid genus because cordons do not appear in any of the stages, including the adults. *P. adunca* is a common parasite of the gizzard of piscivorous birds in many parts of the world. Its transmission and development were studied in a colony of ring-billed gulls (*Larus delawarensis*) in Lake Ontario, Canada, by Anderson and Wong (1982).

Eggs were oval. First-stage larvae were 176–210 µm in length with a tapered tail. Eggs hatched and larvae developed to the infective stage in the haemocoel of the amphipods *Crangonyx laurentianus*, *Gammarus duebeni* and *Hyalella azteca*. Most larvae were found in the head region. At 18–20°C the first moult occurred 10 days and the second 18 days postinfection. Larvae were found encapsulated in the haemocoel in 31 days. Infective larvae were 2.2–2.8 mm in length and had well-developed pseudolabia. The tail was short with a rounded tip. Larvae had unusually well-developed genital primordia and were easily sexed.

In experimentally infected juvenile gulls, infective larvae invaded the mucosa of the proventriculus and rapidly developed into subadults; the third moult occurred 2.5 and the fourth 5 days postinfection. In 8 days subadults were found under the gizzard lining near the proventriculus and by 16 days females had larvated eggs.

Although about 65% of adult birds at the colony in early spring (2 April) were infected with *P. adunca*, intensities were low. Both prevalence and intensity rose as the nesting season progressed and larval stages were found frequently in the birds. Worms began to appear in young birds in the colony 1 week after hatching and both prevalence and intensity increased markedly up to the time the birds fledged, when prevalence was about 85%.

Infective larvae of *P. adunca* became encapsulated on the mesentery of experi-mentally infected goldfish (*Carassius auratus*) and larvae were found on the mesentery of the fish *Culaea inconstans*, *Notropis hudsonius* and *Semotilus atromaculatus* collected in waters adjacent to the gull colony. These, and other species of fish, dying naturally in the lake, were presumably the source of *P. adunca* in gulls.

Ellis and Williams (1973) concluded that *P. adunca* lives for less than a year since prevalence and intensity in captive naturally infected *Larus fuscus* declined markedly in 1 year. Anderson and Wong (1982) noted that prevalences and especially intensities were considerably lower in birds arriving in the spring than those leaving in the autumn, suggesting an annual loss of the parasite.

Skrjabinocerca

S. americana Wong and Anderson, 1992

S. americana (syn. *S. prima* in Bartlett *et al.*, 1989 not Schikhobalova, 1930) is a common parasite of the oesophagus of American avocets (*Recurvirostra americana*) and other Charadriiformes migrating through western Canada. Its transmission and development in avocets in Alberta were studied by Bartlett *et al.* (1989).

Eggs hatched in the gut; larvae (about 180 µm in length) invaded and developed in the haemocoel of the freshwater amphipod *Hyalella azteca*; larvae failed to develop in *Gammarus fasciatus*. Development at 21°C was slow and asynchronous in the amphipod. The first moult occurred 21 days and the second 31 days postinfection. The fully developed infective third-stage larva (Fig. 7.8C) was present 40 days postinfection. Infective larvae (1.9–2.4 mm in length) were free in the haemocoel in posterior segments of the abdomen. Larvae tended to lie semi-rigid and parallel to the longitudinal axis of the amphipod. They had lateral alae and longitudinal ridges, which may have reduced flexibility and contributed to the rigid posture of larvae in the haemocoel. Bartlett *et al.* (1989) suggested that these semi-rigid larvae may have hindered the amphipod host from rapidly extending the abdomen and made it susceptible to predation. Pseudolabia were well developed in larvae and delicate non-recurrent and non-anastomosing cordons were present. The tail was conical. Genital primordia were highly developed in male and female third-stage larvae from *H. azteca*.

In experimentally infected avocets, the third moult occurred in the oesophagus 2 days and the fourth 4 days postinfection. Cordons in the fourth stage were similar to those in the third stage and the adult. Females had unembryonated eggs 7 and 10 days postinfection but by 18 days larvated eggs were present. Various stages were found with their cephalic ends embedded mainly in the upper third of the oesophagus and a few were found in the buccal cavity.

Transmission in avocets probably occurs on the prairies where *H. azteca* is common in marshes and sloughs and where avocets nest and rear their young. The birds leave the nesting area as soon as the young are fledged (one month). It is not known if worms found in adult birds in spring were acquired the previous summer, since longevity of *S. americana* in avocets is unknown. Nevertheless, adult avocets might acquire new infections from ingesting infected amphipods which have overwintered. Development in the avocet was rapid (18 days) and eggs could be dispersed into the environment and be available to amphipods during the nesting period. Development in the amphipods was extremely slow (40 days) and newly infected amphipods might not have third-stage larvae until after the birds have left the nesting area. However, amphipods would have third-stage larvae the following spring when the birds return.

A godwit (*Limosa fedoa*) and a western willet (*Catoptrophorus semipalmatus inornatus*) were successfully infected experimentally.

S. prima Schikhobalova, 1930

S. prima was described from the oesophagus of rooks *Corvus frugileus pastinator* (syn. *Trypanocorax pastinator*) on the Island of Sakhalin off the Pacific coast of the CIS. Tsimbalyuk and Kulikov (1966) reported third-stage larvae in the marine semi-terrestrial amphipod *Orchestia ochotensis* (beach fleas) on Komandor Island in the Bering Sea. The parasite has been reported in a variety of birds on islands in the Bering Sea, including members of Charadriidae, Alcedinidae, Emberrizidae, Montacillidae, Troglodytidae and Cuculidae, which presumably acquire infections from feeding on beach fleas (for review see Wong *et al.*, 1987).

Skrjabinoclava

S. inornatae Wong and Anderson, 1987

This is a common parasite of the posterior half of the oesophagus and proventriculus of western willets (*Catoptrophorus semipalmatus inornatus*) in North America. Its development was studied by Wong *et al.* (1989).

Larvae developed in fiddler crabs (*Uca longisignalis*, *U. panacea* and *U. virens*) collected in Louisiana, where many willets winter after nesting in freshwater habitats on the western prairies; fiddler crabs are prey of willets in Louisiana. Larvae were most commonly associated with pigmented tissues lining the carapace. Infective larvae (1.2–1.4 mm in length) were distinctive in that the posterior half of the body was bowed dorsally, the tail tip had two prominent dorsally directed spines and the genital primordia were precociously developed. Pseudolabia were offset from the body by grooves but cordons were absent.

In experimentally infected willets, the third moult occurred before 3 days and the fourth 5 days postinfection. Subadult worms were recovered in 10 days and females contained larvated eggs 15 days postinfection. Larval and subadult worms occurred in the mucosa of the proventriculus but most adult worms were found with their heads embedded in the posterior half of the oesophagus. Cordons appeared for the first time in the fourth stage and were triangular in shape and less complex than in the adult.

This parasite was transmitted experimentally to marbled godwits (*Limosa fedoa*) although this host has never been found naturally infected. *S. inornatae* has been found rarely in ring-billed gulls (*Larus delawarensis*) and ruddy turnstones (*Arenaria interpres*), indicating that these birds occasionally feed on fiddler crabs on their wintering grounds in marine habitats.

The longevity of *S. inornatae* in willets has not been determined. Experimentally, infections persisted for up to 2 months and birds arriving on the breeding grounds in the Canadian prairies retained infections at least into June.

S. morrisoni Wong and Anderson, 1987

This parasite occurs in the proventriculus of semi-palmated sandpipers (*Calidris pusilla*) which breed in the Arctic from May to July and winter in South America. During spring and autumn the birds stop at two major staging areas, namely the Gulf of Mexico and the Bay of Fundy, Canada.

S. morrisoni developed to the infective stage in the haemocoel of the marine amphipod *Corophium volutator*, which is the important prey of semi-palmated sandpipers in the Bay of Fundy. At 20°C the first moult occurred 8 days and the second 11 days postinfection (Wong and Anderson, 1988). Third-stage larvae (for descriptions see Wong and Anderson, 1988, and Wong *et al.*, 1989) were similar to those of *S. inornatae* but smaller (560–590 μm in length) and the genital primordia were not as precociously developed (Fig. 7.8D).

Juvenile semi-palmated sandpipers collected in Ontario during the autumn migration were not infected with *S. morrisoni*, suggesting that transmission does not occur on the breeding grounds in the Arctic. The presence of moulting fourth-stage larvae in birds collected on Grand Manan Island in the Bay of Fundy and the high prevalence and intensity of infections in birds collected in Nova Scotia indicated that the Bay of Fundy was a major site of transmission of this parasite. The presence of larvae in birds arriving in western Canada in spring from the Gulf of Mexico indicated the latter is also a site for transmission (Wong and Anderson, 1990a). Birds in captivity retained infections for up to 4 months.

Syncuaria

S. squamata (Linstow, 1883)

S. squamata is a common gizzard parasite of cormorants (Phalacrocoracidae) in various parts of the world. Kurochkin (1958) gave eggs from nematodes from cormorants in Russia to *Cypris pubera* and observed, without providing details, that the larvae developed. Subsequently its development and transmission were studied in double-crested cormorants (*Phalacrocorax a. auritus*) from Lake Erie, Ontario, by Wong and Anderson (1987a).

Eggs hatched and larvae developed in the haemocoel of ostracods (*Cyclocypris ovum* and *Cypridopsis vidua*). First-stage larvae were small (165–200 μm in length). The tail was attenuated to a point. At 19–20°C the first moult occurred 11 days and the second 18 days postinfection. Infective larvae (0.9–1.4 mm in length) (Fig. 7.8E) could be sexed although the genital primordia (in a posterior position) were weakly developed. Pseudolabia were well developed. The tail was short and rounded.

In experimentally infected young cormorants, larval stages of *S. squamata* occurred between folds of the proventriculus, where they moulted 3.5 days and 8 days postinfection. Mature males and gravid females were found under the gizzard lining 29 days postinfection. Some female worms were embedded in the gizzard musculature, with their tails protruding into the lumen. Cordons first made their appearance during development to the fourth stage. Cordons in the fourth stage were similar to those in the adults but they only extended to the region of the nerve ring.

Larvae were found in granulomas on the intestinal serosa of goldfish (*Carassius auratus*) allowed to feed on infected ostracods. Young cormorants became infected when fed the intestines of the goldfish. In Lake Erie, juvenile cormorants as young as 1 week acquired infections from ingesting fish regurgitated by their parents. It was estimated that *S. squamata* lives for about 4 months in the final host and there was evidence that females became senescent and passed out of the host sooner than males.

Moravec and Scholz (1994) followed development in *Notodromas monacha* in the Czech Republic. They recorded moults at 9–11 and 13–15 days at 20–22°C. The following fishes were suitable paratenic hosts: *Alburnoides bipunctatus*, *Noemacheilus barbatulus*, *Oncorhynchus mykiss*, *Poecilia reticulata* and *Tinca tinca* (*T. tinca* was naturally infected).

Subfamily Seuratiinae

Streptocara

S. crassicauda (Creplin, 1829)

S. crassicauda (syn. *S. pectinifera*) is a cosmopolitan parasite of the gizzard of waterfowl (Anatidae) (Boughton 1969). Garkavi (1949b, 1950, 1953, 1956) established that the amphipod *Gammarus lacustris* was a suitable intermediate host. Larvae developed in the haemocoel and reached the infective stage in 19–25 days, depending on the temperature. Larvae persisted in nodules on the intestine of fish paratenic hosts (e.g. *Carassius carassius* and *Phoxinus perenurus*) and would then be available to piscivorous ducks. Ducks could also get infected from ingesting amphipods with larvae. Development in the final host was extremely rapid. There were two moults and the females were ready to oviposit in 9–10 days. Kovalenko (1960, 1963) found larvae in the fishes *Capialosa brashnikovi*, *Neogobius fluviatilis*, *N. melanostomus* and *Scardius erythrophalmus*.

Richter (1960) found larval *S. crassicauda* in naturally infected *Gammarus triacanthus* in Yugoslavia. Infective larvae were given to ducks and chickens, and adult worms were found in the oesophagus and gizzard by 12 days postinfection.

Klesov and Kovalenko (1967) found larval *S. crassicauda* in marine amphipods (*Gammarus locusta* and *G. maeoticus*) in a duck farm off the Azov coast in the southern Ukraine. Infected amphipods survived winter and were able to transfer the parasite to ducks in spring. At 15–25°C larvae took 23–24 days to attain the infective stage. The prepatent period in goslings was 9–10 days.

In Canada, Denny (1969) found larval *S. crassicauda* in *Gammarus lacustris* and Laberge and McLaughlin (1989) studied development in *Hyalella azteca*. First-stage larvae were small with conical tails. Larvae developed mainly in the cephalic haemocoel. At 18–20°C the first moult occurred 11 days and the second 15 days postinfection; the fully developed infective larva was found 19 days postinfection. Infective larvae (Fig. 7.8F) were 1.7–3.7 mm in length. The buccal cavity was short and thick-walled. The tail was short with a rounded tip. Mature female worms with larvated eggs were found in domestic ducks 9–21 days postinfection and eggs were first found in faeces on the 26th day. The longevity of the parasite in the duck is probably limited, since females recovered from two ducks 42 days postinfection did not contain eggs.

McLaughlin and McGurk (1987) found *S. crassicauda* in 12 species of dabbling and diving ducks in Manitoba in autumn. Of the dabbling ducks the blue-winged teal (*Anas discors*) had the highest prevalence and intensity. Of the diving species (and also among all duck species), the highest prevalence and intensity were found in the lesser scaup (*Aythya affinis*). Prevalence of *S. crassicauda* in the various duck species generally reflected the importance of amphipods in the diet. Certain duck species which do not feed extensively on amphipods (e.g. *Anas clypeata*, *A. strepera*, *Aythya valisineria*) were rarely infected. However, Laberge and McLaughlin (1991) gave equal numbers of larvae of *S. crassicauda* to blue-winged teal, gadwall (*A. strepera*) and lesser scaup ducklings. Highest intensities were present in teal and gadwall.

Subfamily Schistorophinae

Ancyracanthopsis

Wong and Anderson (1993) associated (and described) third-stage larvae (Fig. 7.8G) with adult *Ancyracanthopsis schikhobalovi* in whimbrels (*Numenius p. phaeopus*) collected in Iceland. Similar larvae were associated with adult *A. heardi* Wong and Anderson, 1990 in clapper rails (*Rallus longirostris*) from Louisiana. These larvae were morphologically similar to those of *A. winegardi* found in fiddler crabs (*Uca* sp.) in Louisiana (see below). The larvae have characteristics found in adults, i.e. a long narrow buccal cavity and muscular oesophagus, large single pointed deirids immediately anterior to the nerve ring and the posterior position of the vulva primordium. Larvae were about 6–7 mm in length and had complex oral structures and precociously developed genital primordia.

Members of the genus occur in larger shorebirds and it is likely that intermediate hosts are decapods such as various types of crabs.

A. winegardi Wong and Anderson, 1990

A. winegardi occurs under the gizzard lining of shorebirds, namely black-bellied plover (*Pluvialis squatarola*), long-billed curlew (*Numenius americanus*), marbled godwit (*Limosa fedoa*), spotted sandpiper (*Acitits macularia*), western willet (*Catoptrophorus semipalmatus inornatus*), Wilson's plover (*Charadrius wilsonia*) and whimbrel (*Numenius phaeopus*). Wong and Anderson (1990b) found an adult female *A. winegardi* in a laboratory-raised willet given large acuarioid-type third-stage larvae found encapsulated in the haemocoel of naturally infected fiddler crabs (*Uca* spp.) collected near Champagne Bay, Louisiana.

7.8

The Superfamily Filarioidea

Filarioids are parasites of the tissues and tissue spaces of all classes of vertebrates other than fishes (Anderson and Bain, 1976). Cephalic structures are generally rather simple and pseudolabia (or vestiges of them) are absent; in some groups there may be cuticular elevations or spines. Cephalic papillae are well developed and there are usually four submedian pairs. The buccal cavity is generally much reduced. Spicules are variable in length but consistently dissimilar in morphology (cf. Aproctoidea). The filarioids probably had a common ancestry with other Spirurida but branched off early during evolution. Two families are recognized: the Filariidae and the Onchocercidae.

Since adult filarioids occur in tissues of the host, they have evolved specialized means of transmission. They are all transmitted by haematophagous arthropods. The Filariidae elicit skin lesions and release eggs and/or larvae which attract arthropod vectors, mainly Muscidae. The Onchocercidae, on the other hand, have evolved blood- or skin-inhabiting microfilariae and are transmitted by arthropods which create lesions or pierce the skin and then suck blood. Anderson (1957b) suggested that the specialized life cycles of onchocercids evolved from those of the orbit-inhabiting *Thelazia* and the subcutaneous filariids (*Filaria, Parafilaria*). Other authors suggested a relationship between some onchocercids and habronematoids like *Draschia* and *Habronema* (Chandler *et al.*, 1941; Bain, 1981c).

Family Filariidae

The filariids are small to medium-sized subcutaneous parasites of certain mammals. The vulva in the five genera assigned to the family is always anterior to the nerve ring and in some genera (e.g. *Filaria, Parafilaria*) is beside the oral opening. The anterior position of the vulva is undoubtedly related to the methods these worms use to release eggs and/or larvae. Species studied elicit cutaneous lesions which attract dipteran intermediate hosts to eggs or larvae. There are only two subfamilies, namely the Filariinae and the Stephanofilariinae.

Subfamily Filariinae

The subfamily includes *Filaria* of carnivores and rodents, *Suifilaria* of swine, *Pseudofilaria* of African antelope and *Parafilaria* of horses and cattle. Transmission of two species in the latter genus has been elucidated.

Filaria

Filaria taxidae Keppner, 1969

Keppner (1971), O'Toole *et al.* (1993, 1994a,b) and Saito and Little (1997) described skin lesions in badgers (*Taxidea taxus*) and a skunk (*Mephitis mephitis*) infected with *Filaria taxidae*. According to O'Toole *et al.* (1993) eggs were deposited between the epidermis and the cutaneous basement membrane. Dermoepidermal separation resulted in fibrin-rich blisters containing eggs. Necrosis of the separated epidermis resulted in exposure of eggs on to the surface of the skin, where they might be available to intermediate hosts (perhaps muscid dipterans attracted to lesions would be the most likely vectors – R.C.A.). The theoretical interest of the observations is that the method used by *Filaria taxideae* to release eggs to the outside of the host seems to bridge the gap between the methods used by species of *Parafilaria* and *Stephanofilaria* to release eggs or larvae (see below) which were regarded as possible steps in the evolution of life cycles with blood- and skin-inhabiting microfilariae (Anderson, 1957b, 1958; Anderson *et al.*, 1998).

Parafilaria

There are only four species in the genus, including *P. bovicola* of cattle and *P. multipapillosa* of horses. Adult worms are found in the subcutaneous connective tissue, usually in haemorrhagic nodules in the upper parts of the body (Metianu, 1949). The female pierces the skin of the nodule, causing bleeding. She oviposits eggs containing larvae in the blood, which attracts dipterans in which development to the infective third stage takes place. Infective larvae migrate to the mouthparts of the vector and escape when the vector feeds on lesions or orbits of the final host. Infective larvae of *Parafilaria* are similar to those found in *Thelazia*, which reinforces the presumption of a phylogenetic relationship between the two genera (Anderson, 1957; Bain, 1981a). One species, *P. bassoni* Ortlepp 1962, has been reported in the orbit of its host, the springbuck (*Antidorcas marsupialis*), in southwest Africa (Ortlepp, 1962).

P. bovicola Tubangui, 1934

P. bovicola occurs in cattle mainly in Europe and Africa (Metianu, 1949; Fain and Herin, 1955; Nevill, 1975; Bech-Nielsen *et al.*, 1982). Metianu (1949) and Fain and Herin (1955) noted that the female made a small hole in the skin and released eggs in blood. Larvae were 215–230 μm in length, with attenuated tails. Fain and Herin (1955) suspected that *Musca domestica* was a possible vector. Nevill (1975) reviewed the literature in detail up to that time and examined many wild-caught flies around cattle in South Africa. He found third-stage larvae of *Parafilaria* in *Musca* (*Eumusca*) *lusoria* and *M.* (*Eumusca*) *xanthomelas*. Prevalence was usually less than 1% and most larvae were in the heads of the flies. He successfully infected *Musca* spp. by allowing them to feed on blood containing eggs. The main period of transmission was apparently in summer from August to February, when most lesions were noted on cattle and when the prevalence of larvae in flies was highest. Calves born during this period became infected and lesions appeared 7–10 months later. Nevill (1975) gave the length of the third-stage larvae as 2.0–4.3 mm. Bain (1981a) described infective larvae which were 3.4–4.4 mm in length. The oesophagus was short and entirely muscular.

Nevill (1979) showed that cattle became infected when infected *M. lusoria* were allowed to feed on a fresh incision or when infective larvae were placed directly on fresh incisions, inoculated subcutaneously into the jugular vein or placed in the orbit. Thus, there were various possible routes whereby larvae could gain entry to the body of the host. Worms usually migrated and matured some distance from the point of entry. The prepatent period was 242–319 days in experimentally infected cattle. Lesions tended to bleed when exposed to sunlight but not when kept in shade, suggesting that heat and/or light are needed to stimulate females to oviposit. This activity is probably correlated with vector activity. Also, infective larvae were stimulated to leave the mouthparts of *Musca* spp. when the latter were fed warm (38–40°C) ox blood but not when given warm saline or 15% sucrose solution.

Bech-Nielsen *et al.* (1982) reported that the face fly, *Musca autumnalis*, was a natural vector of *P. bovicola* in Sweden. Peak prevalence of infective larvae in the flies was 30% in June. Two calves inoculated by the intraconjunctival route became infected, whereas three calves inoculated subcutaneously did not. The prepatent period was 301 days in 26 cattle with natural infections.

P. multipapillosa (Condamine and Drouilly, 1879)

This is a parasite of the subcutaneous and intermuscular connective tissues of horses in various parts of the world, including Eurasia, Africa and South America. Infection with the parasite results in the condition known as 'bloody sweat' or 'summer bleeding'. Female worms were found in small haemorrhagic nodules under the skin. The female pierced the nodule and released larvated eggs in blood which flowed onto the skin of the host. First-stage larvae were unsheathed at 220–230 μm in length according to Supperer (1953) and 176–220 μm according to Gibson *et al.* (1964). Baumann (1946) observed oviposition and thought muscoid flies were likely vectors because he saw them feeding on blood which trickled from nodules. Gnedina and Osipov (1960) showed that the intermediate host was the female of the biting dipteran *Haematobia atripalpis*. Larvae developed in the body cavity and fat body of the fly and reached the infective stage in 10–15 days at 20–36°C. Infective larvae were 1.67–2.67 mm in length. Lesions on horses were seasonal, occurring in spring and summer and disappearing in winter in temperate regions.

Osipov (1962) showed that *H. atripalpis* became infected when it fed on blood containing eggs on the skin of horses and concluded that the infective stage left the fly and invaded the break in the skin made by the latter when feeding. Osipov (1962) failed to infect horses by feeding them infected flies or by inoculating larvae subcutaneously. Lesions made by ovipositing female worms appeared 281 and 387 days after horses had been bitten by infected *H. atripalpis*. There was evidence that worms migrated extensively in the subcutaneous tissues of the horse.

Baumann (1946) reported that sunlight was necessary to stimulate lesions to bleed.

Subfamily Stephanofilariinae

The subfamily contains the single genus *Stephanofilaria* with about eight nominal species found in subcutaneous tissues of bovids (especially cattle), rhinoceros and elephants. The genus is readily recognized because the worms are small, the oral opening is surrounded by numerous spines and they are associated with characteristic skin lesions on the host. Species of *Stephanofilaria* in cattle tend to occur in specific regions of the body (Ivashkin *et al.*, 1971). Johnson (1987) has published a useful review of certain aspects of stephanofilariasis.

Stephanofilaria

S. assamensis Pande, 1936

This species causes 'hump sore' in cattle in India. Srivastava and Dutt (1963) found infective larvae of *S. assamensis* in *Musca conducens* feeding on hump sores of cattle in Assam and Orissa and Patnaik and Roy (1966) infected *M. conducens* experimentally. Patnaik (1973) noted that only female flies engorged while feeding on lesions. At 25.5°C, 23–25 days were required for development of infective larvae and their appearance in mouthparts of the vector. The first moult occurred 5–6 days and the second 13–14 days postinfection. Larvae then moved to the thoracic muscles and grew. In the head of the fly two sizes of larvae were found: a short one 0.75–0.97 mm and a long one 1.08–1.26 mm in length. In longer larvae, the genital primordium was at the end of the first third of the body while in the shorter larvae the genital primordium was in the middle third of the body (these were presumably sexual differences).

Minor surgical wounds were made on humps of four calves and infected flies were allowed to feed on these sites on two calves. One of the latter became infected, a small circular lesion appearing on the hump about 6 weeks postinfection.

Sultanov *et al.* (1979) and Kabilov (1980b) studied *S. assamensis* in Uzbekistan. They collected flies from sores on the hump of cattle and concluded that the vector and intermediate host was *Lyperosia titillans*. Larvae reached the infective stage in thoracic muscles in 21–24 days. Larvae were not found in other flies (*Musca domestica*, *Musca* spp., *Stomoxys calcitrans*) collected from sores. *L. titillans* is widely distributed in Uzbekistan and attacks mainly the upper parts of the body of cattle, including the head. The flies are most active from June to July.

S. kaeli Buckley, 1937

S. kaeli causes 'filarial sore' on the legs of cattle on the west coast of peninsular Malaysia. Fadzil (1973, 1975) obtained microfilariae from the lesions; some were enclosed in a sheath, others free of it. They were 109–117 µm in length, with conical tails. He allowed *Musca conducens* to ingest eggs from lesions and reported that complete development in the fat body was reached in 10 days at 26–30°C. The first moult occurred in 4 and the second in 8 days. Third-stage larvae from the proboscis of the vector were 0.98–1.32 mm in length.

S. stilesi Chitwood, 1934

S. stilesi causes a circumscribed dermatitis along the midventral line of the body of cattle in parts of continental USA, Hawaii and the CIS (Alicata, 1947; Gnedina, 1950a; Hibler, 1966). Dikmans (1934) suggested that the horn fly, *Haematobia irritans*, was possibly an intermediate host since, in the southwestern USA, it is a fairly host-specific pest of cattle and commonly found on stephanofilarial lesions. Ivashkin *et al.* (1963) found larvae in the European horn fly, *Lyperosia titillans*, similar to larvae of *S. stilesi* occurring in the skin of cattle. *H. irritans* is also an important vector of stephanofilariasis in the CIS (Ivashkin *et al.*, 1971).

Hibler (1964, 1966) finally elucidated in detail the development and transmission of *S. stilesi* in cattle in New Mexico. Lesions developed in skin along the mid-ventral line of the body between the brisket and navel when cattle were 8–10 months old. Lesions were raw and bloody or covered with serous exudate for the next 2–3 years (Dikmans, 1934, 1948; Maddy, 1955; Hibler, 1966). Adult worms occurred in the dermis 1–2 mm below the epidermis, and microfilariae occurred in the dermal papillae (Smith and Jones, 1957; Jensen and Mackey, 1965; Hibler, 1966).

Microfilariae were 45–60 μm long. The anterior end was blunt and had a digitiform process and the posterior end was conical. Each microfilaria was enclosed in a 'thick, semirigid, fluid-filled membrane that was oval in shape and 67 (58–72) μm long by 50 (42–55) μm thick' (Hibler, 1966); because of this structure microfilariae apparently cannot migrate from the site where they are deposited.

According to Hibler (1966) the intermediate host and vector of *S. stilesi* is the horn fly, *H. irritans*, the immature stages of which develop in manure. Both sexes fed on stephanofilarial lesions but the females more so than the males. In the horn fly, microfilariae invaded and developed in the abdominal haemocoel to the infective stage in 18–21 days at room temperature, the first moult occurring 8–10 days and the second 14–16 days postinfection. Larvae migrated to the head and proboscis of the fly 2–5 days after the second moult. They were 695–900 μm long with an elevated oral opening, a single ventral lateral spine near the oral opening, large amphids, and a rounded tail with delicate ventroterminal serrations.

Hibler (1966) allowed infected horn flies to feed on calves. The latter developed lesions after 2 weeks of daily exposure. In three calves lesions continued to increase in size. He allowed uninfected flies to feed on lesions. Nine of 62 female flies became infected but none of 122 male flies. Infective third-stage larvae grew in the dermis of cattle to a length of 950–1200 μm and moulted to the fourth stage. When larvae were 1.75–2.15 mm in length they moulted to the fifth stage.

According to Hibler (1966) cattle rarely became infected with *S. stilesi* until they were 8–10 months old, apparently because horn flies prefer to feed on more mature animals. Once an individual animal has become infected and has an established population of horn flies, the system becomes self-perpetuating. The flies flourish when manure is abundant, as on rangeland, where cattle may have more than 2000 flies each. Horn flies are relatively scarce in dry lots when manure is removed or scattered.

Hibler (1964) mentioned that Morgan (1964) reported that horn flies preferred temperatures of 22.7–26.6°C and that when temperatures rose above these levels flies moved to the shaded parts of the host, including the belly, where lesions were found.

S. zaheeri Singh, 1958

S. zaheeri is responsible for 'ear-sore' in India and Pakistan. Dutt (1970) collected numerous species of dipterans feeding on ear-sore lesions in India and examined them for larvae. Infective larvae (946–1097 μm in length) found in *Musca planiceps* were identified as *S. zaheeri*.

Family Onchocercidae

The Onchocercidae includes a diverse array of nematodes (about 70–80 genera in eight subfamilies). Unlike the Filariidae, larval stages are generally found in the definitive host distant from their site of origin in the gravid female. The vulva in the latter is, with few exceptions, in the anterior part of the body, a position which may facilitate oviposition. In developing the microfilaria, onchocercids have freed themselves from the food chain and the viscera of the definitive host and been able to radiate extensively throughout the tissues of all the vertebrate classes other than the fishes – where, incidentally, the tissue-inhabiting dracunculoids have flourished. Microfilaria-like larvae occur in blood of some highly specialized Dracunculoidea of fishes (e.g. *Ichthyofilaria*, *Molnaria*, *Skrjabillanus*) and one species is known to be transmitted by ectoparasitic, blood-sucking branchiurans. Also, microfilaria-like larvae occur in some Filariidae. However, it is only in the Onchocercidae that the microfilaria has attained its full potential as a key element in the transmission of a major group of nematodes. Thus, onchocercids have been reported from all the organ systems in the body and from most tissues. Each species has a preferred site in which it lives in the host and one of the mysteries is how they manage to find highly specific locations in the body. Microfilariae oviposited by female worms live in the blood or skin, where they are available to haematophagous vectors, which include most of the major arthropod groups known to suck the blood of higher vertebrates, i.e. biting midges, blackflies, fleas, horse and deer flies, mosquitoes, lice, louse flies, mites, and ticks.

Microfilariae develop in the uteri of the female worm. Surrounded by the egg membrane, which consists of a vitelline membrane and a thin chitinous shell (Christenson, 1950; McFadzean and Smiles, 1956; Rogers *et al.*, 1976; Fuhrman and Piessens, 1985; Zaman, 1987), the microfilaria may be long (up to 500 μm) and slender, or short (less than 100 μm) and stout. The anterior end is rounded; the tail is blunt and rounded, conical, attenuated or filamentous. The microfilariae of many species hatch *in utero* and are said to be unsheathed when they appear in the blood or skin of the host. Microfilariae of other species, however, retain the egg envelopes and are deposited into the tissues of the host as sheathed microfilariae; they are in a sense fully embryonated eggs which hatch (i.e. exsheath) only in the arthropod intermediate host. The sheath is usually stretched by the microfilaria. It may be delicate and closely applied to the body of the microfilaria or loose and extend well beyond the extremities of the body. In some species the sheath is oval and does not permit the microfilaria to extend itself fully. Such forms are generally found in the skin of the host and may not be able to move in capillaries and lymphatics. Not all microfilariae occurring in the skin are of this type, however. Studies of periodicity have shown that fully extended microfilariae, with or without sheaths, are capable of responding to physiological changes in the host and they may flood into the peripheral circulation at a certain time of the day or night and disappear from the peripheral circulation at other times.

Microfilariae have a unique internal anatomy which is a combination of primitive and specialized features. Certain structures normally found in the first-stage larva of Spirurida (e.g. a fully developed alimentary tract) do not exist in the microfilaria whereas other structures are similar to those found in the most fully developed first-stage larva of the Spirurida. Buckley (1955) noted that 'the positions of the primordia (see below) in relation to one another give the impression that the narrow elongate shape of the microfilaria is the result of it having been stretched out. Whatever may have been its origin, this character is of much importance for the survival of the microfilaria as it facilitates its passage through the fine capillaries and also through the narrow mouthparts and the gut wall of the arthropod vectors.'

Early workers including Manson (1906), Rodenwaldt (1908) and Fülleborn (1913), by skilled staining and optical microscopy, recognized a nerve ring, an excretory vesicle and cell, an inner body ('*innen Körper*') slightly behind the middle of the body, a large, deeply staining cell (G cell) behind the inner body, a row of three smaller interconnected cells (R2-4) staining like the G cell and connected to a clear area called the anal vesicle. The G and the R2-4 cells were thought to be the genital primordium. Fülleborn (1913) discerned, but did not identify, amphids ('*Mundgebilde*') and phasmids ('*Schwanzgebilde*') in specially stained (AzurII–Eosin) microfilariae. The various structures mentioned above (nerve ring, excretory vesicle, etc.) are used as fixed points in giving measurements of microfilariae.

More recent studies have confirmed many of the early observations and added substantially to our understanding of microfilarial structure (Fig. 7.10). A ventral cephalic hook seems to be characteristic of many, if not all, microfilariae. In addition, rows of minute spines behind the cephalic extremity have been observed in some microfilariae (Laurence and Simpson, 1968; McLaren, 1972) as well as eight minute cephalic papillae (McLaren, 1969, 1972) with nerve axons from the nerve ring to innervate their cilia. Cilia have also been found in the '*Mundgebilde*' and '*Schwanzgebilde*', confirming them as amphids and phasmids (Kozek, 1968, 1971; McLaren, 1972; Martinez-Palomo and Martinez-Baez, 1977). A hypodermis is present with muscle cells (McLaren, 1972). The oesophagus (pharynx) has been shown to be represented by a delicate cuticularized thread to which are attached elongated cells (Laurence and Simpson, 1971; McLaren, 1972; Singh *et al.*, 1974). The thread extends to the anterior region of the inner body ('*innen Körper*'). The inner body is a sac-like structure containing refractile granules or spheres (Laurence and Simpson, 1971) and has been suggested as a food reserve which can be charged by way of the oral opening and the oesophageal thread (McLaren, 1972; Singh *et al.*, 1974). The concept of the excretory system has not been changed significantly by more recent studies and its precise function is still unclear (Kozek, 1971; McLaren, 1972).

The function of the G and the R2-4 cells is still obscure although the former is unusually prominent in most microfilariae. Kanagasuntheram *et al.* (1974a,b) suggested a sensory role for the excretory cell and the R2-4 cells and a motor (neurosecretory) role for the G cell in the microfilaria. It seems well established that the R2-4 cells are eventually involved in the formation of the rectum during development in the intermediate host (Feng, 1936; Abe, 1937; Kobayashi, 1940; Anderson, 1956b; Schacher, 1962; Bain, 1972; Vincent *et al.*, 1979). Bain (1970c) and Petit (1981) concluded that

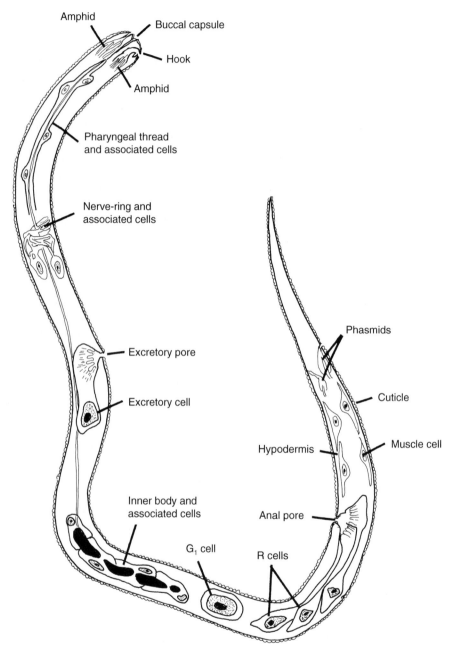

Fig. 7.10. Diagram of microfilaria. (After D.J. McLaren, 1972 – courtesy *Parasitology.*)

the G cell eventually formed the musculature of the adult nematode but this was considered doubtful by McLaren (1972).

Manson (1878) first showed that microfilariae were taken up in the blood meal of the arthropod intermediate host. In the latter, the microfilariae exsheathed, penetrated the gut wall, migrated to the haemocoel and developed to an infective stage in a certain tissue. Manson made his discovery when studying *Wuchereria bancrofti* in humans in Amoy, China. Since the time of Manson the development and transmission of 108 species in 37 genera and eight subfamilies has been investigated. Some species of medical importance have been the subject of a voluminous literature; knowledge of other species is sometimes limited to the observation that they will develop in certain arthropods.

Of the species studied in which the precise site of development in the intermediate host is indicated, about 32 develop in fat body tissues, 36 in muscles (usually the flight muscles), eight in the haemocoel and seven in the Malpighian tubules. Generally, species in the same genus develop in similar locations in the arthropod host. For example, most species of *Dirofilaria* prefer the Malpighian tubules and members of *Brugia* and *Wuchereria* the thoracic muscles.

In the intermediate host, microfilariae of most species shorten and thicken into the so-called sausage stage (see *Wuchereria bancrofti* below). Anderson (1957) suggested that the sausage stage was necessary to bring together the scattered primordial cells of the elongate microfilaria so that development could proceed normally. However, a few filarioids do not undergo a sausage stage during their development in the intermediate host (e.g. *Acanthocheilonema viteae*). During the sausage stage the G cell divides and the R2-4 cells coalesce near the anus to form the rectum and secrete an anal plug. The fully developed oesophagus, intestine and genital primordia make their appearance, and after continued growth, the larva undergoes the first moult. The second-stage larva becomes long and slender and eventually moults to the infective third stage. After a short period of further growth the latter moves to the haemocoel and migrates to the head and mouthparts of the arthropod. When an infected dipteran intermediate host is feeding, the proboscis is bent and larvae break out of the proboscis and wriggle on to the surface of the skin of the final host. If moisture conditions are satisfactory, larvae will survive and invade the puncture wound made by the vector and enter the tissues of the definitive host. Infective larvae in hard ticks which remain attached for prolonged periods apparently invade the salivary glands and are inoculated into the host.

Infective larvae vary from about 0.60 to 5.00 mm in length, depending on the species; the latter length is unusual, however. The oesophagus is generally long and divided into muscular and glandular parts. The genital primordia are poorly developed but sexual differences in their positions in the body are usually evident. The tail is short or long and generally blunt. The end of the tail often has rounded papillae-like elevations or tongue-like structures. Comparative studies of the morphology of infective larvae have been made by Nelson (1959), Yen *et al.* (1982) and Bain and Chabaud (1986). The latter contains a useful illustrated key to different infective filarioid larvae which have been described in the literature.

The Onchocercidae is divided into eight subfamilies. The Oswaldofilariinae is confined to reptiles and the Icosiellinae and Waltonellinae to amphibians. The Setariinae is confined to large mammals. The Dirofilariinae and Onchocercinae contain mainly mammal parasites although a few genera occur in reptiles and birds. The Splendidofilariinae and the Lemdaninae contain parasites of birds, reptiles and mammals; most of the bird filarioids occur in these two subfamilies.

Subfamily Oswaldofilariinae

Members of the Oswaldofilariinae are found in lacertilians and crocodilians and are distinguished morphologically from other onchocercids by the posterior position of the vulva, which is generally in the middle or posterior region of the body.

Conispiculum

C. flavescens (Castellani and Willey, 1905)

C. flavescens (syn. *C. guindiensis*) is a parasite of the connective tissues of the common garden lizard (*Calotes versicolor*) in India and Sri Lanka. Microfilariae were 86–100 μm in length and sheathed. Pandit *et al.* (1929) allowed *Culex fatigans* to feed on infected lizards. Microfilariae exsheathed in 18–24 h in the stomach of the mosquito. In 2 days, most larvae were present in thoracic muscles. Development was completed by 12 days and infective larvae were 1.00–1.25 mm in length. The authors claimed that young worms were found in four of 21 lizards exposed to mosquitoes 1–2 months previously but microfilariae were not detected in the blood.

Menon *et al.* (1944) inoculated larvae subcutaneously into the limbs of *C. versicolor* and followed their development. Larvae migrated to lymph vessels between muscle fibres. Later (in 16–21 days) they migrated into the pelvic tissues and from there to the mesentery, where they matured. They came to lie in a dilated lymphatic vessel between the layers of the peritoneum. Microfilariae usually appeared in the blood 41–72 days postinfection.

Oswaldofilaria

O. bacillaris (Molin, 1858)

This is a parasite of the thoracic muscles and lungs of *Caiman crocodilus, C. sclerops* and *Champsa nigra* (Crocodylidae) in Brazil. Microfilariae were short (84–117 μm in length) and sheathed; they had pointed tails, and occurred in blood. According to Prod'hon and Bain (1972) microfilariae developed in the fat body of *Anopheles stephensi*. At 23–26°C two moults occurred, in 8 and 14 days, and infective larvae appeared in the head in 13–14 days. The infective larva was 1.6 mm in length with a long divided oesophagus. The cuticle had numerous (45) longitudinal ridges and the tail was rounded with a terminal protuberance and two subterminal, lateroventral tongue-like structures. In female larvae the genital primordium was in the region of the glandular oesophagus and in male larvae it was in the region of the intestine.

O. belemensis Bain and Sulahian, 1975

This is a parasite of the heart, aorta and vena cava of the lizard *Dracaena guyanensis* in South America. Microfilariae, which occurred in blood, were 68–78 μm in length and unsheathed; the tail was rounded. According to Bain and Chabaud (1975) microfilariae developed to the infective stage in *Aedes aegypti*. The infective stage was 870 μm in length. Its cuticle had numerous longitudinal ridges and the tail was rounded with a pair of lateral, subterminal tongue-like structures.

O. chlamydosauri (Breinl, 1912)

This is a parasite of the subcutaneous tissue, body cavity and lungs of Agamidae (*Amphibolurus barbatus*, *A. muricatus* and *Chlamydosaurus kingi*) in Australia. Sheathed microfilariae occurring in the blood were 80–105 µm in length (Mackerras, 1953). Microfilariae developed in the abdominal fat body of *Culex annulirostris* and *C. fatigans* (see Johnston and Bancroft, 1920; Johnston and Mawson, 1943; Mackerras, 1953, 1962). According to Mackerras (1953) development to the infective stage was completed in 23 days at 19–32°C in *C. annulirostris*; infective larvae were 1.05–1.16 mm in length. Infective larvae were inoculated into two lizards and in one of them adult worms and microfilariae were found 203 days later. Another three lizards were inoculated with infective larvae and microfilariae appeared in the blood of all three about 4.5 months later.

O. petersi Bain and Sulahian, 1974

This is a parasite of the mesentery, intestinal wall and thigh muscles of *Tupinambis nigropunctatus* (Teiidae) in Brazil. Microfilariae occurred in the blood and were 66–72 µm in length, apparently unsheathed and with rounded tails (Bain and Sulahian, 1974). According to Bain and Chabaud (1975) microfilariae developed to the infective stage in the fat body of *Culex fatigans* and *C. pipiens* in 16 days at 25°C. The first moult occurred in 7 days and the second in 16 days. Infective larvae were 0.92–1.20 mm in length and had numerous longitudinal ridges; the tail was rounded with a pair of lateral, subterminal tongue-like structures.

O. spinosa Bain and Sulahian, 1974

This is a parasite of the armpit ('aisselle'), aponeuroses and, rarely, the body cavity of the lizard *Mabuya mabouia* (Scincidae) in Brazil. Sheathed microfilariae occurred in the blood and were 55–59 µm in length with attenuated tails. Microfilariae were said to develop in *Culex fatigans* and *C. pipiens* (see Bain and Chabaud, 1975). The infective larva was 980 µm in length and had numerous longitudinal ridges. The tail had two lateral, subterminal tongue-like structures.

Subfamily Icosiellinae

The subfamily consists of the single genus *Icosiella*, the species of which occur in the subcutaneous aponeuroses of amphibians. Sheathed microfilariae occur in the blood.

Icosiella

I. neglecta (Diesing, 1851)

I. neglecta has been found in frogs of the genera *Diploglossus* and *Rana* in Europe and North Africa. Desportes (1941, 1942) discovered that the biting midge *Forcipomyia velox* (Ceratopogonidae) fed naturally on the head of green frogs (*Rana esculanta*) in France and was a suitable vector of *I. neglecta*. Microfilariae developed in the thoracic muscles and in 22 days infective larvae appeared in the head. The lesion made by the fly was apparently necessary to allow infective larvae to enter the final host. *Sycorax silacea*

(Psychodidae) also fed on green frogs and a few contained larvae that were identified as belonging to *I. neglecta*.

Subfamily Waltonellinae

Members of the subfamily are widely distributed parasites of frogs and toads (Bufonidae, Leptodactyglidae, Racophoridae, Ranidae) (Bain and Prod'hon, 1974). Only three genera are recognized in the subfamily and the transmission of only a few species in the genus *Waltonella* has been studied.

Waltonella

Members of the genus are parasites of the body cavity and mesentery of frogs and toads. Microfilariae, which occur in the blood, are sheathed and develop to the infective stage in culicine mosquitoes, generally in the abdominal fat body.

W. brachyoptera (Wehr and Causey, 1939)

W. brachyoptera (syn. *Foleyella brachyoptera*) is a parasite of the abdominal mesentery of *Rana sphenocephala* in Florida. Microfilariae were about 120 μm in length, with tapered tails ending in a rounded point. Kotcher (1941) allowed *Culex quinquefasciatus* and *C. pipiens* to feed on infected frogs. Microfilariae were reported to develop in the body cavity. The first moult took place on day 9 and the second 13–16 days postinfection, after which larvae migrated to the head of the mosquito. Infective larvae were 700–886 μm in length.

W. flexicauda (Schacher and Crans, 1973)

W. flexicauda (syn. *Foleyella flexicauda*) is a parasite of the mesentery and body cavity of the bullfrog (*Rana catesbeiana*) in the eastern USA. Microfilariae were about 91 μm in length and their tails ended in a rounded point. *Culex territans*, which feeds predominantly on frogs (Crans, 1970), has been shown to be a suitable intermediate host (Benach and Crans, 1973). In the mosquito the exsheathed microfilariae invaded the fat body, where development took place. Second-stage larvae were found on day 7 at $24 \pm 1°C$ and by 13 days moulting second-stage larvae were observed as well as third-stage larvae. Third-stage larvae at 15 days were 0.84–1.01 mm in length and had migrated to the head of the mosquitoes. Bullfrogs were infected by allowing mosquitoes to feed on them. The prepatent period was 228–232 days. *Rana clamitans* and *R. pipiens* were refractory to infection.

W. ranae (Walton, 1929)

W. ranae (syn. *Foleyella ranae*) is a parasite of the mesentery (generally occurring in capsules) of *Rana catesbeiana* and *R. clamitans* in the USA. The microfilaria was 114 μm in length and the tail tapered to a rounded point. Microfilariae developed in the muscles of the abdomen or thorax of *Aedes aegypti*, *Culex quinquefasciatus* and *C. pipiens* (see Causey, 1939a,b,c; Kotcher, 1941). According to Kotcher (1941) the first moult occurred 9 days and the second 13–16 days postinfection at approximately 23°C. Infective larvae were 0.70–1.06 mm in length.

W. duboisi Witenberg and Gerichter, 1944 not Gedoelst, 1916

This species was found in lymphatic sinuses and the mesentery of *Rana esculenta ridibunda* in North Palestine by Witenberg and Gerichter (1944). Microfilariae had tapered posterior extremities and were 95–145 μm in length (on films). Microfilariae developed in the haemocoel of *Culex molestus* (see Witenberg and Gerichter, 1944). A moult (probably the second) was noted 11–12 days postinfection. Infective larvae (640–840 μm in length) were found in 14 days at 25–28°C. *C. molestus* was apparently rare in the locality where *W. duboisi* was found in frogs and probably some other culicid is the natural vector in Palestine.

W. sp. (in Bain and Prod'hon, 1974)

Bain and Prod'hon (1974) reported without details that an undescribed species of *Waltonella* in *Bufo marinus* in Guadeloupe developed in the fat body of *Aedes polynesiensis*. The infective larva was 530 μm in length (see also Bain and Chabaud, 1986).

Subfamily Setariinae

The subfamily includes *Setaria* with some 43 species found normally in the abdominal cavity of artiodactyls (especially Bovidae), hyracoids and equines. Species are medium to large in size and generally have rather complex cephalic structures in the form of spines, median or lateral cuticular elevations associated with the oral opening, or shield-like thickenings. Certain members of the genus have worldwide distributions in horses and cattle. Sheathed microfilariae occur in blood and known vectors are haematophagous insects. Unfortunately, little is known about the behaviour of *Setaria* spp. in the definitive host even though at least three species have been associated with the central nervous system or eyes of large mammals. Yeh (1959) and Desset (1966) described or redescribed many of the species in the genus *Setaria*, and Sonin (1977) provided an excellent review of the literature on the history, taxonomy and biology of the Setariinae.

Setaria

Weinmann and Shoho (1975) cited limited evidence that prenatal infection occurred in *Setaria yehi* of *Odocoileus hemionus* in California. Recently there is convincing evidence that *Setaria marshalli* is transmitted prenatally in cattle.

S. cervi (Rudolphi, 1819)

S. cervi (syn. *S. altaica*) is a parasite of *Alces alces*, *Capreolus capreolus*, *Cervus axis*, *C. dama*, *C. elaphus*, *C. nippon* and *Muntiacus muntjak* (Cervidae) in Europe and Asia. The mature parasites are generally found in the abdominal and thoracic cavities of the host but they also may be found in the central nervous system (Lubimov, 1945, 1948, 1959; Shol', 1964; Blažek *et al.*, 1968). Neural setariasis is often complicated by concurrent infections of *Elaphostrongylus cervi* (Metastrongyloidea) which also frequently invades the central nervous system and causes neurological disease in deer.

Microfilariae were 205–231 μm long. Osipov (1966b) determined that the vector of *S. cervi* in the southern Altai of the CIS was the stable fly *Haematobia stimulans*. Microfilariae exsheathed in the stomach of the fly within 90 min and invaded the haemocoel; after 48 h they were found in the fat body, where development took place. The second moult occurred 11 days postinfection. Larvae left the fat body in 12 days and grew in the haemocoel for an additional 5–11 days, when the infective stage was reached. Infective larvae were 1.65–2.32 mm in length and numerous tubercles were present on the tail end. In deer, the third moult took place when larvae were about 5.0 mm in length. Full development in the final host required 224–235 days and the longevity of the worms was 1.5 years. Shol' and Drobishchenko (1973) and Drobishchenko and Shol' (1978) compared the biology of *S. cervi* and *S. labiatopapillosa* and concluded that both were valid species.

S. digitata (Linstow, 1906)

S. digitata is a parasite of the abdominal cavity of Asian cattle. As noted by Yeh (1959), Shoho (1959) and Desset (1966), its distribution is strictly Asian (although it has been introduced to Mauritius). It frequently occurs with the cosmopolitan species *S. labiatopapillosa*.

The following mosquitoes have been reported as natural intermediate hosts: *Aedes togoi* (Japan: Itagaki and Taniaguchi, 1954); *Armigeres obturbans* (Sri Lanka: Shoho and Nair, 1960); *Aedes vitattus* and *Armigeres obturbans* (India: Varma *et al.*, 1971). Niles (1961) experimentally infected *Anopheles nigerrimus*.

Varma *et al.* (1971), following methods of Nelson (1962a), transferred adult *S. digitata* to rabbits and calves. *A. vittatus* and *A. obturbans* were allowed to feed on the blood of the inoculated animals. Microfilariae, which were 190 μm in length, exsheathed in the stomach of the mosquitoes, entered the haemocoel and migrated to the thoracic muscles in 4–5 h. Infective larvae, 1.95–2.52 mm in length, appeared in the mouthparts of the mosquitoes 11–13 days after the latter had fed.

Infections of *S. digitata* are evidently silent in cattle. However, immature stages of *S. digitata* invade the central nervous system of horses, sheep and goats. The larval worms inflict traumatic damage on the neuropil, resulting in neurological disease (Innes and Shoho, 1953; Ishii *et al.*, 1953; Innes and Pillai, 1955; Shoho and Nair, 1960). The factors responsible for the curious behaviour of *S. digitata* in unusual hosts are not understood but Anderson (1968a) wondered if *S. digitata* might be normally neurotropic in its usual host. Hagiwara *et al.* (1992) collected mosquitos in the Ibaratci Prefecture in central Japan which feed on cattle, sheep and goats and transmit *S. digitata*, which causes lumbar paralysis in the ovids. *Culex tritaeniorhynchus* was the most abundant, followed by *Culex pipiens pallens* and *Anopheles sinensis*. However, *A. sinenis* was the most abundant mosquito during the *Setaria* transmission seasons in June and July. Four of 312 *A. sinensis* had third-stage larvae of *S. digitata*. Only one of 408 *C. tritaeniorhynchus* had larvae.

S. equina (Abildgaard, 1789)

S. equina is essentially a parasite of the peritoneal cavity of equines (horses, mules, donkeys) although it has been reported from bovids and camelids. Heisch *et al.* (1959) first showed that this species developed in mosquitoes. Nelson (1959) confirmed this by feeding *Aedes aegypti* and *A. pembaensis* on an infected donkey and recovering and describing the infective stage. The parasite developed in the thoracic muscles of the mosquitoes. Infective larvae were 1.28–1.72 mm in length and the tail ended in a terminal papilla and two small subterminal papillae. Larvae identical to those described by Nelson were found in *Aedes* spp. on the island of Pate in Kenya.

Platonov (1966) established that *Aedes communis* and *A. maculatus* were intermediate hosts of *S. equina* in western CIS. Microfilariae appeared in thoracic muscles 10–12 h after ingestion. At 18.0–25.5°C, larvae became sausage-shaped in 5 days. The first moult occurred in 10–12 days postinfection, when larvae were 788 µm in length. The second moult took place 14–15 days postinfection, when larvae commenced to arrive in the mouthparts of the vector. Fully formed infective larvae were 1.37–1.77 mm in length.

S. javensis Vevera, 1923

This is a parasite of the body cavity of the mouse-deer (*Tragulus* sp.) (Tragulidae) in Malaysia. Yen *et al.* (1982) allowed *Aedes togoi* to feed on an infected deer and recovered third-stage larvae identified as *S. javensis* 13–16 days later in mosquitoes held at 27°C. Larvae were 1.55–1.96 mm in length. The tail was rounded, with a sharp point and two ventrolateral, knob-like papillae.

S. labiatopapillosa (Alessandrini, 1848)

This is a common cosmopolitan parasite of cattle. It has also been reported in buffalo, old world antelope, deer, sheep, goats, camels, pigs and horses (Sonin, 1977). Itagaki (1958, in Sonin, 1977) reported that *Aedes togoi*, *Anopheles hyrcanus* and *Armigeres obturbans* were suitable intermediate hosts of *Setaria* in cattle in Japan but there is some question as to the species he was studying. Nelson (1962a) used various techniques, including surrogate hosts, to infect *Aedes aegypti* with *S. labiatopapillosa* from cattle in Kenya. The infective stage was reached in 8–10 days at 23–24°C. There were two moults and most larvae were in the head and proboscis by 11 days. Infective larvae were 2.2 mm in length and the short tail ended in a single knob. Nelson (1962a) felt that, in Kenya, *Mansonia* spp. were the most likely natural vectors. Pietrobelli *et al.* (1998) reported that *Aedes caspius* was a natural vector in Italy.

Osipov (1966a) showed that in the western CIS the intermediate hosts of *S. labiatopapillosa* in cattle were *A. caspius*, *A. flavescens*, *A. cinereus*, *A. venax* and *Anopheles bifurcatus*; the first two were the most important. The infective stage was reached in 16–24 days at 19–26°C. On entering the skin of the final host, larvae underwent a tissue migration to the peritoneal cavity, where they matured in 6 months.

Bain (1970a) studied the development in detail in *A. aegypti* at 26–28°C. Microfilariae, which were 240–310 µm in length, invaded the thoracic muscles. The first moult took place in 4.5 days and the second in 9.5 days after mosquitoes had ingested microfilariae. At the final moult, larvae were 1.8 mm in length. Further growth occurred after the second moult and infective larvae in the mouthparts and head of the mosquitoes were 2.2–2.6 mm in length. The tail of the infective larva ended in a blunt point and there were two small lateral, subterminal papillae-like structures.

Setaria marshalli (Boulinger, 1921)

S. marshalli is a parasite of cattle. Fujii *et al.* (1995) examined 65 bovine fetuses 2–9 months old. *S. marshalli* (39–55 mm in length) was found in the peritoneal cavity of some fetuses 7–9 months old during October to December but not in fetuses during January–September. Microfilariae were noted in the blood of two fetuses. No worms were found in any of the dams. *S. marshalli* has not been detected in cattle older than 2 years and so prenatal infection is probably the common type of transmission, while the postnatal type is uncommon. The most likely time for prenatal infection is June–August, when mosquitos are active and a fetus is 4–5 months of age.

Wee *et al.* (1996) reported 50 *S. marshalli* from the peritoneal cavity of three neonatal calves. Males were 41–52 mm and females 68–98 mm in length. See also Shoho (1965).

Subfamily Dirofilariinae

Representatives of the subfamily (with ten genera) occur in reptiles (one genus only), birds (one genus only) and mammals. Males have highly developed caudal alae which distinguish them from most other members of the Onchocercidae.

Dirofilaria

Members of the genus which have been studied generally occur in the subcutaneous tissues of primates and carnivores (*D. immitis* occurs in the heart). Microfilariae, which occur in the blood, are long with tapered tails and unsheathed. Intermediate hosts are usually mosquitoes (*D. ursi* is transmitted by simuliids). Development generally takes place in the Malpighian tubules (*D. corynodes* develops in the fat body).

D. corynodes (Linstow, 1899)

D. corynodes (syns *D. aethiops*, *D. schoutendeni*) is a parasite of the subcutaneous tissues of African monkeys (Cercopithecidae) (e.g. *Cercopithecus aethiops*, *Colobus* sp., *Erythrocebus patas*). Microfilariae were 256–288 µm in length and markedly nocturnal in the peripheral blood (Hawking and Webber, 1955; Orihel, 1969). Development in mosquitoes was first demonstrated by Webber (1955) who found that microfilariae would develop in *Aedes aegypti* and *Anopheles maculipennis* to the infective stage in 18–20 days at 26°C; the former species seemed most suitable for development. Development took place in 'connective tissue in all parts of the... body' (presumably the fat body). A sausage stage was present and the first moult occurred about 12 days and the second about 16 days postinfection. Third-stage larvae were found in the proboscis of the mosquitoes in 18–20 days. Infective larvae were 0.70–0.80 mm in length. Hawking and Webber (1955) inoculated several monkeys with infective larvae from *A. aegypti*. Six *C. aethiops* and one *Macaca mulatta* were infected successfully. The prepatent period was about 49–51 weeks.

Nelson (1959) described infective larvae in *A. aegypti* and *Aedes pembaensis* which had fed on an infected monkey. He gave the length of the infective larvae as 0.68–1.00 mm. The tail was 'cigar-shaped' and three small papillae were on the terminal end.

Orihel (1969) found microfilariae of *D. corynodes* in patas monkeys (*E. patas*) from northern Nigeria and studied development in *A. aegypti* and monkeys. Microfilariae, which were 250–288 μm in length and nocturnally periodic in the blood, developed in the fat body of *A. aegypti*. At 28°C the first moult occurred in 9 days and the second about 14 days postinfection. Infective larvae were approximately 1.1 mm in length and could be sexed by the position of the genital primordia. Infective larvae were inoculated subcutaneously into monkeys. All three patas monkeys and four of seven rhesus monkeys (*M. mulatta*) became infected. The prepatent period was 260–290 days.

D. immitis (Leidy, 1856)

D. immitis occurs in the right ventricle, pulmonary artery, right atrium and the vena cava mainly of dogs. It has also been reported in wolves, dingoes, coyotes, foxes, various felines (including the domestic cat), sea-lions, harbour seals, mustelids, bears, pandas, beavers, coatis, rabbits, deer, horses, non-human primates and humans (Abraham, 1988). Most of these infections were sporadic and few exhibited microfilariaemias other than those in canids, cats, California sea-lions and some of the mustelids (e.g. ferrets). *D. immitis* has elicited a voluminous literature because of its importance as a dog pathogen and its use as a model for the study of various aspects of filariasis. For a recent review of dirofilariasis, refer to Boreham and Atwell (1988).

Noè (1900, 1901) and Grassi and Noè (1900) showed that microfilariae of this filarioid developed in the Malpighian tubules of *Anopheles maculipennis*. Bancroft (1901, 1903) reported that infective larvae left the proboscis during blood-sucking and he successfully transmitted *D. immitis* by allowing mosquitoes to feed on a puppy; microfilariae appeared in the blood 9 months later and adult worms were found in the heart. Fülleborn (1908) concluded, as a result of experiments, that larvae entered the lesions made by the feeding mosquitoes. Fülleborn (1929) and Feng (1937) inoculated dogs with infective larvae and reported a prepatent period of about 8 months.

The literature was considerably confused by reports that *D. immitis* developed in fleas (Breinl, 1921; Brown and Sheldon, 1940; Summers, 1940, 1943; Bradley, 1952; Stueben, 1954a,b). It is now known that these reports refer to *Acanthocheilonema reconditum*, a subcutaneous parasite of dogs which frequently coexists in this host with heartworm (Rosen, 1954; Newton and Wright, 1957).

Microfilariae were 258 ± 7 μm (SD) in length according to Webber and Hawking (1955) and 300 ± 15 μm (SD) according to Taylor (1960). They were most numerous in the blood at about 1800 h and least numerous at about 0600 h in England (Webber and Hawking, 1955). Microfilariae have been shown to develop to the infective stage in a great variety of mosquitoes of the genera *Aedes*, *Anopheles*, *Culex* and *Mansonia* (for reviews see Kartman, 1957; Lok, 1988). Most reports are the results of feeding mosquitoes on infected dogs and it is often not clear which serve as vectors under natural conditions. Rosen (1954) believed that *Culex annulirostris* and *A. polynesiensis* were natural vectors of heartworm in French Oceania. Symes (1960) found larvae in naturally infected *A. fijiensis*, *A. polynesiensis*, *A. pseudoscutellaris*, *Culex annulirostris* and *C.*

fatigans in Fiji. Christenson (1977) provided strong evidence that *Aedes trivittatus* was a natural vector in Iowa.

Taylor (1960) followed development in *Aedes aegypti* at 26°C and reported a sausage stage in 3–4 days, the first moult in about 10 days and the second in about 13 days. Infective larvae from the head of the mosquito were an average of 1.30 mm in length. Kartman (1953) noted that larvae grew somewhat after attaining the third stage, e.g. from 0.75 to 1.09 mm. Yen *et al.* (1982) gave the length of the infective larvae as 0.82–1.12 mm and stated that the short, cigar-shaped tail had three variable, indistinct, knob-like papillae. Christenson (1977) followed development in the natural vector *A. trivittatus* in Iowa at 26.5 ± 1°C. The first moult occurred 7–8 days and the second 10–11 days postinfection. Lichtenfels *et al.* (1985) reported that infective larvae from *A. aegypti* were 0.70–1.00 mm in length and had two small swellings near the conical tail end. Sexes were distinguished by the position of the genital primordia; spicular primordia were present in males.

Kume and Itagaki (1955) infected dogs and examined them at various times postinfection for developing *D. immitis*. Development occurred in subcutaneous tissues and the muscles. Worms migrated to the heart apparently via the veins between 85 and 120 days postinfection, when they were 3.2–11.0 cm in length.

Orihel (1961) inoculated dogs with infective larvae and examined them 5–278 days later for developing worms. Larvae were found in the subcutaneous tissues and muscles for the first 80 days, from the heart and the tissues at 67–80 days and from the heart after 90 days. The third and fourth moults were noted in the tissues 9–12 and 60–70 days postinfection, respectively. After the final moult, juvenile worms migrated to the right ventricle of the heart and matured over a period of several months.

Kotani and Powers (1982) inoculated 36 beagles and examined them 3–196 days later. On day 3, 93% of larvae were from the site of inoculation and 5% from the abdomen. On day 21, 87% of larvae were from the abdomen and 8% from the thorax. After day 21, larvae in the abdomen decreased to 46%; by day 41, 41% of larvae were from the thorax. Young adults were first found in the heart and pulmonary arteries on day 70.

Campbell and Blair (1978) and Campbell *et al.* (1979) showed that ferrets (*Mustela putorius*) were suitable hosts of *D. immitis*. Supakorndej *et al.* (1994) infected female ferrets each with 60 infective larvae inoculated subcutaneously and examined them at intervals of 3–140 days. Most worms were found in the subcutaneous tissues up to and including day 91. Worms, however, first appeared in the heart on day 70 (i.e. 3.8%). By 119 and 140 days most worms had reached the heart (i.e. 59% and 65.8% respectively). The third and fourth moults occurred at 3 and 56 days, respectively. On day 91 after worms had reached the heart, the length of males was 43–75 mm and females 55–69 mm. By 140 days, males from the heart were 62–146 mm and females 105–168 mm in length.

The precise route taken to reach the heart is still not fully understood. Third- and fourth-stage larvae occur in muscle fibres during their movements whereas adult worms seem able to penetrate muscles and could invade veins to reach the heart or penetrate directly into the heart muscle. According to Hayasaki (1996) young fifth-stage *D. immitis* recovered from pulmonary arteries of dogs migrated to the pulmonary arteries after they had been transplanted into the subcutaneous tissues of uninfected dogs, cats and rabbits. The mean percentage recoveries of the transplanted worms were

45.3% in dogs, 60.9% in cats and 18% in rabbits. Female worms from four transplanted dogs produced microfilariae.

Lichtenfels *et al.* (1985) described various stages collected from experimentally infected dogs and reported that the third moult occurred within 3 days and that the fourth-stage larvae were 1.50 mm in length 3–6 days postinfection. The fourth moult occurred 50–58 days postinfection when worms were 12.0–14.8 mm in length. Palmer *et al.* (1986) and Lichtenfels *et al.* (1987) described the fine structure of the cuticle and body wall of developing *D. immitis*.

The prepatent period of *D. immitis* was reported as about 6–9 months (Bancroft, 1904; Webber and Hawking, 1955; Newton, 1957; Orihel, 1961). Christenson (1977) allowed infected *A. trivittatus* to engorge on a dog, which developed a patent infection about 210 days later.

Genchi *et al.* (1995) infected beagles with *D. immitis* and *D. repens* to determine if there was any interaction between them. The comparison between cross-infections revealed a lower intensity of heartworm in dogs first infected with *D. repens*. This apparent interaction may explain why *D. immitis* is rare in southern Italy, where *D. repens* is common. There might be a similar explanation for the failure of *D. repens* to spread in North America, where heartworm is common.

D. magnilarvatum Price, 1959

This species was found by Price (1959) in the subcutaneous connective tissues of *Macaca irus* in Malaysia. It was distinguished by the remarkable size of its microfilaria (580 ± 10 μm) which was described in detail by Taylor (1959). Hawking (1959) reported that microfilariae showed no periodicity and tended to accumulate in arterioles especially of the skin, including that of the tail. Wharton (1959) showed that microfilariae developed in *Mansonia longipalpis* and *M. uniformis*, the third stage being reached about 10.5 days postinfection (temperature not specified). Infective larvae were 0.96–1.09 mm in length; the tail ended in a cone-shaped structure and had a pair of lateroterminal papilla-like structures. Yen *et al.* (1982) described infective larvae of *D. magnilarvatum* as 0.74–1.01 mm in length with a group of 'three papillae at the end of the cigar-shaped tail'.

D. repens Railliet and Henry, 1911

D. repens is a common parasite of the subcutaneous connective tissues mainly of dogs in southern Europe, Africa, southeast Asia and the CIS. There are reports from domestic cats, genet cats (*Genetta tigrina*), lions and foxes. Sporadic cases have been reported in humans (Faust, 1957; O'Grady *et al.*, 1962); see also Genchi *et al.* (1995).

Microfilariae were about 290 ± 20 μm in length and most numerous in blood about midnight and least numerous about midday (Webber and Hawking, 1955). Fülleborn (1908) discovered that microfilariae developed in *Anopheles maculipennis*; Bernard and Bauche (1913) infected *Aedes fasciata* and noted that at 31–35°C the infective stage was reached in 9 days. Webber and Hawking (1955) infected *Aedes aegypti*, *Anopheles maculipennis* and *A. stephensi* and at 24–27°C infective larvae appeared in the proboscis in 10–11 days. Gunewardene (1956) reported development in *A. aegypti*, *A. albopictus*, *Armigeres obturbans*, *Mansonia annulifera* and *M. uniformis* in Sri Lanka. Nair *et al.* (1961) infected *A. albopictus* in India and Nelson *et al.* (1962) infected *A. aegypti*, *A. pembaensis*, *Mansonia africanus* and *M. uniformis* in Africa.

Coluzzi (1964) in Italy, infected *Aedes geniculatus, A. vexans, Anopheles atroparvus, A. claviger, A. maculipennis, A. petragnanii* and *Culex pipiens* and claimed that microfilariae also developed to the infective stage in the tabanid *Haematopota variegata*. Mantovani and Restani (1965) believed that *A. petragnanii* was the most important vector and that *A. atroparvus* and *A. maculipennis* were of minor importance as vectors in Italy.

Bain (1978) infected *Aedes caspius* and *A. detritus* in France and described the developmental stages. Microfilariae were said to be 370 μm in length with vital staining. At 27°C the first moult occurred in 7 days and the infective stage appeared in 13 days. At 22°C infective larvae appeared in 20 days.

Gunewardene (1956) gave the length of the infective larvae as 0.90–1.01 mm; Nelson (1959) gave their length as 0.64–1.08 mm and noted that larvae had a simple terminal swelling and two minute subcutaneous papillae on the ventral surface near the tip of the tail. Yen *et al.* (1982) stated that infective larvae were 0.79–1.02 mm in length.

According to Webber and Hawking (1955), who inoculated dogs subcutaneously with infective larvae, the prepatent period was 27–34 weeks.

D. striata (Molin, 1858)

D. striata is a parasite of the subcutaneous connective tissues and intermuscular fascia mainly of the thigh muscles of *Felis concolor, F. macroura, F. pardalis, F. tigrina* and *Lynx rufus* (Felidae) in North and South America. Microfilariae occurred in the blood and were 235–270 μm in length (Anderson and Diaz-Ungria, 1959).

Orihel and Ash (1964) found this species in bobcats (*Lynx rufus*) in Louisiana. Microfilariae on thick blood films were 230–240 μm in length; they invaded the Malpighian tubules of *Anopheles quadrimaculatus* and developed, the first moult occurring in about 7 days and the second in 11–12 days postinfection at 26–27°C. Infective larvae, which appeared in the head and mouthparts in 14 days, were 0.95–1.14 mm in length. Infective larvae were injected into kittens and a puppy but the parasite did not establish itself in these animals.

D. tenuis Chandler, 1942

D. tenuis is a parasite of the subcutaneous tissues in various regions of the body of raccoons (*Procyon lotor*) in the USA. Microfilariae occurred in the blood. Pistey (1958) allowed a variety (16 species) of mosquitoes to feed on infected raccoons. Development to the infective stage took place in the Malpighian tubules of *Aedes taeniorhynchus* and *Anopheles quadrimaculatus*. Infective larvae were also found in one of 30 *Psorophora confinnis* and one of seven *Aedes sollicitans* which fed on an infected raccoon. The first moult occurred in 7–8 days and the second in 9–10 days postinfection at ambient temperatures; infective larvae appeared in the mouthparts in 10–12 days and were of two sizes: 0.78–0.99 mm and 1.00–1.15 mm. The tail was short and rounded. Two young raccoons were exposed to the bites of infected mosquitoes but microfilariae did not appear in their blood. Two female worms lacking microfilariae and two small males were found at necropsy of one raccoon 14 months postinfection. Other attempts to infect raccoons were inconclusive and attempts to infect dogs failed.

D. ursi Yamaguti, 1941

D. ursi is a common subcutaneous parasite of Ursidae in Japan, the CIS and North America. Microfilariae occurred in blood and were 185–292 µm in length with attenuated, filamentous tails (Anderson, 1952). Addison (1980) studied biting flies attracted to immobilized black bears (*Ursus americanus*) in Ontario and determined that the vector of *D. ursi* was the blackfly *Simulium venustum* (Simuliidae); microfilariae failed to develop in culicids, tabanids and ceratopogonids which also fed on bears. Ingested microfilariae apparently moved directly from the midgut of the blackfly to the Malpighian tubules, where development took place. At 23°C the first moult occurred 7–8 days and the second 9–11 days postinfection; infective larvae first appeared in the head in 10 days. At 27°C larvae appeared in the head in 5 days. Infective larvae were about 650 µm in length and the tail had one terminal and two subventral papillae. Larvae were inoculated into young uninfected bears and the prepatent period was 210–271 days.

Foleyella

Members of the genus occur in the subcutaneous and intermuscular connective tissues and body cavity of agamid and chameleonid reptiles. Short microfilariae with loose sheaths occur in the blood of the host and development takes place in the fat body of mosquitoes (Bartlett, 1986).

F. candezei (Fraipont, 1882)

This is a parasite of the subcutaneous tissues of chameleons, including *Agama colonorum* and *Uromastix acanthinurus*. Microfilariae were 83–96 µm in length with tapered rounded tails. Bain (1970b) allowed *Anopheles stephensi* to feed on an infected *Chameleo senegalensis* from the Upper Volta in Africa. Microfilariae developed in the fat body of *A. stephensi*. At 24–26°C the first moult occurred 6 days and the second commenced 9 days postinfection, when larvae were 420–540 µm in length. The cuticle of the first moult was retained throughout the second stage. The infective stage was reached by day 10 and was 750 µm in length.

F. furcata (Linstow, 1899)

F. furcata is a parasite of the subcutaneous tissues and body cavity of chameleons in Madagascar. Microfilariae were 125–157 µm in length. Brygoo (1960) reported that *F. furcata* in *Chamaeleo oustaleti* developed in *Culex pipiens fatigans*. At 12–20°C the infective stage was reached in 20 days. Bain (1969a) reported that this species developed in the fat body of experimentally infected *Anopheles stephensi* which had fed on an infected *Chameleo verrucosus*. At 24–26°C the first moult took place in 6 days but the cuticle was not shed. The second moult took place when larvae were 400–600 µm in length. Infective larvae observed on day 10 were 780–1250 µm in length. The anterior part of the infective larva was wider than the body behind it and the tail had a terminal lobe and a pair of ventrolateral papilla-like structures.

F. philistinae Schacher and Khalil, 1967

This is a parasite of the subcutaneous tissues and muscle fascia of the lizard *Agama stellio* in Lebanon. Fixed and stained microfilariae were on average 108 μm in length with a tapered but blunt tail. Schacher and Khalil (1968) showed that *F. philistinae* developed in the fat body of *Culex molestus*. At 26°C the first moult took place in 7 days. Infective larvae appeared as early as 8 days but were most common 12–13 days postinfection. Infective larvae were 525–975 μm in length, with elongate buccal capsules and rounded tails with a terminal bulb and two ventrolateral papilla-like structures.

Loa

L. loa (Cobbold, 1864)

This species occurs in the subcutaneous connective tissues of humans in Africa, where it is confined to the tropical rain forest and gallery forest of West Africa, the Congo basin and, to a lesser extent, the southern Sudan and Uganda (Nelson, 1965). Adult worms (3–7 cm in length) wander extensively in the subcutaneous tissues and elicit transient, oedematous swellings on the body known as Calibar swellings (named after Calibar, where loiasis is common). In addition, the parasite commonly passes across the surface of the eye under the conjunctiva during peregrinations in the body. Microfilariae (250–300 μm long) appear in the blood, are sheathed with long attenuated tails and exhibit a marked diurnal periodicity in the blood.

Leiper (1913), following Manson's suggestions, implicated *Chrysops dimidiata* and *C. silacea* as vectors of *L. loa* in Nigeria. Kleine (1915) provided more convincing evidence that *C. dimidiata* and *C. silacea* were intermediate hosts in Cameroon; he correctly noted that development occurred in the fat body and that infective larvae left the fat body and migrated to the head, where he saw them emerge from the proboscis. Connal and Connal (1921) dissected wild flies in Nigeria and also found larvae in *C. dimidiata* and *C. silacea*. This was followed by much more detailed studies in Calibar by Connal and Connal (1922) who outlined development in flies and the behaviour of the latter as vectors.

Lavoipierre (1958) showed histologically that *L. loa* developed in fat body tissue of the head and thorax but the fat body of the abdomen was the main site. Soon after being ingested, microfilariae penetrated oenocyte-like cells of the fat body and Lavoipierre concluded that the parasite was initially intracellular. By 3 days larvae had outgrown the confines of the cells harbouring them. Fat body tissues had become somewhat disorganized and places with larvae appeared syncytial. By 10 days most larvae had left the fat body and migrated in the haemocoel to the head and mouthparts.

Williams (1960) studied larval development in *C. silacea*. The first moult occurred 3–4 days and the second 6 days postinfection at 30°C. In six days larvae were an average 1.70 mm in length. By day 7 infective larvae were 1.34–2.30 mm in length and many were found in the head. The tip of the tail ended in a digitiform process, two ventrolateral papillae, and a pair of subterminal papillae (phasmids?) near the bases of the latter.

Various authors (Kleine, 1915; Connal and Connal, 1922; Gordon and Crewe, 1953) observed infective larvae of *L. loa* leaving the mouthparts of *Chrysops* spp. Lavoipierre (1958) felt that most larvae left the mouthparts of the fly by rupturing the labiohypopharyngeal membrane. This occurred mainly when the labium was kinked during the actual feeding process. Gordon and Crewe (1953) had noted earlier that no larvae left the mouthparts until the labium was bent and then the larvae emerged 'with monotonous regularity'. It was concluded, therefore, that the mechanism of uptake of blood was necessary for the release of larvae from the head of the fly. Gordon (1948) and Gordon and Crewe (1953) showed that infective larvae of *L. loa* cannot penetrate intact skin and are dependent on the presence of the lesion made by the fly for their entry into the body of the final host.

C. *dimidiata* and *C. silacea*, the main vectors of *L. loa*, are daytime feeders and the biting population consists of females. *C. zahrai* also bites humans and can serve as an intermediate host but is considered less important than *C. dimidiata* and *C. silacea* (see Duke, 1954; Gordon, 1955). *Chrysops* spp. lay their eggs in wet areas, preferably 'sluggishly moving water impeded by rotted vegetation' (Gordon, 1955). Larvae of the flies are found in mud a few centimetres from the surface (Chwatt *et al.*, 1948). For details on the biting behaviour of *Chrysops* spp., refer to Duke (1958).

Gordon *et al.* (1950) found in Cameroon a type of *Loa* in monkeys (*Cercopithecus mona*, *C. nictitans* and *Madrillus leucophaeus*) the microfilariae of which, it was subsequently discovered (Duke, 1958), have a nocturnal periodicity in contrast with the human strain. The monkey form was transmitted by *Chrysops centurionis* and *C. langi*, which fed on monkeys at night in the canopy (Duke, 1958, 1972; Duke and Wijers, 1958). Duke and Wijers (1958) transmitted the human form to drills (*M. leucophaeus*) and after 4–5 months discovered microfilariae in the blood which exhibited a diurnal periodicity, as in humans. Duke (1964) then hybridized the two strains in monkeys and concluded that they were segregating on simple Mendelian lines with respect to microfilarial periodicity and body size (the simian strain is larger than the human strain). Duke (1972), in a review of his work, concluded that there was only limited transfer of the two strains of *L. loa* between monkeys and humans.

Orihel and Moore (1975) successfully infected the Afro-Asian primates *Erythrocebus patas* (patas monkeys) and *Papio anubis* (baboon) with *L. loa* larvae from *C. silacea* which had fed on humans. In the monkeys the prepatent period was 135–148 days and microfilariae exhibited a diurnal periodicity. Orihel and Lowrie (1975) also successfully infected *Chrysops atlanticus* from North America. Microfilariae reached the fat body within 30 min after they had been taken up in blood from infected monkeys. At 26.6°C the first moult occurred in 5 days and the second in 9 days. Nine days later infective larvae were found free in various regions of the fly, including the mouthparts. Infective larvae were about 2.0 mm in length.

Eberhard and Orihel (1981) obtained larvae of *L. loa* from *C. atlanticus* and *C. silacea* which had engorged on infected humans or experimentally infected baboons and patas monkeys. Larvae were inoculated subcutaneously into the lateral abdominal wall of primates and the latter were examined for various stages of *L. loa* 10–1100 days postinfection. The third moult took place about 18 days postinfection, when larvae were 2.4–2.6 mm in length. By 30 days the fourth-stage male larvae were 4.4 mm and female 5.3 mm in length. The final moult to the subadult stage took place in about 50 days. Inseminated females observed at 90 days were 23–30 mm in length; males at this time were 16–20 mm in length. Thereafter, males grew only slightly and were 2.7 cm in length whereas females grew markedly and reached a maximum of 6.4 cm. The prepatent period was 120–150 days.

Bain *et al.* (1998) injected infective larvae of *L. loa* into mice and jirds. The third moult took place in the rodents on day 8 postinfection. The immature adult stage was attained in 25 days when the worms were 3–3.5 mm in length. The fourth moult apparently occurred some time between 19 and 22 days.

Loaina

L. uniformis (Price, 1957)

L. uniformis (syn. *Dirofilaria uniformis*) is a parasite of the subcutaneous connective tissues of the cottontail rabbit (*Sylvilagus floridanus*) in the USA (Bartlett, 1983). The sheathed microfilaria occurred in the blood and was 285 μm in length with an attenuated tail (Price, 1957; Eberhard and Orihel, 1984). Bray and Walton (1961), Duxbury *et al.* (1961) and Price *et al.* (1963) showed that *Anopheles quadrimaculatus* was a suitable intermediate host. Microfilariae shed their sheaths in the stomach of the mosquitoes and entered the abdominal haemocoel. They reached the sausage stage in 72 h at about 24°C. The first moult occurred in 4–5 days and the second in about 7–8 days. Third-stage larvae left the abdomen and reached the head and mouthparts of the mosquitoes in 9 days. Development did not occur in *Aedes* spp. or *Culex pipiens*.

Bray and Walton (1961) inoculated infective larvae into ten wild cottontail rabbits and microfilariae appeared in the blood 97–222 days later. Eight laboratory rabbits (*Oryctolagus cuniculus*) were infected by inoculation of infective larvae. The prepatent period in these animals was 101–136 days but there was evidence that the laboratory rabbit was a poor host (only 1% of larvae inoculated were recovered as adults).

Microfilariaemias increased in intensity for several months after the prepatent period and infections could be retained for more than 2 years in wild rabbits. Microfilariae tended to be most numerous in the blood between 1600 and 2400 h. Duxbury *et al.* (1961) reported that the optimum temperature for development in *Anopheles quadrimaculatus* was 27°C and that more and longer larvae were found in older mosquitoes than in younger mosquitoes.

Pelecitus

Microfilariae of *Pelecitus* spp. are usually fairly long, with tight-fitting sheaths. Adult worms usually live among tendons and muscles near joints of the legs and feet of birds and mammals (Bartlett and Greiner, 1986; Bartlett and Anderson, 1987a,b). *Pelecitus* is the only filarioid genus which contains species in both birds and mammals. Bartlett and Greiner (1986) concluded that *P. scapiceps* of North American lagomorphs and *P. roemeri* of Australian marsupials belong to the genus and suggested that both species are captures from birds.

P. fulicaeatrae (Diesing, 1861)

This parasite occurs among tendons near the ankle of various species of birds (Bartlett and Greiner, 1986) and is common in coots (*Fulica americana*) and red-necked grebes (*Podiceps grisegena*) in western Canada (Bartlett and Anderson, 1987a). Bartlett and Anderson (1989a) considered the parasite from coots as the subspecies *P. f. americanae* and the parasite from red-necked grebes as *P. f. grisegenae*. Microfilariae were about 92–122 μm in length in coots and 94–115 μm in grebes and occurred in skin of the feathered portions of the lower leg. They were common in the dermis around feather follicles. Microfilariae were also abundant in fluid around gravid female worms. The microfilaria was surrounded by a delicate, loose-fitting sheath and had a pointed tail.

Bartlett and Anderson (1987a) found *Pseudomenopon pilosum* (Mallophaga: Amblycera) on 85% of wild-caught coots and *Pseudomenopon dolium* on all wild-caught juvenile red-necked grebes examined. They reported microfilariae as well as developing stages in lice from coots. They found infective third-stage larvae of *P. fulicaeatrae* in 9.5% of female lice and 3.5% of male lice taken from naturally infected coots maintained in captivity; intensity was 1.4 in female lice and 1.2 in male lice. Bartlett and Anderson (1989b) found infective third-stage larvae in 2.3% of female lice and in no male lice taken from two of five naturally infected coots collected in the wild; intensity was 1. Bartlett and Anderson (1987a) examined adult lice from one captive coot for all stages, including microfilariae, and 31.9% of females and 15.7% of males were infected; about two-thirds contained third-stage larvae. Microfilariae were found within the crop and free in the abdominal haemocoel of the lice. Other larvae were found in the abdomen, apparently free. However, a few first- and second-stage larvae were found in the fat body and some third-stage larvae were in delicate, clear sacs presumed to be remnants of fat body. Infective male larvae were 580–689 μm and female larvae 610–745 μm in length. The tail was short and terminated by a rounded lobe and a pair of terminolateral lobes. Third-stage larvae were found only in adult *P. pilosum*, although microfilariae and developing larvae were found in nymphs. This suggested that development to the third stage was physiologically related to louse stage.

Bartlett and Anderson (1989b) reported some preliminary observations on the biology of *P. pilosum* on coots. *P. pilosum* were highly mobile and probably transferred from adult coots to coot chicks shortly after the latter hatched. *P. pilosum* probably acquired microfilariae when feeding on the legs of coots, especially around the neck ('collar') of the feather follicles, where microfilariae were common.

Bartlett and Anderson (1989a) recovered adult *P. fulicaeatrae* from laboratory-reared coots inoculated with infective larvae and from laboratory-reared coots housed with wild-caught coots infected with *P. fulicaeatrae* and infested with lice. At 20 days postinfection, immature adult worms were present in the ankles of experimentally infected coots; microfilariae were found in fluid adjacent to adult worms 210–265 days postinfection.

Bartlett (1993) pointed out that *P. fulicaetrae*, after producing microfilariae, undergoes **reproductive senescence** (rather than death) in the leg joints of coots. This ensures that the host does not experience tenosynovitis or arthritis, which could occur if the worms died. Coots must run vigorously on water to become airborne and health of their legs is of paramount importance to their survival.

P. roemeri (Linstow, 1905)

P. roemeri (syn. *Dirofilaria roemeri*) is a common parasite of the subcutaneous and intermuscular tissues of the knee region of certain Australian marsupials, namely *Macropus giganteus* (eastern grey kangaroo), *M. robustus* (eastern wallaroo), *M. rufogrisea* (red-necked wallaby) and *Megaleia rufa* (red kangaroo) (Spratt, 1970, 1972a, 1975). Spratt (1970) identified larval stages found in Tabanidae in Australia as *P. roemeri*. He collected infective larvae from wild-caught flies (*Dasybasis hebes*), inoculated them into a young grey kangaroo and recovered adult *P. roemeri* 159 days later. In the wallaroo, microfilariae occurred abundantly in the blood and were 179–220 µm in length, with closely fitting sheaths (Spratt, 1972b). In this host, microfilariae exhibited a diurnal subperiodicity. Spratt (1972a) experimentally infected grey kangaroos, wallabies and wallaroos. Microfilariae appeared in the peripheral blood of wallaroos 256–272 days postinfection and adult worms were found in 'highly vascular capsules' in the knees. On the other hand, microfilariae did not appear in the blood of kangaroos or wallabies apparently because they were destroyed in the tough, infiltrated capsule surrounding the adult worms in the legs of the host (Spratt, 1972a). Spratt (1975) experimentally infected a red kangaroo; microfilariae first appeared in its blood 241 days later and reached a peak 118 days after their first detection. It was concluded that the wallaroo is the primary reservoir of infection although the red kangaroo may serve as a secondary reservoir. Examination of wild-caught macropodids supported the conclusion that the wallaroo is the main host.

Spratt (1972c) found *P. roemeri* larvae in wild *Dasybasis acutipalpis* as well as *D. hebes* in an enzootic area in southeastern Queensland; 12–1978 larvae per individual fly were found. Development was studied in experimentally infected flies. By 24 h most larvae were in the abdominal fat body. Cells around larvae had the appearance of a syncytium. The sausage stage was attained in 3 days. The first moult occurred in 5 days and the second in 10–11 days. Infective larvae were 1.94–2.67 mm in length. The tail was rounded and possessed two small lateroterminal papillae and a pair of minute phasmids.

Spratt (1974) collected tabanids off bait animals in two enzootic areas in Queensland from October to May and examined them for larvae. *D. hebes* was the most frequently infected fly caught. Other tabanids carrying larvae were *Dasybasis acutipalpis/circumdata*, *D. dubiosa*, *D. moretonensis*, *D. neobasalis*, *D. oculata*, *Mesomyia fuliginosa*, *Scaptia testaceomaculata*, *Tabanus australicus*, *T. pallipennis*, *T. particaecus*, *T. parvicallus* and *T. townesvilli*. Averages of 646 and 865 larvae (all stages) were found per infected fly in the two localities studied (Allan and Durakai, Queensland). Apparently, *D. hebes* was the most significant vector, since 41.4–60.3% of all larvae (all stages) found in the tabanids examined from the two localities were in this species. According to Spratt (1974) moderate levels of transmission occurred in November and December, low to moderate levels in January through March and maximum levels in April, associated with peaks of *D. hebes*. Spratt (1974) also noted that peak activity of tabanids occurred between 1000 and 1500 h during clear, hot weather. This corresponded to the diurnal activity of macropodids and peak numbers of microfilariae in the blood of wallaroos (Spratt, 1972a). Highest prevalence occurred in commercially harvested animals in southcentral and southwest Queensland, where *Tabanus particaecus*, *T. strangmenii*, *T. townsvilli* and *Mesomyia fuliginosa* are the only tabanids known to occur (Spratt, 1975).

P. scapiceps (Leidy, 1886)

P. scapiceps (syn. *Dirofilaria scapiceps*, *Loaina scapiceps*) is a common parasite of the area between the synovial sheath and tendons of the ankles of the hindlegs of cottontail rabbits (*Sylvilagus floridanus*) and snowshoe hares (*Lepus americanus*) in North America

(Bartlett, 1983). They are rather short nematodes, helically coiled along their entire length. Microfilariae occur in the blood and are sheathed.

Highby (1938, 1943a) showed that *P. scapiceps* was transmitted by mosquitoes and made some preliminary observations on its development. Bartlett (1984a,b,c) carried out a detailed study of the development and transmission of this nematode in Ontario. She collected mosquitoes which had fed on infected rabbits and discovered that microfilariae invaded the fat body and developed to the infective stage in *Aedes canadensis, A. euedes, A. excrucians, A. provocans, A. punctor, A. stimulans/fitchii, A. vexans* and *Mansonia perturbans*. Intensities of third-stage larvae were greatest in *A. euedes* and *A. excrucians* (Bartlett, 1984a). Rabbits placed outside during the mosquito season in July and August acquired natural infections. The parasite also developed in laboratory-raised *Aedes aegypti*.

In the most suitable intermediate hosts, larvae developed in syncytia (containing hypertrophied adipocyte nuclei) which seemed to serve as a nutrient source necessary for larval development. Unsheathed microfilariae were found in the fat body as early as 1 h after mosquitoes had fed. Microfilariae were 262–300 µm in length. At 26°C the sausage stage was reached in 4–6 days. The first moult occurred 7–8 days and the second 9–10 days postinfection. Third-stage larvae appeared free in the abdomen in 10 days and in the head in 12 days. Third-stage larvae were 1.02–1.31 mm in length and the tail had a terminal swelling or cap, two small subventral papillae and two terminal phasmids. Spicular primordia were present in male larvae.

Bartlett (1984b) infected experimentally 34 cottontail rabbits. Intensity was 8–76% of the numbers of larvae inoculated. The third and fourth moults occurred 6 and 12 days postinfection. Development to subadults occurred in subcutaneous tissues in various regions of the body; the specific region was influenced by the site of inoculation. Subadults migrated through the subcutaneous tissues and reached the ankles as early as 16 days postinfection; they were mature by 67 days. The prepatent period was 137–234 days and microfilariae were non-periodic. Few worms in cottontails degenerated or died and rabbits with 1-year-old infections could be reinfected. *P. scapiceps* was recovered from 12 of 14 experimentally inoculated snowshoe hares and two of four New Zealand white domestic rabbits (*Oryctolagus cuniculus*). However, intensity in hares and domestic rabbits was only 0.4–15% and 6–7%, respectively, of the numbers of larvae given and many worms in these hosts degenerated or died.

Bartlett (1984c) undertook a detailed histopathological study of *P. scapiceps* in the ankle region of wild-caught lagomorphs. In cottontail rabbits, tendons and sheaths generally appeared normal and all worms were adults. In many snowshoe hares, however, a chronic proliferative tenosynovitis developed and led to encapsulation of worms. Thus, 46% of infected hares contained only a few viable or only dead worms. Few microfilariae reached the peripheral blood and microfilariaemias were extremely low. Microfilariae apparently became trapped and were destroyed in the lesions in the tendon sheaths. Thus, although *P. scapiceps* can be maintained within snowshoe hare populations, the latter are relatively poor hosts and Bartlett (1984c) suggested that the parasite may have spread from cottontail rabbits to snowshoe hares relatively recently.

P. ceylonensis Dissanaike, 1967

This species was found in the tendons and muscles of the legs of 'ash-doves' (Columbidae), chickens and crows in Sri Lanka (Dissanaike, 1967; Dissanaike and

Niles, 1967). Infective larvae of *P. ceylonensis* were found in wild-caught *Mansonia crassipes* (Culicidae) in Sri Lanka. Larvae were inoculated into chickens and doves and adult worms were recovered from the legs of these birds 4–5 months later. Infective larvae were 1.12–1.18 mm in length.

Subfamily Onchocercinae

The subfamily is characterized by its long non-alate tail and markedly dissimilar spicules. With the exception of *Macdonaldius*, which occurs in reptiles, all genera occur in mammals. The subfamily contains many important genera, such as *Brugia*, *Wuchereria*, *Onchocerca* and *Elaeophora*. In some species microfilariae occur in the skin producing a **microfiladerma**. Bain *et al.* (1994) compared certain features of development in rodents of six species of the subfamily in the genera *Litomosoides*, *Acanthocheilonema*, *Molinema*, *Monanema* and *Brugia* which migrate in the lymphatics.

Acanthocheilonema

Species of *Acanthocheilonema* occur in a variety of hosts (e.g. rodents, canids, seals, insectivores) and are transmitted by several types of vectors (e.g. ticks, fleas, hippoboscids, sucking lice). Microfilariae occur in the blood and are unsheathed, with attenuated, pointed tails. Most species develop in the fat body of the intermediate host.

A. mansonbahri (Nelson, 1961)
This is a parasite of the subcutaneous tissues and fascia of the rodent *Pedetes surdaster* (Pedetidae) in Kenya. Microfilariae were about 262 μm in length, with pointed tails. Infective larvae were found in two of 29 fleas, *Delopsylla crassipes* (Siphonaptera), found on *P. surdaster* (see Nelson, 1961). *Ctenocephalides felis* and *Xenopsylla cheopis* were allowed to feed on infected *P. surdaster* and both proved to be suitable hosts. Microfilariae (which had an unusually prominent cephalic hook) developed in large fat body cells and there was apparently no sausage stage. The first moult took place about 7 days and the second about 10 days after fleas had fed. Infective larvae appeared in fleas in about 15 days; most were free in the haemocoel. Infective larvae were 1.15–1.43 mm in length; the tail had a terminal conical projection and two ventrolateral appendages.

A. reconditum (Grassi, 1889)
A. reconditum (syn. *Dipetalonema reconditum*) is a parasite of the subcutaneous tissues and fascia of canids, including dogs, jackals and hyenas. It has been found in Europe, Africa, India and North America. Microfilariae were 269–283 μm in length (Newton and Wright, 1956, 1957). Grassi and Calandruccio (1890) showed that *A. reconditum* developed in fat body cells of the fleas *Ctenocephalides canis*, *C. felis* and *Pulex irritans* (Siphonaptera). Stueben (1954a,b) and others reported that *Dirofilaria immitis* developed in fleas but it is clear that they confused the microfilariae of *A. reconditum* in dogs with those of *D. immitis*. Newton and Wright (1956) showed that *C. canis* and *C. felis* were suitable vectors of *A. reconditum*. Farnell and Faulkner (1978) reported that infective larvae appeared in fleas as early as 7 days after they had fed; they gave the

prepatent period as 61–68 days. Pennington (1971) reported that *Echidnophaga gallinacea* and *Pulex simulans* were suitable intermediate hosts. Nelson (1962b) reported that *A. reconditum* was common in dogs, jackals and hyenas in Kenya, where it is transmitted mainly by *C. felis*. He also reported that the mallophagan louse *Heterodoxus spiniger* (Amblycera) was a suitable intermediate host. Infective larvae were 1.08–1.64 mm in length. The tail ended in a conical projection and had two ventrolateral ear-like appendages.

A. spirocauda (Leidy, 1858)

A. spirocauda (syn. *Dipetalonema spirocauda*) was first described from the heart and pulmonary arteries of harbour seals (*Phoca vitulina*). Measures *et al.* (1997) reported this species in four of six seal species examined in Atlantic Canada (*Phoca vitulina*, *P. groenlandica*, *P. hispida* and *Cystophora cristata*). Worms were usually in the right ventricle but some were found in the pulmonary artery and deep within the lungs. None was found in *Halichoerus grypus* and *Erignathus barbatus*. Microfilariae were 266–302 µm in length. Geraci *et al.* (1981) found larval stages of a filarioid in 70 of 102 anopluran lice (*Echinophthirius horridus*) commonly found on the seals. Larvae were usually in the fat body with an average of three third-stage larvae found in 54% of the lice. Infective larvae were 1.17–1.97 mm in length; the tail ended in a small knob and a pair of finger-like ventrolateral papillae.

A. viteae (Krepkogorskaya, 1933)

A. viteae (syn. *Dipetalonema blanci*) is a parasite of the subcutaneous tissues of *Meriones libycus* and other Gerbillidae of the genera *Jaculus* and *Rhombomys* in Iran, the CIS and North Africa. Baltazard *et al.* (1952) and Chabaud (1954a) have elucidated its development. Microfilariae were 170–205 µm in length and developed to the infective stage in *Ornithodorus tartakovskyi* and *Rhipicephalus* sp. (Argasidae, Ixodidae) in about 20 days at 28°C (Chabaud, 1954a). The exact site of early development in the tick was not determined but in 14–18 days larvae were in muscles. The first moult occurred about 12 days and the second about 20 days postinfection. Larvae attained the infective stage in 22 days, when they were 1.30–1.60 mm in length; the tail had a conical point and two ventrolateral appendages. Four gerbils (*M. libycus*) were inoculated with infective larvae and examined 5–20 days later (Chabaud, 1954a). The developing worms were found in the aponeuroses or subcutaneous tissues. The third moult probably took place 7 days postinfection; by 21 days all were in the fifth stage but microfilariae did not appear in the blood of the gerbils until about 50 days.

Baltazard *et al.* (1952) and Sullivan and Chernin (1976) found it difficult or impossible to infect jirds (*Meriones unguiculatus*) with *A. viteae* by the oral route.

Johnson *et al.* (1974) reported that *A. viteae* completed the third moult 7 days and the fourth moult 23 days postinfection in jirds. Insemination usually took place in 25–28 days. Microfilariae appeared in the uterus in 42 days and in the peripheral blood in 47 days. Mated and unmated worms lived for up to 2 years and frequent matings were needed for continuous fertility.

Beaver *et al.* (1974) reported that microfilariae were first detected in blood when worms were about 8 weeks old, regardless of the numbers of females present. Maximum levels of microfilaraemias were similar with two to five pairs of worms but were greater when ten pairs of worms were present in the host. Infections with one to five pairs of worms produced substantial and persistent microfilaraemias for up to 2 years. Once-mated females produced microfilaraemias of about 100/20 mm^3 of blood which lasted at diminishing levels for up to 15 months. The findings indicated that the most stable and enduring infections were produced by one to five pairs of worms and that periodic mating was essential.

According to Mössinger and Barthold (1987) the embryonic development of fertilized eggs in the uteri of *A. viteae* commenced 29–32 days postinfection and microfilariae were released about 20 days later. In *M. unguiculatus* the prepatent period was 45–59 days and worms were in subcutaneous tissues (65.3%), fascia (26.9%) and body cavities (7.8%). It was estimated that a female worm produced 7000 microfilariae each day.

Barthold and Wenk (1992) gave increasing doses (5–90) of infective larvae of *A. viteae* to jirds and reported that the number of adults recovered increased steadily with smaller doses but declined markedly with doses of 60 and 90 infective larvae. Increasing numbers of worms led to smaller females (Schacher, 1962a). Trickle infections at intervals of 2–6 days, throughout the prepatency period, resulted in the recovery of seven to ten worms in each jird. Larvae inoculated into patent jirds were destroyed by the immune system.

Wenk *et al.* (1994) studied the turnover of microfilariae inoculated into jirds and contrasted location and survival of the microfilariae with those in animals with patent infections produced by adult worms in the tissues. The results indicated a low turnover of microfilariae.

Votava *et al.* (1974) noted that unfed ticks (*O. tartakovskyi*) which fed on engorged ticks containing microfilariae of *A. viteae* became infected. Stein *et al.* (1997) showed that repeated feeding of ticks on hosts with *A. viteae* did not increase the number of infective larvae beyond that found in single feeding. Some mechanism must be operating to prevent overcrowding. Reproduction in infected ticks was significantly reduced.

A. weissi (Seurat, 1915)

A. weissi (syn. *Dipetalonema weissi*) is a parasite of the peritoneum and subcutaneous tissues of the head, abdomen and back of the insectivore *Elephantulus rozeti* (Macroscelidae) in North Africa. Microfilariae were 270–318 μm long, with markedly attenuated tails. Bain and Quentin (1977) reported that microfilariae developed in the ixodid tick *Ornithodorus erraticus*. One of two infective larvae found in the tick was 1.58 mm long and the tail had a terminal and two ventrolateral papillae-like structures.

Breinlia

Species which have been studied occur in the thoracic or abdominal cavities of rodents and primates. In these species microfilariae are unsheathed, occur in the blood and have long tapering tails. Intermediate hosts are mosquitoes and development takes place in the fat body.

B. booliati Singh and Ho, 1973

B. booliati was found in the thoracic and abdominal cavities of the forest rat (*Rattus sabanus*) in West Malaysia (Singh and Cheong, 1971; Singh and Ho, 1973; Lim *et al.*, 1975). It is also known from Sarawak, East Malaysia (Mak and Lim, 1974) and the Philippines (Miyata and Tsukamoto, 1975). Microfilariae were 188–206 μm in length (Singh and Ho, 1973). Ho *et al.* (1973) reported that *B. booliati* developed in the fat body, mainly the abdomen, of *Aedes aegypti*, *A. togoi* and *Armigeres subalbatus*. In *A. togoi*, at 26–30°C, the first moult occurred in 7 days and the second in 11 days postinfection. Parts of the infected fat body took on a syncytial appearance. By 11 days larvae entered the haemocoel and appeared in the mouthparts of the mosquitoes in 12 days. Infective larvae were 0.84–1.28 mm in length and the tail had two 'prominent ear-like posterior (lateral) papillae'.

Singh *et al.* (1976) infected *R. sabanus* and laboratory albino rats with larvae of *B. booliati* from *A. togoi* and *A. subalbatus*. In rats the third moult occurred 6–8 days and the final moult 24–28 days postinfection. Larvae were found initially in the skin and carcass but after 35 days the developing worms were found only in the thoracic and abdominal cavities where they matured in 11–12 weeks and continued to grow. According to Yap *et al.* (1975) the prepatent period was 11–14 weeks. They also reported that microfilariae of *B. booliati* exhibited nocturnal subperiodicity in *R. sabanus* but not in laboratory albino rats.

B. manningi Bain, Petit, Ratanaworabhan, Yenbutra and Chabaud, 1981

This is a parasite of the thoracic cavity of the rodent *Menetes berdmorei* (Sciuridae) in Thailand. Microfilariae occurred in the blood and were about 235 μm in length, with a filamentous tail. Microfilariae developed in the fat body of *Aedes aegypti* and *A. togoi* (see Bain *et al.*, 1981a). At 27°C the first moult took place in 6 days and the infective stage was attained in 10 days. The infective stage was 0.97–1.20 mm in length and had an unusually long tail, with one terminal papilla and two lateroterminal papillae.

B. sergenti (Mathis and Léger, 1909)

B. sergenti occurs in the peritoneum of the slow loris (*Nycticebus coucang*) (Lorisidae) in Malaysia. Microfilariae were 240–250 μm in length, with a long, attenuated, filamentous tail (Ramachandran and Dunn, 1968). They developed in the haemocoel of *Aedes aegypti*, *A. togoi* and *Armigeres subalbatus* (Ramachandran and Dunn, 1968; Zaman and Chellappah, 1968). At 24–30°C the first moult started in 6.5 days and was completed in 10.5 days and the second occurred 10–12 days postinfection (Ramachandran and Dunn, 1968). Infective larvae at 15.5–26.5 days were 0.94–1.28 mm in length. The tail had two prominent ear-like structures and, in some specimens, a tiny protuberance between them. Infective larvae were inoculated into a slow loris. Microfilariae were found in the heart but not in the peripheral blood. Adult specimens were found in the peritoneum. The authors suggested that about 5 months may be required for maturation of the worms.

Brugia and *Wuchereria*

Species of *Brugia* and *Wuchereria* are delicate nematodes mainly of the lymphatic system, including the nodes, as well as testes of primates, carnivores, shrews (Tupaidae) and lagomorphs. *B. buckleyi* of *Lepus nigricollis* is reported from the blood vascular system but it may not belong to *Brugia*. The development and transmission of species of the two genera are similar. Microfilariae occur in the blood and have prominent sheaths extending well beyond the extremities of the microfilariae. A dark-staining nucleus is present at the extremity of the tail of the microfilariae of *Brugia* spp., a feature which, along with a reduced number of male caudal papillae, helps to distinguish this genus from the closely related *Wuchereria* (see Buckley, 1960). Microfilariae frequently exhibit a more or less marked periodicity in the blood, presumably associated with the feeding behaviour of vectors. Hawking (1975, 1976) related periodicity to the accumulation of microfilariae in lung capillaries during the day and their more or less even distribution throughout all the circulation by night, when the microfilariae were in a passive phase. The periodicity is synchronized with the sleeping and waking habits of the host and seems to be controlled by the difference in oxygen tension between venous and arterial blood during day and night.

Development takes place in thoracic muscles of mosquitoes of the genera *Aedes, Anopheles, Culex* and *Mansonia*. There are the usual two moults in the intermediate host leading to the infective stage, which migrates to the mouthparts. Larvae leave the mouthparts of the feeding mosquito. Under moist conditions some larvae can survive for sufficient time on the skin of the host to enable them to locate and invade the puncture wound made by the mosquito. Larvae then enter lymphatic vessels and lymph nodes, where they mature.

The infective larvae of *Brugia* spp. and *Wuchereria bancrofti* can be distinguished with difficulty on the basis of certain morphological and morphometric characters. The tail of the infective larvae has three rounded papillae, said to be more prominent in *W. bancrofti* than in *Brugia* spp. (Nelson, 1959; Yen *et al.*, 1982; Bain and Chabaud, 1986).

Brugia malayi infects humans in southeast Asia but *Wuchereria bancrofti* has been spread by humans to many parts of the tropical world (Laurence, 1989). Both species are associated in humans with inflammation of the lymphatics (lymphangitis) and blockage of lymph flow, which frequently leads to elephantiasis, mainly of the groin, genitalia and lower limbs. Genital elephantiasis is common in bancroftian filariasis. Brugian filariasis mainly involves the lower limbs.

A remarkable feature of some species of *Brugia* is the extremely wide spectrum of definitive hosts which are found infected in nature and which can be infected experimentally (Ahmed, 1966). *W. bancrofti* is generally considered to occur only in humans but it has now been transmitted experimentally to leaf monkeys (*Presbytis* spp.). *W. kalimantani*, which is clearly related to *W. bancrofti*, occurs commonly in leaf monkeys in Kalimantan, Indonesia. *W. lewisi* Schacher is known only from microfilariae.

Xie *et al.* (1994) carried out a molecular comparison of six of the ten known species of *Brugia* and identified two clades: *pahangi–beaveri* and *malayi–timori–buckleyi*. The first included parasites of carnivores in Asia and America and the second included parasites of primates and lagomorphs in Asia. *B. malayi* and *B. timori* are very close, suggesting recent separation.

The literature on human filariasis is voluminous. For recent reviews of many aspects of filariasis, refer to Denham and McGreavy (1977), Mak (1983) and Mak and Yong (1986). Orihel and Beaver (1989) reviewed zoonotic *Brugia* infections in North and South America.

B. beaveri Ash and Little, 1964

B. beaveri was found in the lymph nodes, skin and carcass of the raccoon (*Procyon lotor*) in Louisiana. Formalin-fixed microfilariae were 285–325 μm in length; those fixed in alcohol and stained with haematoxylin were 200–258 μm (Ash and Little, 1964). Similar microfilariae were noted in a bobcat (*Lynx rufus*) and a mink (*Mustela vison*). According to Harbut (1976) and Harbut and Orihel (1995) microfilariae developed in 10 days to the infective stage in the flight muscles of *Aedes aegypti*. Infective larvae were 1.30–1.69 mm in length. Three of three raccoons, one cat and nine of 21 jirds (*Meriones unguiculatus*) developed patent infections after subcutaneous inoculation of larvae from *A. aegypti*. The prepatent period was 69–74 days in the raccoons, 70 days in the cat and an average of 107 days in jirds.

B. ceylonensis Jayewardene, 1962

This filarioid was found in the lymph nodes of dogs and cats in Sri Lanka; its relationship to *B. patei* needs clarification. Microfilariae were 220–275 μm in length. Jayewardene (1963) studied development in *Aedes aegypti* at 26.6–30.0°C. In 4 days the cuticle was 'entirely loose' and the first moult was completed in about 5 days (123 h). A moulting second-stage larva was observed at 138 h. Infective third-stage larvae, first observed in 6 days 17 h, were 1.33–2.20 mm in length; the tail had a prominent terminal papilla and two subterminal ear-like papillae.

Other intermediate hosts which have been reported are *Anopheles stephensi*, *Mansonoides annulifera* and *M. uniformis* (Abdulcader *et al.*, 1966; Pattanayak, 1968; Lahiri *et al.*, 1972).

B. malayi (Brug, 1927)

Lichtenstein (1927) first recognized that brugian filariasis was clinically distinct from bancroftian filariasis. Brug (1927) first described the microfilaria of *B. malayi* in humans in Indonesia and distinguished it from that of *W. bancrofti*. Rao and Maplestone (1940) found and described adults taken from a human in India. The parasite was subsequently reported in humans in parts of Malaysia, Vietnam, Korea, China, Indonesia, Papua New Guinea, the Philippines, Thailand, India and Sri Lanka where in some places it coexists with *W. bancrofti* (for a review see Denham and McGreavy, 1977). The parasite occurs in various wild and domesticated animals, including monkeys (*Macaca* spp., *Presbytis* spp.), cats (*Felis catus, F. bengalensis, F. planiceps*), dogs, viverids (*Arctogalidia trivergata, Paradoxurus hermaphroditis*) and pangolins (*Manis javanica*) (see Laing *et al.*, 1960; Edeson and Wilson, 1964; Orihel and Pacheco, 1966). It has been experimentally transmitted to domestic cats (Edeson and Wharton, 1957), monkeys (*Macaca irus, M. rhesus*), the slow loris (*Nycticebus coucang*) and the civet cat (*Viverra tangalunga*) (Edeson and Wharton, 1958).

Two strains of *B. malayi* have been recognized in Malaysia (Turner and Edeson, 1957; Wilson *et al.*, 1958). In the nocturnally subperiodic strain, microfilariae are always present in the blood but tend to be more prevalent at night than during the day. In the nocturnally periodic strain, microfilariae are absent or in very low numbers during the day but relatively abundant in the blood at night. The subperiodic strain occurs in swamp-forest areas, where it is transmitted mainly between leaf monkeys (*Presbytis* spp.) and other forest-dwelling animals by species of *Mansonia*. The nocturnally periodic strain is found predominantly in coastal rice-growing regions and open swamps, where it is transmitted between humans and domesticated carnivores by species of *Anopheles* (e.g. *A. barbirostris, A. campestris, A. donaldi*) and *Aedes*. Some species of *Mansonia* are vectors in both environments, however. The two strains of *B. malayi* are not necessarily fixed, since the periodic strain from humans becomes subperiodic in experimentally infected cats while the subperiodic strain in humans and cats becomes periodic in monkeys (Denham and McGreavy, 1977).

There is considerable evidence (from attempts to control brugian filariasis in villages, by the use of anthelmintics to remove worms from humans and insecticides to eliminate mosquitoes) that domesticated and wild animals serve as reservoirs for the reinfection of humans (Wharton *et al.*, 1958; Wharton, 1962).

Development of *B. malayi* was studied in detail by Feng (1936) using *Anopheles hyrcanus* as an intermediate host. At 29–32°C the first moult occurred in 4 days and the second in 6 days postinfection. Infective larvae were 1.30 mm long and active, being found in palpi and antennae as well as the abdomen. Infective larvae were 1.16–1.83 mm in length according to Mak (1985, in Mak and Yong, 1986). According to Zahedi (1991) microfilariae were found in the thorax of *Armigeras subalbatus* held at 28°C in 15 min after feeding and most reached this site in 6 h. The first moult occurred in 96 h and the second in 120 h. By 144 h larvae had migrated into the head and abdomen.

In leaf monkeys (*Presbytis melalophos*) the final moult took place 35–40 days postinfection. The prepatent period was about 67 days in this host and 70–116 days in cats and jirds (Edeson and Wharton, 1957; Edeson and Buckley, 1959; Ash and Riley, 1970a). In humans the prepatent period is said to be about 3.5 months (Edeson and Wharton, 1957).

Ahmed (1966) listed the various locations in different hosts where species of *Brugia* were located. In addition to the lymphatic system, *B. malayi* has been found in the lungs and testes of rats and jirds.

B. malayi (subperiodic) developed to the infective stage in *Culex tarsalis* and *C. erthrothorax* from California (Bangs *et al.*, 1995). This is the first evidence that *Culex* spp. can serve as intermediate hosts of *B. malayi*.

B. pahangi (Buckley and Edeson, 1956)

B. pahangi was originally found in dogs in Asia (Malaysia) but was subsequently reported in Cebidae (*Presbytis obscura*), Erinaceidae (*Echinosorex gymnurus*), Felidae (*Felis bengalensis, F. planiceps, F. senegalensis, Panthera tigris, Paradoxurus hermaphroditis*), Lorisidae (*Nycticebus coucang*), Manidae (*Manis javanica*), Sciuridae (*Ratufa bicolor*) and Viveridae (*Arctictis binturong, Arctogalidia trivergata, Viverra zibetha*) (Laing *et al.*, 1960; Nelson, 1965).

The parasite is transmissible to jirds (*Meriones unguiculatus*) and ferrets (*Mustela putorius*) (Ash and Riley, 1970b; Schacher, 1973; Campbell *et al.*, 1979). Adult worms localize in the lymphatics of ferrets but occur in aberrant locations in jirds, including the heart, lungs and testes, even when a microfilariaemia has developed (for comments on this phenomenon, see Ash and Riley, 1970b). *B. pahangi* is transmissible to humans but does not always become patent in this host (Buckley, 1958a,b; Edeson *et al.*, 1960a).

Microfilariae were 280 μm (wet mount) or 186–200 μm (thick films) in length according to Schacher (1962a). For an ultrastructural account of the microfilaria, see Laurence and Simpson (1974). Development to the infective stage took place in *Mansonia annulatus* and *M. longipalpus*, which are natural vectors in Malaysia (Edeson *et al.*, 1960a; Laing *et al.*, 1960). *Anopheles barbirostris* and *Armigeres obturbans* were also suitable intermediate hosts. Schacher (1962) infected *Anopheles quadrimaculatus*, *A. crucians* and *Psorophora confinnis* found in the southeastern USA. At 26.6°C microfilariae reached the haemocoel within 30 min of being ingested and were found in flight muscles in 1 h. The first moult occurred in 4–5 days; by day 8 most larvae had completed the second moult and some had started migrating to the head and labium. In 9–11 days infective larvae from *A. quadrimaculatus* were 1.55 mm in length. The tail had three papilliform elevations (see also Aoki *et al.*, 1980). According to Yen *et al.* (1982) infective larvae were 1.36–1.85 mm in length. Zahedi (1990a,b) colonized *Armigerus subalbatus*, a known vector of *Brugia* spp., and studied certain aspects of the development of *B. pahangi* in the mosquitoes (Zahedi, 1991, 1994; Zahedi *et al.*, 1992a,b, 1993). Microfilariae appeared in the haemocoel within 5 min of feeding. Most microfilariae lost their sheaths during migration through the gut and some had already migrated to the thorax. Migration continued for 6 h at 28°C.

Schacher (1962b) followed in detail the development of *B. pahangi* in experimentally infected cats. Third-stage larvae were commonly found in lymphatic vessels near nodes, where they underwent the third moult in 8–9 days postinfection. Male fourth-stage larvae moulted to the adult stage in about 23 days whereas females began the final moult in about 27 days and completed it by 33 days. About 50% of female worms had mated in 33 days (males were about 18 mm and females 15 mm in length). By 55 days (end of the prepatent period) numerous microfilariae appeared in female worms but microfilariae were not noted in blood smears until 74 days. Edeson *et al.* (1960a) found microfilariae in smears as early as 59 days, however.

Ash and Riley (1970b) followed development in the jird (*Meriones unguiculatus*). Early developmental stages were found in the pelt, carcass, lymph nodes, fat, testes, heart and lungs. The third moult occurred 6–9 days and the fourth 18–24 days postinfection. Males moulted 1–2 days before females. Twenty-five of 33 jirds developed patent infections (in 57–84 days) and microfilariaemias persisted for at least 13 weeks. The most stable microfilariaemias occurred in animals with the majority of worms in the testes. Infective larvae were inoculated into the jird *M. libycus*, the wood rat *Neotoma lepida*, the kangaroo rat *Dipodomys merriami* and the golden hamster *Mesocricetus auratus*. Developing worms were found in all rodents except the kangaroo rat but microfilariae appeared only in jirds. Worms were found almost entirely in the heart and lungs.

B. patei (Buckley, Nelson and Heisch, 1958)

B. patei is a parasite of dogs, cats (*Felis domestica*) and *Genetta tigrina* (Viveridae) in Kenya (Buckley *et al.*, 1958). Larvae were found in naturally infected *Aedes pembaensis*, *Mansonia africanus* and *M. uniformis* (Heisch *et al.*, 1959; Nelson *et al.*, 1962).

Laurence and Pester (1961a,b, 1967) experimentally infected *Aedes togoi*, *Anopheles gambiae* and *M. uniformis*. In *M. uniformis* at 26–30°C the first moult occurred in 4 days and the second in 6–8 days. Larvae were fully infective and in the head and proboscis of the mosquito in 9–10 days; larvae were about 1.1 mm in length. Laurence and Pester (1967) showed that the microfilariae of *B. patei* could be adapted better to *A. togoi* by repeated passage through this mosquito and cats.

B. timori Partono, Dennis, Atmosoedjono, Oemijeti and Cross, 1977

B. timori is a parasite of the lymphatics of humans on Flores Island, southeast Indonesia (David and Edison, 1964, 1965). Microfilariae were 265–323 µm in length in Giemsa-stained films and 332–383 µm in formalin-fixed material. *Aedes togoi* proved to be a suitable intermediate host (Purnomo *et al.*, 1976). Larvae from *Aedes togoi* which had fed on the blood of infected individuals were inoculated into jirds (*Meriones unguiculatus*), which were killed and examined 65 and 142 days later. Adult worms found in testes, lungs, heart and some 'major vessels' at the base of the heart (Partono *et al.*, 1976, 1977) were later described as *B. timori* and distinguished from *B. malayi*. In humans, the parasite behaves as *B. malayi*.

According to Partono *et al.* (1976) and Janz and Sing (1977) *B. timori* is transmitted by *Anopheles barbirostris* in East Timor. Domestic cats can be infected. In *A. togoi* at 27–29°C, the first moult occurred in 3–5 days (Purnomo *et al.*, 1976). A silver leaf monkey, *Presbytis cristata*, was experimentally infected and was patent in 95 days. The microfilaraemia reached a plateau 151 days postinfection (Sartono *et al.*, 1990). The periodicity was nocturnal.

B. tupaiae Orihel, 1966

This species was found in tree shrews (*Tupaia glis* and *T. tana* – Tupaidae) in Malaysia, Thailand and Vietnam (Orihel, 1967a). Microfilariae were 283–322 µm in length and developed to the infective stage in *Aedes aegypti* in 8.5–9.0 days at 22.7°C (Orihel, 1967a). The first moult occurred 5 days and the second 7–8 days postinfection. At 8–10 days, most larvae were in the head and mouthparts of the mosquito. In 10 days infective larvae were 0.97–1.50 mm in length. The tail had a terminal and two subventral papillae. Tree shrews were inoculated subcutaneously with infective larvae and two *T. glis* and one *T. tana* developed patent infections in 62–80 days.

W. bancrofti (Cobbold, 1877)

W. bancrofti is a parasite of humans from latitudes of about 41° N to about 28° S in the eastern hemisphere and from about 30° N to about 30° S in the western hemisphere (Faust, 1949). As indicated earlier, no other naturally infected hosts are known. There have been numerous unsuccessful attempts to establish *W. bancrofti* in certain primates (see for example Nelson, 1965; Cross *et al.*, 1979). However, *W. bancrofti* has now been successfully transmitted to leaf monkeys (*Presbytis cristatus*) (Harinasuta *et al.*, 1981; Palmieri *et al.*, 1982; Sucharit *et al.*, 1982) and it has been concluded that the disease produced in this host is essentially indistinguishable from the human infection (Palmieri *et al.*, 1983). This raises the possibility that *W. bancrofti* is of simian origin.

    W. bancrofti is thought to have originated in southeast Asia, from which it spread to the Pacific islands, Australia, central China, Japan, India, Sri Lanka, Africa and eventually, by means of the slave trade, to the Caribbean and Central and South America. It once existed in the southeastern USA but has disappeared from there as well as from Australia. Laurence (1989) suggested that *W. bancrofti* originated in nomadic Malay-speaking humans in southeast Asia, where it was transmitted by *Anopheles* and *Aedes* mosquitoes. These sea-faring people carried the parasite to Polynesia and across the Indian Ocean to Africa. The slave trade introduced the parasite to the Americas in more recent times.

    Studies of *W. bancrofti* in the 19th century are regarded as landmarks in the history of tropical medicine. The important contributions of Bancroft, Cobbold, Demarquay, Lewis, Low, Wucherer and, especially, Manson have been detailed by Manson-Bahr (1959). More recent comments and reviews are those of Chernin (1983), Grove (1990), Laurence (1989, 1990) and Wenk (1990).

    Microfilariae of *W. bancrofti* in thick films were 244–296 μm in length, with pointed tails devoid of the terminal nucleus found in microfilariae of *Brugia* spp. The morphological development of microfilariae in mosquitoes was studied by Abe (1937), Kobayashi (1940) and Bain (1972). Kobayashi (1940) followed in detail the development of *W. bancrofti* in *Culex fatigans* at 22.8–32.0°C. Microfilariae exsheathed in 2–6 h and established themselves in the thoracic muscles in 4–17 h. The first moult occurred in 5–6 days; during this period the larvae were in the sausage stage. The second moult occurred 9–10 days postinfection and the infective larvae were 0.93–1.78 mm in length 12 days postinfection. According to Bain (1972), who studied development in *Anopheles gambiae* at 26–28°C (Fig. 7.11A–K), the first moult took place in 5 days and the second in 10 days. Larvae were infective in 11 days. Infective larvae were 1.17–1.57 mm in length according to Nelson (1959) and 1.50–2.00 mm according to Yen *et al.* (1982). The caudal end of the infective larva had one terminal and two subterminal bulbous papillae.

    Manson (1881) first demonstrated that the microfilariae of *W. bancrofti* in China exhibited a marked nocturnal periodicity in the blood (maximum numbers between 2200 h and 0200 h). This periodic form occurs in many parts of the world and is transmitted mainly by night-biting mosquitoes. However, strains of *W. bancrofti* exist in the south Pacific (Fiji, Samoa, Philippines, Tahiti) which are not periodic and are transmitted by day-biting mosquitoes (Bahr, 1912). A great variety of species and races of *Aedes, Anopheles, Culex* and *Mansonia* are known intermediate hosts. It has been noted that the current success of bancroftian filariasis can be attributed to its ability to develop in *Culex* spp. (e.g. *C. pipiens fatigans*) which flourish in the poor drainage systems of overcrowded tropical towns (Laurence, 1989).

    Lardeux and Cheffort (1996) noted that the percentage reduction of infective larvae from *W. bancrofti* from *Aedes polynesiensis* ranged from 66 to 100% when the mosquitoes fed on mice. They make the point that although prevalence and intensity of infected larvae in some *Aedes polynesiensis* may be low, the 'adaptive strategy leads in fact to an increase in the parasite yield (i.e. one ingested microfilariae may give one infective

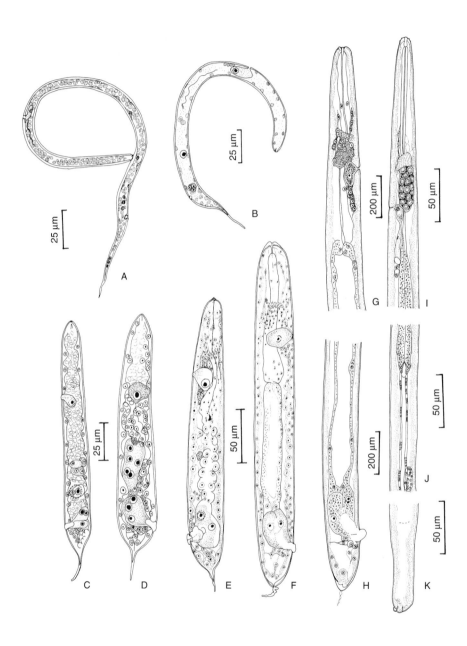

Fig. 7.11. Developmental stages of *Wuchereria bancrofti* in the mosquito host:
(A) sheathed microfilaria; (B) larva at 60 h; (C) larva at 70 h; (D) larva at 4 days; (E)
larva at 5 days, beginning the first moult; (F) second-stage larva with cuticle of first
stage; (G) late second-stage larva, anterior end (10 days); (H) late second-stage larva,
posterior end (10 days); (I) infective third-stage larva, anterior end; (J) infective
third-stage larva, postoesophageal region; (K) infective third-stage larva, posterior end.
(After O. Bain, 1972 – courtesy *Annales de Parasitologie Humaine et Comparée*.)

larva in most cases and the parasite may be transmitted even in low prevalence or low density areas (Pichon, 1974)'. Southgate and Bryan (1992) reviewed recent ideas concerning the factors involved in the transmission of *W. bancrofti* by anopheline mosquitos and discuss at length the concepts of limitations, facilitation and proportionality.

Larvae which have invaded the skin of humans apparently move about in the lymphatics and tend to concentrate in the groin glands and in the gland tissue of the scrotum, especially that around the epididymis. Worms mature in these locations and microfilariae appear in the blood about 1 year after the mosquito bite (Faust, 1949). In leaf monkeys the prepatent period was 206–285 days (Harinasuta *et al.*, 1981; Palmieri *et al.*, 1982; Sucharit *et al.*, 1982). It has been estimated that the fecundic longevity of female worms in humans is 10.2 years without chemotherapy (Vanamail *et al.*, 1990).

W. kalimantani Palmieri, Purnomo, Dennis and Marwoto, 1980
This parasite was found recently in the inguinal lymphatics and testes of the silvered leaf monkey (*Presbytis cristatus*) in South Kalimantan (Borneo), Indonesia. Microfilariae in the blood were nocturnally periodic (Palmieri *et al.*, 1980; Campbell *et al.*, 1986) and *Aedes togoi* and *Anopheles balabacensis* were suitable intermediate hosts (Campbell *et al.*, 1986). At 23–26°C infective larvae were present in the mosquitoes in about 14–17 days. Microfilariae were present in the blood of an experimentally infected *P. cristatus* 532 days postinfection (it is not clear if the blood was examined earlier).

Cercopithifilaria

The approximately 14 species in the genus have a wide host and geographical distribution in Bovidae, Cervidae, Canidae, Didelphidae, Hystricidae, Lagomorpha, Primates, Muridae and Sciuridae (Bain *et al.*, 1982b). They are subcutaneous parasites in which the microfilariae occur in the skin. Microfilariae are slender, apparently sheathed or unsheathed, with attenuated tails. Transmission of three species has been investigated and the intermediate hosts were Ixodidae.

C. grassii (Noè, 1907)
C. grassii (syn. *Acanthocheilonema grassii*) is a parasite of dogs in Europe and Africa. Microfilariae were about 570 µm long and unsheathed. According to Noè (1908) the microfilariae developed in nymphs of *Rhipicephalus sanguineus*. Bain *et al.* (1982a) identified as *C. grassii* larvae found in two *R. sanguineus* in France; these third-stage larvae were 1.05–1.62 mm in length and the tail ended in a conical point and had two lateral tongue-like structures.

C. johnstoni (Mackerras, 1954)
C. johnstoni occurs in Australian rodents and marsupials. Adults are found in subcutaneous tissues. The microfilariae are sheathed and inhabit the lymphatic vessels of the skin, mainly the ears. The parasite occurs in the following animals in Australia: Muridae – *Rattus fuscipes, R. lutriolus, Uromys candimaculatus*; Marsupialia; Paramelidae – *Perameles gunni, P. nasuta, Isooson macrourus, I. obsulus*; Petauridae – *Petauroides volans*; Dasyuridae – *Sarcophilus harrisii*.

Intermediate hosts are ixodid ticks, mainly *Ixodes trichosura* but also *I. facialis*, *I. holocyclus* and *I. tasmani* (Spratt and Haycock, 1988). Development occurs after engorged ticks leave the definitive host and during ecdysis from larva to nymph or from nymph to adult (Spratt and Haycock, 1988). Transmission from *Rattus fuscipes* occurred in summer and winter and was associated with peaks of larvae and/or nymphal ticks on the animals. *R. fuscipes* was infected by inoculation of infective larvae or by allowing infected ticks to feed on the host. The prepatent period was about 3 months and the microfiladerma persisted for more than 25 months.

According to Vuong *et al.* (1993) the parasite induces skin and ocular lesions in infected animals. Microfilariae live in lymphatic vessels and their exit gives rise to localized inflammatory reactions, resulting in fibrosis. Microfilariae were found in the limbus, cornea and eyelids and in the stroma of the mucosal region of the eyelid. They were inside and outside lymphatic vessels. Inflammation was most pronounced in the limbus, with dilation of lymphatics and blood vessels, vascularization and infiltration of melanophages and mast cells. Acute inflammatory lesions of the limbus frequently spread to the periphery of the cornea. Corneal fibrosis was mild to severe in the various animals examined.

C. roussilhoni Bain, Petit and Chabaud, 1986

This species was found in the African rodent *Atherurus africanus* (Hystricidae). Microfilariae were unsheathed, about 130–162 μm in length and dorsoventrally flattened (Bain *et al.*, 1986b). They developed in larval *Rhipicephalus sanguineus*. At 25–27°C the first moult occurred in 13 days and the second in 19 days. Male third-stage larvae were 1.20–1.48 mm in length. The tail ended in a conical point and had two lateral tongue-like appendages as well as a swelling or a pair of tiny points. Microfilariae were not found in the ears of three *A. africanus* inoculated with infective larvae 4 months earlier. Microfilariae were found, however, when the animals were examined 7 months postinfection.

Petit *et al.* (1988b) described in detail the development of *C. roussilhoni* in *R. sanguineus*. The parasite causes the formation of an epidermal syncytium (cf. *Monanema martini*).

C. rugosicauda (Böhm and Supperer, 1953)

C. rugosicauda (syn. *Wehrdikmansia rugosicauda*) is a parasite of roedeer (*Capreolus capreolus*) in Europe. Microfilariae were sheathed, 212–222 μm long, with pointed tails (Böhm and Supperer, 1953). Microfilariae were found mainly in the deeper layer of the skin of the ears. They developed in the nymphs of *Ixodes ricinus* to the infective stage in 56–67 days at 20–22°C (Winkhardt, 1980). During this period nymphs moulted to adulthood. Infected adult ticks in southern Germany each contained up to 12 infective larvae; most ticks (43%) were infected in spring and microfilariae appeared in 46% of roedeer by autumn. The prepatent period was about 6 months. Infective larvae were 1.97–2.18 mm in length and the tail had a terminal swelling with tiny spines and two large lateral appendages. Winkhardt (1979) studied different methods of infecting *I. ricinus* artificially using glass capillaries and various media.

Cherylia

C. guyanensis Bain, Petit, Jacquet-Viallet and Houin, 1985

C. guyanensis occurs in the subcutaneous and intermuscular connective tissues of the opossum *Metachirops opossum* in French Guyana, South America. Sheathed microfilariae occurred in the skin and were 140–170 μm in length, with pointed tails (Bain *et al.*, 1985a). The sheath extended beyond the length of the microfilaria and an oblong clear body was present within the sheath. Larval *Ixodes ricinus* were allowed to engorge on an infected opossum and kept at 29°C; the larval ticks moulted 15 days later. About 28 days later 50 of the nymphs were examined and six infective larvae were found in them. Larvae were 1.50–1.86 mm in length, with long tails. The rounded tip of the tail was bent ventrally and had a few tiny spines.

Deraiophoronema

D. evansi (Lewis, 1882)

This species (syn. *Dipetalonema evansi*) is a parasite of the pulmonary and internal spermatic arteries, lymph spaces and nodes of camels (*Camelus bactrianus*, *C. dromedarius*) in Egypt, the Far East and the eastern republics of the CIS. It has been introduced to Australia. Microfilariae in the blood were sheathed and 200–315 μm in length, with slightly tapered tails (Nagaty, 1947). Kataitseva (1968) examined many arthropods associated with camels in Turkmenia. Larvae were found only in *Aedes caspius*. This mosquito was allowed to feed on an infected camel; third-stage larvae appeared in the head and proboscis 10 days later. Development took place in the flight muscles. The first moult occurred in 2 and the second in 9 days postinfection at 26–28°C according to Kataitseva (1968), but Sonin (1975) pointed out that the first moult probably occurred 5–6 days postinfection, after the sausage stage had been passed. Infective larvae were 0.96–1.64 mm in length.

Dipetalonema

D. caudispina Molin, 1858

D. caudispina is a parasite of the abdominal and thoracic cavities of neotropical Cebidae (*Ateles ater*, *A. paniscus*, *Cebus albifrons*, *Saimiri boliviensis* and *S. sciureus*) (Sonin, 1975). Microfilariae were about 200 μm in length, with closely fitting sheaths and tapered, pointed tails. Eberhard *et al.* (1979) allowed *Culicoides hollensis* (Ceratopogonidae) from Mississippi to feed on an infected squirrel monkey (*S. sciureus*). Microfilariae shed their sheaths and invaded the abdominal fat body of the fly. The first moult occurred in 5 days and the second in about 8 days postinfection (temperature not indicated). Infective larvae were 455–670 μm in length and the tail ended in three rounded papillae.

D. dracunculoides (Cobbold, 1870)

D. dracunculoides is a parasite of the peritoneal cavity of canids. Cobbold (1870) collected it from an aard wolf (*Protales cristatus*) from South Africa. It was reported in domestic dogs in North Africa by Railliet *et al.* (1912) and Baylis (1929) and in the spotted hyena (*Crocuta crocuta*) in Kenya by Leger (1911) and Nelson (1963). It is common in dogs in Spain and Portugal (Azevedo, 1943; Ortega-Mora and Rojo-Vázquez, 1988; Rojo-Vázquez *et al.*, 1990; Olmeda-Garcia *et al.*, 1993). Azevedo (1943) noted that microfilariae occurred in the blood, were unsheathed and $189 \pm 4.48 \, \mu m$ in length; microfilariae of *D. immitis* were longer ($234.7 \pm 2.76 \, \mu m$) and more slender (see also Leger, 1911).

Nelson (1963) reported third-stage larvae (2.4 mm in length) in the house fly *Hippobosca longipennis* removed from a dog infected with *D. dracunculoides* in Kenya. Bain (1972) described third-stage larvae from a tick, *Rhipicephalus sanguineus*, which had fed on a dog from Algeria with *D. dracunculoides*. The larva was 2.3 mm in length and the tail ended in a terminal cone-shaped structure flanked by two small dorsoventral projections. Davadan (1988) and Rodriguez-Rodriguez *et al.* (1989) believe that *R. sanguineus* is the vector in Spain and Portugal. Olmeda-Garcia *et al.* (1993) allowed nymphs of *R. sanguineus* to feed on an infected dog. After engorging, the ticks were kept at 30°C until they had moulted to adults and were then allowed to feed on dogs. The prepatent period in two dogs was 69 and 76 days.

D. gracile (Rudolphi, 1805)

D. gracile is a common parasite of the body cavity of neotropical monkeys of the families Callitrichidae (*Callitrix, Leontideus, Saguineus*), Cebidae (*Aotes, Ateles, Brachyteles, Callicebus, Cebus, Lagothrix* and *Saimiri*) and Cercopithecidae (*Cercopithecus*) (Sonin, 1975). Microfilariae were about 130 μm long, with voluminous sheaths and blunt tails. Eberhard *et al.* (1979) allowed *Culicoides hollensis* (Ceratopogonidae) from Mississippi to feed on an infected moustached marmoset (*Sanguineus mystax*). Microfilariae exsheathed in the midgut of the fly and invaded the abdominal fat body. The first moult took place 7 days and the second 12 days postinfection (temperature not indicated). Infective larvae were 670–740 μm in length and the tail had three prominent rounded papillae.

Elaeophora

Species of *Elaeophora* are medium-sized worms which occur in the heart and arteries of artiodactyls. *E. schneideri* has been investigated extensively in the USA.

E. schneideri Wehr and Dikmans, 1935

This is a parasite mainly of the common carotid arteries and their branches in mule deer, *Odocoileus hemionus* (Cervidae), in western North America. Adult worms are relatively large and have a characteristic coiled appearance, which is undoubtedly related to the need to resist the strong flow of blood in the arteries in which they live. Microfilariae were 239–279 μm in length, broad and sheathed, with blunt tails. Microfilariae were found most commonly in skin of the forehead and face of the host (Hibler and Adcock, 1968); they also occurred in capillaries supplying the eyes, brain and other internal organs (Hibler and Adcock, 1971). For a recent review see Pence (1991).

Microfilariae developed to the infective stage in various species of Tabanidae (Hibler *et al.*, 1971; Hibler and Metzger, 1974). Microfilariae invaded the fat body lining the abdomen of flies. They developed into first-stage larvae and, when about 350–400 µm long, left the fat body and continued development to the third stage in the haemocoel (Hibler and Metzger, 1974). Infective larvae were 3.29–5.03 mm in length; the tail had a terminal cone-shaped structure flanked by two ventrolateral tongue-like structures.

Hibler *et al.* (1970, 1974), Hibler and Gates (1974) and Hibler and Metzger (1974) inoculated mule deer and sheep with infective larvae. They suggested that larvae remained at the sites of inoculation for a few hours and then entered the circulation, which eventually carried them to the lungs and arterial system. It is not known how larvae entered the carotid arteries or cephalic arterial system. Development took place in the leptomeningeal arteries. Growth was rapid and immature adults (10–13 mm in length) were present in about 2 weeks. In 3.5–4 weeks, worms migrated into the carotids, and matured in about 4.5 months. Infected mule deer exhibited no clinical signs and were believed to be the normal host of *E. schneideri*.

Horsefly vectors were found in the Gila National Forest, New Mexico, only above elevations of about 2200 m. They were active between 0900 and 1700 h and especially from 1100 to 1400 h at high temperatures (Hibler and Adcock, 1971). Flies preferred to feed on the forehead and face of deer and wapiti (*Cervus canadensis*) and complete engorgement required about 3 min. Tabanids emerged in the last week in May; they reached maximum numbers in June and disappeared beginning the first week of July. *Hybomitra laticornis*, *H. phaenops*, *H. tetrica*, *Tabanus abditus*, *T. eurycerus* and *T. gilanus* were found naturally infected in New Mexico (Hibler *et al.*, 1971; Clark, 1972; Clark and Hibler, 1973). In southwest New Mexico *H. laticornis* appeared to be the most important vector, since it was the most abundant and most often infected (16% prevalence).

E. schneideri has been reported in low numbers in white-tailed deer (*Odocoileus virginianus*) but this cervid is not considered a particularly suitable host (Prestwood and Ridgeway, 1972; Titche *et al.*, 1979). Couvillion *et al.* (1984) and Couvillion (1985) reported that the parasite occurred in low prevalence in white-tailed deer in parts of southeastern USA, where it is apparently transmitted by *Tabanus lineola* and *T. nigrovittatus*.

E. schneideri has gained considerable attention because it is an important pathogen of wapiti (*Cervus canadensis*) in southwestern USA (Adcock *et al.*, 1965; Adcock and Hibler, 1969; Hibler *et al.*, 1969; Hibler and Metzger, 1974). The presence of worms in the arteries results in vascular lesions leading to occlusion and ischaemic necrosis involving the ears, muzzle, eyes and brain. It is also a well-known pathogen of domestic sheep (Kemper, 1938; Hibler *et al.*, 1970). More recently it has been reported in moose (*Alces alces*) (Worley *et al.*, 1972), sika deer (*Cervus nippon*) (Robinson *et al.*, 1978) and barbary sheep (*Ammotragus lervia*) (Pence and Gray, 1981). For a detailed review of the pathology of arterial worm the reader is referred to Adcock and Hibler (1969) and Hibler and Adcock (1971).

Litomosoides

Species of *Litomosoides* occur in the body cavities of rodents, bats and opossums. Members of the genus are easily recognized by their highly developed, thick-walled buccal capsules. Microfilariae are short with pointed tails; they are sheathed and occur in the blood of the host. They develop in the haemocoel of mites (Acarina – Dermanyssidae).

L. galizai Bain, Petit and Diagne, 1989

This is a parasite of *Oecomys trinitatis tapajinus* in Brazil. Diagne *et al.* (1989) reported on its development in *Bdellonyssus bacoti* (syn. *Ornithonyssus bacoti*). Microfilaria developed mainly in fat body cells. The first and second moults occurred 5 and 7 days post-feeding, respectively.

L. legerae Bain, Petit and Berteaux, 1980

L. legerae was found in the abdominal cavity of *Oxymycterus quaestor* (Cricetidae) in Brazil. Uterine microfilariae were 80–83 µm in length. According to Bain *et al.* (1980) microfilariae developed to the infective stage in the tropical rat mite *Bdellonyssus bacoti*. Infective larvae were 0.86–0.98 mm in length.

L. petteri Bain, Petit and Berteaux, 1980

L. petteri was found in the pleural and abdominal cavities of the marsupial *Marmosa cinerea* (Didelphidae) in Brazil. Uterine microfilariae were 78–83 µm in length. According to Bain *et al.* (1980) microfilariae developed to the infective stage in the tropical rat mite *Bdellonyssus bacoti*. Two infective larvae were 0.73 and 0.98 mm in length.

L. sigmodontis Chandler, 1931

The species in the cotton rat (*Sigmodon hispidus*) used for a great deal of laboratory studies is *L. sigmodontis* Chandler, 1931 and not *L. carninii* as indicated by the original literature (for a review of the problem, see Bain *et al.*, 1989). *L. sigmodontis* is a parasite of the pleural cavity and, less commonly, body cavity of cotton rats in North America. It is transmissible to laboratory rats and mice (Scott, 1960), *Mastomys natalensis*, *Microtus arvalis* and hamsters (*Mesocricetus auratus*) (Hawking and Burroughs, 1946; Wenk, 1967; Wenk and Heimburger, 1967; Lammler *et al.*, 1968). *L. sigmodontis* is easily maintained in laboratory rodents and has been used extensively in studies of anthelmintics, immunology and biochemistry since the discovery of the details of its transmission and development in the late 1940s and 1950s.

Microfilariae were 70–90 µm in length (Williams, 1948). McFadzean and Smiles (1956) showed that the sheath of the microfilaria was a stretched egg membrane. Microfilariae did not exhibit a periodicity (Bell and Brown, 1945). Williams and Brown (1945, 1946) showed convincingly that microfilariae hatched in the gut of the tropical rat mite *Bdellonyssus bacoti* (syns *Liponyssus*, *Ornithonyssus bacoti*) and entered the haemocoel, where development to the infective stage took place. According to Scott *et al.* (1951) the first moult in mites occurred about 9 days and the second about 13 days postinfection at 18–24°C. Third-stage larvae reached their maximum size (0.51–0.96 mm) in about 15 days (Bertram, 1947; Scott, 1946; Scott *et al.*, 1951); the shorter forms were probably the males and the longer forms the females (Williams and Brown, 1946; Williams, 1948).

Cross and Scott (1947) noted that females were mature when they were only 40% of their eventual lengths and that the moult from the fourth to the adult stage took place when worms were only 7.0 mm in length.

Infection of rats took place during the feeding of mites on rats and not through the ingestion of mites by rats. In cotton rats, infective larvae apparently migrated to the pleural cavity, where development occurred. The third moult occurred 9–15 days and the final moult about 23–24 days postinfection. The prepatent period was about 50–80 days (Scott *et al.*, 1946; Williams and Brown, 1946). Wenk (1967) infected cotton rats and white mice and reported that larvae migrated from the site of the bite of an infected mite via the lymphatics to the regional lymph nodes within 24 h. Larvae then followed the lymphatic system to the venous system, the right heart and the lungs, which they reached in 5–10 days in rats and in 1–2 days in mice.

According to Webber (1954a) spermatozoa developed by 25 days and females were inseminated by 28–32 days. Webber (1954b) transferred inseminated females to uninfected cotton rats and studied the longevity of microfilariae production, concluding that repeated insemination of females was unnecessary. Scott *et al.* (1946) and Bertram (1953) reported that cotton rats were susceptible to reinfections, but microfilariaemias declined in intensity and duration and microfilariae even disappeared in the blood although many adults remained in the pleural cavity. Hawking (1954) estimated from *in vitro* studies that female worms produced 4000–43,000 microfilariae in 2 h and that development from unfertilized ovum to the microfilaria required 5–6 days.

Hoffmeister and Wenk (1991) inoculated cotton rats with infective larvae and showed that the latter migrated from the inoculation site via the lymphatics to the heart, lungs and pleural cavity within 5 days. After doses of 60 larvae per rat, 19.9 worms reached the pleural cavity. The recovery of worms in the pleural cavity decreased from 53%, 43% and 26% when doses were six, 12 and 60, respectively. However, when the 60 larvae were distributed to the four extremities, the recovery was 59% whereas when 60 larvae were inoculated into one leg, recovery was only 26%.

Wenk and Mösinger (1991) studied the establishment and the microfilariaemias in cotton rats inoculated with low numbers of larvae. A single dose of only five larvae resulted in long-lasting patent infections in one of eight animals. With doses of six, 12 and 60 larvae, the recovery of adult worms declined from 75%, 43% and 26%, respectively. Short patent periods were noted in many animals given three to 60 larvae, indicating that defence reactions limit the number of worms maturing.

Macdonaldius

M. innisfailensis (Mackerras, 1962)

This parasite (syn. *Saurofilaria innisfailensis*) was found in the subperitoneal tissues of the lizard *Physignathus lesueurii* (Agamidae) in Australia. According to Mackerras (1962) microfilariae in the peripheral blood were 115–133 µm in length and unsheathed. The tail was attenuated. Microfilariae developed in the fat body of *Culex fatigans*; infective third-stage larvae were 0.93–1.06 mm in length.

M. oschei Chabaud and Frank, 1961

M. oschei was first described from the major arteries of *Python molurus* from India and *P. reticulatus* from Indonesia held in the Zoological Garden in Stuttgart, Germany (Chabaud and Frank, 1961a). Subsequently the parasite was found in *Constrictor constrictor* in the same zoo and microfilariae were redescribed (Chabaud and Frank,

1961b). It was suggested that *M. oschei* originated from some New World reptile and was being transmitted in the zoo to reptiles from other parts of the world (Chabaud and Frank, 1961c).

Microfilariae were about 207 μm long, with cone-shaped tails and oval sheaths shorter than the microfilariae. Frank (1962, 1964a,b) reported that microfilariae invaded and developed in the Malpighian tubules of the New World argasid tick *Ornithodoros talaje*, which had been introduced to the zoo. At 28–29°C the first moult occurred in 14–20 days and the second in 20–25 days after the tick had fed. No sausage stage was noted. Infective larvae were 1.80–2.10 mm in length. The caudal end of the larva was rounded and had a tiny terminal papilla. Infective larvae left the Malpighian tubules and migrated to the head region of the tick, where they were found in the subpharyngeal musculature and between the chelicerae and the hypostome.

Frank (1962, 1964c) described necrotic lesions in the skin and muscles of *Python* spp. caused by occlusion of arteries by adult *M. oschei*. In contrast, *Epicrates cenchria* of the New World had high microfilariaemias but displayed no clinical signs. These observations support the view that *M. oschei*, like other members of the genus *Macdonaldius*, is normally a parasite of New World reptiles and its appearance in other hosts in the Stuttgard Zoological Garden was the result of the introduction to the zoo of infected reptiles along with suitable New World vectors.

Mansonella

Species of *Mansonella* occur in subcutaneous tissues and, less commonly, in the body cavities mainly of primates. Unsheathed microfilariae occur in the blood or skin (*M. streptocerca*) and develop in the thoracic muscles of Ceratopogonidae and Simuliidae.

M. llewellyni (Price, 1962)
This is a parasite of the subcutaneous connective tissues and fascia of raccoons (*Procyon lotor*) in the USA. Microfilariae, which occurred in the blood, were long (275–304 μm) and sinuous with attenuated tails. Yates *et al.* (1982) collected *Culicoides hollensis* (Ceratopogonidae) in tidal marshes in Mississippi and allowed them to feed on an infected raccoon. Larval stages, presumably of *M. llewellyni*, were found in the thoracic muscles of the midges. The first moult occurred 6–7 days postinfection and the second in 9 days. Infective larvae collected from the head and mouthparts of the midges 10 days postinfection were 619–720 μm in length. The caudal end terminated in four papillae.

M. marmosetae (Faust, 1935)
This is a parasite of the intermuscular connective tissues of the neck and spine of monkeys in South America, namely: Cebidae – *Aotes trivigratus*, *A. zonali*, *Ateles dariensis*, *A. geoffroyi*, *A. paniscus*, *Cebus capucinus*, *Saimiri bolivensis*, *S. oerstedi* and *S. sciureus*; Callithricidae – *Saguinus geoffroyi*, *S. fuscicollis*, *S. oedipus* and *S. tamarin*. Microfilariae, which occurred in the blood, were about 330 μm long, with rounded tails with nuclei. Lowrie *et al.* (1978) described the development of *M. marmosetae* in *Culicoides furens* and *C. hollensis* (Ceratopogonidae). Microfilariae migrated from the midgut of the fly to the flight muscles within 24 h at 26.6°C. The first moult

occurred 6 days postinfection. Third-stage infective larvae found in the midges 8 days postinfection were 670–780 µm in length. The posterior half of the body was broader than the anterior half and the oesophagus was long. The tail was rather long and ended in four terminal papillae.

M. ozzardi (Manson, 1897)

M. ozzardi (syn. *Filaria demarquayi*) is a parasite of the body cavity of humans in Latin America from northern Argentina to Yucatan and the West Indies (Faust *et al.*, 1975). It was first studied by Manson in blood from Carib Indians in British Guyana. Microfilariae, which occurred in the blood, were 185–200 µm in length, with long attenuated tails free of nuclei. Microfilariae were non-periodic in the peripheral blood. The biting midge *Culicoides furens* (Ceratopogonidae) was shown convincingly by Buckley (1933, 1934) to be the vector of *M. ozzardi* in St Vincent, West Indies. Microfilariae in *C. furens* migrated within 24 h to the thorax, where development took place as the larvae lay 'stretched out straight between the muscle fibres'. The sausage stage was reached in 24 h at temperatures of about 28–32°C. The first moult took place in about 3–4 days and the second about 6 days postinfection. The infective larva was a maximum of 0.78 mm in length and the tail possessed four small, terminal papillae-like swellings. Buckley (1934) suggested that *Culicoides paraensis* is probably also a vector. Cerqueira (1959) showed that *M. ozzardi* developed successfully in the blackfly *Simulium amazonicum* in Brazil. Tidwell *et al.* (1980) concluded that the *Simulium sanguineum* group was the main vector in the Mitú region of Colombia. Orihel (1967b) believed that *M. ozzardi* is especially well adapted to aboriginal Indians, in whom prevalences are frequently extremely high. Marenkelle and German (1970) reported that 96.2% of adult Indians were infected in a region in southern Colombia; they associated infection with eosinophilia and severe articular pain and regarded it as a major health concern.

M. perstans (Manson, 1891)

M. perstans is a parasite of the pleural and abdominal cavities, perirenal and retroperitoneal tissues, and the pericardium of humans in tropical Africa, Algeria, Tunis and the east coast of South America, from Panama to Argentina. It may occur in gorillas and chimpanzees in Africa (Nelson, 1965). Microfilariae occurred in the blood and were about 200 µm in length, with a slightly tapered tail having a rounded extremity containing nuclei. Microfilariae, first discovered by Daniels (1897) in Demeraran Indians from British Guyana, were described by Manson, who also discovered them in humans in Africa. Sharp (1928) showed that the biting midge *Culicoides austeni* (Ceratopogonidae) was a suitable intermediate host in Cameroon. Microfilariae developed in thoracic muscles. There were two moults and infective larvae appeared in the mouthparts of the flies 7–10 days postinfection. Infective larvae were 600–900 µm in length.

Orihel (1967b) pointed out that *M. perstans* and *M. ozzardi* frequently occurred together in British Guyana, that their prepatent periods may be as short as 2–4 months and that the microfilariae of both species may persist in the blood for long periods in the absence of adult worms.

Chardome and Peel (1949) and Henrard and Peel (1949) noted that, in mixed infections, *Culicoides grahami* ingested microfilariae of *Mansonella streptocerca* from the skin of humans in preference to *M. perstans* from the blood. They doubted, therefore, that *C. grahami* was a vector of *M. perstans*, as suggested by Sharp (1928).

M. streptocerca (Macfie and Corson, 1922)

The adult of this parasite is found in the dermis of humans in certain regions of Africa, especially Zaire. Microfilariae, which were 180–240 µm in length, also occurred in the dermis but mainly near the epidermis (Meyers *et al.*, 1972). The tail of the microfilaria was slightly narrower than the body and its termination was rounded. Also, the caudal end of the microfilaria was bent in the form of a shepherd's crook. Chardome and Peel (1949) studied the transmission of *M. streptocerca* in and around Mbandaka (= Coquilhatville) in Zaire. In *Culicoides grahami* (Ceratopogonidae) microfilariae passed rapidly to the thoracic muscles, where they developed to the infective stage in 7 days. The infective stage was about 574 µm in length. *C. grahami* ingested microfilariae of *M. streptocerca* in preference to those of *M. perstans* in mixed human infections (see also *M. perstans*). *M. streptocerca* has been reported in chimpanzees (*Pan* spp.) and gorillas (*Gorilla gorilla*) (Peel and Chardome, 1946a,b) but Nelson (1965) believed that transmission to humans was mainly from humans.

Molinema

Species of this genus occur in the peritoneal cavity of large rodents, including porcupines and beavers. Microfilariae occur in the blood and are unsheathed with pointed, attenuated tails. Microfilariae develop in the fat body of mosquitoes of the genera *Aedes*, *Anopheles* and *Taeniorhynchus*.

M. arbuta (Highby, 1943)

M. arbuta (syn. *Dipetalonema arbuta*) is a common parasite of the peritoneal cavity of porcupines (*Erethizon dorsatum*, Erethizontidae) in North America. Microfilariae were 280–297 µm in length according to Highby (1943b) and 251–282 µm according to Anderson (1953). Highby (1943c) reported that the following mosquitoes were suitable intermediate hosts: *Aedes aegypti*, *A. canadensis*, *A. cinereus*, *A. excrucians*, *A. fitchii*, *A. stimulans*, *A. vexans* and *Taeniorhynchus perturbans*. The most suitable hosts were considered to be *A. stimulans* and *T. perturbans*. The timing of the two moults was not reported but a sausage stage was attained by day 3 and infective third-stage larvae appeared in 9 days. Infective larvae were 0.88–1.16 mm in length and the tail ended in three papilla-like structures.

   Bartlett and Anderson (1985b) reported that *M. arbuta* in Ontario, developed in *Aedes canadensis*, *A. euedes* and *A. stimulans/fitchii*. Infective larvae, which appeared in the head in 14–18 days postinfection, were 1.00–1.20 mm in length. The tail was long and ended in one terminal and two lateral digitiform appendages. As a result of examining two experimentally infected young porcupines, it was suggested that the third moult occurred 20 days postinfection and the fourth later than 28 days in subcutaneous tissues. Fifth-stage worms probably migrated from subcutaneous tissues to the body cavity of the host.

M. dessetae (Bain, 1973)

This (syn. *Dipetalonema dessetae*) is a parasite of the abdominal cavity of *Proechimys guyanensis* (Echimyidae) from Brazil. Microfilariae were 280–310 μm in length. Bain (1974) followed the development in *Anopheles stephensi* at 25°C. The first moult occurred in 11–13 days and the second moult in 16–20 days postinfection. Infective larvae migrated to the head and mouthparts of the mosquitoes in about 27 days. Infective larvae were 0.76–0.95 mm in length; the tail was long and ended in three prominent tongue-like structures.

M. sprenti (Anderson, 1953)

M. sprenti (syn. *Dipetalonema sprenti*) is a common parasite of the peritoneal and, less commonly, pleural and pericardial cavities of beaver (*Castor canadensis*) in Canada. Microfilariae were 322–450 μm in length (Anderson, 1953). Addison (1973) determined that in Ontario the vectors were *Aedes abserratus* and *A. intrudens* (Culicidae). *M. sprenti* did not develop in *Anopheles earlei*, which was abundant in beaver lodges in the study area. Microfilariae invaded the haemocoel as early as 4 min after being ingested. Development took place in the abdomen and thorax. At 27°C the first moult occurred 4–6 days and the second 7–9 days after the blood meal had been taken. Infective larvae were first observed in the mouthparts 8 days postinfection. At 23°C development was slower; after 9 days larvae were still in the second stage and, in contrast to development at 27°C, a sausage stage did not appear. Third-stage larvae 18 days postinfection were 0.85–1.18 mm in length. The tail ended in a central dorsal protuberance and two ventrolateral digitiform appendages. Six uninfected beavers were inoculated with infective larvae and microfilariae appeared in their blood 116–135 days later. *A. abserratus* and *A. intrudens* fed on beavers from the first week of June until about the middle of July. Addison (1973) suggested that transmission of *M. sprenti* may occur mainly within beaver lodges.

Addison (1973) failed to infect porcupines (*Erethizon dorsatum*) with *M. sprenti*. It seems that each of *M. arbuta* of porcupines and *M. sprenti* of beavers is specific to a single host. *M. sprenti* failed to develop in *Aedes stimulans*, a suitable intermediate host of *M. arbuta*.

Monanema

Species of *Monanema* are parasites of the lymphatics, pulmonary vessels and the walls of the caecum and colon (depending on the species) of rodents. Microfilariae, which occur mainly in the skin (especially the ears) of the host, are short with slightly attenuated tails and are enveloped by rather voluminous sheaths which are shorter than the body of the microfilaria. Intermediate hosts are hard ticks (Ixodidae). The tail of the infective larva ends in three prominent processes.

M. globulosa (Muller and Nelson, 1975)

This species (syn. *Ackertia globulosa*) was found in the pulmonary arteries of rodents (*Aethomys kaiseri*, *Lemniscomys striatus*, *Otomys angoensis* and *Tatera robusta*) in Kenya (Muller and Nelson, 1975). *L. striatus* was most commonly infected (82%). Microfilariae (135–150 μm in length) were found in various regions of the body but were more numerous in the ears than elsewhere. The sheath of the microfilaria contained 10–11 small refractile globules. A nymph of *Haemaphysalis leachi* (Ixodidae) removed from inside the ear of *L. striatus* contained 11 infective filarioid larvae; identical larvae were obtained from laboratory-bred ticks 30 days after feeding on infected rodents (Muller and Nelson, 1975). Bianco and Muller (1977) and Bianco *et al.* (1983) transmitted *M. globulosa* to Mongolian jirds (*Meriones unguiculatus*) by subcutaneous inoculation of 28–50 infective larvae from *H. leachi*. At 5 days third-stage larvae were found in the liver, kidney and pelt of the jirds. By 10 days most larvae had moulted to the fourth stage and were found in the heart, lungs, and peritoneal cavity, in addition to the liver, kidney and pelt. In 18 and 25 days all fourth-stage larvae (which were tiny) were in the peritoneal cavity, free around the gut; females were undergoing the final moult. In 80 days mature worms were in the pulmonary blood vessels and by 105 days worms had grown considerably. The prepatent period was 69–88 days.

Bianco (1984) reported that, of several species or strains of commercially available rodents, only jirds (*M. unguiculatus*) were susceptible to infection. Infections were produced by subcutaneous inoculations and by exposure to bites of infected *H. leachi* but not by intravenous, intraperitoneal or oral routes of administering larvae. The mean prepatent period in jirds was 74 days and in *L. striatus* 88 days. Microfilariae persisted longer in jirds than in *L. striatus*.

M. marmotae (Webster, 1967)

This species (syn. *Ackertia marmotae*) occurs in lymphatic vessels associated with the extrahepatic bile duct, the cystic duct and the gall bladder of the groundhog (*Marmota monax* – Sciuridae). Development and transmission were studied by Ko (1972) in Ontario, Canada. Microfilariae were rarely found in blood but were numerous in the skin (reticular layers of the dermis), especially of the ears. They also occurred in the skin of the neck and legs; they were rare in skin of the belly. Microfilariae were 111–160 μm in length. One or two conspicuous round bodies (polar bodies?) were present in the sheath. Ko (1972) showed that *Ixodes cookei* ingested microfilariae as it fed on groundhogs. In the tick the microfilaria lost its sheath and invaded the fat body cell or, less commonly, an epidermal cell, where development took place. Seven days after detachment of the tick, larvae moulted; they were about 300 μm in length at that time. Larvae did not pass through a typical sausage stage during this period of development. Growth was rapid after the first moult. The second moult occurred 14 days after detachment and larvae (631 μm in length) were still in individual fat body cells. Third-stage larvae, recovered 14–25 days after detachment of ticks, were 620–940 μm in length. Infective larvae collected 36, 43, 50 days after detachment were 1.1–1.5 mm in length, with long tails. At 30°C about 1 month was required for larvae to reach the infective stage in ticks, if one takes into account the fact that ticks, remained attached to the final host for about 1 week.

Infective larvae were inoculated into a number of young groundhogs. An immature female was found in one groundhog and a male worm in a second groundhog examined 317 and 307 days postinfection.

Infected ticks were examined histologically 1.5–3 h after they had attached to a groundhog. Sections of larvae (presumably infective larvae) were found in the haemocoel near the gut and between alveoli of salivary glands. Since salivary secretions are injected continuously into the host during tick feeding, it is likely that infective larvae are inoculated with salivary secretions into the host during the long period of attachment of the tick.

I. *cookei* inhabits the deep, extensive burrows of groundhogs. Interburrow activities of groundhogs are high in spring, which helps to spread ticks between groundhogs (Ko, 1972).

M. martini Bain, Bartlett and Petit, 1986

Bain *et al.* (1985b) assumed that microfilariae in the skin of the ears of *Lemniscomys striatus* from the Central African Republic and from *Arvicanthas niloticus* from Mali belonged to *M. nilotica*, although adult worms were not found in naturally and experimentally infected murids. Microfilariae were 250–288 μm in length. They developed to the infective stage in *Hyalomma truncatum* and *Rhipicephalus sanguineus*. Infective larvae were 0.84–1.08 mm in length and had a long terminal process on the tail and two shorter dorsolateral processes. Later, Bain *et al.* (1986a) found and described worms from colonic lymphatics in *A. niloticus* as *M. martini* and concluded that material studied earlier (Bain *et al.*, 1985b) referred to this species.

Petit *et al.* (1988a) reported that hexapod larvae of *Hyalomma truncatum*, *R. sanguineus* and *R. turanicus* gorged on an infected *L. striatus*. Early larvae were found in and under the epidermis and sometimes at the level of the epidermal glands, where they elicited a syncytium of epidermal cells which bulged into the haemocoel.

Wanji *et al.* (1990) concluded that *M. martini* is mainly a lymphatic dweller. The authors infected *L. striatus* and *Meriones inguiculatus* experimentally. In *L. striatus*, larvae entered the peripheral lymphatics and migrated to lumbar and mesenteric lymph nodes. In 5 days most larvae were found in the mesentery and in 21 days most were in the intestinal wall. The third and fourth moults occurred on days 10 and 21 postinfection. Wanji *et al.* (1994) also carried out a detailed study of the microfiladerma produced by *M. martini* in *L. striatus* and *A. niloticus*. The level of the microfiladerma increased in the ear pinnae with the increase in number of larvae inoculated in *L. striatus*. *A. niloticus* was a less suitable host than *L. striatus*, microfiladermas were much reduced and adults had shorter life spans.

Onchocerca

Species of *Onchocerca* are medium-sized filarioids which usually inhabit subcutaneous tissues, ligaments and aponeuroses of large mammals; one species (*O. armillata*) occurs in the wall of the aorta. The various species are found in more or less specific sites in the host and many elicit the formation of nodules.

The genus consists of about 27 species reported from Equidae (four species), Cervidae (seven), Camelidae (one), Suidae (one), Bovidae (13) and humans (one). In the Bovidae, species occur in members of *Bos* (seven species of *Onchocerca*), *Cephalophus* (three), *Hippotragus* (one) and *Kobus* (1). The genus has a worldwide distribution. Species in horses and cattle have been distributed widely throughout the world whereas those in wild ungulates tend to have more restricted distributions, depending on that of their hosts.

Microfilariae deposited by female worms are unsheathed, with pointed, attenuated tails, and generally invade the skin of the host rather than the blood – a discovery first made by Montpellier and Lacroix (1920). The exception is *O. armillata*, in which microfilariae occur in the blood and have a nocturnal periodicity. There is some evidence that female worms in some species (e.g. *O. flexuosa*) pierce the nodule to release larvae into tissues around the capsule (Schulz-Key, 1975) but there is apparently no evidence of this in other species (Beveridge *et al.*, 1980a). Microfilariae tend to occur in specific locations in the skin, depending to some degree on the location of adult worms. For example, microfilariae of *O. lienalis* occur mainly near the umbilicus whereas those of *O. cervicalis* are found mainly in skin of the head and neck. In some instances it seems that the position of microfilariae in skin is related to vector feeding behaviour.

Vectors of *Onchocerca* spp. are species of blackflies (Simuliidae) (Blacklock, 1926) and biting midges (Ceratopogonidae) (Steward, 1935) which ingest microfilariae with their blood meals. Microfilariae invade the haemocoel of the fly and migrate to the thoracic muscles, where the third and infective stage is attained. Larvae leave the mouthparts of the fly and enter the final host.

The fate of infective larvae after entering the definitive host is not well understood. According to Beveridge *et al.* (1980a) juvenile female *O. gibsoni* provoke encapsulation and males only invade the capsule and remain there when females are mature. Schulz-Key (1975) reported that males of *O. flexuosa* probably moved from nodule to nodule; individual males were not found in nodules in *O. flexuosa* but they have been found in nodules in *O. volvulus* (Schulz-Key and Albiez, 1977). Worms may live for a long time. It is estimated, for example, that *O. volvulus* in humans can live for up to 16 years (Roberts *et al.*, 1967). Schulz-Key *et al.* (1980) and Schulz-Key and Karam (1986) concluded that each female *O. volvulus* of humans had three or four reproductive cycles of 2–4 months per year and that reinsemination was necessary for the development of each cycle. Vankan and Copeman (1988) reported cyclic phases of reproduction, each lasting 14.5 weeks, in *O. gibsoni* of cattle.

O. volvulus, a major human pathogen in parts of Africa and Central and South America, has been the subject of an extensive literature. Muller and Horsburgh (1987) published a bibliography (containing 5449 references) on animal and human onchocerciasis.

O. cebei Galliard, 1937

O. cebei (syn. *O. sweetae*) was found in nodules over the pectoral region of water buffalo (*Bubalus bubalis*) in southern Asia (India, Malaysia). It has been introduced to Australia. According to Spratt *et al.* (1978) microfilariae were 222–258 µm in length, with moderately filamentous tails. Most microfilariae were found in the superficial layers of the dermis of the skin of the flank, sternum and shoulder. Nearly all microfilariae (98%) were at a depth of less than 0.60 mm from the skin surface. Various biting insects were collected from water buffalo but only one species, *Culicoides* sp. 'M', ingested microfilariae. Developing larvae were found in the thorax of three of seven *Culicoides* which remained alive. A sausage stage was present in 2 days and a late second-stage larva 8 days postinfection.

O. cervicalis Railliet and Henry, 1910
O. reticulata Diesing, 1841

Both species are cosmopolitan parasites of horses, mules and donkeys and it is convenient to consider them together. *O. cervicalis* occurs in the *ligamentum nuchae* and

O. reticulata is found in the suspensory ligament of the fetlock, chiefly of the forelegs. Microfilariae of *O. cervicalis* were 210–250 µm in length, whereas those of *O. reticulata* were 310–395 µm in length (Bain, 1981b). Microfilariae occurred in skin near the site of adult worms. *O. cervicalis* has been associated with ocular lesions in horses (Böhn and Supperer, 1954).

Steward (1935, 1937) showed that *O. cervicalis* developed in *Culicoides nubeculosus* in England. Microfilariae apparently stayed for 3–4 days in the midgut of the fly before migrating to the thoracic muscles. Larvae were in the sausage stage in about 7 days. Infective larvae appeared in the head and proboscis 22–25 days postinfection and were 600–700 µm in length. Steward (1935, 1937) suggested that *Culicoides parroti* and *C. obsoletus* might also be vectors in England. Mellor (1971, 1975) successfully infected both *C. nubeculosus* and *C. variipennis*; microfilariae reached the infective stage in 14–15 days at 21–23°C. Mellor (1974) noted that *C. nubeculosus* had two peaks of activity in England: the evening peak occurred just before sunset and was 2–2.5 times as great as the morning peak. Most flies attacked the ventral mid-line from the front legs to the sheath or mammae.

Collins and Jones (1978), using membrane feeding, successfully infected three strains of *C. variipennis*. In Louisiana, Foil *et al.* (1984) determined that of the six species of *Culicoides* collected from ponies with *O. cervicalis*, only *C. variipennis* proved to be a suitable intermediate host.

Bain and Petit (1978) briefly described the development of *O. cervicalis* in *C. nubeculosus*. The infective stage, attained in 10 days at 26°C, was 680–870 µm in length.

O. reticulata has been associated with summer itch or cutaneous onchocerciasis in the Far East and Australia. Riek (1954) suggested that lesions on the legs of horses in Australia were the result of the bites of *Culicoides robertsi* rather than microfilariae of *O. reticulata*. Moignoux (1952) believed that *O. reticulata* developed to the infective stage in *C. nubeculosus*; the optimal temperature for development was 25°C. Other authors believed that Moignoux (1952) was dealing with *O. cervicalis*.

O. gibsoni (Cleland and Johnston, 1910)

O. gibsoni occurs in nodules in the brisket and intercostal spaces and less commonly in the shoulder of *Bos indicus* and *B. taurus*. Found mainly in southeast Asia, it has been introduced to Australia. Microfilariae were 230–270 µm in length and were found by Buckley (1938) immediately beneath the epidermis at a depth of 0.05–0.20 mm in the skin. Buckley (1938) incriminated *Culicoides buckleyi, C. orientalis, C. oxystoma, C. pungens* and *C. shortti* as vectors in Malaysia. The largest infective larvae that he found in the midges were 540–750 µm in length.

In Australia, Ottley and Moorhouse (1980) showed that *Forcipomyia townsvillensis* was a suitable intermediate host. Development in the midge to the infective stage took 6 days at 30°C, but there was continued growth of the infective larva on day 7 after the midges had fed. Beveridge *et al.* (1981) assessed the biting midges likely to be vectors of *O. gibsoni* and considered *Culicoides marksi* as a likely possibility. It was the most abundant species in North Queensland; it fed on sites on calves where microfilariae are known to occur; and it was found naturally infected with a filarioid larvae. Uterine microfilariae of *O. gibsoni* migrated into the skin of all parts of the body when inoculated into rats and mice and persisted for 21–70 days (Beveridge *et al.*, 1980a,b).

Beveridge *et al.* (1980a,b) reported that, as nodule size and female worm size increased, the numbers of females with microfilariae and the number of nodules with males increased to reach almost 100% in nodules weighing more than 3 g. This suggested that females normally become encapsulated when immature and that males enter the nodule later and remain there after fertilizing the female.

O. gutturosa Neumann, 1910
O. lienalis (Stiles, 1892)

Because of confusion in the literature, these two species from cattle will be reviewed together. Some authors regarded the two species as synonyms, but the validity of both has been established by Bain *et al.* (1978). *O. gutturosa* is a widely distributed parasite of the *ligamentum nuchae* and the femurotibial ligaments, and its microfilariae occur in skin of the neck and back of the host. *O. lienalis* is a common parasite of the gastrosplenic region between the spleen and the rumen and its microfilariae occur in the skin of the belly, especially around the umbilicus. Eichler and Nelson (1971) in England reported as *O. gutturosa* worms from the *ligamentum nuchae* (presumably *O. gutturosa*) and from the gastrosplenic region (presumably *O. lienalis*). They noted that most microfilariae occurred in the ventral belly skin, especially near the umbilicus. Thus, these microfilariae probably belonged to *O. lienalis* and not *O. gutturosa*.

Steward (1937), Gnedina (1950b), Supperer (1952), Mikhailyuk (1967), Eichler (1971) and Bain (1972) reported that *O. gutturosa* was transmitted by *Simulium* (*Adagmia*) *ornatum* in Europe. Bain (1979) pointed out, however, that *S. ornatum* feeds preferentially on the ventral belly of cattle, as noted earlier by Eichler (1971), Eichler and Nelson (1971) and others. It feeds much less commonly on the areas where microfilariae of *O. gutturosa* are found. Therefore, it is likely that the species studied by Steward (1937) and others mentioned above refers to *O. lienalis* and not to *O. gutturosa*. This conclusion is supported by Bain (1979), who described the development of the microfilariae of *O. gutturosa* in *Culicoides nubeculosus*. Ivashkin and Golovanov (1974) reported development in simuliids and midges and one suspects that they were working with mixed infections.

Steward (1937) gave the length of the microfilariae of *O. lienalis* as 235–266 µm. He collected *Simulium ornatum* that had fed on a cow and noted sausage stages in 10 days and infective larvae in the head of the flies 19–22 days postinfection (temperature not stated).

Supperer (1952) noted that microfilariae of *O. gutturosa* were 190–221 µm in length and that blackflies took up most microfilariae during the final part of the feeding; this observation suggested that microfilariae were in some way attracted to the site of the bite of the fly. A sausage stage was found in the thoracic muscles of the fly in 3–4 days. The first moult occurred in 11–13 days and the second in 15–18 days. Infective larvae were 540–563 µm in length. In Austria in May, development to the infective stage took 20–23 days whereas during warmer weather in the latter part of June the infective stage was reached in only 8 days. Bain (1972) also described the early development of *O. lienalis* in *S. ornatum*.

In the CIS, Gnedina (1950b) and Mikhailyuk (1967) reported that *Boophthora sericata*, *Simulium galeratum*, *S. ornatum* and *S. tuberosum* were intermediate hosts and that the flies would travel little more than 6 km from stream breeding sites to feed on cattle. Eichler (1971) noted that most *S. ornatum* alighted on the ventrum of cattle and that the umbilicus was both the preferred site and the site with the heaviest concentration of microfilariae. Infective larvae apparently leave flies when they are feeding and migrate to the gastrosplenic region. In Nakhicheran, CIS, *O. lienalis* developed in *Simulium* (*Adagmia*) *variegata* and *Friesia condica*, according to Khudaverdiev (1977). Most infected flies were found in June–July in the mountains but cattle may get infected from May to December. Infective larvae were found in the head of flies 8–9 days after the latter had fed. Larvae did not develop in flies at 21–22°C.

Lok *et al.* (1983a) in the eastern USA collected *Simulium jenningsi* from the umbilical area of cattle infected with *O. lienalis*. Third-stage larvae were recovered from 22% of the flies examined 8–13 days after they had fed. Ovarian dissections of 304 *S. jenningsi* attacking cattle indicated a parous rate of 56%. Larvae were found in 7.3% of the parous flies. Lok *et al.* (1983b) also inoculated blackflies with microfilariae of *O. lienalis* and showed that *Simulium decorum*, *S. pictipes* and *S. vittatum* were suitable intermediate hosts. *S. pictipes* could readily be infected by membrane feeding.

According to Bremner (1955) the fixed and stained microfilariae of *O. gutturosa* were 197–259 µm in length, in contrast to those of *O. gibsoni*, which were 242–286 µm.

Bain (1979) allowed *Culicoides nubeculosus* to feed on a bull infected with *O. gutturosa* and found infective larvae in some flies 6.5 days later at 27°C. Infective larvae were 770–900 µm in length and the tail had a terminal and two subterminal inconspicuous papillae. Mwaiko (1981) gave the length of the infective larva as 521 µm.

O. ochengi Bwangamoi, 1969

O. ochengi was found in intradermal nodules in *Bos indicus* and *B. taurus* in Africa (Togo). Denke and Bain (1978) reported circumstantial evidence that infective larvae, found in two of 82 *Simulium damnosum* which fed on a cow infected with four species of *Onchocerca*, belonged to *O. ochengi*. Infective larvae were 540–680 µm in length.

Wahl *et al.* (1998a) concluded that in North Cameroon members of the *Simulium damnosum* complex (especially *S. squamosum* and *S. damnosum*) are the only important vectors of *O. ochengi*. The infective larva of *O. ochengi* was longer and more slender and with a relatively shorter tail than the larva of *O. volvulus* and the tail was thick and rounded (Wahl and Schibel, 1998). The morphological distinction between the infective larvae of the two species was supported by DNA probes. Wahl *et al.* (1998b) studied the relationship between the transmission in North Cameroon of *O. volvulus* (in humans), *O. ramachandrini* (in warthogs) and *O. ochengi* (in cattle). These data led to the hypothesis that live infective larvae of *O. ochengi* (and possibly *O. ramachandrini*) transmitted in high numbers by *S. damnosum* s.l. to humans can result in partial protection from infection with *O. volvulus*.

O. ramachandrini Bain Wahl and Renz, 1993

O. ramachandrini of the subcutaneous tissue of the feet of the warthog (*Phacochoerus aethiopicus*) of North Cameroon had microfilariae evenly distributed across the body. Microfilariae injected into the thorax of *Simulium squamosum* and *S. damnosum* developed to the third stage, which were 955 µm in length. Similar larvae were found in wild *S. damnosum* in the Cameroon and Liberia (Wahl and Bain, 1995).

O. tarsicola Bain and Schulz-Key, 1974

O. tarsicola was found in the subcutaneous tissues surrounding the abductor tendons of the radiocarpal and tibiotarsal joints of European cervids (*Cervus elaphus, Rangifer tarandus*). Microfilariae occurred in skin near female worms but were most abundant in the ears. Schulz-Key and Wenk (1981) gave the length of the microfilariae as 340–395 μm and noted that two other *Onchocerca* spp. occurred in red deer in addition to *Cutifilaria wenki*, but the microfilariae of *O. tarsicola* were the only ones found in the ears. Forty per cent of *Simulium ornatum* induced to feed on the ears of a deer became infected and 30% of those which survived more than 18 days contained infective larvae. *Prosimulium nigripes* appeared also to be a natural vector. In *S. ornatum* the first moult was completed in 12–14 days; the cuticle of the first moult was retained until the early third stage was reached. Infective larvae which appeared in the head of the flies 23–25 days postinfection were 523–800 μm in length (unfixed). Prevalence of larvae in wild-caught flies was about 40%.

O. volvulus (Leuckart, 1893)

O. volvulus is a common and widespread parasite of humans in tropical Africa, especially in the rain forest regions and the savannah belt from Senegal to Tanzania (Nelson, 1970). It occurs also in Yemen as well as in the New World (Guatemala, Mexico, Venezuela, Columbia), where it was presumably introduced by Africans. Although *O. volvulus* has been reported in the spider monkey (*Ateles geoffroyi*) in Guatemala (Caballero and Barrera, 1958) and the gorilla (*Gorilla gorilla*) in Africa (van den Berghe *et al.*, 1964), and is transmissible to the chimpanzee (*Pan paniscus*) (see Duke, 1962), the parasite is not regarded as zoonotic. Authoritative reviews of onchocerciasis have been published by Nelson (1970), Duke (1971, 1981) and Godoy *et al.* (1986).

Onchocerciasis or river blindness is one of the most important tropical diseases. The most obvious sign of the disease is the presence of small to large tumours containing a variable number of adult worms in subcutaneous tissues. In Africa, nodules tend to occur most commonly around the pelvic region, and in Central America in the upper parts of the body, including the head. In early stages of infection in children and in lightly infected individuals, worms may be free in subcutaneous tissues and not grossly detectable. According to Nelson (1970) the presence of microfilariae in skin results in pruritus, thickening, discoloration, lichenification, pachydermia and loss of elasticity. The latter may lead to hanging groin, hernia and elephantiasis. The most serious consequence of the infection is the migration of microfilariae from the face through the conjunctiva to the eyes. Robles (1919) in Central America was the first to show that microfilariae were responsible for the chronic inflammation which often leads to blindness.

Blacklock (1926) discovered in Sierra Leone that *Simulium damnosum* was an intermediate host of *O. volvulus*. *S. damnosum* is now recognized as a complex of some 26 species (Service, 1982), many of which are known to be important vectors in Africa (e.g. Garms and Cheke, 1985). *S. neavei*, a slowly reproducing vector, the larvae and pupal stages of which have an obligatory association with freshwater crabs (*Potomonautes* spp.), is an important vector in East Africa and the related species *S. woodi* (also associated with crabs) is important in Tanzania. For recent articles on the vectors of *O. volvulus* in Africa, refer to Fain *et al.* (1981), Renz (1987), Renz and Wenk (1987), Traoré-Lamizana and Lemasson (1987) and Quillevere *et al.* (1988). Basánez *et al.*

(1995) analysed the relationship between the number of microfilariae ingested and successful larval development in the simuliid vectors in Guatemala, Venezuela and northern Brazil. A maximum number of one to three third-stage larvae per fly was observed in the four simuliid vectors, leading to the conclusion that development to the infective stage is dependent on density-dependent factors acting in the early microfilarial phase in the blackfly.

In Central and South America, *Simulium callidum*, *S. metallicum* and especially *S. ochraceum* are major vectors. Other species which have been implicated as possible vectors in Latin America are *S. haematopotum* (see Takaoka *et al.*, 1984b), *S. horacioi* (see Takaoka, 1983) and *S. pintoi* (see Takaoka *et al.*, 1984a). In Venezuela the *S. amazonicum–sanguineum* complex, consisting of four species, is important in transmission (Ramirez Perez and Peterson, 1981). Schiller *et al.* (1984) reported that *S. quadrivittatum* was a vector in Panama.

Simuliids require fast-flowing water for their larvae and pupae and it is in such places that human onchocerciasis thrives in Africa and Central America. *S. damnosum* s.l. tends to feed on the legs whereas *S. ochraceum* in the New World prefers to feed on the upper parts of the body. It is conjectured that this may account for differences in the distribution of nodules in humans in Africa and the New World (Duke, 1971). If nodules are in the head, the microfilariae may attain the eyes sooner than if nodules are in other parts of the body.

In the blackfly vector, the infective stage is reached in 6–12 days, depending on temperature. In humans, the time from infection to the time when microfilariae can first be detected in the skin is '10–20 months (commonly 15–18)' (Duke, 1971). It is estimated that adult worms can live up to 15 years.

According to Collins *et al.* (1982) a female worm initiates nodule formation and later males and other female worms may enter the nodule as well. Occasionally mature females may not become encapsulated but remain free in the subcutaneous connective tissue or between muscles. Nodules often occur in deeper tissue, such as that attached to the capsules of hip joints, and are difficult to detect. Duke (1991) pointed out the paucity of information on the development of larvae to adulthood in *O. volvulus* and described an immature male and an immature female worm from nodules in humans from Guatemala. Schulz-Key and Soboslay (1994) reviewed the reproduction of *O. volvulus* and concluded that, if the mean life span of a microfilaria was 1.0–1.5 years in the human host, then in a stable parasitdermia $22–33 \times 10^3$ microfilariae would have to be produced each day. It was estimated that 30 female worms would be needed to produce that number of microfilariae (see also Schulz-Key and Karam, 1986; Schulz-Key, 1990).

Skrjabinofilaria

S. skrjabini Travassos, 1925

This is a parasite of the subcutaneous tissues and body cavity of opossums (Didelphidae) in South America (e.g. *Caluromys philander*, *Didelphis marsupialis*, *Marmosa demarerae*, *M. murina*, *Metachirops opossum*). Sheathed microfilariae occurred in the blood and were 180–205 µm in length, with attenuated tails. Bain and Durette-Desset (1973) allowed *Culex pipiens* to feed on an infected *M. opossum* and studied the development of the larvae. Development took place in the flight muscles of the mosquitoes. At 26–32°C the first moult took place on day 8 and the second started about 12 days and lasted until 18 days postinfection, when the infective stage was reached. Infective larvae were 1.47–2.10 mm in length. The tail had two large, triangular lateral appendages a short distance from the terminal end.

Yatesia

Y. hydrochoerus (Yates and Jorgenson, 1983)

This species (syn. *Dipetalonema hydrochoerus*) is a common parasite of the skeletal muscle fascia of capybaras (*Hydrochaeris hydrochaeris*) in Colombia, South America. Microfilariae appearing in the blood were unsheathed, had attenuated tails and were 250–305 µm in length. According to Yates and Lowrie (1984) the intermediate hosts were ticks (Ixodidae), specifically *Amblyomma americanum* and *A. cajennense*. Microfilariae invaded the muscles of the tick. At 25°C development was slow during the first 2 weeks but on days 14–15 the first moult occurred. The second moult took place 26 days postinfection. Third-stage larvae were 2.0–2.7 mm in length. The tail was rather long and slightly tapered, with one terminal and two subterminal papillae.

Subfamily Splendidofilariinae

Members of the subfamily occur in reptiles, mammals and birds. They are widespread in the latter, especially Passeriformes, and some species are not highly host specific. Species are small and often difficult to locate in their hosts.

Aproctella

The seven species in the genus are generally found in the body cavity of birds. *A. stoddardi* has been found in Passeriformes and Galliformes in North America. Microfilariae are moderately long, with pointed tails.

A. alessandroi Bain, Petit, Kozek and Chabaud, 1981

A. alessandroi was found in the body cavity of the blue-grey tanager (*Thraupis episcopus*) (Passeriformes – Thraupidae) in Colombia, South America. According to Bain *et al.* (1981b) microfilariae were 153–155 µm in length in the uterus and 147–157 µm when immobilized by heat. Microfilariae were unsheathed and had pointed tails. They invaded the thoracic muscles of *Aedes togoi* and developed to the infective stage in 13 days. Infective larvae were 1.33–1.60 mm long. The oesophagus was long and divided and the tail ended in two small lateroterminal papillae.

Cardiofilaria

The genus consists of about 12 species found mainly in the body and pericardial cavity of birds. Microfilariae, which occur in the blood, are unusually long (approximately 300 μm), unsheathed and with long tapered tails. One species (*C. pavlovskyi*) has been reported in birds as diverse as Falconiformes, Charadriiformes and Passeriformes.

C. nilesi Dissanaike and Fernando, 1965

Niles (1962) found unidentified third-stage larvae in the fat body of naturally infected *Mansonia crassipes* (syn. *Coquillettidia crassipes*) in Sri Lanka. Larvae were inoculated into chickens (Niles *et al.*, 1965) and adults were subsequently found in the body cavity and described by Dissanaike and Fernando (1965). Microfilariae were 330–390 μm in length. Adult worms were successfully transplanted to uninfected chickens which subsequently developed microfilaraemias (Gooneratne, 1968, 1969; Niles and Kulasiri, 1970).

Niles and Kulasiri (1970) noted that *M. crassipes* was a most efficient intermediate host and in one experiment 11 mosquitoes had 859 infective larvae when examined 10.5 days after they had fed on an infected chicken. Numerous (32) chicks (1 day and 1 week old) were inoculated with infective larvae. Microfilariae appeared in the blood 21–36 days postinfection. The patent period in the chickens was at least 5 months. There was a tendency for microfilariae to increase in numbers in the blood from 1600 and 2000 h; *M. crassipes* is crepuscular in its feeding habits. Worms were almost always found in the body cavity of experimentally infected birds.

Cheong and Omar (1970) and Mak *et al.* (1984) found larval *C. nilesi* in *M. crassipes* in Malaysia; Cheong *et al.* (1981) reported infective larvae of *Cardiofilaria* in *Armigeres subalbatus*, *Culex annulus*, *C. tritaeniorhyncus* and *M. crassipes* near Kuala Lumpur, Malaysia.

Unfortunately the wild hosts of *C. nilesi* have not been identified. *C. nilesi* resembles *C. pavlovskyi* described by Strom (1937) but probably not *C. inornata* (Anderson, 1956a) which Sonin (1966) and Bartlett and Anderson (1980a) placed into synonymy with *C. pavlovskyi*. *C. inornata* has a dilated cephalic extremity and a clearly defined anus in the female, and the microfilariae are considerably longer (403–445 μm) than those reported in *C. pavlovskyi* and *C. nilesi* (see Anderson, 1956a). The latter may be a synonym of *C. pavlovskyi*, however, which has been reported in a wide range of birds in France, the CIS and southeast Asia.

Chandlerella

Twenty-six species are known. They are commonly found in connective tissues around arteries in passerine birds but may also occur in other sites. Microfilariae occur in the blood, are long and sheathed and have blunt, very slightly tapered caudal ends. Vectors are ornithophilic ceratopogonids which are crepuscular or nocturnal in their feeding habits.

C. chitwoodae Anderson, 1961

C. chitwoodae (syn. *C. flexivaginalis*) was found in connective tissues around the vena cava, the splenic, hepatic, mesenteric, pulmonary and brachial arteries and the adrenal glands and in subcutaneous connective tissues of the neck of various passerine and gallinaceous birds (Bartlett and Anderson, 1980a,b). Microfilariae (183–208 μm in length) occurred in the blood.

Microfilariae developed to the third stage in the thoracic muscles of *Culicoides travisi* and *C. stilobezzioides* which engorged on crows (*Corvus brachyrhynchos*) in Ontario. At 32°C microfilariae were found in the thorax of the flies in 12 h after they had fed on infected crows. First-stage larvae prepared to moult as early as 3 days, when they were 135–185 μm in length. Third-stage larvae, present in the mouthparts of midges by 5 days, were 370–450 μm in length.

C. stilobezzioides was considered a better potential vector than *C. travisi* because it ingested microfilariae more consistently than the latter species (cf. *Chandlerella quiscali*). *C. chitwoodae* seems to be a parasite of birds sharing woodland habitats in Ontario, where it is transmitted by ornithophilic *Culicoides* spp. to a variety of woodland birds.

C. quiscali (Linstow, 1904)

C. quiscali is a parasite of the pia mater of the cerebrum of grackles (*Quiscalus quiscula versicolor*) in the USA. The microfilaria had a blunt posterior end (Odetoyinbo and Ulmer, 1959). Robinson (1971) collected biting midges (*Culicoides crepuscularis, C. travisi*) from infected grackles in Minnesota and examined them for larval stages of *C. quiscali*. Exsheathing microfilariae were found in *C. crepuscularis* 15 min after midges were collected. By 12 h microfilariae were found lying parallel to the thoracic muscles. Larvae passed through a sausage stage. The first moult apparently occurred in 72–84 h, the second in 120 h, and in 120–132 h larvae migrated to the head and mouthparts of the midges. An infective larva was 475 μm in length.

C. crepuscularis appeared to be a more suitable vector than *C. travisi*, based on the number of flies feeding on grackles which became infected.

C. striatospicula Hibler, 1964

C. striatospicula was found commonly in connective tissues around the splenic artery, oesophagus and mesentery of the American black-billed magpie (*Pica pica hudsonia*) in Colorado. Microfilariae were 138–181 μm in length. According to Hibler (1963) microfilariae were nocturnally periodic in the blood and reached a peak at 2300–2400 h. Microfilariae developed to the infective stage in the thoracic muscles of *Culicoides haematopotus*. At 29°C a sausage stage was present in 3–5 days. The first moult occurred in 6–8 days and the second in 8–10 days. Infective larvae appeared in the head of the midges in 10–12 days. The prepatent period of 37–53 days was determined by the regular examination of the blood of birds exposed to wild biting midges.

Splendidofilaria

The genus contains a number (31) of small nematode species found in the tissues of birds, including leg joints, subcutaneous tissues and artery walls. Microfilariae are short and unsheathed and occur in the blood. Vectors are ornithophilic ceratopogonids and simuliids.

S. *californiensis* (Wehr and Herman, 1956)

S. californiensis (syn. *Lophortofilaria californiensis*) was found in the ventricles of the heart of California quail (*Lophortyx californicus*) in California (Wehr and Herman, 1956). It was subsequently reported to be a common parasite of quail. Worms were encapsulated in the wall of the aorta near the heart (Weinmann *et al.*, 1979). Microfilariae were 126 μm long and occurred in the blood. From June to September, 68 nocturnally active *Culicoides multidentatus* were collected and kept for 2–15 days after feeding on infected quails. Forty-five per cent of the flies contained developing larvae. Larvae appeared to be in fat body cells. Infective larvae were 385–480 μm in length. Field data suggested the prepatent period was about 6 months and birds remained patent for at least 15 months. *C. multidentatus* was active from June to September.

S. *fallisensis* (Anderson, 1954)

S. fallisensis (syn. *Ornithofilaria fallisensis*) is a parasite of the subcutaneous connective tissues of wild black ducks (*Anas rubripes*) and domestic ducks (*A. platyrhynchos domesticus*) in North America. Microfilariae occurred in the blood and were 90–121 μm in length, with blunt tails. In the peripheral blood, microfilariae exhibited a marked diurnal periodicity. Anderson (1956b, 1968b) reported that the diurnal, **ornithophilic** simuliids *Simulium anatinum* and *S. rugglesi* were vectors of *S. fallisensis* in Algonquin Park, Ontario. These **nettaphilic** (duck-loving) and **chenophilic** (goose-loving) simuliids attacked ducks on water but not ducks on land or elevated above the water level. Flies crawled under the feathers and quickly engorged on blood; the two simuliids are also vectors of the pathogenic sporozoan *Leucocytozoon simondi* of ducks and geese. *S. fallisensis* also developed in mammal-biting simuliids such as *Simulium parnassum* and *S. venustum*, but these flies were not attracted to birds and do not feed naturally on waterfowl.

S. *anatinum* emerged from streams in Algonquin Park during the early part of May and disappeared in about 2 weeks whereas *S. rugglesi* emerged the end of May and persisted to the middle of July with decreasing numbers. Domestic ducks exposed to the feeding of blackflies became infected, the source of the parasite being wild ducks in the area.

Microfilariae penetrated the stomach wall and entered the haemocoel of the flies within 15 min of ingestion. Development took place in the haemocoel and the third-stage larvae appeared in the mouthparts of the blackfly in 7–14 days, depending on the temperature. Larvae passed through a sausage stage; the first moult occurred when larvae were 210 μm in length and the second when they were 350 μm in length. At an average ambient temperature of 17.1°C in the third week of June, the first moult occurred in 4 days, and the second in 7 days; larvae appeared in the head in 8 days. Infective larvae were 389–486 μm in length and the tail had two prominent lateroterminal swellings. The oesophagus was short, narrow and undivided.

Microfilariae appeared in the blood of ducklings 30–36 days after they were inoculated with third-stage larvae. In contrast to naïve ducklings, which consistently became infected when inoculated with larvae or exposed to blackflies in the field, re-exposure failed to alter declining microfilariaemias in ducks, or to result in the reappearance of microfilariae in ducks which had been previously infected.

S. picacardina Hibler, 1964

S. picacardina was found in the myocardium behind the aortic and pulmonary semi-lunar valves of 81% of American black-billed magpies (*Pica pica hudsonia*) examined in northern Colorado. Microfilariae were 120–148 µm in length, with blunt caudal extremities, and exhibited a marked nocturnal periodicity in the blood, with a peak in 2200–2300 h. Microfilariae invaded the abdominal fat body of *Culicoides crepuscularis* which readily fed on the magpies. Developing first-stage larvae left the fat body and developed free in the abdominal haemocoel. According to Hibler (1963) the first moult occurred in 4–6 days and the second in 6–8 days at 29°C. Larvae were found in the head in 7–9 days postinfection. Third-stage larvae were 465–580 µm in length. The caudal end was truncated, with two terminolateral swellings, and the cephalic end had a constriction behind the cephalic extremity. Five fourth-stage larvae were found in the aorta of magpies exposed 2 weeks earlier to the bites of wild midges, and ten early fifth-stage worms were found in three birds which had been exposed to wild biting midges 3 weeks earlier. Microfilariae appeared in the blood of exposed magpies in 42–73 days.

Thamugadia

Members of the genus occur in subcutaneous tissues of geckoes. Microfilariae are short and sheathed and occur in the blood.

T. ivaschkini Annaev, 1976

T. ivaschkini is a parasite of Gekkonidae in southern CIS. Reznik (1982) captured infected geckoes in Turkmenia and allowed sandflies (Psychodidae) to feed on them. Microfilariae in the blood of the geckoes were 122–140 µm in length, with tapered, rounded extremities. Microfilariae invaded the flight muscles of the flies (*Phlebotomus caucasicus*, *P. papatasii* and *Sergentomyia arpaclensis*) and at 24–27°C the sausage stage was found in 3–4 days. The first moult occurred in 6–7 days and the second in 8–9 days; infective larvae appeared in the head and haemocoel of the thorax in 9–10 days.

Madathamugadia

M. ineichi Bain, Wanji, Petit, Paperna and Finkelman, 1993

This species was found in *Pseudocordylus microlepidotus melanotus* (Cordylidae) in South Africa. The microfilariae were 112–125 µm in length and attained the infective stage in the thorax of *Phlebotomus dubosqui* (Bain *et al.*, 1993). The infective larvae were 490–530 µm in length.

Subfamily Lemdaninae

Eufilaria

There are about 15 species in the genus. They are delicate parasites found in the subcutaneous connective tissues of birds, especially of the neck region. Microfilariae are short and unsheathed and occur in the blood. Species studied are transmitted by ornithophilic ceratopogonids.

E. bartlettae Bain, 1980

This species was found, along with *E. delicata*, in a blackbird (*Turdus merula*) in France. Its microfilariae were 110–122 µm when immobilized by heat and 116 µm with vital staining (Bain, 1980). Microfilariae apparently developed in *Culicoides nubeculosus*. Infective larvae found in the same biting midges were distinguished from those of *E. delicata* on the basis of frequency, i.e. larval types found most frequently in the flies were assigned to *E. bartlettae*. The latter were 410–485 µm in length.

E. delicata Supperer, 1958

E. delicata was found in the subcutaneous connective tissues of the blackbird (*Turdus merula*) in Europe. Microfilariae in the uterus of fixed female worms were 107–155 µm in length, in blood immobilized by heat 146–170 µm, and after vital staining 170 and 180 µm. They were unsheathed, with attenuated tails. Bain (1980) reported development of the microfilariae to the infective stage in *Culicoides nubeculosus* in 8 days at 26°C. Infective larvae were 568 µm in length. The oesophagus was long and divided; the tail was blunt and rounded.

E. kalifai Millet and Bain, 1984

E. kalifai is a parasite of the subcutaneous connective tissues of magpies (*Pica pica pica*) in France. Microfilariae were 193–198 µm in length when immobilized with heat and stained vitally, and 120–155 µm from the uteri of fixed females. The tail was attenuated. Microfilariae reached the infective stage in *Culicoides nubeculosus* in 8–10 days at 25°C (Millet and Bain, 1984). Infective larvae were 440–580 µm in length and had a long divided oesophagus. The tail had four rounded papilla-like structures (Millet and Bain, 1984).

E. longicaudata Hibler, 1964

This species was found embedded in the wall of the pulmonary arteries (usually in the tunica media) of American black-billed magpies (*Pica pica hudsonia*) in Colorado. Microfilariae, which occurred in blood, were 88–122 µm in length, with long attenuated tails. According to Hibler (1963) microfilariae developed in the thoracic muscles of *Culicoides crepuscularis* and *C. haematopotus*. At 29°C a sausage stage was found in 1–3 days. The first moult occurred in 3–5 days and the second in 5–7 days. Infective larvae appeared in the head of midges in 6–8 days. Microfilariae appeared in the blood of magpies in 34–76 days after they had been exposed to bites of wild *Culicoides* spp.

Eulimdana

Members of the genus are found in birds, particularly those in the order Charadriiformes, where there are 11 known and likely many undescribed species. Adult worms occur mainly in the neck region of the host, both subcutaneously and on the serosa of the oesophagus. Microfilariae are sheathed and most occur in skin. Skin-inhabiting microfilariae in birds were first discovered in African house swifts infected with *E. cypseli*. Unfortunately, this discovery went largely unnoticed, since it was published as a brief statement within the context of a larger manuscript (Nelson, 1962b) on an unrelated filarioid. Dutton (1905) found that *E. cypseli* was transmitted by lice (Mallophaga); this was the first demonstration of a louse vector among the avian filarioids. Bartlett *et al.* (1989) suggested and Bartlett (1992b) showed that lice transmit species in Charadriiformes.

Bartlett *et al.* (1989), Bartlett and Anderson (1990) and Bartlett (1992a) provided convincing evidence that, in Charadriiformes, post-reproductive adult worms are **ephemeral** and that microfilariae are long-lived (for review, see Anderson and Bartlett, 1994). In most infected adult birds the only evidence of infection was the presence of microfilariae in the skin. Transmission was presumed generally to be from adult birds to the newly hatched precocial chicks. Juvenile birds were, therefore, the best prospect in which to find the combination of adult worms and microfilariae required for taxonomic study.

E. bainae Bartlett, 1992

This species occurs in whimbrels (*Numenius phaeopus*) in Iceland. Microfilariae in 2% formalin in saline were 96–120 μm in length. They had an oval sheath and a short pointed tail and were common in skin of the wings. Bartlett (1992b) collected lice from a wild-caught adult whimbrel harbouring skin-inhabiting microfilariae and examined the lice for filarioid larvae. Eleven sausage- and one moulting first-stage larvae presumed to be those of *E. bainae* were found in three of four adult males and in two of six adult females of the amblyceran louse *Austromenopon phaeopodis*. One third-stage larva was found in each of two of 17 adult females of the ischnoceran louse *Lunaceps numenii phaeopi*. The tail of the third-stage larvae had two subventral, sublateral tongue-like structures and a terminal pointed structure. One third-stage larva was 775 μm in length. Larvae were free in the abdominal haemocoel of lice.

E. cypseli (Annett, Dutton and Elliot, 1901)

E. cypseli (syns *Filaria cypseli*, *Eufilaria cypseli*, *Lemdana cypseli*) is a parasite of the subcutaneous connective tissues of the head and neck of the house swift, *Apus affinis* (syn. *Cypselus affinis*) (Apodidae) in West Africa (Gambia and Nigeria). Dutton (1905) found filarioid larvae in the fat body and haemocoel of an unidentified mallophagan (Leiothinae), generally 'on the undersurfaces of the long feathers of the wing on either side of the quill'. He noted that the lice fed not only on the feathers of the bird but also on blood and lymph. Microfilariae were rarely found in the peripheral or heart blood but numerous microfilariae were found in lymph from the swollen feet of many birds; the relationship between the swollen feet and the presence of infection with *E. cypseli* was not specified. Microfilariae in lymph were 84.7 μm in length and ensheathed; the sheath was somewhat shorter than the microfilariae. In fixed and stained preparations, the microfilariae were 75.0–84.7 µm in length. A cephalic spine was noted as well as 'a short, rather broad, highly refractile tubercle, which was always in contact with the sheath'.

Various stages in the development of *E. cypseli* were found in infected mallophagans, including a sausage stage. The lengths of the most advanced stages, which were active in the haemocoel, were 368.6 µm and 612.88 µm.

Nelson (1962b) reported that the observations of Dutton (1905) had been confirmed and that the louse concerned was *Dennysus hirundis*. He goes on to say that 'in Kenya *D. hirundis* collected from the house swift, showed full development of a species belonging to the subfamily Eufilariinae. Adult worms were found in the connective tissues of the neck. The short, sheathed microfilariae are confined to the skin.'

E. wongae Bartlett, 1992

This species occurs in marbled godwits (*Limosa fedoa*) in North America. Microfilariae in 2% formalin in saline were 128–155 µm in length. They had an oval sheath and an attenuated, pointed tail and were common in skin of the wings. Bartlett (1992b) collected lice from a wild-caught juvenile godwit harbouring skin-inhabiting microfilariae and examined the lice for filarioid larvae. Microfilariae and one sausage-stage, one moulting first-stage and one third-stage larvae (all presumed to be *E. wongae*) were found in four of 33 adult females and three of 35 adult males of the ischnoceran louse *Carduiceps clayae*. Microfilariae and one second-stage larva were found in two of eight adult females of the amblyceran louse *Austromenopon limosae*. One third-stage larva was found in one of ten adult females of the amblyceran louse *Actornithophilus limosae*. The tail of the third-stage larva had two subventral, sublateral tongue-like structures, a terminal point, and two small subdorsal, sublateral tongue-like structures. Third-stage larvae were 635 and 690 µm in length. Larvae were free in the abdominal haemocoel of lice.

Sarconema

The genus consists of *S. eurycerca* of the myocardium of swans and geese and *S. pseudolabiata* of the subcutaneous tissues of ducks.

S. eurycerca Wehr, 1939

Originally described from the whistling swan (*Cygnus columbianus*) in the USA, this species has since been reported in other species of swans as well as in geese (*Anser* spp.) in the USA, England and the CIS (Sonin, 1966). Sheathed microfilariae appeared in the blood and were 329–331 µm in length (Seegar *et al.*, 1976)

Seegar *et al.* (1976) found larval stages of *S. eurycerca* in the mallophagan *Trinoton anserinum* collected from infected whistling swans. Infective larvae from lice were inoculated into two naïve mute swans. One swan died 27 days after inoculation and an immature *S. eurycerca* was found in its myocardium. Microfilariae appeared in the blood of the other swan in 98 days.

Saurositus

S. agamae hamoni Bain, 1969

S. agamae hamoni was found in the mesentery of the lizard *Agama agama* in the savannah of Upper Volta (Bain, 1969b). Microfilariae in the blood were short (89–96 μm) with a conical tail and surrounded by sheaths closely applied to the body. Microfilariae developed to the infective stage in flight muscles of *Anopheles stephensi* in 7.5 days at 23–25°C. A sausage stage was formed. The first moult took place in 4.5 days, when larvae were 100–130 μm in length. The second moult took place in larvae which were 210–250 μm in length. Infective larvae were 250–350 μm in length. The long oesophagus of the third-stage larva consisted of a filamentous muscular part and an equally long expanded glandular part.

7.9

The Superfamily Aproctoidea

Aproctoids are small to medium-sized worms found in air sacs, nasal cavities, subcutaneous tissues of the head and neck, and orbits of birds (Anderson and Bain, 1976). They produce thick-shelled eggs with a fully developed first-stage larva. The superfamily is divided into the Aproctidae and the Desmidocercidae. Little is known about the transmission of any of the species in these two families. The aproctoids occur mainly in terrestrial birds whereas the desmidocercids occur in piscivorous birds (e.g. herons, cormorants, albatrosses).

Family Aproctidae

Aproctids living in the orbits may release eggs which pass down the lachrymal ducts, although this has not been demonstrated. Reports of specimens in subcutaneous tissues may have been in error because the worms may occur in extensions of the cervical air sacs. On the other hand, if some species do live in the subcutaneous tissues of the head and neck, they could presumably migrate to the orbits to release their eggs as suggested by Anderson and Chabaud (1958).

Aprocta

A. cylindrica Linstow, 1883

This species has been found in ploceids (*Quelea quelea, Ploceus capitalis, P. cucullatus, Euplectes orix*) in Africa by Quentin *et al.* (1976a). They infected *Locusta migratoria* with eggs from worms found in these birds. Eggs were oval and each contained a larva similar in morphology to that of the diplotriaenoids (cf. *Serratospiculum tendo*). Developing larvae were found in the haemocoel 6–15 days postinfection. At 28°C the first moult took place 6–8 days and the second 8–11 days postinfection. Larvae were not encapsulated and had completed their development by 15 days, when they were 408 μm in length.

Family Desmidocercidae

Some species have been reported in sites other than the air sacs (e.g. kidney, liver, lungs, gall bladder, on the intestine and gizzard). Air sacs in birds are complex and easily disrupted during necropsy and thus the location of worms has probably sometimes been incorrectly identified. Also, worms may migrate from damaged air sacs to abnormal sites. There is some evidence that desmidocercids reach fish-eating birds through the ingestion of fish paratenic hosts. Given the systematic affinities of the family, some aquatic arthropods may serve as intermediate hosts.

Desmidocercella

D. numidica (Seurat, 1920)

D. numidica is a common parasite of the air sacs of herons (Ardeidae) in Africa, North America and the CIS. Female worms are frequently non-gravid. Anderson (1959) found gravid females and described the oval, rather thin-shelled eggs (56 × 24 μm in size) and the first-stage larva (170–176 μm in length). The latter was differentiated and had a short tail ending in a sharp point. Dubinin (1949) found larvae in the eyes of small fish (including *Abramis brama, Rutilus frisi, Scardinius erythrophthalmus, Perca fluviatilis*) in the Volga region of the CIS. He recovered adult worms identified as *D. numidica* in herons and cormorants fed eyes containing larvae.

7.10

The Superfamily Diplotriaenoidea

Diplotriaenoids are large worms which inhabit the air sacs of reptiles and birds (Anderson and Bain, 1976). The common genus *Diplotriaena* contains many species found in birds throughout the world. In birds, diplotriaenoids usually occur in abdominal and thoracic air sacs but they are often erroneously reported in earlier literature as being in the body cavity. Some early authors (e.g. Blanc, 1919) correctly reported them from air sacs of birds, however, and Chabaud (1955) noted that eggs passed through the respiratory system and out in faeces. Anderson (1957a) confirmed these observations, demonstrated experimentally that grasshoppers were suitable intermediate hosts and described various developmental stages for the first time. These observations led eventually to the recognition that diplotriaenoids were unrelated to the Filarioidea, with which they had been associated in the classification of the nematodes.

The female produces oval, smooth, thick-shelled eggs containing a fully developed first-stage larva. Eggs are probably deposited in vents leading from the lungs to the air sacs and make their way through the respiratory system to the throat, where they are swallowed and passed in faeces. Eggs hatch in the gut of grasshoppers and locusts. The first-stage larva is short and stout and has a prominent dorsal spine near the oral opening. The anterior end is encircled by rows of minute, posteriorly directed spines. The tail tip is rounded and encircled by a row of spines. In the proper intermediate host, first-stage larvae are extremely active and they penetrate the gut of the insect and lodge in the fat body, where development with the usual two moults occurs. Larvae may become surrounded by delicate transparent capsules. The third-stage larva is short and stout with a voluminous glandular oesophagus which, in the genera *Diplotriaena* and *Quadriplotriaena* (but not in *Serratospiculum*), compresses the intestine into a confined space anterior to the rectum.

Development of the diplotriaenoids in the final host has been studied in only one species (*Diplotriaena tricuspis*). The evidence suggests that species of *Diplotriaena* develop in the hepatic portal system and migrate to the lungs as late fourth-stage or subadult worms via the heart and the pulmonary arteries. They break out of the arterial system in the lungs and invade the air sacs.

Family Diplotriaenidae

Subfamily Diplotriaeninae

Diplotriaena

D. agelaeus (Walton, 1927)

D. agelaeus (syn. *Diplotriaenoides translucidus*) is a parasite of the air sacs of Icteridae in North America. Anderson (1957) collected eggs from worms found in *Quiscalus quiscula* in Canada and exposed various insects to them, including ants, beetles (carabids, histerids and tenebrionids), camel crickets (Rhaphidophorinae) and field crickets (Gryllidae) as well as the grasshopper *Camnula pellucida*. Development occurred only in *C. pellucida*. Larvae (250 μm in length) hatched in the gut of grasshoppers kept at 25°C and were exceedingly active. They penetrated the gut wall and entered the fat body. The first moult occurred 24 days and the second 27 days postinfection. At the time of the second moult the oesophagus occupied about half the length of the body. During development of the third-stage larva to the infective stage 31 days postinfection, the oesophagus grew posteriorly and compressed the intestine. Infective larvae had prominent lateral alae and were about 710 μm in length.

D. bargusinica Skrjabin, 1917

This species occurs in thrushes (Turdidae) in North America. Anderson (1962) took eggs (55 × 36 μm in size) from worms found in air sacs of the veery (*Catharus fuscescens*, syn. *Hylocichla fuscescens*) in Canada and followed development in grasshoppers (*Melanoplus bilituratus*, *M. fasciatus*, *Camnula pellucida*) maintained at 30–33°C. First-stage larvae were 220 μm in length. In 2 days larvae were found free in the haemocoel of the grasshoppers. In 9–11 days larvae were in delicate capsules in the fat body and undergoing the first moult. The second moult took place 14–16 days postinfection. Growth of the third-stage larva (533–624 μm in length) was completed 18–19 days postinfection. The glandular oesophagus grew posteriorly after the second moult and compressed the intestine. The infective larva had prominent lateral alae.

Attempts were made to infect wild-caught birds. Infective larvae were given to two veeries and at 55 and 84 days immature adult worms were found in air sacs. A single immature male was found in air sacs of an olive-backed thrush (*Catharus ustulata*, syn. *Hylocichla ustulata*) 301 days postinfection. Attempts to infect a robin (*Turdus migratorius*) and a grackle (*Quiscalus quiscula*) were unsuccessful.

Worms identified as late fourth-stage larvae preparing to moult were described from the heart and aorta of a 3-week-old olive-backed thrush captured in the wild and it was suggested that species of *Diplotriaena* may spend time in the circulatory system before taking up their final location in the respiratory system (cf. *Diplotriaena tricuspis*).

D. tricuspis (Fedchenko, 1874)

The development of this species was studied by Cawthorn and Anderson (1980) in the American crow (*Corvus brachyrhynchos*) (Fig. 7.12). Eggs were 51 ± 3.9 × 36 ± 2.7 μm in size and first-stage larvae were 230 ± 18 μm in length. Larvae developed in the fat body of grasshoppers (*Melanoplus sanguinipes*) and locusts (*Schistocerca gregaria*). At 30°C the first moult occurred 4 days and the second 8 days postinfection. Third-stage

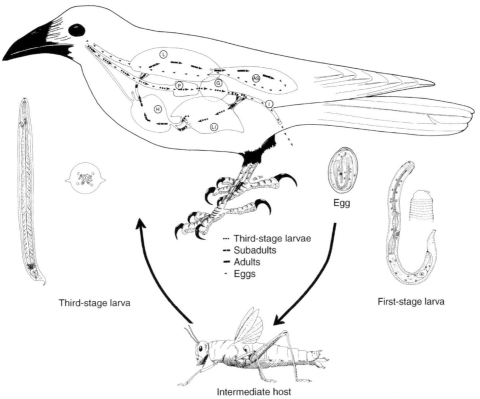

Fig. 7.12. Development and transmission of *Diplotriaena tricuspis*. AS = air sac; G = gizzard; H = heart; I = intestine; L = lung; Ll = liver; P = proventriculus.

30°C the first moult occurred 4 days and the second 8 days postinfection. Third-stage larvae (641 ± 46 μm in length) did not increase significantly in size after the second moult but the glandular oesophagus grew posteriorly and almost obliterated the intestine. The genital primordium was present behind the excretory pore and lateral alae were prominent.

Larvae were given to hand-reared crows, which were examined at intervals thereafter. Washes were made of the air sacs and body cavity. Larvae undergoing the third moult were recovered in washes 4 days postinfection. Sexes were indistinguishable. Fourth-stage larvae were recovered in washes and the sexes could be distinguished 7 days postinfection. In 10 days trident primordia were recognizable. The fourth moult occurred 20 days postinfection. All third- and fourth-stage larvae were recovered after major blood vessels had been broken and this suggested that development was occurring in blood vessels. Subadult worms were recovered from the air sacs and the right heart 25–26 days postinfection. They were also found in lungs at 25 days. They were found in the pulmonary arteries 30–33 days postinfection and by 45 days most were in the air sacs. At 60 days all worms found were in the air sacs but one was found in the mesenteric vein. Sperm appeared in male worms in 60 days and a crow began to pass eggs 117 days postinfection. The evidence suggested that *D. tricuspis* develops in the vascular system, probably the hepatic portal system; histological evidence supported this conclusion (Cawthorn *et al.*, 1980). When the subadult stage is reached, worms apparently migrate to the heart and follow the pulmonary arteries to the lungs, where they invade the channels leading to the air sacs.

D. tridens (Molin, 1858)

D. tridens was collected from *Sylvia atricapilla* in France by Bain and Vaucher (1973). Eggs were given to *Locusta migratoria* and development in the fat body of this host was described. At 29°C the first moult occurred 7 days and the second 16 days postinfection. Infective larvae were fully developed in 19 days. During development of the third stage, the glandular oesophagus grew posteriorly and compressed the intestine. The infective larva had prominent lateral alae.

Quadriplotriaena

Q. hypsokysta Crites, 1964

This species occurs in meadowlarks (*Sturnella* spp.). Anderson (1989) infected nymphs of the grasshopper *Aeropedellus clavatus* with eggs from worms found in the air sacs of *Sturnella neglecta* in western Canada. Infective larvae were found in delicate capsules in the fat body 30 days later. Infective larvae were similar in size and morphology to those of *Diplotriaena*.

Subfamily Dicheilonematinae

Dicheilonema

D. rheae (Owen, 1843)

D. rheae, a parasite of the air sacs of *Rhea americana*, has been reported commonly in zoos. Vakarenko (1994) found infective larvae in *Acrida turrita* and *Calliptamus barbatus* in a zoo harbouring rheas. Larvae in infected crickets eaten by predaceous insects entered and persisted in the tissues of the latter as well as in snakes and lizards. These latter hosts may be regarded as significant paratenic hosts, in zoo situations at least.

Serratospiculum

S. tendo (Nitzsch, 1857)

This species is a parasite of falcons. Worms were collected from *Falco peregrinus* in Europe and Bain and Vassiliades (1969) used eggs from the parasite in an attempt to infect *Geotrupes sylvaticus*, *Locusta migratoria*, *Periplaneta americana* and *Tenebrio molitor*. Larvae were subsequently found in *G. sylvaticus* and *L. migratoria*. In locusts at 25–28°C the first moult occurred 6 days and the second about 12 days postinfection. At

48 days worms were found in delicate capsules in the fat body. The first-stage larva was similar in morphology to that of *Diplotriaena*. The infective third-stage larva was similar to that of *Diplotriaena* but the glandular oesophagus was shorter and did not compress the intestine posteriorly. The caudal end of the larva had a group of about six to eight finger-like projections.

References (Spirurina)

Abdulcader, M.H.M., Tharumarajah, K. and Rajendran, K. (1966) Filarial infection in domestic cats in Ceylon. *Bulletin of the Indian Society of Malaria and Communicable Diseases* 3, 155–158.

Abe, S. (1937) Development of *Wuchereria bancrofti* in the mosquito, *Culex quinquefasciatus*. *Journal of Medical Association of Formosa* 36, 483–519. (In Japanese.)

Abraham, D. (1988) Biology of *Dirofilaria immitis*. In: Boreham, P.F.L. and Atwell, R.B. (eds) *Dirofilariasis*. CRC Press, London, pp. 29–46.

Adcock, J.L. and Hibler, C.P. (1969) Vascular and neuro-ophthalmic pathology of elaeophorosis in elk. *Pathological Veterinary* 6, 185–213.

Adcock, J.L., Hibler, C.P., Abdelbaki, Y.Z. and Davis, R.W. (1965) Elaeophoriasis in elk (*Cervus canadensis*). *Bulletin of Wildlife Disease Association* 1, 48.

Addison, E.M. (1973) Life cycle of *Dipetalonema sprenti* Anderson (Nematoda: Filarioidea) of beaver (*Castor canadensis*). *Canadian Journal of Zoology* 51, 403–416.

Addison, E.M. (1980) Transmission of *Dirofilaria ursi* Yamaguti, 1941 (Nematoda: Onchocercidae) of black bear (*Ursus americanus*) by blackflies (Simuliidae). *Canadian Journal of Zoology* 58, 1913–1922.

Addison, E.M. and Anderson, R.C. (1969) A review of eyeworms of the genus *Oxyspirura* (Nematoda: Spiruroidea). *Wildlife Diseases* (microfiche), pp. 55, 58 (W.D. 69–6).

Africa, C.M., Refuerzo, P.G. and Garcia, E.Y. (1936a) Observations on the life cycle of *Gnathostoma spinigerum*. *Philippine Journal of Science* 59, 513–521.

Africa, C.M., Refuerzo, P.G. and Garcia, E.Y. (1936b) Further observations on the life cycle of *Gnathostoma spinigerum*. *Philippine Journal of Science* 61, 221–225.

Ahmed, S.S. (1966) Location of developing and adult worms of *Brugia* sp. in naturally and experimentally infected animals. *Journal of Tropical Medicine and Hygiene* 69, 291–293.

Akahane, H., Iwata, K. and Miyazaki, I. (1982) Studies on *Gnathostoma hispidum* Fedchenko, 1872 parasitic in loaches imported from China. *Japanese Journal of Parasitology* 31, 507–516.

Akahane, H., Iwata, K. and Miyazaki, I. (1983) Studies on the life cycle of *Gnathostoma hispidum* Fedchenko, 1872. I. Experimental studies on susceptibility of various vertebrates to the early third-stage larvae from loaches. *Japanese Journal of Parasitology* 32, 459–464.

Alicata, J.E. (1935) Early developmental stages of nematodes occurring in swine. *United States Department of Agriculture Technical Bulletin* No. 489, 96 pp.

Alicata, J.E. (1937) Larval development of the spirurid nematode *Physaloptera turgida* in the cockroach, *Blattella germanica*. *Papers on Helminthology Published in Commemoration of the 30th year Jubileum K.I. Skrjabin and 15th Anniversary*. All-Union Institute of Helminthology, pp. 11–14.

Alicata, J.E. (1938) The life history of the gizzard worm (*Cheilospirura hamulosa*) and its mode of transmission to chickens with special reference to Hawaiian conditions. *Livro Jubilar do Professor Lauro Travassos. Editado para Commemorar o 25 Anniversario de suas Actividades Scientificas (1913–1938) Rio de Janeiro*, pp. 11–19.

Alicata, J.E. (1947) Parasites and parasitic diseases of domestic animals in the Hawaiian Islands. *Pacific Science* 1, 69–84.

Allen, R.W. and Spindler, L.A. (1949) Note on the natural occurrence in farm-raised chickens of encysted third-stage larvae of *Physocephalus sexalatus*, a spirurid stomach worm of swine. *Proceedings of the Helminthological Society of Washington* 16, 1–3.

Anantaraman, M. and Jayalakshmi, N. (1963) On the life history of *Spirocerca lupi* (Rudolphi, 1809), a nematode of dogs in India. *Proceedings of the Indian Academy of Science* 58, 137–147.

Anderson, R.C. (1952) Description and relationships of *Dirofilaria ursi* Yamaguti, 1941, and a review of the genus *Dirofilaria* Railliet and Henry, 1911. *Transactions of the Royal Canadian Institute* Part II, 29, 35–65.

Anderson, R.C. (1953) *Dipetalonema sprenti* n.sp. from *Castor canadensis* Kuhl. *Parasitology* 43, 215–221.

Anderson, R.C. (1956a) Two new nematodes from Ontario birds. *Canadian Journal of Zoology* 34, 213–218.

Anderson, R.C. (1956b) The life cycle and seasonal transmission of *Ornithofilaria fallisensis* Anderson, a parasite of domestic and wild ducks. *Canadian Journal of Zoology* 34, 485–525.

Anderson, R.C. (1957a) Observations on the life cycle of *Diplotriaenoides translucidus* Anderson and members of the genus *Diplotriaena*. *Canadian Journal of Zoology* 35, 15–24.

Anderson, R.C. (1957b) The life cycles of the dipetalonematid nematodes (Filarioidea; Dipetalonematidae): the problem of their evolution. *Journal of Helminthology* 31, 203–224.

Anderson, R.C. (1958) Possible steps in the evolution of filarial life cycles. *Proceedings of the Sixth International Congress on Tropical Medicine and Malaria, Lisbon* 2, 444–449.

Anderson, R.C. (1959) The egg and first-stage larvae of *Desmidocercella numidica* (Seurat, 1920) with remarks on the affinities of the Desmidocercidae. *Canadian Journal of Zoology* 37, 407–413.

Anderson, R.C. (1962) On the development, morphology and experimental transmission of *Diplotriaena bargusinica* (Filarioidea: Diplotriaenidae). *Canadian Journal of Zoology* 40, 1175–1186.

Anderson, R.C. (1968a) The pathogenesis and transmission of neurotropic and accidental nematode parasites of the central nervous system of mammals and birds. *Helminthological Abstracts* (Review Article) 37, 191–210.

Anderson, R.C. (1968b) The simuliid vectors of *Splendidofilaria fallisensis* of ducks. *Canadian Journal of Zoology* 46, 610–611.

Anderson, R.C. (1988) Nematode transmission patterns. *Journal of Parasitology* 74, 30–45.

Anderson, R.C. (1989) The development of *Quadriplotriaena hypsokysta* (Nematoda: Diplotriaenoidea) in grasshoppers (Orthoptera). *Proceedings of the Helminthological Society of Washington* 56, 199–201.

Anderson, R.C. and Bain, O. (1976) No. 3. Keys to genera of the Order Spirurida. Part 3. Diplotriaenoidea, Aproctoidea and Filarioidea. In: Anderson, R.C., Chabaud, A.G. and Willmott, S. (eds) *CIH Keys to the Nematode Parasite of Vertebrates*. Commonwealth Agricultural Bureaux, Farnham Royal, UK, pp. 59–116.

Anderson, R.C. and Bartlett, C.M. (1994) Ephemerality and reproductive senescence in avian filarioids (Nematoda). *Parasitology Today* 10, 33–35.

Anderson, R.C. and Chabaud, A.G. (1958) Taxonomie de la filaire *Squamofilaria sicki* (Strachan, 1957) n.comb., et place du genre *Squamofilaria* Schmerling, 1925 dans la sousfamille Aproctinae. *Annales de Parasitologie Humaine et Comparée* 33, 254–266.

Anderson, R.C. and Diaz-Ungria, C. (1959) Nematodes de Venezuela, VI. *Dirofilaria striata* (Molin, 1858) Railliet y Henry, 1911, en felinos suramericanos, con comentarios sobre las *Dirofilaria* en carnivoros. *Boletin Venezolano de Laboratorio Clínico* 4, 3–15.

Anderson, R.C. and Wong, P.L. (1982) The transmission and development of *Paracuaria adunca* (Creplin, 1846) (Nematoda: Acuarioidea) of gulls (Laridae). *Canadian Journal of Zoology* 60, 3092–3104.

Anderson, R.C., Gustafson, B.W. and Williams, E.S. (1998) Acute filariasis in a springhaus. *Journal of Wildlife Diseases* 34, 145–149.

Anderson, R.C., Barnes, E.T. and Bartlett, C.M. (1993) Restudy of *Spirura infundibuliformis* McLeod, 1933 (Nematoda: Spiruroidea) from *Spermophilus richardsoni*, with observations on its development in insects. *Canadian Journal of Zoology* 71, 1869–1873.

Anderson, R.C., Wong, P.L. and Bartlett, C.M. (1996) The acuarioid and habronematoid nematodes (Acarioidea, Habronematoidea) of the upper digestive tract of waders. A review of observations on their host and geographic distributions and transmission in marine environments. *Parasite* 4, 303–312.

Ando, K., Tokura, H., Matsuoka, H., Taylor, D. and Chinzei, Y. (1992) Life cycle of *Gnathostoma nipponicum* Yamaguti, 1941. *Journal of Helminthology* 66, 53–61.

Ando, K., Sato, Y., Miura, K. and Chinzei, Y. (1994) Migration and development of the larvae of *Gnathostoma nipponicum* in the rat, second intermediate or paratenic host, and the weasel, definitive host. *Journal of Helminthology* 68, 13–17.

Annaev, D. and Meredov, M. (1990) Life cycle of the nematode *Harteria obesa* Seurat, 1915. *Izvestiya Akademii Nauk Turkmenskoi SSR. Seriya Biologicheskikh Nauk* 6, 61–63.

Annaev, D. and Mushkambarova, M.G. (1975) The life cycle of *Abbreviata* (*Abbreviata*) *turkomanica* Andrushko and Markov, 1956. *Izvestiya Akademii Nauk Turkmenskoi SSR, Biologicheskie Nauki* 5, 81–87. (In Russian.)

Aoki, Y., Vincent, A.L., Ash, L.R. and Katamine, D. (1980) Scanning electron microscopy of third- and fourth-stage larvae and adults of *Brugia pahangi* (Nematoda: Filarioidea). *Journal of Parasitology* 66, 449–457.

Appleby, E.C., Gibsons, L.M. and Georgiou, K. (1995) Distortion of the gizzard in Cyprus pigeons (*Columba livia*) associated with *Hadjelia truncata* infestation. *Veterinary Record* 136, 561–564.

Appy, R.G. and Butterworth, E. (1983) Direct development of *Ascarophis* sp. (Nematoda: Cystidicolida) in an amphipod. *Annual Meeting, Canadian Society of Zoologists, May 15–18, University of Ottawa*, p. 37 (Abstract).

Appy, R.G. and Dadswell, M.J. (1978) Parasites of *Acipenser brevirostrum* LeSueur and *Acipenser oxyrhynchus* Mitchill (Osteichthyes: Acipenseridae) in the Saint John River Estuary, N.B., with a description of *Cabelleronema pseudoargumentosus* sp.n. (Nematoda: Spirurida). *Canadian Journal of Zoology* 56, 1382–1391.

Appy, R.G. and Dadswell, M.J. (1983) Transmission and development of *Capillospirura pseudoargumentosa* (Appy and Dadswell, 1978) (Nematoda: Cystidicolidae). *Canadian Journal of Zoology* 61, 848–859.

Arita, M. (1953) Studies on two species of *Gnathostoma* parasitic in weasels. *Acta Medica* 23, 1729–1749. (In Japanese.)

Ash, L.R. (1962a) Development of *Gnathostoma procyonis* Chandler, 1942 in the first and second intermediate hosts. *Journal of Parasitology* 48, 298–305.

Ash, L.R. (1962b) Migration and development of *Gnathostoma procyonis* Chandler, 1942, in mammalian hosts. *Journal of Parasitology* 48, 306–313.

Ash, L.R. and Little, M.D. (1964) *Brugia beaveri* sp.n. (Nematoda: Filarioidea) from the raccoon (*Procyon lotor*) in Louisiana. *Journal of Parasitology* 50, 119–123.

Ash, L.R. and Riley, J.M. (1970a) Development of subperiodic *Brugia malayi* in the jird, *Meriones unguiculatus*, with notes on infection in other rodents. *Journal of Parasitology* 56, 969–973.

Ash, L.R. and Riley, J.M. (1970b) Development of *Brugia pahangi* in the jird, *Meriones unguiculatus* with notes on infections in other rodents. *Journal of Parasitology* 56, 962–968.

Austin, F.G. and Welch, H.E. (1972) The occurrence, life cycle, and pathogenicity of *Echinuria uncinata* (Rudolphi, 1819) Soloviev, 1912 (Spirurida, Nematoda) in waterfowl at Delta, Manitoba. *Canadian Journal of Zoology* 50, 385–393.

Awachie, J.B.E. (1973) Ecological observations on *Metabronema truttae* Baylis, 1935, and *Cystidicola farionis* Fischer and Waldheim, 1798 (Nematoda, Spiruroidea) in their intermediate and definitive hosts in Afon Terrig. *Acta Parasitologica Polonica* 21, 661–670.

Azevendo, F. de J. (1943) On the presence of *Dipetalonema dracunculoides* (Cobbald, 1870) among dogs in Portugal, a contribution to the study of its morphology. *Anais Instituto de Medicina Tropical, Lisbon* 1, 105–114.

Bahr, P. (1912) *Filariasis and Elephantiasis in Fiji, Being a Report to the London School of Tropical Medicine.* London.

Bailey, W.S. (1963) Parasites and cancer: sarcoma in dogs associated with *Spirocerca lupi. Annals of the New York Academy of Sciences* 108, 890–923.

Bailey, W.S., Cabrera, D.J. and Diamond, D.L. (1963) Beetles of the family Scarabaeidae as intermediate hosts for *Spirocerca lupi. Journal of Parasitology* 49, 485–488.

Bailey, W.S., Morgan, D.H. and Cabrera, D.J. (1964) Observations on the epidemiology of *Spirocerca lupi* in the southeastern USA. *Auburn Veterinarian* 20, 124–128.

Bain, O. (1969a) Etude morphologique du développement larvaire de *Foleyella furcata* chez *Anopheles stephensi. Annales de Parasitologie Humaine et Comparée* 44, 165–172.

Bain, O. (1969b) Développement larvaire de *Saurositus agamae hamoni* n.s.sp., Eufilariinae parasite d'agame en Haute-Volta, chez *Anopheles stephensi. Annales de Parasitologie Humaine et Comparée* 44, 581–594.

Bain, O. (1970a) Morphologie larvaire de *Setaria labiatopapillosa* (Nematoda, Filarioidea) chez *Aedes aegypti. Annales de Parasitologie Humaine et Comparée* 45, 431–439.

Bain, O. (1970b) Etude morphologique de développement larvaire de *Foleyella candezei* chez *Anopheles stephensi* et *Aedes aegypti. Annales de Parasitologie Humaine et Comparée* 45, 21–30.

Bain, O. (1970c) La cellule R1 des microfilaires (Nematoda) initiale du mesenchyme. *Annales de Parasitologie Humaine et Comparée* 45, 227–235.

Bain, O. (1972) Recherches sur le morphogénèse des filaires chez l'hôte intermédiaire. *Annales de Parasitologie Humaine et Comparée* 47, 251–303.

Bain, O. (1974) Développement larvaire de *Dipetalonema dessetae*, filaire de rongeur entretenue au laboratoire. *Annales de Parasitologie Humaine et Comparée* 49, 457–466.

Bain, O. (1978) Développement en Camargue de la filaire du chien, *Dirofilaria repens* Railliet et Henry, 1911, chez les *Aedes* halophile. *Bulletin du Muséum National d'Histoire Naturelle* 351, 19–27.

Bain, O. (1979) Transmission de l'onchocerque bovine, *Onchocerca gutturosa*, par *Culicoides. Annales de Parasitologie Humaine et Comparée* 54, 483–488.

Bain, O. (1980) Deux filaires du genre *Eufilaria* chez le merle: développement chez *Culicoides nubeculosus. Annales de Parasitologie Humaine et Comparée* 55, 583–590.

Bain, O. (1981a) Redescription du stade infestant de la filaire *Parafilaria bovicola*: affinités du genre avec les *Thelazia. Annales de Parasitologie Humaine et Comparée* 56, 527–530.

Bain, O. (1981b) Le genre *Onchocerca*: hypothèses sur son évolution et clé dichotomique des espèces. *Annales de Parasitologie Humaine et Comparée* 56, 503–526.

Bain, O. (1981c) Filariids and their evolution. In: Willmott, S. (ed.) *Evolution of Helminths.* Workshop Proceedings EMOP3. *Parasitology* 82, 167–168.

Bain, O. and Chabaud, A.G. (1975) Développement chez des moustiques de trois filaires de Lézards sud-américains du genre *Oswaldofilaria. Annales de Parasitologie Humaine et Comparée* 50, 209–221.

Bain, O. and Chabaud, A.G. (1986) Atlas des larves infestantes de filaires. *Tropical Medicine and Parasitology* 37, 301–340.

Bain, O. and Durette-Desset, M.C. (1973) Cycle de *Skrjabinofilaria skrjabini*, filaire de marsupial sud-américain position systématique (1). *Annales de Parasitologie Humaine et Comparée* 48, 61–79.

Bain, O. and Petit, G. (1978) Redescription du stade infestant d'*Onchocerca cervicalis* R. et H., 1910. *Annales de Parasitologie Humaine et Comparée* 53, 315–318.

Bain, O. and Prod'hon, J. (1974) Homogéneité des filaire de genres *Waltonella*, *Ochoterenella* und *Madochotera*; création des Waltonellinae n.subfam. *Annales de Parasitologie Humaine et Comparée* 49, 721–739.

Bain, O. and Quentin, J.C. (1977) Développement de *Dipetalonema* (*A.*) *weissi*, filaire de Macroscélide, chez un ornithodore. *Annales de Parasitologie Humaine et Comparée* 52, 569–575.

Bain, O. and Sulahian, A. (1974) Trois nouvelles filaires du genre *Oswaldofilaria* chez des lézards sud-américains; essai de classification des Oswaldofilariinae. *Bulletin du Muséum National d'Histoire Naturelle* No. 232, 156, 827–841.

Bain, O. and Vassiliades, G. (1969) Cycle évolutif d'un Dicheilonematinae, *Serratospiculum tendo* filaire parasite du faucon. *Annales de Parasitologie Humaine et Comparée* 44, 595–604.

Bain, O. and Vaucher, C. (1973) Développement larvaire de *Diplotriaena tridens* (Nematoda: Filarioidea) chez *Locusta migratoria*. *Annales de Parasitologie Humaine et Comparée* 48, 81–89.

Bain, O., Petit, G. and Poulain, B. (1978) Validité des deux espèces *Onchocerca lienalis* et *O. gutturosa*. *Annales de Parasitologie Humaine et Comparée* 53, 421–430.

Bain, O., Petit, G. and Berteaux, S. (1980) Description de deux nouvelles filaires du genre *Litomosoides* et de leurs stades infestants. *Annales de Parasitologie Humaine et Comparée* 55, 225–237.

Bain, O., Petit, G., Ratanaworabhan, N., Yenbutra, S. and Chabaud, A.G. (1981a) Une nouvelle filaire d'ecureuil en Thaïlande, *Breinlia* (*B.*) *manningi* n.sp., et son développement chez *Aedes*. *Annales de Parasitologie Humaine et Comparée* 56, 193–201.

Bain, O., Petit, G., Kozek, W.J. and Chabaud, A.G. (1981b) Sur les filaires Splendidofilariinae du genre *Aproctella*. *Annales de Parasitologie Humaine et Comparée* 56, 95–105.

Bain, O., Aeschlimann, A. and Chatelanat, P. (1982a) Présence, chez des tiques de la région de Genève, de larves infestantes qui pourraient se rapporter a la filaire de chien *Dipetalonema grassii*. *Annales de Parasitologie Humaine et Comparée*. 57, 643–646.

Bain, O., Baker, M. and Chabaud, A.G. (1982b) Nouvelles données sur la lignée *Dipetalonema* (Filarioidea, Nematoda). *Annales de Parasitologie Humaine et Comparée* 57, 593–620.

Bain, O., Petit, G., Jacquet-Viallet, P. and Houin, R. (1985a) *Cherylia guyanensis* n.gen., n.sp., filaire d'un marsupial sud-américain, transmise par tique. *Annales de Parasitologie Humaine et Comparée* 60, 727–737.

Bain, O., Petit, G. and Gueye, A. (1985b) Transmission experimentale de *Monanema nilotica* El Bihari et coll. 1977. Filaire à microfilaire dermigue parasite de muridés africains. *Annales de Parasitologie Humaine et Comparée* 60, 83–89.

Bain, O., Bartlett, C.M. and Petit, G. (1986a) Une filaire de muridés Africains dans la paroi colon, *Monanema martini* n.sp. *Annales de Parasitologie Humaine et Comparée* 61, 465–472.

Bain, O., Petit, G. and Chabaud, A.G. (1986b) Une nouvelle filaire, *Cercopithifilaria roussilhoni* n.sp., parasite de l'Athérure au Gabon, transmise par tiques; hypothèse sur l'évolution du genre. *Annales de Parasitologie Humaine et Comparée* 61, 81–93.

Bain, O., Petit, G. and Diagne, M. (1989) A morphological study of some *Litomosoides* species from rodents; taxonomic conclusions. *Annales de Parasitologie Humaine et Comparée* 64, 268–289.

Bain, O., Wahl, G. and Renz, A. (1993a) *Onchocerca ramachandrini* n.sp. From the warthog in Cameroon. *Annales de Parasitologie Humain et Comparée* 68, 139–143.

Bain, O., Wanji, S., Petit, G., Paperna, I. and Finkelman, S. (1993b) Splendidofilariinae of lizards: new species, redescription, and development in the Phlebotomine host. *Systematic Parasitology* 26, 97–115.

Bain, O., Wanji, S., Vuong, P.N., Maréchal, P., LeGoff, L. and Petit, G. (1994) Larval biology of six Filariae of the subfamily Onchocercinae in a vertebrate host. *Parasite* 1, 241–254.

Bain, O., Wanji, S., Enyong, P., Petit, G., Noireau, F., Eberhard, M.I. and Wahl, G. (1998) New features on the moults and morphogenesis of the human filaria *Loa loa* by using rodent hosts. Consequences. *Parasite* 5, 37–46.

Baker, M.R. (1987) *Synopsis of the Nematoda Parasitic in Amphibians and Reptiles*. Memorial University of Newfoundland Occasional Papers in Biology, No. 11, 325 pp.

Baltazard, M., Chabaud, A.G. and Minou, A. (1952) Cycle évolutif d'une filaire parasite de mérion. *Comptes Rendu de l'Académie des Sciences* 234, 2115–2117.

Bancroft, T.L. (1901) Preliminary note on the intermediary host of *Filaria immitis* Leidy. *Journal and Proceedings of the Royal Society of New South Wales* 35, 41–46.

Bancroft, T.L. (1903) On some further observations on the life history of *Filaria immitis* Leidy. *Journal and Proceedings of the Royal Society of New South Wales* 37, 254–257.

Bancroft, T.L. (1904) Some further observations of the life history of *Filaria immitis* Leidy. *British Medical Journal* 1, 822–823.

Bangs, M.J., Ash, L.R. and Barr, A.R. (1995) Susceptibility of various mosquitoes of California to subperiodic *Brugia malayi*. *Acta Tropica* 59, 323–332.

Barger, M.A. and Janovy, J., Jr (1994) Host specificity of *Rhabdochona canadensis* (Nematoda: Rhabdochonidae) in Nebraska. *Journal of Parasitology* 80, 1032–1035.

Barthold, E. and Wenk, P. (1992) Dose-dependent recovery of adult *Acanthocheilonema viteae* (Nematoda: Filarioidea) after single and trickle inoculations in jirds. *Parasitology Research* 78, 229–234.

Bartlett, C.M. (1983) Zoogeography and taxonomy of *Dirofilaria scapiceps* (Leidy, 1886) and *D. uniformis* Price, 1957 (Nematoda: Filarioidea) of lagomorphs in North America. *Canadian Journal of Zoology* 61, 1011–1022.

Bartlett, C.M. (1984a) Development of *Dirofilaria scapiceps* (Leidy, 1886) (Nematoda: Filarioidea) in *Aedes* spp. and *Mansonia perturbans* (Walker) and responses of mosquitoes to infection. *Canadian Journal of Zoology* 62, 112–129.

Bartlett, C.M. (1984b) Development of *Dirofilaria scapiceps* (Leidy, 1886) (Nematoda: Filarioidea) in lagomorphs. *Canadian Journal of Zoology* 62, 965–979.

Bartlett, C.M. (1984c) Pathology and epizootiology of *Dirofilaria scapiceps* (Leidy, 1886) (Nematoda: Filarioidea) in *Sylvilagus floridanus* (J.A. Allen) and *Lepus americanus* Erxleben. *Journal of Wildlife Diseases* 20, 197–206.

Bartlett, C.M. (1986) The reptilian filarioid genus *Foleyella* Seurat, 1917 (Onchocercidae: Dirofilariinae) and its relationship to other dirofilariine genera. *Systematic Parasitology* 9, 43–56.

Bartlett, C.M. (1991) A new hypothesis concerning attachment of parasitic nematodes (Spirurida: Acuarioidea) to the upper alimentary tract of birds. *Canadian Journal of Zoology* 69, 1829–1833.

Bartlett, C.M. (1992) New, known and unidentified species of *Eulimdana* (Nematoda): additional information on biologically unusual filarioids of charadriiform birds. *Systematic Parasitology* 23, 209–230.

Bartlett, C.M. (1993) Lice (*Amblycerra* and *Ischnocera*) as vectors of *Eulimdana* spp. (Nematoda: Filarioidea) in charadriform birds and the necessity of short reproductive periods in adult worms. *Journal of Parasitology* 79, 85–91.

Bartlett, C.M., Crawshaw, J. and Appy, R.G. (1984) Epizootiology, development, and pathology of *Geopetitia aspiculata* Webster, 1971 (Nematoda: Habronematoidea) in tropical birds at the Assiniboine Park Zoo, Winnipeg, Canada. *Journal of Wildlife Disease* 20, 289–299.

Bartlett, C.M. and Anderson, R.C. (1980a) Filarioid nematodes (Filarioidea: Onchocercidae) of *Corvus brachyrhynchos brachyrhynchos* Brehm in southern Ontario, Canada and a consideration of the epizootiology of avian filariasis. *Systematic Parasitology* 2, 77–102.

Bartlett, C.M. and Anderson, R.C. (1980b) Development of *Chandlerella chitwoodae* Anderson, 1961 (Filarioidea: Onchocercidae) in *Culicoides stilobezzioides* Foote and Pratt and *C. travisi* Vargas (Diptera: Ceratopogonidae). *Canadian Journal of Zoology* 58, 1002–1006.

Bartlett, C.M. and Anderson, R.C. (1985a) Larval nematodes (Ascaridida and Spirurida) in the aquatic snail, *Lymnaea stagnalis*. *Journal of Invertebrate Pathology* 46, 153–159.

Bartlett, C.M. and Anderson, R.C. (1985b) The third-stage larva of *Molinema arbuta* (Highby, 1943) (Nematoda) and development of the parasite in the porcupine (*Erethizon dorsatum*). *Annales de Parasitologie Humaine et Comparée* 60, 703–708.

Bartlett, C.M. and Anderson, R.C. (1987b) *Pelecitus fulicaeatrae* (Nematoda: Filarioidea) of coots (Gruiformes) and grebes (Podicipediformes): skin-inhabiting microfilariae and development in Mallophaga. *Canadian Journal of Zoology* 65, 2803–2812.

Bartlett, C.M. and Anderson, R.C. (1987b) Additional comments on species of *Pelecitus* (Nematoda: Filarioidea) from birds. *Canadian Journal of Zoology* 65, 2813–2814.

Bartlett, C.M. and Anderson, R.C. (1989a) Mallophagan vectors and the avian filarioids: new subspecies of *Pelecitus fulicaeatrae* (Nematoda: Filarioidea) in sympatric North American hosts, with development, epizootiology, and pathogenesis of the parasite in *Fulica americana* (Aves). *Canadian Journal of Zoology* 67, 2821–2833.

Bartlett, C.M. and Anderson, R.C. (1989b) Some observations on *Pseudomenopon pilosum* (Amblycera: Menoponidae), the louse vector of *Pelecitus fulicaeatrae* (Nematoda: Filarioidea) of coots, *Fulica americana* (Aves: Gruiformes). *Canadian Journal of Zoology* 67, 1328–1331.

Bartlett, C.M. and Anderson, R.C. (1990) *Eulimdana florencae* n.sp. (Nematoda: Filarioidea) from *Micropalama himantopus* (Aves: Charadriiformes): evidence for neonatal transmission, ephemeral adults, and long-lived microfilariae among filarioids of shorebirds. *Canadian Journal of Zoology* 68, 986–992.

Bartlett, C.M. and Greiner, E.C. (1986) A revision of *Pelecitus* Railliet and Henry, 1910 (Filarioidea, Dirofilariinae) and evidence for the 'capture' by mammals of filarioids from birds. *Bulletin du Muséum National d'Histoire Naturelle, Series 4, Section A* 8, 47–99.

Bartlett, C.M., Anderson, R.C. and Bush, A.O. (1989) Taxonomic descriptions and comments on the life history of new species of *Eulimdana* (Nematoda: Filarioidea) with skin-inhabiting microfilariae in the Charadriiformes (Aves). *Canadian Journal of Zoology* 67, 612–629.

Bartlett, C.M., Bush, A.O. and Anderson, R.C. (1987) Unusual finding of encapsulated nematode larvae (Spiruroidea) in *Bartramia longicauda* and *Numenius americanus* (Charadriiformes) in western Canada. *Journal of Wildlife Diseases* 23, 591–595.

Bartlett, C.M., Anderson, R.C. and Wong, P.L. (1989) Development of *Skrjabinocerca prima* (Nematoda: Acuarioidea) in *Hyalella azteca* (Amphipoda) and *Recurvirostra americana* (Aves: Charadriiformes), with comments on its precocity. *Canadian Journal of Zoology* 67, 2883–2892.

Bartlett, C.M., Anderson, R.C. and Bush, A.O. (1989) Taxonomic descriptions and comments on the life history of new species of *Eulimdana* (Nematoda: Filarioidea) with skin-inhabiting microfilariae in the Charadriiformes (Aves). *Canadian Journal of Zoology* 67, 612–629.

Baruš, V., Moravec, F. and Prokopic, J. (1970) Two new intermediate hosts of the nematode *Physocephalus sexalatus* (Molin, 1860). *Folia Parasitologica* 17, 94.

Basáñez, M.G., Remme, J.H.F., Alley, E.S., Bain, O., Shelley, A.J., Medley, G.F. and Anderson, R.M. (1995) Density dependent processes in the transmission of human onchocerciasis: relationship between the numbers of microfilariae ingested and successful larval development in the simuliid vector. *Parasitology* 110, 409–427.

Basir, M.A. (1948) On the *Physaloptera* larvae from an insect. *Canadian Journal of Research* 26, 197–200.

Bauer, O.N. and Nikolskaya, N.P. (1952) New information about the intermediate hosts of the parasites of *Coregonus lavaretus lodogae*. *Doklady Akademii Nauk SSSR* 84, 1109–1112. (In Russian.)

Baumann, R. (1946) Beobachtungen beim parasitären Sommerbluten der Pferde. *Wiener Tierärztliche Monastsschrift* 32, 52–55.

Baylis, H. (1929) *A Manual of Helminthology*. Ballière, Tindall and Cox, London.

Baylis, H.A. (1928) On a collection of nematodes from Nigerian mammals (chiefly rodents). *Parasitology* 20, 280–304.

Baylis, H.A. (1931) *Gammarus pulex* as an intermediate host for trout parasites. *Annales and Magazine of Natural History* 7, 431–433.

Beaucournu, J.C. and Chabaud, A.G. (1963) Infestation spontanée de puces par le spiruride *Mastophorus muris* (Gmelin). *Annales de Parasitologie Humaine et Comparée* 38, 931–934.

Beaver, P.C., Orihel, T.C. and Johnson, M.H. (1974) *Dipetalonema viteae* in the experimentally infected jird, *Meriones unguiculatus*. II. Microfilaraemia in relation to worm burden. *Journal of Parasitology* 60, 310–315.

Bech-Nielsen, S., Bornstein, S., Christensson, D., Wallgren, T.B., Zakrisson, G. and Chirico, J. (1982) *Parafilaria bovicola* (Tubangui, 1934) in cattle; epizootiology–vector study and experimental transmission of *Parafilaria bovicola* to cattle. *American Journal of Veterinary Research* 43, 948–954.

Beglinger, R., Illgen, R., Pfister, K. and Heider, K. (1988) The parasite *Trichospirura leptostoma* associated with wasting disease in a colony of common marmosets, *Callithrix jacchus*. *Folia Primatology* 51, 45–51.

Bell, S.D. and Brown, H.W. (1945) Studies on the microfilarial periodicity of *Litomosoides carinii*, filariid parasite of the cotton rat. *American Journal of Tropical Medicine and Hygiene* 25, 137–140.

Benach, J.L. and Crans, W.J. (1973) Larval development and transmission of *Foleyella flexicauda* Schacher and Crans, 1973 (Nematoda: Filarioidea) in *Culex territans*. *Journal of Parasitology* 59, 797–800.

Berghe, L. van den, Chardome, M. and Peel, E. (1964) The filariid parasite of the Eastern Gorilla in the Congo. *Journal of Helminthology* 38, 349–368.

Bernard, P.N. and Bauche, J. (1913) Conditions de propagation de la filariose sous-cutanée du chien, *Stegomiyia fasciata* hôte intermédiaire de *Dirofilaria repens*. *Bulletin de la Société de Pathologie Exotique* 4, 482–485.

Bertram, D.S. (1947) The period required by *Litomosoides carinii* to reach the infective stage in *Liponyssus bacoti*, and the duration of the mite's infectivity. *Annales of Tropical Medicine and Parasitology* 41, 253–261.

Bertram, D.S. (1953) Laboratory studies on filariasis in the cotton rat. *Transactions of the Royal Society of Tropical Medicine and Hygiene* 47, 85–106.

Bethel, W.M. (1973) The life cycle and notes on the developmental stages of *Microtetrameres corax* Schell, 1953 (Nematoda: Tetrameridae). *Proceedings of the Helminthological Society of Washington* 40, 22–26.

Beveridge, I. (1987) *Echinocephalus overstreeti* Deardorff and Ko, 1983 (Nematoda: Gnathostomatoidea) from elasmobranchs and molluscs in South Australia. *Transactions of the Royal Society of South Australia* 111, 79–92.

Beveridge, I., Kummerow, E.L. and Wilkinson, P. (1980a) Observations on *Onchocerca gibsoni* and nodule development in naturally infected cattle in Australia. *Tropenmedizin und Parasitologie* 31, 75–81.

Beveridge, I., Kummerow, E.L. and Wilkinson, P. (1980b) Experimental infection of laboratory rodents and calves with microfilariae of *Onchocerca gibsoni*. *Tropenmedizin und Parasitologie* 31, 82–86.

Beveridge, I., Kummerow, E.L., Wilkinson, P. and Copeman, D.B. (1981) An investigation of biting midges in relation to their potential as vectors of bovine onchocerciasis in North Queensland. *Journal of the Australian Entomological Society* 20, 39–45.

Bianco, A.E. (1984) Laboratory maintenance of *Monanema globulosa*, a rodent filaria with skin-dwelling microfilariae. *Zeitschrift für Parasitenkunde* 70, 255–264.

Bianco, A.E. and Muller, R. (1977) A hard tick as vector of a new rodent filaria. *Transactions of the Royal Society of Tropical Medicine and Hygiene* 71, 383.

Bianco, A.E., Muller, R. and Nelson, G.S. (1983) Biology of *Monanema globulosa*, a rodent filaria with skin-dwelling microfilariae. *Journal of Helminthology* 57, 259–278.

Birova, V., Macko, J.K. and Espaine, L. (1974a) The life cycle of *Dispharynx nasuta* (Rudolphi, 1819) in experimentally infected intermediate hosts in Cuba. *Helminthologia* 15, 693–713.

Birova, V., Macko, J.K. and Ovies, D. (1974b) The life cycle of *Dispharynx nasuta* (Rudolphi, 1819) in experimentally infected chickens in Cuba. *Helminthologia* 15, 715–740.

Black, G.A. (1983a) Taxonomy of a swimbladder nematode, *Cystidicola stigmatura* (Leidy) and evidence of its decline in the Great Lakes. *Canadian Journal of Fisheries and Aquatic Sciences* 40, 643–647.

Black, G.A. (1983b) Origin, distribution, and postglacial dispersal of a swimbladder nematode, *Cystidicola stigmatura*. *Canadian Journal of Fisheries and Aquatic Sciences* 40, 1244–1253.

Black, G.A. (1983c) *Cystidicola farionis* (Nematoda) as an indicator of lake trout (*Salvelinus namaycush*) of Bering ancestry. *Canadian Journal of Fisheries and Aquatic Sciences* 40, 2034–2040.

Black, G.A. (1984) Morphology of *Cystidicola stigmatura* (Nematoda) in relation to its glacial history. *Journal of Parasitology* 70, 967–974.

Black, G.A. (1985) Reproductive output and population biology of *Cystidicola stigmatura* (Leidy) (Nematoda) in Arctic char, *Salvelinus alpinus* (L.) (Salmonidae). *Canadian Journal of Zoology* 63, 617–622.

Black, G.A. and Lankester, M.W. (1980) Migration and development of swimbladder nematodes, *Cystidicola* spp. (Habronematoidea), in their definitive hosts. *Canadian Journal of Zoology* 58, 1997–2005.

Black, G.A. and Lankester, M.W. (1981) The transmission, life span, and population biology of *Cystidicola cristivomeri* White, 1941 (Nematoda: Habronematoidea) in char, *Salvelinus* spp. *Canadian Journal of Zoology* 59, 498–509.

Blacklock, D.B. (1926) The development of *Onchocerca volvulus* in *Simulium damnosum*. *Annales of Tropical Medicine and Parasitology* 20, 1–48.

Blanc, G.R. (1919) Sur quelques espèces du genre *Diplotriaena* Railliet et Henry. *Archives de Parasitologie* 22, 546–556.

Blažek, K., Dyková, J. and Páv, J. (1968) The occurrence and pathogenicity of *Setaria cervi* Rud., in the central nervous system of deer. *Folia Parasitologica* 15, 123–130.

Böhn, L.K. and Supperer, R. (1953) Beobachtungen über eine neue Filarie (Nematoda), *Wehrdikmansia rugosicauda* Böhn und Supperer, 1953, aus dem subkutanen Bindegewebe des Rehes. *Sitzungsberichte der Österreichischen Akademie der Wissenschaften. Abteilung I. Mathematisch-Naturwissenschaftliche Klasse* 162, 95–104.

Böhn, L.K. and Supperer, R. (1954) Weitere Untersuchungen über Mikrofilarien als Erreger der periodischen Augenentzündung der Pferde. *Wiener Tierärztliche Monatsschrift* 41, 129–139.

Boreham, P.F.L. and Atwell, R.B. (eds) (1988) *Dirofilariasis*. CRC Press, London.

Boughton, R.V. (1935) Endoparasitic infestations in grouse, their pathogenicity and correlation with meteoro-topographical conditions. *University of Michigan Agriculture Experiment Station Technical Bulletin* No. 121, 50 pp.

Boughton, E. (1969) On the occurrence of the oesophageal worm, *Streptocara crassicauda*, in ornamental ducks in Hampshire. *Journal of Helminthology* 43, 273–280.

Bradley, R.E. (1952) Observations on the development of *Dirofilaria immitis* in certain insects. *Journal of the Tennessee Academy of Science* 27, 206.

Branch, G.J. and Stoffolano, J.G. (1974) Face fly: invertebrate vector and host of a mammalian eyeworm in Massachusetts. *Journal of Economical Entomology* 67, 304–305.

Bray, R.L. and Walton, B.C. (1961) The life cycle of *Dirofilaria uniformis* Price and transmission to wild and laboratory rabbits. *Journal of Parasitology* 47, 13–22.

Breinl, A. (1921) Preliminary note on the development of the larvae of *Dirofilaria immitis* in the dog fleas *Ctenocephalus felis* and *canis*. *Annals of Tropical Medicine and Parasitology* 14, 389–392.

Bremner, K.C. (1955) Morphological studies on the microfilariae of *Onchocerca gibsoni* Cleland and Johnston and *Onchocerca gutturosa* Neumann (Nematoda: Filarioidea). *Australian Journal of Zoology* 49, 324–330.

Brenes, R.R., Arroyo, G. and Delgado Flores, E. (1962) Helmintos de la Republica de Costa Rica XIX. Nematoda 5. Algunos nematodos parasitos de *Gallus gallus domesticus* (L.). *Revista Biologia Tropical Universidad de Costa Rica* 10, 183–197.

Brown, H.W. and Sheldon, A.J. (1940) Natural infection of fleas with the dog heartworm (*Dirofilaria immitis*). *North American Veterinarian* 21, 230–231.

Brug, S.L. (1927) Een nieuwe *Filaria* – soort (*Filaria malayi*) parasiteerende bij den meësch (voorloopige mededealing). *Geneeskindig Tijdschrift voor Nederlandsch-Indië* 67, 750–754.

Brumpt, E. (1931) Némathelminthes parasites des rats sauvage (*Epimys norvegicus*) de Caracas. I. *Protospirura bonnei*. Infections expérimentales et spontanées. Formes adultes et larvaires. *Annales de Parasitologie Humaine et Comparée* 9, 344–358.

Brygoo, E.R. (1960) Evolution de *Foleyella furcata* (von Linstow, 1899) chez *Culex fatigans* Wiedemann, 1828. *Archives de l'Institut Pasteur de Madagascar* 28, 129–138.

Buckley, J.J.C. (1933) A note on the development of *Filaria ozzardi* in *Culicoides furens* Poey. *Journal of Helminthology* 11, 257–258.

Buckley, J.J.C. (1934) On the development, in *Culicoides furens* Poey, of *Filaria* (= *Mansonella*) *ozzardi* Manson, 1897. *Journal of Helminthology* 12, 99–118.

Buckley, J.J.C. (1938) On *Culicoides* as a vector of *Onchocerca gibsoni* (Cleland and Johnston, 1910). *Journal of Helminthology* 16, 121–158.

Buckley, J.J.C. (1955) Symposium on loiasis. V. The morphology of the larval stages in the vector: some of the problems involved. *Transactions of the Royal Society of Tropical Medicine and Hygiene* 49, 122–126.

Buckley, J.J.C. (1958a) Tropical pulmonary eosinophilia in relation to filarial infections (*Wuchereria* spp.) in animals. *Transactions of the Royal Society of Tropical Medicine and Hygiene* 52, 335–336.

Buckley, J.J.C. (1958b) Occult filarial infections of animal origin as a cause of tropical pulmonary eosinophilia. *East African Medical Journal* 35, 493–500.

Buckley, J.J.C. (1960) On *Brugia* gen.nov. for *Wuchereria* spp. of the 'malayi' group i.e. *W. malayi* (Brug, 1927), *W. pahangi* Buckley and Edeson, 1956, *W. patei* Buckley, Nelson and Heisch, 1958. *Annals of Tropical Medicine and Parasitology* 54, 75–77.

Buckley, J.J.C., Nelson, G.S. and Heisch, R.B. (1958) On *Wuchereria patei* n.sp. from the lymphatics of cats, dogs and genet cats on Pate Island, Kenya. *Journal of Helminthology* 32, 73–80.

Bull, L.B. (1919) A contribution to the study of habronemiasis: a clinical, pathological, and experimental investigation of a granulomatous condition of the horse – habronemic granuloma. *Transactions and Proceedings of the Royal Society of South Australia* 43, 85–141.

Burnett, H.S., Parmelee, W.E., Lee, R.D. and Wagner, E.D. (1957) Observations on the life cycle of *Thelazia californiensis* Price, 1930. *Journal of Parasitology* 43, 433.

Butler, J.M. and Grundmann, A.W. (1954) The intestinal helminths of the coyote, *Canis latrans* Say in Utah. *Journal of Parasitology* 40, 1–4.

Byrne, P.J. (1992a) *Rhabdochona rotundicaudatum* n.sp. and a redescription of *R. cascadilla* Wigdon, 1918 (Nematoda: Thelazioidea) from minnows in southern Ontario, Canada. *Canadian Journal of Zoology* 70, 476–468.

Byrne, P.J. (1992b) On the biology of *Rhabdochona rotundicaudatum* and *R. cascadilla* (Nematoda: Thelazioidea) in stream fishes from southern Ontario, Canada. *Canadian Journal of Zoology* 70, 485–493.

Caballero, C.E. and Berrera, A. (1958) Estudios helmintologicos de la region oncocercosa de México y de la Republica de Quatemala. Nematoda IIa Parte. Filarioidea. V. Halazgo de un nodulae oncocercosa en un monoarana, *Ateles geoffroyi vellerosus* Gray, del Estado de Chiapas. *Revista Latino-americana de Microbiología* 1, 79–94.

Campbell, H. and Lee, L. (1953) *Studies on Quail Malaria in New Mexico and Notes on Other Aspects of the Quail Population.* New Mexico Department of Game and Fish, 79 pp.

Campbell, J.R., Soekartono, Purnomo, Atmosoedjono, S. and Marwoto, H.A. (1986) Experimental *Wuchereria kalimantani* infection in the leaf monkey, *Presbytis cristatus*. *Annals of Tropical Medicine and Parasitology* 80, 141–142.

Campbell, W.C. and Blair, L.S. (1978) *Dirofilaria immitis*: experimental infections in the ferret (*Mustula putorius furo*). *Journal of Parasitology* 64, 119–122.

Campbell, W.C., Blair, L.S. and McCall, J.W. (1979) *Brugia pahangi* and *Dirofilaria immitis*: experimental infections in the ferret, *Mustela putorius furo*. *Experimental Parasitology* 47, 327–332.

Cancrini, G., Pietrobelli, M., Frangipane Di Regalbobo, A. and Tampieri, M.P. (1997) Mosquitos as vectors of *Setaria labiatopapillosa*. *International Journal for Parasitology* 27, 1061–1064.

Causey, O.R. (1939a) *Aedes* and *Culex* mosquitoes as intermediate hosts of frog filaria, *Foleyella* sp. *American Journal of Hygiene Section C* 29, 79–81.

Causey, O.R. (1939b) The development of frog filaria larvae, *Foleyella ranae* in *Aedes* and *Culex* mosquitoes. *American Journal of Hygiene Section D* 29, 131–132.

Causey, O.R. (1939c) Development of the larval stages of *Foleyella brachyoptera* in mosquitoes. *American Journal of Hygiene Section D* 30, 69–71.

Cawthorn, R.J. and Anderson, R.C. (1976a) Development of *Physaloptera maxillaris* (Nematoda: Physalopteroidea) in skunk (*Mephitis mephitis*) and the role of paratenic and other hosts in its life cycle. *Canadian Journal of Zoology* 54, 313–323.

Cawthorn, R.J. and Anderson, R.C. (1976b) Effects of age, temperature, and previous infection on the development of *Physaloptera maxillaris* (Nematoda: Physalopteroidea) in field crickets (*Acheta pennsylvanicus*). *Canadian Journal of Zoology* 54, 442–448.

Cawthorn, R.J. and Anderson, R.C. (1976c) Seasonal population changes of *Physaloptera maxillaris* (Nematoda: Physalopteroidea) in striped skunk (*Mephitis mephitis*). *Canadian Journal of Zoology* 54, 522–525.

Cawthorn, R.J. and Anderson, R.C. (1977) Cellular reactions of field crickets (*Acheta pennsylvanicus* Burmeister) and German cockroaches (*Blattella germanica* L.) to *Physaloptera maxillaris* Molin (Nematoda: Physalopteroidea). *Canadian Journal of Zoology* 55, 368–375.

Cawthorn, R.J. and Anderson, R.C. (1980) Development of *Diplotriaena tricuspis* (Nematoda: Diplotriaenoidea) a parasite of Corvidae, in intermediate and definitive hosts. *Canadian Journal of Zoology* 58, 95–108.

Cawthorn, R.J., Anderson, R.C. and Barker, I.K. (1980) Lesions caused by *Diplotriaena tricuspis* (Nematoda: Diplotriaenoidea) in the American crow, *Corvus brachyrhynchos* Brehm. *Canadian Journal of Zoology* 58, 1892–1898.

Cebotarev, R.S. and Poliscuk, V.P. (1959) Gongylonematosis of domestic animals under conditions of Ukrainian Polesie and forest-steppe areas. *Acta Parasitologica Polonica* 7, 549–558.

Cebotarev, R.S. and Poliscuk, V.P. (1961) New data on *Gongylonema pulchrum* Molin, 1857. *Zoologicheski Zhurnal* 40, 976–982. (In Russian.)

Cerqueira, M.L. (1959) Sôbre a transmissao da *Mansonella ozzardi. Journal Brasileiro de Medicina* 1, 900–914.

Chabaud, A.G. (1950) Cycle évolutif de *Synhimantus* (*Desportesius*) *spinulatus* (Nematoda Acuariidae). *Annales de Parasitologie Humaine et Comparée* 25, 150–166.

Chabaud, A.G. (1951) Cycle évolutif chez des coléoptères ténébrionides de deux especes de nématodes Habronematinae (Genre *Sicarius* et genre *Hadjelia*) parasite de *Upupa epops* L. à Banyuls. *Comptes Rendus des Séances de l'Academie des Sciences* 232, 564–565.

Chabaud, A.G. (1954a) Sur le cycle évolutif des spirurides et de nématodes ayant une biologie comparable. Valeur systématics des caractères biologique suites. *Annales de Parasitologie Humaine et Comparée* 29, 206–249.

Chabaud, A.G. (1954b) Sur le cycle évolutif des spirurides et de nématodes ayant une biologie comparable. Valeur systématique des caractères biologiques. *Annales de Parasitologie Humaine et Comparée* 29, 42–88, 206–249, 358–425.

Chabaud, A.G. (1954c) Sur le cycle évolutif des spirurides et de nématodes ayant une biologie comparable. *Annales de Parasitologie Humaine et Comparée* 29, 206–249.

Chabaud, A.G. (1955) Remarques sur le cycle évolutif des filaires du genre *Diplotriaena* et redescription de *D. monticelliana* (Stossich, 1890). *Vie et Milieu* 6, 342–347.

Chabaud, A.G. (1975a) No. 3. Keys to genera of the order Spirurida. Part I. Camallanoidea, Dracunculoidea, Gnathostomatoidea, Physalopteroidea, Rictularioidea and Thelazioidea. In: Anderson, R.C., Chabaud, A.G. and Willmott, S. (eds) *CIH Keys to the Nematode Parasites of Vertebrates.* Commonwealth Agricultural Bureaux, Farnham Royal, UK, pp. 1–27.

Chabaud, A.G. (1975b) No. 3. Keys to genera of the order Spirurida. Part 2. Spiruroidea, Habronematoidea and Acuarioidea. In: Anderson, R.C., Chabaud, A.G. and Willmott, S. (eds) *CIH Keys to the Nematode Parasites of Vertebrates.* Commonwealth Agricultural Bureaux, Farnham Royal, UK, pp. 29–58.

Chabaud, A.G. and Frank, W. (1961a) Nouvelle filaire parasite des artères de pythons: *Maldonaldius oschei* n.sp. (Nematodes, Onchocercidae). *Zeitschrift für Parasitenkunde* 20, 434–439.

Chabaud, A.G. and Frank, W. (1961b) Description de la microfilaire de *Macdonaldius oschei.* Note additionnelle. *Annales de Parasitologie Humaine et Comparée* 36, 133–134.

Chabaud, A.G. and Frank, W. (1961c) Nouvelle filaire parasite des artères de l'*Heloderma suspectum* cope: *Macdonaldius andersoni* n.sp. (Nematodes, Onchocercidae). *Annales de Parasitologie Humaine et Comparée* 36, 127–133.

Chabaud, A.G. and Mahon, J. (1958) Cycle évolutif du nématode *Spirura talpae* (Gmelin, 1790). *Comptes Rendus des Séances de la Société de Biologie* 152, 474–476.

Chandler, A.C. (1925) A contribution to the life history of a gnathostome. *Parasitology* 17, 237–244.

Chandler, A.C. (1946) Helminths of armadillos, *Dasypus novenicinctus*, in eastern Texas. *Journal of Parasitology* 32, 237–241.

Chandler, A.E., Alicata, J.E. and Chitwood, M.B. (1950) Life history (Zooparasitica): parasites of vertebrates. In: Chitwood, B.G. and Chitwood, M.B. (eds) *Introduction to Nematology.* University Park Press, Baltimore, pp. 267–301.

Chardome, M. and Peel, E. (1949) La répartition des filaires dans la region de Coquilhatville et la transmission de *Dipetalonema streptocerca* par *Culicoides grahami. Annales de la Société Belge de Médicine Tropicale* 29, 99–119.

Cheong, W.H. and Omar, A.H. (1970) A new bird filarioid isolated from *Mansonia crassipes* in Malaysia. *Southeast Asian Journal of Tropical Medicine and Public Health* 1, 302–304.

Cheong, W.H., Mak, J.W., Naidu, S. and Mahadevan, S. (1981) *Armigeres subalbatus* incriminated as an important vector of the dog heartworm *Dirofilaria immitis* and the bird *Cardiofilaria* in urban Kuala Lumpur. *Southeast Asian Journal of Tropical Medicine and Public Health* 12, 611.

Chernin, E. (1983) Sir Patrick Manson's studies on the transmission and biology of filariasis. *Reviews of Infectious Diseases* 5, 148–163.

Chhabra, R.C. and Singh, K.S. (1972a) On the life cycle of *Spirocerca lupi*: preinfective stages in the intermediate host. *Journal of Helminthology* 46, 125–137.

Chhabra, R.C. and Singh, K.S. (1972b) On *Spirocerca lupi* infection in some paratenic hosts infected experimentally. *Indian Journal of Animal Sciences* 42, 297–304.

Chhabra, R.C. and Singh, K.S. (1972c) Study of the life-history of *Spirocerca lupi*: histotropic juveniles in dogs. *Indian Journal of Animal Sciences* 42, 628–636.

Chhabra, R.C. and Singh, K.S. (1972d) On the adult *Spirocerca lupi* in experimentally infected dogs. *Indian Journal of Animal Sciences* 42, 636–641.

Chhabra, R.C. and Singh, K.S. (1973) A study of the life-history of *Spirocerca lupi*; intermediate hosts and their biology. *Indian Journal of Animal Sciences* 43, 49–54.

Chhabra, R.C. and Singh, K.S. (1977) The life-history of *Spirocerca lupi*: development and biology of infective juvenile. *Indian Journal of Animal Sciences* 47, 178–184.

Chitwood, B.G. and Chitwood, M.B. (1950) *Introduction to Nematology*. Monumental Printing Co., Baltimore, Maryland.

Choquette, L.P.E. (1955) The life history of the nematode *Metabronema salvelini* (Fujita, 1920) parasitic in the speckled trout, *Salvelinus fontinalis* (Mitchill), in Quebec. *Canadian Journal of Zoology* 33, 1–4.

Chowdhury, N. and Pande, B.P. (1969) Intermediate hosts of *Ascarops strongylina* in India and the development of the parasite in rabbit and guinea pig. *Indian Journal of Animal Sciences* 39, 139–148.

Christenson, B.M. (1977) Laboratory studies on the development and transmission of *Dirofilaria immitis* by *Aedes trivittatus*. *Mosquito News* 37, 367–372.

Christenson, R.O. (1950) Significance of the embryonic sheath. In: Chitwood, B.G. and Chitwood, M.B. (eds) *Introduction to Nematology*. Monumental Printing Co., Baltimore, Maryland, pp. 179–180.

Chwatt, L.J., Gordon, R.M. and Jones, C.M. (1948) The breeding places of *Chrysops silacea*. *Annals of Tropical Medicine and Parasitology* 42, 251.

Clark, G.G. (1972) The role of horse flies (Diptera: Tabanidae) in the transmission of *Elaeophora schneideri* Wehr and Dikmans, 1935, in the Gila National Forest, New Mexico. Unpublished PhD thesis, Colorado State University, Fort Collins, Colorado.

Clark, G.G. and Hibler, C.P. (1973) Horse flies and *Elaeophora schneideri* in the Gila National Forest, New Mexico. *Journal of Wildlife Diseases* 9, 21–25.

Cobbold, T.S. (1870) Description of a new generic type of entozoon from the aard wolf (*Proteles*) with remarks on its affinities, specially in reference to the question of parthenogenesis. *Proceedings of the Zoological Society of London* 1, 9–14.

Collins, R.C. and Jones, R.H. (1978) Laboratory transmission of *Onchocerca cervicalis* with *Culicoides variipennis*. *American Journal of Tropical Medicine and Hygiene* 27, 46–50.

Collins, R.C., Lujan, R., Figueroa, M.H. and Campbell, C.C. (1982) Early formation of the nodule in Guatemalan onchocerciasis. *American Journal of Tropical Medicine and Hygiene* 31, 267–269.

Coluzzi, M. (1964) Osservazioni sperimentali sul comportamento di *Dirofilaria repens* in diversi gruppi di artropodi ematofagi. *Parassitologia* 6, 57–63.

Connal, A. and Connal, S.L.M. (1921) A preliminary note on the development of *Loa loa* Guyot, in *Chrysops silacea* Austen. *Transactions of the Royal Society of Tropical Medicine and Hygiene* 15, 131–134.

Connal, A. and Connal, S.L.M. (1922) The development of *Loa loa* (Guyot) in *Chrysops silacea* (Austen) and in *Chrysops dimidiata* (Van der Wulp.). *Transactions of the Royal Society of Tropical Medicine and Hygiene* 16, 64–89.

Cosgrove, G.E., Humason, G. and Lushbaugh, C.C. (1970) *Trichospiruira leptostoma*, a nematode of the pancreatic duct of marmosets (*Saguinus* spp.). *Journal of the American Veterinary Medical Association* 157, 696–698.

Couvillion, C.E. (1985) Studies on the arterial worm (*Elaeophora schneideri* Wehr and Dikmans, 1935) in white-tailed deer in the southeastern USA. Unpublished PhD thesis, University of Georgia, Athens, Georgia.

Couvillion, C.E., Sheppard, D.C., Nettles, V.F. and Bannaga, O.M. (1984) Intermediate hosts of *Elaeophora schneideri* Wehr and Dikmans, 1935 on South Island, South Carolina. *Journal of Wildlife Diseases* 20, 59–61.

Cram, E.B. (1924) Certain phases of the life history of a spirurid. *Journal of Parasitology* 11, 117.

Cram, E.B. (1926) A new nematode from the rat and its life history. *Proceedings of the USA National Museum* 68, 1–7.

Cram, E.B. (1928) Observations on the life history of the swine stomach worm, *Physocephalus sexalatus* in the USA. *Journal of Parasitology* 15, 136.

Cram, E.B. (1930) Aberrant larvae of *Physocephalus sexalatus* in birds. *Journal of Parasitology* 17, 56.

Cram, E.B. (1931) Developmental stages of some nematodes of the Spiruroidea parasitic in poultry and game birds. *USA Department of Agriculture Technical Bulletin* No. 227, pp. 1–27.

Cram, E.B. (1932) Recent findings in connection with parasites of game birds. *Transactions of the 18th American Game Conference*, pp. 243–247.

Cram, E.B. (1933a) Observations on the life history of *Seurocyrnea colini*. *Journal of Parasitology* (Suppl.) 20, 98.

Cram, E.B. (1933b) Observations on the life history of *Tetrameres pattersoni*. *Journal of Parasitology* (Suppl.) 20, 97–98.

Cram, E.B. (1934a) Recent records of the gizzard worm, *Acuaria anthuris* (Rudolphi, 1819) (Nematoda: Acuariidae), with observations on its life history. *Proceedings of the Helminthological Society of Washington* 1, 48–49.

Cram, E.B. (1934b) Orthopterans and pigeons as secondary and primary hosts, respectively, for the crow stomach-worm *Microtetrameres helix* (Nematoda: Spiruridae). *Proceedings of the Helminthological Society of Washington* 1, 50.

Cram, E.B. (1935) New avian and insect hosts for *Gongylonema ingluvicola* (Nematoda: Spiruroidea). *Proceedings of the Helminthological Society of Washington* 2, 59.

Cram, E.B. (1937) A species of Orthoptera serving as intermediate host of *Tetrameres americana* of poultry in Puerto Rico. *Proceedings of the Helminthological Society of Washington* 4, 24.

Crans, W.J. (1970) The blood feeding habits of *Culex territans* Walker. *Mosquito News* 30, 445–447.

Crites, J.L. and Overstreet, R.M. (1991) *Heliconema brooksi* n.sp. (Nematoda: Physalopteridae) from the ophichthid eel (*Ophichthus gomesi*) in the Gulf of Mexico. *Journal of Parasitology* 77, 42–50.

Crook, J.R. and Grundmann, A.W. (1964) The life history of *Protospirura numidica* Seurat, 1914 (Nematoda: Spiruroidea). *Proceedings of the Helminthological Society of Washington* 31, 225–229.

Cross, J.H. and Scott, J.A. (1947) The developmental anatomy of the fourth-stage larvae and adults of *Litomosoides carinii*, a filarial worm of the cotton rat. *Transactions of the American Microscopical Society* 66, 1–21.

Cross, J.H., Partono, F., Hsu, M.K., Ash, L.R. and Oemijati, S. (1979) Experimental transmission of *Wuchereria bancrofti* to monkeys. *American Journal of Tropical Medicine and Hygiene* 28, 56–66.

Cuvillier, E. (1934) Notes on the life history of *Cheilospirura hamulosa*, the chicken gizzard worm. *Proceedings of the Helminthological Society of Washington* 1, 14–15.

Cuvillier, E. (1937) Observations on the biological and morphological relationships of *Dispharynx spiralis* in bird hosts. *Papers on Helminthology Published in Commemoration of the 30-year Jubileum of K.I. Skrjabin and the 15th Anniversary of All-Union Institution of Helminthology*, pp. 97–104.

Cvetaeva, N.P. (1960) Pathomorphological changes in the proventriculus of ducks by tetrameriasis. *Helminthologia* 2, 143–150. (In Russian.)

Daengsvang, S. (1968) Further observations on the experimental transmission of *Gnathostoma spinigerum*. *Annals of Tropical Medicine and Parasitology* 62, 88–94.

Daengsvang, S. (1971) Infectivity of *Gnathostoma spinigerum* larvae in primates. *Journal of Parasitology* 57, 476–478.

Daengsvang, S. (1980) *A Monograph on the Genus* Gnathostoma *and Gnathostomiasis in Thailand*. Southeast Asia Medical Information Center, Tokyo, 85 pp.

Daengsvang, S. and Tansurat, P. (1938) A contribution to the knowledge of the second intermediate host of *Gnathostoma spinigerum* Owen, 1836. *Annals of Tropical Medicine and Parasitology* 32, 137–140.

Daengsvang, S., Sirichakwal, P., Yingyourd, P. and Machimasatha, R. (1971) Studies of new experimental hosts, life cycles and modes of transmission of gnathostomes. *Annual Progress Report SEATO Medical Research Laboratory*, 156 pp.

Daiya, G.G. (1969) New data on the reservoir hosts of *Gnathostoma hispidum* (Nematoda: Gnathostomatidae). *Zoologicheskii Zhurnal* 48, 1730–1732. (In Russian.)

Daniels, C.W. (1897) The *Filaria sanguinis hominis perstans* found in the aboriginals of British Guyana. *British Guinea Medical Annales* 9, 24–41.

Davadan, A. (1988) Contribution á l'etude de l'infestation parasitaire du chien par *Dipetalonema dracunculoides*. Doctoral thesis, Veterinary Medicine, Toulouse, France, 149 pp.

David, H.L. and Edeson, J.F.B. (1964) The microfilaria of man in Portuguese Timor. *Transactions of the Royal Society of Tropical Medicine and Hygiene* 56, 6.

David, H.L. and Edeson, J.F.B. (1965) Filariasis in Portuguese Timor, with observations on a new microfilaria found in man. *Annals of Tropical Medicine and Parasitology* 59, 193–204.

De, N.C. and Moravec, F. (1979) Some new data on the morphology and development of the nematode *Cystidicoloides tenuissima* (Zeder, 1800). *Folia Parasitologica* 26, 231–238.

Denham, D.A. and McGreavy, P.B. (1977) Brugian filariasis: epidemiological and experimental studies. *Advances in Parasitology* 15, 243–309.

Denke, A.M. and Bain, O. (1978) Notes et informations. Données sur le cycle d'*Onchocerca ochengi* chez *Simulium damnosum* s.l. au Togo. *Annales de Parasitologie Humaine et Comparée* 53, 757–760.

Denny, M. (1969) Life cycles of helminth parasites using *Gammarus lacustris* as an intermediate host in a Canadian lake. *Parasitology* 59, 795–827.

Desportes, C. (1941) Nouvelles recherches sur la morphologie et sur l'évolution d'*Icosiella neglecta* (Diesing, 1851) filaire commune de la grenouille verte. *Annales de Parasitologie Humaine et Comparée* 18, 46–67.

Desportes, C. (1942) *Forcipomyia velox* Winn. et *Sycorax silacea* Curtis, vecteurs d'*Icosiella neglecta* (Diesing) filaire commune de la grenouille verte. *Annales de Parasitologie Humaine et Comparée* 19, 53–68.

Desportes, C., Chabaud, A.G. and Campana, Y. (1949) Sur les gongylonèmes de Muridae et leurs formes larvaires. *Annales de Parasitologie Humaine et Comparée* 24, 447–459.

Desset, M.C. (1966) Contribution à la systématique des filaires du genre *Setaria*; valeur des diérides. *Mémoires du Muséum National d'Histoire Naturelle Serie A, Zoologie* 39, 257–287.

Diagne, M., Petit, G., Seureau, C. and Bain, O. (1989) Développement de la filaire *Litomosoides galizai* chez l'acarien vecteur. *Annales de Parasitologie Humaine et Comparée* 64, 478–488.

Dikmans, G. (1934) Observations on stephanofilariasis in cattle. *Proceedings of the Helminthological Society of Washington* 1, 42–43.

Dikmans, G. (1948) Skin lesions of domestic animals in the USA due to nematode infestation. *Cornell Veterinarian* 38, 2–23.

Dimitrova, E. (1965a) New intermediary hosts of *Simondia paradoxa* Cobb 1864 – nematode of swine. *Compte Rendu de l'Académie Bulgare des Sciences* 18, 355–357.

Dimitrova, E. (1965b) Reservoir parasitism in *Simondia paradoxa* (Spirurida: Nematoda). *Izvestiya na Tsentralnata Khelmintologichna Laboratoriya, Sofia* 10, 35–45. (In Russian.)

Dimitrova, E. (1966) Distribution of *Ascarops strongylina* and *Physocephalus sexalatus* in Bulgaria and their intermediate and reservoir hosts in the Strandzha Mountain area. *Izvestiya na Tsentralnata Khelmintologichna Laboratoriya, Sofia* 11, 43–56. (In Russian.)

Dissamarn, R., Thirapat, K., Aranyakanada, P. and Chai-anan, P. (1966) Studies on morphology and life history of *Gnathostoma doloresi* and *G. hispidum* in Thailand. *Journal of the Thai Veterinary Medical Association* 17, 1–10.

Dissanaike, A.S. (1967) *Pelecitus ceylonensis* n.sp., from the chick and ash-dove experimentally infected with larvae from *Mansonia crassipes*, and from naturally infected crows in Ceylon. *Ceylon Journal of Biological Sciences* 7, 96–104.

Dissanaike, A.S. and Fernando, M.A. (1965) *Cardiofilaria nilesi* n.sp., recovered from a chicken experimentally infected with infective larvae from *Mansonia crassipes*. *Journal of Helminthology* 39, 151–158.

Dissanaike, A.S. and Niles, W.J. (1967) On two infective filarial larvae in *Mansonia crassipes* with a note on other infective larvae in wild-caught mosquitoes in Ceylon. *Journal of Helminthology* 41, 291–298.

Dittrich, L. (1963) *Gongylonema neoplasticum* (Fibiger und Ditlevsen, 1914) bei Ratten und Schaben. *Berliner und Münchener Tierärztliche Wochenschrift* 76, 12–13.

Dixon, J.M. and Roberson, J.M. (1967) Case report: aberrant larvae of *Physaloptera* sp. in a quail (*Colinus virginianus*). *Avian Diseases* 11, 41–44.

Dobrynin, M.I. (1972) The discovery of the intermediate host of *Thelazia leesei* Railliet and Henry, 1910. *Izvestiya Akademii Nauk Turkmenskoi SSR* 3, 73–77. (In Russian.)

Dobrynin, M.I. (1974) Development of *Thelazia leesei* in the intermediate host (*Musca lucidula*). *Izvestiya Akademii Nauk Turkmenskoi SSR* 5, 39–45. (In Russian.)

Dotsenko, T.K. (1953) Life cycle of *Cheilospirura hamulosa*, parasite of gallinaceous birds. *Doklady Akademii Nauk SSSR* 88, 583–584. (In Russian.)

Drobishchenko, N.I. and Shol', V.A. (1978) Features of interrelationships between *Haematobia stimulans* and the filaria of *Setaria cervi*, *S. equina* and *S. labiatopapillosa*. *Zhiznennye Tsikly, Ekologiya i Morfologiya gel'Mintov Zhivotnykh Kazakhstana. Alma-Ata, USSR*, *'Nauka'*, pp. 151–156. (In Russian.)

Dubinin, V.B. (1949) Experimental studies on the life cycles of some parasitic worms in animals of the Volga delta. *Parazitologicheskii Sbornik* 11, 145–151. (In Russian.)

Duke, B.O.L. (1954) The transmission of loiasis in the forest-fringe area of the British Cameroons. *Annals of Tropical Medicine and Parasitology* 48, 349–355.

Duke, B.O.L. (1958) Studies on the biting habits of *Chrysops*. V. The biting cycle and infection rates of *C. silacea*, *C. dimidiata*, *C. langi* and *C. centurionis* at canopy level in the rain-forest of Bombe, British Cameroons. *Annals of Tropical Medicine and Parasitology* 52, 24–35.

Duke, B.O.L. (1962) Experimental transmission of *Onchocerca volvulus* to chimpanzee. *Transactions of the Royal Society of Tropical Medicine and Hygiene* 56, 271.

Duke, B.O.L. (1964) Studies on loiasis in monkeys. IV. Experimental hybridization of the human and simian strains of *Loa. Annals of Tropical Medicine and Parasitology* 58, 390–408.

Duke, B.O.L. (1971) The ecology of onchocerciasis in man and animals. In: Fallis, A.M. (ed.) *Ecology and Physiology of Parasites. A Symposium.* University of Toronto Press, Toronto pp. 213–222.

Duke, B.O.L. (1972) Behaviourial aspects of the life cycle of *Loa.* In: Canning, E.U. and Wright, C.A. (eds) *Behaviour Aspects of Parasite Transmission.* Academic Press, London, pp. 97–107.

Duke, B.O.L. (1981) The six diseases of WHO: onchocerciasis. *British Medical Journal* 283, 961–962.

Duke, B.O.L. (1991) Observations and reflections of the immature stages of *Onchocerca volvulus* in the human host. *Annals of Tropical Medicine and Parasitology* 85, 103–110.

Duke, B.O.L. and Wijers, D.J.B. (1958) Studies on loiasis in monkeys. I. The relationship between human and simian *Loa* in the rain-forest zone of the British Cameroons. *Annals of Tropical Medicine and Parasitology* 52, 158–175.

Dutt, S.C. (1970) Preliminary studies on the life history of *Stephanofilaria zaheeri* Singh, 1958. *Indian Journal of Helminthology* 22, 139–143.

Dutton, J.E. (1905) The intermediary host of *Filaria cypseli* (Annett, Dutton, Elliot), the filaria of the African swift *Cypselus affinis. Thompson Yates and Johnston Laboratories Report n.s.* 6, 137–147.

Duxbury, R.E., Moon, A.P. and Sadun, E.H. (1961) Susceptibility and resistance of *Anopheles quadrimaculatus* to *Dirofilaria uniformis. Journal of Parasitology* 47, 687–691.

Dyer, W.G. and Olsen, O.W. (1967) Biology of *Mastophorus numidica* (Seurat, 1914) Read and Millemann, 1953 (Nematoda: Spiruridae) with a description of the juvenile stages. *Proceedings of the Helminthological Society of Washington* 34, 98–103.

Eberhard, M.L. and Orihel, T.C. (1981) Development and larval morphology of *Loa loa* in experimental primate hosts. *Journal of Parasitology* 67, 557–564.

Eberhard, M.L. and Orihel, T.C. (1984) *Loaina* gen.n (Filarioidea: Onchocercidae) for the filariae parasitic in rabbits in North America. *Proceedings of the Helminthological Society of Washington* 51, 49–53.

Eberhard, M.L., Lowrie, R.C. and Orihel, T.C. (1979) Development of *Dipetalonema gracile* and *D. caudispina* to the infective stage in *Culicoides hollensis. Journal of Parasitology* 65, 89–95.

Edeson, J.F.B. and Buckley, J.J.C. (1959) Studies on filariasis in Malaya: on the migration and rate of growth of *Wuchereria malayi* in experimentally infected cats. *Annals of Tropical Medicine and Parasitology* 53, 113–119.

Edeson, J.F.B. and Wharton, R.H. (1957) The transmission of *Wuchereria malayi* from man to the domestic cat. *Transactions of the Royal Society of Tropical Medicine and Hygiene* 51, 366–370.

Edeson, J.F.B. and Wharton, R.H. (1958) The experimental transmission of *Wuchereria malayi* from man to various animals in Malaya. *Transactions of the Royal Society of Tropical Medicine and Hygiene* 52, 25–45.

Edeson, J.F.B. and Wilson, T. (1964) The epidemiology of filariasis due to *Wuchereria bancrofti* and *Brugia malayi. Annual Review of Entomology* 9, 245–268.

Edeson, J.F.B., Wharton, R.H. and Laing, A.B.G. (1960a) A preliminary account of the transmission, maintenance and laboratory vectors of *Brugia pahangi. Transactions of the Royal Society of Tropical Medicine and Hygiene* 54, 439–449.

Edeson, J.F.B., Wilson, T., Wharton, R.H. and Laing, A.B.G. (1960b) Experimental transmission of *Brugia malayi* and *Brugia pahangi* to man. *Transactions of the Royal Society of Tropical Medicine and Hygiene* 54, 229–234.

Eichler, D.A. (1971) Studies on *Onchocerca gutturosa* (Neumann, 1910) and its development in *Simulium ornatum* (Meigen, 1818). II. Behaviour of *S. ornatum* in relation to the transmission of *O. gutturosa*. *Journal of Helminthology* 45, 259–270.

Eichler, D.A. and Nelson, G.S. (1971) Studies on *Onchocerca gutturosa* (Neumann, 1910) and its development in *Simulium ornatum* (Meigen, 1818). I. Observations on *O. gutturosa* in cattle in southeast England. *Journal of Helminthology* 45, 245–258.

Ellis, C.J. (1969a) Life history of *Microtetrameres centuri* Barus, 1966 (Nematoda: Tetrameridae). I. Juveniles. *Journal of Nematology* 1, 84–93.

Ellis, C.J. (1969b) Life history of *Microtetrameres centuri* Barus, 1966 (Nematoda: Tetrameridae). II. Adults. *Journal of Parasitology* 55, 713–719.

Ellis, E. and Williams, J.C. (1973) The longevity of some species of helminth parasites in naturally acquired infections of the lesser black-headed gull, *Larus fuscus* L. in Britain. *Journal of Helminthology* 47, 329–338.

Fadzil, M. (1973) *Musca conducens* Walker, 1859 – its prevalence and potential as a vector of *Stephanofilaria kaeli* in Peninsular Malaysia. *Kajian Veterinaire* 5, 27–37.

Fadzil, M. (1975) The development of *Stephanofilaria kaeli* Buckley, 1937 in *Musca conducens* Walker, 1859. *Kajian Veterinaire* 1, 1–7.

Fagerholm, H.P. and Butterworth, E. (1988) *Ascarophis* sp. (Nematoda: Spirurida) attaining sexual maturity in *Gammarus* spp. (Crustacea). *Systematic Parasitology* 12, 123–139.

Fain, A. and Herin, V. (1955) Filarioses des bovidés au Ruanda-Urundi. III. Etude parasitologique. *Annales de la Société Belge de Médecine Tropicale* 35, 535–554.

Fain, A., Wery, M. and Tilkin, J. (1981) Transmission of *Onchocerca volvulus* by *Simulium albivirgulatum* in the onchocerciasis focus of Cuvette Centrale, Zaire. *Annales de la Société Belge de Médecine Tropicale* 61, 307–309.

Farnell, D.R. and Faulkner, D.R. (1978) Prepatent period of *Dipetalonema reconditum* in experimentally infected dogs. *Journal of Parasitology* 64, 565–567.

Faust, E.C. (1927) Migration route of *Spirocerca sanguinolenta* in its definitive host. *Proceedings of the Society for Experimental Biology and Medicine* 25, 192–195.

Faust, E.C. (1928) The life cycle of *Spirocerca sanguinolenta* – a natural nematode parasite of the dog. *Science* 68, 407–409.

Faust, E.C. (1929) The egg and first-stage (rhabditiform) larva of the nematode, *Spirocerca sanguinolenta*. *Transactions of the American Microscopical Society* 48, 62–65.

Faust, E.C. (1949) *Human Helminthology. A Manual for Physicians, Sanitarians and Medical Zoologists*, 3rd edn. Lea and Febiger, Philadelphia.

Faust, E.C. (1957) Human infection with species of *Dirofilaria*. *Zeitschrift für Tropenmedizin und Parasitologie* 9, 59–68.

Faust, E.C., Beaver, P.C. and Jung, R.C. (1975) *Animal Agents and Vectors of Human Diseases*. Lea and Febiger, Philadelphia.

Feng, L.C. (1936) The development of *Microfilaria malayi* in *A. hyrcanus* var. *sinensis* Wied. *Chinese Medical Journal* (Suppl.) 1, 345–367.

Feng, L.C. (1937) Attempt to immunize dogs against infection with *Dirofilaria immitis* Leidy, 1856. *Festschrift Bernhard Nocht zum 80. Geburtstag, Hamburg*, pp. 140–142.

Fibiger, J.A.G. (1913) Untersuchungen über eine Nematode (*Spiroptera* sp. n.) und deren Fähigkeit papillomatöse und carcinomatöse Geschwulstbildung im Magen der Ratte hervorzurufen. *Zeitschrift für Krebsforschung* 13, 217–280.

Fibiger, J.A.G. (1919) On *Spiroptera carcinomata* and their relationship to true malignant tumors with some remarks on cancer age. *Journal of Cancer Research* 4, 367–387.

Fibiger, J.A.G. and Ditlevsen, H. (1914) A contribution to the biology and morphology of *Spiroptera* (*Gongylonema*) *neoplastica* n.sp. *Mindeskrift i Anledning af Hundredaaret, Københaven*, No. 25, 28 pp.

Fielding, J.W. (1926) Preliminary note on the transmission of the eyeworms of Australian poultry. *Australian Journal of Experimental Biology and Medical Science* 3, 225–232.

Fielding, J.W. (1927) Further observations on the life history of the eyeworm of poultry. *Australian Journal of Experimental Biology and Medical Science* 4, 273–281.

Fielding, J.W. (1928) Additional observations on the development of the eyeworm of poultry. *Australian Journal of Experimental Biology and Medical Science* 5, 1–8.

Fincher, G.T., Stewart, T.B. and Davis, R. (1969) Beetle intermediate hosts for swine spirurids in southern Georgia. *Journal of Parasitology* 55, 355–358.

Fitzsimmons, W.M. (1960) Observations on the incidence, pathology and aetiology of *Spirocerca lupi* infestations in Nyasaland. *British Veterinary Journal* 116, 272–275.

Foil, L., Stage, D. and Klei, T.R. (1984) Assessment of wild-caught *Culicoides* (Ceratopogonidae) species as natural vectors of *Onchocerca cervicalis* in Louisiana. *Mosquito News* 44, 204–206.

Foster, A.O. and Johnson, C.M. (1939) A preliminary note on the identity, life cycle and pathogenicity of an important nematode parasite of captive monkeys. *American Journal of Tropical Medicine* 19, 265–277.

Frank, W. (1962) Biologie von *Macdonaldius oschei* (Filarioidea, Onchocercine), zugleich ein Beitrag über die Wirtsspezifität von *Ornithodoros talaje* (Ixodoidea) Argasidae. *Zeitschrift für Parasitenkunde* 22, 107–108.

Frank, W. (1964a) Die Entwicklung von *Macdonaldius oschei* Chabaud et Frank 1961 (Filarioidea, Onchocercidae) in der Lederzecke *Ornithodoros talaje* Guérin-Méneville (Ixodoidea, Argasidae). *Zeitschrift für Parasitenkunde* 24, 319–350.

Frank, W. (1964b) Die Übertragung der Filarien-infektionsstadien von *Macdonaldius oschei* Chabaud et Frank 1961 (Filarioidea, Onchocercidae) durch *Ornithodoros talaje* (Ixodoidea, Argasidae) auf den Endwirt; Zugleich ein Beitrag zur Biologie des Überträgers. *Zeitschrift für Parasitenkune* 24, 415–441.

Frank, W. (1964c) Die pathogenen Wirkungen von *Macdonaldius oschei* Chabaud et Frank 1961 (Filarioidea, Onchocercidae) bei verschiedenen Arten von Schlangen (Reptilia, Ophidia). *Zeitschrift für Parasitenkunde* 24, 249–275.

French, R.A., Todd, K.S., Meehan, T.P. and Zachary, J.F. (1994) Parasitology and the pathogenesis of *Geopetitia aspiculata* (Nematoda: Spirurida) in zebra finches (*Taeniopygia guttata*): experimental infection and new host records. *Journal of Zoo and Wildlife Medicine* 25, 403–422.

Fuhrman, J.A. and Piessens, W.F. (1985) Chitin synthesis and sheath morphologenesis in *Brugia malayi* microfilariae. *Molecular Biochemistry and Parasitology* 17, 93–104.

Fujii, T., Hayashi, T., Ishimoto, A., Takahashi, S., Asano, H. and Kato, T. (1995) Prenatal infection with *Setaria marshalli* (Boulenger, 1921) in cattle. *Veterinary Parasitology* 56, 303–309.

Fülleborn, F. (1908) Ueber Versuche an Hundefilarien und deren Uebertragung durch Mücken. *Archiv für Schiffs- und Tropen-Hygiene, Pathologie und Therapie Exotischer Krankheiten* 12, 313–351.

Fülleborn, F. (1913) Beiträge zur Morphologie und Differentialdiagnose der Mikrofilarien. *Archiv für Schiffs- und Tropen-Hygiene, Pathologie und Therapie Exotischer Krankheiten* 1, 7–72.

Fülleborn, F. (1929) Filariosen des Menschen. *Handbuch der Pathogenen Mikroorganismen. Begründet von W. Kolle und A.V. Wassermann* 6, 1043–1224.

Gafurov, A. (1969) Life-cycle of *Vigisospirura potekhina* (Petrov and Potekhina, 1953) Chabaud, 1959. *Materialy Nauchnoi Konferentsii Vsesoyuznogo Obshchestva Gel'mintologov*, Part II, pp. 171–180. (In Russian.)

Gafurov, A.K. and Lunkina, E.P. (1970) A study of the developmental cycle of *Abbreviata kazachstanica* Markov and Paraskiv, 1956 (Nematoda: Spirurata). *Izvestiya Akademii Nauk Tadzhikskoi SSR Otdelenie Biologicheskikh Nauk* 4, 100–104. (In Russian.)

Galeb, O. (1878) Observations sur les migrations du *Filaria rytipleurites*, parasite des blattes et des rats. *Compte Rendu Hebdomadaire des Séances de l'Académie des Sciences, Paris* 88, 75–77.

Garkavi, B.L. (1949a) Elucidation of the life cycle of the nematode *Tetrameres fissispina* parasite of domestic and wild ducks. *Doklady Akademii Nauk SSSR* 66, 1215–1218. (In Russian.)

Garkavi, B.L. (1949b) A study of the life cycle of the nematode *Streptocara crassicauda* (Creplin, 1829) parasitic in domestic and wild ducks. *Doklady Akademii Nauk SSSR* 65, 421–424. (In Russian.)

Garkavi, B.L. (1950) Reservoir hosts of *Streptocara crassicauda* (Creplin, 1829), a parasite of domestic ducks. *Trudi Vsesoyuznogo Instituta Gelmintologii* 4, 5–7. (In Russian.)

Garkavi, B.L. (1953) The life cycle of the nematode *Streptocara crassicauda*. Diagnosis and epizootiology of streptocariasis of ducks. *Trudi Vsesoyuznogo Instituta Gelmintologii* 5, 5–22. (In Russian.)

Garkavi, B.L. (1956) The propagation and natural foci of the *Streptocara* nematodes of ducks. *Zoologicheskii Zhurnal* 35, 376–378. (In Russian.)

Garkavi, B.L. (1960) Observations on the biology of *Echinuria uncinata* and epizootiology of the infection in ducks in the Krasnodar Territory I. (Abstract). *Tezigy Doklady Nauchnoi Konferentsii Vsesoyuznogo Obshchestva Gelminthologeskii, Moscow*, Dec. 15–20 1960, pp. 28–29. (In Russian.)

Garms, R. and Cheke, R.A. (1985) Infections with *Onchocerca volvulus* in different members of the *Simulium damnosum* complex in Togo and Benin. *Zeitschrift für Angewandte Zoologie* 72, 479–495.

Geden, C.J. and Stoffolano, J.G. (1981) Geographic range and temporal patterns of parasitization of *Musca autumnalis* De Geer by *Thelazia* sp. in Massachusetts, with observations on *Musca domestica* L. as an unsuitable intermediate host. *Journal of Medical Entomology* 18, 449–456.

Geden, C.J. and Stoffolano, J.G. (1982) Development of the bovine eyeworm, *Thelazia gulosa* (Railliet and Henry), in experimentally infected female *Musca autumnalis* de Geer. *Journal of Parasitology* 68, 287–292.

Genchi, C., Basano, F.S., Bandi, C., DiSacco, B., Venco, L., Vezzoni, A. and Crancrini, G. (1995) Factors influencing the spread of heartworms in Italy; interaction between *Dirofilaria immitis* and *Dirofilaria repens*. *Proceedings of the Heartworm Symposium* 95, Auburn, Alabama, USA, pp. 65–71.

Geraci, J.R., Fortin, J.F., St Aubin, D.J. and Hicks, B.D. (1981) The seal louse, *Echinophthirius horridus*, an intermediate host of the seal heartworm, *Dipetalonema spirocauda* (Nematoda). *Canadian Journal of Zoology* 59, 1457–1459.

Gibson, T.E., Pepin, G.A. and Pinsent, P.J.N. (1964) *Parafilaria multipapillosa* in the horse. *Veterinary Record* 76, 774–777.

Gnedina, M.P. (1950a) Discovery of *Stephanofilaria stilesi* Chitwood, 1934, in cattle. *Trudi Vsesoyuznogo Instituta Gel'mintolgii im Akademika K.I. Skryabina* 4, 82–84. (In Russian.)

Gnedina, M.P. (1950b) Biology of the nematode *Onchocerca gutturosa* Neumann (1910) parasitic in cattle. *Doklady Akademii Nauk SSSR* 70, 169–171. (In Russian.)

Gnedina, M.P. and Osipov, A.N. (1960) The biology of the causative agent of parafilariasis in horses. *Veterinariya* 37, 49–50. (In Russian.)

Godoy, G.A., Volcan, G., Medrano, C. and Guevara, R. (1986) The parasitology of onchocerciasis in America with special reference to Venezuela. *Boletín de la Oficina Saniaria Panamericana* 101, 1–18.

Golovin, O.V. (1956) The biology of *Gnathostoma hispidum*. *Doklady Akademii Nauk SSSR* 111, 242–244. (In Russian.)

Golvan, Y.J. and Chabaud, A.G. (1963) Infestation spontanée de phlébotomes par le spiruride *Mastophorus muris* (Gmelin). *Annales de Parasitologie Humaine et Comparée* 38, 934.

Gooneratne, B.W.M. (1968) On a successful peritoneal transplant of *Cardiofilaria nilesi* into an experimental chicken. *Journal of Helminthology* 42, 279–282.

Gooneratne, B.W.M. (1969) On *Cardiofilaria nilesi* in experimentally infected chickens with a note on the morphology and periodicity of the microfilariae. *Journal of Helminthology* 43, 311–317.

Gordon, R.M. (1948) The mechanisms by which mosquitoes and tsetse flies obtain their blood-meal, the histology of the lesions produced, and the subsequent reactions of the mammalian host; together with some observations on the feeding of *Chrysops* and *Cimex*. *Annals of Tropical Medicine and Parasitology* 42, 334.

Gordon, R.M. (1955) A brief review of recent advances in our knowledge of loiasis and some of the still outstanding problems. *Transactions of the Royal Society of Tropical Medicine and Hygiene* 49, 98–105.

Gordon, R.M. and Crewe, W. (1953) The deposition of the infective stage of *Loa loa* by *Chrysops silacea* and the early stages of its migration to the deeper tissues of the mammalian host. *Annals of Tropical Medicine and Parasitology* 47, 74–85.

Gordon, R.M., Kershaw, W.E., Crewe, W. and Oldroyd, H. (1950) The problems of loiasis in west Africa with special reference to recent investigations at Kumba in the British Cameroons and at Sapele in southern Nigeria. *Transactions of the Royal Society of Tropical Medicine and Hygiene* 44, 11–41.

Gorshkov, I.P. (1948) Biology of *Draschia megastoma* from horses. In: *Collected Papers on Helminthology Dedicated to K.I. Skrjabin on the 40th Anniversary of his Scientific Activity and on the 25th Anniversary of the All-Union Institute of Helminthology*, pp. 98–108. (In Russian.)

Grassi, G.B. (1888) Ciclo evolutivo della *Spiroptera* (*Filaria*) *sanguinolenta*. *Giornale di Anatomia Fisiologia e Patologia delgi Animali, Pisa* 20, 99–101.

Grassi, B. and Calandruccio, S. (1890) Ueber Haematozoon Lewis: Entwicklungscyklus einer Filaria (*Filaria reconditum* Grassi) des Hundes. *Zentralblatt für Bakteriologie, Parasitenkunde und Infektionskrankheiten* 7, 8–26.

Grassi, B. and Noè, G. (1900) The propagation of the Filariae of the blood exclusively by means of the puncture of peculiar mosquitoes. *British Medical Journal* 2, 1306–1307.

Gray, J.B. (1981) Biology of *Turgida turgida* (Rudolphi) (Physalopteroidea: Nematoda) in the intermediate and final host. Unpublished MS thesis, University of Guelph, Guelph, Ontario, Canada.

Gray, J.B. and Anderson, R.C. (1982a) Observations on *Turgida turgida* (Rudolphi, 1819) (Nematoda: Physalopteroidea) in the American opossum (*Didelphis virginiana*). *Journal of Wildlife Diseases* 18, 279–286.

Gray, J.B. and Anderson, R.C. (1982b) Development of *Turgida turgida* (Rudolphi, 1819) (Nematoda: Physalopteroidea) in the opossum (*Didelphis virginiana*). *Canadian Journal of Zoology* 60, 1265–1274.

Gray, J.B. and Anderson, R.C. (1982c) Development of *Turgida turgida* (Rudolphi, 1819) in the common field cricket (*Acheta pennsylvanicus* Burmeister). *Canadian Journal of Zoology* 60, 2134–2142.

Greenberg, B. (1971) *Flies and Diseases,* Vol. II, *Biology and Disease Transmission*. Princeton University Press, Princeton, 447 pp.

Grétillat, S. and Touré, S. (1970) Premières recherches concernant l'épidémiologie et la détermination de vecteur de la thélaziose bovine en Afrique de l'Quest. *Comptes Rendus de l'Academie des Sciences Série D* 270, 239–241.

Grove, D.I. (1990) *A History of Human Helminthology*. CAB International, Wallingford, UK.

Guilhon, J., Saratsiotis, A. and Jolivet, G. (1971) Experimental echinuriasis in ducks. *Recueil de Médecine Vétérinaire de l'Ecole d'Alfort* 147, 1–21.

Gunewardene, K. (1956) Observations on the development of *Dirofilaria repens* in *Aedes* (*Stegomyia*) *albopictus* and other common mosquitoes in Ceylon. *Ceylon Journal of Medical Science* 9, 45–53.

Gupta, V.P. (1969) Pond-snakes and ground squirrels as paratenic hosts of *Ascarops strongylina*. *Current Science* 22, 548–549.

Gupta, V.P. (1970) Experimental development of the oesophageal worm of cattle in rabbits. *Current Science* 39, 237–238.

Gupta, V.P. and Pande, B.P. (1970a) *Hemidactylus flaviviridis*, a paratenic host of *Rictularia cahirensis*. *Current Science* 23, 535–536.

Gupta, V.P. and Pande, B.P. (1970b) Natural intermediaries and paratenic hosts of *Simondsia paradoxa* – an important stomach worm of pigs. *Current Science* 39, 536.

Gupta, V.P. and Pande, B.P. (1981) *Cyathospirura chabaudi* n.sp. from pups infected with re-encapsulated larvae from a paratenic host. *Indian Journal of Animal Science* 51, 526–534.

Gustafson, P.V. (1939) Life cycle studies on *Spinitectus gracilis* and *Rhabdochona* sp. (Nematoda: Thelaziidae). *Journal of Parasitology* (Suppl.) 25, 12–13.

Gustafson, P.V. (1942) A peculiar larval development of *Rhabdochona* spp. (Nematoda: Spiruroidea). *Journal of Parasitology* (Suppl.) 28, 30.

Hagiwara, S., Suzuki, M., Shirasaka, S. and Kurihara, T. (1992) A survey of the vector mosquitoes of *Setaria digitata* in Ibaraki Prefecture, central Japan. *Japanese Journal of Sanitary Zoology* 43, 291–295.

Hagstrom, T. (1979) Population ecology of *Triturus cristatus* and *T. vulgaris* (Urodela) in SW Sweden. *Holarctic Ecology* 2, 108–114.

Hamann, O. (1893) Die Filarienseuche der Enten und der Zwischenwirt von *Filaria uncinata* R. *Zentralblatt für Bakteriologie und Parasitenkunde* 14, 555–557.

Harbut, C.L. (1976) *Brugia beaveri*: development in the intermediate and final hosts. *Dissertation Abstracts International* 36B (12), 6088.

Harbut, C.L. and Orihel, T.C. (1995) *Brugia beaveri*: microscopic morphology in host tissues and observations on its life history. *Journal of Parasitology* 81, 239–243.

Harinasuta, C., Sucharit, S. and Choochote, W. (1981) The susceptibility of leaf monkeys to Bancroftian filariasis in Thailand. *Southeast Asian Journal of Tropical Medicine and Public Health* 12, 581–589.

Hasegawa, H. and Otsuru, M. (1977) Life cycle of a frog nematode, *Spinitectus ranae* Morishita, 1926 (Cystidicolidae). *Japanese Journal of Parasitology* 26, 336–344.

Hasegawa, H. and Otsuru, M. (1978) Notes on the life cycle of *Spiroxys japonica* Morishita, 1926 (Nematoda: Gnathostomatidae). *Japanese Journal of Parasitology* 27, 113–122.

Hasegawa, H. and Otsuru, M. (1979) Life history of an amphibian nematode, *Hedruris ijimai* Morishita, 1926 (Hedruridae). *Japanese Journal of Parasitology* 28, 89–97. (In Japanese.)

Hawking, F. (1954) The reproductive system of *Litomosoides carinii*, a filarial parasite of the cotton rat. III. The number of microfilariae produced. *Annals of Tropical Medicine and Parasitology* 48, 382–385.

Hawking, F. (1959) *Dirofilaria magnilarvatum* Price, 1959 (Nematoda: Filarioidea) from *Macaca irus* Cuvier. III. The behaviour of the microfilariae in the mammalian host. *Journal of Parasitology* 45, 511–512.

Hawking, F. (1975) Circadian and other rhythms of parasites. *Advances in Parasitology* 13, 123–182.

Hawking, F. (1976) The 24 h periodicity of microfilariae: biological mechanism responsible for its production. *Proceedings of the Royal Society* 169, 59–76.

Hawking, F. and Burroughs, A.M. (1946) Transmission of *Litomosoides carinii* to mice and hamsters. *Nature* (*London*) 158, 98.

Hawking, F. and Webber, W.A.F. (1955) *Dirofilaria aethiops* Webber, 1955, a filarial parasite of monkeys. II. Maintenance in the laboratory. *Parasitology* 45, 378–387.

Hayasaki, M. (1996) Re-migration of fifth-stage juvenile *Dirofilaria immitis* into pulmonary arteries after subcutaneous transplantation in dogs, cats, and rabbits. *Journal of Parasitology* 82, 835–837.

Healey, M.C. and Grundmann, A.W. (1974) The influence of intermediate hosts on the infection pattern of *Protospirura numidica criceticola* Quentin, Karimi and Rodriquez de Almeida, 1968 (Nematoda: Spiruridae) in the Bonneville Basin, Utah. *Proceedings of the Helminthological Society of Washington* 41, 50–63.

Hedrick, L.R. (1935) The life history and morphology of *Spiroxys contortus* (Rudolphi); Nematoda: Spiruridae. *Transactions of the American Microscopical Society* 54, 307–335.

Heisch, R.B., Nelson, G.S. and Furlong, M. (1959) Studies on filariasis in East Africa. I. Filariasis on the island of Pate, Kenya. *Transactions of the Royal Society of Tropical Medicine and Hygiene* 53, 41–53.

Henrard, C. and Peel, E. (1949) *Culicoides grahami* Austen. Vecteur de *Dipetalonema streptocerca* et non de *Acanthocheilonema perstans*. *Annales de la Société Belge de Médicine Tropicale* 29, 127–143.

Heydon, G.M. (1929) Creeping eruption or larva migrans in North Queensland and a note on the worm *Gnathostoma spinigerum* (Owen). *Medical Journal of Australia* 1, 583–591.

Hibler, C.P. (1963) Onchocercidae (Nematoda: Filarioidea) of the American magpie, *Pica pica hudsonia* (Sabine) in northern Colorado. Unpublished PhD thesis, Colorado State University, Fort Collins, Colorado.

Hibler, C.P. (1964) The life history of *Stephanofilaria stilesi* Chitwood, 1934. *Journal of Parasitology* (Suppl.) 50, 34.

Hibler, C.P. (1966) Development of *Stephanofilaria stilesi* in the horn fly. *Journal of Parasitology* 52, 890–898.

Hibler, C.P. and Adcock, J.L. (1968) Redescription of *Elaeophora schneideri* Wehr and Dikmans, 1935 (Nematoda: Filarioidea). *Journal of Parasitology* 54, 1095–1098.

Hibler, C.P. and Adcock, J.L. (1971) Elaeophorosis. In: Davis, R.W. and Anderson, R.C. (eds) *Parasitic Diseases of Wild Mammals*. Iowa State University Press, Ames, Iowa, pp. 263–268.

Hibler, C.P. and Gates, G.H. (1974) Experimental infection of immature mule deer with *Elaeophora schneideri*. *Journal of Wildlife Diseases* 10, 44–46.

Hibler, C.P. and Metzger, C.J. (1974) Morphology of the larval stages of *Elaeophora schneideri* in the intermediate and definitive hosts with some observations on their pathogenesis in abnormal definitive hosts. *Journal of Wildlife Diseases* 10, 361–369.

Hibler, C.P., Adcock, J.L., Davis, R.W. and Abdelbaki, Y.Z. (1969) Elaeophorosis in deer and elk in the Gila Forest, New Mexico. *Bulletin of Wildlife Disease Association* 5, 27–30.

Hibler, C.P., Adcock, J.L., Gates, G.H. and White, R. (1970) Experimental infection of domestic sheep and mule deer with *Eleophora schneideri* Wehr and Dikmans, 1935. *Journal of Wildlife Diseases* 6, 110–111.

Hibler, C.P., Gates, G.H., White, R. and Donaldson, B.R. (1971) Observations on horseflies infected with larvae of *Elaeophora schneideri*. *Journal of Wildlife Diseases* 7, 43–45.

Hibler, C.P., Gates, G.H. and Donaldson, B.R. (1974) Experimental infection of immature mule deer with *Elaeophora schneideri*. *Journal of Wildlife Diseases* 10, 44–46.

Highby, P.R. (1938) Development of the microfilaria of *Dirofilaria scapiceps* (Leidy, 1886) in mosquitoes of Minnesota. *Journal of Parasitology* (Suppl.) 24, 36.

Highby, P.R. (1943a) Vectors, transmission, development, and incidence of *Dirofilaria scapiceps* (Leidy, 1886) (Nematoda) from the snowshoe hare in Minnesota. *Journal of Parasitology* 29, 253–259.

Highby, P.R. (1943b) *Dipetalonema arbuta* n.sp. (Nematoda) from the porcupine, *Erethizon dorsatum* (L.). *Journal of Parasitology* 29, 239–242.

Highby, P.R. (1943c) Mosquito vectors and larval development of *Dipetalonema arbuta* Highby (Nematoda) from the porcupine, *Erethizon dorsatum*. *Journal of Parasitology* 29, 243–252.

Hill, G.F. (1918) Relationship of insects to parasitic diseases of stock. *Proceedings of the Royal Society of Victoria* 31, 11–107.

Hill, G.F. (1920) The life histories of *Habronema muscae* (Carter), *Habronema microstoma* (Schneider) and *Habronema megastoma* (Rudolphi). *Veterinary Journal* 76, 155–156.

Hitchcock, C.R. and Bell, E.T. (1952) Studies on the nematode parasite, *Gongylonema neoplasticum* (*Spiroptera neoplasticum*) and avitaminosis A in the forestomach of rats: comparison with Fibiger's results. *Journal of the National Cancer Institute* 12, 1345–1387.

Hiyeda, K. and Faust, E.C. (1929) Aortic lesions in dogs caused by infection with *Spirocerca sanguinolenta*. *Archives of Pathology* 7, 253–272.

Ho, B.C., Singh, M. and Lim, B.L. (1973) Observations on the development of a new filaria (*Breinlia booliati* Singh and Ho, 1973) of a rat *Rattus sabanus* in the mosquito *Aedes togoi*. *Journal of Helminthology* 47, 135–140.

Hobmaier, M. (1941) Extramammalian phase of *Physaloptera maxillaris* Molin, 1860 (Nematoda). *Journal of Parasitology* 27, 233–235.

Hoffmeister, K. and Wenk, P. (1991) Experiments on the regulation of worm load in the rodent filariid *Litomosoides carinii* (Nematoda, Filarioidea) in *Sigmodon hispidus*. *Mitteilungen der Österreichischen Gesellschaft für Tropenmedizin und Parasitologie* 13, 119–124.

Hsu, H.F. and Chow, C.Y. (1938) On the intermediate host and larva of *Habronema mansioni* Seurat, 1914 (Nematoda). *Chinese Medical Journal* (Suppl.) 2, 419–422. (In Chinese.)

Hu, C.H. and Hoeppli, R.J.C. (1936) The migration route of *Spirocerca sanguinolenta* in experimentally infected dogs. *Chinese Medical Journal* (Suppl.) 1, 293–311.

Hu, C.H. and Hoeppli, R.J.C. (1937) Further study on the migration route of *Spirocerca sanguinolenta* in experimentally infected dogs. *Chinese Medical Journal* 51, 489–495.

Illgen-Wilcke, B., Beglinger, R., Pfister, R. and Heider, K. (1992) Studies on the developmental cycle of *Trichospirura leptostoma* (Nematoda: Thelaziidae). Experimental infection of the intermediate hosts *Blatella germanica* and *Supella longipalpa* and the definitive host *Callithrix jacchus* and development in the intermediate hosts. *Parasitology Research* 78, 509–512.

Imai, J.I., Akahane, H., Horiuchi, S., Maruyama, H. and Nawa, Y. (1989) *Gnathostoma doloresi*: development of the larvae obtained from snake, *Agkistrodon halys*, to adult worm in a pig. *Japanese Journal of Zoology* 38, 221–225.

Innes, J.R.M. and Pillai, C.P. (1955) Kumri – so called lumbar paralysis – of horses in Ceylon (India and Burma) and its identification with cerebrospinal nematodiasis. *British Veterinary Journal* 111, 223–235.

Innes, J.R.M. and Shoho, C. (1953) Cerebrospinal nematodiasis. Focal encephalomyelomala of animals caused by nematodes (*Setaria digitata*), a disease which may occur in man. *Archives of Neurology and Psychiatry* 70, 325–349.

Ishii, S., Yagima, A., Sugawa, Y., Ishiwara, T., Ogata, T. and Hashiguchi, Y. (1953) The experimental reproduction of so called lumbar paralysis–epizootic cerebrospinal nematodiasis in goats in Japan. *British Veterinary Journal* 107, 160–167.

Itagaki, S. and Taniaguchi, M. (1954) Pathogenicity of *S. digitata* in domestic animals (sheep, goats, and horses) and its life cycle. *Japanese Journal of Sanitary Zoology IV Special Number Commemorating the 70th Birthday of Dr H. Kobayashi*. (In Japanese.)

Ivashkin, V.M. (1956) Elucidation of the life cycle of the nematode *Parabronema skrjabini* of ruminants. *Doklady Academii Nauk SSSR* 107, 773–775. (In Russian.)

Ivashkin, V.M. (1958) Observations on the ecology of the fly, *Lyperosia titillans* and its significance in the biology of *Parabronema skrjabini*. In: *Work of the Expeditions of the Helminthological Laboratory of the Academy of Science of USSR (1945–1957)*. Gel'mintologicheskoi Laboratorii, Akademiya Nauk SSR, Moscow, pp. 109–119. (In Russian.)

Ivashkin, V.M. (1959) Parabronemiasis in ruminants. *Veterinariya* 36, 26–28. (In Russian.)

Ivashkin, V.M. and Golovanov, V.I. (1974) The biology of *Onchocerca gutturosa*, the causative organism of onchocerciasis in cattle. *Trudi Gel'mintologicheskoi Laboratorii (Ekologiya i Geografiya gel'Mintov)* 24, 40–46. (In Russian.)

Ivashkin, V.M. and Khromova, L.A. (1961a) Epizootiology of habronemiasis in domestic ungulates. *Trudi Gelmintologicheskoi Laboratorii. Adademiya Nauk SSSR* 11, 105–108. (In Russian.)

Ivashkin, V.M. and Khromova, L.A. (1961b) Intermediate hosts of *Gongylonema pulchrum* in the Uzbek SSR. *Trudi Gelminthologicheskoi Laboratorii Akademiya Nauk SSSR* 11, 102–104. (In Russian.)

Ivashkin, V.M., Khromova, L.A. and Shmytova, G.A. (1963) Stephanofilariasis of cattle. *Veterinariya* 40, 36–39. (In Russian.)

Ivashkin, V.M., Khromova, L.A. and Baranova, N. (1979) The development cycle of *Thelazia lacrymalis*. *Veterinariya Moscow* 7, 46–47. (In Russian.)

Ivashkin, V.M., Shmitova, G.Y. and Koishibaev, G.K. (1971) Stephanofilariasis of herbivorous animals. *Veterinariya* 48, 66–68. (In Russian.)

Jackson, C.J., Marcogliese, D.J. and Burt, M.D.B. (1997) Precociously developed *Ascarophis* sp. (Nematoda, Spirurata) and the *Hemiarus levinseni* (Digenea, Hemiuridae) in their crustacean intermediate hosts. *Acta Parasitologica* 42, 31–35.

Janz, J.G. and Sing, C.K. (1977) Filariase linfatica em Timor Lest. *Anais do Instituto de Higiene e Medicina Tropical* 5, 257–280.

Jayewardene, L.G. (1963) Larval development of *Brugia ceylonensis* Jayewardene, 1962, in *Aedes aegypti*, with a brief comparison of the infective larvae with those of *Brugia* spp., *Dirofilaria repens* and *Artionema digitata*. *Annals of Tropical Medicine and Parasitology* 57, 359–370.

Jensen, R. and Mackey, D.R. (1965) *Diseases of Feedlot Cattle*. Lea and Febiger, Philadelphia.

Jilek, R. and Crites, J.L. (1980) Pathological implications of *Spinitectus carolini* (Spirurida: Nematoda) infections to survival of mayflies and dragonflies. *Journal of Invertebrate Pathology* 36, 144–146.

Jilek, R. and Crites, J.L. (1981) Observations on the lack of specificity of *Spinitectus carolini* and *Spinitectus gracilis* (Spirurida: Nematoda) for their intermediate hosts. *Canadian Journal of Zoology* 59, 476–477.

Jilek, R. and Crites, J.L. (1982a) The life cycle and development of *Spinitectus carolini* Holl, 1928 (Nematoda: Spirurida). *American Midland Naturalist* 107, 100–106.

Jilek, R. and Crites, J.L. (1982b) The life cycle and development of *Spinitectus gracilis* (Nematoda: Spirurida). *Transactions of the American Microscopical Society* 101, 75–83.

Johnson, S. (1966) On a *Spinitectus* larva (Spiruroidea: Nematoda) from a shrimp (Crustacea) in India. *Indian Journal of Helminthology* 18, 49–52.

Johnson, M.H., Orihel, T.C. and Beaver, P.C. (1974) *Dipetalonema viteae* in the experimentally infected jird, *Meriones unguiculatus*. I. Insemination, development from egg to microfilariae, reinsemination and longevity of mated and unmated worms. *Journal of Parasitology* 60, 302–309.

Johnson, S.J. (1987) Stephanofilariasis – a review. *Helminthological Abstracts, Series A, Animal and Human Helminthology* 56, 287–299.

Johnston, T.H. and Bancroft, M.J. (1920) Experiments with certain Diptera as possible transmitters of bovine onchocerciasis. *Proceedings of the Royal Society of Queensland* 32, 31–57.

Johnston, T.H. and Mawson, P.M. (1943) Remarks on some nematodes of Australian reptiles. *Transactions of the Royal Society of South Australia* 67, 183–186.

Jones, H.J. (1995) Pathology associated with physalopterid larvae (Nematoda: Spirurida) in the gastric tissues of Australian reptiles. *Journal of Wildlife Diseases* 31, 710–715.

Jurášek, V., Alonso, M. and Ovies, D. (1970) Nota sobre el coleoptero *Dermestes ater* (Deg.), nuevo hospedero intermediario de los nematodos *Tropisurus confusus*, Travassos, 1919 y *Cheilospirura hamulosa* (Diesing, 1851) Spirurida en las condiciones de Cuba. *Revista Cubana de Ciencia Avicola* 3–4, 64.

Kabilov, T.K. (1980a) The life cycle of *Abbreviata kazachstanica*. *Parazitologiya* 14, 263–270. (In Russian.)

Kabilov, T.K. (1980b) New data on biology of *Stephanofilaria assamensis* (Nematoda, Filariata). *Helminthologia* 17, 191–196.

Kabilov, T.K. and Siddikov, B.Kh. (1978) Finding the intermediate hosts of *Abbreviata kazachstanica* Markov et Paraskiv, 1956. *Doklady Akademii Uzbekskoi SSR*, No. 2, 67–68. (In Russian.)

Kalyanasundaram, R. (1977) Life cycles of some spirurid parasitic of fowl. *First National Congress of Parasitology, Baroda 24–26 Feb, 1977 India.* Indian Society of Parasitology, pp. 38–39.

Kanagasuntheram, R., Singh, M., Ho, B.C. and Chan, H.L. (1974a) Some ultrastructural observations on the microfilaria of *Breinlia sergenti* – the excretory complex, rectal cells and anal vesicle. *International Journal for Parasitology* 4, 7–15.

Kanagasuntheram, R., Singh, M., Ho, B.C., Yap, E.H. and Chan, H.L. (1974b) Some ultrastructural observations on the microfilaria of *Breinlia sergenti* – nervous system. *International Journal for Parasitology* 4, 489–495.

Kartman, L. (1953) On the growth of *Dirofilaria immitis* in the mosquito. *American Journal of Tropical Medicine and Hygiene* 2, 1062–1069.

Kartman, L. (1957) The vectors of canine filariasis: a review with special reference to factors influencing susceptibility. *Publicacoes Avulsas. Revista Brasileira de Malariologia e Doencas Tropicais* 5, 1–41.

Kataitseva, T.Y. (1968) The life cycle of *Dipetalonema evansi* Lewis, 1882. *Doklady Akademii Nauk SSSR* 180, 1262–1264. (In Russian.)

Kawamata, K. (1961) Studies on parasitism by nematodes unknown as human parasites. The development of the larval stage of *Habronema megastoma* (Rudolphi, 1819) in the intermediate host. *Medical Journal of Kagoshima University* 13, 281–298.

Keiserovskaya, M.A. (1975) The biology of *Thelazia rhodesi* in Azerbaijzhan. In: *Issledovaniya po gelmintologii v Azerbaidzhane Baku USSR "Elm"*, pp. 66–69. (In Russian.)

Kemper, H.E. (1938) Filarial dermatosis in sheep. *North American Veterinarian* 19, 36–41.

Kennedy, M.J. (1993) Prevalence of eyeworms (Nematoda: Thelazoidea) in beef cattle grazing different range pasture zones in Alberta, Canada. *Journal of Parasitology* 79, 866–869.

Kennedy, M.J. and MacKinnon, J.D. (1994) Site segregation of *Thelazia skrjabini* and *Thelazia gulosa* (Nematoda: Thelazioidea) in the eyes of cattle. *Journal of Parasitology* 80, 501–504.

Keppner, E.J. (1971) The pathology of *Filaria taxideae* (Filarioidea, Filariidae) from the badger *Taxidea taxus taxus* from Wyoming. *Transactions of the American Microscopic Society* 88, 581–588.

Keppner, E.J. (1975) Life cycle of *Spinitectus micracanthus* Christian, 1972 (Nematoda: Rhabdochonidae) from the bluegill, *Lepomis macrochirus* Rafinesque, in Missouri with a note on *Spinitectus gracilis* Ward and Magath, 1917. *American Midland Naturalist* 93, 411–423.

Khromova, L.A. (1969) Spontaneous infection of molluscs by spirurid larvae. *Problemy Parazitologie* Part 1, 261–263. (In Russian.)

Khromova, L.A. (1971) Molluscs as reservoir hosts of *Spiroxys contortus* (Rudolphi, 1819) (Nematoda: Gnathostomatidae). *Trudi Gelmintologicheskoii Laboratorii* 21, 126–128. (In Russian.)

Khudaverdiev, T.P. (1977) The biology of *Onchocerca lienalis* Stiles, 1892, the agent of onchocerciasis in cattle in the Nakhichevan ASSR. *Tezisy Dokladov Nauchnoi Konferentsii, Posvyashchennoi 75-letiyu so dnya Osnovaniya Azerbaidzhanskogo Nauchno-Issledovatel'skogo Veterinarnogo Instituta,* pp. 99–101. (In Russian.)

Kleine, F.K. (1915) Die Uebertragung von Filarien durch *Chrysops. Zeitschrift für Hygiene und Infektionskrankheiten* 80, 345–349.

Klesov, M.D. (1949) The biology of the nematode *Thelazia rhodesi* (Desmarest, 1827). *Doklady Akademii Nauk SSSR* 66, 309–311. (In Russian.)

Klesov, M.D. (1950) Contribution to the biology of two nematodes of the genus *Thelazia* Bosc, 1819 parasites of cattle. *Doklady Akademii Nauk SSSR* 75, 591–594. (In Russian.)

Klesov, M.D. and Kovalenko, I.I. (1967) The biology, epizootiology and prophylaxis of helminths of ducks on the Azov coast. *Veterinariya Kiev* 11, 3–7. (In Russian.)

Ko, R.C. (1972). The transmission of *Ackertia marmotae* Webster, 1967 (Nematoda: Onchocercidae) of groundhogs (*Marmota monax*) by *Ixodes cookei. Canadian Journal of Zoology* 50, 437–450.

Ko, R.C. (1975) *Echinocephalus sinensis* n.sp. (Nematoda: Gnathostomatidae) from the ray (*Aetabatus flagellum*) in Hong Kong, Southern China. *Canadian Journal of Zoology* 53, 490–500.

Ko, R.C. (1976) Experimental infection of mammals with larval *Echinocephalus sinensis* (Nematoda: Gnathostomatidae) from oysters (*Crassostrea gigas*). *Canadian Journal of Zoology* 54, 597–609.

Ko, R.C. (1977) Effects of temperature acclimation on infection of *Echinocephalus sinensis* (Nematoda: Gnathostomatidae) from oysters to kittens. *Canadian Journal of Zoology* 55, 1129–1132.

Ko, R.C. (1986) *A preliminary review of* Ascarophis *(Nematoda) of fishes*. Occasional Publication, Department of Zoology, Hong Kong, 54 pp.

Ko, R.C. and Anderson, R.C. (1969) A revision of the genus *Cystidicola* Fischer, 1798 (Nematoda: Spriruroidea) of the swimbladder of fishes. *Journal of the Fisheries Research Board of Canada* 26, 849–864.

Ko, R.C., Morton, B. and Wong, P.S. (1975) Prevalence and histopathology of *Echinocephalus sinensis* (Nematoda: Gnathostomatidae) in natural and experimental hosts. *Canadian Journal of Zoology* 53, 550–559.

Kobayashi, H. (1927) On the life history of *Oxyspirura mansoni* and the pathological changes in the conjunctiva and the *ductus lacrymalis*, caused by this worm, with further observations on the structures of the adult worm. *Journal of the Medical Association of Formosa* 29, 491–532.

Kobayashi, H. (1940) On the development of *Microfilaria bancrofti* in the body of the mosquito (*Culex fatigans*). *Acta Japonica Medicinae Tropicalis* 2, 63–88. (In Japanese.)

Koga, M. and Ishii, Y. (1981) Larval gnathostomes found in reptiles in Japan and experimental life cycle of *Gnathostoma nipponicum. Journal of Parasitology* 67, 565–570.

Koga, M. and Ishii, Y. (1990) Persistence of *Gnathostoma hispidum* in chronically infected rats. *Journal of Helminthology* 64, 46–50.

Koshida, T. (1905) On a species of Nematoda parasitic in salmonoid fishes. *Hokkaido Suisan Zasshi* 5, 7–9. (In Japanese.)

Koshida, T. (1910) Survey of parasitic nematodes in fish farms. *Third Annual Report of the Hokkaido Fisheries Experimental Station*, pp. 516–519. (In Japanese.)

Kotani, T. and Powers, K.G. (1982) Developmental stages of *Dirofilaria immitis* in the dog. *American Journal of Veterinary Research* 43, 2199–2206.

Kotcher, E. (1941) Studies on the development of frog filariae. *American Journal of Hygiene* 34, 36–65.

Kotelnikov, G.A. (1961) Biology of *Echinuria uncinata* from ducks. *Sbornik Nauk Informastsii Vsesoyuznogo Instituta Gelmintologii K.I. Skrjabin* 7/8, 30–33. (In Russian.)

Kovalenko, I.I. (1960) Study of the life cycles of some helminths of domestic ducks from farms on Azov coast. *Doklady Academii Nauk SSSR* 133, 1259–1261. (In Russian.)

Kovalenko, I.I. (1963) Endemic of mixed infections with *Streptocara, Tetrameres* and *Polymorphus* in fowls. *Trudi Ukrainskogo Respublikanskogo Nauchnogo Obshchestva Parazitologov* 2, 137–140. (In Russian.)

Kozek, W.J. (1968) Unusual cilia in the microfilaria of *Dirofilaria immitis. Journal of Parasitology* 54, 838–844.

Kozek, W.J. (1971) Ultrastructure of the microfilaria of *Dirofilaria immitis. Journal of Parasitology* 57, 1052–1067.

Kozlov, D.P. (1962) The life cycle of the nematode, *Thelazia callipaeda*, parasitic in the eye of man and carnivores. *Doklady Akademii Nauk SSSR* 142, 732–733. (In Russian.)

Kozlov, D.P. (1963) Biology of *Thelazia callipaeda* Railliet and Henry, 1910. *Trudi Gelmintologischeskoi Laboratorii* 13, 330–346. (In Russian.)

Krahwinkel, D.J. and McCue, J.F. (1967) Wild birds as transport hosts of *Spirocerca lupi* in the southeastern USA. *Journal of Parasitology* 53, 650–651.

Krastin, N.I. (1949a) Elucidation of the life cycle of *Thelazia rhodesi* (Desmarest, 1827) parasitic in the eyes of cattle. *Doklady Akademii Nauk SSSR* 64, 885–887. (In Russian.)

Krastin, N.I. (1949b) Epizootiology of thelaziasis in cattle and biology of *Thelazia rhodesi* (Desmarest, 1827). *Veterinariya* 26, 6–8. (In Russian.)

Krastin, N.I. (1950a) Elucidation of the biological cycle of a second vector of thelaziasis. *Veterinariya* 27, 20–21. (In Russian.)

Krastin, N.I. (1950b) A study of the developmental cycle of the nematode, *Thelazia gulosa* (Railliet and Henry, 1910) a parasite of the eyes of cattle. *Doklady Akademii Nauk SSSR* 70, 549–551. (In Russian.)

Krastin, N.I. (1952) The life cycle of the nematode *Thelazia skrjabini* Ershov, 1928 ocular parasite in cattle. *Doklady Akademii Nauk SSSR* 82, 829–831. (In Russian.)

Krastin, N.I. (1958) Development of *Thelazia rhodesi* (Desmarest, 1827) in the final host. *Collected Papers on Helminthology Presented to Prof. R.S. Shults on his 60th Birthday.* Kazakhskoe Gosudarstvennogo Izdatelstvo, Alma-Ata, pp. 236- 244. (In Russian.)

Kudo, N., Oyamada, T., Okutsu, M. and Kinoshita, M. (1996) Intermediate hosts of *Gongylonema pulchrum* Molin, 1857, in Aomori Prefecture, Japan. *Japanese Journal of Parasitology* 45, 222–229.

Kume, S. and Itagaki, S. (1955) On the life-cycle of *Dirofilaria immitis* in the dog as the final host. *British Veterinary Journal* 111, 16–24.

Kurochkin, Y.V. (1958) Study of the nematodes of the genus *Skrjabinocara* Kuraschwili, 1941. *Trudi Astrakhanskogo Gosudarstvennago Zap.* 43, 325–336. (In Russian.)

Laberge, R.J.A. and McLaughlin, J.D. (1989) *Hyalella azteca* (Amphipoda) as an intermediate host of the nematode *Streptocara crassicauda*. *Canadian Journal of Zoology* 67, 2335–2340.

Laberge, R.J.A. and McLaughlin, J.D. (1991) Susceptibility of blue-winged teal, gadwall, and lesser scaup ducklings to experimental infection with *Streptocara crassicauda*. *Canadian Journal of Zoology* 69, 1512–1515.

Lahiri, S.K., Menon, P.K.M. and Vijayan, C.P. (1972) Development of *Brugia ceylonensis* Jayewardene infection in *Anopheles stephensi*. *Journal of Communicable Diseases* 4, 201–207.

Laing, A.B.G., Edeson, J.F.B. and Wharton, R.H. (1960) Studies on filariasis in Malaya: the vertebrate hosts of *Brugia malayi* and *B. pahangi*. *Annals of Tropical Medicine and Parasitology* 54, 92–99.

Lambertsen, R.H. (1985) Taxonomy and distribution of a *Crassicauda* species (Nematoda: Spirurida) infecting the kidney of the common fin whale (*Balaenoptera physalus* Linné, 1758). *Journal of Parasitology* 71, 485–488.

Lambertsen, R.H. (1986) Disease of the common fin whale (*Balaenoptera physalus*) crassicaudiosis of the urinary system. *Journal of Mammology* 67, 353–366.

Lamberston, R.H. (1992) Crassicaudosis: a parasitic disease threatening the health and population recovery of large baleen whales. *Revue Scientifique et Technique – Office International des Épizooties* 11, 1131–1141.

Lammler, G., Saupe, E. and Herzog, H. (1968) Infectionsversuche mit der Baumwollrattenfilarie *Litomosoides carinii* bei *Mastomys natalensis* (Smith, 1834). *Zeitschrift für Parasitenkunde* 30, 281–290.

Lankester, M.W. and Smith, J.D. (1980) Host specificity and distribution of the swimbladder nematodes, *Cystidicola farionis* Fischer, 1798 and *C. cristivomeri* White, 1941

(Habronematoidea), in salmonid fishes of Ontario. *Canadian Journal of Zoology* 58, 1298–1305.

Lardeux, F. and Cheffort, J. (1996) Behavior of *Wuchereria bancrofti* (Filarioidea: Onchocercidae) infective larvae in the vector *Aedes polynesiensis* (Diptera: Culicidae) in relation to parasite transmission. *Journal of Medical Entomology* 33, 516–524.

Laurence, B.R. (1989) The global dispersal of bancroftian filariasis. *Parasitology* 5, 260–264.

Laurence, B.R. (1990) Filariasis discovery (Letter to Editor). *Parasitology Today* 6, 153–154.

Laurence, B.R. and Pester, F.R.N. (1961a) The behaviour and development of *Brugia patei* (Buckley, Nelson and Heisch, 1958) in a mosquito host, *Mansonia uniformis* (Theobald). *Journal of Helminthology* 35, 285–300.

Laurence, B.R. and Pester, F.R.N. (1961b) The ability of *Anopheles gambiae* to transmit *Brugia patei* (Buckley, Nelson and Heisch). *Journal of Tropical Medicine and Hygiene* 64, 169–171.

Laurence, B.R. and Pester, F.R.N. (1967) Adaptation of a filarial worm, *Brugia patei*, to a new mosquito host, *Aedes togoi*. *Journal of Helminthology* 41, 365–392.

Laurence, B.R. and Simpson, M.G. (1968) Cephalic and pharyngeal structures in microfilariae revealed by staining. *Journal of Helminthology* 42, 309–330.

Laurence, B.R. and Simpson, M.G. (1971) The microfilaria of *Brugia*: a first stage nematode larva. *Journal of Helminthology* 45, 23–40.

Laurence, B.R. and Simpson, M.G. (1974) The ultrastructure of the microfilaria of *Brugia*, Nematoda: Filarioidea. *International Journal for Parasitology* 4, 523–536.

Lavoipierre, M.M.J. (1958) Studies on the host–parasite relationships of filarial nematodes and their arthropod hosts. I. The sites of development and migration of *Loa loa* in *Chrysops silacea*, the escape of the infective forms from the head of the fly and the effect of the worm on its insect host. *Annals of Tropical Medicine and Parasitology* 52, 103–121.

Lee, S.H. (1955) The mode of egg dispersal in *Physaloptera phrynosoma* Ortlepp (Nematoda: Spiruroidea), a gastric nematode of Texas horned toads, *Phrynosoma cornutum*. *Journal of Parasitology* 41, 70–74.

Lee, S.H. (1957) The life cycle of *Skrjabinoptera phrynosoma* (Ortlepp) Schulz, 1927 (Nematoda: Spiruroidea) a gastric nematode of Texas horned toads, *Phrynosoma cornutum*. *Journal of Parasitology* 43, 66–75.

Leger, A. (1911) Filaire à embryons sanquicoles de *Hyaena crocuta* Erxleben. *Bulletin Société Pathologé Exotique* 4, 629–631.

Leiper, R.T. (1913) Reports to the Colonial Office, London School of Tropical Medicine. *Report of the Helminthologist for the Half Year Ending 30th April 1913.*

Leuckart, K. (1865) Helminthologische Experimentaluntersuchungen. Vierte Reihe. *Nachrichten von der K. Gesellschaft der Wissenschaften und der Georg-Augusts Universität. Göttingen* 8, 219–232.

Leuckart, K. (1876) *Die menschlichen Parasiten und die von ihnen herrührenden Krankheiten*, Vol. 2. Leipzig, pp. 513–882.

Lichtenfels, J.R., Pilitt, P.A., Kotani, T. and Powers, K.G. (1985) Morphogenesis of developmental stages of *Dirofilaria immitis* (Nematoda) in the dog. *Proceedings of the Helminthological Society of Washington* 52, 98–113.

Lichtenfels, J.R., Pilitt, P.A. and Wergin, W.P. (1987) *Dirofilaria immitis*: fine structure of cuticle during development in dogs. *Proceedings of the Helminthological Society of Washington* 54, 133–140.

Lichtenstein, A. (1927) *Filaria* – onderzoek te Bireuën. *Geneeskundig Tijdschrift voor Nederlandsch-Indië* 67, 742–749.

Lim, C.W. (1975) The fowl (*Gallus domesticus*) and a lepidopteran (*Setomorpha rutella*) as experimental hosts for *Tetrameres mohtedai* (Nematoda). *Parasitology* 70, 143–148.

Lim, B.L., Mak, J.W., Ho, B.C. and Yap, E.H. (1975) Distribution and ecological considerations of *Breinlia booliati* infecting wild rodents in Malaysia. *Southeast Asian Journal of Tropical Medicine and Public Health* 6, 241–246.

Lincoln, R.C. and Anderson, R.C. (1973) The relationship of *Physaloptera maxillaris* (Nematoda: Physalopteroidea) to skunk (*Mephitis mephitis*). *Canadian Journal of Zoology* 51, 437–441.

Lincoln, R.C. and Anderson, R.C. (1975) Development of *Physaloptera maxillaris* (Nematoda) in the common field cricket (*Gryllus pennsylvanicus*). *Canadian Journal of Zoology* 53, 385–390.

Linstow, O. (1894) Helminthologische Studien. *Jenaische Zeitschrift für Naturwissenschaft, Jena* 28, 328–342.

Lok, J.B. (1988) *Dirofilaria* sp.: taxonomy and distribution. In: Boreham, P.F.L. and Atwell, R.B. (eds) *Dirofilariasis*. CRC Press, London, pp. 1–28.

Lok, J.B., Cupp, E.W. and Bernardo, M.J. (1983a) *Simulium jenningsi* Malloch (Diptera: Simuliidae): a vector of *Onchocerca lienalis* Stiles (Nematoda: Filarioidea) in New York. *American Journal of Veterinary Research* 44, 2355–2358.

Lok, J.B., Cupp. E.W., Bernardo, M.J. and Pollack, R.J. (1983b) Further studies on the development of *Onchocerca* spp. (Nematoda: Filarioidea) in Nearctic blackflies (Diptera: Simuliidae). *American Journal of Tropical Medicine and Hygiene* 32, 1298–1305.

Lowrie, R.C., Eberhard, M.L. and Orihel, T.C. (1978) Development of *Tetrapetalonema marmosetae* to the infective stage in *Culicoides hollensis* and *C. furens*. *Journal of Parasitology* 64, 1003–1007.

Lubimov, M.P. (1945) New worm diseases of the brain of deer with unossified antlers. *Sbornik Nauchno-issledovatel'skikh rabot (Laboratoriyi Panto-vogo Olenvodstva Ministerstva Sovkhozov SSSR)* 1, 225–232. (In Russian.)

Lubimov, M.P. (1948) New helminths of the brain of maral deer. *Trudi Gel'mintologicheskoi Laboratorii* 1, 198–201. (In Russian.)

Lubimov, M.P. (1959) The season dynamics of elaphostrongylosis and setariosis in *Cervus elaphus maral*. *Trudy Gel'mintologicheskoi Laboratorii* 9, 155–156. (In Russian.)

Lucker, J.T. (1932) Some cross transmission experiments with *Gongylonema* of ruminant origin. *Journal of Parasitology* 19, 134–141.

Lyons, E.T., Drudge, J.H. and Tolliver, S.C. (1980) Experimental infections of *Thelazia lacrymalis*: maturation of third-stage larvae from face flies (*Musca autumnalis*) in eyes of ponies. *Journal of Parasitology* 66, 181–182.

Machida, M., Araki, J., Koyama, T., Kumada, M., Horii, Y., Imada, I., Takasaka, M., Honjo, S., Matsubayashi, K. and Tiba, T. (1978) The life cycle of *Streptopharagus pigmentatus* (Nematoda, Spiruroidea) from the Japanese monkey. *Bulletin of the National Science Museum, Series A (Zool.)* 4, 1–9.

Mackerras, M.J. (1953) Lizard filaria: transmission by mosquitoes of *Oswaldofilaria chlamydosauri* (Breinl) (Nematoda: Filarioidea). *Parasitology* 43, 1–3.

Mackerras, M.J. (1962) Filarial parasites (Nematoda: Filarioidea) of Australian animals. *Australian Journal of Zoology* 10, 400–457.

Maddy, K.T. (1955) Stephanofilarial dermatitis of cattle. *North American Veterinarian* 36, 275–278.

Mak, J.W. (ed.) (1983) Filariasis. *Institute for Medical Research, Kuala Lumpur, Malaysia, Bulletin* No. 19.

Mak, J.W. and Lim, B.L. (1974) New hosts of *Breinlia booliati* in wild rats from Sarawak with further observations on its morphology. *Southeast Asian Journal of Tropical Medicine and Public Health* 5, 22–28.

Mak, J.W. and Yong, H.S. (eds) (1986) Control of brugian filariasis. In: *Proceedings of the W.H.O. Regional Seminar*, July 1–5, 1985, Kuala Lumpur, Malaysia.

Mak, J.W., Chiang, G.L. and Cheong, W.H. (1984) *Cardiofilaria nilesi* (Filarioidea: Onchocercidae) infection in chickens in Peninsular Malaysia. *Tropical Biomedicine* 1, 115–120.

Mamaev, Y.L. (1971) Helminth larvae in freshwater crustaceans in the Primorsk Region. In: *Parazity Zhivotnykh i Rastenii dal'Nego Vostoka*. Dal'nevostochnoe Knizhnoe Izdatel'stvo, Vladivostok, pp. 120–132. (In Russian.)

Manson, P. (1878) On the development of *Filaria sanguinis hominis* and on the mosquito considered as a nurse. *Journal of Linnaean Society (Zoology)* 14, 304–311.

Manson, P. (1881) On the periodicity of filarial migrations to and from the circulation. *Journal of Quekett Microscopy Club* 6, 239–248.

Manson, P. (1906) *Tropical Diseases: A Manual of Diseases of Warm Climates*, 3rd edn. William Wood and Co., New York.

Manson-Bahr, P. (1959) The story of *Filaria bancrofti*. A critical review. *Journal of Tropical Medicine and Hygiene* 62, 53–61, 85–94, 106–117, 138–145, 160–173.

Mantovani, A. and Restani, R. (1965) Ricerche sui possibili artropodi vettori di *Dirofilaria repens* in alcune provincie dell'Italia centrale. *Parassitologia* 7, 109–116.

Mar, P.H. and Fei, A.C.Y. (1995) Epizootiological and preventive study of *Seteria digitata* in Taiwan. *Asia Seasonly Report of Environmental Microbiology* 4, 27–35.

Marchi, P. (1871) Monografia sulla storia genetica e sulla anatomia della *Spiroptera obtusa* Rud. *Memorie della Reale Accademia delle Scienze di Torino* 25, 1–30.

Marenkelle, C.J. and German, E. (1970) Mansonelliasis in the Comisaria del Vaupes of Colombia. *Tropical and Geographical Medicine* 22, 101–111.

Margolis, L. and Moravec, F. (1982) *Ramellogammarus vancouverensis* Bousfield (Amphipoda) as an intermediate host for salmonid parasites in British Columbia. *Canadian Journal of Zoology* 60, 1100–1104.

Martinez-Palomo, A. and Martinez-Baez, M. (1977) Ultrastructure of the microfilaria of *Onchocerca volvulus* from Mexico. *Journal of Parasitology* 63, 1007–1018.

McAllister, C.T., Goldberg, S.R. and Holshuh, H.J. (1993a) *Spiroxys contorta* (Nematoda: Spirurida) in gastric granulomas of *Apalone spinifera pallida* (Reptilia: Testudines). *Journal of Wildlife Diseases* 29, 509–511.

McAllister, C.T., Goldberg, S.R., Bursey, C.R., Freed, P.S. and Holshuh, H.J. (1993b) Larval *Ascarops* sp. (Nematoda: Spirurida) in introduced Mediterranean geckos, *Hemidactylus furcicus* (Sauria: Gekkonidae) from Texas. *Journal of the Helminthological Society of Washington* 60, 280–282.

McFadzean, J.A. and Smiles, J. (1956) Studies of *Litomosoides carinii* by phase-contrast microscopy: the development of the larvae. *Journal of Helminthology* 30, 25–32.

McLaren, D.J. (1969) Ciliary structures in the microfilaria of *Loa loa*. *Transactions of the Royal Society of Tropical Medicine and Hygiene* 63, 290–291.

McLaren, D.J. (1972) Ultrastructural studies on microfilariae (Nematoda: Filarioidea). *Parasitology* 65, 317–332.

McLaughlin, J.D. and McGurk, B.P. (1987) An analysis of gizzard worm infections in fall migrant ducks at Delta, Manitoba, Canada. *Canadian Journal of Zoology* 65, 1470–1477.

Measures, L., Gosselin, J.F. and Bergeron, F. (1997) Heartworm, *Acanthocheilonema spirocauda* (Leidy, 1858) infections in Canadian phocid seals. *Canadian Journal of Fisheries and Aquatic Sciences* 54, 842–846.

Mello, M.J. and Cuocolo, R. (1943) Alguns aspetos das relacões do *Habronema muscae* (Carter, 1861) com a mosca domestica. *Arquivos do Instituto Biologico, Saõ Paulo* 14, 227–234.

Mellor, P.S. (1971) A membrane feeding technique for the infection of *Culicoides nubeculosus* Mg. and *Culicoides variipennis sonorensis* Coq. with *Onchocerca cervicalis* Rail. and Henry. *Transactions of the Royal Society of Tropical Medicine and Hygiene* 65, 199–201.

Mellor, P.S. (1974) Studies on *Onchocerca cervicalis* Raillet and Henry, 1910. IV. Behaviour of the vector *Culicoides nubeculosus* in relation to the transmission of *Onchocerca cervicalis*. *Journal of Helminthology* 48, 283–288.

Mellor, P.S. (1975) Studies on *Onchocerca cervicalis* Railliet and Henry, 1910: V. The development of *Onchocerca cervicalis* larvae in the vectors. *Journal of Helminthology* 49, 33–42.

Menon, T., Ramamurti, B. and Rao, D.S. (1944) Lizard filariasis. An experimental study. *Transactions of the Royal Society of Tropical Medicine and Hygiene* 37, 373–386.

Metianu, T. (1949) Considerations sur la parafilariose hémorragique des bovins. *Parafilaria bovicola* en Roumanie. *Annales de Parasitologie Humaine et Comparée* 24, 54–59.

Meyers, W.M., Conne, D.H., Harman, L.E., Fleshman, K., Moris, R. and Neafie, R.C. (1972) Human streptocerciasis. A clinical-pathologic study of 40 Africans (Zairians) including identification of the adult filaria. *American Journal of Medicine and Hygiene* 21, 528–545.

Mikhailyuk, A.P. (1967) Biology of *Onchocerca gutturosa* and *O. lienalis* in the forest steppe zone of the Ukrainian SSR. *Veterinariya* 2, 62–67. (In Ukrainian.)

Millemann, R.E. (1951) *Echinocephalus pseudouncinatus* n.sp., a nematode of the abalone. *Journal of Parasitology* 37, 435–439.

Millemann, R.E. (1963) Studies on the taxonomy and life history of echinocephalid worms (Nematoda: Spiruroidea) with a complete description of *Echinocephalus pseudouncinatus* Millemann, 1951. *Journal of Parasitology* 49, 754–764.

Millet, P. and Bain, O. (1984) Une nouvelle filaire de la pie, *Eufilaria kalifai* n.sp. (Lemdaninae) et son développement chez *Culicoides nubeculosus*. *Annales de Parasitologie Humaine et Comparée* 59, 177–187.

Minailova, N.M. (1977) The development cycle of *Gongylonema ivaschkini* n.sp. *Izvestiya Akademii Nauk Turkmenskoi SSR, Biologicheskie Nauki* 3, 31–35. (In Russian.)

Misiura, M. (1970) Development of *Echinuria uncinata* (Rud., 1819) larvae (Nematoda) in Cladocera and Ostracoda. *Acta Parasitologica Polonica*. 17, 247–251.

Miyamoto, K., Shinonaga, S. and Kano, R. (1981) Experimental studies on the development of *Thelazia rhodesi* larvae in the intermediate and definitive hosts. *Japanese Journal of Parasitology* 30, 15–21.

Miyata, J. (1939) Study of the life cycle of *Protospirura muris* (Gmelin) of rats and particularly of its intermediate hosts. *Volumen Jubilare Pro Professore Sadao Yoshida* 1, 101–136. (In Japanese.)

Miyata, A. and Tsukamoto, M. (1975) Blood parasites from small mammals in Palawan Islands, the Philippines. *Tropical Medicine* 16, 113–130.

Miyazaki, I. (1952a) Studies on the life history of *Gnathostoma spinigerum* Owen, 1836 in Japan (Nematoda: Gnathostomidae). *Igaku Kenkyuu Kyushu University Acta Medica* 22, 1135–1144.

Miyazaki, I. (1952b) On the second-stage larvae of three species of *Gnathostoma* occurring in Japan (Nematoda: Gnathostomidae). *Igaku Kenkyuu Kyushu University Acta Medica* 22, 1433–1441.

Miyazaki, I. (1954) Studies on *Gnathostoma* occurring in Japan (Nematoda: Gnathostomidae). II. Life history of *Gnathostoma* and morphological comparison of its larval forms. *Kyushu Memoirs of Medical Science* 5, 123–143.

Miyazaki, I. (1960) On the genus *Gnathostoma* and human gnathostomiasis, with special reference to Japan. *Experimental Parasitology* 9, 338–370.

Miyazaki, I. (1966) *Gnathostoma* and gnathostomiasis in Japan. In: *Progress in Medical Parasitology in Japan*. Meguro Parasitology Museum, Tokyo, Japan 3, 531–586.

Miyazaki, I. and Ishii, Y. (1952) On a gnathostome larva encysted in the muscles of salamanders, *Hynobius*. *Acta Medica* 22, 467–473. (In Japanese.)

Miyazaki, I. and Kawashima, K. (1962) On the larval *Gnathostoma doloresi* Tubangui found in a snake from Ishigaki-jima, the Ryukyu Island (Nematoda: Gnathostomatidae). *Kyusku Journal of Medical Science* 13, 165.

Moignoux, J.B. (1952) Les onchocerques des équides. *Acta Tropica* 9, 125–150.

Montali, R.J., Gardiner, C.H., Evans, R.E. and Bush, M. (1983) *Pterygodermatites nycticebi* (Nematoda: Spirurida) in golden lion tamarins. *Laboratory Animal Science* 33, 194–197.

Montpellier, J. and Lacroix, A. (1920) Le craw-craw ou gale filarienne; son origine dans les kystes sous-cutanées à *Onchocerca volvulus*. *Bulletin de la Société de Pathologie Exotique et de ses Filiales* 13, 305–315.

Moolenbeek, W.J. and Surgeoner, G.A. (1980) Southern Ontario survey of eyeworms, *Thelazia gulosa* and *Thelazia lacrymalis* in cattle and larvae of *Thelazia* spp. in the face fly *Musca autumnalis*. *Canadian Veterinary Journal* 21, 50–52.

Moravec, F. (1971a) Studies on the development of the nematode *Cystidicoloides tenuissima* (Zeder, 1800). *Véstnik Československe Společnosti Zoologické* 35, 43–55.

Moravec, F. (1971b) On the life history of the nematode *Cystidicoloides tenuissima* (Zeder, 1800) in the River Bystrice, Czechoslovakia. *Folia Parasitologica* 18, 107–112.

Moravec, F. (1972) Studies on the development of the nematode *Rhabdochona* (*Filochona*) *ergensi* Moravec, 1968. *Folia Parasitologica* 19, 321–333.

Moravec, F. (1975) Reconstruction of the nematode genus *Rhabdochona* Railliet, 1916 with a review of species parasitic in fishes in Europe and Asia. *Studies CSAV (Prague)* 8, 1–104.

Moravec, F. (1976) Observations on the development of *Rhabdochona phoxini* Moravec, 1968 (Nematoda: Rhabdochonidae). *Folia Parasitologica* 23, 309–320.

Moravec, F. (1977) Life history of the nematode *Rhabdochona phoxini* Moravec, 1968 in the Rokytka Brook, Czechoslovakia. *Folia Parasitologica* 24, 97–105.

Moravec, M.F. (1995) Trichopteran larvae (Insecta) as the intermediate hosts of *Rhabdochona hellichi* (Nematoda: Rhabdochonidae) a parasite of *Barbus barbus* (Pisces). *Parasitology Research* 81, 268–270.

Moravec, F. and De, N.C. (1982) Some new data on the bionomics of *Cystidicoloides tenuissima* (Nematoda: Cystidicolidae). *Véstnik Československe Společnosti Zoologické* 46, 100–108.

Moravec, F. and Nagasawa, K. (1986) New records of amphipods as intermediate hosts for salmonid nematode parasites in Japan. *Folia Parasitologica* 33, 45–49.

Moravec, F. and Scholz, T. (1994) Observations on the development of *Syncuaria squamata* (Nematoda: Acuariidae) a parasite of cormorants, in the intermediate and paratenic hosts. *Folia Parasitologica* 41, 183–192.

Moravec, F., Nagasawa, K. and Urushibara, Y. (1998) Observations on the seasonal maturation of the nematode *Rhabdochona zacconis* in Japanese dace, *Tribolodon hakonensis*, of the Okitsu River, Japan. *Acta Societatis Zoologica Bohemia* 62, 45–50.

Morgan, N.O. (1964) Autecology of the adult horn fly, *Haematobia irritans* (L.) (Diptera: Muscidae). *Ecology* 45, 728–736.

Morozov, Yu.F. (1960) Life cycle of *Rictularia amurensis* (Nematoda, Rictulariidae). *Uchenye Zapiski Gorkovskogo Gosudarstvennogo Pedagogicheskogo Instituta Gelmintologicheskoi Sbornik* 27, 17–28. (In Russian.)

Mössinger, J. and Barthold, E. (1987) Fecundity and localization of *Dipetalonema viteae* (Nematoda, Filarioidea) in the jird *Meriones unguiculatus*. *Parasitology Research* 74, 84–87.

Moya, R.A. and Ovies, Y.D. (1982) The biology of *Cheilospirura hamulosa* under the sub-tropical conditions of Cuba. *Revista Avicultura* 26, 181–204.

Mozgovoi, A.A., Popova, T.I. and Semenova, M.K. (1965) Study of the life cycle of *Synhimantus brevicaudatus* (Duj. 1845), a parasite of ciconiform birds and freshwater fish. *Doklady Akademii Nauk SSSR* 162, 719–721. (In Russian.)

Muller, R.L. and Horsburgh, R.C.R. (1987) *Bibliography of Onchocerciasis (1841–1985).* CAB International, Wallingford, UK.

Muller, R.L. and Nelson, G.S. (1975) *Ackertia globulosa* sp.n. (Nematoda: Filarioidea) from rodents in Kenya. *Journal of Parasitology* 61, 606–609.

Mwaiko, G.L. (1981) The development of *Onchocerca gutturosa* Neumann to the infective stage in *Simulium vorax* Pomeroy. *Tropenmedizin und Parasitologie* 32, 276–277.

Nagaty, H.F. (1947) *Dipetalonema evansi* (Lewis, 1882) and its microfilariae from *Camelus dromedarius. Parasitology* 38, 86–92.

Nair, C.P., Roy, R. and Raghavan, N.G.S. (1961) Susceptibility of *Aedes albopictus* to *Dirofilaria repens* infection in cats. *Indian Journal of Malariology* 15, 49–52.

Nawa, Y., Imai, J.I., Horii, Y., Ogata, K. and Otsuka, K. (1993) *Gnathostoma doloresi* found in *Lepomis macrochirus* Rafinesque captured in the central part of Migazaki Prefecture, Japan. *Japanese Journal of Parasitology* 42, 40–43.

Nazarova, N.S. (1964) Migration of *Spirocerca lupi* in the final host. *Trudi Gel'mintologicheskoi Laboratorii Akademiya Nauk SSR* 14, 131–135. (In Russian.)

Nelson, G.S. (1959) The identification of infective filarial larvae in mosquitoes: with a note on the species found in 'wild' mosquitoes on the Kenya coast. *Journal of Helminthology* 33, 233–256.

Nelson, G.S. (1961) On *Dipetalonema mansonbahri* n.sp., from the spring-hare, *Pedetes surdaster larvalis*, with a note on its development in fleas. *Journal of Helminthology* 35, 143–160.

Nelson, G.S. (1962a) Observations on the development of *Setaria labiatopapillosa* using new techniques for infecting *Aedes aegypti* with this nematode. *Journal of Helminthology* 36, 281–296.

Nelson, G.S. (1962b) *Dipetalonema reconditum* (Grassi, 1889) from the dog with a note on its development in the flea, *Ctenocephalides felis* and the louse, *Heterodoxus spiniger. Journal of Helminthology* 36, 297–308.

Nelson, G.S. (1963) *Dipetalonema dracunculoides* (Cobbold, 1870), from the dog in Kenya: with a note on its development in the louse-fly, *Hippobosca longipennis. Journal of Helminthology* 37, 235–240.

Nelson, G.S. (1965) Filarial infections as zoonoses. *Journal of Helminthology* 39, 229–250.

Nelson, G.S. (1970) Onchocerciasis. *Advances in Parasitology* 8, 173–224.

Nelson, G.S., Heisch, R.B. and Furlong, M. (1962) Studies in filariasis in East Africa. II. Filarial infections in man, animals and mosquitoes on the Kenya coast. *Transactions of the Royal Society of Tropical Medicine and Hygiene* 56, 202–217.

Nevill, E.M. (1975) Preliminary report on the transmission of *Parafilaria bovicola* in South Africa. *Onderstepoort Journal of Veterinary Research* 42, 41–48.

Nevill, E.M. (1979) The experimental transmission of *Parafilaria bovicola* to cattle in South Africa using *Musca* species (subgenus *Eumusca*) as intermediate hosts. *Onderstepoort Journal of Veterinary Research* 46, 51–57.

Newton, W.L. (1957) Experimental transmission of the dog heartworm, *Dirofilaria immitis* by *Anopheles quadrimaculatus. Journal of Parasitology* 43, 589.

Newton, W.L. and Wright, W.H. (1956) The occurrence of a dog filariid other than *Dirofilaria immitis* in the USA. *Journal of Parasitology* 42, 246–258.

Newton, W.L. and Wright, W.H. (1957) A reevaluation of the canine filariasis problem in the USA. *Veterinary Medicine* 52, 75–78.

Niles, W.J. (1961) *Anopheles nigerrimus* (*hyrcanus* group) as a laboratory vector of some filariae of man and animals. *Annals of Tropical Medicine and Parasitology* 55, 379–380.

Niles, W.J. (1962) Natural infections of developing animal filariae in the fat-body of *Mansonia crassipes. Transactions of the Royal Society of Tropical Medicine and Hygiene* 56, 437–438.

Niles, W.J. and Kulasiri, C. de S. (1970) Studies on *Cardiofilaria nilesi* in experimental chickens. *Ceylon Journal of Medical Science* 19, 18–28.

Niles, W.J., Fernando, M.A. and Dissanaike, A.S. (1965) *Mansonia crassipes* as the natural vector of filarioids, *Plasmodium gallinaceum* and other plasmodia of fowls in Ceylon. *Nature, London* 205, 411–412.

Nishikubo, K. (1963) Studies on experimental gnathostomiasis with special reference to host–parasite relationship in *Gnathostoma spinigerum*. III. An investigation on the development and migration route of the larval *G. spinigerum* in the gastrectonized cat. *Endemic Diseases Bulletin Nagasaki University* 5, 199–207.

Noè, G. (1900) Propagazione delle filarie del sangue, esclusivamente per mezzo della puntura della zanzara. 2. Nota preliminare. *Atti della Reale Accademia dei Lincei. Rendiconti della Classe di Scienze Fisiche, Matematiche e Naturali* 9, 357–362.

Noè, G. (1901) Propangazione delle filarie del sangue unicamente per la puntura delle zanzare. 3. Nota preliminare. *Atti della Reale Accademia dei Lincei. Rendiconti della Classe di Scienze Fisiche, Matematiche e Naturali* 10, 317–319.

Noè, G. (1908) Il ciclo evoluto della *Filaria grassii* mihi, 1907. *Atti dell' Accademia Nazionale dei Lincei Rendiconti. Classe di Scienze Fisiche Matematiche e Naturali.* 17, 282–283.

Odetoyinbo, J.A. and Ulmer, M.J. (1959) Studies on avian filarial worms of the subfamily Splendidofilariinae (Nematoda: Dipetalonematidae). *Journal of Parasitology* (Suppl.) 45, 58.

O'Grady, F., Fawcett, A.N. and Buckley, J.J.C. (1962) A case of human infection with *Dirofilaria* (*Nochtiella*) sp. probably of African origin. *Journal of Helminthology* 36, 309–312.

Oguz, T. (1970) *Gongylonema pulchrum* Molin, 1857'nin morfolojisi ile Ankara civarindaki arakonakcilarina dair arastirmalar. *Veteriner Fakültesi Dergisi. Anakara Universitesi* 17, 136–155. (In Turkish.)

O'Hara, J.E. and Kennedy, M.J. (1991) Development of the nematode eyeworm, *Thelazia skrjabini* (Nematoda: Thelazioidea), in experimentally infected face flies, *Musca autumnalis* (Diptera: Muscidae). *Journal of Parasitology* 77, 417–425.

Olmeda-Garcia, A.S., Rodríguez-Rodríguez, J.A. and Roja-Vázquez, F.A. (1993) Experimental transmission of *Dipetalonema dracunculoides* (Cobbold, 1870) by *Rhipicephalus sanguineus* (Latreille, 1806). *Veterinary Parasitology* 47, 339–342.

Olsen, J.L. (1971) Life history of *Physaloptera rara* Hall and Wigdor, 1918, in definitive, intermediate, and paratenic hosts. Unpublished PhD thesis, Colorado State University, Fort Collins, Colorado.

Ono, S. (1929) *Gymnopleurus sinnatus* as the intermediate host of Spiruridae found in the vicinity of Mukden, South Manchuria. I. *Gymnopleurus sinnatus* as the intermediate host of *Spirocerca sanguinolenta* and inquiry into Grassi's experiment with *Blatta orientalis*. *Journal of the Japanese Society of Veterinary Science* 8, 233–237. (In Japanese.)

Ono, S. (1932) *Gymnopleurus* sp. as the intermediate host of Spiruridae found in the vicinity of Murden, South Manchuria. II. Report. Studies on the life history of *Arduenna strongylina*. *Journal of the Japanese Society of Veterinary Science* 11, 105–117. (In Japanese.)

Ono, S. (1933) Studies on the life history of Spiruridae in Manchuria. I. The morphologic studies on the encysted larvae found in two species of dung beetles, dragonfly, hedgehog, domestic fowl and duck as well as their infestation experiment with rabbits and dogs. *Journal of the Japanese Society of Veterinary Science* 12, 165–184. (In Japanese.)

Orihel, T.C. (1961) Morphology of the larval stages of *Dirofilaria immitis* in the dog. *Journal of Parasitology* 47, 251–262.

Orihel, T.C. (1967a) Development of *Brugia tupaiae* in the intermediate and definitive hosts. *Journal of Parasitology* 53, 376–381.

Orihel, T.C. (1967b) Infections with *Dipetalonema perstans* and *Mansonella ozzardi* in the aboriginal Indians of Guyana. *American Journal of Tropical Medicine and Hygiene* 16, 628–635.

Orihel, T.C. (1969) *Dirofilaria corynodes* (von Linstow, 1899): morphology and life history. *Journal of Parasitology* 55, 94–103.

Orihel, T.C. and Ash, L.R. (1964) Occurrence of *Dirofilaria striata* in the bobcat (*Lynx rufus*) in Louisiana with observations on its larval development. *Journal of Parasitology* 50, 590–591.

Orihel, T.C. and Beaver, P.C. (1989) Zoonotic *Brugia* infections in North and South America. *American Journal of Tropical Medicine and Hygiene* 40, 638–647.

Orihel, T.C. and Lowrie, R.C. (1975) *Loa loa*: development to the infective stage in the American deer fly, *Chrysops atlanticus*. *American Journal of Tropical Medicine and Hygiene* 24, 610–615.

Orihel, T.C. and Moore, P.J. (1975) *Loa loa*: experimental infection in two species of African primates. *American Journal of Tropical Medicine and Hygiene* 24, 606–609.

Orihel, T.C. and Pacheco, G. (1966) *Brugia malayi* in the Philippine macaque. *Journal of Parasitology* 52, 394.

Orihel, T.C. and Seibold, H.R. (1971) Trichospirurosis in South American monkeys. *Journal of Parasitology* 57, 1366–1368.

Ortega-Mora, L.M. and Rojo-Vázquez, F.A. (1988) Sobre la presentia de *Dipetalonema dracunculoides* (Cobbold, 1870) en el perro en Epaña. *Revista Ibérica de Parasitologia, Spain* 48, 187–188.

Ortlepp, R.J. (1962) *Parafilaria bassoni* spec. nov. from the eyes of springbuck (*Antidorcas marsupialis*). *Onderstepoort Journal of Veterinary Research* 29, 165–168.

Osipov, A.N. (1962) The development of *Parafilaria* in the final host. *Tezisy Dokladov Nauchnoi Konferentsii Vsesoyuznogo Obshchestva Gelmintologov AN SSSR* Part I, 129–131. (In Russian.)

Osipov, A.N. (1966a) Study of the development of *Setaria labiatopapillosa* (Alessandrini, 1848) in the body of experimentally infected cattle. *Materialy Nauchnoi Konferentsii Vses Ob-va Gel'mintologii. 3, 214–216. (In Russian.)*

Osipov, A.N. (1966b) Life cycle of *Setaria altaica* (Rajewskaja, 1928), a parasite of the brain of Siberian deer. *Doklady Akademii Nauk SSSR* 168, 247–248. (In Russian.)

Oswald, V.H. (1958a) Studies on *Rictularia coloradensis* Hall, 1916 (Nematoda: Thelaziidae). I. Larval development in the intermediate host. *Transactions of the American Microscopical Society* 77, 229–240.

Oswald, V.H. (1958b) Studies on *Rictularia coloradensis* Hall, 1916 (Nematoda: Thelaziidae). II. Development in the definitive host. *Transactions of the American Microscopical Society* 77, 413–422.

O'Toole, D., Welch, V. and Williams, B. (1994a) Immuno-histochemistry of parasitic subepidermal vesiculobullous disease in American badgers (*Taxidea taxus*). *Journal of Veterinary Diagnostic Investigation* 6, 72–76.

O'Toole, D., Welch, V. and Williams, E.S. (1994b) Vesiculobullous skin disease in free-ranging badgers (*Taxidea taxus*). *Veterinary Pathology* 30, 343–351.

Ottley, M.L. and Moorhouse, D.E. (1980) Laboratory transmission of *Onchocerca gibsoni* by *Forcipomyia* (*Lasiohelea*) *townsvillensis*. *Australian Veterinary Journal* 56, 559–560.

Oyamada, T., Kawagoe, T., Matsunaga, T., Kudo, N., Yoshikawa, H. and Yoshikawa, T. (1995a) Larval *Gnathostoma nipponicum* found in catfish, *Silurus asotus*, in Aomori Prefecture, Japan. *Japanese Journal of Parasitology* 44, 283–289.

Oyamada, T., Kudo, N., Sano, T., Narai, H. and Yoshikawa, T. (1995b) Prevalence of the advanced third-stage larvae of *Gnathostoma nipponicum* in loaches (*Misgurnus anguillicaudatus*) in Aomori Prefecture, northern part of Honshu, Japan. *Japanese Journal of Parasitology* 44, 222–227.

Oyamada, T., Esaka, Y., Kudo, N. and Yoshikawa, T. (1996a) Epidemiological survey of *Gnathostoma nipponicum* larvae in fishes as the source of human infection in northern Japan. *Journal of the Japan Veterinary Medical Association* 49, 574–578. (In Japanese.)

Oyamada, T., Esaka, Y., Kudo, N. and Yoshikawa, T. and Kamiya, H. (1996b) Prevalence of *Gnathostoma nipponicum* larvae in *Oncorhynchus masou* (Salmonidae) and *Tribolodon hakonensis* (Cyprinidae) collected from eastern Aomori Prefecture, Japan. *Japanese Journal of Parasitology* 45, 201–206.

Oyamada, T., Kobayashi, H., Kindou, T., Kudo, N., Yoshikawa, H. and Yoshikawa, T. (1996c) Discovery of mammalian hosts to *Gnathostoma nipponicum* larvae and prevalence of the larvae in rodents and insectivores. *Journal of Veterinary Medical Science* 58, 839–843.

Oyamada, T., Ohta, Y., Nogutti, S., Kudo, N. and Yoshikawa, T. (1996d) Assessment of three species of copepods as the first intermediate host of *Gnathostoma nipponicum*, in Aomori, Prefecture, Japan. *Japanese Journal of Parasitology* 45, 234–237.

Oyamada, T., Kudo, N., Yoshikawa, H., Oyamada, T., Yoshikaw, T. and Suzuki, N. (1997) Survey for *Gnathostoma nipponicum* larvae in gobiid freshwater fish and infectivity of the larvae to a bobiid fish (*Chaenogobius urotaenia*). *Journal of Veterinary Medical Science* 59, 671–675.

Oyamada, T., Hirata, T., Hara, M., Kudo, N., Oyamada, T., Yoshikawa, H., Yoshikawa, T. and Suzuki, N. (1998) Spontaneous larval *Gnathostoma nipponicum* infection in frogs. *Journal of Veterinary Medical Science* 60, 1029–1031.

Palmer, C.A., Wittrock, D.D. and Christenson, B.M. (1986) Ultrastructure of the body wall of larval stages of *Dirofilaria immitis* (Nematoda: Filarioidea) in the mosquito host. *Proceedings of the Helminthological Society of Washington* 53, 224–231.

Palmieri, J.R., Purnomo, Dennis, D.T. and Marwoto, H.A. (1980) Filariid parasites of South Kalimantan (Borneo) Indonesia. *Wuchereria kalimantani* sp.n. (Nematoda: Filarioidea) from the silvered leaf monkey, *Presbytis cristatus* Eschscholtz, 1821. *Journal of Parasitology* 66, 645–651.

Palmieri, J.R., Connor, D.H., Purnomo, Dennis, D.T. and Marwoto, H. (1982) Experimental infection of *Wuchereria bancrofti* in the silvered leaf monkey *Presbytis cristatus* Eschscholtz, 1821. *Journal of Helminthology* 56, 243–245.

Palmieri, J.R., Connor, D.H., Purnomo, and Marwoto, H.A. (1983) Bancroftian filariasis: *Wuchereria bancrofti* infection in the silvered leaf monkey (*Presbytis cristatus*). *American Journal of Pathology* 112, 383–386.

Pandit, C.G., Pandit, S.R. and Iyer, P.V.S. (1929) The development of the filaria *Conispiculum guindiensis* (1929) in *C. fatigans*, with a note on the transmission of the infection. *Indian Journal of Medical Research* 17, 421–429.

Partono, F., Oemijati, S., Dennis, D.T., Purnomo, Atmosoedjono, S. and Cross, J.H. (1976) The Timor filaria on Flores and experimental transmission of the parasite. *Transactions of the Royal Society of Tropical Medicine and Hygiene* 70, 354–355.

Partono, F., Purnomo, Dennis, D.T., Atmosoedjono, S., Oemijati, S. and Cross, J.H. (1977) *Brugia timori* sp.n. (Nematoda: Filarioidea) from Flores Island, Indonesia. *Journal of Parasitology* 63, 540–546.

Patnaik, B. (1973) Studies on stephanofilariasis in Orissa. III. Life cycle of *S. assamensis* Pande, 1936. *Zeitschrift für Tropenmedizin und Parasitologie* 24, 457–466.

Patnaik, B. and Roy, S.P. (1966) On the life cycle of the filariid *Stephanofilaria assamensis* Pande, 1936, in the arthropod vector *Musca conducens* Walker, 1859. *Indian Journal of Animal Health* 5, 91–101.

Pattanayak, S. (1968) Some observations on the animal filariasis in India. *Bulletin of the National Society of India for Malaria and Other Mosquito Borne Diseases* 5, 262–266.

Pearse, J.S. and Timm, R.W. (1971) Juvenile nematodes (*Echinocephalus pseudouncinatus*) in the gonads of sea urchins (*Centrostephanus coronatus*) and their effect on host gameto-genesis. *Biological Bulletin Marine Biological Laboratory, Woods Hole, Massachusetts* 140, 95–103.

Peel, E. and Chardome, M. (1946a) Sur des filarides de chimpanzés *Pan paniscus* et *Pan satyrus* au Congo belge. *Annales de la Société Belge de Médicine Tropicale* 26, 117–156.

Peel, E. and Chardome, M. (1946b) Note préliminaire sur des filarides de chimpanzés, *Pan paniscus* et *Pan satyrus* au Congo belge. *Recueil de Travaux de Sciences Médicales au Congo Belge* 5, 244–247.

Pence, D.B. (1991) Elaephorosis in wild ruminants. *Bulletin of the Society for Vector Ecology* 16, 149–160.

Pence, D.B. and Gray, G.G. (1981) Elaeophorosis in barbary sheep and mule deer from the Texas Panhandle. *Journal of Wildlife Diseases* 17, 49–56.

Pennington, N.E. (1971) Arthropod vectors, cyclodevelopment and prepatent period of *Dipetalonema reconditum* (Grassi) and the incidence of canine filariasis and ectoparasites in north-central Oklahoma. Unpublished PhD thesis, Oklahoma State University, Stillwater, Oklahoma.

Petit, G. (1981) Cellule R_1 et musculature des filaires; analyse ultrastructurale (1). *Annales de Parasitologie Humaine et Comparée* 56, 81–95.

Petit, G., Bain, O., Carrat, C. and Marval, F. de (1988a) Développement de la filaire *Monanema martini* dans l'épiderme des tiques Ixodidae. *Annales de Parasitologie Humaine et Comparée* 63, 54–63.

Petit, G., Bain, O., Cassone, J. and Seureau, C. (1988b) La filaire *Cercopithofilaria roussilhoni* chez la tique vectrice. *Annales de Parasitologie Humaine et Comparée* 63, 296–302.

Petri, L.H. (1950) Life cycle of *Physaloptera rara* Hall and Wigdor, 1918 (Nematoda: Spiruroidea) with the cockroach, *Blattella germanica*, serving as intermediate host. *Transactions of the Kansas Academy of Science* 53, 331–337.

Petri, L.H. and Ameel, D.J. (1950) Studies on the life cycle of *Physaloptera rara* Hall and Wigdor, 1918 and *Physaloptera praeputialis* Linstow, 1889. *Journal of Parasitology* (Suppl.) 36, 40.

Petrov, Y.F. (1970) Intermediate hosts of the nematode *Tetrameres fissispina* from domestic aquatic birds. *Sbornik Rabot Molodykh Uchenykh, Vsesoyuznogo Nauchno-issledovatel'skogo i Technologicheskogo Instiuta Ptitsevodstva* 12, 222–225. (In Russian.)

Petter, A.J. (1960) La blatte germanique (*Blattella germanica* L.), hôte intermédiaire probable d'*Abbreviata caucasica* (Linstow, 1902), nématode parasite des primates et de l'homme. *Comptes Rendus des Séances de la Société de Biologie* 154, 87–90.

Petter, A.J. (1970) Quelques spirurides de poissons de la region nantaise. *Annales de Parasitologie Humaine et Comparée* 45, 31–46.

Petter, A.J. (1971) Redescription d'*Hedruris androphora* Nitzsch, 1821 (Nematoda, Hedruridae) et étude de son developpement chez l'hôte intermediare. *Annales de Parasitologie Humaine et Comparée* 46, 479–495.

Pfister, R., Heider, K., Illgen, R. and Beglinger, R. (1990) *Trichospirura leptostoma*: a possible cause of wasting disease in the marmoset. *Zeitschrift für Versuchstierkd* 33, 157–161.

Piana, G.P. (1897) Osservazioni sul *Dispharagus nasutus* Rud. dei polli e sulla larve Nematodelmintiche delle mosche e dei porcellioni. *Atti Societa Italiana delle Scienze Naturelle (Milano)* 36, 239–262.

Pichon, G. (1974) Relations mathematique entre le nombre des microfilariae ingérées et le nombre des parasites chez différent vecteurs natural on expérimentaux de filarioses. *Cahiers ORSTROM, Série Entomologie Médical et Parasitologie* 12, 199–216.

Pietrobelli, M., Cancrina, G., Frangipane di Regalbono, A., Gallupi, R. and Tampieri, M.P. (1998) Development of *Setaria labiatopapillosa* in *Aedes caspius*. *Medical and Veterinary Entomology* 12, 106–108.

Pistey, W.R. (1958) Studies on the development of *Dirofilaria tenuis* Chandler, 1942. *Journal of Parasitology* 44, 613–626.

Platonov, A.V. (1966) Study of the cycle of development of *Setaria equina* Abild., 1789 – a parasite of horses. *Doklady Akademii Nauk SSSR* 169, 982–984. (In Russian.)

Poinar, G.O. Jr and Kannangara, D.W.W. (1972) *Rhabdochona praecox* sp.n. and *Proleptus* sp. (Spiruroidea: Nematoda) from freshwater crabs in Ceylon. *Annales de Parasitologie Humaine et Comparée* 47, 121–129.

Poinar, G.O. Jr and Hess, R. (1974) An ultrastructural study of the response of *Blattella germanica*, (Orthoptera: Blattidae) to the nematode *Abbreviata caucasica* (Spirurida: Physalopteridae). *International Journal for Parasitology* 4, 133–138.

Poinar, G.O. Jr and Kuris, A.M. (1975) Juvenile *Ascarophis* (Spirurida: Nematoda) parasitizing intertidal decapod Crustacea in California: with notes on prevalence and effects on host growth and survival. *Journal of Invertebrate Pathology* 26, 375–382.

Poinar, G.O. Jr and Quentin, J.C. (1972) The development of *Abbreviata caucasica* (von Linstow) (Spirurida: Physalopteridae) in an intermediate host. *Journal of Parasitology* 58, 23–28.

Poinar, G.O. Jr and Thomas, G.M. (1976) Occurrence of *Ascarophis* (Nematoda: Spiruridea) in *Callianassa californiensis* Dana and other decapod crustaceans. *Proceedings of the Helminthological Society of Washington* 43, 28–33.

Popova, Z.G. (1960) Further studies on the biology of *Gongylonema* in farm animals and the diagnosis of this disease. *Nauchnie Trudi, Ukrainski Nauchno-Issledovatelski Institut Eksperimentalnoi Veterinarii* 27, 28–33. (In Russian.)

Porter, D.A. (1939) Some new intermediate hosts of the swine stomach worms, *Ascarops strongylina* and *Physocephalus sexalatus*. *Proceedings of the Helminthological Society of Washington* 6, 79–80.

Potekhina, L.F. (1968) Epizootiology of echinuriasis of birds. *Trudi Vsesoyuznogo Instituta Gelmintologii* 14, 263–271. (In Russian.)

Prestwood, A.K. and Ridgeway, T.R. (1972) Elaeophorosis in white-tailed deer of the southeastern USA: case report and distribution. *Journal of Wildlife Diseases* 8, 233–236.

Price, D.L. (1957) *Dirofilaria uniformis* n.sp. (Nematoda: Filarioidea) from *Sylvilagus floridanus mallurus* (Thomas) in Maryland. *Proceedings of the Helminthological Society of Washington* 24, 15–19.

Price, D.L. (1959) *Dirofilaria magnilarvatum* n.sp. (Nematoda: Filarioidea) from *Macaca irus* Cuvier. I. Description of the adult filarial worms. *Journal of Parasitology* 45, 499–504.

Price, D.L., Sprinz, H., Duxbury, R.E. and Smith, R.W. (1963) Development of *Dirofilaria uniformis* Price in *Anopheles quadrimaculatus* Say. *Journal of Parasitology* (Suppl.) 49, 48.

Prod'hon, J. and Bain, O. (1972) Développement larvaire chez *Anopheles stephensi* d'*Oswaldofilaria bacillaris*, filaire de caiman sud-américain, et redescription des adultes. *Annales de Parasitologie Humaine et Comparée* 47, 745–758.

Prommas, C. and Daengsvang, S. (1933) Preliminary report of a study on the life cycle of *Gnathostoma spinigerum*. *Journal of Parasitology* 19, 287–292.

Prommas, C. and Daengsvang, S. (1936) Further report of a study on the life cycle of *Gnathostoma spinigerum*. *Journal of Parasitology* 22, 180–186.

Prommas, C. and Daengsvang, S. (1937) Feeding experiments on cats with *Gnathostoma spinigerum* larvae obtained from the second intermediate host. *Journal of Parasitology* 23, 115–116.

Purnomo, Partono, F., Dennis, D.T. and Atmosoedjono, S. (1976) Development of the Timor filaria in *Aedes togoi*: preliminary observations. *Journal of Parasitology* 62, 881–885.

Quentin, J.C. (1969a) Cycle biologique de *Protospirura muricola* Gedoelst, 1916 Nematoda Spiruridae. *Annales de Parasitologie Humaine et Comparée* 44, 485–504.

Quentin, J.C. (1969b) Infestation spontanée d'un dermaptère par des larves de *Pseudophysaloptera vincenti* n.sp., parasite du lemurien *Galagoides demidovii* (Fischer, 1808). *Annales de Parasitologie Humaine et Comparée* 44, 749–755.

Quentin, J.C. (1969c) Cycle biologique de *Pterygodermatites desportesi* (Chabaud et Rousselot, 1956) Nematoda, Rictulariidae. *Annales de Parasitologie Humaine et Comparée* 44, 47–58.

Quentin, J.C. (1969d) Essai de classification des nématodes rictulaires. *Mémoires du Muséum National d'Histoire Naturelle, Nouvelle Série, A. Zoologie* 54, 55–115.

Quentin, J.C. (1970a) Morphogénèse larvaire du spiruride *Mastophorus muris* (Gmelin, 1790). *Annales de Parasitologie Humaine et Comparée* 45, 839–855.

Quentin, J.C. (1970b) Cycle biologique de *Rictularia proni*, Seurat, 1915 Nematoda Rictulariidae. Ontogénèse des structures cephaliques. *Annales de Parasitologie Humaine et Comparée* 45, 89–103.

Quentin, J.C. (1970c) Cycle biologique de *Pterygodermatites* (*Mesopectines*) *taterilli* (Baylis, 1928) Nematoda Rictulariidae. *Annales de Parasitologie Humaine et Comparée* 45, 629–635.

Quentin, J.C. (1973) Présence de *Spirura guianensis* (Ortlepp, 1924) chez des marsupiaux néotropicaux. Cycle évolutif. *Annales de Parasitologie Humaine et Comparée* 48, 117–133.

Quentin, J.C. (1975) Nématodes *Spirura* parasites des *Tupaia* et du nycticèbe en Malaisie. *Annales de Parasitologie Humaine et Comparée* 50, 795–812.

Quentin, J.C. and Barre, N. (1976) Description et cycle biologique de *Tetrameres* (*Tetrameres*) *cardinalis* n.sp. *Annales de Parasitologie Humaine et Comparée* 51, 65–81.

Quentin, J.C. and Gunn, T. (1981) Morphologie et biologie larvaire de *Gongylonema soricis* Fain, 1955. *Annales de Parasitologie Humaine et Comparée* 56, 167–172.

Quentin, J.C. and Poinar, G.O. Jr (1973) Comparative study of the larval development of some heteroxenous subulurid and spirurid nematodes. *International Journal for Parasitology* 3, 809–827.

Quentin, J.C. and Seguignes, M. (1979) Cycle biologique de *Gongylonema mucronatum* Seurat, 1916 parasite du hérisson d'Afrique du Nord. *Annales de Parasitologie Humaine et Comparée* 54, 637–644.

Quentin, J.C. and Seureau, C. (1974). Cycle biologique de *Pterygodermatites hispanica* Quentin, 1973 (Nematoda Rictulariidae). *Annales de Parasitologie Humaine et Comparée* 49, 701–719.

Quentin, J.C. and Seureau, C. (1983) Cycle biologique d'*Acuaria gruveli* (Genre, 1913), nématode acuaride parasite du francolin au Togo. *Annales de Parasitologie Humaine et Comparée* 58, 43–56.

Quentin, J.C., Karimi, Y. and Rodriguez de Almeida, C. (1968) *Protospirura numidica criceticola* n. subsp. parasite de rongeurs Cricetidae du Brésil. Cycle évolutif. *Annales de Parasitologie Humaine et Comparée* 43, 583–596.

Quentin, J.C., Seureau, C. and Gabrion, C. (1972) Cycle biologique d'*Acuaria anthuris* (Rudolphi, 1819), nématode parasite de la pie. *Zeitschrift für Parasitenkunde* 39, 103–126.

Quentin, J.C., Troncy, P.M. and Barre, H. (1976a) *Aprocta cylindrica* Linstow, 1883, filaire ovipare parasite d'oiseaux plocéidés au Tchad. Morphogénèse larvaire du nématode. *Annales de Parasitologie Humaine et Comparée* 51, 83–93.

Quentin, J.C., Seureau, C. and Vernet, R. (1976b) Cycle biologique du nématode rictulaire *Pterygodermatites* (*Multipectines*) *affinis* (Jagerskiold, 1904). *Annales de Parasitologie Humaine et Comparée* 51, 51–64.

Quentin, J.C., Seureau, C. and Railhac, C. (1983) Cycle biologique de *Cyrnea* (*Procyrnea*) *mansioni* Seurat, 1914 nématode habronème parasite des rapaces au Togo. *Annales de Parasitologie Humaine et Comparée* 58, 165–175.

Quentin, J.C., Seureau, C. and Kulo, S.D. (1986a) Cycle biologique de *Tetrameres* (*Microtetrameres*) *inermis* (Linstow, 1879) nématode Tetrameridae parasite du Tisserin *Ploceus aurantius* au Togo. *Annales de Parasitologie Humaine et Comparée* 61, 321–332.

Quentin, J.C., Seureau, C. and Sapin, J.M. (1986b) *Gongylonema congolense* Fain, 1955, (Nematoda, Spirurida) synonymie et biologie larvaire, réactions cellulaires de l'insecte hôte intermédiare. *Zeitschrift für Parasitenkunde* 72, 227–239.

Quillevere, D., Prod'hon, J. and Traore, S. (1988) Transmission of *Onchocerca* infections by female *Simulium damnosum. Bobo-Dioulasso, Burkina Faso; Organisation de Coordination et de Cooperation pour la Lutte Contre les Grandes Endemies*, pp. 25–34.

Railliet, A., Henry, A. and Langeron, M. (1912) Le genre *Acanthocheilonema* Cobbold et les filaires peritoneales des carnivores. *Bulletin de la Société Pathologie Exotique* 5, 292–395.

Ramachandran, C.P. and Dunn, F.L. (1968) The development of *Breinlia sergenti* (Dipetalonematidae) in *Aedes* mosquitoes. *Annals of Tropical Medicine and Parasitology* 62, 441–449.

Ramirez-Perez, J. and Peterson, B.V. (1981) Study of the complex of *Simulium amazonicum–sanguineum* in Venezuela. Description of three new species. *Boletín de la Dirección de Malariología y Saneamiento Ambiental* 21, 151–160.

Ramishvili, N.D. (1973) Study of the distribution and life-cycle of *Gongylonema pulchrum. Parazitologicheskii Sbornik, Tbilisi* 3, 112–136. (In Russian.)

Ramaswamy, K. and Sundaram, R.K. (1979) *Porcellio laevis* (Isopoda) as a new intermediate host for *Tetrameres mohtedai* Bhalerao and Rao, 1944. *Indian Veterinary Journal* 56, 363–366.

Ransom, B.H. (1904) A new nematode (*Gongylonema ingluvicola*) parasitic in the crop of chicken. *United States Department of Agriculture, Bureau of Animal Industry*, Circular (64) 3 pp.

Ransom, B.H. (1911) The life history of a parasitic nematode *Habronema muscae. Science* 34, 690–692.

Ransom, B.H. (1913) The life history of *Habronema muscae* (Carter), a parasite of the horse transmitted by the housefly. *USA Department of Agriculture Bureau of Animal Industry Bulletin* No. 163, pp. 1–36.

Ransom, B.H. and Hall, M.C. (1915) A further note on the life-history of *Gongylonema scutatum. Journal of Parasitology* 3, 177–181.

Rao, S.S. and Maplestone, P.A. (1940) The adult of *Microfilaria malayi* Brug, 1927. *Indian Medical Gazette* 75, 159–160.

Refuerzo, P.G. and Garcia, E.Y. (1938) The crustacean intermediate hosts of *Gnathostoma spinigerum* in the Philippines and its pre- and intracrustacean development. *Philippine Journal of Animal Industry* 5, 351–362.

Refuerzo, P.G. (1940) Arthropod intermediate hosts of *Acuaria hamulosa* in the Philippines, I. *Natural and Applied Science Bulletin University of Philippines* 7, 407–414.

Renz, A. (1987) Studies on the dynamics of transmission of onchocerciasis in a Sudan-savanna area of North Cameroon. II. Seasonal and diurnal changes in the biting densities and in the age-composition of the vector population. *Annals of Tropical Medicine and Parasitology* 81, 229–237.

Renz, A. and Wenk, P. (1987) Studies on the dynamics of transmission of onchocerciasis in a Sudan-savanna area of North Cameroon. I. Prevailing *Simulium* vectors, their breeding sites. *Annals of Tropical Medicine and Parasitology* 81, 215–228.

Reznik, E.P. (1982) Development of filaria *Thamugadia ivaschkini* Annaev, 1976 (Splendidofilariidae) in mosquitoes. *Helminthologia* 19, 141–149.

Ribelin, W.F. and Bailey, W.S. (1958) Esophageal sarcomas associated with *Spirocerca lupi* infection in the dog. *Cancer* 11, 1242–1246.

Richter, S. (1960) Intermediate host of *Streptocara pectinifera* (Neumann, 1900), life cycle and infestation mode of this parasite. *Veterinarski Arhiv* 30, 86–92. (In Russian.)

Riek, R.F. (1954) Studies on allergic dermatitis (Queensland itch) of the horse: the aetiology of the disease. *Australian Journal of Agricultural Research* 5, 109–129.

Rietschel, G. (1973) Untersuchungen zur Entwicklung einiger in Krähen (Corvidae) vorkommenden Nematoden. *Zeitschrift für Parasitenkunde* 42, 243–250.

Roberts, J.M.D., Neumann, E., Göckel, C.W. and Highton, R.B. (1967) Onchocerciasis in Kenya 9, 11 and 18 years after elimination of the vector. *Bulletin of the World Health Organization* 37, 195–212.

Robinson, E.J. Jr (1971) *Culicoides crepuscularis* (Malloch) (Diptera: Ceratopogonidae) as a host for *Chandlerella quiscali* (von Linstow, 1904) comb.n. (Filarioidea: Onchocercidae). *Journal of Parasitology* 57, 772–776.

Robinson, R.M., Jones, L.P., Galvin, T.J. and Harwell, G.M. (1978) Elaeophorosis in sika deer in Texas. *Journal of Wildlife Diseases* 14, 137–141.

Robles, R. (1919) Onchocercose humaine au Quatémala produisant la cécité et l'érysipèle du littoral (erisipela de la costa). *Bulletin de la Société de Pathologie Exotique* 12, 442–460.

Rodenwaldt, E. (1908) Die Verteilung der Mikrofilarien im Körper und die Ursachen des Turnus bei Mikrofilaria nocturna und diurna Studien zur Morphologie der Mikrofilarien. *Archiv für Schiffs- und Tropen-Hygiene, Pathologie und Therapie Exotischer Krankheiten* 10, 18–30.

Rodriguez-Rodriguez, J.A., Omelda-Garcia, A.S., Valcarcel-Sancho, F. and Gómes-Bautista, M. (1989) *Rhipicephalus sanguineus* (Letrilla, 1908) vector potential de *Dipetalonema dracunculoides* (Cobbold, 1870) VI Congress Nacional y I Congreso Ibérico de Parasitologia. Caja de Ahorros Salamanca, Caceres, p. 229.

Rogers, R., Ellis, D.S. and Denham, D.A. (1976) Studies with *Brugia pahangi*. 14. Intrauterine development of the microfilaria and a comparison with other filarial species. *Journal of Helminthology* 40, 251–257.

Rojo-Vázquez, F.A., Valcárcel, F., Guerro, J. and Gómez Bautista, M. (1990) Prevalencia de la dirofilariosis canina en cuartro áreas geográficas de España. *Medicina Veterinaria* 7, 297–305.

Romanova, N.P. (1947) A study of the development cycle of *Echinuria uncinata* Rud., 1819 a nematode parasitic of the stomach of natatores. *Comptes Rendus (Doklady) de l'Academie des Sciences de l'URSS* 55, 371–372. (In Russian.)

Rosen, L. (1954) Observations on *Dirofilaria immitis* in French Oceania. *Annals of Tropical Medicine and Parasitology* 48, 318–328.

Roubaud, E. and Descazeaux, J. (1921) Contribution à l'histoire de la mouche domestique comme agent vecteur des habronèmoses d'équides. Cycle évolutif et parasitisme de l'*Habronema megastoma* (Rudolphi, 1819) chez la mouche. *Bulletin de la Société Pathologique Exotique* 14, 471–506.

Roubaud, E. and Descazeaux, J. (1922a) Evolution de l'*Habronema muscae* Carter chez la mouche domestique et de l'*H. microstomum* Schneider chez le stomoxe. (Note préliminaire). *Bulletin de la Société Pathologique Exotique* 15, 572–574.

Roubaud, E. and Descazeaux, J. (1922b) Deuxième contribution à l'étude des mouches dans leurs rapports avec l'évolution des habronèmes d'équides. *Bulletin de la Societé Pathologique Exotique* 15, 978–1001.

Rust, W. (1908) Entenerkrankung durch *Tropidocerca fissispina* (Abstract). *Veröffentlichungen aus den Jahres-Veterinär-Berichten der beamteten Tierärzte Preussens, Berlin* 6, 30.

Ryzhikov, K.M. and Nazarova, N.S. (1959) On the reservoir parasitism in *Physocephalus sexalatus* and *Spirocerca lupi*. *Trudi Gel'mintologicheskoi Laboratorii* 9, 249–252. (In Russian.)

Saito, E.K. and Little, S.E. (1997) Filaria taxideae-induced dermatitis in a striped skunk (*Mephitis mephitis*). *Journal of Wildlife Diseases* 34, 873–876.

Sanders, D.A. (1929) Manson's eyeworms of poultry. *Technical Bulletin Agricultural Experimental Station, University of Florida* 206, 567–585.

Sartono, E., Purnomo, Bahang, Z.B. and Partono, F. (1990) Experimental infection and periodicity studies of *Brugia timori* in the silvered leaf monkey. *Mosquito Borne Diseases Bulletin* 6, 65–67.

Schacher, J.F. (1962a) Morphology of the microfilaria of *Brugia pahangi* and the larval stages in the mosquito. *Journal of Parasitology* 48, 679–692.

Schacher, J.F. (1962b) Developmental stages of *Brugia pahangi* in the final host. *Journal of Parasitology* 48, 693–703.

Schacher, J.F. (1973) Laboratory models in filariasis: a review of filarial life cycle patterns. *Southeast Asian Journal of Tropical Medicine and Public Health* 4, 336–349.

Schacher, J.F. and Khalil, G.M. (1968) Development of *Foleyella philistinae* Schacher and Khalil, 1967 (Nematoda: Filarioidea) in *Culex pipiens molestus* with notes on pathology in the arthropod. *Journal of Parasitology* 54, 869–878.

Schell, S.C. (1952a) Studies on the life cycle of *Physaloptera hispida* Schell (Nematoda: Spiruroidea), a parasite of the cotton rat (*Sigmodon hispidus littoralis* Chapman). *Journal of Parasitology* 38, 462–472.

Schell, S.C. (1952b) Tissue reactions of *Blattella germanica* L. to the developing larva of *Physaloptera hispida* Schell, 1950 (Nematoda: Spiruroidea). *Transactions of the American Microscopical Society* 71, 293–302.

Schiller, E.L., Petersen, J.L., Shirazian, D. and Marroquin, H.F. (1984) Morphogenesis of larval *Onchocerca volvulus* in the Panamanian black fly, *Simulium quadrivittatum. American Journal of Tropical Medicine and Hygiene* 33, 410–413.

Schtein, G.A. (1959) The life cycle and ecology of the nematode *Rhabdochona denudata* (Dujardin, 1845). *Doklady Akademii Nauk SSSR* 127, 1320–1321. (In Russian.)

Schulz-Key, H. (1975) Untersuchungen über die Filarien der Cerviden in Süddeutschland. I. Knotenbildung, Geschlechterfindung und Mikrofilarienausschüttung bei *Onchocerca flexuosa* (Wedl, 1856) im Rothirsch (*Cervus elaphus*). *Tropenmedizin und Parasitologie* 26, 60–69.

Schulz-Key, H. (1990) Observations on the reproductive biology of *Onchocerca volvulus. Acta Leidensia* 59, 27–44.

Schulz-Key, H. and Albiez, E.J. (1977) Worm burden of *Onchocerca volvulus* in a hyperendemic village of the rainforest in West Africa. *Tropenmedizin und Parasitologie* 28, 431–438.

Schulz-Key, H. and Karam, M. (1986) Periodic reproduction of *Onchocerca volvulus. Parasitology Today* 2, 284–286.

Schulz-Key, H. and Soboslay, P.T. (1994) Reproductive biology and population dynamics of *Onchocerca volvulus* in the vertebrate host. (In *3rd CEC Filariasis Network Meeting. Lisbon – IHMT, 4–7 September 1993*). *Parasite* 1 (Suppl. 1) 53–55.

Schulz-Key, H. and Wenk, P. (1981) The transmission of *Onchocerca tarsicola* (Filarioidea: Onchocercidae) by *Odagmia ornata* and *Prosimulium nigripes* (Diptera: Simuliidae). *Journal of Helminthology* 55, 161–166.

Schulz-Key, H., Jean, B. and Albiez, E.J. (1980) Investigations on female *Onchocerca volvulus* for the evaluation of drug trials. *Tropenmedizin und Parasitologie* 31, 34–40.

Schwabe, C.W. (1950) Studies on *Oxyspirura mansoni* the tropical eyeworm of poultry. III. Preliminary observations on eyeworm pathogenicity. *American Journal of Veterinary Research* 11, 286–290.

Schwabe, C.W. (1951) Studies on *Oxyspirura mansoni* the tropical eyeworm of poultry. II. Life history. *Pacific Science Honolulu* 5, 18–35.

Scott, J.A. (1946) Observations on the rate of growth and maturity of *Litomosoides carinii*, a filarial worm of the cotton rat. *Journal of Parasitology* 32, 570–573.

Scott, J.A. (1960) A description of certain details of growth and development of the filarial worm of cotton rats and the value of the observations in the study of immunity. *Libro Homenaje al Dr. Eduardo Caballero y Caballero, Jubileo* 1930–1960, pp. 501–510.

Scott, J.A., Sisley, N.M. and Stembridge, V.A. (1946) The susceptibility of cotton rats and white rats to *Litomosoides carinii* in relation to the presence of previous infections. *Journal of Parasitology* (Suppl.) 32, 17.

Scott, J.A., MacDonald, E.M. and Terman, B. (1951) A description of the stages in the life cycle of the filarial worm *Litomosoides carinii. Journal of Parasitology* 37, 425–432.

Seegar, W.S., Schiller, E.L., Sladen, W.J.L. and Trpis, M. (1976) A mallaghaga, *Trinoton anserinum*, as a cyclodevelopmental vector for a heartworm parasite of waterfowl. *Science* 194, 739–741.

Seibold, H.R., Bailey, W.S., Hoerlein, B.F., Jordan, E.M. and Schwabe, C.W. (1955) Observations on the possible relation of malignant esophageal tumors and *Spirocerca lupi* lesions in the dog. *American Journal of Veterinary Research* 16, 5–14.

Sen, K. and Anantaraman, M. (1971) Some observations on the development of *Spirocerca lupi* in its intermediate and definitive hosts. *Journal of Helminthology* 45, 123–131.

Service, M.W. (1982) Importance of vector ecology in vector disease control in Africa. *Bulletin of the Society of Vector Ecologists* 7, 1–13.

Seurat, L.G. (1911) Sur l'habitat et les migrations de *Spirura talpae* Gmel. (= *Spiroptera strumosa* Rud.). *Compte Rendu des Séances de la Société de Biologie* 71, 606–608.

Seurat, L.G. (1913a) Sur l'evolution du *Spirura gastrophila* (Müll.). *Compte Rendu des Séances de la Société de Biologie* 74, 286–289.

Seurat, L.G. (1913b) Sur l'évolution du *Physocephalus sexalatus* (Molin). *Compte Rendu des Séances de la Société de Biologie* 75, 517–520.

Seurat, L.G. (1914) Sur un nouveau spiroptère du chat ganté. *Compte Rendu des Séances de la Société de Biologie* 77, 344–347.

Seurat, L.G. (1916) Contributions à l'études des formes larvaires des nématodes parasites hétéroxènes. *Bulletin Scientifique de la France et de la Belgique* 49, 299–377.

Seurat, L.G. (1919) Contributions nouvelle a l'étude des formes larvaires des nématodes parasites hétéroxènes. *Bulletin Biologique de la France et de la Belgique, Fondé par Alfred Giard, Paris* 52, 344–378.

Seureau, C. (1973) Réactions cellulaires provoquées par les nématodes subulures et spirurides chez *Locusta migratoria* (Orthoptère): localisation et structure des capsules. *Zeitschrift für Parasitenkunde.* 41, 119–138.

Seureau, C. and Quentin, J.C. (1983) Sur la biologie larvaire de *Cyrnea* (*Cyrnea*) *eurycerca* Seurat, 1914 nématode habronème parasite du francolin au Togo. *Annales de Parasitologie Humaine et Comparée* 58, 151–164.

Sharp, N.A.D. (1928) *F. perstans*, its development in *Culicoides austeni. Transactions of the Royal Society of Tropical Medicine and Hygiene* 21, 371–396.

Shimazu, T. (1996) Mayfly larvae – *Ephemera japonica*, a natural intermediate host of salmonid nematodes *Sterliadochona ephemeridarum* and *Rhabdochona oncorzynchi* in Japan. *Japanese Journal of Parasitology* 45, 167–172.

Shmitova, G.Y. (1963) Experimental study of reservoir parasitism in *Ascarops strongylina. Helminthologia* 4, 456–463. (In Russian.)

Shmitova, G.Y. (1964) Study of the ontogenetic development of *Ascarops strongylina. Trudi Gel'mintologicheskoi Laboratorii* 14, 288–301. (In Russian.)

Shogaki, Y., Mizuno, S. and Itoh, H. (1972) On *Protospirura muris* (Gmelin), a parasitic nematode of the brown rat in Nagoya city. *Japanese Journal of Parasitology* 21, 28–38. (In Japanese.)

Shoho, C. (1959) Sur les filaires chez les équidés et les bovidés. *Revue d'Élevage et de Médecine Vétérinaire des Pays Tropicaux* 12, 42–53.

Shoho, C. (1965) *Setaria marshalli* Boulenger 1921, et *Setaria marshalli pandei* n. s.sp. *Annales de Parasitologie (Paris)* 40(3), 285–302.

Shoho, C. and Nair, V.K. (1960) Studies on cerebrospinal nematodiasis in Ceylon 7. Experimental production of cerebrospinal nematodiasis by the inoculation of infective larvae of *Setaria digitata* into susceptible goats. *Ceylon Veterinary Journal* 8, 2–12.

Shol', A.V. (1964) Diagnosis of *Setaria* infections in deer. In: Boev, S.N. (ed.) *Parasites of Farm Animals in Kazakhstan*. Alma Ata: Izdatel Akademii Nauk Kazakhstan SSR 31, 101–103. (In Russian.)

Shol', A.V. and Drobishchenko, N.I. (1973) Species validity of *Setaria cervi* (Rudolphi, 1819), a parasite of deer. *Zhiznennye tsikly gel'mintov zhivotnykh Kazakhstana (Sbornik).* Alma-Ata Akademiya Nauk Kazakhskoi SSR Institut Zoologii, pp. 124–133. (In Russian.)

Singh, M. and Cheong, C.H. (1971) On a collection of nematode parasites from Malaysian rats. *Southeast Asian Journal of Medicine and Public Health* 2, 516–522.

Singh, M. and Ho, B.C. (1973) *Breinlia booliati* sp.n. (Filarioidea: Onchocercidae), a filaria of the Malayan forest rat, *Rattus sabanus* (Thos.). *Journal of Helminthology* 47, 127–133.

Singh, M., Kanagasuntheram, R., Ho, B.C., Yap, E.H. and Chan, H.L. (1974) Some ultrastructural observations on the microfilaria of *Breinlia sergenti* – the pharyngeal thread and innenkorper. *International Journal for Parasitology* 4, 375–382.

Singh, M., Yap, E.H., Ho, B.C., Kang, K.L., Lim, E.P.C. and Lim, B.L. (1976) Studies on the Malayan forest rat filaria *Breinlia booliati* (Filarioidea: Onchocercidae): course of development in the rat host. *Journal of Helminthology* 50, 103–110.

Skrjabin, K.I. and Sobolev, A.A. (1963) *Principles of Nematology, Vol. II, Spirurida in Animals and Man and their Control.* Academy of Science of the USSR, Moscow, 511 pp. (In Russian.)

Skrjabin, K.I., Sobolev, A.A. and Ivashkin, V.M. (1967) *Essentials of Nematology, Vol. XVI, Spirurata of Animals and Man and the Diseases Caused by Them. Part 4. Thelazioidea.* (Israel Program for Scientific Translations, Jerusalem 1971.)

Skvortsov, A.A. (1937) Some observations on the biology of the nematode *Habronema megastoma* (Rud. 1819). In: *Papers on Helminthology published in Commemoration of the 30 year Jubileum of K.I. Skrjabin and of the 15th Anniversary of All-Union Institute of Helminthology,* pp. 653–662. (In Russian.)

Slepnev, N.K. (1971) The development of *Ascarops strongylina* (Rudolphi, 1819) in the intermediate host. *Trudi (Nauchnye Trudi) NauchnoIssledovatel'skogo Veterinarnogo Instituta Belorusskoi SSR* 9, 106–108. (In Russian.)

Smith, H.A. and Jones, T.C. (1957) *Veterinary Pathology.* Lea and Febiger, Philadelphia.

Smith, J.D. and Lankester, M.W. (1979) Development of swimbladder nematodes (*Cystidicola* spp.) in their intermediate hosts. *Canadian Journal of Zoology* 57, 1736–1744.

Sohn, W.M., Kho, W.G. and Lee, S.H. (1993) Larval *Gnathostoma nipponicum* found in the imported Chinese loaches. *Korean Journal of Parasitology* 31, 347–352.

Sohn, W.M. and Lee, S.H. (1996) Identification of larval *Gnathostoma* obtained from imported Chinese loaches. *Korean Journal of Parasitology* 34, 161–167.

Sonin, M.D. (1966) *Filariata of Animals and Man and Diseases Caused by Them. Fundamentals of Nematology.* Vol. 17, *Part I. Aproctoidea.* (Israel Program for Scientific Translations, Jerusalem, 1974.)

Sonin, M.D. (1968) *Filariata of Animals and Man and the Diseases Caused by Them. Fundamentals of Nematology.* Vol. 24, *Part II. Diplotriaenoidea.* (Israel Program for Scientific Translations, Jerusalem, 1975.)

Sonin, M.D. (1975) *Filariata of Animals and Man and the Diseases Caused by Them. Fundamentals of Nematology.* Vol. 24, *Part III. Filariidae and Onchocercinae.* (Translated by Amerind Publishing, New Delhi, India, 1985.)

Sonin, M.D. (1977) *Filariata of Animals and Man and the Diseases Caused by Them. Fundamentals of Nematology.* Vol. 28, *Part IV. Onchocercidae.* (Translated by Amerind Publishing, New Delhi, India, 1985.)

Southgate, B.A. and Bryan, J.H. (1992) Factors affecting transmission of *Wuchereria bancrofti* by anopheline mosquitos. 4. Facilitation, limitation, proportionality and their epidemiological significance. *Transactions of the Royal Society of Tropical Medicine and Hygiene* 86, 523–530.

Spratt, D.M. (1970) The synonymy of *Agamofilaria tabanicola* and *Dirofilaria roemeri*. *Journal of Parasitology* 56, 622–623.

Spratt, D.M. (1972a) Aspects of the life-history of *Dirofilaria roemeri* in naturally and experimentally infected kangaroos, wallaroos and wallabies. *International Journal for Parasitology* 2, 139–156.

Spratt, D.M. (1972b) Histological morphology of adult *Dirofilaria roemeri* and anatomy of the microfilaria. *International Journal for Parasitology* 2, 193–200.

Spratt, D.M. (1972c) Natural occurrence, histopathology and developmental stages of *Dirofilaria roemeri* in the intermediate host. *International Journal for Parasitology* 2, 201–208.

Spratt, D.M. (1974) Comparative epidemiology of *Dirofilaria roemeri* infection in two regions of Queensland. *International Journal for Parasitology* 4, 481–488.

Spratt, D.M. (1975) Further studies of *Dirofilaria roemeri* (Nematoda: Filarioidea) in naturally and experimentally infected Macropodidae. *International Journal for Parasitology* 5, 561–564.

Spratt, D.M. and Haycock, P. (1988) Aspects of the life history of *Cercopithifilaria johnstoni* (Nematoda: Filarioidea). *International Journal for Parasitology* 18, 1087–1092.

Spratt, D.M., Dyce, A.L. and Standfast, H.A. (1978) *Onchocerca sweetae* (Nematoda: Filarioidea): notes on the intermediate host. *Journal of Helminthology* 52, 75–81.

Srivastava, H.D. and Dutt, S.C. (1963) Studies on the life history of *Stephanofilaria assamensis*. *Indian Journal of Veterinary Science* 33, 173–177.

Stefanski, W. (1934) Sur le developpement et les caractères specifique de *Spirura rytipleurites* (Deslongchamps, 1824). *Annales de Parasitologie Humaine et Comparée* 12, 203–217.

Stein, G., Hinz, E., Textor-Schneider, G. and Petney, T.N. (1997) Quantitative untersuchungen zur Enwicklung von *Acanthocheilonema vitae* in *Ornithodoros moubata* nach ein-und mehrmaliger Infektion. *Mitteilungen der Österreichischen Gesellschaft für Tropenmedizin und Parasitologie* 19, 75–82.

Steward, J.S. (1935) Fistulous withers and poll-evil. Equine and bovine onchocerciasis compared, with an account of the life histories of the parasite concerned. *Veterinary Record* 15, 1563–1595.

Steward, J.S. (1937) The occurrence of *Onchocerca gutturosa* Neumann in cattle in England, with an account of its life history and development in *Simulium ornatum* Mg. *Parasitology* 29, 212–219.

Stewart, T.B. and Kent, K.M. (1963) Beetles serving as intermediate hosts of swine nematodes in southern Georgia. *Journal of Parasitology* 49, 158–159.

Strom, J. (1937) A new nematode of birds: *Cardiofilaria pavlovskyi* n.g. n.sp.. *Trudi Soveta po Izucheniyu Proizvoditel'nykh Sil, Seriya Turkmenskaya* 9, 217–221. (In Russian.)

Stueben, E.B. (1954a) Incidence of infection of dogs and fleas with *Dirofilaria immitis* in Florida. *Journal of American Veterinary Medical Association* 125, 57–60.

Stueben, E.B. (1954b) Larval development of *Dirofilaria immitis* (Leidy) in fleas. *Journal of Parasitology* 40, 580–589.

Sucharit, S., Harinasuta, C. and Choochote, W. (1982) Experimental transmission of subperiodic *Wuchereria bancrofti* to the leaf monkey (*Presbytis melalophos*), and its periodicity. *American Journal of Tropical Medicine and Hygiene* 31, 599–901.

Sullivan, J.J. and Chernin, E. (1976) Oral transmission of *Brugia pahangi* and *Dipetalonema viteae* to adult and neonatal jirds. *International Journal for Parasitology* 6, 75–78.

Sultanov, M.A. and Kabilov, T. (1969) Intermediate hosts of *Gongylonema pulchrum* in the Fergansk valley (Uzbek SSR). *Doklady Akademii Nauk UzSSR* 8, 44–45. (In Russian.)

Sultanov, M.A., Kabilov, T.K. and Siddikov, B.K. (1979) The development cycle of the nematode *Stephanofilaria assamensis* Pande, 1936 in Uzbekistan (USSR). *Doklady Akademii Nauk Uzbekskoi SSR* 9, 73–74. (In Russian.)

Summers, W. (1940) Fleas as acceptable intermediate hosts of the dog heartworm, *Dirofilaria immitis*. *Proceedings of the Society for Experimental Biology and Medicine* 43, 448–450.

Summers, W. (1943) Experimental studies on the larval development of *Dirofilaria immitis* in certain insects. *American Journal of Hygiene* 37, 173–178.

Sundaram, R.K., Radhakrishnan, C.V., Sivasubramaniyam, M.S., Padmanabhan, R. and Peter, C.T. (1963) Studies on the life cycle of *Tetrameres mohtedai* Bhale Rao and Rao, 1944. *Indian Veterinary Journal* 40, 7–15.

Supakorndej, P., McCall, J.W. and Jun, J.J. (1994) Early migration and development of *Dirofilaria immitis* in the ferret, *Mustela putorius furo*. *Journal of Parasitology* 80, 237–244.

Supperer, R. (1952) Über das Vorkommen der Filarie (s.l.) *Onchocerca gutturosa* Neumann in Rindern in Österreich und ihre Entwicklung in der Kriebelmücke *Odagmia ornata* Mg. *Wiener Tierärztliche Monatsschrift* 39, 173–179.

Supperer, R. (1953) Filariosen der Pferde in Österreich. *Wiener Tierärztliche Monatsschrift* 40, 193–220.

Swales, W.E. (1936) *Tetrameres crami* Swales, 1933, a nematode parasite of ducks in Canada. Morphological and biological studies. *Canadian Journal of Research, D*, 14, 151–164.

Symes, C.B. (1960) A note on *Dirofilaria immitis* and its vectors in Fiji. *Journal of Helminthology* 34, 39–42.

Tada, I., Sato, A. and Nagano, K. (1969) On the larval *Gnathostoma doloresi* found in snakes, *Trimeresurus flavoviridis flavoviridis*, from Amami-oshima Is., Kagoshima, Japan. *Japanese Journal of Parasitology* 18, 289–293.

Takaoka, H. (1983) Entomological aspects of onchocerciasis in Guatemala. *Actas del seminario para CAICET, Puerto Ayacucho, Territoria Federal Amazonas, 13–15 Oct. 1982*, pp. 155–159.

Takaoka, H., Suzuki, H., Noda, S., Onofre Ochoa, A.J. and Tada, I. (1984a) The intake, migration and development of *Onchocerca volvulus* microfilariae in *Simulium haematopotum* in Guatemala. *Japanese Journal of Sanitary Zoology* 35, 121–127.

Takaoka, H., Suzuki, H., Noda, S., Tada, I., Basanez, M.G. and Yarzabal, L. (1984b) Development of *Onchocerca volvulus* larvae in *Simulium pintoi* in the Amazonas region of Venezuela. *American Journal of Tropical Medicine and Hygiene* 33, 414–419.

Taylor, A.E.R. (1959) *Dirofilaria magnilarvatum*, Price, 1959 (Nematoda: Filarioidea) from *Macaca irus* Cuvier. II. Microscopical studies on the microfilariae. *Journal of Parasitology* 45, 505–509.

Taylor, A.E.R. (1960) The development of *Dirofilaria immitis* in the mosquito *Aedes aegypti*. *Journal of Helminthology* 34, 27–38.

Theiler, A. (1919) A new nematode in fowls, having a termite as an intermediary host (*Filaria gallinarum* (nova species)). *Fifth and Sixth Report of the Director of Veterinary Research, Department of Agriculture, Union South Africa*, pp. 695–707.

Théodoridès, J. (1952) Le coléoptère scarabéide hôte intermédiaire naturel de *Spirocerca lupi* (Rud.) [= *S. sanguinolenta* (Rud.)], en Chine, n'est pas un *Canthon* mais un *Paragymnopleurus*. *Annales de Parasitologie Humaine et Comparée* 27, 571–572.

Tidwell, M.A., Tidwell, M.A. and Munoz de Hoyos, P. (1980) Development of *Mansonella ozzardi* in a black fly species of the *Simulium sanguineum* group from eastern Vaupés, Colombia. *American Journal of Tropical Medicine and Hygiene* 29, 1209–1214.

Titche, A.R., Prestwood, A.K. and Hibler, C.P. (1979) Experimental infection of white-tailed deer with *Elaeophora schneideri*. *Journal of Wildlife Diseases* 15, 273–280.

Traoré-Lamizana, M. and Lemasson, J.J. (1987) Participation in a feasibility study for onchocerciasis control in the Logone Basin area. Distribution of *Simulium damnosum* complex species in the Cameroonian area of the project. *Cahiers ORSTOM, Entomologie Médicale et Parasitologie* 25, 171–186.

Trifonov, T. (1962) The biology of *Simondsia paradoxa* Cobbold, 1864. Part I. *Izvestiya na Tsentralniya Veterinaren Institut za Zarazni i Parazitni Bolesti, Sofia* 5, 151–153. (In Bulgarian.)

Tromba, F.G. (1952) *Myotis lucifugus lucifugus*; a new aberrant host for the third stage larva of *Physocephalus sexalatus* (Molin, 1860). *Journal of Parasitology* 38, 497.

Tsimbalyuk, A.K. and Kulikov, V.V. (1966) The biology of *Skrjabinocerca prima* Schikhobalova, 1930 (Nematoda: Acuariidae) – a parasite of birds. *Zoologicheskii Zhurnal* 45, 1565–1569. (In Russian.)

Tsimbalyuk, E.M., Kulikov, V.V. and Tsimbalyuk, A.K. (1970) A contribution to the biology of *Ascarophis pacificus* (Nematoda, Ascarophididae). *Zoologicheskii Zhurnal* 49, 1874–1875. (In Russian.)

Turner, L.H. and Edeson, J.F.B. (1957) Studies on filariasis in Malaya: the periodicity of the microfilariae of *Wuchereria malayi*. *Annals of Tropical Medical Parasitology* 51, 271–277.

Ueki, T. (1957) Experimental studies on the third-stage larva of *Gnathostoma spinigerum*. *Egaku Kenkyu Fukuoka* 27, 1162–1196.

Uspenskaya, A.V. (1953) Life cycle of the nematodes belonging to the genus *Ascarophis* van Beneden. *Zoologicheskii Zhurnal* 32, 828–832. (In Russian.)

Uspenskaya, A.V. (1954) The parasite fauna of deep water Crustacea in East Murmansk. *Trudi Problemnykh i Tematicheskikh Soveshchanii Zoologicheskii Institut, Akademiya Nauk SSSR* 4, 123–127. (In Russian.)

Uzmann, J.R. (1967) Juvenile *Ascarophis* (Nematoda: Spiruroidea) in the American lobster: *Homarus americanus*. *Journal of Parasitology* 53, 218.

Vakarenko, E.G. (1994) On the occurrence of paratenic parasitism in nematodes of the suborder Filariata. *Vestnik Zoologii* 94, 78–80.

Vanamail, P., Subramanian, S., Das, P.K., Pani, S.P. and Rajagopalan, P.K. (1990) Estimation of fecundic life span of *Wuchereria bancrofti* from longitudinal study of human infection in an endemic area of Pondcherry (south India). *Indian Journal of Medical Research, Section A Infectious Diseases* 91, 293–297.

Vankan, D.M. and Copeman, D.B. (1988) Reproduction in female *Onchocerca gibsoni*. *Annals of Tropical Medicine and Parasitology* (Suppl.) 39, 469–471.

Varma, A.K., Sahai, B.N., Singh, S.P., Lakra, P. and Shrivastava, V.K. (1971) On *Setaria digitata*, its specific characters, incidence and development in *Aedes vittatus* and *Armigeres obturbans* in India with a note on its ectopic occurrence. *Zeitschrift für Parasitenkunde* 36, 62–72.

Varma, S., Malik, P.D. and Lal, S.S. (1976) White mice as reservoirs of swine stomach worm *Ascarops strongylina*. *Haryana Agricultural University Journal of Research* 6, 246–247.

Velikanov, V.P. (1988) The life cycle of *Procyrnea zorillae* (Nematoda, Spirurina). *Parazitologiya* 22, 408–416.

Velikanov, V.P. (1992) Life-cycle of *Thubunea baylisi* (Nematoda: Spirurina) *Parazitologiya* 26, 436–440.

Velikanov, V.P. and Sharpilo, V.P. (1984) Experimental identification of *Physaloptera praeputialis* and *Pterygodermatites cahirensis* larvae (Nematoda, Spirurata) from paratenic hosts. *Vestnik Zoologii* 6, 25–29. (In Russian.)

Vilagiova, I. (1967) Results of experimental studies on the development of preinvasive stages of worms of the genus *Thelazia* Bosc., 1819 (Spirurata: Nematoda), parasitic in the eye of cattle. *Folia Parasitologica* 14, 275–280.

Vincent, A.L., Frommes, S.P. and Ash, L.R. (1979) Ultrastructure of the rectum of infective stage *Wuchereria bancrofti* (Nematoda: Filarioidea). *Journal of Parasitology* 65, 246–252.

Vojtkova, L. (1971) Beitrag zur Kenntnis der Helminthfauna der Wasserwirbellosen III. Cestoda, Nematoda, Acanthocephala. *Věstník Československe Společnosti Zoologické* 35, 146–155.

Votava, C.L., Rabalais, F.C. and Ashley, D.C. (1974) Transmission of *Dipetalonema viteae* by hyperparasitism in *Ornithodorus tartakovskyi*. *Journal of Parasitology* 60, 479.

Vuong, P.N., Spratt, D., Wanjii, S., Aimard, L. and Bain, O. (1993) Onchocerca-like lesions induced by the filarioid nematode *Cercopithifilaria johnstoni* in its natural hosts and in the laboratory rat. *Annales de Parasitologie Humaine et Comparée* 68, 176–181.

Wahl, G. and Bain, O. (1995) Development by injection in *Simulium damnosum* s.l. of two *Onchocerca* species from the wart hog to infective larvae resembling type D larvae (Duke, 1967). *Parasite* 2, 55–62.

Wahl, G. and Schibel, J.M. (1998) *Onchocerca ochengi*: morphological identification of the L₃ in wild *Simulium damnosum* s.l. verified by DNA probes. *Parasitology* 116, 337–348.

Wahl, G., Ekale, D. and Schmitz, A. (1998a) *Onchocerca ochengi*: assessment of the simulian vectors in North Cameroon. *Parasitology* 116, 327–336.

Wahl, G., Enyong, P., Ngosso, J.M., Schibel, R., Moyou, F., Tubbesing, H., Ekale, D. and Renz, A. (1998b) *Onchocerca ochengi* epidemiological evidence of cross-protection against *Onchocerca volvulus* in man. *Parasitology* 116, 349–362.

Wang, P., Sun, Y. and Zhao, Y. (1976) On the development of *Gnathostoma hispidum* in the intermediate host with special reference to its transmission route in pigs. *Acta Zoologica Sinica* 22, 45–52. (In Chinese.)

Wanji, S., Cabaret, J., Gantier, J.C., Bonnand, N. and Bain, O. (1990) The fate of the filaria *Monanema martini* in two rodent hosts: recovery rate, migration, and localization. *Annales de Parasitologie Humaine et Comparée* 65, 80–88.

Wanji, S., Gantier, J.C., Petit, G., Rapp, J. and Bain, O. (1994) *Monanema martini* in its murid hosts: microfiladermia related to infective larvae and adult filariae. *Tropical Medicine and Parasitology* 45, 107–111.

Webber, W.A.F. (1954a) The reproductive system of *Litomosoides carinii*, a filarial parasite of the cotton rat. I. Development of gonads and initial insemination. *Annals of Tropical Medicine and Parasitology* 48, 367–374.

Webber, W.A.F. (1954b) The reproductive system of *Litomosoides carinii*, a filarial parasite of the cotton rat. II. The frequency of insemination. *Annals of Tropical Medicine and Parasitology* 48, 375–381.

Webber, W.A.F. (1955) *Dirofilaria aethiops* Webber, 1955, a filarial parasite of monkeys. III. The larval development in mosquitoes. *Parasitology* 45, 388–400.

Webber, W.A.F. and Hawking, F. (1955) Experimental maintenance of *Dirofilaria repens* and *D. immitis* in dogs. *Experimental Parasitology* 4, 143–164.

Wee, S.H., Hwan, J., HooDon, J., Joo, H.D., Kang, Y.B. and Lee, C.G. (1996) *Setaria mashalli* infection in neonatal calves. *Korean Journal of Parasitology* 34, 207–210.

Wehr, E.E. and Herman, C.M. (1956) *Lophortofilaria californiensis* n.g. n.sp. (Filarioidea, Dipetalonematidae) from California quail, *Lophortyx californicus*, with notes on its microfilaria. *Journal of Parasitology* 42, 42–44.

Weinmann, C.J. and Shoho, C. (1975) Abdominal worm infection in newborn deer (Filarioidea: Setariidae). *Journal of Parasitology* 61(2), 317.

Weinmann, C.J., Anderson, J.R., Longhurst, W.M. and Connolly, G. (1973) Filarial worms of Columbian black-tailed deer in California. I. Observations in the vertebrate host. *Journal of Wildlife Diseases* 9, 213–220.

Weinmann, C.J., Murphy, K., Anderson, J.R., DeMartini, J.C., Longhurst, W.M. and Connolly, G. (1979) Seasonal prevalence, pathology, and transmission of the quail heartworm, *Splendidofilaria californiensis* (Wehr and Herman, 1956), in northern California (Nematoda: Filarioidea). *Canadian Journal of Zoology* 57, 1871–1877.

Weller, T.H. (1938) Description of *Rhabdochona ovifilamenta* n.sp. (Nematoda: Thelaziidae) with a note on the life history. *Journal of Parasitology* 24, 403–408.

Wenk, P. (1967) Der Invasionweg der metazyklischen Larven von *Litomosoides carinii* Chandler, 1931 (Filariidae). *Zeitschrift für Parasitenkunde* 28, 240–263.

Wenk, P. (1990) Filariasis discovery (Letter to Editor). *Parasitology Today* 6, 153.

Wenk, P. and Heimburger, L. (1967) Infectionsversuche mit der Baumwollrattenfilarie *Litomosoides carinii* Chandler, 1931 (Filariidae) bei inadäquaten Wirten. *Zeitschrift für Parasitenkunde* 29, 282–298.

Wenk, P. and Mössinger, J. (1991) Recovery of adult stages and microfilaraemia after low dose inoculation of third stage larvae of *Litomosoides carinii* in *Sigmodon hispidus*. *Journal of Helminthology* 65, 219–225.

Wenk, P., Kellermann, E. and Hafner, C. (1994) Turnover of microfilariae in small mammals. 2. Disintegration of microfilariae (*Acanthocheilonema viteae*) (Filarioidea: Nematoda) after intravenous injection into the jiird, *Meriones unguiculatus. Parasitology* 109, 201–207.

Wertheim, G. (1962) A study of *Mastophorus muris* (Gmelin, 1790) (Nematoda: Spiruridae). *Transactions of the American Microscopical Society* 81, 274–279.

Wharton, R.H. (1959) *Dirofilaria magnilarvatum* Price (1959) (Nematoda: Filarioidea) from *Macaca irus* Cuvier. IV. Notes on larval development in *Mansonioides* mosquitoes. *Journal of Parasitology* 45, 513–518.

Wharton, R.H. (1962) The biology of *Mansonia* mosquitoes in relation to the transmission of filariasis in Malaya. *Institute for Medical Research, Kuala Lumpur, Bulletin* No. 11, 114 pp.

Wharton, R.H., Edeson, J.F.B., Wilson, T. and Reid, J.A. (1958) Studies on filariasis in Malaya: pilot experiment in the control of filariasis due to *Wuchereria malayi* in East Pahang. *Annals of Tropical Medicine and Parasitology* 52, 191–205.

Widmer, E.A. (1967) Helminth parasites of the prairie rattlesnake, *Crotalus viridis* Rafinesque, 1818, in Weld County, Colorado. *Journal of Parasitology* 53, 362–363.

Widmer, E.A. (1970) Development of third-stage *Physaloptera* larvae from *Crotalus viridis* Rafinesque, 1818 in cats with notes on pathology of the larvae in the reptile (Nematoda, Spiruroidea). *Journal of Wildlife Diseases* 6, 89–93.

Williams, P. (1960) Studies on Ethiopian *Chrysops* as possible vectors of loiasis. II. *Chrysops silacea* Austen and human loiasis. *Annals of Tropical Medicine and Parasitology* 54, 439–459.

Williams, R.W. (1948) Studies on the life cycle of *Litomosoides carinii*, filariid parasite of the cotton rat, *Sigmodon hispidus litoralis. Journal of Parasitology* 34, 24–43.

Williams, R.W. and Brown, H.W. (1945) The development of *Litomosoides carinii* filariid parasite of the cotton rat in the tropical rat mite. *Science* 102, 482–483.

Williams, R.W. and Brown, H.W. (1946) The transmission of *Litomosoides carinii* filariid parasite of the cotton rat by the tropical rat mite *Liponyssus bacoti. Science* 103, 224.

Wilson, T., Edeson, J.F.B., Wharton, R.H., Reid, J.A., Turner, L.H. and Laing, A.B.G. (1958) The occurrence of two forms of *Wuchereria malayi* in man. *Transactions of the Royal Society of Tropical Medicine and Hygiene* 52, 480–481.

Winkhardt, H.J. (1979) Untersuchungen über den Enwicklungszyklus von *Dipetalonema rugosicauda* (syn. *Wehrdikmansia.*) (Nematoda: Filarioidea). I. Experimentelle Infektion der Zecke *Ixodes ricinus* mit Mikrofilarien durch künstliche Fütterung. *Tropenmedizin und Parasitologie* 30, 455–462.

Winkhardt, H.J. (1980) Untersuchungen über den Entwicklungszyklus von *Dipetalonema rugosicauda* (syn. *Wehrdikmansia rugosicauda*) (Nematoda: Filarioidea). II. Die Entwicklung von *Dipetalonema rugosicauda* im Zwischenwirt *Ixodes ricinus* und Untersuchungen über das Vorkommen der Mikrofilarien im Reh (*Capreolus capreolus*). *Tropenmedizin und Parasitologie* 31, 21–30.

Witenberg, G.G. (1928) Reptilienals Zwischenwirte parasitischer Würmer von Katze und Hund. *Tierärztliche Rundschau* 34, 603.

Witenberg, G. and Gerichter, C. (1944) The morphology and life history of *Foleyella duboisi* with remarks on allied filariids of amphibians. *Journal of Parasitology* 30, 245–256.

Wong, P.L. and Anderson, R.C. (1982) The transmission and development of *Cosmocephalus obvelatus* (Nematoda: Acuarioidea) of gulls (Laridae). *Canadian Journal of Zoology* 60, 1426–1440.

Wong, P.L. and Anderson, R.C. (1987a) Development of *Syncuaria squamata* (Linstow, 1883) (Nematoda: Acuarioidea) in ostracods (Ostracoda) and double-crested cormorants (*Phalacrocorax auritus auritus*). *Canadian Journal of Zoology* 65, 2524–2531.

Wong, P.L. and Anderson, R.C. (1987b) New and described species of *Skrjabinoclava* Sobolev, 1943 (Nematoda: Acuarioidea) of the proventriculus of nearctic waders (Aves:

Charadriiformes) with a review of the genus and a key to species. *Canadian Journal of Zoology* 65, 2760–2779.

Wong, P.L. and Anderson, R.C. (1988) Transmission of *Skrjabinoclava morrisoni* Wong and Anderson, 1988 (Nematoda: Acuarioidea) to semipalmated sandpipers (*Calidris pusilla* (L.)) (Charadriiformes: Scolopacidae). *Canadian Journal of Zoology* 66, 2265–2269.

Wong, P.L. and Anderson, R.C. (1990a) Host and geographic distribution of *Skrjabinoclava* spp. (Nematoda: Acuarioidea) in nearctic shorebirds (Aves: Charadriiformes) and evidence for transmission in marine habitats in staging and wintering areas. *Canadian Journal of Zoology* 68, 2539–2552.

Wong, P.L. and Anderson, R.C. (1990b) *Ancyracanthopsis winegardi* n.sp. (Nematoda: Acuarioidea) from *Pluvialis squatarola* (Aves: Charadriidae) and *Ancyracanthopsis heardi* n.sp. from *Rallus longirostris* (Aves: Rallidae), and a review of the genus. *Canadian Journal of Zoology* 68, 1297–1306.

Wong, P.L. and Anderson, R.C. (1993) New and described acuarioids (Nematoda: Acuarioidea) from Icelandic shorebirds (Charadriiformes). *Systematic Parasitology* 25, 187–202.

Wong, P.L. and Lankester, M.W. (1984) Revision of the genus *Schistorophus* Raillet, 1916 (Nematoda: Acuarioidea). *Canadian Journal of Zoology* 62, 2527–2540.

Wong, P.L., Bush, A.O. and Anderson, R.C. (1987) Redescription of *Skrjabinocerca prima* Shikhobalova, 1930 (Nematoda: Acuarioidea) from the American Avocet (*Recurvirostra americana* Gmelin). *Canadian Journal of Zoology* 65, 1569–1573.

Wong, P.L., Watson, T. and Anderson, R.C. (1980) *Vigisospirura potekhina* (Petrov and Potekhina, 1953) (Nematoda: Spiruroidea) from the bobcat, *Lynx rufus* (Schreber), in the southeastern USA. *Canadian Journal of Zoology* 58, 1612–1616.

Wong, P.L., Anderson, R.C. and Bartlett, C.M. (1989) Development of *Skrjabinoclava inornatae* (Nematoda: Acuarioidea) in fiddler crabs (*Uca* spp.) (Crustacea) and western willets (*Catoptrophorus semipalmatus inornatus*) (Aves: Scolopacidae). *Canadian Journal of Zoology* 67, 2893–2901.

Worley, D.E., Anderson, C.K. and Greer, K.R. (1972) Elaeophorosis in moose from Montana. *Journal of Wildlife Diseases* 8, 242–244.

Xie, H., Bain, O. and Williams, S.A. (1994) Molecular phylogenetic studies on *Brugia filariae* using HHA 1 repeat sequences. *Parasite* 1, 255–260.

Yamaguchi, T., Nishimoto, M. and Murakami, K. (1956) Studies on *Gnathostoma* in Shikoku. *Schikoku Acta Medica* 9, 78–88.

Yamaguti, S. and Nishimura, H. (1944) One nematode and two trematode larvae from *Caridina denticulata* de Haan. *Hokuoka Acta Medica* 37, 411. (In Japanese.)

Yap, E.H., Ho, B.C., Singh, M. and Kang, K.L. (1975) Studies on the Malayan forest rat filaria *Breinlia booliati*: periodicity and microfilaraemic patterns during the course of infection. *Journal of Helminthology* 49, 263–269.

Yates, J.A. and Lowrie, R.C. (1984) Development of *Yatesia hydrochoerus* (Nematoda: Filarioidea) to the infective stage in ixodid ticks. *Proceedings of the Helminthological Society of Washington* 51, 187–190.

Yates, J.A., Lowrie, R.C. and Eberhard, M.L. (1982) Development of *Tetrapetalonema llewellyni* to the infective stage in *Culicoides hollensis*. *Journal of Parasitology* 68, 293–296.

Yeh, L.S. (1959) A revision of the nematode genus *Setaria* Viborg, 1795, its host–parasite relationship, speciation and evolution. *Journal of Helminthology* 33, 1–98.

Yen, P.K.F., Zaman, V. and Mak, J.W. (1982) Identification of some common infective filarid larvae in Malaysia. *Journal of Helminthology* 56, 69–80.

Yoshida, S. (1934) Observations on *Gnathostoma spinigerum* Owen, 1836, cause of esophageal tumour of the Japanese mink *Lutreola itatsi itatsi* (Temminck, 1844) with special reference to its life history. *Japanese Journal of Zoology* 6, 113–122.

Yue, M.Y. and Jordan, H.E. (1986) Studies of the life cycle of *Pterygodermatites nycticebi* (Monnig, 1920) Quentin 1969. *Journal of Parasitology* 72, 788–790.

Zago Filho, H. (1957) Contribuição para o conhecimento de hospedeiros intermediarios e definitivos da *Physaloptera praeputialis* Linstow, 1889 (Nematoda, Spiruroidea). *Revista Brasileira de Biologia* 17, 513–520.

Zago Filho, H. (1958) Contribuição para o conhecimento de hospedeiros intermediarios e definitivos da *Turgida turgida* (Rud., 1819) Travassos, 1920 (Nematoda, Spiruroidea). *Revista Brasileira de Biologia* 18, 41–46.

Zago Filho, H. (1962a) Contribuição para o conhecimento do ciclo evolutivo da *Physaloptera praeputialis* von Linstow, 1889 (Nematoda: Spiruroidea). *Arquivos de Zoologia do Estado de Sao Paulo (Years 1958–1962)* 11, 59–98.

Zago Filho, H. (1962b) Contribuição para o conhecimento do ciclo evolutivo da *Turgida turgida* (Rud., 1819) Trav., 1920 (Nematoda: Spiruroidea). *Arquivos de Zoologia do Estado de Sao Paulo (Years 1958–1962)* 11, 99–119.

Zago Filho, H. and Pereira Barretto, M. (1962) Contribuicão para o conhecimento do ciclo evolutivo da *Tetrameres confusa* Trâv., 1917 (Nematoda: Spiruroidea). *Papeis Avulsos do Departmento de Zoologia, São Paulo* 15, 111–122.

Zahedi, M. (1990a) The development of *Brugia pahangi* and *B. malayi* in *Armigeres subalbatus*. *Malaysian Journal of Medical Laboratory Sciences* 7, 5–11.

Zahedi, M. (1990b) Laboratory studies on the biology and colonization of *Armigeres subalbatus* (Coquillett, 1898) and some preliminary observations on their susceptibility to *Brugia pahangi* and *B. malayi*. *Malaysian Journal of Medical Laboratory Sciences* 7, 34–43.

Zahedi, M. (1991) Histological observation on the development of *Brugia malayi* and *B. pahangi* in *Armigeres subalbatus*. *Tropical Biomedicine* 8, 7–15.

Zahedi, M. (1994) The fate of *Brugia pahangi* microfilariae in *Armigeres subalbatus* during the first 48 h post ingestion. *Tropical Medicine and Parasitology* 45, 33–35.

Zahedi, M., Jeffery, J. and Abdullah, W.O. (1992a) The migration and distribution of *Brugia malayi* microfilariae in a susceptible strain of *Armigeres subalbatus*. *Mosquito Borne Diseases Bulletin* 9, 81–85.

Zahedi, M., Omar, A.W. and Jeffery, J. (1992b) Variation in peripheral microfilarial density influences microfilarial uptake by mosquitoes. *Journal of Bioscience* 3, 91–96.

Zahedi, M., Denham, D.A. and Ham, P.J. (1993) Exsheathment of *Brugia pahangi* microfilariae in *Armigeres subalbatus*. *Tropical Biomedicine* 10, 100–106.

Zajicek, D. and Pav, J. (1972) The intermediate hosts of *Ascarops strongylina* (Rud.) and *Physocephalus sexalatus* (Molin) in Bohemia. *Folia Parasitologica* 19, 121–127.

Zaman, V. (1987) Ultrastructure of *Brugia malayi* egg shell and its comparison with microfilarial sheath. *Parasitology Research* 73, 281–283.

Zaman, V. and Chellappah, W.T. (1968) Studies on vector susceptibility and larval morphology of a filaria of the slow loris, *Nycticebus coucang*. *Annals of Tropical Medicine and Parasitology* 62, 450–454.

Part II
Class Adenophorea

Chapter 8

Order Enoplida – Suborder Dioctophymina

8.1

The Superfamily Dioctophymatoidea

The four genera of the superfamily are divided into two morphologically distinct families (Anderson and Bain, 1982). In the Soboliphymatidae the anterior end is modified into a globular sucker (*Soboliphyme*) which enables the worm to remain attached to the gut wall. Members of the family are found mainly in mustelids and insectivores. The cephalic end of the Dioctophymatidae is normal. Some members of this family (*Eustrongylides*, *Hystrichis*) occur in large tumours in the stomach wall of piscivorous birds; one species (*Dioctophyme renale*) occurs in the kidney, mainly of mustelids and canids. All members of the superfamily produce thick-shelled eggs, often with roughened surfaces. One or both poles of the egg shells have clear areas commonly referred to as opercula. Eggs are passed from the host in a one- or two-cell stage. The egg embryonates in moist conditions and forms a characteristic first-stage larva. The latter is short with a rounded or pointed tail and a prominent dorylaimid-type stylet but is otherwise poorly differentiated. Fully embryonated eggs can survive for prolonged periods in moist conditions. Eggs hatch in the intestine of terrestrial or aquatic oligochaetes. First-stage larvae invade the tissues of the oligochaete and develop into third-stage larvae, which may be infective to the definitive host. Paratenesis is important in transmission of species found in piscivorous hosts. In one genus (*Eustrongylides*) precocious development to the fourth stage takes place in a fish host before the parasites are infective to the definitive host.

Karmanova (1968) has published an important monograph on the biology of the dioctophymatoids and reviewed a great deal of research conducted in the CIS.

Family Soboliphymatidae

Soboliphyme

The family contains one genus (*Soboliphyme*). Members are medium-sized nematodes found in the stomach, and less commonly the intestine, of mustelids and insectivores, including shrews and moles. The parasites use the anterior sucker to attach themselves firmly to the gut wall. The transmission and development of only one species, *S. baturini* (in mustelids) have been investigated; it develops to the infective stage in terrestrial oligochaetes. Presumably the species in shrews (i.e. *S. soricis* and *S. jamesoni*) and moles (i.e. *S. hirudiniformis* and *S. caucasica*) also develop in earthworms and are acquired by their hosts through the ingestion of the intermediate host. Mustelids may acquire *S. baturini* from eating oligochaetes but the possibility of paratenesis has not been ruled out (Karmanova, 1968).

S. baturini Petrov, 1930

This is a parasite mainly of holarctic terrestrial mustelids (*Martes* spp. and *Mustela* spp.) although it apparently occurs sporadically in canids and felids, usually in the larval stage. Eggs are barrel-shaped and the surface of the shell is roughened by semi-circular depressions. There are clear areas (opercula) at each pole. Karmanova (1963) elucidated the development and transmission of this helminth. Eggs passed from the host were unsegmented; they were fully larvated in 28–30 days at 17–19°C in moist soil. At this time, larvae were active inside the eggs. This was followed by a period of dormancy, during which eggs retained their viability for over 3 years. The first-stage larva was stout with a short pointed tail. The tissues were packed with granules and the larva apparently was poorly differentiated. Eggs hatched in the intestine of hydrophilic soil oligochaetes of the family Enchytraidae. Larvae penetrated the intestine of the oligochaete and entered the coelom, where they became surrounded by a transparent capsule attached to the outer wall of the intestine. In the capsule, larvae grew and moulted twice to attain the infective third stage. The first moult took place 8–11 days and the second 20–60 days postinfection. Third-stage larvae, which remained in the capsule, were 1.5–2.2 mm in length; they had the characteristic oral sucker found in adults and were, therefore, readily identifiable to genus. The genital primordia were in the form of simple rudimentary tubes.

Mink (*Mustela vison*) readily consumed infected oligochaetes. Larvae attached to the gut wall of the mink and developed to adulthood. The first moult occurred 19 days and the second about 28–30 days postinfection. Females apparently began to lay eggs soon after the final moult and continued to do so for about 20 months. Karmanova (1968) suggested that mink and marten acquire infections from eating terrestrial oligochaetes but she did not rule out the possibility that rodents and shrews might serve as paratenic hosts.

Family Dioctophymatidae

Dioctophyme

D. renale (Goeze, 1782)

The giant kidney worm is essentially a parasite of the kidney of wild mustelids, especially mink (*Mustela vison*) and less commonly wild canids (especially *Canis lupus*) in temperate regions of the world (Fig. 8.1).

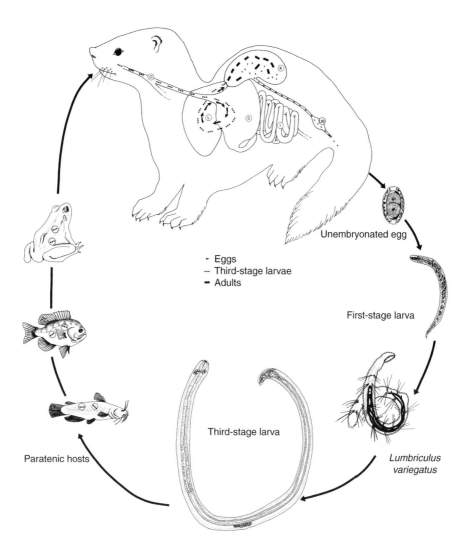

Unembryonated egg

- Eggs
··· Third-stage larvae
- Adults

First-stage larva

Third-stage larva

Paratenic hosts

Lumbriculus variegatus

Fig. 8.1. Development of transmission of the giant kidney worm, *Dioctophyme renale*. L = liver; S = stomach; K = kidney; I = intestine; UB = urinary bladder. (Original, U.R. Strelive.)

The parasite has been reported (Karmanova, 1968; Mace, 1974, 1976b) in other wild hosts, including mustelids (*Lutra* spp., *Martes* spp., *Mustela itatsi, Galictis vittata*), canids (*Chrysocyon brachyurus, Vulpes fulva, Speothos venaticus*), bears (*Ursus ursus*) and raccoons (*Procyon lotor, Nasua nasua*). It has also been reported rarely in rats and seals. Most reports of dioctophymatosis are in dogs living in enzootic areas and being fed fish (Mace, 1974). There are sporadic reports of the parasite in humans and some domesticated animals.

In mink, which is regarded as the usual host in North America (Mace and Anderson, 1975), the parasite is almost invariably found only in the right kidney (rarely, both kidneys may be infected). The left kidney is hypertrophied in infected mink and the right kidney is modified into an enlarged capsule containing the worms and usually a bony staghorn-like structure (McNeill, 1948; Mace, 1976b; Measures and Anderson, 1985). The capsule is filled with fluid which, in mature infections, contains huge numbers of eggs which are, along with fluid, ingested by the parasite. The opening to the ureter remains open and eggs in fluid pass readily to the bladder and are excreted. In other hosts, such as wolves, dogs, martens and otters, the worms are sometimes found free in the body cavity. Female worms in such locations are generally unfertilized and they oviposit enormous numbers of unfertilized eggs, which contribute to an extensive peritonitis. Also, in wolves and dogs both kidneys are sometimes infected.

Eggs passed in the urine of the host have thick, pitted shells with a plug-like, clear area at each pole (Mace and Anderson, 1975). Fertilized eggs were in the two-cell stage in the uterus of the female worm and when passed in the urine of the final host. First-stage larvae developed in eggs incubated in water between 14 and 30°C and were active for about 1 week after development; the rate of development varied according to temperature (Mace and Anderson, 1975). At 20°C larvae were fully developed in about 35 days.

Karmanova (1959a, 1960, 1962) and Mace and Anderson (1975) reported that eggs hatched in, and larvae invaded, the tissues of the aquatic oligochaete *Lumbriculus variegatus*, in which development to the third stage took place. Eggs hatched in some other species of oligochaetes but larvae did not invade the tissue of these hosts. Thus, *L. variegatus* is the only known intermediate host of *D. renale*. The report of Woodhead (1950) that branchiobdellid ectoparasitic oligochaetes found on crayfish were intermediate hosts of kidney worm is unfounded; the larvae he described in branchiobdellids were horsehair worms (Gordiacea). The first-stage larva of *D. renale* is short and poorly differentiated but its stylet is clearly defined. The tail is tapered and ends in a point. Larvae invaded the coelom of *L. variegatus* and remained there for about 9 days (Mace and Anderson, 1975). By 10 days larvae were found in the ventral blood vessel. There was little growth or development for the first 30 days. This was followed by rapid growth leading to the first moult, which occurred about 50 days postinfection. The second-stage larva retained the first-stage cuticle and was about 2 mm in length. The first moult was followed by little growth for about 20 days and there was rapid growth leading to the second moult, which took place 100 days postinfection. Third-stage larvae, still in the ventral blood vessel of *L. variegatus*, were large (about 6–12 mm in length) in relation to the size of the oligochaete and usually only one or two fully developed larvae could be sustained by the host. Larvae were red in colour and retained cuticles of the first and second moults. The genital primordia were transparent but sufficiently well developed to determine that females were monodelphic and opisthodelphic and males were diorchic and opisthorchic.

Third-stage larvae have been found encapsulated in tissues of various species of fish, including *Ictalurus melas*, *I. nebulosus* and *Lepomis gibbosus* in North America (Woodhead, 1950; Hallberg, 1953; Mace and Anderson, 1975; Measures and Anderson, 1985), *Leuciscus idus* in Romania (Ciurea, 1921) and *Esox lucius*, *Perca fluviatilis*, *Aspius aspius*, *Gambusia affinis*, *Rutilus rutilus*, *Chalcaeburnus chalcoides*, *Pelecus cultratus*, *Barbus branchicephalus*, *Gobio gobio*, *Alburus taeniatus* and *Pseudoscaphirynchus kaufmanni* in various other parts of the world (Karmanova, 1961, 1968). Karmanova (1968) and Mace and Anderson (1975) also found larvae in frogs (*Rana ridibunda* in Russia and *R. catesbeiana*, *R. clamitans* and *R. septentrionalis* in Canada) and the latter authors also infected frogs by feeding them oligochaetes with larvae. Larvae were generally found encapsulated on the stomach and intestinal serosa and on the mesentery. Usually one to 12 larvae were found per host. They must be distinguished from third-stage larvae of *Eustrongylides*, which are similar in appearance. Measures and Anderson (1985) found larvae encapsulated in the hypaxial musculature of *L. gibbosus* and used them successfully to infect ferrets.

Larvae do not grow in paratenic hosts (cf. *Eustrongylides* spp.) and they remain in the third stage. It is possible, therefore, to infect susceptible definitive hosts with larvae directly from oligochaetes (Karmanova, 1968; Mace and Anderson, 1975) or from fish and frogs (Ciurea, 1921; Woodhead, 1950; Karmanova, 1968; Mace and Anderson, 1975; Measures and Anderson, 1985). Karmanova (1968) and Measures and Anderson (1985) reported that dogs, mink and ferrets (*Mustela putorius*) vomited violently 30 min after they were given larvae. Vomition continued for as long as 1 h and was presumably related to larval penetration of the stomach. Mace and Anderson (1975) and Mace (1976a) found moulting larvae in the stomach wall of mink 5 days postinfection. They also found lesions in the liver of a mink 50 days postinfection and this led them to believe that larvae spend some time in this organ before invading the kidney. They noted the proximity of the right lobe of the liver to both the stomach and the right kidney and wondered if this had anything to do with the fact worms almost invariably invade the right kidney of mink and ferrets. However, they reported a larva in the body cavity of the same mink, which indicated that other factors may be involved in directing worms to the right kidney. Nevertheless, worms apparently become disoriented rather frequently in hosts other than mink and ferrets and are not uncommonly found in the abdominal cavity.

Karmanova (1968) reported a moult at 30 days and one at 2 months. She gave the prepatent period in dogs as about 135 days. Mace and Anderson (1975) found eggs in the urine of a mink infected 154 days earlier. Measures and Anderson (1985) found mature worms in the right kidney of two ferrets examined 108 and 134 days postinfection.

Eustrongylides

The genus contains about 11 nominal species which inhabit large tumours in the wall of the proventriculus of piscivorous birds. Measures (1988a) believed that only three species should be considered valid, namely *E. ignotus*, *E. excisus* and *E. tubifex*.

Development and transmission of the latter two have been studied (Karmanova, 1965, 1968; Measures, 1988b,c,d). Tumours containing worms communicate with the lumen of the proventriculus by means of small apertures through which the nematodes project their tails and heads. Within tumours, which bulge prominently into the body cavity, the parasites are generally encased in fibrous tubes. Eggs released by female worms are thick-shelled with a pitted surface and unsegmented. Eggs embryonate in the usual way in water in the external environment until each contains a small, poorly differentiated, stylet-bearing larva with a bluntly pointed tail bent to one side of the body. Eggs hatch in aquatic oligochaetes and first-stage larvae invade the ventral blood vessel, where development takes place to the third stage. Development is prolonged but in about 2–4 months larvae are about 9–10 mm in length and highly conspicuous by their bright red colour and relatively large size. Fish which consume oligochaetes serve as paratenic hosts. Larvae become encapsulated mainly on the mesentery and the intestinal serosa. In these locations they develop to the fourth stage, in which the reproductive system is highly developed. Piscivorous birds acquire their infections through the ingestion of fish containing fourth-stage larvae. Because of the precocious development of the ingested larvae, worms form tumours and mature in the final host in a very short time and eggs are produced in large numbers as early as 10 days. Worms live for only a few weeks after they have produced eggs and tumours are gradually resolved. Species of *Eustrongylides* are uniquely adapted for transmission to migratory water birds which spend only short periods of time in specific locations, where they acquire fresh infections from ingesting fish. It is necessary for the parasite to mature quickly in order to contaminate the area with eggs before the departure of the birds to either their nesting or wintering areas.

E. excisus Jagerskiold, 1909

This is a parasite of Pelecaniformes, Ciconiiformes and Anseriformes in Europe, southeast Asia, the Middle East and Australia (Measures, 1988a). Karmanova (1968) reviewed the contributions of Ciurea (1921) and various CIS authors to our understanding of the development and transmission of *E. excisus*. Karmanova (1965) showed that fully developed larvae appeared in eggs in 21 days at 26–37°C. Eggs hatched and invaded the coelom of the aquatic oligochaetes *Lumbriculus variegatus*, *Tubifex tubifex* and *Limnodrilus* sp. Larvae eventually invaded the abdominal blood vessels and developed to the third stage (which were 5.0–6.2 mm in length) in 60–70 days. Third- and fourth-stage larvae have been reported in *Perca fluviatilis*, *Aspius aspius*, *Silurus glanis*, *Esox lucius*, *Lucioperca volgensis*, *L. lucioperca*, *Acerina cernua*, *Chalcalburnus chalcoides*, *Bentophilus macrocephalus*, *Huso huso*, *Acipenser ruthenus*, *A. guldenstadti*, *Leucisus idus* and *Barubus brachycephalus*, as well as amphibians (*Rana ridibunda*) and grass snakes (*Natrix tesselata*). Karmanova (1968) pointed out that perch (*P. fluviatilis*) were the most important source of *E. excisus* in cormorants.

In the final host, fourth-stage larvae acquired from the ingestion of fish invaded the wall of the proventriculus and in 7–8 days underwent the final moult; in 12–14 days females commenced egg laying. Karmanova showed that cormorants returning to the Volga Delta in spring were uninfected and acquired their infections in the Delta. The parasites were mature by the second half of June. Young birds were infected early in their lives. Shedding of eggs into the environment started about the end of June and extended throughout the summer. In late summer and autumn, cormorants migrated to the Caspian Sea and infections in the birds become senescent and eventually disappeared.

E. ignotus Jagerskiold, 1909

E. ignotus is a common parasite of the proventriculus of herons and egrets in Florida where it occurs mainly in the great blue heron (*Ardea herodius*) and egrets (*Casmerodius albus* and *Egretta thula*) but has been found in other ardeids as well (Measures, 1988a; Spalding *et al.*, 1993). Tubificid oligochaetes are presumably the first intermediate hosts (Lichtenfels and Stroup, 1985). Various fish species have been found serving as paratenic hosts in Florida, namely *Lepisosteus platyrhincus, Fundulus chrysotus, F. confluentus, Gambusia holbrooki, Heterandria formosa, Poecillia latipinna, Lepomis gulosus, L. macrochirus, L. microlophus* and *L. punctatus*. Larvae were usually coiled within spherical capsules attached to the mesentery. In small fish, the worm caused distension of the stomach wall, resulting in abnormal swimming (Spalding *et al.*, 1993; Frederick *et al.*, 1996).

I. ignotus is highly pathogenic in the young of herons and egrets (Spalding, 1990; Spalding and Forrester, 1993; Spalding *et al.*, 1993, 1994).

E. tubifex (Nitzsch in Rudolphi, 1819)

This parasite has been reported in mergansers and other Anseriformes, Gaviiformes, Ciconiiformes and Podicipediformes in Europe, CIS, Brazil and North America (Measures, 1988a). In Canada it is most commonly a parasite of the common merganser (*Mergus merganser*) (Fig. 8.2). Measures (1988b) cultured eggs of *E. tubifex* from common mergansers. Eggs contained motile first-stage larvae in 31–33 days at 22°C and in 23 days at 25°C. Larvated eggs retained their infectivity for at least 2.5 years when stored at 4°C. Eggs hatched and larvae developed in the ventral blood vessel of the oligochaetes *Limnodrilus hoffmeisteri* and *Tubifex tubifex*. The first moult took place in less than 30 days and the second in 70–109 days postinfection at 27.5°C. Third-stage larvae were 9.6–10.1 mm in length; the genital primordia were well developed and worms remained in the shed cuticle of the second stage. During the early stages of development in oligochaetes, larvae were found anywhere in the ventral blood vessel; but as early as 30 days postinfection they moved anteriorly, especially near the gonads, where the more advanced larvae were always found.

Measures (1988c) found third- and fourth-stage larvae encapsulated in various fish (*Lepomis gibbosus, Ambloplites rupestris* and *Perca flavescens*) in Guelph Lake, Ontario which is visited by migrating mergansers for short periods in spring and autumn. Overall prevalence in the three fish species was 12.9% and the overall mean intensity was 1.8. Larvae occurred only in *L. gibbosus* and *A. rupestris* that were at least 9.0 and 8.1 cm in length, respectively. Male and female third-stage larvae in the fish were 11.5–34.8 mm in length. Fourth-stage larvae were considerably larger, i.e. males 32.2–66.0 mm and females 45.1–83.4 mm. Fourth-stage larvae retained the shed cuticle of the third moult and the genital primordia were highly developed. Rosinski *et al.* (1997) reported larvae in yellow perch (*Perca flavescens*) in Lake Huron.

Measures (1988d) gave fourth-stage larvae from fish to various laboratory-raised birds. The experiments confirmed field data that common and red-breasted mergansers (*Mergus serrator*) were important hosts of *E. tubifex* in Ontario. Fourth-stage larvae invaded the stomach wall of birds and the fourth and final moult took place in 2 days. Worms developed in the tunica muscularis and rapidly elicited large, raised, oval tumours. The females produced eggs 10–17 days postinfection and then degenerated. Tumours resolved rapidly and the proventriculus returned to normal in about 30

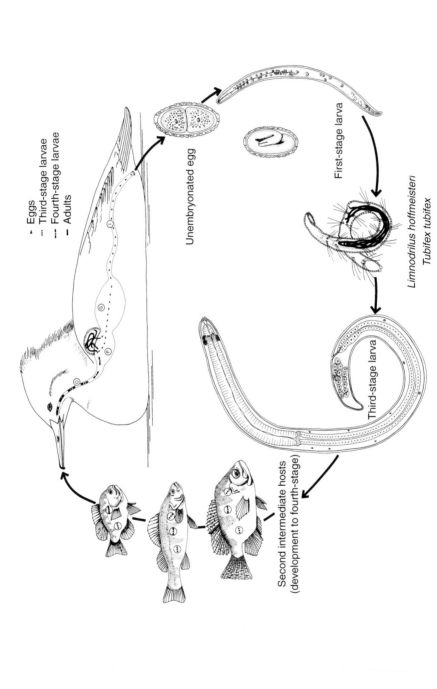

Fig. 8.2. Development and transmission of *Eustrongylides tubifex* to the common merganser. O = oesophagus; P = proventriculus; G = gizzard; I = intestine. (Illustrations of fish adapted from *Freshwater Fishes of Canada*, W.B. Scott and E.J. Crossman, 1998, Fisheries Research Board of Canada. Original, U.R. Strelive.)

days. In addition to the two above-mentioned mergansers, Measures (1988d) also experimentally infected hooded mergansers (*Mergus cucullatus*), a great blue heron (*Ardea herodias*), double-crested cormorants (*Phalacrocorax auritus*) and domestic ducks (*Anas platyrhynchos*).

In summarizing her data, Measures (1988d) concluded that worms developed to an advanced stage in fish and development to adulthood in the avian host was correspondingly rapid. Thus, mergansers visiting Guelph Lake for about 3–4 weeks during the spring and also during the autumn migrations acquired infections and shed eggs before they departed. *E. tubifex* is, therefore, highly adapted to hosts which remain in enzootic areas for only brief periods of time. Transmission is probably dependent on high populations of suitable oligochaete intermediate hosts and organically enriched or polluted bodies of water can be important places for epizootics.

Hystrichis

Species of *Hystrichis* are found in small to medium-sized tumours in the proventriculus of anatids and Ciconiiformes. These nematodes are readily recognized to genus because they have a dilated cephalic end and the anterior part of the body has numerous regularly arranged spines, particularly numerous in the head region. The development and transmission of one species have been well studied in the CIS by Karmanova (1956, 1959b, 1968).

H. tricolor Dujardin, 1845

This is a common parasite mainly of waterfowl in Europe, including many regions of the CIS. It occurs in dabbling ducks such as mallards (*Anas platyrhynchos*), teal (*Anas crecca*) and shovelers (*Anas clypeata*) as well as species which are piscivorous (*Mergus serrator*). It has been reported in geese, coots, egrets and some waders (Karmanova, 1968). The cephalic and caudal ends of the parasites project into the lumen of the proventriculus; the oval, unsegmented, thick-shelled eggs with pitted surfaces are passed into the lumen of the gut and out in the faeces of the host. In water at 22–30°C, eggs embryonated and active larvae appeared 21 days later. After 7 days, larvae ceased to move and in 28–30 days they had reached the infective stage. Karmanova (1956, 1959b) and other Soviet authors demonstrated that lumbricoid oligochaetes (*Criodrilus lacuum, Allolobophora dubiosa, Eophilia leoni*) living in shallow water or moist shoreline soil served as intermediate hosts of *H. tricolor*. Eggs hatched in the oligochaetes and the first-stage larvae migrated to the abdominal blood vessel, where development to the infective stage occurred. Development in oligochaetes was prolonged and, according to Karmanova, the first moult did not take place until 5.5 months postinfection, when larvae were about 4.9 mm in length. The third stage (29–31 mm in length) is enormous relative to the size of the oligochaete host and is readily recognized to genus by the presence of the characteristic body spines. The sexes are distinguishable by the precociously developed genital primordia.

According to CIS authors, ducks readily consume oligochaetes which are especially abundant in reservoirs known as important foci of infection. Infective larvae invaded the wall of the proventriculus of the final host and moulted 3–4 days postinfection (Gumenshchikova, 1963). The final moult occurred 8–9 days postinfection and ovipositing began in 12–13 days. Egg-laying continued for 30–55 days. In experimental infections most worms died 85 days postinfection. Karmanova (1968) estimated that 1 month was essential for development of the eggs in the environment and 6 months for development in the oligochaetes, and that the worms would live about 3 months in the definitive host. According to Karmanova (1968), larvae of *H. tricolor* can survive winter in the oligochaetes. Also, larvated eggs can survive for prolonged periods in the environment, where they are available to the intermediate hosts. Heavy adult infections occurred in domestic ducks in one area in Georgia, CIS, from April to May and from August to October. The parasite is considered a significant pathogen in some areas. According to Karmanova (1968), the highest prevalence of infected oligochaetes occurred in lakes where large flocks of migrating waterfowl spent time.

References (Dioctophymina)

Anderson, R.C. and Bain, O. (1982) No. 9. Keys to genera of the superfamilies Rhabditoidea, Dioctophymatoidea, Trichinelloidea, and Muspiceoidea. In: Anderson, R.C., Chabaud, A.G. and Willmott, S. (eds) *CIH Keys to the Nematode Parasites of Vertebrates*. Commonwealth Agricultural Bureaux, Farnham Royal, UK, pp. 1–26.

Ciurea, J. (1921) Sur la source d'infestation par l'eustrongyle géant (*Eustrongylus gigas* Rud.). *Compte Rendu de la Société de Biologie, Paris* 85, 532–534.

Frederick, P.C., McGehee, S.M. and Spalding, M.G. (1996) Prevalence of *Eustrongylides ignotus* in mosquito fish (*Gambusia holbrooki*) in Florida: historical and regional comparisons. *Journal of Wildlife Diseases* 32, 552–555.

Gumenshchikova, V.P. (1963) Dynamics of pathological and morphological changes during experimental hystrichosis of ducks. *Trudi Vsesoyuznogo Instituta Gelmintologii im Akademii K.I. Skrjabini* 10, 126–141. (In Russian.)

Hallberg, C.W. (1953) *Dioctophyme renale* (Goeze, 1782). A study of the migration routes to the kidneys of mammals and resultant pathology. *Transactions of the American Microscopical Society* 72, 351–363.

Karmanova, E.M. (1956) An interpretation of the biological cycle of the nematode *Hystrichis tricolor* Dujardin, 1845, a parasite of domestic and wild ducks. *Doklady Akademii Nauk SSSR* 111, 245–247. (In Russian.)

Karmanova, E.M. (1959a) The life cycle of the nematode, *Dioctophyme renale*. *Doklady Akademii Nauk SSSR* 127, 1317–1319. (In Russian.)

Karmanova, E.M. (1959b) Biology of the nematode *Hystrichis tricolor* Dujardin, 1845 and some data on the epizootiology of hystrichosis in ducks. *Trudi Gelmintologicheskoi Laboratorii Akademiya Nauk SSSR* 9, 113–125. (In Russian.)

Karmanova, E.M. (1960) The life cycle of the nematode, *Dioctophyme renale* (Goeze, 1782) parasitic in the kidneys of carnivorous animals and man. *Doklady Akademii Nauk SSSR* 132, 1219–1221. (In Russian.)

Karmanova, E.M. (1961) First case of the larval nematode, *Dioctophyme renale* (Goeze, 1782) found in fish in the USSR. *Trudi Gelmintologicheskoi Laboratorii Akademiya Nauk SSSR.* 11, 118–121. (In Russian.)

Karmanova, E.M. (1962) The life cycle of *Dioctophyme renale* in its intermediate and definitive hosts. *Trudi Gelmintologicheskoi Laboratorii Akademiya Nauk SSSR* 12, 27–36. (In Russian.)

Karmanova, E.M. (1963) Development of *Soboliphyme baturini* Petrov, 1930 (Nematoda, Dioctophymata) in the body of an intermediate host. *Gel'minty Cheloveka, Zhivotnykh i Rastenii i Bor'ba s Nimi. Izd-vo Akademii Nauk SSSR*, pp. 241–243. (In Russian.)

Karmanova, E.M. (1965) Detection of intermediate hosts of the nematode *Eustrongylides excisus*, parasite of waterfowl. *Trudi Gelmintologicheskoi Laboratorii Akademiya Nauk SSSR* 15, 86–87. (In Russian.)

Karmanova, E.M. (1968) *Dioctophymidea of Animals and Man and Diseases Caused by Them. Fundamentals of Nematology*, Vol. 20, 381 pp. (Translated by Amerind Publishing, New Delhi, 1985.)

Lichtenfels, R. and Stroup, C.F. (1985) *Eustrongylides* sp. (Nematoda: Dioctophymatoidea): first report of an invertebrate host (Oligochaeta: Tubificidae) in North America. *Proceedings of the Helminthological Society of Washington* 52, 320–323.

Mace, T.F. (1974) Studies on the biology of the giant kidney worm, *Dioctophyme renale* (Goeze, 1782) (Nematoda: Dioctophymatoidea). Unpublished PhD thesis, University of Guelph, Guelph, Ontario.

Mace, T.F. (1976a) Lesions in mink (*Mustela vison*) infected with giant kidney worm (*Dioctophyme renale*). *Journal of Wildlife Diseases* 12, 88–92.

Mace, T.F. (1976b) Bibliography of giant kidney worm *Dioctophyme renale* (Goeze, 1782). *Wildlife Diseases* (microfiche), pp. 36. (WD-76-1.)

Mace, T.F. and Anderson, R.C. (1975) Development of the giant kidney worm, *Dioctophyme renale* (Goeze, 1782) (Nematoda: Dioctophymatoidea). *Canadian Journal of Zoology* 53, 1552–1568.

McNeill, C.W. (1948) Pathological changes in the kidney of mink due to infection with *Dioctophyme renale* (Goeze, 1782), the giant kidney worm of mammals. *Transactions of the American Microscopical Society* 67, 257–261.

Measures, L.N. (1988a) Revision of the genus *Eustrongylides* Jagerskiold 1909 (Nematoda: Dioctophymatoidea) of piscivorous birds. *Canadian Journal of Zoology* 66, 885–895.

Measures, L.N. (1988b) The development of *Eustrongylides tubifex* (Nematoda: Dioctophymatoidea) in oligochaetes. *Journal of Parasitology* 74, 294–304.

Measures, L.N. (1988c) Epizootiology, pathology, and description of *Eustrongylides tubifex* (Nematoda: Dioctophymatoidea) in fish. *Canadian Journal of Zoology* 66, 2212–2222.

Measures, L.N. (1988d) The development and pathogenesis of *Eustrongylides tubifex* (Nematoda: Dioctophymatoidea) in piscivorous birds. *Canadian Journal of Zoology* 66, 2223–2232.

Measures, L.N. and Anderson, R.C. (1985) Centrarchid fish as paratenic hosts of the giant kidney worm, *Dioctophyme renale* (Goeze, 1782). *Journal of Wildlife Diseases* 21, 11–19.

Rosinski, J.L., Muzzall, P.M. and Haas, R.C. (1997) Nematodes of yellow perch from Saginaw Bay, Lake Huron, with emphasis on *Eustrongylides tubifex* (Dioctophymatidae) and *Philometra cylindracea* (Philometridae). *Journal of the Helminthological Society of Washington* 64, 96–101.

Spalding, M.G. (1990) Antemortem diagnosis of eustrongylidosis in wading birds (Ciconiiformes). *Colonial Waterbirds* 13, 75–77.

Spalding, M.G. and Forrester, D.J. (1993) Pathogenesis of *Eustrongylides ignotus* (Nematoda: Dioctophymatoidea) in ciconiiformes. *Journal of Wildlife Diseases* 29, 250–260.

Spalding, M.G., Bancroft, G.T. and Forrester, D.J. (1993) The epizootiology of eustrongylidosis in wading birds (Ciconiiformes) in Florida. *Journal of Wildlife Diseases* 29, 237–249.

Spalding, M.G., Smith, J.P. and Forrester, D.J. (1994) Natural and experimental infections of *Eustrongyloides ignotus*: effect on growth and survival of nestling wading birds. *The Auk* 111, 328–336.

Woodhead, A.E. (1950) Life-history cycle of the giant kidney worm, *Dioctophyme renale* (Nematoda), of man and many other mammals. *Transactions of the American Microscopical Society* 69, 21–46.

Chapter 9

Order Enoplida – Suborder Trichinellina

9.1

The Superfamily Trichinelloidea

The structure of the oesophagus distinguishes trichinelloids from other Nematoda. It consists of a short muscular anterior part and a long glandular posterior portion. The latter consists of a narrow tube composed of myofilaments around cuticle-secreting epithelial cells. Attached to the tube are one to three rows of large, naked gland cells (stichocytes), each of which communicates by a single pore with the cuticle-lined lumen of the oesophagus (Wu, 1955; Wright *et al.*, 1985). The entire glandular structure is known as a stichosome. The anterior part of the body generally contains only the stichosome and is usually much narrower than the posterior part of the body containing the reproductive organs; the extreme condition is found in the whipworms (*Trichuris*).

The Trichinelloidea is divided into three families (Anderson and Bain, 1982). The Trichuridae includes: (i) the Trichurinae (whipworms) of the large intestine of mammals; (ii) the Capillariinae, which are extremely common in the gut, urinary tract (especially the bladder), respiratory system and, less commonly, the liver, spleen and skin of all vertebrate groups (but especially birds and mammals); and (iii) the Trichosomoidinae, which includes a few highly specialized species associated with the buccal, nasal, gut and bladder mucosal epithelium of mammals. *Trichinella* spp. are placed in their own family, the Trichinellidae. The little understood *Cystoopsis* of sturgeons and *Dioctowittus* of reptiles are placed in their own family, the Cystoopsidae.

Trichinelloids occur in a wide range of hosts. Although a few species occupy spleen or liver tissue (e.g. *Calodium hepaticum* syn. *Capillaria hepatica*), the vast majority are associated with epithelial surfaces, as Wright (1989) pointed out in an excellent review. Species of the Trichurinae bury only the anterior (stichosome) part of the body in the caecal epithelium whereas members of the Capillariinae and Trichosomoidinae generally are found totally embedded in tunnels in epithelium or, less commonly, in liver or spleen tissue.

Eggs are generally thick-shelled with polar, plug-like structures (sometimes with filaments) giving them a lemon or barrel shape. Eggs are generally oviposited and passed from the host in an unembryonated state (e.g. *Trichuris*, *Capillaria*) but in other species the eggs embryonate *in utero* (*Trichinella*, Trichosomoidinae). First-stage larvae, as well as some other stages, including adults, have a stylet which relates the group, along with the Dioctophymatoidea, to soil-dwelling dorylaimid ancestors. Trichinelloids always infect the final host in the first stage even if a sojourn in an intermediate host (e.g. earthworms) is necessary for larvae to become infective. Even larvae of *Trichinella spiralis*, which grow and develop markedly in the muscles of the intermediate host, remain in the first stage. In all the trichinelloids which have been well investigated, there are the usual four moults leading to the reproducing adults.

The development and transmission of *Trichinella spiralis* (*sensu lato*) and the whipworms (*Trichuris* spp.) have been investigated intensively because of their medical importance and the ease with which they can be maintained in the laboratory. The development and transmission of a number of species in the Capillariinae have also been studied, as well as some of the trichosomoidines.

Family Trichuridae

Subfamily Trichurinae

Trichuris

The some 60–70 species of whipworms found in the large intestine (caecum and colon) of mammals (primates, pigs, sheep, goats, deer, rodents, lagomorphs, African antelopes, opossums, shrews, felids and foxes) are among the most readily recognized of nematodes (Skrjabin *et al.*, 1957). The anterior (stichosome) part of the body is long, narrow, tapered and whip-like. The posterior part of the body, containing the ample reproductive organs and their products, is broad and handle-like. *In situ* the anterior end of the worm is buried in the mucosal epithelium of the gut of the host and the posterior part is free in the lumen. Lee and Wright (1978) have shown convincingly that the cephalic end of the *Trichuris muris* induces the formation of a syncytium of epithelial origin. This syncytium is apparently used as food by the parasite while it threads its way through the epithelium. Presumably these observations pertain to all members of the genus.

Eggs of *Trichuris* spp. are barrel-shaped with thick, smooth shells and have an elevated, plug-like operculum at each pole. Davaine, Leuckart and Railliet in the latter half of the 19th century first showed that eggs embryonated outside the body of the host and were directly infective to the final host (Grove, 1990).

More recent studies have provided information on the rate of development and longevity of eggs under various conditions and development in the final host. It can probably be assumed that all species behave similarly. Eggs hatch in the small intestine of the host and larvae migrate rapidly to the large intestine, where they invade the mucosal epithelium and develop. There are undoubtedly four moults during development to the adult stage.

T. muris Schrank, 1788

This is a common cosmopolitan parasite of the large intestine of mice, rats and other rodents of the genera *Arvicola*, *Arvicanthis*, *Citellus*, *Cricetulus*, *Echinomys*, *Epimys*, *Georychus*, *Holochilus*, *Loncheres*, *Meriones*, *Mesocricetus*, *Microtus*, *Mus*, *Opodemus*, *Rattus* and *Trychomys* (see Skrjabin *et al.*, 1957). Eggs are passed from the host unsegmented. Fusion of pronuclei took place in 24–30 h at 30°C (Shikhobalova, 1937); in 16–20 days active larvae were present in the eggs. Eggs given to mice appeared in the colon and caecum 2.5 h later and some had hatched. Shikhobalova (1937) reported two moults, at 20–21 days and at 31 days postinfection. Efremov and Shikhobalova (1939) reported that 10-day-old larvae in mice invaded the mucosal glands, mainly of the caecum. The anterior oesophageal region of adult worms (40–55 days postinfection) had penetrated into the mucosa and the posterior end of the parasite was free in the lumen. Worms commenced laying eggs 36–43 days postinfection. The life span of the parasite in the host was 76–100 days.

Fahmy (1954) reported that at 25–26°C eggs were fully embryonated in 30–31 days. He reported two moults in the caecum of mice at 11 and 23 days postinfection. Wakelin (1969) incubated eggs at 18°C ± 2°C and obtained fully developed larvae in 43 days in 75% of incubated eggs, although he believed that 56–63 days were required for eggs to become infective to mice. He orally inoculated mice with larvated eggs and reported that they hatched in the small intestine within 60 min and that some larvae had reached the caecum at this time. He observed three moults at 9, 10 and 11 days.

Shmytova (1971) reported that eggs took 20–30 days to embryonate fully at 27–28°C. In mice, mass hatching of eggs occurred in the intestine 10–20 cm behind the stomach within 30 min of ingestion; in 18–20 h most larvae had reached the caecum. The nematodes moulted four times during the 29–30-day period of development to adulthood.

Panesar (1981) injected eggs directly into the caeca of mice and observed that hatched larvae entered gland openings before penetrating into the epithelium. Panesar and Croll (1980) reported that larvae established themselves in the epithelium of the ileum in mice in which the caecum had been removed. Larvae transferred from the caecum of one mouse to another mouse at 5 and 10 days posthatching, established themselves in the host, but larvae transferred at later times failed to do so, presumably because of the inability of older worms to re-establish the association with the epithelium. Panesar (1989) reported the first moult in 9–11 days postinfection and the second in 20–21 days, at which time the sexes could be recognized. The third moult took place in 25 days in males and 28–30 days in females. The fourth moult occurred on day 28 in males and 32–34 days in females. The worms were quite active during moulting. The stylet remained extruded in many of the worms and may help to break the moulted cuticle before it is shed.

T. ovis (Abildgaard, 1795)

This is a common cosmopolitan parasite of the caecum and colon of sheep, cattle, camels, cervids, giraffes, antelope and porcupines (Skrjabin *et al.*, 1957). Leuckart (1876) was the first to transmit a species of *Trichuris*: he gave larvated eggs of *T. ovis* to sheep and 16 days later found many immature worms in their caeca. Artyukh (1936) reported that eggs developed completely in water in 16 days at 30°C and gave the prepatent period in sheep as 48–50 days. Pukhov (1939) reported that eggs contained motile larvae in 24–25 days at 25–30°C but claimed that eggs were not infective until 50 days. He gave the prepatent period as 46–52 days.

According to Thapar and Singh (1954) first-stage larvae remained viable in eggs for 24 weeks. Larvae were 125–150 μm in length. Deo (1960) gave the length of first-stage larvae as 134–160 μm. He gave eggs to kids and lambs and recovered worms from their caeca at various times later. The prepatent period was said to be 85–135 days.

None of the above authors reported moults during development in the final host.

T. skrjabini (Baskakov, 1924)

This is a parasite of the caecum and colon of camels, sheep, goats, cattle, deer, gazelles and ibex in the CIS (Skrjabin *et al.*, 1957). Magomedbekov (1953) reported that at 30°C eggs reached the infective stage in 41–42 days. In faeces at 25–30°C, eggs reached the infective stage in 40–45 days. In sheep, the prepatent period was 42–47 days and the life span was 125–180 days. In the Dagestan Republic, sheep were mainly infected from March to May and from September to October. Thus, adult worms appeared in sheep in two periods, June–July and November–December.

T. suis (Schrank, 1788)

This is a cosmopolitan nematode of the caecum and colon of pigs. Alicata (1935) reported that eggs embryonated in 210 days at ambient temperatures of 6–24°C and eggs remained alive for several years in moist conditions. Beer (1973a,b) determined that the optimum temperature for development of the eggs was 34°C, at which the infective larval stage was reached in 19 days. Hill (1957) reported that eggs survived in soil for 6 years.

Powers *et al.* (1960) infected piglets and reported that larvae which hatched from eggs mainly entered the wall of the caecum and colon. They were said to remain there for 3–10 days and to return to the lumen when mature. The prepatent period was 41–45 days and the life span in the final host was 4–5 months.

Beer (1973a,b) reported that *T. suis* hatched within 9 h in the distal region of the small intestine and in the caecum and colon. During hatching, both polar plugs were dissolved and the larva used its stylet (5–7 μm in length) to pierce the vitelline membrane in order to escape from the egg shell. Larvae were 136–207 μm in length and in 18 h had penetrated the epithelium lining the crypts of Lieberkühn. The first moult occurred 10 days and the second 16 days postinfection. The third moult occurred 20 days and the fourth 32–37 days postinfection. Eggs appeared in faeces in 41 days.

T. trichiura (Linnaeus, 1771)

T. trichiura is a parasite of the caecum and colon of humans and other primates. It is regarded as the second most common parasitic infection in humans in the tropics and it also occurs elsewhere (Bundy and Cooper, 1989). Ooi *et al.* (1993) described specimens from *Macaca fuscata* and *Papio papio* as *T. trichiura*. They noted that the species in pigs (*T. suis*) is morphologically distinct from *T. trichiura* in that the latter has a pair of pericloacal papillae and a cluster of small papillae near the caudal end not found in the latter species.

The discovery of the typical direct transmission of the parasite was claimed by Calandruccio, who ingested embryonated eggs and found eggs in his stool 27 days later.

According to Grove (1990), Grassi attempted to take credit for the discovery. It is highly improbable, however, that Calandruccio's observations were accurate because the prepatent period of *T. trichiura* is never likely to be as short as 27 days. Presumably both Italian workers were aware of Leuckart's (1876) prior discovery of the transmission of *T. ovis* in sheep.

According to Fülleborn (1923) eggs became fully embryonated in 3.5 weeks at 26°C but only remained infective for 4–4.5 weeks at this temperature. Hasegawa (1924) reported that eggs were fully developed in 28 days at 28–30°C. Spindler (1929) reported the rate of development as 21 days at 30°C.

Artemenko (1937) claimed that the most rapid development was 14 days at 37°C. Dinnik and Dinnik (1937, 1939) gave the following rates of development at various temperatures: 120 days at 20°C; 57 days at 25°C; 17.5 days at 30°C; and 11 days at 35°C.

Little is known of the development of *T. trichiura* in humans but it is undoubtedly similar to that of other species in mammals. Fülleborn (1923) gave eggs to guinea pigs and rabbits and reported that larvae reached the caecum in the former in 24 h and in the latter in 48 h.

The prepatent period of *T. trichiura* in humans has apparently not been accurately determined but is probably 2–3 months (Faust *et al.*, 1975; Bundy and Cooper, 1989). Estimates of the number of eggs each female produces per day vary from 3000 to 20,000 (Faust *et al.*, 1975). The average parasitic life span of the worms in humans is not known but may be 1–2 years and individual worms may live as long as 4 years (Bundy and Cooper, 1989).

For a review of the epidemiological significance of *T. trichiura* in human health, refer to Bundy and Cooper (1989).

T. vulpis (Froelich, 1789)

This is a cosmopolitan parasite of the large intestine, particularly the caecum, of dogs, foxes and polecats (*Mustela putorius*). According to Rubin (1954) eggs reached the infective stage in 25–26 days at 19–25°C or 9–10 days at 33–38°C. Eggs hatched in the intestine of dogs, especially in the duodenum and the upper half of the jejunum (Miller, 1947). Larvae migrated into the crypts of Lieberkühn and then invaded the mucosa of the large intestine. The prepatent period was given as 70–107 days, which is much longer than the 30–41 days reported earlier by Tochihara (1922), Hasegawa (1924) and Fernan-Nunez (1927). According to Whitney (1938) the longevity of *T. vulpis* does not exceed 5 months.

Subfamily Capillariinae

The classification of the Capillariinae is one of the most difficult in the Nematoda. There are some 300 described species in *Capillaria* s.l. parasitizing a wide range of hosts from fishes to mammals and birds. There have been numerous attempts to define genera in this mass of species but none of them was well received or seemed biologically sound. Skrjabin *et al.* (1957), in the most comprehensive account of the Capillariinae, recognized five genera, namely *Capillaria*, *Hepaticola*, *Thominx*, *Skrjabinocapillaria* and *Eucoleus*. Anderson and Bain (1982), noting the confusion in the generic classification of the subfamily, placed four genera in Skrjabin *et al.* (1957) into synonymy with *Capillaria* along with 16 other genera which had been proposed up to 1978. They expressed dissatisfaction with the characters which had been used to define genera (e.g. spicules present or absent, cirrus with or without spines or tubercles).

Moravec (1982) recognized that the subfamily was in a confused state and provisionally proposed 16 genera and five subgenera to which he provided descriptions and a key. Not all species known in the Capillariinae could be assigned to genera by Moravec; nevertheless, we shall follow his system whenever possible, as well as Skrjabin *et al.* (1957) for species not mentioned by Moravec. Unfortunately, such important and familiar species as *Capillaria hepatica*, *C. philippinesis* and *C. aerophilus* are assigned to genera unfamiliar to most readers.

Eggs of species of the Capillariinae living in the epithelium are probably released during the normal turnover of epithelial cells. They pass out in the excretory products of the host, depending on the location of the parasite in the host (i.e. in faeces in species living in the viscera or lungs and in urine in species in the urinary system). Species living in the liver parenchyma do not have a direct outlet for their eggs, which must, therefore, be released into the environment through the agency of carnivorous or scavenging animals. Eggs are usually deposited unembryonated; *Aonchotheca philippinensis* also lays embryonated autoinfective eggs. Under suitable conditions of humidity and temperature, eggs embryonate until each contains a small first-stage larva with a well-defined stylet. In some species (e.g. *Baruscapillaria obsignata*, *Eucoleus contortus*) eggs are directly infective to the final host whereas in other species (e.g. *Aonchotheca caudinflata*, *Eucoleus aerophilus*) eggs must first be ingested by earthworms, in which they hatch. One species (*Aonchotheca putorii*) can be transmitted directly or indirectly through the agency of earthworms. Larvae invade the tissues of earthworms and, although remaining in the first stage, become infective to the final host after a period of modest growth. Heteroxeny in the Capillariinae was first discovered by Wehr (1936) in *Eucoleus annulatus* of fowl.

Aonchotheca

A. caudinflata (Molin, 1858)

This is a cosmopolitan parasite of the small intestine of gallinaceous and anatid birds, including chickens, grouse, pheasants, turkeys, pigeons, ducks and geese. Eggs embryonate in 11–12 days under favourable conditions of temperature and humidity. Allen and Wehr (1942), Morehouse (1944), Wehr and Allen (1945) and Nickel (1953) showed that earthworms (*Allolobophora caliginosa*, *Eisenia foetida*) were necessary inter-mediate hosts. Eggs readily hatched *in vitro* in earthworm digestive juices (Morehouse, 1944). Larvae were infective to the definitive host in about 9 days after they were ingested by earthworms. Worms matured in chickens 22–24 days postinfection.

A. putorii (Rudolphi, 1819)

A. putorii is found in the stomach and small intestine of Mustelidae in western Europe, the CIS and the USA. Skarbilovich (1945) was able to infect sables, marten, mink and cats by giving them fully embryonated eggs. She reported that eggs also hatched in earthworms; in 2–3 days first-stage larvae invaded the body cavity, where they reached the infective stage in 30–38 days. Larvae were located immediately behind the genital girdle of the earthworm. Infected earthworms were given to sable (*Martes zibellina*), marten (*Martes* sp.), mink (*Mustela vison*) and a cat. Sable began to pass eggs 26 days and marten at 32 days postinfection. Mink and the cat failed to pass eggs; original sources of eggs for the experiments were from sable.

A. philippinensis (Chitwood, Valesquez and Salazar, 1968)

This parasite (syn. *Capillaria philippinensis*) was found originally in the small intestine of humans in northern Luzon of the Philippines, where it was associated with a wasting disease characterized by intractable diarrhoea, loss of protein and electrolytes, and malabsorption of sugars and fats (Chitwood *et al.*, 1968; Detels *et al.*, 1969; Fresh *et al.*, 1972; Cross and Bhaibulaya, 1983). The disease was often fatal and involved hundreds of individuals; it has also been reported in Thailand and Japan (Bhaibulaya *et al.*, 1979).

The female worm in humans produced two types of eggs, namely a typical *Capillaria*-type egg with a thick shell and polar plugs, which passed out unembryonated in the host's faeces, and an egg with only a vitelline membrane, which embryonated and hatched *in utero* or in the lumen of the host's gut. The latter larvae were autoinfective, giving rise to a second generation of adult worms.

The mode of transmission to humans has been subject to intense investigation. Cabrera *et al.* (1969) reported eggs in the freshwater fishes *Ambassis commersoni* and *Elotris melanosoma* and first raised the possibility that humans acquired the parasite from eating raw fish harbouring eggs. The latter were noted around households (Cross *et al.*, 1970) and would presumably be common in water bodies near places where people defaecated. However, attempts to infect a variety of vertebrates, including humans, with the embryonated, thick-shelled eggs were never successful (Cross *et al.*, 1970).

Cross *et al.* (1970, 1972) collected eggs from human faeces and fed them to the freshwater fishes *A. commersoni*, *E. melanosoma* and *Hypselotris bipartita* (= 'Apogon sp.'). Larvae hatched from the eggs in the gut of the fish and entered the gut epithelium, where they grew from 130–150 μm to about 250–300 μm in length in 18–21 days. Most larvae occurred in the gastric mucosa. *H. bipartita* was by far the most suitable fish host, since 37% harboured larvae after exposure to eggs. Larvae from fish were given to *Macaca* spp., many of which became infected and passed eggs 22–96 (average 46) days postinfection. The most susceptible monkey was *M. mulatta*, followed by *M. cyclopis* and *M. irus*. At necropsy, all stages of the parasite were found but only in the small intestine. None of the monkeys developed the disease signs noted in humans, however.

Cross *et al.* (1978) gave embryonated eggs of *A. philippinensis* from humans and monkeys to *H. bipartita*. They collected larvae from the fish 3 weeks later or more and orally inoculated them into gerbils (*Meriones unguiculatus*). All gerbils became infected. Larvae developed to adulthood within 10–11 days and females produced larvae 13–14 days postinfection. These larvae developed into second-generation adults in 22–24 days and second-generation females produced eggs which appeared in the faeces of gerbils 24–35 days postinfection. Most female worms were oviparous but a few with larvae were always present. Gerbils died 31–84 (average 46) days postinfection without evidence of diarrhoea or other signs, although they lost weight. At necropsy, 852–5353 worms were recovered in the gerbils, showing the importance of autoinfection, since the animals were inoculated with only 5–100 larvae. Three gerbils became infected when fed *H. bipartita* from a lagoon in the endemic area in Luzon, Philippines.

Bhaibulaya *et al.* (1979) successfully infected six species of freshwater fishes in Thailand with *A. philippinensis*, namely *Aplocheilus panchax, Cyprinus carpio, Gambusia holbrookii, Puntius gonionotus, Rasbora borapetensis* and *Trichopsis vittatus.* Bhaibulaya and Indra-Ngarm (1979) showed that in Thailand the fish-eating birds *Amaurornis phoenicurus* and *Ardeola bacchus* could readily be infected by feeding them the fish *G. holbrookii* infected with eggs of *A. philippinensis* from experimentally infected gerbils. The prepatent period in three *A. phoenicurus* was 22–30 days and all birds became ill. One bird which survived was given challenge infections but was evidently immune. One of the three *A. bacchus* became infected and passed eggs 16 days postinfection; it died 28 days later. Some 19,000 adult and larval *C. philippinensis* were found in this bird.

Cross and Basaca-Sevilla (1983) confirmed the findings of Bhaibulaya and Indra-Ngarm (1979) by successfully infecting fish-eating birds from Taiwan, namely *A. phoenicurus, Bubulcus ibis, Gallinula chloropus, Ixobrychus sinensis* and *Nycticorax nycticorax. A. philippinensis* also developed in pigeons, greater painted snipe (*Rostratula benghalensis*) and a few ducks and chickens. Birds which survived infections could not be reinfected.

There is considerable evidence, therefore, that fish-eating birds are the natural hosts of *A. philippinensis* in the Far East and that freshwater fishes serve as intermediate hosts. Humans and other animals acquire their infections undoubtedly from the consumption of raw fish containing larvae.

Baruscapillaria

B. obsignata (Madsen, 1945)

B. obsignata (syn. *Capillaria columbae*) is found in the mucosa of the small intestine of chickens, partridges (*Perdix perdix*), pigeons and turkeys. Eggs of this species developed to the infective stage in 6–8 days at room temperature (Graybill, 1924; Levine, 1937, 1938; Wehr, 1937, 1939). Eggs were markedly resistant to low temperature and remained infective after 14 days exposure to temperatures of −6.7 to 12.2°C (Wehr, 1939). First-stage larvae were 123–150 µm in length and had a distinct stylet (Wehr, 1939). Eggs were directly infective to pigeons. In the latter, the first moult occurred in the second week of infection; the second moult took place before 14 days and the third 2–3 weeks postinfection. Mature worms were found 19 days after pigeons were given eggs. Eggs appeared in the droppings of the birds 26 days postinfection. Worms lived for at least 7 months.

Wakelin (1965) confirmed many of the earlier findings using chickens as hosts. He reported the prepatent period as usually 20–21 days and that output of eggs ceased 60 days postinfection, although worms (mainly males) persisted in the birds for up to a year.

Calodium

C. hepaticum (Bancroft, 1893)

C. hepaticum (syn. *Hepaticola hepatica; Capillaria hepatica*) is a cosmopolitan parasite of the liver parenchyma mainly of rodents and lagomorphs of the genera *Actomys,*

Apodemus, Arvicanthis, Arvicola, Castor, Citellus, Clethrionomys, Cricetulus, Cynomys, Dasymys, Ellobius, Lemniscomnis, Lepus, Marmota, Microtus, Mus, Myopotamus, Napaeozapus, Ondatra, Oryctolagus, Otomys, Peromyscus, Rattus, Sciurus, Sigmodon, Synaptomys and *Tatera*. It has been reported sporadically in dogs, cats, hyraxes, peccaries and primates, including humans (Lubinsky, 1956; Skrjabin *et al.*, 1957; Freeman and Wright, 1960). Wright (1974) demonstrated that *C. hepaticum* adults were enclosed by multinucleate cytoplasmic masses originating from liver cells and he concluded that they fed on the cytoplasm surrounding the anterior end.

The female worm moves in the liver and deposits small groups of uncleaved eggs, which become encapsulated by host tissue (Luttermoser, 1938a,b; Pavlov, 1955). Eggs may embryonate to the four-cell or, rarely, the eight-cell stage after being deposited (Wright, 1961). Since eggs are encapsulated they cannot normally be released from the living host, a fact noted by Bancroft (1893) and Fülleborn (1924). In addition, Shorb (1931), Luttermoser (1938a), Orlov (1948) and Pavlov (1955) showed that viable eggs rarely resulted from decaying livers. The consensus is that the infected liver must be eaten by animals during cannibalism, predation or scavenging. Unembryonated eggs pass through the gut of disseminator animals and are dispersed into the environment with the faeces. Given the appropriate temperature and humidity, eggs will develop to the infective stage.

Eggs are susceptible to desiccation but in a moist environment at room temperature or 25°C they will develop fully in 35–45 days (Luttermoser, 1938a; Pavlov, 1955; Wright, 1961). Luttermoser (1938a) reported that unembryonated eggs from rats from Baltimore, Maryland, died after exposure to temperatures of −1 to 5°C for 19 days and that temperatures of −7 to 1°C killed most eggs in 16 days. Wright (1961), on the other hand, found that eggs from *Peromyscus maniculatus* from Algonquin Park, Ontario, could withstand temperatures below 0°C for considerable periods (i.e. −15°C for 60 days); however, a temperature of −40°C killed the eggs.

Freeman and Wright (1960) and Wright (1961) gave eggs to mice to follow development of *C. hepaticum*. First-stage larvae (140–190 µm in length) were first found in the liver 2 days postinfection and Wright (1961) suggested that they had travelled by the hepatic portal system. Moulting first-stage larvae were observed 3–4 days postinfection. No moulting second-stage larvae were found but third-stage larvae appeared in 5 days and fourth-stage larvae in 9 days. By day 13 lesions were grossly visible on the liver. Males were considered to have reached the adult stage in 18 days and females in 20 days. In 21 days few eggs appeared in the liver parenchyma. All males died by 40 days but females remained alive for up to 59 days after having deposited groups of eggs.

Pavlov (1955) studied infections in wild rodents in the Voronezh Reserve in the CIS. The common red-backed vole (*Clethrionomys glareolus*) and the striped field mouse (*Apodemus agrarius*), which live in damp floodplain sites, were most often infected. Pavlov (1955) believed that the floodplain habitat was highly suitable for the development of eggs. Highest prevalence in *C. glareolus* and *A. agrarius* was in mid-July (10.1%), leading to the conclusion that eggs become infective by late June.

Freeman and Wright (1960) studied the epizootiology of *C. hepaticum* in a wild population of white-footed mice (*Peromyscus maniculatus*) in Algonquin Park, Ontario. Prevalence varied directly with host density and could not be correlated with predation. Most infections were acquired during winter and it was concluded that the winter nest was the primary focus of infection through cannibalism. Eggs of *C. hepaticum* are long-lived both when embryonated or unembryonated and can probably withstand winter temperatures within the nests of mice.

Shimatani (1961) infected mice and reported that eggs appeared in female worms 23 days postinfection. Some mice became infected after he had injected eggs into the abdominal cavity and the caudal vein, but more liver lesions were associated with abdominal injection of eggs than by intravenous injection.

Eucoleus

E. aerophilus (Creplin, 1839)

E. aerophilus is a cosmopolitan parasite of the mucosa of the respiratory system of carnivores (Canidae, Felidae, Mustelidae). Under optimal conditions of temperature and humidity, eggs will reach the infective stage in 35–45 days and attempts to infect cats and foxes with embryonated eggs were unsuccessful (Christensen, 1938). Borovkova (1947) gave eggs to the earthworms *Allolobophora caliginosa*, *Bimestus tenuis*, *Eisenia foetida*, *Lumbricus rubellus* and *L. terrestris* and examined them 2–50 days later. Larvae were found in all earthworms except *E. foetida*. Borovkova (1947) infected cats, dogs and foxes by feeding them infected earthworms. Young foxes passed eggs 25–26 days postinfection. She suggested that larvae of *E. aerophilus* attained the lungs via the lymph and blood and that the life span of the parasite in foxes was about 10–11 months.

E. annulatus (Molin, 1858)

This is a cosmopolitan parasite of the oesophagus, crop and, rarely, the oral cavity of chickens and wild game birds. Wehr (1936) showed that transmission involved earthworm intermediate hosts (*A. caliginosa* and *E. foetida*). Zucchero (1942) reported that eggs were fully developed in 18 days at 27–28°C and 50 days at 20.5–21°C. Allen (1949) noted that eggs of *E. annulatus* embryonated in tap water in 32 days at room temperature. Unembryonated eggs held for 21 days at either 37°C or 4–6°C did not develop appreciably. Eggs hatched *in vitro* when placed in an extract of intestinal contents of earthworms. The longitudinal muscles were the main sites for larvae in earthworms and the infective stage was reached in 14–21 days in *A. caliginosa* and 21–28 days in *L. terrestris*. Larvae increased in size in earthworms and changed morphologically. The prepatent period in turkeys was 25 days and in chickens 16–26 days.

E. contortus (Creplin, 1839)

This parasite (syn. *Capillaria oesphagealis* Rietschel, 1973) has been reported in the oesophagus, crop and oral cavity of a great variety of birds throughout the world, including Anseriformes, Charadriiformes, Falconiformes, Galliformes and Passeriformes (Cram, 1936). The parasite has been reported in the submucosa as well as in the mucosa. Cram (1936) reported that eggs developed to the infective stage in 27–40 days in tap water at room temperature. The fully embryonated eggs were directly infective to ducks, bobwhite quail (*Colinus virginianus*) and turkeys. The prepatent period was 45–54 days. *E. contortus* has been associated with severe damage to the crop in turkeys (Wehr, 1948).

According to Rietschel (1973) first-stage larvae from *Corvus corone* grew from 180 to 370 mm in the earthworms *L. rubellus* and *L. terrestris*.

E. frugilegi (Czaplinski, 1962)

First-stage larvae of *E. frugilegi* of the oesophagus of Corvidae grew from 180 to 240 μm in length in the earthworm *Allolobophora chlorotica* (see Rietschel, 1973). The prepatent period was 38–45 days in *Corvus corone*.

Pearsonema

P. mucronata (Molin, 1856)

This is a parasite of the urinary bladder of Mustelidae in Europe and the CIS. According to Skarbilovich (1945) earthworms (*L. rubellus*) are necessary for transmission. Eggs embryonated in 30–43 days at summer temperatures and in 20–25 days at 25°C. The prepatent period was 42–45 days. The parasite lived for less than 12–14 months.

P. plica (Rudolphi, 1819)

This is a cosmopolitan parasite of the urinary bladder of canids and mustelids. According to Petrov and Borovkova (1942) eggs embryonated in 20–21 days at 26–28°C. Eggs hatched in *L. rubellus* and *L. terrestris* and larvae invaded the tissues of the earthworms. Infected earthworms were given to foxes and dogs and adult worms were recovered from the urinary bladder in 49–63 days. Enigk (1950) reported that larvae were infective to dogs and foxes within a day after eggs were ingested by earthworms. First-stage larvae invaded the intestinal wall of the final host and developed to the second stage in 8–10 days. These larvae then entered blood vessels and migrated to the bladder, where they attained the third stage in about 30 days and the adult stage in 58–63 days postinfection.

Pseudocapillaria

P. tomentossa (Dujardin, 1843)

Although mainly a parasite of the intestine of Cyprinidae in Europe, Asia and North America, this species (syn. *P. brevispicula*) has also been found in Cobitidae, Perciformes, Gadiformes, Anguilliformes, Esociformes and Siluriformes). It occurs also in aquarium-reared fishes. Eggs passed from the host develop in 7 days at 20–22°C (Moravec, 1983). Lomakin and Trofimenko (1982) infected *Tubifex* sp., *Limnodrilus hoffmeisteri* and *Lumbriculus variegatus*. *Lebistes reticulatus* was infected after being fed oligochaetes allowed to ingest eggs. Guppies also became infected when given larvated eggs.

According to Moravec (1985), fishes (*Cyprinus carpio* and *Tinca tinca*) in the Czech Republic acquire new infections all year round but gravid females occurred only during the warmer period from April to October.

Schulmanela

S. petruschewskii (Shulman, 1948)

S. petruschewskii (syn. *Hepaticola petruschewskii, Hepatospina acerinae*) is a parasite of the liver of a variety of fishes (Salmonidae, Cyprinidae, Cobitidae, Centrarchidae and Percidae) in Europe. As in *Calodium hepaticum*, unembryonated eggs, released in the gut of predators and scavengers, pass into the environment in faeces, where they embryonate (Thieme, 1961, 1964). Kutzer and Otte (1966) infected *Eiseniella tetraedra* but suggested that the real intermediate hosts were tubificids and small crustaceans. The newly hatched larvae were 240–350 µm in length and grew to about 760–1250 µm in the intermediate host. They gave infected oligochaetes to *Oncorhynchus mykiss* and estimated that worms would mature within 6 months. According to Thieme (1964) and Shulman (1948) transmission begins early in spring and adult worms appear by late autumn.

Subfamily Trichosomoidinae

There are only three genera (including *Huffmonella*: see Moravec *et al.*, 1998) and eight species in the subfamily. Only one species has been studied in any detail.

Eggs are typical of the trichurids but are larvated when laid. Eggs of species of *Anatrichosoma* and *Trichosomoides*, deposited in tunnels in epithelium, are presumably released during the normal sloughing of epithelial cells (Orihel, 1970; Pence and Little, 1972). This process is dependent upon little or no local inflammation which might encapsulate the eggs. The eggs pass out of the body of the host in excretory products and are presumably already infective to the final host.

Anatrichosoma

The following species have been assigned to *Anatrichosoma*:

- *A. buccalis* Pence and Little, 1972 in the buccal mucosal epithelium of opossums.
- *A. cutaneum* (Swift, Boots and Miller, 1922) in the nasal mucosal epithelium of *Macaca mulatta*.
- *A. cynamolgi* Smith and Chitwood, 1954 in the nasal mucosal epithelium of *Macaca philippinensis*.
- *A. gerbillis* (Bernard, 1964) in the gastric mucosa of gerbils.
- *A. haycocki* Spratt, 1982 in the paracloacal glands of *Antechinus* spp. (Dasyuridae).
- *A. ocularis* File, 1974 in the squamous epithelium of the sclera, cornea and conjunctiva of the tree shrew (*Tupaia glis*).

An unusual feature of *Anatrichosoma* spp. is that males, although often as long as females, are extremely slender. During copulation the male inserts the posterior part of his body, often up to half its length, into the vagina and uterus of the female (Little and Orihel, 1972).

Trichosomoides

Only two species of *Trichosomoides* have been described, namely *T. nasicola* Biocca and Aurizi, 1961 from the nasal mucosal epithelium of rats and *T. crassicauda* (Bellingham, 1965) from the urinary tract (usually the bladder) of rats (especially *Rattus norvegicus*). Males of species of *Trichosomoides* are much smaller than females and are typically found within the vagina and uterus of females.

T. crassicauda (Bellingham, 1965)

T. crassicauda is associated generally with papillomas in the bladder wall (Thomas, 1924). The route that larvae take to reach the bladder has not been convincingly demonstrated, presumably because of the difficulty in locating a few larvae in the tissue and tissue spaces of the host. Linstow (1874), who first noted the stylet in the first-stage larva, speculated that larvae reached the kidneys and then followed ureters to the bladder. Löwenstein (1910, 1912) was interested in *T. crassicauda* because of the possibility that the parasite would elicit tumours. He concluded that eggs were directly infective to the rat and that larvae reached the urinary tract from the blood. Yokogawa (1920) infected a rat with oral inoculations of larvated eggs over a period of several days and examined it after the final dose on day 4. He found four larvae in the body cavity, two in the pleural cavity and three in the lungs. He suggested that larvae hatched in the gut, invaded the body cavity, then penetrated the diaphragm and attained the lungs.

Thomas (1924) infected rats and examined them later at various times. The first-stage larva was 287 μm in length and had a stylet; the stylet was present also in the adult male but not in the female. Thomas (1924) concluded that eggs hatched in the stomach and that larvae passed into veins and reached the heart by way of the hepatic portal system. Only larvae subsequently reaching the urinary system developed to adulthood. There was little growth of worms during the first 8 days but in 3–6 weeks females began to lay eggs. According to Thomas (1924) copulation can take place in various regions of the urogenital tract and some males may leave females after fertilizing them, although generally they remained within the female.

Hasslinger and Schwarzler (1980) infected 40 rats orally with 50–70 eggs of *T. crassicauda* and killed and examined three each week for larvae. Larvae passed from the gut to the lungs and finally to the urogenital tract via the thoracic and abdominal cavities and also by the blood. Prenatal infection did not occur. The prepatent period was 8–10 weeks. The authors successfully infected *Mastomys natalensis*, which passed eggs in urine 50–62 days later.

Family Trichinellidae

Trichinella

Trichinella spiralis s.l. is probably the most intensively studied of all helminths. Investigations on the parasite reach back to the early half of the 19th century and attracted the attention of such luminaries as Leuckart, Leidy, von Siebold, Zenker, Virchow, Herbst and Küchenmeister as well as James Paget, a freshman medical student, who took muscle samples with larvae to the distinguished plant physiologist

Robert Brown at the British Museum. Brown had the good fortune to own a microscope but when asked if he knew anything about intestinal worms replied, 'No, thank heavens. Nothing whatever.' Nevertheless, he and Paget managed to extract and study the *Trichinella* larvae under the microscope for the first time. For an account of the elucidation of the transmission of *T. spiralis*, refer to Schwartz (1960), Kean *et al.* (1978) and Grove (1990). Important reviews of *Trichinella* and trichinellosis have been published by Gould (1945), Campbell (1983) and Tanner *et al.* (1989).

In recent years understanding of the systematics of *Trichinella* has been considerably complicated by evidence that there exist differences in isolates from animals throughout the world where the parasite is known to occur. There was considerable reluctance to recognize additional species, mainly because of a lack of distinguishing morphological characters and clear evidence of reproductive isolation. For a review of the controversy, refer to Dick (1983a,b), Dick and Chadee (1983), Lichtenfels *et al.* (1983) and Pozio *et al.* (1989).

Pozio *et al.* (1989) and La Rosa *et al.* (1989), in examining isoenzymic patterns in isolates of *Trichinella* from numerous hosts and localities, found seven distinct electrophoretic clusters, four of which accorded with species proposed earlier and three that did not fit any of the others. The genus now contains five recognized species (La Rosa *et al.*, 1989, 1992; Pozio *et al.*, 1989, 1992a,b,c) as follows:

1. *T. spiralis* (Owen) is the common, cosmopolitan and well-studied species found in domesticated pigs and which develops readily in rats and mice. It has six unique allozymes, 13–16 allozyme differences and ribosomal DNA fragments 4.8, 2.2 and 1.8 kb. It has no resistance to freezing.
2. *T. nativa* Britov and Boev, 1972 is a northern species responsible for sylvatic trichinosis in arctic and subarctic bears, canids and felids. It is holarctic in distribution (Shaikenov, 1992; Handeland *et al.*, 1995; Zarnke *et al.*, 1995; Christian *et al.*, 1996). *T. nativa* has two unique allozymes, 4–20 allozyme differences and ribosomal DNA fragments 2.2 and 0.7 kb. It does not develop readily in pigs and rats and is highly resistant to freezing.
3. *T. pseudospiralis* Garkavi, 1972, is a cosmopolitan species that is unique in that larvae do not elicit nurse cells and it is infectious to birds as well as mammals. Originally discovered in *Procyon lotor* in the northern Causasus by Garkavi (1972a,b), this species is much smaller than *T. spiralis* and the larvae invades both fast and slow twitch muscles (Bagheri *et al.*, 1986). Pozio *et al.* (1992b) reported this species in *Aquila rapax* and *Vulpes cosae* in Kazakhstan and *Dasyurus maculatus* in Tasmania and listed previous reports in birds and mammals. More recently Andrews *et al.* (1994) reported human cases and Lindsay *et al.* (1995) described the species in the black vulture (*Coragyps atratus*) and transmitted it to pigs, chickens and mice. *T. pseudospiralis* has 12 unique allozymes, 16–20 allozyme differences and ribosomal DNA fragments 4.8, 2.2, 1.8 and 0.7 kb. Pigs can be infected only with difficulty but rats are readily infected. The species has no resistance to freezing.
4. *T. nelsoni* Britov and Boev, 1972 occurs in wildlife in tropical Africa and is regarded as the species studied by Nelson (1970) in Kenya (Pozio *et al.*, 1997). The species has four unique allozymes and 9–17 allozyme differences and only the ribosomal DNA fragment 1.8 kb is present. It is not readily transmitted to pigs and has no resistance to freezing.

5. *T. britovi* Pozia *et al.*, 1992 is found in wildlife in the temperate zone of the Palaearctic. According to Pozio *et al.* (1996) this species has been found from the Iberian Peninsula and France to Kazakhstan and Japan. The parasite has little resistance to freezing and seems to occur below the isotherm of −6°C in January but its southern extent is still unknown. The species has one unique allozyme, four to 17 allozyme differences and ribosomal DNA fragments 2.2 and 1.8 kb. It has been reported in canids, felids, bears, a horse and humans (Pozio *et al.*, 1992a). Rats and pigs can be infected with some difficulty.

The general account which follows is based on our knowledge of *Trichinella spiralis* and will apply in most respects to other species except *T. pseudospiralis*.

T. spiralis (Owen, 1835)

Adult *T. spiralis* are tiny worms (males about 1.4–1.6 mm in length, females about twice the size of males) which occupy the intestinal mucosal epithelium at the base of the villi, the glandular crypts and, less commonly, the tips of villi; the entire body of the worm is embedded in epithelium (Gardiner, 1976). Females lay active first-stage larvae which invade veins or lymphatics of the intestine and are carried in the circulation. It is estimated that each female will produce 500 larvae in light infections (Chirasak Khamboonruang, 1971). Larvae emerge from blood vessels in striated muscles, especially of the diaphragm, larynx, tongue, intercostal, biceps, abdomen, psoas, pectoral, gastrocnemius and deltoid. The newborn larva is capable of invading the striated muscle fibres of most mammal species (Gould, 1945) and of surviving for prolonged periods as an intracellular parasite. The infected cell is modified into what is commonly referred to as a nurse cell, which functions to nourish the parasite in some way or protect it from host response (Purkenson and Despommier, 1974; Stewart and Giannini, 1982). Despommier (1975) injected newborn larvae directly into thigh muscles, which made it possible to study synchronously developing infections. No changes occurred in the first 2 days. However, in 3–4 days, spaces developed between the plasma membrane and the myofilament and near the regions of triads. Myofilaments were disrupted. Nuclei enlarged and migrated to the central portion of the infected cytoplasm. By day 5, sarcomeres were disorganized and sarcoplasmic reticulum increased. By day 8, only the extreme periphery of the fibre contained Z bands with attached actin filaments and there was proliferation of the tubule system. On day 10, myofilaments were entirely replaced by sarcoplasmic reticulum. The plasma membrane became hyperinvoluted and was associated with a 36-fold increase in the thickness of the glycocalyx. Finally, a host-derived double membrane developed around the larvae. The nurse cell remains alive for the life of the larva and the two can live for long periods, in many species for the entire life of the host.

The larva in the nurse cell grew substantially. Despommier *et al.* (1975) divided the growth of the larva into three phases: (i) an initial growth phase from 0–1 day; (ii) a lag phase from 1–3 days; and (iii) an exponential growth phase from 3–19 days, after which there was no significant growth. Larvae arriving in muscles were 69–99 μm in length and at the end of the growth phases in 19 days were 745–975 μm in length. Growth was similar in rats and mice. A remarkable feature of the fully developed muscle larva is the precocious development of the reproductive primordia, which makes it easy to differentiate the sexes (Kozek, 1971a,b). The precocity is probably related to the extreme rapidity of maturation in the final host (see below) which allows the parasite to reproduce before the host's immune system has been effectively mobilized.

T. spiralis is transmitted to a new host when muscle larvae are ingested along with the flesh containing them. Thus, trichinellosis is mainly a disease of flesh-eating animals. In humans, most cases are the result of eating the flesh of wild or domesticated Suidae. However, *T. spiralis* will develop in most mammals and has recently been discovered in horses and an ox which had apparently ingested infected rodents with their food (Knapen and Franchimont, 1989; Dick *et al.*, 1990; Murrell, 1994). In horses, the larvae are most abundant in head and neck muscles and relatively uncommon in the diaphragm, tongue and masseter (Pozio *et al.*, 1998).

Larvae, digested out of the muscle capsules in the stomach and intestine of the mammal host, invaded the region between the lamina propria and the columnar epithelium of the small intestine 10–60 min postinfection (Gould, 1945; Despommier *et al.*, 1978) and reached adulthood within 36 h. During the rapid period of development the worms undergo four moults (Ali Khan, 1966; Kozek, 1971a,b). According to Ali Khan (1966) the moults took place at approximately 10, 17, 24 and 29 h in males and 12, 19, 26 and 36 h in females. Kozek (1971a,b) reported the final moults at 26 h in males and 29 h in females.

Gould *et al.* (1957) reported that the seminal receptacle of females contained increasing amounts of sperm from 32 h, when they were first inseminated, until 144 h postinfection. This indicated that more than one insemination had taken place. Thomas (1965) gave varying ratios of males and females to mice and compared the numbers of mated females with time. He concluded that females could be mated several times. Chirasak Khamboonruang (1971) believed that there was only one mating. Kozlov (1972) placed inseminated females in mice without males and found that in 19 days developing larvae were absent or few in the females, in contrast to the large numbers of larvae in females in control mice in which males accompanied females. Gardiner (1976) believed that mating occurred in the epithelium of the host, since males were not found outside this site. He estimated that one male could inseminate four females. Larvae deposited by female worms in the epithelium migrated through the stroma to venules and lymphatics and gained access to the general circulation, which carried them to the muscles. Adult worms were rarely found in mice later than 28 days postinfection and the life span of the *T. spiralis* in the gut of the host was probably about 1 month (Rappaport, 1943; Coker, 1955; Larsh, 1963; Chirasak Khamboonruang, 1971).

Bruschi *et al.* (1992) labelled newborn *T. spiralis* with ^{131}I and noted that labelled parts of the cuticle detached and were retained in the lungs of rats; they suggested that modifications in the cuticle of larvae during their passage through the lungs may increase resistance to non-specific defence mechanisms of the host. In recent years efforts have been made to understand better the processes which transform the myofibre containing a larval *T. spiralis* into a nurse cell. Wright (1989) confirmed the presence of capillary networks around the nurse cell and suggested that these were the result of appression of existing capillaries to the greatly enlarged muscle cell. Lee *et al.* (1991), Ko *et al.* (1992, 1994) and Mak and Ko (1999) explored the effects of excretory/secretory products from larvae in the transformation of the muscle cell and concluded that invasion of the cell elicited a degenerative/regenerative response and specific changes in the genomic expression of the myonuclei. Capó *et al.* (1998) reviewed the literature on the formation of the nurse cell and stressed the probable importance of vascular endothelial growth factor in the formation and maintenance of the cell. Despommier (1998) suggested that secreted tyrelosylated proteins played a central role in cell formation.

Poirier *et al.* (1995) reported that *Peromyscus maniculatus* infected with *T. spiralis* had 'activity deficits' proportional to the number of larvae recovered from their tissue whereas mice infected with *T. pseudospiralis* did not display 'activity deficits'. Saumier *et al.* (1988) reported that *Falco sparverius* infected with *T. pseudospiralis* exhibited mild behavioural changes related to the adult phase in the first 5 days postinoculation. More severe effects on mobility were noted as larvae invaded the muscles; these signs persisted for at least 5 weeks.

Family Cystoopsidae

Subfamily Cystoopsinae

Cystoopsis

C. accipenseri Wagner, 1867
This, the only species in the genus, lives in the subcutaneous connective tissue of the ganoid *Accipenser ruthensus* (sturgeon) and elicits tumours on the ventral wall. Janicki and Rasin (1930) stated that eggs were released into water and developed in different species of *Gammarus*. Larvae hatched in the intestine of the amphipod, penetrated the gut wall and became localized under the tegument especially of the appendages, where they were coiled and encapsulated. It is presumed that sturgeons became infected by eating the parasitized gammarids.

Subfamily Dioctowittinae

Dioctowittus

D. wittei Chabaud and Le Van Hoa, 1960
D. wittei was found in a snake, *Psammophis sibilans*, in Africa (Chabaud and Le Van Hoa, 1960). The precise location in the host was not determined but it may have been the air sac. Eggs are typical of the trichurids in that they are barrel-shaped with polar plugs. Tufts of filaments, however, are attached to the polar plugs, indicating that the parasite likely has an aquatic intermediate host, although nothing is known about its transmission. Eggs were apparently embryonated *in situ*.

9.2

The Superfamily Muspiceoidea

The few known species were found in the tissues of bats, rodents, deer, kangaroos, humans and crows. Their wide host and geographic distribution, diversity, extremely small size and often cryptic locations in the host, suggest that numerous undescribed species may have been overlooked. Members of the superfamily are highly modified (Brumpt, 1930; Chabaud and Campana, 1950; Bain and Chabaud, 1968, 1979; Chabaud and Bain, 1974; Bain and Nikander, 1983; Spratt and Speare, 1982; Spratt *et al.*, 1999) but have been assigned to the Enoplida of the Adenophorea along with the Trichinelloidea and the Dioctophymatoidea (Anderson and Bain, 1982); it has been suggested that they have distinct affinities with the mermithoids. The placement of the superfamily would be strengthened if a dorylaimid-type stylet could be found in any of the adults and larvae. Their mode of transmission has not been elucidated.

The superfamily is divided into the families Muspiceidae and Robertdollfusidae.

Family Muspiceidae

Muspicea

M. borreli Sambon, 1925

M. borreli of domestic mice was studied by Brumpt (1930) who found various stages in subcutaneous tissues, in muscles (especially near the shoulder blades) and less commonly in the abdominal and pleural cavities and salivary glands. Brumpt (1930) concluded that the worms were protandrous hermaphrodites. When filled with larvae, worms were 3.5–4.4 mm in length and 220–240 μm in width. Larvae (300–350 μm in length) left the parent worm by a rupture at the anterior extremity. Larvae were deposited in muscles and connective tissue but also on the surface of the skin through a small ulcer associated with the head of the female worm. He ruled out the possibility of an intermediate host being involved in transmission and finally concluded that mice probably ingested infective larvae crawling on the skin when the mice groomed themselves or each other, particularly after bathing. He did not rule out the possibility that cannibalism might be a method for parasite transmission from mouse to mouse.

Riouxgolvania, Lukonema and *Pennisia*

Five species belonging to these genera are found in the skin of the wings of bats (i.e. *R. beveridgei* Bain and Chabaud, 1979; *R. nyctali* Bain and Chabaud, 1978; *R. rhinolophi* Bain and Chabaud, 1968; *L. lukoschusi* Chabaud and Bain, 1974; *P. nagorseni* Bain and Chabaud, 1979). Members of *Riouxgolvania* and *Lukonema* are, like *Muspicea borreli*, protandrous hermaphrodites. *P. nagorseni*, however, is dioecious (but only known from immature specimens). Larvae apparently leave the parent worm by migrating between layers of the cuticle of the head end, where they emerge. It is suspected that the parent worm elicits a tiny ulcer, which would allow larvae to attain the surface of the wing of the bat. Infection of new hosts would presumably require free larvae to be ingested during individual or mutual grooming.

Family Robertdollfusidae

The transmission of members of the family is still not clear. Bain and Renz (1993) found larvae with characteristics of the family in 25 simuliids dissected near the rainforest in Cameroon, Africa. The main group of infected flies was caught near a national game park. The larvae were 1.2–1.3 mm long and 8–9.5 mm in width. The cephalic end was rounded, with a cephalic cap through which the elongate papillae possibly passed. Beaver and Burgdorfer (1984) found a long microfilaria-like nematode in *Ixodes damnini* in New York which Beaver and Burgdorfer later (1987) believed was a member of the family. These tentative observations might suggest that arthropod vectors are involved in the transmission of the nematodes.

Robertdollfusa

R. paradoxa Chabaud and Campana, 1950
This species was found in the anterior chamber of the eye of *Corvus corone*. It is not known how it is transmitted although Chabaud and Campana (1950) suggested that it might be by the ingestion of infected flesh. Males are unknown.

Durikainema

D. macropi Spratt and Speare, 1982
This species was found in mesenteric and hepatic veins of Macropodidae (Spratt and Speare, 1982). It is dioecious. Larvae in females were 848–950 μm long. Larvae were reported in peripheral blood, in a lactating mammary gland, and in the capillaries of thigh skin. Males were present.

D. phascolarcti Spratt and Gill 1998

Adults were found in the pulmonary arteries and arterioles and larvae were observed in blood vessels in lungs, brain, liver, kidney, uterus, cervix and bladder of koalas (*Phascolarctos cinerius*) in New South Wales, Australia (Spratt and Gill, 1998). Females are ovoviviparous and larvae were 588–634 μm in length. Males were present.

Lappnema

L. auras Bain and Nikander, 1983

This parasite was found in large nodules in the ears of reindeer (Bain and Nikander, 1983). Males were not found. The female is strikingly similar to that of *Durikainema macropi*.

Haycocknema

H. perplexum Spratt, Beveridge, Andrews, and Dennett, 1999

This species was found within individual myofibres in a human in Tasmania, Australia (Spratt *et al.*, 1999). Larvae were found within and between myofibres. Males were 295–307 μm and females 277–509 μm in length. Larvae were 147–152 μm in length. Males had a single spicule.

References (Trichinellina)

Ali Khan, Z. (1966) The postembryonic development of *Trichinella spiralis* with special reference to ecdysis. *Journal of Parasitology* 52, 248–259.

Alicata, J.E. (1935) Early developmental stages of nematodes occurring in swine. *USA Department of Agriculture Technical Bulletin* No. 489, pp. 47–51.

Allen, R.W. (1949) Studies on the life history of *Capillaria annulata* (Molin, 1858) Cram, 1926. *Journal of Parasitology* (Suppl.) 35, 35.

Allen, R.W. and Wehr, E.E. (1942) Earthworms as possible intermediate hosts of *Capillaria caudinflata* of chickens and turkeys. *Proceedings of the Helminthological Society of Washington* 9, 72–73.

Anderson, R.C. and Bain, O. (1982) No. 9. Keys to genera of the superfamilies Rhabditoidea, Dioctophymatoidea, Trichinelloidea and Muspiceoidea. In: Anderson, R.C., Chabaud, A.G. and Willmott, S. (eds) *CIH Keys to the Nematode Parasites of Vertebrates*. Commonwealth Agricultural Bureaux, Farnham Royal, UK, pp. 1–26.

Andrews, J.R.H., Ainsworth, R. and Abernethy, D. (1994) *Trichinella pseudospiralis* in humans: description of a case and its treatment. *Transactions of the Royal Society of Tropical Medicine and Hygiene* 88, 200–203.

Artemenko, V.D. (1937) The biology of eggs of *Trichocephalus trichiurus*. *Pratsy Protoz. Parazyt. viddilu Odes' Bakt. Int.*, pp. 120–134 Kyyiv. (In Russian.) (Reviewed in Skrjabin *et al.*, 1957.)

Artyukh, E.S. (1936) Helminth fauna of sheep in the Ukrainian SSR. *Uchenye Zapiski Vitebskogo Veterinaro-Zootekhnickeskogo Instituta* 4, 115–122. (In Russian.) (Reviewed in Skrjabin *et al.*, 1957.)

Bagheri, Ali, Ubelaker, J.E., Stewart, G.L. and Wood, B. (1986) Muscle fiber selectivity of *Trichinella spiralis* and *Trichinella pseudospiralis. Journal of Parasitology* 72, 277–282.

Bain, O. and Chabaud, A.G. (1968) Description de *Riouxgolvania rhinolophi* n.g., n.sp., nématode parasite de Rhinolophe, montrant les affinités entre Muspiceoidea et Mermithoidea. *Annales de Parasitologie Humaine et Comparée* 43, 45–50.

Bain, O. and Chabaud, A.G. (1979) Sur les Muspiceidae (Nematoda–Dorylaimina). *Annales de Parasitologie Humaine et Comparée* 54, 207–225.

Bain, O. and Nikander, S. (1983) Un nématode aphasmidien dans les capillaires de l'oreille du renne, *Lappnema auris* n.gen., n.sp. (Robertdollfusidae). *Annales de Parasitologie Humaine et Camparée* 58, 383–390.

Bain, O. and Renz, A. (1993) Infective larvae of a new species of Robertdollfusidae (Adenophorea, Nematoda) in the gut of *Simulium damnosum* in Cameroon. *Annales de Parasitologia Humaine et Comparie* 68, 182–184.

Bancroft, T.L. (1893) On the whipworm of the rat's liver. *Journal and Proceedings of the Royal Society of New South Wales* 27, 86–90.

Beaver, P.C. and Burgdorfer, W. (1984) A microfilaria of exceptional size from the ixodic tick *Ixodes damnini* from Shelter Island, New York. *Journal of Parasitology* 70, 963–966.

Beaver, P.C. and Burgdorfer, W. (1987) Critical comments. Letter to the Editor. *Journal of Parasitology* 73, 389.

Beer, R.J.S. (1973a) Morphological description of the egg and larval stages of *Trichuris suis* Schrank, 1788. *Parasitology* 67, 263–278.

Beer, R.J.S. (1973b) Studies on the biology of the life-cycle of *Trichuris suis* Schrank, 1788. *Parasitology* 67, 253–262.

Bhaibulaya, M. and Indra-Ngarm, S. (1979) *Amaurornis phoenicurus* and *Ardeola bacchus* as experimental definitive hosts for *Capillaria philippinensis* in Thailand. *International Journal for Parasitology* 9, 321–322.

Bhaibulaya, M., Indra-Ngarm, S. and Ananthapruti, M. (1979) Freshwater fishes of Thailand as experimental intermediate hosts for *Capillaria philippinensis. International Journal for Parasitology* 9, 105–108.

Bober, C.M. and Dick, T.A. (1983) A comparison of the biological characteristics of *Trichinella spiralis* var. *pseudospiralis* between mice and bird hosts. *Canadian Journal of Zoology* 61, 2110–2119.

Boev, S.N., Britov, V.A. and Orlov, I.V. (1979) Species composition of Trichinellae. *Wiadomosci Parazytologicnze* 25, 495–503.

Borovkova, A.M. (1947) Life cycle of the causative agent of thominxiasis infection in silver foxes, epizootiology and prophylaxis of the disease. Thesis (reviewed in Skrjabin *et al.*, 1957).

Brumpt, E. (1930) *Muspicea borreli* Sambon, 1925 et cancers des souris. *Annales de Parasitologie Humaine et Comparée* 8, 309–343.

Bruschi, F., Solfanelli, S. and Binaghi, R.A. (1992) *Trichinella spiralis*: modifications of the cuticle of the newborn larva during passage through the lung. *Experimental Parasitology* 75, 1–9.

Bundy, D.A.P. and Cooper, E.S. (1989) *Trichuris* and trichuriasis in humans. *Advances in Parasitology* 28, 107–173.

Cabrera, B.D., Baltazar, R., Caviles, A.P., Domingo, E., Gonzaga, A., Lingao, A. and Campos, P.C. (1969) *Capillaria philippinensis* eggs demonstrated in fish from Pudos, Ilocos Sur. *Journal of the Philippine Medical Association* 45, 275.

Campbell, W.C. (ed.) (1983) Trichinella *and Trichinosis*. Plenum Press, New York.

Capó, V.A., Despommier, D.D. and Polvere, R.I. (1998) *Trichinella spiralis*: vascular endothelial growth factor is up-regulated within the nurse cell during the early phase of its formation. *Journal of Parasitology* 84, 209–214.

Chabaud, A.G. and Bain, O. (1974) Données nouvelles sur la biologie des Nématodes Muspicéides fournies par l'étude d'un parasite de Chiroptères: *Lukonema lukoschusi* n.gen. n.sp. *Annales de Parasitologie Humaine et Comparée* 48, 819–834.

Chabaud, A.G. and Campana, Y. (1950) Nouveau parasite remarquable par l'atrophie de ses organes: *Robertdollfusa paradoxa* (Nematoda, incertae sedis). *Annales de Parasitologie Humaine et Comparée* 25, 325–334.

Chabaud, A.G. and Le Van Hoa (1960) Adaptation à la vie tissulaire d'un nématode aphasmidien. *Compte Rendus des Séances de l'Académie des Séances* 251, 1837–1839.

Chirasak Khamboonruang (1971) Output of larvae and life span of *Trichinella spiralis* in relation to worm burden and superinfection in the mouse. *Journal of Parasitology* 57, 289–297.

Chitwood, M.B., Velasquez, C. and Salazar, N.G. (1968) *Capillaria philippinensis* sp.n. (Nematoda: Trichinellida), from the intestine of man in the Philippines. *Journal of Parasitology* 54, 368–371.

Christensen, R.O. (1938) Life history and epidemiological studies on the fox lungworm *Capillaria aerophila* (Creplin, 1839). *Livro Jubilar Lauro Travassos*, pp. 119–136.

Christian, M.O., Kapel, M.O., Hendriksen, S.A., Berg, T.B. and Nansen, P. (1996) Epidemiologic and zoogeographic studies on *Trichinella nativa* in arctic fox, *Alopes lagopus* in Greenland. *Journal of the Helminthological Society of Washington* 63, 226–232.

Coker, C.M. (1955) Effects of cortisone on *Trichinella spiralis* infection in non-immunized mice. *Journal of Parasitology* 41, 498–504.

Cram, E.B. (1936) Species of *Capillaria* parasites in the upper digestive tract of birds. *USA Department of Agriculture Technical Bulletin* No. 516, 27 pp.

Cross, J.H. and Basaca-Sevilla, V. (1983) Experimental transmission of *Capillaria philippinensis* to birds. *Transactions of the Royal Society of Tropical Medicine and Hygiene* 77, 511–514.

Cross, J.H. and Bhaibulaya, M. (1983) Intestinal capillariasis in the Philippines and Thailand. In: Croll, N.A. and Cross, J.H. (eds) *Human Ecology and Infectious Diseases*. Academic Press, London, pp. 104–136.

Cross, J.H., Banzon, T., Murrell, K.D., Watten, R.H. and Dizon, J.J. (1970) A new epidemic diarrheal disease caused by the nematode *Capillaria philippinensis*. *Industry and Tropical Health* 7, 124–131.

Cross, J.H., Banzon, T., Clarke, M.D., Basaca-Servilla, V., Watten, R.H. and Dizon, J.J. (1972) Studies on the experimental transmission of *Capillaria philippinensis* in monkeys. *Transactions of the Royal Society of Tropical Medicine and Hygiene* 66, 819–827.

Cross, J.H., Banzon, T. and Singson, C. (1978) Further studies on *Capillaria philippinensis*: development of the parasite in the Mongolian gerbil. *Journal of Parasitology* 64, 208–213.

Deo, P.G. (1960) Studies on the biology and life-history of *Trichuris ovis* (Abildgaard) Smith. II. Development of infective embryonated eggs of *Trichuris ovis* (Abildgaard) in sheep and goats. *Indian Journal of Veterinary Science and Animal Husbandry* 30, 165–177.

Despommier, D. (1975) Adaptive changes in muscles fibers infected with *Trichinella spiralis*. *American Journal of Pathology* 78, 477–496.

Despommier, D.D. (1998) How does *Trichinella spiralis* make itself at home? *Parasitology Today* 14, 318–323.

Despommier, D., Aron, L. and Turgeon, L. (1975) *Trichinella spiralis*: growth of the intracellular (muscle) larva. *Experimental Parasitology* 37, 108–116.

Despommier, D., Sukhdeo, M. and Maerovitch, E. (1978) *Trichinella spiralis*: site selection by the larva during the enteral phase of infection in mice. *Experimental Parasitology* 44, 209–215.

Detels, R., Gutman, L., Jaramillo, J., Zerrudo, E., Banzon, T., Valera, J., Murrell, K.D., Cross, J.H. and Dizon, J.J. (1969) An epidemic of intestinal capillariasis in man. A study in a barrio in Northern Luzon. *American Journal of Tropical Medicine and Hygiene* 18, 676–682.

Dick, T.A. (1983a) Species and intra specific variation. In: Campbell, W.C. (ed.) Trichinella *and* Trichinosis. Plenum Press, New York, pp. 33–73.

Dick, T.A. (1983b) The species problem in *Trichinella*. In: Stone, A.R., Platt, H.M. and Khali, L.F. (eds) *Concepts in Nematode Systematics*. Academic Press, London, pp. 351–360.

Dick, T.A. and Chadee, K. (1983) Interbreeding and gene flow in the genus *Trichinella*. *Journal of Parasitology* 69, 176–180.

Dick, T.A., deVos, T. and Dupauy-Camet, J. (1990) Identification of two isolates of *Trichinella* recovered from humans in France. *Journal of Parasitology* 76, 41–44.

Dinnik, Y.A. and Dinnik, N.N. (1937) The effects of temperature, oxygen-free environment and desiccation on eggs of *Trichocephalus trichiurus* (L.). *Meditsinskaya Parazitologiya i Parazitarnie Bolezni* 6, 103–118. (In Russian.) (Reviewed in Skrjabin *et al.*, 1957.)

Dinnik, Y.A. and Dinnik, N.N. (1939) Observations on the development of eggs of *Trichocephalus trichiurus* in the soil. *Meditsinskaya Parazitologiya i Parazitarnie Bolezni* 8, 221–229. (In Russian.) (Reviewed in Skrjabin *et al.*, 1957.)

Efremov, V.V. and Shikhobalova, N.P. (1939) Histopathogenesis of experimental trichocephaliasis in white mice. (In Russian.) *Meditsinskaya Parazitologiya i Parazitarnie Bolezni* 8, 81–88.

Enigk, K. (1950) Die Biologie von *Capillaria plica* (Trichuroidea, Nematodes). *Zeitschrift für Tropenmedizin und Parasitologie* 1, 560–571.

Fahmy, M.A.M. (1954) An investigation on the life cycle of *Trichuris muris*. *Parasitology* 44, 50–57.

Faust, F.C., Beaver, P.C. and Jung, R.C. (1975) *Animal Agents and Vectors of Human Disease*. Lea and Febiger, Philadelphia.

Fernan-Nunez, M. (1927) The pathogenic role of *Trichocephalus dispar*. *Archives of Internal Medicine* 40, 46–57.

Freeman, R.S. and Wright, K.A. (1960) Factors concerned with the epizootiology of *Capillaria hepatica* (Bancroft, 1893) (Nematoda) in a population of *Peromyscus maniculatus* in Algonquin Park. *Journal of Parasitology* 46, 373–382.

Fresh, J.W., Cross, J.H., Reyes, V., Whalen, G.E., Uylangco, C.V. and Dizon, J.J. (1972) Necropsy findings in intestinal capillariasis. *American Journal of Tropical Medicine and Hygiene* 21, 169–173.

Fülleborn, F. (1923) Über den Kopfstachel der Trichocephaliden und Bemerkungen über die jüngsten Stadien von *Trichocephalus trichiurus*. *Archiv für Schiffs- und Tropen-Hygiene, Pathologie und Therapie Exotischer Krankheiten* 27, 421.

Fülleborn, F. (1924) Ueber den Infectionsweg bei *Hepaticola hepatica*. *Archiv für Schiffs- und Tropen-Hygiene, Pathologie und Therapie Exotischer Krankheiten* 28, 48–61.

Gardiner, C.H. (1976) Habitat and reproductive behaviour of *Trichinella spiralis*. *Journal of Parasitology* 62, 865–870.

Garkavi, B.L. (1972a) *Trichinella* species from a raccoon. *Materials of the All-Union Conference on the Problem of Trichinellosis in Man and Animals, May 30–June 1, 1972, Vilnius, Akademia Nauk Litovskoi SSSR*, pp. 53–54. (In Russian.)

Garkavi, B.L. (1972b) Species of *Trichinella* from wild carnivores. *Veterinariya* 49, 90–91. (In Russian.)

Gould, S.E. (1945) *Trichinosis*. Charles C. Thomas, Springfield, Illinois.

Gould, S.E., Villella, J.B. and Hertz, C.S. (1957) Studies on *Trichinella spiralis*. VI. Effects of cobalt-60 and X-ray on morphology and reproduction. *American Journal of Pathology* 33, 79–105.

Graybill, H.W. (1924) *Capillaria columbae* (Rud.) from the chicken and turkey. *Journal of Parasitology* 10, 205–207.

Grove, D.I. (1990) *A History of Human Helminthology*. CAB International, Wallingford, UK.

Haehling, E., Niederkorn, J.Y. and Stewart, G.L. (1995) *Trichinella spiralis* and *Trichinella pseudospiralis* induce collagen synthesis by host fibroblasts *in vitro* and *in vivo*. *International Journal for Parasitology* 25, 1393–1400.

Handeland, K., Slettbakk, T. and Helle, O. (1995) Freeze-resistant *Trichinella* (*Trichinella nativa*) established on the Scandinavian peninsula. *Acta Veterinaria Scandinavica* 36, 149–151.

Hasegawa, T. (1924) Beitrag zur Entwicklung von *Trichocephalus* im Wirte. *Archiv für Schiffs- und Tropen-Hygiene, Pathologie und Therapie Exotischer Krankheiten* 28, 337–340.

Hasslinger, M.A. and Schwarzler, C. (1980) Der Blasenwurm der Ratte, *Trichosomoides crassicauda* Untersuchungen zur Entwicklung, Übertragung und Diagnose. *Berliner und Münchener Tierärztliche Wochenschrift* 93, 132–135.

Hill, C.H. (1957) The survival of swine whipworm eggs in hog lots. *Journal of Parasitology* 43, 104.

Janicki, C. and Rasin, K. (1930) Bemerkungen über *Cystoopsis accipenseri* des Wolga-Sterlets, sowie über die Entwicklung dieses Nematoden im Zwischenwirt. *Zeitschrift für Wissenschaftliche Zoologie* 136, 1–37.

Kapel, C.M.O., Henriksen, S.A., Berg, T.B. and Nansen, P. (1996) Epidemiologic and zoogeographic studies on *Trichinella nativa* in arctic fox, *Alopex lagopus*, in Greenland. *Journal of the Helminthological Society of Washington* 63, 226–232.

Kean, B., Mott, K.E. and Russell, A.J. (1978) *Tropical Medicine and Parasitology. Classic Investigations.* Cornell University Press, Ithaca, New York. pp. 458–472.

Khamboonruang, C. (1971) Output of larvae and life span of *Trichinella spiralis* in relation to worm burden and superinfection in the mouse. *Journal of Parasitology* 57, 289–297.

Knapen, F. and Franchimont, J.H. (1989) *Trichinella spiralis* infection in horses. In: Tanner, C.E., Martinez-Fernandez, R. and Bolas-Fernandez, F. (eds) *Trichinellosis. Proceedings of the 7th International Conference on Trichinellosis, Alicante, Spain*, pp. 376–381.

Ko, R.C. and Fan, L. (1996) Heat shock response of *Trichinella spiralis* and *Trichinella pseudospiralis*. *Parasitology* 112, 89–95.

Ko, R.C., Fan, L. and Lee, D.L. (1992) Experimental reorganization of host muscle cells by excretory/secretory products of infective *Trichinella spiralis* larvae. *Transcripts of the Royal Society of Tropical Medicine and Hygiene* 86, 77–78.

Ko, R.C., Fan, L., Lee, D.L. and Compton, H. (1994) Changes in host muscle induced by excretory/secretory product of larval *Trichinella spiralis* and *Trichinella pseudospiralis*. *Parasitology* 108, 195–205.

Kozek, W.J. (1971a) The molting pattern in *Trichinella spiralis*. I. A light microscope study. *Journal of Parasitology* 57, 1015–1028.

Kozek, W.J. (1971b) The molting pattern in *Trichinella spiralis*. II. An electron microscope study. *Journal of Parasitology* 57, 1029–1038.

Kozlov, D.P. (1972) Repeated fertilization in *Trichinella spiralis*. *Parasitologiya* 6, 360–363. (In Russian.)

Kutzer, E. and Otte, E. (1966) *Capillaria petruschewskii* (Schulman, 1948), Morphologie, Biologie und Pathogene Bedeutung. *Zeitschrift für Parasitenkunde* 28, 16–30.

La Rosa, G., Pozio, E., Rossi, P. and Murrell, K. (1992) Allozyme analysis of *Trichinella* isolates from various host species and geographical regions. *Journal of Parasitology* 78, 641–646.

La Rosa, G., Pozio, E. and Rossi, P. (1989) New taxonomic contribution to the genus *Trichinella* Railliet, 1895. II. Multivariate analysis on genetic and biological data. In: Tanner, C.E., Martinez-Fernandez, R. and Bolas-Fernandez, F. (eds), *Trichinellosis. Proceedings of the 7th International Conference on Trichinellosis, Alicante, Spain*, pp. 83–88.

Larsh, J.E. (1963) Experimental trichiniasis. *Advances in Parasitology* 1, 213–286.

Lee, D.L., Ko, R.C., Yi, X.Y. and Yeung, M.H.F. (1991) *Trichinella spiralis*: antigenic epitopes from the stichocytes detected in the hypertrophic nuclei and cytoplasm of the parasitized muscle fibre (nurse cell) of the host. *Parasitology* 102, 117–123.

Lee, T.D.G. and Wright, K.A. (1978) The morphology of the attachment and probable feeding site of the nematode, *Trichuris muris* (Schrank, 1788) Hall, 1916. *Canadian Journal of Zoology* 56, 1889–1905.

Leuckart, K. (1876) Die menschlichen Parasiten und die von ihnen herrührenden Krankheiten. *Ein Hand- und Lehrbuch für Naturforscher und Aerzte*, Vol. 2. Leipzig and Heidelberg, pp. 513–882.

Levine, P.P. (1937) The effects of various environmental conditions on the viability of the ova of *Capillaria columbae. Journal of Parasitology* 23, 427–428.

Levine, P.P. (1938) Infection of the chicken with *Capillaria columbae* (Rudolphi). *Journal of Parasitology* 24, 45–52.

Lichtenfels, J.R., Murrell, K.D. and Pilitt, P.A. (1983) Comparison of three subspecies of *Trichinella spiralis* by scanning electron microscopy. *Journal of Parasitology* 69, 1131–1140.

Lindsay, D.S., Zarlenga, D.S., Gamble, H.R., Alyaman, F., Smith, P.C. and Blagburn, B.L. (1995) Isolation and characterization of *Trichinella pseudospiralis* Garkavi, 1972 from a black vulture (*Coragyps atratos*). *Journal of Parasitology* 81, 920–923.

Linstow, O. von (1874) Beobachtungen an *Trichodes crassicauda* Bell. *Archiv für Naturgeschichte* 1, 271–286.

Little, M.D. and Orihel, T.C. (1972) The mating behaviour of *Anatrichosoma* (Nematoda: Trichuroidea). *Journal of Parasitology* 58, 1019–1020.

Lomakin, V.V. and Trofimenko, V.Y. (1982) Capillariids (Nematoda: Capillariidae) of the freshwater fish fauna of the USSR. *Trudy Gel'mintologicheski Laboratorii* 31, 60–87. (In Russian.)

Löwenstein, S. (1910) Epithelwucherungen und Papillombildungen der Rattenblase, verursacht durch die *Trichosoma* (*Tr. crassicauda?*). Vorläufige Mitteilung. *Beiträge zur Klinischen Chirurgie* 69, 533–546.

Löwenstein, S. (1912) *Trichodes crassicauda* specifica als Erreger von Papillomen der Blase und Niere. *Verhandlungen der Gesellschaft Deutscher Naturforscher und Aerzte* 2. Teil, 2. Hälfte pp. 162–166.

Lubinsky, G. (1956) On the probable presence of parasitic liver cirrhosis in Canada. *Canadian Journal of Comparative Medicine* 20, 457–465.

Luttermoser, G.W. (1938a) Factors influencing the development and viability of eggs of *Capillaria hepatica. American Journal of Hygiene* 27, 275–289.

Luttermoser, G.W. (1938b) An experimental study of *Capillaria hepatica* in the rat and the mouse. *American Journal of Hygiene* 27, 321–340.

Magomedbekov, V.A. (1953) Biology of the nematode *Trichocephalus skrjabini* (Baskakov, 1924) and study of certain problems of the epizootiology of trichocephaliasis in sheep in Dagestan. *Trudi Gel'mintologicheskoi Laboratorii AN SSR* 8. (In Russian.) (Reviewed in Skrjabin *et al.*, 1957.)

Mak, C. and Ko, R.C. (1999) Characterization and endonuclease activity from excretory/secretory products of a parasitic nematode, *Trichinella spiralis. European Journal of Biochemistry* 259, 1–6.

Miller, M.J. (1947) Studies on the life-history of *Trichocephalus vulpis*, the whipworm of dogs. *Canadian Journal of Research* 25, 1–11.

Moravec, F. (1982) Proposal of a new systematic arrangement of nematodes of the family Capillariidae. *Folia Parasitologica* 29, 119–132.

Moravec, F. (1983) Observations on the bionomy of the nematode *Pseudocapillaria brevispicula* (Linstow, 1973). *Folia Parasitologica* (Praha) 30, 229–241.

Moravec, F. (1985) Occurrence of the endoparasitic helminths in tench (*Tinca tinca*) from the Mácha Lake fish pond system. *Věstnik československé Společnosti Zoologicke* 49, 32–50.

Moravec, F., Koudela, B., Ogawa, K. and Nagasawa, K.C. (1998) Two new *Huffmanela* species, *H. japonica* n.sp. and *H. shikokuensis* n.sp. (Nematoda: Trichosomoididae), from marine fishes in Japan. *Journal of Parasitology* 84, 589–593.

Morehouse, N.F. (1944) Life cycle of *Capillaria caudinflata*, a nematode parasite of the common fowl. *Iowa State College Journal of Science* 18, 217–253.

Murrell, K.D. (1994) Beef as a source of Trichinellosis. *Parasitology Today* 10, 434.

Nelson, G.S. (1970) Trichinosis in Africa. In: Gould, S.E. (ed.) *Trichinellosis in Man and Animals.* C.C. Thomas Publishers, Springfield, Illinois, pp. 473–492.

Nickel, E.A. (1953) Ein Beitrag zur Biologie und Pathogenität des Geflügelhaarwurms *Capillaria caudinflata* (Molin, 1858). *Berliner und Münchener Tierärztliche Wochenschrift* 66, 245–248.

Ooi, H.K., Tenora, F., Itoh, K. and Kamiya, M. (1993) Comparative study of *Trichuris trichiura* from non-human primates and from man and their difference with *T. suis*. *Journal of Veterinary Medical Science* 55, 363–366.

Orihel, T.C. (1970) Anatrichosomiasis in African monkeys. *Journal of Parasitology* 56, 982–985.

Orlov, I.V. (1948) Study of the helminth fauna of beaver. Parasitic fauna and diseases of wildlife. *Glavnoe Upravlenie po Zapovednikam* pp. 114–125. (In Russian.) (Reviewed in Skrjabin *et al.*, 1957.)

Panesar, T.S. (1981) The early phase of tissue invasion by *Trichuris muris* (Nematoda: Trichuroidea). *Zeitschrift für Parasitenkunde* 66, 163–166.

Panesar, T.S. (1989) The moulting pattern in *Trichuris muris* (Nematoda: Trichuroidea). *Canadian Journal of Zoology* 67, 2340–2343.

Panesar, T.S. and Croll, N.A. (1980) The location of parasites within their hosts: site selection by *Trichuris muris* in the laboratory mouse. *International Journal for Parasitology* 10, 261–273.

Pavlov, A.V. (1955) Biology of the nematode *Hepaticola hepatica* and features of the epizootiology of the disease caused by it. (In Russian.) (Thesis reviewed in Skrjabin *et al.*, 1957).

Pence, D.B. and Little, M.D. (1972) *Anatrichosoma buccalis* sp.n. (Nematoda: Trichosomoididae) from the buccal mucosa of the common opossum, *Didelphis marsupialis* L. *Journal of Parasitology* 58, 767–773.

Petrov, A.M. and Borovkova, A.M. (1942) Study of the developmental cycle of *Capillaria plica* (Rudolphi, 1819) the causative agent of helminth diseases of the bladder of dogs and foxes. *Doklady AN SSSR* 38, 175–176. (In Russian.)

Poirier, S.R., Rau, M.E. and Wang, X. (1995) Diel locomotory activity of deer mice (*Peromyscus maniculatus*) infected with *Trichinella nativa* or *Trichinella pseudospiralis*. *Canadian Journal of Zoology* 73, 1323–1334.

Powers, K.G., Todd, A.C. and McNutt, S.H. (1960) Experimental infections of swine with *Trichuris suis*. *American Journal of Veterinary Research* 21, 262–268.

Pozio, E., La Rosa, G., Rossi, P. and Murrell, K.D. (1989) New taxonomic contribution to the genus *Trichinella* (Owen, 1835). I. Biochemical identification of seven clusters by gene–enzyme systems. In: Tanner, C.E., Martinez-Fernandez, R. and Bolas-Fernandez, F. (eds) *Trichinellosis. Proceedings of the 7th International Conference on Trichinellosis, Alicante, Spain*, pp. 76–82.

Pozio, E., La Rosa, G., Murrell, K.D. and Lichtenfels, J.R. (1992a) Taxonomic revision of the genus *Trichinella*. *Journal of Parasitology* 78, 654–659.

Pozio, E., Shaikenov, B., La Rosa, G. and Obendorf, D.L. (1992b) Allozymic and biological characters of *Trichinella pseudospiralis* isolates from free-ranging animals. *Journal of Parasitology* 78, 1087–1090.

Pozio, E., La Rosa, G., Rossi, P. and Murrell, K.D. (1992c) Biological characterizations of *Trichinella* isolates from various host species and geographic regions. *Journal of Parasitology* 78, 647–653.

Pozio, E., La Rosa, G., Yamaguchi, T. and Saito, S. (1996) *Trichinella britovi* from Japan. *Journal of Parasitology* 82, 847–849.

Pozio, E., de Meneghi, D., Roelke-Parker, M.E. and La Rosa, G. (1997) *Trichinella nelsoni* in carnivores from the Serengeti ecosystem, Tanzania. *Journal of Parasitology* 83, 1195–1198.

Pozio, E., Celano, G.V., Sacchi, L., Pavia, C., Rossi, P., Tamburni, A., Corona, S. and La Rosa, G. (1998) Distribution of *Trichinella spiralis* larvae in muscles from a naturally infected horse. *Veterinary Parasitology* 74, 19–27.

Pukhov, V.I. (1939) The biology of *Trichocephalus ovis* (Abildgaard, 1795). *Trudi Rostovskoi Oblastnoi Nauchno-Issledovatel'skoi Veterinarnoi Opytnoi Stantsii* 7, 70–76. (In Russian.) (Reviewed in Skrjabin *et al.*, 1957.)

Purkenson, M. and Despommier, D.D. (1974) Fine structure of the muscle phase of *Trichinella spiralis* in the mouse. In: Kim, C. (ed.) *Trichinellosis*. Intext Publishers, New York.

Rappaport, I. (1943) A comparison of three strains of *Trichinella spiralis*. II. Longevity and sex ratio of adults in the intestine and rapidity of larval development in the musculature. *American Journal of Tropical Medicine* 23, 351–161.

Rietschel, G. (1973) Untersuchungen zur Entwicklung einiger in Krähen (Corvidae) vorkommender Nematoden. *Zeitschrift für Parasitenkunde* 42, 243–250.

Rubin, R. (1954) Studies on the common whipworm of the dog, *Trichuris vulpis*. *Cornell Veterinarian* 44, 36–49.

Saumier, M.D., Rau, M.E. and Bird, D.M. (1988) The influence of *Trichinella pseudospiralis* infection on the behaviour of captive, non-breeding American kestrels (*Falco sparverius*). *Canadian Journal of Zoology* 66, 1685–1692.

Schwartz, B. (1960) Discovery of trichinae and determination of their life history and pathogenicity. *Proceedings of the Helminthological Society of Washington* 27, 261–268.

Shaikenov, B.S. (1992) Ecological border of distribution of *Trichinella nativa* Britov et Boev, 1972 and *Trichinella nelsoni* Britov et Boev, 1972. *Wiadomosci Parazytologiczne* 38, 85–91.

Shikhobalova, N.P. (1937) Experimental study of the chemotherapy of trichocephalosis. I. Trichocephalosis of white mice. *Meditsinskaya Parazitologiya i Parazitarnie Bolezni* 6, 389–400. (In Russian.)

Shimatani, T. (1961) Studies on the ecology of *Capillaria hepatica* eggs. *Journal of Kyoto Prefectural Medical University* 64, 1063–1083. (In Japanese.)

Shmytova, G.Y. (1971) The embryonic and postembryonic development of nematodes of the genus *Trichuris*. *Trudi Gel'mintologicheskoi Laboratorii Voprosy Biologii Fiziologii i Biokhimii Gel'mintov Zhivotnykh i Rastenii* 21, 157–166. (In Russian.)

Shorb, D.A. (1931) Experimental infestation of white rats with *Hepaticola hepatica*. *Journal of Parasitology* 17, 151–154.

Shulman, S.S. (1948) A new species of roundworm parasite in the liver of fish. *Izvestiya Vsesoyuznogo Nauchno-Issedovatel'skogo Instituta Ozernoga i Rechnogo Rybnogo Khozgaistva* 27, 235–238. (In Russian.)

Skarbilovich, T.S. (1945) Determination of two different types of developmental cycles in *Capillaria putorii* (Rud. 1819). *Doklady AN SSR* 50, 553–554.

Skrjabin, K.E., Shikhobalova, N.P. and Orlov, I.V. (1957) *Essentials of Nematology. Trichocephalidae and Capillariidae of Animals and Man and the Diseases Caused by Them.* Vol. VI. Academy of Sciences of the USSR. (Israel Program for Scientific Translations, Jerusalem, 1970.)

Spindler, L.A. (1929) The relation of moisture to the distribution of human *Trichuris* and *Ascaris*. *American Journal of Hygiene* 10, 476–496.

Spratt, D.M. and Gill, P.A. (1998) *Durikainema phascolarcti* n.sp. (Nematoda: Muspiceoidea: Robertdollfusidae) from the pulmonary arteries of the koala *Phascolarctus cinereus* with associated pathological changes. *Systematic Parasitology* 39, 101–106.

Spratt, D.M. and Speare, R. (1982) *Durikainema macropi* gen. et sp. nov. (Muspiceoidea: Robertdollfusidae), a remarkable nematode from Macropodidae (Marsupialia). *Annales de Parasitologie Humaine et Comparée* 57, 53–62.

Spratt, D.M., Beveridge, I., Andrews, J.R.H. and Dennett, X. (1999) *Haycocknema perplexa* n.g. n.sp. (Nematoda: Robertdollfusidae) an intramyofibre parasite of man. *Systematic Parasitology* 43, 123–131.

Stewart, G.L. (1989) Biological and immunological characteristics of *Trichinella pseudospiralis*. *Parasitology Today* 5, 344–349.

Stewart, G.L. and Giannini, S.Z. (1982) *Sarcocystis, Trypanosoma, Toxoplasma, Brugia, Ancylostoma* and *Trichinella* spp. A review of the intracellular parasites of striated muscles. *Experimental Parasitology* 53, 406–447.

Tanner, C.E., Martinez-Fernandez, R. and Bolas-Fernandez, F. (eds) (1989) *Trichinellosis. Proceedings of the 7th International Conference on Trichinellosis, Alicante, Spain.*

Thapar, G.S. and Singh, Kr.S. (1954) Studies on the life-history of *Trichuris ovis* (Abildgaard, 1795) (Family: Trichuridae: Nematoda). *Proceedings of the Indian Academy of Science* 40, 69–88.

Thieme, H. (1961) *Capillaria acerinae* nov. spec. (Nematoda, Trichuroidea), ein neuer Haawurm aus der Leber vom Kaulbarsch (*Acerina cernua*). *Zoologischer Anzeiger* 166, 135–139.

Thieme, H. (1964) The development and life cycle of *Hepatospina acerinae* Thieme 1961 from the liver of *Acerina cernua*. In: Ergens, R. and Ryšavý, B. (eds) *Parasitic Worms and Aquatic Conditions. Proceedings of Symposium, Prague, Oct. 29–Nov. 2, 1962,* pp. 115–119.

Thomas, H. (1965) Beitrag zur Biologie und mikroskopischen Anatomie von *Trichinella spiralis* (Owen, 1835). *Zeitschrift für Tropenmedizin und Parasitologie* 16, 148–180.

Thomas, L.J. (1924) Studies on the life history of *Trichosomoides crassicauda* (Bellingham). *Journal of Parasitology* 10, 105–135.

Tochihara, J. (1922) Studies on the development of the whipworm. *Seikingaku Zasshi, Tokyo* 327, 809–819.

Wakelin, D. (1965) Experimental studies on the biology of *Capillaria obsignata* Madsen, 1945, a nematode of the domestic fowl. *Journal of Helminthology* 39, 399–412.

Wakelin, D. (1969) The development of the early larval stages of *Trichuris muris* in the albino laboratory mouse. *Journal of Helminthology* 43, 427–436.

Wehr, E.E. (1936) Earthworms as transmitters of *Capillaria annulatus*, the cropworm of chickens. *North American Veterinarian* 17, 18–20.

Wehr, E.E. (1937) Studies on the development of the pigeon capillarid, *Capillaria columbae* (Abstract). *Journal of Parasitology* 23, 573.

Wehr, E.E. (1939) Studies on the development of the pigeon capillarid, *Capillaria columbae*. *United States Department of Agriculture Technical Bulletin* No. 679, 19 pp.

Wehr, E.E. (1948) A cropworm, *Capillaria contorta*, the cause of death in turkeys. *Proceedings of the Helminthological Society of Washington* 15, 80.

Wehr, E.E. and Allen, R.W. (1945) Additional studies on the life cycles of *Capillaria caudinflata*, a nematode of chickens and turkeys. *Proceedings of the Helminthological Society of Washington* 12, 12–14.

Whitney, L.F. (1938) Longevity of the whipworm. *Veterinary Medicine* 33, 69–70.

Wright, K.A. (1961) Observations on the life cycle of *Capillaria hepatica* (Bancroft, 1893) with a description of the adult. *Canadian Journal of Zoology* 38, 167–182.

Wright, K.A. (1974) The feeding site and probable feeding mechanism of the parasitic nematode *Capillaria hepatica* (Bancroft, 1893). *Canadian Journal of Zoology* 52, 1215–1220.

Wright, K.A. (1989) Parasites in peril – the trichuroid nematodes. In: Ko, R. (ed.) *Current Concepts in Parasitology.* Hong Kong University Press, Hong Kong, pp. 65–80.

Wright, K.A., Lee, D.L. and Shivers, R.R. (1985) A freeze–fracture study of the digestive tract of the parasitic nematode *Trichinella*. *Tissue and Cell* 17, 189–198.

Wright, K.A., Matta, I., Hong, H.P. and Flood, N. (1989) *Trichinella* larvae and the vasculature of the murine diaphragm. In: Tanner, C.E., Martinez-Fernandez, A.R. and Bolas-Fernandez, F. (eds) *Trichinellosis. Proceedings of the Seventh International Conference on*

Trichinellosis, Alicante, Spain, 2–6 October, 1988. Consejo Superior de Investigaciones Cientificas Press, Madrid, Spain, pp. 70–75.

Wu, L.Y. (1955) The development of the stichosome and associated structures in *Trichinella spiralis. Canadian Journal of Zoology* 33, 440–446.

Wu, Z., Nagano, I. and Takahashi, Y. (1998) Differences and similarities between *Trichinella spiralis* and *T. pseudospiralis* in morphology of stichocyte granules, peptide maps of excretory and secretory (E–S) products and messenger RNA of stichosomal glycoproteins. *Parasitology* 116, 61–66.

Xu, D., Wu, Z., Nagano, I. and Takahashi, Y. (1997) A muscle larva of *Trichinella pseudospiralis* is intracellular, but does not form a typical cyst wall. *Parasitology International* 46, 1–5.

Yokogawa, S. (1920) On the migratory course of *Trichosomoides crassicauda* (Bellingham) in the body of the final host. *Journal of Parasitology* 7, 80–84.

Zarnke, R.L., Gajadhar, A.A., Tiffin, G.B. and Hoef, J.M. (1995) Prevalence of *Trichinella nativa* in Lynx (*Felis lynx*) from Alaska, 1988–1993. *Journal of Wildlife Diseases* 31, 314–318.

Zucchero, P.J. (1942) Notes on the life cycle of *Capillaria annulata. Proceedings of the West Virginia Academy of Science* 15, 96–106.

Index

Main reference in **bold** type